THE OXFORD HANDBOOK OF
THE HISTORY OF MATHEMATICS

THE OXFORD HANDBOOK OF

THE HISTORY OF MATHEMATICS

Edited by

Eleanor Robson and Jacqueline Stedall

OXFORD
UNIVERSITY PRESS

Great Clarendon Street, Oxford OX2 6DP

Oxford University Press is a department of the University of Oxford.
It furthers the University's objective of excellence in research, scholarship,
and education by publishing worldwide in

Oxford New York

Auckland Cape Town Dar es Salaam Hong Kong Karachi
Kuala Lumpur Madrid Melbourne Mexico City Nairobi
New Delhi Shanghai Taipei Toronto

With offices in

Argentina Austria Brazil Chile Czech Republic France Greece
Guatemala Hungary Italy Japan Poland Portugal Singapore
South Korea Switzerland Thailand Turkey Ukraine Vietnam

Oxford is a registered trade mark of Oxford University Press
in the UK and in certain other countries

Published in the United States
by Oxford University Press Inc., New York

© Eleanor Robson and Jacqueline Stedall 2009

First published 2009
First published in paperback 2010

British Library Cataloguing in Publication Data
Data available

Library of Congress Cataloging-in-Publication Data
The Oxford handbook of the history of mathematics / edited by Eleanor
Robson & Jacqueline Stedall.
p. cm.
Includes index.
ISBN 978–0–19–921312–2
1. Mathematics—History—Handbooks, manuals, etc. I. Robson,
Eleanor. II. Stedall,Jacqueline A.
QA21.O94 2008
510.9—dc22 2008031793

Typeset by Newgen Imaging Systems (P) Ltd, Chennai, India
Printed in Great Britain
on acid-free paper by
Ashford Colour Press Ltd, Gosport, Hants

ISBN 978–0–19–921312–2 (Hbk.)
ISBN 978–0–19–960319–0 (Pbk.)

2 4 6 8 10 9 7 5 3 1

TABLE OF CONTENTS

INTRODUCTION

Eleanor Robson and Jacqueline Stedall

We hope that this book will not be what you expect. It is not a textbook, an encyclopedia, or a manual. If you are looking for a comprehensive account of the history of mathematics, divided in the usual way into periods and cultures, you will not find it here. Even a book of this size is too small for that, and in any case it is not what we want to offer. Instead, this book explores the history of mathematics under a series of themes which raise new questions about what mathematics has been and what it has meant to practise it. The book is not descriptive or didactic but investigative, comprising a variety of innovative and imaginative approaches to history.

The image on the front cover captures, we hope, the ethos of the *Handbook* (Chapter 1.2, Fig. 1.2.5). At first glance it has nothing to do with the history of mathematics. We see a large man in a headdress and cloak, wielding a ceremonial staff over a group of downcast kneeling women. Who are they, and what is going on? Who made this image, and why? Without giving away too much—Gary Urton's chapter has the answers—we can say here that the clue is in the phrase written in Spanish above the women's heads: *Repartición de las mugeres donzellas q[ue] haze el ynga* 'categorization (into census-groups) of the maiden women that the Inka made'. As this and many other contributions to the book demonstrate, mathematics is not confined to classrooms and universities. It is used all over the world, in all languages and cultures, by all sorts of people. Further, it is not solely a literate activity but leaves physical traces in the material world: not just writings but also objects, images, and even buildings and landscapes. More often, mathematical practices are ephemeral and transient, spoken words or bodily gestures recorded and preserved only exceptionally and haphazardly.

A book of this kind depends on detailed research in disparate disciplines by a large number of people. We gave authors a broad remit to select topics and approaches from their own area of expertise, as long as they went beyond straight 'what-happened-when' historical accounts. We asked for their writing to be exemplary rather than exhaustive, focusing on key issues, questions, and methodologies rather than on blanket coverage, and on placing mathematical content into context. We hoped for an engaging and accessible style, with striking images and examples, that would open up the subject to new readers and challenge those already familiar with it. It was never going to be possible to cover every conceivable approach to the material, or every aspect we would have liked to include. Nevertheless, authors responded to the broad brief with a stimulating variety of styles and topics.

We have grouped the thirty-six chapters into three main sections under the following headings: geographies and cultures, people and practices, interactions and interpretations. Each is further divided into three subsections of four chapters arranged chronologically. The chapters do not need to be read in numerical order: as each of the chapters is multifaceted, many other structures would be possible and interesting. However, within each subsection, as in the book as a whole, we have tried to represent a range of periods and cultures. There are many points of cross-reference between individual sections and chapters, some of which are indicated as they arise, but we hope that readers will make many more connections for themselves.

In working on the book, we have tried to break down boundaries in several important ways. The most obvious, perhaps, is the use of themed sections rather than the more usual chronological divisions, in such a way as to encourage comparisons between one period and another. Between them, the chapters deal with the mathematics of five thousand years, but without privileging the past three centuries. While some chapters range over several hundred years, others focus tightly on a short span of time. We have in the main used the conventional western BC/AD dating system, while remaining alert to other world chronologies.

The *Handbook* is as wide-ranging geographically as it is chronologically, to the extent that we have made geographies and cultures the subject of the first section. Every historian of mathematics acknowledges the global nature of the subject, yet it is hard to do it justice within standard narrative accounts. The key mathematical cultures of North America, Europe, the Middle East, India, and China are all represented here, as one might expect. But we also made a point of commissioning chapters on areas which are not often treated in the mainstream history of mathematics: Russia, the Balkans, Vietnam, and South America, for instance. The dissemination and cross-fertilization of mathematical ideas and practices between world cultures is a recurring theme throughout the book.

The second section is about people and practices. Who creates mathematics? Who uses it and how? The mathematician is an invention of modern Europe. To limit the history of mathematics to the history of mathematicians is to lose much of the subject's richness. Creators and users of mathematics have included cloth weavers, accountants, instrument makers, princes, astrologers, musicians, missionaries, schoolchildren, teachers, theologians, surveyors, builders, and artists. Even when we can discover very little about these people as individuals, group biographies and studies of mathematical subcultures can yield important new insights into their lives. This broader understanding of mathematical practitioners naturally leads to a new appreciation of what counts as a historical source. We have already mentioned material and oral evidence; even within written media, diaries and school exercise books, novels and account books have much to offer the historian of mathematics. Further, the ways in which people have chosen to express themselves—whether with words, numerals, or symbols, whether in learned languages or vernaculars—are as historically meaningful as the mathematical content itself.

From this perspective the idea of mathematics itself comes under scrutiny. What has it been, and what has it meant to individuals and communities? How is it demarcated from other intellectual endeavours and practical activities? The third section, on interactions and interpretations, highlights the radically different answers that have been given to these questions, not just by those actively involved but also by historians of the subject. Mathematics is not a fixed and unchanging entity. New questions, contexts, and applications all influence what count as productive ways of thinking or important areas of investigation. Change can be rapid. But the backwaters of mathematics can be as interesting to historians as the fast-flowing currents of innovation. The history of mathematics does not stand still either. New methodologies and sources bring new interpretations and perspectives, so that even the oldest mathematics can be freshly understood.

At its best, the history of mathematics interacts constructively with many other ways of studying the past. The authors of this book come from a diverse range of backgrounds, in anthropology, archaeology, art history, philosophy, and literature, as well as the history of mathematics more traditionally understood. They include old hands alongside others just beginning their careers, and a few who work outside academia. Some perhaps found themselves a little surprised to be in such mixed company, but we hope that all of them enjoyed the experience, as we most certainly did. They have each risen wonderfully and good-naturedly to the challenges we set, and we are immensely grateful to all of them.

It is not solely authors and editors who make a book. We would also like to thank our consultants Tom Archibald and June Barrow-Green, as well as the team at OUP: Alison Jones, John Carroll, Dewi Jackson, Tanya Dean, Louise Sprake, and Jenny Clarke.

GEOGRAPHIES AND CULTURES

1. Global

What was mathematics in the ancient world? Greek and Chinese perspectives

G E R Lloyd

Two types of approach can be suggested to the question posed by the title of this chapter. On the one hand we might attempt to settle a priori on the criteria for mathematics and then review how far what we find in different ancient cultures measures up to those criteria. Or we could proceed more empirically or inductively by studying those diverse traditions and then deriving an answer to our question on the basis of our findings.

Both approaches are faced with difficulties. On what basis can we decide on the essential characteristics of mathematics? If we thought, commonsensically, to appeal to a dictionary definition, which dictionary are we to follow? There is far from perfect unanimity in what is on offer, nor can it be said that there are obvious, crystal clear, considerations that would enable us to adjudicate uncontroversially between divergent philosophies of mathematics. What mathematics is will be answered quite differently by the Platonist, the constructivist, the intuitionist, the logicist, or the formalist (to name but some of the views on the twin fundamental questions of *what* mathematics studies, and *what* knowledge it produces).

The converse difficulty that faces the second approach is that we have to have some prior idea of what is to count as 'mathematics' to be able to start our cross-cultural study. Other cultures have other terms and concepts and their interpretation poses delicate problems. Faced with evident divergence and heterogeneity, at what point do we have to say that we are not dealing with a different concept of mathematics, but rather with a concept that has nothing to do with mathematics at all? The past provides ample examples of the dangers involved in legislating that certain practices and ideas fall beyond the boundaries of acceptable disciplines.

My own discussion here, which will concentrate largely on just two ancient mathematical traditions, namely Greek and Chinese, will owe more to the second than to the first approach. Of course to study the ancient Greek or Chinese contributions in this area—their theories and their actual practices—we have to adopt a provisional idea of what can be construed as mathematical, principally how numbers and shapes or figures were conceived and manipulated. But as we explore further their ancient ideas of what the studies of such comprised, we can expect that our own understanding will be subject to modification as we proceed. We join up, as we shall see, with those problems in the philosophy of mathematics I mentioned: so in a sense a combination of both approaches is inevitable.

Both the Greeks and the Chinese had terms for studies that deal, at least in part, with what we can easily recognize as mathematical matters, and this can provide an entry into the problems, though the lack of any exact equivalent to our notion in both cases is obvious from the outset. I shall first discuss the issues as they relate to Greece before turning to the less familiar data from ancient China.

Greek perspectives

Our term 'mathematics' is, of course, derived from the Greek *mathēmatikē*, but that word is derived from the verb *manthanein* which has the quite general meaning of 'to learn'. A *mathēma* can be any branch of learning, anything we have learnt, as when in Herodotus, *Histories* 1.207, Croesus refers to what he has learnt, his *mathēmata*, from the bitter experiences in his life. So the *mathēmatikos* is, strictly speaking, the person who is fond of learning in general, and it is so used by Plato, for instance, in his dialogue *Timaeus* 88c, where the point at issue is the need to strike a balance between the cultivation of the intellect (in general) and that of the body—the principle that later became encapsulated in the dictum *mens sana in corpore sano* 'a healthy mind in a healthy body'. But from the fifth century BC certain branches of study came to occupy a privileged position as the *mathēmata* par excellence. The terms mostly look familiar enough, *arithmētikē*, *geōmetrikē, harmonikē, astronomia*, and so on, but that is deceptive. Let me spend a little time explaining first the differences between the ancient Greeks' ideas and our own, and second some of the disagreements among Greek authors themselves about the proper subject-matter and methods of certain disciplines.

Arithmētikē is the study of *arithmos*, but that is usually defined in terms of positive integers greater than one. Although Diophantus, who lived at some time in late antiquity, possibly in the third century AD, is a partial exception, the Greeks did not normally think of the number series as an infinitely divisible continuum, but rather as a set of discrete entities. They dealt with what we call fractions as ratios between integers. Negative numbers are not *arithmoi*. Nor is the number one, thought of as neither odd nor even. Plato draws a distinction, in the *Gorgias* 451bc, between *arithmētikē* and *logistikē*, calculation, derived from the verb *logizesthai*, which is often used of reasoning in general. Both studies focus on the odd and the even, but *logistikē* deals with the pluralities they form while *arithmētikē* considers them—so Socrates is made to claim—in themselves. That, at least, is the view Socrates expresses in the course of probing what the sophist Gorgias was prepared to include in what he called 'the art of rhetoric', though in other contexts the two terms that Socrates thus distinguished were used more or less interchangeably. Meanwhile a different way of contrasting the more abstract and practical aspects of the study of *arithmoi* is to be found in Plato's *Philebus* 56d, where Socrates distinguishes the way the many, *hoi polloi*, use them from the way philosophers do. Ordinary people use units that are unequal, speaking of two armies, for instance, or two oxen, while the philosophers deal with units that do not differ from one another in any respect; abstract ones in other words.[1]

At the same time, the study of *arithmoi* encompassed much more than we would include under the rubric of arithmetic. The Greeks represented numbers by letters, where α represents the number 1, β the number 2, γ 3, ι 10, and so on. This means that any proper name could be associated with a number. While some held that such connections were purely fortuitous, others saw them as deeply significant. When in the third century AD the neo-Pythagorean Iamblichus claimed that 'mathematics' is the key to understanding the whole of nature and all its parts, he illustrated this with the symbolic associations of numbers, the patterns they form in magic squares and the like, as well as with more widely accepted examples such as the identification of the main musical concords, the octave, fifth, and fourth, with the ratios 2:1, 3:2, and 4:3. The beginnings of such associations, both symbolic and otherwise, go back to the pre-Platonic Pythagoreans of the fifth and early fourth centuries BC, who are said by Aristotle to have held that in some sense 'all things' 'are' or 'imitate' numbers. Yet this is quite unclear, first because we cannot be sure what 'all things' covers, and secondly because of the evident discrepancy between the claim that they *are* numbers and the much weaker one that they merely *imitate* them.

---

1. Cf. Asper, Chapter 2.1 in this volume, who highlights divergences between practical Greek mathematics and the mathematics of the cultured elite. On the proof techniques in the latter, Netz (1999) is fundamental.

What about 'geometry'? The literal meaning of the components of the Greek word *geōmetria* is the measurement of land. According to a well-known passage in Herodotus, 2 109, the study was supposed to have originated in Egypt in relation, precisely, to land measurement after the flooding of the Nile. Measurement, *metrētikē*, still figures in the account Plato gives in the *Laws* 817e when his spokesman, the Athenian Stranger, specifies the branches of the *mathēmata* that are appropriate for free citizens, though now this is measurement of 'lengths, breadths and depths', not of land. Similarly, in the *Philebus* 56e we again find a contrast between the exact *geometria* that is useful for philosophy and the branch of the art of measurement that is appropriate for carpentry or architecture.

Those remarks of Plato already open up a gap between practical utility—mathematics as securing the needs of everyday life—and a very different mode of usefulness, namely in training the intellect. One classical text that articulates that contrast is a speech that Xenophon puts in the mouth of Socrates in the *Memorabilia*, 4 7 2–5. While Plato's Socrates is adamant that mathematics is useful primarily because it turns the mind away from perceptible things to the study of intelligible entities, in Xenophon Socrates is made to lay stress on the usefulness of geometry for land measurement and on the study of the heavens for the calendar and for navigation, and to dismiss as irrelevant the more theoretical aspects of those studies. Similarly, Isocrates too (11 22–3, 12 26–8, 15 261–5) distinguishes the practical and the theoretical sides of mathematical studies and in certain circumstances has critical remarks to make about the latter.

The clearest extant statements of the opposing view come not from the mathematicians but from philosophers commenting on mathematics from their own distinctive perspective. What mathematics can achieve that sets it apart from most other modes of reasoning is that it is exact and that it can demonstrate its conclusions. Plato repeatedly contrasts this with the merely persuasive arguments used in the law-courts and assemblies, where what the audience can be brought to believe may or may not be true, and may or may not be in their best interests. Philosophy, the claim is, is not interested in persuasion but in the truth. Mathematics is repeatedly used as the prime example of a mode of reasoning that can produce certainty: and yet mathematics, in the view Plato develops in the *Republic*, is subordinate to dialectic, the pure study of the intelligible world that represents the highest form of philosophy. Mathematical studies are valued as a propaedeutic, or training, in abstract thought: but they rely on perceptible diagrams and they give no account of their hypotheses, rather taking them to be clear. Philosophy, by contrast, moves from its hypotheses up to a supreme principle that is said to be 'unhypothetical'.

The exact status of that principle, which is identified with the Form of the Good, is highly obscure and much disputed. Likening it to a mathematical axiom immediately runs into difficulties, for what sense does it make to call an axiom 'unaxiomatic'? But Plato was clear that both dialectic and the mathematical sciences deal with independent intelligible entities.

Aristotle contradicted Plato on the philosophical point: mathematics does not study independently existing realities. Rather it studies the mathematical properties of physical objects. But he was more explicit than Plato in offering a clear definition of demonstration itself and in classifying the various indemonstrable primary premises on which it depends. Demonstration, in the strict sense, proceeds by valid deductive argument (Aristotle thought of this in terms of his theory of the syllogism) from premises that must be true, primary, necessary, prior to, and explanatory of the conclusions. They must, too, be indemonstrable, to avoid the twin flaws of circular reasoning or an infinite regress. Any premise that can be demonstrated should be. But there have to be *ultimate* primary premises that are evident in themselves. One of Aristotle's examples is the equality axiom, namely if you take equals from equals, equals remain. That cannot be shown other than by circular argument, which yields no proof at all, but it is clear in itself.

It is obvious what this model of axiomatic-deductive demonstration owes to mathematics. I have just mentioned Aristotle's citation of the equality axiom, which figures also among Euclid's 'common opinions',[2] and most of the examples of demonstrations that Aristotle gives, in the *Posterior analytics*, are mathematical. Yet in the absence of substantial extant texts before Euclid's *Elements* itself (conventionally dated to around 300 BC) it is difficult, or rather impossible, to say how far mathematicians before Aristotle had progressed towards an explicit notion of an indemonstrable axiom. Proclus, in the fifth century AD, claims to be drawing on the fourth century BC historian of mathematics, Eudemus, in reporting that Hippocrates of Chios was the first to compose a book of 'Elements', and he further names a number of other figures, Eudoxus, Theodorus, Theaetetus, and Archytas among those who 'increased the number of theorems and progressed towards a more epistemic or systematic arrangement of them' (*Commentary on Euclid's Elements I* 66.7–18).

That is obviously teleological history, as if they had a clear vision of the goal they should set themselves, namely the Euclidean *Elements* as we have it. The two most substantial stretches of mathematical reasoning from the pre-Aristotelian period that we have are Hippocrates' quadratures of lunes and Archytas' determining two mean proportionals (for the sake of solving the problem of the duplication of the cube) by way of a complex kinematic diagram involving the intersection of three surfaces of revolution, namely a right cone, a cylinder, and a torus. Hippocrates' quadratures are reported by Simplicius (*Commentary on Aristotle's Physics* 53.28–69.34), Archytas' work by Eutocius (*Commentary on*

2. Often translated as 'common notions'.

Archimedes' On the sphere and cylinder II, vol. 3, 84.13–88.2), and both early mathematicians show impeccable mastery of the subject-matter in question. Yet neither text confirms, nor even suggests, that these mathematicians had defined the starting-points they required in terms of different types of indemonstrable primary premises.

Of course the principles set out in Euclid's *Elements* themselves do not tally exactly with the concepts that Aristotle had proposed in his discussion of strict demonstration. Euclid's three types of starting-points include definitions (as in Aristotle) and common opinions (which, as noted, include what Aristotle called the equality axiom) but also postulates (very different from Aristotle's hypotheses). The last included especially the parallel postulate that sets out the fundamental assumption on which Euclidean geometry is based, namely that non-parallel straight lines meet at a point. However, where the philosophers had demanded arguments that could claim to be incontrovertible, Euclid's *Elements* came to be recognized as providing the most impressive sustained exemplification of such a project. It systematically demonstrates most of the known mathematics of the day using especially reductio arguments (arguments by contradiction) and the misnamed method of exhaustion. Used to determine a curvilinear area such as a circle by inscribing successively larger regular polygons, that method precisely did *not* assume that the circle was 'exhausted', only that the difference between the inscribed rectilinear figure and the circumference of the circle could be made as small as you like. Thereafter, the results that the *Elements* set out could be, and were, treated as secure by later mathematicians in their endeavours to expand the subject.

The impact of this development first on mathematics itself, then further afield, was immense. In statics and hydrostatics, in music theory, in astronomy, the hunt was on to produce axiomatic-deductive demonstrations that basically followed the Euclidean model. But we even find the second century AD medical writer Galen attempting to set up mathematics as a model for reasoning in medicine—to yield conclusions in certain areas of pathology and physiology that could claim to be incontrovertible. Similarly, Proclus attempted an *Elements of theology* in the fifth century AD, again with the idea of producing results that could be represented as certain.

The ramifications of this development are considerable. Yet three points must be emphasized to put it into perspective. First, for ordinary purposes, axiomatics was quite unnecessary. Not just in practical contexts, but in many more theoretical ones, mathematicians and others got on with the business of calculation and measurement without wondering whether their reasoning needed to be given ultimate axiomatic foundations.[3]

---

3. Cuomo (2001) provides an excellent account of the variety of both theoretical and practical concerns among the Greek mathematicians at different periods.

Second, it was far from being the case that all Greek work in arithmetic and geometry, let alone in other fields such as harmonics or astronomy, adopted the Euclidean pattern. The three 'traditional' problems, of squaring the circle, the duplication of the cube, and the trisection of an angle were tackled already in the fifth century BC without any explicit concern for axiomatics (Knorr 1986). Much of the work of a mathematician such as Hero of Alexandria (first century AD) focuses directly on problems of mensuration using methods similar to those in the traditions of Egyptian and Babylonian mathematics by which, indeed, he may have been influenced.[4] While he certainly refers to Archimedes as if he provided a model for demonstration, his own procedures sharply diverge, on occasion, from Archimedes'.[5] In the *Metrica*, for instance, he sometimes gives an arithmetized demonstration of geometrical propositions, that is, he includes concrete numbers in his exposition. Moreover in the *Pneumatica* he allows exhibiting a result to count as a proof. Further afield, I shall shortly discuss the disputes in harmonics and the study of the heavens, on the aims of the study, and the right methods to use.

Third, the recurrent problem for the model of axiomatic-deductive demonstration that the *Elements* supplied was always that of securing axioms that would be both self-evident and non-trivial. Moreover, it was not enough that an axiom set should be internally consistent: it was generally assumed that they should be true in the sense of a correct representation of reality. Clearly, outside mathematics they were indeed hard to come by. Galen, for example, proposed the principle that 'opposites are cures for opposites' as one of his indemonstrable principles, but the problem was to say what counted as an 'opposite'. If not trivial, it was contestable, but if trivial, useless. Even in mathematics itself, as the example of the parallel postulate itself most clearly showed, what principles could be claimed as self-evident was intensely controversial. Several commentators on the *Elements* protested that the assumption concerning non-parallel straight lines meeting at a point should be a theorem to be proved and removed from among the postulates. Proclus outlines the controversy (*Commentary on Euclid's Elements I* 191.21ff.) and offers his own attempted demonstration as well as reporting one proposed by Ptolemy (365.5ff., 371.10ff.): yet all such turned out to be circular, a result that has sometimes been taken to confirm Euclid's astuteness in deciding to treat this as a postulate in the first place. In time, however, it was precisely the attack on the parallel postulate that led to the eventual emergence of non-Euclidean geometries.

These potential difficulties evidently introduce elements of doubt about the ability of mathematics, or of the subjects based on it, to deliver exactly what

---

4. Cf. Robson (Chapter 3.1), Rossi (Chapter 5.1), and Imhausen (Chapter 9.1) in this volume.

5. Moreover Archimedes himself departed from the Euclidean model in much of his work, especially, for example, in the area we would call combinatorics; cf. Saito (Chapter 9.2) in this volume and Netz (forthcoming).

some writers claimed for it. Nevertheless, to revert to the fundamental point, mathematics, in the view both of some mathematicians and of outsiders, was superior to most other disciplines, precisely in that it could outdo the merely persuasive arguments that were common in most other fields of inquiry.

It is particularly striking that Archimedes, the most original, ingenious, and multifaceted mathematician of Greek antiquity, insisted on such strict standards of demonstration that he was at one point led to consider as merely heuristic the method that he invented and set out in his treatise of that name. He there describes how he discovered the truth of the theorem that any segment of a parabola is four-thirds of the triangle that has the same base and equal height. The method relies on two assumptions: first that plane figures may be imagined as balanced against one another around a fulcrum and second that such figures may be thought of as composed of a set of line segments indefinitely close together. Both ideas breached common Greek presuppositions. It is true that there were precedents both for applying some quasi-mechanical notions to geometrical issues—as when figures are imagined as set in motion—and for objections to such procedures, as when in the *Republic* 527ab Plato says that the language of mathematicians is absurd when they speak of 'squaring' figures and the like, as if they were doing things with mathematical objects. But in Archimedes' case, the first objection to his reasoning would be that it involved a category confusion, in that geometrical objects are not the types of item that could be said to have centres of gravity. Moreover, Archimedes' second assumption, that a plane figure is composed of its indivisible line segments, clearly breached the Greek geometrical notion of the continuum. The upshot was that he categorized his method as one of discovery only, and he explicitly claimed that its results had thereafter to be demonstrated by the usual method of exhaustion. At this point, there appears to be some tension between the preoccupation with the strictest criteria of proof that dominated one tradition of Greek mathematics (though only one) and the other important aim of pushing ahead with the business of discovery.

The issues of the canon of proof, and of whether and how to provide an axiomatic base for work in the various parts of 'mathematics', were not the only subjects of dispute. Let me now illustrate the range of controversy first in harmonics and then in the study of the heavens.

'Music', or rather *mousikē*, was a generic term, used of any art over which one or other of the nine Muses presided. The person who was *mousikos* was one who was well-educated and cultured generally. To specify what we mean by 'music' the Greeks usually used the term *harmonikē*, the study of harmonies or musical scales. Once again the variety of ways that study was construed is remarkable and it is worth exploring this in some detail straight away as a classic illustration of the tension between mathematical analysis and perceptible phenomena. There were those whose interests were in music-making, practical musicians who were interested in producing pleasing sounds. But there were also plenty of theorists who attempted analyses involving, however, quite different starting assumptions. One approach, exemplified by Aristoxenus, insisted that the unit of measurement should be something identifiable to perception. Here, a tone is defined as the difference between the fifth and the fourth, and in principle the whole of music theory can be built up from these perceptible intervals, namely by ascending and descending fifths and fourths.

But if this approach accepted that musical intervals could be construed on the model of line segments and investigated quasi-geometrically, a rival mode of analysis adopted a more exclusively arithmetical view, where the tone is defined as the difference between sounds whose 'speeds' stand in a ratio of 9:8. In this, the so-called Pythagorean tradition, represented in the work called the *Sectio canonis* in the Euclidean corpus, musical relations are understood as essentially ratios between numbers, and the task of the harmonic theorist becomes that of deducing various propositions in the mathematics of ratios.

Moreover, these quite contrasting modes of analysis were associated with quite different answers to particular musical questions. Are the octave, fifth, and fourth exactly six tones, three and a half, and two and a half tones respectively? If the tone is identified as the ratio of 9 to 8, then you do not get an octave by taking six such intervals. The excess of a fifth over three tones, and of a fourth over two, has to be expressed by the ratio 256 to 243, not by the square root of 9/8.

This dispute in turn spilled over into a fundamental epistemological disagreement. Is perception to be the criterion, or reason, or some combination of the two? Some thought that numbers and reason ruled. If what we heard appeared to conflict with what the mathematics yielded by way of an analysis, then too bad for our hearing. We find some theorists who denied that the interval of an octave plus a fourth can be a harmony precisely because the ratio in question (8:3) does not conform to the mathematical patterns that constitute the main concords. Those all have the form of either a multiplicate ratio as, for example, 2:1 (expressing the octave) or a superparticular one as, for example, 3:2 and 4:3, both of which meet the criterion for a superparticular ratio, namely n+1 : n.

It was one of the most notable achievements of the *Harmonics* written by Ptolemy in the second century AD to show how the competing criteria could be combined and reconciled (cf. Barker 2000). First, the analysis had to derive what is perceived as tuneful from rational mathematical principles. Why should there be any connection between sounds and ratios, and with the particular ratios that the concords were held to express? What hypotheses should be adopted to give the mathematical underpinning to the analysis? But just to select some principles that would do so was, by itself, not enough. The second task the music theorist must complete is to bring those principles to an empirical test, to confirm that the results arrived at on the basis of the mathematical theory did indeed tally with what was perceived by the ear in practice to be concordant—or discordant—as the case might be.

The study of the heavens was equally contentious. Hesiod is supposed to have written a work entitled *Astronomia*, though to judge from his *Works and days* his interest in the stars related rather to how they tell the passing of the seasons and can help to regulate the farmer's year. In the *Epinomis* 990a (whether or not this is an authentic work of Plato) Hesiod is associated with the study of the stars' risings and settings—an investigation that is *contrasted* with the study of the planets, sun, and moon. *Gorgias* 451c is one typical text in which the task of the astronomer is said to be to determine the relative speeds of the stars, sun and moon.

Both *astronomia* and *astrologia* are attested in the fifth century BC and are often used interchangeably, though the second element in the first has *nemo* as its root and that relates to distribution, while *logos*, in the second term, is rather a matter of giving an account. Although genethlialogy, the casting of horoscopes based on geometrical calculations of the positions of the planets at birth, does not become prominent until the fourth century BC, the stars were already associated with auspicious and inauspicious phenomena in, for example, Plato's *Symposium* 188b. Certainly by Ptolemy's time (second century AD) an explicit distinction was drawn between predicting the movements of the heavenly bodies themselves (astronomy, in our terms, the subject-matter of the *Syntaxis*), and predicting events on earth on their basis (astrology, as we should say, the topic he tackled in the *Tetrabiblos*, which he explicitly contrasts with the other branch of the study of the heavens). Yet both Greek terms themselves continued to be used for either. Indeed, in the Hellenistic period the term *mathēmatikos* was regularly used of the astrologer as well as of the astronomer.

Both studies remained controversial. The arguments about the validity of astrological prediction are outlined in Cicero's *De divinatione* for instance, but the Epicureans also dismissed astronomy as speculative. On the other hand, there were those who saw it rather as one of the most important and successful of the branches of mathematics—not that they agreed on how it was to be pursued. We may leave to one side Plato's provocative remarks in the *Republic* 530ab that the *astronomikos* should pay no attention to the empirical phenomena—he should 'leave the things in the heavens alone'—and engage in a study of 'quickness and slowness' themselves (529d), since at that point Plato is concerned with what the study of the heavens can contribute to abstract thought. If we want to find out how Plato himself (no practising astronomer, to be sure) viewed the study of the heavens, the *Timaeus* is a surer guide, where indeed the contemplation of the heavenly bodies is again given philosophical importance—such a vision encourages the soul to philosophize—but where the different problems posed by the varying speeds and trajectories of the planets, sun, and moon are recognized each to need its own solution (*Timaeus* 40b–d).

Quite how the chief problems for theoretical astronomy were defined in the fourth century BC has become controversial in modern scholarship (Bowen 2001). But it remains clear first that the problem of the planets' 'wandering', as their Greek name ('wanderer') implied, was one that exercised Plato. In his *Timaeus*, 39cd, their movements are said to be of wondrous complexity, although in his last work, the *Laws* 822a, he came to insist that each of the heavenly bodies moves with a *single* circular motion. The model of concentric spheres that Aristotle in *Metaphysics lambda (Λ)* ascribes to Eudoxus, and in a modified form to Callippus, was designed to explain *some* anomalies in the apparent movements of the sun, moon, and planets. Some *geometrical* model was thereafter common ground to much Greek astronomical theorizing, though disputes continued over *which* model was to be preferred (concentric spheres came to be replaced by eccentrics and epicycles). Moreover, some studies were purely geometrical in character, offering no comments on how (if at all) the models proposed were to be applied to the physical phenomena. That applies to the books that Autolycus of Pitane wrote *On the moving sphere*, and *On risings and settings*. Even Aristarchus in the one treatise of his that is extant, *On the sizes and distances of the sun and moon*, engaged (in the view I favour) in a purely geometrical analysis of how those results could be obtained, without committing himself to concrete conclusions, although in the work in which he adumbrated his famous heliocentric hypothesis, there are no good grounds to believe he was *not* committed to that as a physical solution.

Yet if we ask *why* prominent Greek theorists adopted *geometrical* models to explain the apparent irregularities in the movements of the heavenly bodies, when most other astronomical traditions were content with purely numerical solutions to the patterns of their appearances, the answer takes us back to the ideal of a demonstration that can carry explanatory, deductive force, and to the demands of a teleological account of the universe, that can show that the movements of the heavenly bodies are supremely orderly.

We may note once again that the history of Greek astronomy is not one of uniform or agreed goals, ideals, and methods. It is striking how influential the contrasts that the philosophers had insisted on, between proof and persuasion or between demonstration and conjecture, proved to be. In the second century AD, Ptolemy uses those contrasts twice over. He first does so in the *Syntaxis* in order to contrast 'mathematics', which here clearly includes the mathematical astronomy that he is about to embark on in that work, with 'physics' and with 'theology'. Both of those studies are merely conjectural, the first because of the instability of physical objects, the second because of the obscurity of the subject. 'Mathematics', on the other hand, can secure certainty, thanks to the fact that it uses—so he says—the incontrovertible methods of arithmetic and geometry. In practice, of course, Ptolemy has to admit the difficulties he faces when tackling such subjects as the movements of the planets in latitude (that is, north and south of the ecliptic): and his actual workings are full of approximations. Yet that is not allowed to diminish the claim he wishes to make for his theoretical study.

Then, the second context in which he redeploys the contrast is in the opening chapters of the *Tetrabiblos*, which I have already mentioned for the distinction it draws between two types of prediction. Those that relate to the movements of the heavenly bodies themselves can be shown demonstratively, *apodeiktikōs*, he says, but those that relate to the fortunes of human beings are an *eikastikē*, conjectural, study. Yet, while some had used 'conjecture' to undermine an investigation's credibility totally, Ptolemy insists that astrology is founded on assumptions that are tried and tested. Like medicine and navigation, it cannot deliver certainty, but it can yield probable conclusions.

Many more illustrations of Greek ideas and practices could be given, but enough has been said for one important and obvious point to emerge in relation to our principal question of what mathematics was in Greece, namely that generalization is especially difficult in the face of the widespread disagreements and divergences that we find at all periods and in every department of inquiry. Some investigators, to be sure, got on with pursuing their own particular study after their own manner. But the questions of the status and goals of different parts of the study, and of the proper methods by which it should be conducted, were frequently raised both within and outside the circles of those who styled themselves mathematicians. But if no single univocal answer can be given to our question, we can at least remark on the intensity with which the Greeks themselves debated it.

Chinese perspectives

The situation in ancient China is, in some respects, very different. The key point is that two common stereotypes about Chinese work are seriously flawed: the first that their concern for practicalities blocked any interest in theoretical issues, and the second that while they were able calculators and arithmeticians, they were weak geometers.

It is true that while the Greek materials we have reviewed may suffer from a deceptive air of familiarity, Chinese ideas and practices are liable to seem exotic. Their map or maps of the relevant intellectual disciplines, theoretical or practical and applied, are very different both from those of the Greeks and from our own. One of the two general terms for number or counting, *shu* 數, has meanings that include 'scolding', 'fate', or 'destiny', 'art' as in 'the art of', and 'deliberations' (Ho 1991). The second general term, *suan* 算, is used of 'planning', 'scheming', and 'inferring', as well as 'reckoning' or 'counting'. The two major treatises that deal with broadly mathematical subjects that date from between around 100 BC and 100 AD, both have *suan* in their title: we shall have more to say on each in due course. The *Zhou bi suan jing* 周 髀 算 經 is conventionally translated 'Arithmetic classic of the gnomon of Zhou'. The second treatise is the *Jiu zhang suan shu* 九 章 算 術, the 'Nine chapters on mathematical procedures'. This draws on an earlier text recently excavated from a tomb sealed in 186 BC, which has both general terms in its title, namely *Suan shu shu* 算 數 書, the 'Book of mathematical procedures', as Chemla and Guo (2004) render it, or more simply, 'Writings on reckoning' (Cullen 2004). But the 'Nine chapters' goes beyond that treatise, both in presenting the problems it deals with more systematically, and in extending the range of those it tackles, notably by including discussing *gou gu* 句股, the properties of right-angled triangles (a first indication of those Chinese interests in geometrical questions that have so often been neglected or dismissed). Indeed, thanks to the existence of the *Suan shu shu* we are in a better position to trace early developments in Chinese mathematics than we are in reconstructing what Euclid's *Elements* owed to its predecessors.

When, in the first centuries BC and AD the Han bibliographers, Liu Xiang and Liu Xin, catalogued all the books in the imperial library under six generic headings, *shu shu* 數 術 'calculations and methods' appears as one of these. Its six subspecies comprise two that deal with the study of the heavens, namely *tian wen* 天 文 'the patterns in the heavens' and *li pu* 曆譜 'calendars and tables', as well as *wu xing* 五 行 'the five phases', and a variety of types of divinatory studies. The five phases provided the main framework within which change was discussed. They are named fire, earth, metal, water, and wood, but these are not elements in the sense of the basic physical constituents of things, so much as processes. 'Water' picks out not so much the substance, as the process of 'soaking downwards', as one text (the Great plan) puts it, just as 'fire' is not a substance but 'flaming upwards'.

This already indicates that the Chinese did not generally recognize a fundamental contrast between what we call the study of nature (or the Greeks called *phusike*) on the one hand and mathematics on the other. Rather, each discipline dealt with the quantitative aspects of the phenomena it covered as and when the need arose. We can illustrate this with harmonic theory, included along with calendar studies in the category *li pu*.

Music was certainly of profound cultural importance in China. We hear of different types of music in different states or kingdoms before China was unified under Qin Shi Huang Di in 221 BC, some the subject of uniform approval and appreciation, some the topic of critical comment as leading to licentiousness and immorality—very much in the way in which the Greeks saw different modes of their music as conducive to courage or to self-indulgence. Confucius is said to have not tasted meat for three months once he had heard the music of *shao* in the kingdom of *Qi* (*Lun yu* 7 14).

But musical sounds were also the subject of theoretical analysis, indeed of several different kinds. We have extensive extant texts dealing with this, starting with the *Huai nan zi*, a cosmological summa compiled under the auspices of Liu An, King of Huainan, in 136 BC, and continuing in the musical treatises contained in the first great Chinese universal history, the *Shi ji* written by Sima Tan and his son Sima Qian around 90 BC. Thus *Huai nan zi*, ch 3, sets out a schema correlating the twelve pitchpipes, that give what we would call the 12-tone scale, with the five notes of the pentatonic scale. Starting from the first pitchpipe, named Yellow Bell (identified with the first pentatonic note, *gong*), the second and subsequent pitchpipes are generated by alternate ascents of a fifth and descents of a fourth—very much in the manner in which in Greece the Aristoxenians thought that all musical concords should be so generated. Moreover, *Huai nan zi* assigns a number to each pitchpipe. Yellow Bell starts at 81, the second pitchpipe, Forest Bell, is 54 —that is 81 times 2/3, the next is 72, that is 54 times 4/3, and so on. The system works perfectly for the first five notes, but then complications arise. The number of the sixth note is rounded from 42 2/3 to 42, and at the next note the sequence of alternate ascents and descents is interrupted by two consecutive descents of a fourth—a necessary adjustment to stay within a single octave.

On the one hand it is clear that a numerical analysis is sought and achieved, but on the other a price has to be paid. Either approximations must be allowed, or alternatively very large numbers have to be tolerated. The second option is the one taken in a passage in the *Shi ji* 25, where the convention of staying within a single octave is abandoned, but at the cost of having to cope with complex ratios such as 32,768 to 59,049. Indeed *Huai nan zi* itself in another passage, 3. 21a, generates the twelve pitchpipes by successive multiplications by 3 from unity, which yields the number 177,147 (that is 3^{11}) as the 'Great Number of Yellow Bell'. That section associates harmonics with the creation of the 'myriad things' from the primal unity. The *Dao* 道 is one, and this subdivides into *yin* 陰 and *yang* 陽, which between them generate everything else. Since *yin* and *yang* themselves are correlated with even and with odd numbers respectively, the greater and the lesser *yin* being identified as six and eight respectively, and the greater and lesser *yang* nine and seven, the common method of divination, based on the hexagrams set out in such texts as the *Yi jing* 易經 'Book of changes', is also given a numerical basis. But, interestingly enough, the 'Book of changes' was not classified by Liu Xiang and Liu Xin under *shu shu*. Rather it was placed in the group of disciplines that dealt with classic, or canonical, texts. Indeed the patterns of *yin* and *yang* lines generated by the hexagrams were regularly mined for insight into every aspect of human behaviour, as well as into the cosmos as a whole.

Similarly complicated numbers are also required in the Chinese studies of the heavens. One division dealt with 'the patterns of the heavens', *tian wen*, and was chiefly concerned with the interpretation of omens. But the other *li fa* included the quantitative analysis of periodic cycles, both to establish the calendar and to enable eclipses to be predicted. In one calendrical schema, called the Triple Concordance System, a lunation is 29 43/81 days, a solar year 365 385/1539 days, and in the concordance cycle 1539 years equals 19,035 lunations and 562,120 days (cf. Sivin 1995). On the one hand, considerable efforts were expended on carrying out the observations needed to establish the data on which eclipse cycles could be based. On the other, the figures for the concordances were also manipulated mathematically, giving in some cases a spurious air of precision—just as happens in Ptolemy's tables of the movements of the planets in longitude and in anomaly in the *Syntaxis*.

Techniques for handling large-number ratios are common to both Chinese harmonics and to the mathematical aspects of the study of the heavens. But there is also a clear ambition to integrate these two investigations—which both form part of the Han category *li pu*. Thus, each pitchpipe is correlated with one of the twelve positions of the handle of the constellation 'Big Dipper' as it circles the celestial pole during the course of the seasons. Indeed, it was claimed that each pitchpipe resonates spontaneously with the *qi* of the corresponding season and that that effect could be observed empirically by blown ash at the top of a half-buried pipe, a view that later came to be criticized as mere fantasy (Huang Yilong and Chang Chih-Ch'eng 1996).

While the calendar and eclipse cycles figure prominently in the work of Chinese astronomers, the study of the heavens was not limited to those subjects. In the *Zhou bi suan jing*, the Master Chenzi is asked by his pupil Rong Fang what his *Dao* achieves, and this provides us with one of the clearest early state-ments acknowledging the power and scope of mathematics.[6] The *Dao*, Chenzi replies, is able to determine the height and size of the sun, the area illuminated by its light, the figures for its greatest and least distances, and the length and breadth of heaven, solutions to each of which are then set out. That the earth is flat is assumed throughout, but one key technique on which the results depend is the geometrical analysis of gnomon shadow differences. Among the observa-tional techniques is sighting the sun down a bamboo tube. Using the figure for the distance of the sun obtained in an earlier study, the dimension of the sun can be gained from those of the tube by similar triangles. Such a result was just one impressive proof of the power of mathematics (here *suan shu*) to arrive at an understanding of apparently obscure phenomena. But it should be noted that although Chenzi eventually explains his methods to his pupil on the whole quite clearly, he first expects him to go away and work out how to get these results on

6. The term *Dao*, conventionally translated 'the Way', can be used of many different kinds of skills, and here the primary reference is to Chenzi's ability in mathematics. But those skills are thought of as subordinate to the supreme principle at work in the universe, which it is the goal of the sage to cultivate, indeed to embody (Lloyd and Sivin 2002).

his own. Instead of overwhelming the student with the incontrovertibility of the conclusion '*quod erat demonstrandum*', the Chinese master does not rate knowledge unless it has been internalized by the pupil.

The major classical Chinese mathematical treatise, the 'Nine chapters', indicates both the range of topics covered and the ambitions of the coverage. Furthermore the first of the many commentators on that text, Liu Hui in the third century AD, provides precious evidence of how he saw the strategic aims of that treatise and of Chinese mathematics as a whole. The 'Nine chapters' deals with such subjects as field measurement, the addition, subtraction, multiplication, and division of fractions, the extraction of square roots, the solutions to linear equations with multiple unknowns (by the rule of double false position), the calculation of the volumes of pyramids, cones, and the like.

The problems are invariably expressed in concrete terms. The text deals with the construction of city-walls, trenches, moats, and canals, with the fair distribution of taxes across different counties, the conversion of different quantities of grain of different types, and much else besides. But to represent the work as just focused on practicalities would be a travesty. A problem about the number of workmen needed to dig a trench of particular dimensions, for instance, gives the answer as 7 427/3064ths labourers. The interest is quite clearly in the exact solution to the equation rather than in the practicalities of the situation. Moreover the discussion of the circle–circumference ratio (what we call π) provides a further illustration of the point. For practical purposes, a value of 3 or 3 1/7 is perfectly adequate, and such values were indeed often used. But the commentary tradition on the 'Nine chapters' engages in the calculation of the area of inscribed regular polygons with 192 sides, and even 3072-sided ones are contemplated (the larger the number of sides, the closer the approximation to the circle itself of course): by Zhao Youqin's day, in the thirteenth century, we are up to 16384-sided polygons (Volkov 1997).

Liu Hui's comments on the chapter discussing the volume of a pyramid illustrate the sophistication of his geometrical reasoning (cf. Wagner 1979). The figure he has to determine is a pyramid with rectangular base and one lateral edge perpendicular to the base, called a *yang ma* 陽馬. To arrive at the formula setting out its volume (namely one third length, times breadth, times height) he has to determine the proportions between it and two other figures, the *qian du* 塹堵 (right prism with right triangular base) and the *bie nao* 鱉臑 (a pyramid with right triangular base and one lateral edge perpendicular to the base). A *yang ma* and a *bie nao* together go to make up a *qian du*, and its volume is simple: it is half its length, times breadth, times depth. That leaves Liu Hui with the problem of finding the ratio between the *yang ma* and the *bie nao*. He proceeds by first decomposing a *yang ma* into a combination of smaller figures, a box, two smaller *qian du*, and two smaller *yang ma*. A *bie nao* similarly can be decomposed into two smaller *qian du* and two smaller *bie nao*. But once so decomposed it can be seen that the box plus two smaller *qian du* in the original *yang ma* are twice the two smaller *qian du* in the original *bie nao*. The parts thus determined stand in a relation of 2:1. The remaining problem is, of course, to determine the ratios of the smaller *yang ma* and the smaller *bie nao*: but an exactly similar procedure can be applied to them. At each stage more of the original figure has been determined, always yielding a 2:1 ratio for the *yang ma* to the *bie nao*. If the process is continued, the series converges on the formula one *yang ma* equals two *bie nao*, and so a *yang ma* is two-thirds of a *qian du*, which yields the requisite formula for the volume of the *yang ma*, namely one third length, times breadth, times height (Fig. 1.1.1).

Two points of particular interest in this stretch of argument are first that Liu Hui explicitly remarks on the uselessness of one of the figures he uses in his decomposition. The *bie nao*, he says, is an object that 'has no practical use'. Yet without it the volume of the *yang ma* cannot be calculated. At this point we have yet another clear indication that the interest in the exact geometrical result takes precedence over questions of practical utility.

Second, we may observe both a similarity and a difference between the procedure adopted by Liu Hui and some Greek methods. In such cases (as in Euclid's determination of the pyramid at *Elements* 12 3) the Greeks used an indirect proof, showing that the volume to be determined cannot be either greater or less than the result, and so must equal it. Liu Hui by contrast uses a direct proof, the technique of decomposition which I have described, yielding increasingly accurate approximations to the volume, a procedure similar to that used in the Chinese determination of the circle by inscribing regular polygons, mentioned above. Such a technique bears an obvious resemblance to the Greek method of exhaustion, though I remarked that in that method the area or volume to be determined was precisely not exhausted. Liu Hui sees that his process of decomposition can be

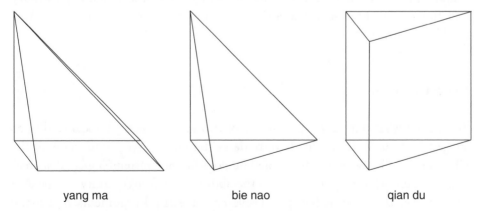

yang ma   bie nao   qian du

Figure 1.1.1 the *yang ma*, *bie nao*, and *qian du*

continued indefinitely, and he remarks on the progressively smaller remainders that this yields. We are dealing evidently with what we would call a converging series, but although Liu Hui has no explicit concept for the limit of such, he ends his investigation with the rhetorical question 'how can there be any remainder?'.

There is no suggestion, however, in any of the texts we have been considering, of giving mathematics an axiomatic base. The notion of axiom is absent from Chinese mathematics until the arrival of the Jesuits in the sixteenth century. Rather the chief aims of Chinese mathematicians were to explore the unity of mathematics and to extend its range. Liu Hui, especially, comments that it is the *same* procedures that provide the solutions to problems in different subject-areas. What he looks for, and finds, in such procedures as those he calls *qi* 齊 'homogenizing' and *tong* 同 'equalizing', is what he calls the *gang ji* 綱紀 'guiding principles' of *suan* 'mathematics'. In his account of how, from childhood, he studied the 'Nine chapters', he speaks of the different branches of the study, but insists that they all have the same *ben* 本 'trunk'. They come from a single *duan* 端 'source'. The realizations and their *lei* 類 'categories', are elaborated mutually. Over and over again the aim is to find and show the connections between the different parts of *suan shu*, extending procedures across different categories, making the whole 'simple but precise, open to communication but not obscure'. Describing how he identified the technique of double difference, he says (92.2) he looked for the *zhi qu* 指趣 'essential characteristics' to be able to extend it to other problems.

While Liu Hui is more explicit in all of this than the 'Nine chapters', the other great Han classic, the *Zhou bi*, represents the goal in very similar terms. We are not dealing with some isolated, maybe idiosyncratic, point of view, but with one that represents an important, maybe even the dominant, tradition. 'It is the ability to distinguish categories in order to unite categories' which is the key according to the *Zhou bi* (25.5). Again, among the methods that comprise the *Dao* 'Way', it is 'those which are concisely worded but of broad application which are the most illuminating of the categories of understanding. If one asks about one category and applies [this knowledge] to a myriad affairs, one is said to know the Way' (24.12ff., Cullen 1996, 177).

Conclusions

To sum up what our very rapid survey of two ancient mathematical traditions suggests, let me focus on just two fundamental points. We found many of the Greeks (not all) engaged in basic methodological and epistemological disagreements, where what was at stake was the ability to deliver certainty—to be able to do better than the merely persuasive or conjectural arguments that many downgraded as inadequate. The Chinese, by contrast, were far more concerned to explore the connections and the unity between different studies, including between those we consider to be mathematics and others we class as physics or cosmology. Their aim was not to establish the subject on a self-evident axiomatic basis, but to expand it by extrapolation and analogy.

Each of those two aims we have picked out has its strengths and its weaknesses. The advantages of axiomatization are that it makes explicit what assumptions are needed to get to which results. But the chief problem was that of identifying self-evident axioms that were not trivial. The advantage of the Chinese focus on guiding principles and connections was to encourage extrapolation and analogy, but the corresponding weakness was that everything depended on perceiving the analogies, since no attempt is made to give them axiomatic foundations. It is apparent that there is no one route that the development of mathematics had to take, or should have taken. We find good evidence in these two ancient civilizations for a variety of views of its unity and its diversity, its usefulness for practical purposes and for understanding. The value of asking the question 'what is mathematics?' is that it reveals so clearly, already where just two ancient mathematical traditions are concerned, the fruitful heterogeneity in the answers that were given.

Bibliography

Barker, A D, *Scientific method in Ptolemy's harmonics*, Cambridge University Press, 2000.

Bowen, A C, 'La scienza del cielo nel periodo pretolemaico', in S Petruccioli (ed), *Storia della scienza*, vol 1, Enciclopedia Italiana, 2001, 806–839.

Chemla, K and Guo Shuchun, *Les Neuf chapitres. Le Classique mathématique de la Chine ancienne et ses commentaires*, Dunod, 2004.

Cullen, C, *Astronomy and mathematics in ancient China: the Zhoubi Suanjing*, Cambridge University Press, 1996.

Cullen, C, *The Suan Shu Shu: writings on reckoning* (Needham Research Institute Working Papers, 1), Needham Research Institute, 2004.

Cuomo, S, *Ancient mathematics*, Routledge, 2001.

Ho Peng-Yoke, 'Chinese science: the traditional Chinese view', *Bulletin of the School of Oriental and African Studies*, 54 (1991), 506–519.

Huang Yilong and Chang Chih-Ch'eng, 'The evolution and decline of the ancient Chinese practice of watching for the ethers', *Chinese Science*, 13 (1996), 82–106.

Knorr, W, *The ancient tradition of geometric problems*, Birkhäuser, 1986.

Lloyd, G E R, and Sivin, N, *The way and the word*, Yale University Press, 2002.

Netz, R, *The shaping of deduction in Greek mathematics*, Cambridge University Press, 1999.

Netz, R, *Ludic proof*, Cambridge University Press, forthcoming.

Sivin, N, 'Cosmos and computation in early Chinese mathematical astronomy', in *Researches and Reflections vol 1: Science in Ancient China*, Variorum, 1995, ch. II.

Volkov, A, 'Zhao Youqin and his calculation of π', *Historia Mathematica*, 24 (1997), 301–331.

Wagner, D B, 'An early Chinese derivation of the volume of a pyramid: Liu Hui, third century AD', *Historia Mathematica*, 6 (1979), 164–188.

CHAPTER 1.2

Mathematics and authority: a case study in Old and New World accounting

Gary Urton

The title of an article published by Alan Bishop, 'Western mathematics: the secret weapon of cultural imperialism' (1990), must surely be one of the most provocative in the recent literature concerning the history of mathematics and the nature and status of mathematical practice.[1] There are several surprises in this title, beginning with the adjective 'western'. According to Platonism, the grounding philosophy that informs the thinking of most mathematicians, mathematical truths lie beyond human experience, in an abstract realm set apart from language, culture, and history. In what sense, then, could mathematics be conceived of as preferentially linked to one or the other of the earthly hemispheres? And how could mathematics—the supposed dispassionate and logical investigation of arrangement, quantity, and related concepts in algebra, analysis, and geometry—be implicated in any meaningful way with such socially and politically loaded objects and concepts as 'weapons', 'culture', and 'imperialism'? Conveniently, Bishop's title provides an answer to this puzzle in the assertion that the association of mathematics with this disturbing set of modifiers is (or was) a 'secret'.

1. Thanks to Carrie Brezine and Julia Meyerson for their critical readings of drafts of this work. I alone am responsible for any errors of fact or logic that remain.

In the article in question, Bishop argues that western European colonizing societies of the fifteenth to nineteenth centuries carried with them to various exotic locales the gifts of rationalism and 'objectism' (that is, a way of conceiving of the world as composed of discrete objects that could be abstracted from their contexts), as well as a number of clearly formulated ways of employing mathematical ideas and procedures, all of which combined to promote western control over the physical and social environments in the colonies. Such regimes of power and control constituted what Bishop (1990, 59) terms a 'mathematico-technological cultural force' embedded in the colonies in institutions related to accounting, trade, administration, and education:

Mathematics with its clear rationalism, and cold logic, its precision, its so-called 'objective' facts (seemingly culture and value free), its lack of human frailty, its power to predict and to control, its encouragement to challenge and to question, and its thrust towards yet more secure knowledge, was a most powerful weapon indeed. (Bishop 1990, 59)

When we look more broadly at the uses to which mathematics has been put, especially in accounting systems and in other administrative projects in ancient and modern states, it becomes clear that what is ideally conceived of as the fine, elegant, and dispassionate art of mathematics has in many times and places been intimately linked to systems and relations of authority in a wide range of ideological, philosophical, and political programs and productions. The central questions that we will address here in relation to this history are: how has the linkage between mathematics and authority come about? And how and why has this relationship evolved in the particular ways it has in different historical settings?

To speak of a relationship between mathematics and authority is by no means to limit the issues to imperialist administrative regimes. It also arises in other settings, from the authority that emerges among mathematicians as a result of the successful execution of mathematical proofs, to the attempt by those steeped in the measurement and quantification of social behaviors to adopt math-based paradigms for ordering society (see Mazzotti, Chapter 3.3 in this volume). In short, what we will be concerned with here are a number of problems connected with the manipulation of numbers by arithmetical procedures and mathematical operations and the ways these activities enhance authority and underlie differences in power between different individuals and/or groups or classes in society—for example, between bureaucrats and commoners, or, as in the particular setting to be discussed below, between conquerors and conquered.

We will address the questions raised above in three different but historically related cultural and social historical contexts. The first concerns mathematical philosophies and concepts of authority in the West in the centuries leading up to the European invasion of the New World. This section will include an overview of the rise of double entry bookkeeping in European mercantile capitalism. Next, we will examine the practice of *khipu* (knotted-string) record-keeping in the Inka empire of the Pre-Columbian Andes. And, finally, we will examine the encounter between Spanish written (alphanumeric) record-keeping practices and Inka knotted-string record-keeping that occurred in the Andes following the European invasion and conquest of the Inka empire, in the sixteenth century.

Accounting, authority, power, and legitimacy

A wealth of literature produced by critical accounting historians over the past several decades has elucidated the role of accounting as a technology of, and a rationality for, governance in state societies. Accounting and its specialized notational techniques are some of the principal instruments employed by states in their attempts to control and manage subjects (Hoskin and Macve 1986; Miller and O'Leary 1987; Miller 1990). As Miller has argued:

Rather than two independent entities, accounting and the state can be viewed as inter-dependent and mutually supportive sets of practices, whose linkages and boundaries were constructed at least in their early stages out of concerns to elaborate the art of statecraft. (Miller 1990, 332)

A focus on accounting is one of the most relevant approaches to take in examining Andean and European (Spanish) mathematical practices, as this was the context of the production of most of the documentation deriving from mathematical activities in these two societies that is preserved in archives and museums. The *khipu* was, first and foremost, a device used for recording information pertaining to state activities, such as census-taking and the assessment of tribute; this was also true of the information recorded by Spanish bureaucrats in written documents in the administration of the crown's overseas holdings. For instance, among the some 34,000 *legajos* (bundles of documents) deriving from Spanish colonial administration in the New World, preserved today in the Archivo de Indias in Seville, the largest collections—other than those labeled *Indiferente* 'miscellaneous/unclassified'—are those categorized under the headings *Contaduría* 'accountancy' (1953 *legajos*) and *Contratación* 'trade contracts' (5873 *legajos*; Gómez Cañedo 1961, 12–13). Focusing on accounting will, therefore, provide us with the best opportunity for investigating the relative complexity of arithmetic and mathematical practices employed in the records of these two states, as well as similarities and differences in their principles of quantification.

Although the focus of this essay is on the relationship between mathematics and authority in the context of accounting, we will not get far in our examination of these concepts and domains of human intellectual activity without first developing a clear sense of the meaning of 'authority' and discussing how this concept relates to the wider field of social and political relations that includes legitimacy, power, and social norms. The principal figure whose work must be engaged on these topics is, of course, Max Weber (1964). Insofar as the question of power is concerned, Weber famously defined this concept as '…the probability that one actor within a social relationship will be in a position to carry out his[/her] own will despite resistance, regardless of the basis on which this probability rests' (cited in Uphoff 1989, 299). It is clear from this definition that power is inextricably linked to authority and legitimacy. Uphoff makes a forceful argument to the effect that authority should be understood as a *claim* for compliance, while legitimacy should be understood as an *acceptance* of such a claim. Thus, different persons are involved in such power relationships; on the one hand there are 'the authorities' and on the other there are those who are subject to and accept the claims of the authorities (Uphoff 1989, 303). Thus, the three central concepts we are concerned with are linked causally in the sense that the power associated with authority depends on the legitimacy accorded to it.

Weber identified three principal types of authority, each having a particular relationship to norms. One type, referred to as 'charismatic authority', which may be embodied by the prophet or the revolutionary, Weber considered the purest form of authority in that, in coming into being, it breaks down all existing normative structures. In 'traditional authority', the leader comes into power by heredity or some other customary route, and the actions of the leader are in turn limited by custom. Thus, in traditional systems of authority, norms generate the leader, and one who comes into such a position of authority—the king, chief, or other hereditary leader—depends on traditional norms for his/her authority. Finally, in what Weber termed 'legal-rational authority', the leader occupies the highest position in a bureaucratic structure and derives authority from the legal norms that define the duties and the jurisdiction of the office he/she occupies (Spencer 1970, 124–5).

In terms of the relationship between types of authority and forms of political rule relevant to our study, both the Inka state under its (possibly dual) dynastic rulers, as well as the Spanish kings of the Hapsburg dynasty, experienced processes of increasing regularization of bureaucratic procedures from traditional to rational-legal authority structures during the century or so leading up to the European invasion of the Andes. Our study will examine ways in which mathematical activities linked to accounting practices in pre-modern states in the Old and New Worlds served to legitimize or empower particular individuals or classes in their claims for compliance of the exercise of their will. Our task will be particularly challenging because we will examine these matters in the context of the Spanish conquest of the Inka empire, a historical conjuncture that brought two formerly completely unrelated world traditions of mathematics and authority into confrontation with each other.

Two almost simultaneous developments in European mathematics and commercialism during the fourteenth and fifteenth centuries are critical to the picture we are sketching here of accounting and record-keeping practices of

Spanish colonial administrators in the sixteenth century. These developments were the invention of double-entry bookkeeping and the replacement of Roman numerals by Hindu-Arabic numerals.

The earliest evidence for double-entry bookkeeping dates from the thirteenth century when the method was put to use by merchants in northern Italy (Yamey 1956; Carruthers and Espeland 1991). The first extended explanation of double-entry bookkeeping appeared in a treatise on arithmetic and mathematics written by the Franciscan monk Luca Pacioli in 1494 (Brown and Johnston 1984). In the double-entry method, all transactions are entered twice, once as a debit and again as a credit (Fig. 1.2.1). Daily entries are posted to a journal, which are later

Hypothetical Medieval Ledger Postings based on Luca Pacioli's Directions

In the Name of God

+Jesus   MCDIII
On this day, Cash shall give to Capital CLI lire in the form of coin.
CLI lire
Cr. ref. page

+Jesus   MCDIII
On this day, Capital shall have from Cash in the form of coin CLI lire.
CLI lire
Dr. ref. page

+Jesus   MCDLXXX
Giovanni Bessini shall give, on This day, CC lire, which he promised to pay to us at our pleasure, for the debt which Lorenzo Vincenti owes us.
CC lire
Cr. ref. page

+Jesus   MCDLXXX
Giovanni Bessini shall have back on Nov. II, the CC lire, which he deposited with us in cash.
CC lire
Dr. ref. page

+Jesus   MCDLXXIV
On this day, Jewels with a value DLXX lire, shall give to Capital
DLXX lire
Cr. ref. page

+Jesus   MCDLXXIV
On this day, Capital shall have of from Jewels, a value of DLXX lire.
DLXX lire
Dr. ref. page

+Jesus   MCDXXX
On this day, Business Expense for office material worth CCC lire Shall give to Cash
CCC lire
Cr. ref. page

+Jesus   MCDXXX
On this day, Cash shall have from Business Expense CCC lire.
CCC lire
Dr. ref. page

Figure 1.2.1 Double entry book-keeping ledger postings based on Luca Pacioli's (1494) directions (Aho 2005, 71, Table 7.2)

transferred to a ledger. The ledger books provide the material for the process of accounting, which relies on the equation: assets = liabilities + equity. For books to remain in balance, a change in one account (a debit or credit) must be matched by an equal change in the other. In the rhetorical form in which Pacioli presented the method, the balancing of accounts by double-entry was constructed as an undertaking that had deep religious and moral implications.

The invention and implementation of double-entry went hand-in-hand with the replacement of Roman numerals by Hindu-Arabic numerals, which had been introduced into western Europe almost five hundred years before their eventual acceptance into accounting practice in the fifteenth century. Ellerman (1985, 232) argues that what is distinctive about double-entry is not that it relates two or more accounts, as that is a characteristic of the transaction itself; rather, the distinction of double-entry is that this is a new system of *recording* transactions. Double-entry required complex mathematics based on an efficient system of numbers—like Hindu-Arabic numerals, rather than the cumbersome Roman numerals. There are extensive literatures documenting (Swetz 1989, 11–13; Durham 1992, 48–49) and demonstrating (Donoso Anes 1994, 106) that the coupling of Hindu-Arabic numerals and double-entry in accounting had a powerful affect in promoting increasing rationality in business, society, and politics. There is controversy over whether capitalism was nurtured initially and primarily by Catholicism, with its emphasis on penance and confession constituting a form of accounting (Sombert 1967; Aho 2005), or by Protestantism (Weber 1958). However, those arguing on both sides of this question agree that the spread of double-entry bookkeeping throughout western Europe was a central component of the increasing rationalization and standardization associated with the rise of mercantile capitalism (Carruthers and Espeland 1991, 32; Aho 2005).

While the centers of development of double-entry bookkeeping were the burgeoning mercantile city-states of northern Italy, the method soon spread to other regions of western Europe, including the Iberian peninsula. From detailed study of accounts pertaining to the sale of gold and silver brought from the Americas kept in the *Casa de Contratación* 'Treasury House', in Seville, Donoso Anes (1994) has shown convincingly that the double-entry method was employed in the central accounts of the Royal Treasury of Castille from as early as 1555. In fact, Spain was the first European country to issue laws (in 1549 and 1552) compelling merchants to apply the double-entry method, as well as the first country in which the method was implemented by a public institution—the *Casa de Contratación* (Donoso Anes 1994, 115). Furthermore, Spanish merchants appear to have taught the method to English traders (Reitzer 1960, 216), and they were instrumental in developing and passing on to French merchants the practice of drawing bills of exchange (Lapeyre 1955, 22; cited in Reitzer 1960, 216). While double-entry was used in Spanish accounting for the sale and minting of gold and silver by the Royal Treasury, single-entry accounts were kept at the same time, primarily as the official accounting procedure controlling the activities of the treasurer of the *Casa de Contratación* (Donoso Anes 1994, 115; cf. Klein and Barbier 1988, 54; Hoffman 1970, 733).

The cities of northern Italy that were the centers of commercial activities from the fourteenth to the sixteenth centuries also became centers of learning in arithmetic and mathematics. It was in these cities—Venice, Bologna, Milan—that Hindu-Arabic numerals were first linked with double-entry to form the basis of modern accounting science. It was here as well that abacus or 'reckoning' schools grew up that were patronized by the sons and apprentices of merchants throughout Europe. The masters of those schools, the *maestri d'abbaco*, taught the new arithmetic, or *arte dela mercadanta*, 'the mercantile art' (Swetz 1989, 10–16). It was in northern Italy as well where, a couple of decades prior to the publication of Pacioli's exposition of double-entry bookkeeping, the first arithmetic textbook, the so-called *Treviso arithmetic*, was published in 1478 (Swetz 1989). While not discussing the double-entry method itself, the *Treviso arithmetic* proclaimed itself from the opening passage as intended for study by those with an interest in commercial pursuits (Swetz 1989, 40).

This, then, was a new kind of authority in mathematics, one that was grounded not in theoretical considerations, but rather with a mathematics that served the practical needs and interests of the merchant. The efficacy of this new mathematics was determined not by how closely it cleaved to some body of theoretical principles or philosophical values, but rather by how well it tracked the debits, credits, and profit fluctuations of merchant capitalists, how well it served in arbitrating disputes, and its overall contribution to the well-being of those who put the methods into practice. This new mathematics of the fifteenth century both stimulated and reflected the development of mercantilism and economic accounting and administration throughout Europe, and it was this mathematical practice that was transplanted to the New World in the fifteenth and sixteenth centuries as the basis of accounting for trade, tribute, and the growth of wealth in the American colonies.

From virtually the earliest years following the invasion of the Andes, European administrators—toting accounting ledgers filled with columns of Hindu-Arabic numerals and alphabetically-written words and organized in complex formats—came into contact with Inka administrative officials wielding bundles of colorful knotted cords. These local administrators—known as *khipukamayuqs* 'knot-keepers/makers/organizers'—were, oddly enough, speaking the language (in Quechua) of decimal numeration and practicing what may have looked for all the world, to any Spaniard trained by the reckoning masters of northern Italy, like double-entry bookkeeping.

A new world of knotted-cord record keeping

Khipus are knotted-string devices made of spun and plied cotton or camelid fibers (Figure 1.2.2).[2] The colors displayed in *khipus* are the result of the natural colors of cotton or camelid fibers or of the dyeing of these materials with natural dyes. The 'backbone' of a *khipu* is the so-called primary cord—usually around 0.5 cm in diameter—to which are attached a variable number of thinner strings, called pendant cords. *Khipus* contain from as few as one up to as many as 1500 pendants (the average of some 450+/- samples studied by the Harvard *Khipu* Database project is 84 cords). Top cords are pendant-like strings that leave the primary cord opposite the pendants, often after being passed through the attachments of a group of pendant strings. Top cords often contain the sum of values knotted on the set of pendant cords to which they are attached. About one-quarter of all pendant cords have second-order cords attached to them; these are called subsidiaries. Subsidiaries may themselves bear subsidiaries, and there are examples of *khipus* that contain up to thirteen levels of subsidiaries, making

Figure 1.2.2 A *khipu* from Museum for World Culture, Göteborg, Sweden (#1931.37.0001 [UR113])

2. According to my own inventory, there are some 780+/- *khipu* samples in museums and private collections in Europe, North America, and South America. While many samples are too fragile to permit study, almost 450 samples have been closely studied to date. Observations on a few hundred *khipus* may be viewed at <http://khipukamayuq.fas.harvard.edu/> and <http://instruct1.cit.cornell.edu/resear4ch/quipu~ascher/>.

the *khipu* a highly efficient device for the display of hierarchically organized information.[3]

The majority of *khipus* have knots tied into their pendant, subsidiary, and top strings (Locke 1923; Pereyra 2001). The most common knots are of three different types, which are usually tied in clusters at different levels in a decimal place system of numerical registry (Fig. 1.2.3).[4] The most thorough treatment to date of the numerical, arithmetic, and mathematical properties of the *khipus* is Ascher and Ascher's *Mathematics of the Incas: code of the quipus* (1997; see also Urton 1997; 2003). The Aschers have shown that the arithmetic and mathematical operations used by Inka accountants included, at a minimum, addition, subtraction, multiplication, and division; division into unequal fractional parts and into proportional parts; and multiplication of integers by fractions (Ascher and Ascher 1997, 151–2).

What kinds of information were registered on the *khipus*? In addressing this question, it is important to stress that, although we are able to interpret the quantitative data recorded in knots on the *khipus*, we are not yet able to read the accompanying nominative labels, which appear to have been encoded in the colors, twist, and other structural features of the cords. The latter would, were we able to read them, presumably inform us as to the identities of the items that were being enumerated by the knots. Thus, in discussing the identities of objects accounted for in the *khipus*, we are forced to rely on the Spanish documents from the early years following the European invasion.

According to the Spanish accounts, records were kept of censuses, tribute assessed and performed, goods stored in the Inka storehouses, astronomical periodicities and calendrical calculations, royal genealogies, historical events, and so on (see Murra 1975; Zuidema 1982; Julien 1988; Urton 2001; 2002; 2006). The overriding interest in the recording, manipulation and eventual archiving of quantitative data in the *khipus* was the attempt to control subject peoples throughout the empire. This meant being able to enumerate, classify, and retain records on each subject group. The most immediate use to which this information was put was the implementation of the labor-based system of tribute. Tribute in the Inka state took the form of a labor tax, which was levied on all married, able-bodied men (and some chroniclers say women as well) between the ages of 18 and 50. In its conception and application to society, Inka mathematics appears to have taken a form remarkably like the political arithmetic of seventeenth-century Europeans.[5] In sum, the decimal place system of recording values—including zero (Urton 1997, 48–50)—of the

---

3. For general works on *khipu* structures and recording principles, see Urton (1994; 2003); Ascher and Ascher (1997); Arellano (1999); Conklin (2002); Radicati di Primeglio (2006).

4. Approximately one-third of *khipu* studied to date do not have knots tied in (decimal-based) tiered arrangements. I have referred to these as 'anomalous *khipu*' and have suggested that their contents may be more narrative than statistical in nature (Urton 2003).

5. See the discussions of Inka arithmetic and mathematics in Ascher (1992); Ascher and Ascher (1997); Pereyra (2001); and Urton (1997).

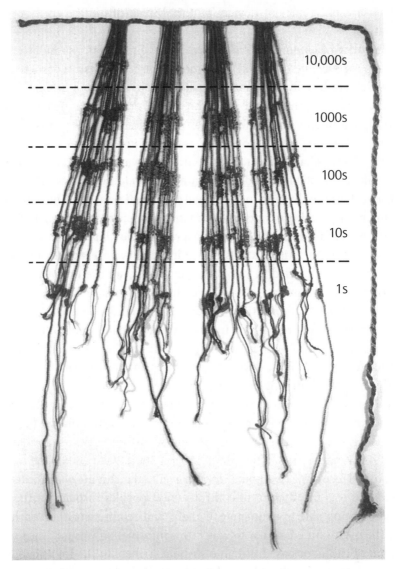

Figure 1.2.3 A clustering of knots on a *khipu* in decimal hierarchy

Inka knotted-cords was as precise and complex a system of recording quantitative data as the written Hindu-Arabic numeral-based recording system of Europeans at the time of the conquest, although the records of the former were not as rapidly produced, nor as easily changeable, as those of the latter.

Richardson (1987, 341) has argued that accounting has long been one of the principal institutions and administrative practices involved in maintaining and legitimizing the status quo in western European nation-states. Can this be said of *khipu* accounting in the pre-Hispanic Andes as well? We gain a perspective on this question by looking at two accounts of how censuses were carried out in the

Inka state. As in other ancient societies, census-taking was a vital practice in the Inka strategy of population control, as well as serving as the basis for the assessment and eventual assignment of laborers in the *mit'a* (taxation by labor) system (Murra 1982; Julien 1988). The first account of census-taking is from the famed mid-sixteenth century soldier and traveller, Cieza de León:

the nobles in Cuzco told me that in olden times, in the time of the Inka kings, it was ordained of all the towns and provinces of Peru that the head men [*señores principales*] and their delegates should [record] every year the men and women who had died and those who had been born; they agreed to make this count for the payment of tribute, as well as in order to know the quantity of people available to go to war and the number that could remain for the defense of the town; they could know this easily because each province, at the end of the year, was ordered to put down in their *quipos*, in the count of its knots, all the people who had died that year in the province, and all those that had been born.[6] (Cieza de León 1967 [1551], 62; my translation)

Some forty years after Cieza wrote down the information cited above, Martín de Murúa gave an account of Inka census-taking that varies somewhat from Cieza's understanding of this process and that contains interesting details concerning the actual procedures involved in local population counts.

They sent every five years *quipucamayos* [*khipu*-keepers], who are accountants and overseers, whom they call *tucuyricuc*. These came to the provinces as governors and visitors, each one to the province for which he was responsible and, upon arriving at the town he had all the people brought together, from the decrepit old people to the newborn nursing babies, in a field outside town, or within the town, if there was a plaza large enough to accommodate all of them; the *tucuyricuc* organized them into ten rows ['streets'] for the men and another ten for the women. They were seated by ages, and in this way they proceeded [with the count]…[7] (Murúa 2004 [1590], 204; my translation)

Late sixteenth-century drawings—what we could term 're-imaginings'—of these male and female accounting events from the chronicle of Martín de Murúa, are shown in Figs. 1.2.4 and 1.2.5.

One would be hard put to find better examples than the two quotations cited above, and the images of census events in Figs. 1.2.4 and 1.2.5, of what Michel

6. …concuerdan los orejones que en el Cuzco me dieron la relación, que antiguamente, en tiempo de los reyes Incas, se mandaba por todos los pueblos y provincias del Perú que los señores principales y sus delegados supiesen cada año los hombres y mugeres que habían sido muertos y todos los que habían nacido; porque, así para la paga de los tributes como para saber la gente que había para la Guerra y la que podia quedar por defensa del pueblo, convenía que se tuviese ésta [cuenta]; la cual fácilmente podían saber porque cada provincia, en fin del año, mandaba asentar en los quipos por la cuenta de sus nudos todos los hombres que habían muerto en ella en aquel año, y por el [con]siguiente los que habían nacido.

7. Enviaba de cinco a cinco años *quipucamayos*, que son contadores y veedores, que ellos llaman *Tucuyricuc*. Estos venían por sus provincias como gobernadores y visitadores, cada uno en las que le cabía, y llegado al pueblo hacía juntar toda la gente, desde los viejos decrépitos hasta los indios niños de teta y en una pampa o plaza, si la había, hacían estos gobernadores, llamados *Tucuyricuc*, señalar diez calles para los indios y otras diez para las indias, con mucho orden y concierto, en que por las edades ponían los dichos indios con mucha curiosidad y concierto…

Figure 1.2.4 Conducting a census count of men, by age-grade (Murúa 2004, 114v)

Figure 1.2.5 Conducting a census count of women, by age-grade (Murúa 2004, 116v)

Foucault characterized as the disciplinary power of state institutions—what he termed power/knowledge structures—as they attend to the work of social surveillance and the control of the bodies of subjects. The result of such procedures was, as Foucault noted, the production of subjects that cooperate and connive in their own subjection (Foucault 1977, 184–187; Hoskin and Macve 1986, 106; Stewart 1992). In Inka census-taking, people were ordered into public spaces to be counted and classified. Although resistance and evasion may have been common in such proceedings, from what the Spanish chroniclers and administrative officials tell us, Inka censuses were accomplished using non-coercive measures— that is, local people apparently were compliant with the claims of authority coming from local officials and state administrators. Such surveillance, reporting, and social control procedures are examples of what Foucault termed a disciplinary, as opposed to sovereign, form of power.

Sovereign power is identified as a diminished form of power. Its ultimate recourse is seizure—of things, of bodies and ultimately of life. Disciplinary power is much richer and entails penetrating into the very web of social life through a vast series of regulations and tools for the administration of entire populations and of the minutiae of people's lives. (cited in Miller and O'Leary 1987, 238)

Thus, as much as an accounting tool, the census *khipu* was an instrument for the performance and display of state authority and power within local communities.[8] The census data collected by local record-keepers were knotted into *khipus*, copies were made of each record, and the data were subsequently reported to higher-level accountants in regional and provincial administrative centers (see Urton and Brezine 2005). Two issues arise with respect to these procedures: one concerns the practice of making one or more copies of *khipu* records, the other concerns the training and education of state record-keepers.

While there are a number of references in the Spanish chronicles to *khipu* copies, the study of such copies in the corpus of extant *khipus* has proceeded slowly. Recent advances have come about, however, following the development of a searchable database—the *Khipu* Database (KDB).[9] From searches of the 450 or so samples included in the KDB, some 12–15 examples of copies of accounts have been identified (Urton 2005). While referred to as duplicate, or 'matching' *khipus*, we could also consider 'pairs' of *khipus* to represent an original and a copy.

Copies (or matching) *khipus* occur in three different forms. First, there are examples in which the numerical values on a sequence of strings on one sample

8. Guevara-Gil and Salomon (1994) have discussed what were similar procedures, and results, in the censuses undertaken by Spanish *visitadores* (administrative 'visitors') who were responsible for counting, classifying, and (re-)organizing local populations in the early colonial Andes.

9. The *Khipu* Database project (KDB), located in the Department of Anthropology, Harvard University, is described fully on the project website <http://khipukamayuq.fas.harvard.edu/>. I gratefully acknowledge the following research grants from the National Science Foundation, which made the creation of the KDB possible: #SBR-9221737, BCS-0228038, and BCS-0408324. Thanks also to Carrie J Brezine, who served as *Khipu* Database Manager from 2002 to 2005.

are repeated exactly on another *khipu*. In some samples of this type, we find that while the pair of *khipus* bears the same knot values, the colors of the strings may vary (see Urton 2005, 150–151). The second type of matching *khipus*, which I have termed 'close matches', involves instances in which two different samples contain not exactly matching sequences of numbers, but rather ones in which the values are similar (for example, those of one sample varying a small amount from those on another sample). And, finally, we have examples in which a numerical sequence recorded on one cord section of a *khipu* are repeated exactly, or closely, on another section of cords of that same *khipu*.

I argued elsewhere (Urton 2005) that duplicate *khipus* may have been produced as a part of a system of 'checks and balances'. However, duplicates seem also to possess most of the requisite elements of double-entry bookkeeping in which 'all transactions were entered twice, once as a debit and once as a credit…The debit side pertained to debtors, while the credit side pertained to creditors' (Carruthers and Espeland 1991, 37). Close matches would be accounts in which the debits and credits sides of the ledger were not in balance. On pairs of *khipus* having identical numerical values on sequences of strings, but in which string colors vary (Urton 2005, 150–151), color could have been used to signal the statuses of credits and debits in the matching accounts.[10] In the Inka state, debit/credit accounting would have been employed primarily in relation to the levying of labor tribute on subject populations.

The principal information that we lack in order to be able to confirm whether or not duplicate *khipus* might have been produced and used as double-entry-like accounts are the identities of the objects recorded on the *khipus*. Since we still cannot read the code of the *khipus*, we are unable to determine whether paired accounts were simply copies made for the purposes of checks-and-balances or if they might represent a relationship between a debit for an item on one account and the credit for that same item on another account. Research into this matter is on-going.[11]

What can we say about the individuals who became *khipu*-keepers for the state? How were these individuals recruited and trained? What role did they play in exercising authority and maintaining social and political control in the Inka state? The late sixteenth-century chronicler Martín de Murúa provided the

10. It is interesting to note that in early Chinese bookkeeping, red rods signified positive numbers while black rods were used for negative numbers. As Boyer noted, '[f]or commercial purposes, red rods were used to record what others owed to you and black rods recorded what you owed to others' (cited in Peters and Emery 1978, 425).

11. Three articles published in the 1960s and 1970s by economists and accounting historians contain a lively debate not only about whether or not the *khipus* contained double-entry bookkeeping, but about the claim made by one of the disputants (Jacobsen) to the effect that the Inkas may in fact have invented the technique (Jacobsen 1964; Forrester 1968; Buckmaster 1974). There is not space here to review the arguments made in these three articles. Suffice it to say that, while interesting for historical purposes, these articles are all poorly informed about the nature of the *khipus*, about what the Spanish documents say about their use, as well as about Inka political and economic organizations.

following account of a school that was set up in the Inka capital of Cusco for the training of *khipu*-keepers.

The Inca…he set up in his house [palace] a school, in which there presided a wise old man, who was among the most discreet among the nobility, over four teachers who were put in charge of the students for different subjects and at different times. The first teacher taught the language of the Inca…and upon gaining facility and the ability to speak and understand it, they entered under the instruction of the next [second] teacher who taught them to worship the idols and the sacred objects [*huacas*]…In the third year the next teacher entered and taught them, by use of *quipus*, the business of good government and authority, and the laws and the obedience they had to have for the Inca and his governors…The fourth and last year, they learned from the other [fourth] teacher on the cords and *quipus* many histories and deeds of the past.[12] (Murúa 2001, 364; my translation).

The curriculum of the young administrators aimed at engendering loyalty to the Inka and adherence to state values, policies, and institutions. The *khipu* studies component of the administrative curriculum fulfilled what Miller and O'Leary (1987) have referred to as accounting education's objective of producing 'governable persons' who themselves went on to administer for the state in the provinces. The curriculum also incorporated what has been described as a process whereby examination, discipline, and accounting are bound together to empower texts, rationalize institutional arrangements for state interests and, ultimately, to transform the bodies of the persons subjected to training (Hoskin and Macve 1986, 107).

The situation outlined above was not to last for long, as less than half a century after the school of administration was set up, a cataclysmic event brought the school, not to mention the entire imperial infrastructure, crashing down; this event was the Spanish conquest.

Conquest, colonization, and the confrontation between knot- and script-based texts

The story of the conquest of the Inka empire by the Spaniards, which was undertaken by Francisco Pizarro and his small force of around 164 battle-hardened *conquistadores*, beginning in 1532, has been told too many times—in all its astonishing

---

12. Dijo el Ynga…puso en su casa una escuela, en la cual presidía un Viejo anciano, de los más discretos *orejones*, sobre cuatro maestros que había para diferentes cosas y diferentes tiempos de los discípulos. El primer maestro enseñaba al principio la lengua del Ynga…Acabado el tiempo, que salían en ella fáciles, y la hablaban y entendían, entraban a la sujeción y doctrina de otro maestro, el cual les enseñaba a adorar los ídolos y sus *huacas*…Al tercer año entraban a otro maestro, que les declaraba en sus *quipus* los negocios pertenecientes al buen gobierno y autoridad suya, y a las leyes y la obediencia que se había de tener al Ynga y a sus gobernadores…El cuarto y postrero año, con otro maestro aprendían en los mismos cordeles y *quipus* muchas historias y sucesos antiguos…

and entrancing/appalling details—for me to add much to the telling in the space available here (see Hemming 1970). The events of the conquest and the processes of colonization that are relevant for our discussion here are the following. The initial battle of conquest, which occurred in November 1532 in the Inka provincial center of Cajamarca, in the northern highlands of what is today Peru, resulted in the defeat of the Inka army and the capture and execution of Atahualpa, one of two contenders for succession to the Inka throne. Pizarro then led his small force southward, arriving in the Inka capital city of Cuzco in 1534. The Spaniards and their native allies were soon forced to defend Cuzco against a rebellion led by the Spanish-installed puppet-king, Manco Inca. This gave rise to a decades-long war of pacification of the rebels, which finally came to an end in 1572 with the execution of the then rebel leader, Tupac Amaru (Hemming 1970).

Three years prior to the capture and execution of Tupac Amaru, a new Viceroy of Peru (the fourth), Francisco de Toledo, had arrived in Peru with a mandate to put down the rebellion and to transform the war- and disease-ravaged land of the former Inka empire into an orderly and productive colony for the benefit of the king of Spain, Philip II. Viceroy Toledo instituted a set of reforms that were in some respects a continuation of certain of the processes of pacification, reorganization, and transformation that had been on-going since the earliest days following the initial conquest. In other ways, Toledo's reforms represented something completely new, different, and profoundly transformative in their effects on Andean ways of life (Stern 1993, 51–79).

The end result of the Toledan reforms, the clear shape of which became manifest by the mid-to-late 1570s, included, most centrally, the following institutions: *encomiendas*—grants of groups of Indians to Spanish *encomenderos* 'overseers' who were charged with the care and religious indoctrination of the natives and who, in exchange, had the right to direct native labor for their personal benefit but without the right (after the Toledan reforms) to levy tribute demands on them; *corregimientos*—territorial divisions for the management and control of civil affairs, including (theoretically) oversight of the *encomenderos*; *reducciones*— newly-formed towns that were laid out in grid-like ground plans to which the formerly dispersed natives were transferred for their surveillance, control, and indoctrination; *doctrinas*—parish districts staffed by clergy who attended to the religious indoctrination of the natives within the reducciones and who received a portion of the tribute for their own maintenance; and *mita*—a form of labor tax based on the Inka-era *mit'a*, which supplemented what was, for Andeans, a new kind of tribute imposed on them by Toledo: specified quantities of agricultural produce, manufactured goods (textiles, sandals, blankets), and coinage (Rowe 1957; Ramírez 1996, 87–102).

The census was a critical institution for reorganizing Andean communities. Spanish censuses were carried out by administrative *visitadores* 'visitors' who produced documents, known as a *visitas*, which were detailed enumerations of the population in the *reducciones* broken down (usually) into household groupings. Each household member was identified by name, age and—in the case of adult males—*ayllu* 'social group' affiliation (Guevara-Gil and Salomon 1994; Urton 2006). The *visitadores* were usually joined in their rounds by the *kurakas* 'local lords' and often by the local *khipukamayuqs*. The *khipu*-keepers could supply historical, corroborating information on population figures and household composition (Loza 1998). It is important to stress that participation by the native record-keepers was not primarily for the benefit of the Spaniards, rather, it was to ensure that the natives would have their own, *khipu*-based accounts of the enumeration in the event—which seems always and everywhere to have come to pass—that a dispute arose over the population count, the amount of tribute levied, or other administrative questions.

There are two contexts in which I will explore native Andean encounters with Old World mathematical principles and practices, each of which was linked to a wholly new relation of authority and power: the manner of collecting information pertaining to the censuses, and the striking and circulation of coinage. These practices were closely linked to new forms of tribute, as well as to what was, for Andean peoples, a completely new form of communication: writing—that is, the inscribing of marks in ordered, linear arrangements on paper, parchment, or some other two-dimensional surface. Such a medium and associated recording technology were unprecedented in the Andean world.

There have been numerous important works published in recent years on the confrontation between *khipu* records and alphabetic texts in the early colonial Andes (Rappaport and Cummins 1994, 1998; Mignolo 1995; Brokaw 1999; Quilter and Urton 2002; Fossa 2006; Quispe-Agnoli 2006). That this body of works responds to what was, in fact, an area of intense interest and concern on both sides of an initially starkly drawn dual—native/Spaniard—world of social interactions and power relations is confirmed by the documentation detailing initial efforts by the Spaniards to establish an orderly colony in the former Inka territories. Central to this process from the 1540s through the 1570s was a program of enumerating the native population, investigating its form(s) of organization, and beginning to sketch out its history. One form that this process took was to call the *khipu*-keepers before colonial officials and have them read the contents of their cords (Loza 1998; Urton 1998). These recitations were made before a *lengua* 'translator'; the Spanish words spoken by the translator were written down by a scribe. This activity resulted in the production of written transcriptions in Spanish alphanumeric script of the census data and other information previously jealously guarded by the *khipu*-keepers in their cords.

Many of the *khipu* transcriptions discovered to date have been assembled in an important collection, entitled *Textos Andinos* (Pärssinen and Kiviharju 2004).

While these documents have been studied in terms of the displacement, and eventual replacement, of *khipu* 'literacy' by alphabetic literacy, what has received virtually no attention to date is the equally striking information they contain with respect to the confrontation between Inka knot-based numeration and Spanish grapheme-based written numerals and mathematics. How and what did individuals on either side of this confrontation think about the translation of quantitative values from knotted-cords to written texts?

Fig. 1.2.6a shows an image of a *khipu* juxtaposed to an *unrelated khipu* transcription, in Fig. 1.2.6b.[13] It is important to stress that we do not have an actual match—such as that suggested in Figs. 1.2.6a and b—between an extant *khipu* and a transcription of that same sample. As for the *khipu* in Fig. 1.2.6a, we are able to read the knot values of this sample and thereby interpret the numerical information encoded on this sample. We assume that the identities of the objects accounted for in this *khipu* were represented in a constellation of elements, including color, structure, and perhaps numbers interpreted as labels (Urton 2003). In the (unrelated) *khipu* transcription in Fig. 1.2.6b, the text is organized line by

CMA-850
LC1-479

Figure 1.2.6 a) A *khipu* from Centro Mallqui, Leymebamba, Amazonas, Peru (#CMA 850/LC1–479 [UR9])

13. The *khipu* sample shown in Fig.1.2.6a is from the site of Laguna de los Cóndores, in the area of Chachapoyas, northern Peru (#CMA 850/LC1–479; in the 'Data table' page of the KDB website, this is sample UR9). The *khipu* transcription shown in Fig. 1.2.6b is from a tribute *khipu* from Xauxa, in the central Peruvian highlands, dating to 1558 (AGI, Lima 205, no. 16 folio 10r; see Pärssinen and Kiviharju 2004, 172–173).

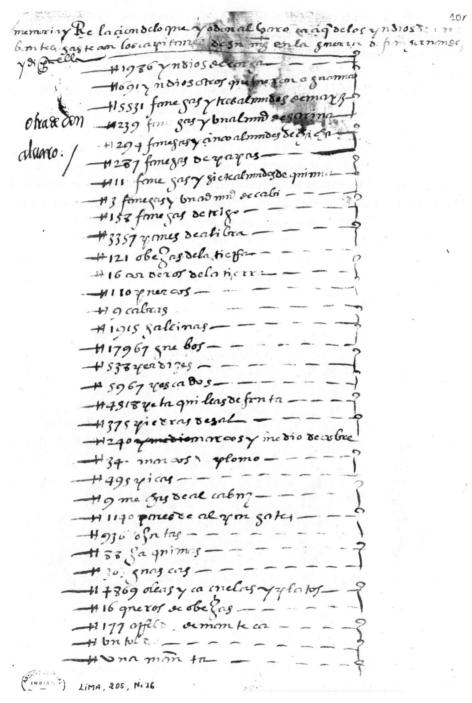

Figure 1.2.6 b) A *khipu* transcription (AGI, Lima 205, no. 16, folio 10r)

line, as the *khipu* itself is organized string by string; each line in the transcription contains a number followed by the name/identity of the object enumerated on that string in the *khipu* from which the transcription was derived.

The numerical values recorded in the *khipu* shown in Fig. 1.2.6a are similar in their range of magnitudes and distribution to those found in early Spanish census accounts in the Andes (Urton 2006). The *khipu* transcription illustrated in Fig. 1.2.6b is a tribute account recorded in the valley of Xauxa, in 1558 (Pärssinen and Kiviharju 2004, 172–173). If Fig. 1.2.6a were the *khipu* from which the transcription in Fig. 1.2.6b was drawn up (which it is not), we assume (but do not know for certain) that there would be a parallelism between number signs and object identity signs that would form a bridge across the semiotic—nominative and quantitative—divide separating these two species of texts.

Not surprisingly, almost all of the information we have in order to address the question of how Andean people thought about *khipus* and their translation and transcription into written texts comes to us from the Spanish side of the equation. The Spaniards were at least initially respectful of the *khipus* and their keepers, as the *khipus* were the primary sources of information on the basis of which Spanish officials began to erect the colonial administration. The most important point that should be made for our interest here concerning the juxtaposition of documents in Figs. 1.2.6a and b is that not only information, but *authority* as well, was located initially in the *khipu* member of the *khipu*/transcription pair juxtaposed in Fig. 1.2.6. However, once the information was transferred from *khipus* to written texts, the locus of textual authority, legitimacy, and power began to shift toward the written documents.

While many native Andeans learnt how to read and write alphabetic script and how to manipulate Hindu-Arabic number signs, only a handful of Spaniards appear to have achieved any degree of familiarity with the *khipus* (Pärssinen 1992, 36–50); it appears that no Spaniard became truly proficient at manipulating and interpreting the cords (Urton 2003, 18–19). What this meant was that, rather than contests over interpretations of information contained in the two sets of documents coming down to reciprocal readings of the two sets of texts, what emerged between the 1540s and the 1570s were separate, contested readings by the keepers of the two different text types before a Spanish judge. As disputes intensified, and as more and more original data were recorded uniquely in the written documents, the *khipu* texts became both redundant and increasingly troublesome for the Spaniards (Platt 2002). By the end of the tumultuous sixteenth century, *khipus* had been declared to be idolatrous objects—instruments of the devil—and were all but banned from official use.[14]

---

14. The *khipus* were declared idolatrous objects and their use was severely proscribed by the Third Council of Lima, in 1583 (Vargas Ugarte 1959). However, the *khipus* continued to be used for local record-keeping purposes—in some cases down to the present day (see Mackey 1970; Salomon 2004).

The circulation of coins is another area in which Andeans were confronted with a completely new and unfamiliar terrain of political relations, economic activity, and shifting relations of authority over the course of the early colonial period. The first mint in South America was formally established in Lima in 1568, just 36 years after the events of Cajamarca. The royal decree that controlled the weights, fineness, and the fractional components of the coins to be struck in Lima—the *real* and the *escudo*—were issued by Ferdinand and Isabella in 1479 and amended by Charles V in 1537 (Craig 1989, 2). The first coins struck in Lima bore a rendering of the Hapsburg coat of arms on the obverse and a cross with castles and lions on the quartered face on the reverse (Craig 1989, 6).

As noted, the two coin types were the *real*, a silver coin, and the gold *escudo*. Each of these coin types was broken down into subunits, each of which was valued in relation to a general, unified standard of valuation known as the *maravedí*. The latter was not a coin but rather it was what Moreyra Paz Soldán (1980, 66) terms *moneda imaginaria y de cuenta* 'imaginary money of account'. The *maravedí* was used to coordinate values between different types of coins as determined by material differences and subdivisions of standard units (for example, the silver *real* = 34 *maravedís*; the gold *escudo* = 350 [from 1537–1566] or 400 [after 1566] *maravedís*). From this primary coordinating function, the *maravedí* served as a common denominator that permitted the interrelating of heterogeneous monetary values pertaining to gold and silver. For example, until 1566, the *maravedí* coordinated the value of silver to gold at 11.5 to 1; after 1566 the ratio was 12.12 to 1 (Moreyra Paz Soldán 1980, 66–67; Craig 1989, 2).

What did any of the above have to do with Andean peoples? How were they to understand the meaning of these words and concepts? To understand the force of these questions, we can begin by imagining how one might go about translating the previous two paragraphs into a language like those spoken by large numbers of people throughout the Andes in the first few decades following the conquest, such as Quechua, Aymara, Puquina, or Yunga. They did not have terms for money or coinage, much less a term like *maravedí*, and had formed such concepts as 'value', 'heterogeneity', and 'account' in the absence of markets and a monetary economy (Murra 1995). It is clear in this case where authority would quickly come to reside in any dispute that might arise over the exchange value of any one of the several coin types in this system that would have begun to circulate through Andean communities by the 1570s. But we are getting ahead of ourselves.

From almost the earliest years following the conquest, Spanish officials in the countryside (the *encomenderos*) had been levying tribute in kind, which in some places included a demand for plates of silver and bars of gold, and translating the value of these items into Spanish currency values (Ramírez 1996, 92–112). Spanish officials regularly produced documents translating the quantities of items of tribute in kind into values in *pesos ensayados* (a unit of value in silver currency). This was the main context within which the *kurakas* 'local lords' in communities would have begun to encounter translations of the use-value of objects, which they were familiar with in their local non-monetized economies, into exchange-values stated in terms of currency equivalents (Spalding 1973). Furthermore, the Viceroy Francisco de Toledo introduced in the mid-1570s a new tribute system, which included not only produce and manufactured goods but also coins; the sum to be given yearly by each tributary was four-to-five *pesos ensayados* (that is, coinage in *plata ensayada* 'assayed silver'). Tribute payers were designated as male heads of households between the ages of 18–50. The native chronicler, Guaman Poma de Ayala, drew several images of native people paying their tribute using what appears to be coinage bearing the quartered reverse face of the *cuatro reales* (Fig. 1.2.7).[15]

People in communities—the newly-built *reducciones*—were able to acquire coins to pay their tribute from forced work in the mines (another component of the Toledan tributary system), as well as from marketing and wage labor. The engagements with currency that resulted from these activities required people to begin to think about the different units of coinage, shifting equivalencies between coinage units, as well as to accommodate themselves to fluctuations in currency values in the periodic currency devaluations and the debasement of coinage that took place during the colonial period. The act of 'devaluing' currency is a claim of authority on the part of some entity (such as the state) over the exchange-value of the coinage one holds in one's own purse. One's subsequent use of that same coinage according to the newly announced rate of exchange represents compliance with the claim by the entity in question to control the value of one's currency. Although we have almost no data on the basis of which to consider how Andean peoples responded to such changes (see Salomon 1991), these were some of the processes that were transpiring on the front lines of the confrontation between Old and New World mathematics and notions and relations of authority in the early colonial Andes.

Conclusions

We began this exploration by asking about the relevance and salience of a characterization of mathematics as 'the secret weapon of cultural imperialism' (Bishop

---

15. See the study by Salomon (1991) of one of the few references in the colonial literature to the engagement with coinage (*la moneda de cuatro reales*) by a native Andean during the colonial period. Salomon argues that the story, which appears in a well-known manuscript from Huarochirí (Salomon and Urioste 1991), is concerned with the internal conflicts of a man due to the competing religious sentiments he experiences over loyalty to a local deity (*huaca*) and the Christian deity. The narrative plays on the precise symbolism of images, as well as the lettering, on a quartered Spanish coin.

Figure 1.2.7 Paying tribute with coin bearing a quartered design (Guaman Poma de Ayala 1980, 521[525])

1990). Having now looked at several aspects of arithmetic, mathematics, and accounting in western Europe and the Andes during the period leading up to, and a century or so beyond, the fateful encounter between Pizarro and Atahualpa in Cajamarca in 1532, we return to ask: in what sense was mathematics linked to authority, power, and legitimacy in this historical conjuncture?

I argue that the answer to the question posed above is found in the same rationale and set of explanations that explain who writes history and who determines truth in history. The answer to both of these questions is: the conqueror. This is not because the conqueror knows what is, in fact, true; rather, it is because the conqueror possesses the power to speak, and to represent and establish the rules of the game as it is to be played from that moment forward. This is the case not only in terms of narrating and writing the events of history and explaining their causes (Urton 1990), but also in taking the measure of the world and accounting for those measurements—geographic, demographic, economic, and so on—for as long as the dominant group holds power.

Power, which is intimately linked to the exercise of authority, takes many forms. In its most extreme and, paradoxically, weakest form, power is maintained by force. As Foucault has shown more clearly than any recent political theorist, the most effective species of power is that which takes shape as individuals and groups become complicit with and participate in institutions of the state, such as in censuses, regulatory and corrective institutions, and accounting (Foucault 1979, 140–141; Stewart 1992; Smart 2002, 102–103). What is the place of mathematics in this Foucauldian, 'genealogical' conception of power and authority? I think that here we must return to the question of the certainty of mathematics, and of how that certainty relates to truth and, ultimately, to power. I suggest that the critical observation on these matters for our purposes here is that mathematics may be made to serve, although it itself is not responsible for giving rise to, regimes of power. A 'regime of power' may be manifested in the trappings of a king's court, in the ministrations of a priestly hierarchy, or in complex 'book-keeping' procedures—such as bundles of knotted cords in the hands of individuals authorized to record information (numerical and otherwise) in the interests of the state.

Bibliography

Aho, James, *Confession and bookkeeping: the religious, moral and rhetorical roots of modern accounting*, State University of New York, 2005.

Arellano, Carmen, 'Quipu y tocapu: sistemas de comunicación incas', in Franklin Pease (ed), *Los incas: arte y símbolos*, Lima: Banco de Crédito del Perú, 1999, 215–261.

Ascher, Marcia, 'Before the conquest', *Mathematics Magazine*, 65 (1992), 211–218.

Ascher, Marcia, and Ascher, Robert, *Mathematics of the Incas: code of the quipu*, Dover, 1997.

Bishop, Alan J, 'Western mathematics: the secret weapon of cultural imperialism', *Race and Class*, 32/2 (1990), 51–65.

Brokaw, Galen, Transcultural intertextuality and quipu literacy in Felipe Guaman Poma de Ayala's *Nueva corónica y buen gobierno*, PhD Dissertation, Indiana University, Ann Arbor: UMI Dissertation Services, 1999.

Brown, R Gene, and Johnston, Kenneth S, *Paciolo on accounting*, McGraw-Hill Book Co, 1984.

Buckmaster, Dale, 'The Incan quipu and the Jacobsen hypothesis', *Journal of Accounting Research*, 12 (1974), 178–181.

Carruthers, Bruce G, and Espeland, Wendy N, 'Accounting for rationality: double-entry bookkeeping and the rhetoric of economic rationality', *The Journal of American Sociology*, 97 (1991), 31–69.

Cieza de León, Pedro de, *El señorio de los Incas*, Lima: Instituto de Estudios Peruanos, 1967.

Conklin, William J, 'A *khipu* information string theory', in Jeffrey Quilter and Gary Urton (eds), *Narrative threads: accounting and recounting in Andean khipu*, University of Texas Press, 2002, 53–86.

Craig, Freeman, Jr, 'Coinage of the viceroyalty of el Perú', in William I Bischoff (ed), *The coinage of el Perú*, The American Numismatic Society, 1989.

Donoso Anes, Rafael, 'The double-entry bookkeeping method applied in Spain to account for transactions related to the minting process of gold and silver in the sixteenth century', *The Accounting Historians Journal*, 21 (1994), 97–116.

Durham, John W, 'The introduction of "Arabic" numerals in European accounting', *The Accounting Historians Journal*, 19 (1992), 25–55.

Ellerman, David P, 'The mathematics of double entry bookkeeping', *Mathematics Magazine*, 58 (1985), 226–233.

Forrester, D A R, 'The Incan contribution to double-entry accounting', *Journal of Accounting Research*, 6 (1968), 283.

Fossa, Lydia, *Narrativas problemáticas: los Inkas bajo la pluma española*, Lima: Instituto de Estudios Peruanos, 2006.

Foucault, Michel, *Discipline and punish*, Allen Lane, 1977.

Foucault, Michel, *The history of sexuality*, vol 1, Allen Lane, 1979.

Gómez Cañedo, Lino, *Los archivos de la historia de America: período colonial español*, Mexico, DF, 1961.

Guaman Poma de Ayala, Felipe, *El primer nueva corónica y buen gobierno*, (trans J L Urioste, eds John V Murra and Rolena Adorno), 3 vols, Siglo Veintiuno, 1980.

Guevara-Gil, Armando, and Salomon, Frank, 'A "personal visit": colonial political ritual and the making of Indians in the Andes', *Colonial Latin American Review*, 3 (1994), 3–36.

Hemming, John, *The conquest of the Incas*, Harcourt Brace Jovanovich, 1970.

Hoffman, Paul E, 'The computer and the colonial treasury accounts: a proposal for a methodology', *The Hispanic American Historical Review*, 50 (1970), 731–740.

Hoskin, Keith W, and Macve, Richard H, 'Accounting and the examination: a genealogy of disciplinary power', *Accounting, Organizations and Society*, 11 (1986), 105–136.

Jacobsen, Lyle E, 'The ancient Inca empire of Peru and the double entry accounting concept', *Journal of Accounting Research*, 2 (1964), 221–228.

Julien, Catherine, 'How Inca decimal administration worked', *Ethnohistory*, 35 (1988), 257–279.

Klein, Herbert S, and Barbier, Jacques A, 'Recent trends in the study of Spanish American colonial public finance', *Latin American Research Review*, 23 (1988), 35–62.

Lapeyre, Henri, *Une famille de marchands, les Ruiz*, Paris, 1955.

Locke, L Leland, *The ancient quipu or Peruvian knot record*, American Museum of Natural History, 1923.

Loza, Carmen Beatriz, 'Du bon usage des quipus face a l'administration coloniale Espagnole, 1553–1599', *Population,* 53 (1998), 139–160.

Mackey, Carol, *Knot records of ancient and modern Peru,* PhD Dissertation, University of California, Berkeley, Ann Arbor: University Microfilms, 1970.

Mignolo, Walter D, *The darker side of the Renaissance: literacy, territoriality, and colonization,* University of Michigan Press, 1995.

Miller, Peter, 'On the Interrelations between Accounting and the State', *Accounting, Organizations and Society,* 15 (1990), 315–338.

Miller, P, and O'Leary, T, 'Accounting and the construction of the governable person', *Accounting, Organizations and Society,* 12 (1987), 235–265.

Moreyra Paz Soldán, Manuel, *La moneda colonial en el Perú,* Lima: Banco Central de la Reserva, 1980.

Murra, John V, 'Las etno-categorías de un *khipu* estatal', in *Formaciones económicas y políti-cas en el mundo andino,* Lima: Instituto de Estudios Peruanos, 1975, 243–254.

Murra, John V, 'The *mit'a* obligations of ethnic groups to the Inka state', in George A Collier, Rosaldo Renato, and John D Wirth (eds), *The Inca and Aztec states, 1400–1800: anthropology and history,* Academic Press, 1982, 237–262.

Murra, John V, 'Did tribute and markets prevail in the Andes before the European invasion?' in Brook Larson and Olivia Harris (eds), *Ethnicity, markets, and migration in the Andes: at the crossroads of history and anthropology,* Duke University Press, 1995, 57–72.

Murúa, Fray Martín de, *Historia general del Perú,* (ed Ballesteros Gaibrois), Dastin Historia, 2001.

Murúa, Fray Martín de, *Códice Murúa - historia y genealogía de los reyes Incas del Perú del Padre Mercenario Fray Martín de Murúa (Códice Galvin),* (ed Juan Ossio), Testimonio Compañía Editorial, SA, 2004.

Pärssinen, Martti, *Tawantinsuyu: the Inca state and its political organization* (Studia Historica, 43), Societas Historica Finlandiae, 1992.

Pärssinen, Martti, and Kiviharju, Jukka (eds), *Textos Andinos: corpus de textos khipu Incaicos y Coloniales,* vol I, Instituto Iberoamericano de Finlandia and Universidad Complutense de Madrid, 2004.

Pereyra, Hugo, 'Notas sobre el descubrimiento de la clave numeral de los quipus incaicos', *Boletín del Museo de Arqueología y Antropología,* 4 (2001), 115–123.

Peters, Richard M, and Emery, Douglas R, 'The role of negative numbers in the development of double entry bookkeeping', *Journal of Accounting Research,* 16 (1978), 424–426.

Platt, Tristan, '"Without deceit or lies": variable *chinu* readings during a sixteenth-century tribute-restitution trial', in Jeffrey Quilter and Gary Urton (eds), *Narrative threads: accounting and recounting in Andean khipu,* University of Texas Press, 2002, 225–265.

Quilter, Jeffrey, and Urton, Gary, (eds), *Narrative threads: accounting and recounting in Andean khipu,* University of Texas Press, 2002.

Quispe-Agnoli, Rocío, *La fe andina en la escritura: resistencia e identidad en la obra de Guaman Poma de Ayala,* Lima: Universidad Nacional Mayor de San Marcos, 2006.

Radicati di Primeglio, Carlos, *Estudios sobre los quipus,* (ed and intro Gary Urton), Fondo Editorial Universidad Nacional Mayor de San Marcos, 2006.

Ramírez, Susan E, *The world upside down: cross-cultural contact and conflict in sixteenth century Peru,* Stanford University Press, 1996.

Rappaport, Joanne, and Cummins, Thomas, 'Literacy and power in colonial Latin America', in George C Bond and Angela Gilliam (eds), *Social construction of the past: representation as power,* London and Routledge, 1994, 89–109.

Rappaport, Joanne, and Cummins, Thomas, 'Between images and writing: the ritual of the king's *quillca*', *Colonial Latin American Review,* 7/1 (1998), 7–32.

Reitzer, Ladislas, 'Some observations on Castilian commerce and finance in the sixteenth century', *The Journal of Modern History*, 32 (1960), 213–231.

Richardson, Alan J, 'Accounting as a legitimating institution', *Accounting, Organizations and Society*, 12 (1987), 341–355.

Rowe, John H, 'The Incas under Spanish colonial institutions', *Hispanic American Historical Review*, 37 (1957), 155–199.

Salomon, Frank, 'La moneda que a don Cristóbal se le cayó: el dinero como elemento simbólico en el texto *Runa Yndio Ñiscap Machoncuna*', in Segundo Moreno and Frank Salomon (eds), *Reproducción y transformación de las sociedades andinas, siglos XVI–XX*, Quito: Ediciones ABYA-YALA, 1991, 481–586.

Salomon, Frank, *The cord keepers: khipus and cultural life in a Peruvian village*, Duke University Press, 2004.

Salomon, Frank, and Urioste, J L, *The Huarochirí manuscript: a testament of ancient and colonial Andean religion*, University of Texas Press, 1991.

Smart, Barry, *Michel Foucault*, Routledge, 2002.

Sombart, Werner, *The quintessence of capitalism* (trans M Epstein), Howard Fertig, 1967.

Spalding, Karen, 'Kurakas and commerce: a chapter in the evolution of Andean society', *The Hispanic American Historical Review*, 53 (1973), 581–599.

Spencer, Martin E, 'Weber on legitimate norms and authority', *The British Journal of Sociology*, 21 (1970), 123–134.

Stern, Steve J, *Peru's Indian peoples and the challenge of Spanish conquest*, University of Wisconsin Press, 1993.

Stewart, Ross E, 'Pluralizing our past: Foucault in accounting history', *Accounting, Organizations and Society*, 5 (1992), 53–73.

Swetz, Frank J, *Capitalism and arithmetic: the new math of the 15th century*, Open Court, 1989.

Uphoff, Norman, 'Distinguishing power, authority and legitimacy: taking Max Weber at his word by using resources-exchange analysis', *Polity*, 22 (1989), 295–322.

Urton, Gary, *The history of a myth: Pacariqtambo and the origin of the Inkas*, University of Texas Press, 1990.

Urton, Gary, 'A new twist in an old yarn: variation in knot directionality in the Inka *khipus*', *Baessler-archiv* Neue Folge, 42 (1994), 271–305.

Urton, Gary, *The social life of numbers: a Quechua ontology of numbers and philosophy of arithmetic*, University of Texas Press, 1997.

Urton, Gary, 'From knots to narratives: reconstructing the art of historical record-keeping in the Andes from Spanish transcriptions of Inka *khipus*', *Ethnohistory*, 45 (1998), 409–438.

Urton, Gary, 'A calendrical and demographic tomb text from northern Peru', *Latin American Antiquity*, 12 (2001), 127–147.

Urton, Gary, 'Recording signs in narrative-accounting *khipu*', in Jeffrey Quilter and Gary Urton (eds), *Narrative threads: accounting and recounting in Andean khipu*, University of Texas Press, 2002, 171–196.

Urton, Gary, *Signs of the Inka khipu: binary coding in the Andean knotted-string records*, University of Texas Press, 2003.

Urton, Gary, '*Khipu* archives: duplicate accounts and identity labels in the Inka knotted string records', *Latin American Antiquity*, 16 (2005), 147–167.

Urton, Gary, 'Censos registrados en cordeles con "amarres". Padrones poblacionales pre-his-pánicos y coloniales tempranos en los *khipu* Inka', *Revista Andina*, 42 (2006), 153–196.

Urton, Gary, and Brezine, Carrie J, '*Khipu* accounting in ancient Peru', *Science*, 309 (2005), 1065–1067.

Vargas Ugarte, Rubén, *Historia de la iglesia en el Perú*, vol 2 (1570–1640), Aldecoa, 1959.

Weber, Max, *The Protestant ethic and the spirit of capitalism* (trans Talcott Parsons), Scribner, 1958.

Weber, Max, *The theory of economic and social organization* (trans A M Henderson and Talcott Parsons), The Free Press, 1964.

Yamey, B S, 'Introduction', in A C Littleton and B S Yamey (eds), *Studies in the history of accounting*, Sweet and Maxwell, 1956, 1–13.

Zuidema, R Tom, 'Bureaucracy and systematic knowledge in Andean civilization', in George A Colliedr, R I Rosaldo, and John D Wirth (eds), *The Inca and Aztec states, 1400–1800*, Academic Press, 1982, 419–458.

Heavenly learning, statecraft, and scholarship: the Jesuits and their mathematics in China

Catherine Jami

The story of the transmission of mathematics from Europe to China in the early modern age is closely linked to that of the Jesuit mission from 1582 to 1773, which spanned the last decades of the Ming dynasty (1368–1644) and the Qing dynasty (1644–1911) from its advent to its apogee in the mid-eighteenth century.[1] For almost two centuries, the Jesuits put the sciences in the service of evangelization: their knowledge enhanced the prestige of their religion and opened the way first to the patronage of individual officials, and then to that of the state. This emphasis on the sciences as a tool for proselytization seems to have been unique at the time, both among the missionary orders present in China, and among Jesuit missions around the world (Standaert 2001, 309–354; Romano 2002). Even within the China mission, most Jesuits devoted their time and effort solely to evangelization, while only a few 'specialists' among them taught and practiced the sciences (Brockey 2007). However, it could be argued that Jesuits' science had a much more pervasive influence than their religion. Christianity remained a minority religion, even a marginal one.[2] On the other hand all scholars interested

---

1. For names and dates of people and dynasties, see Table 2 on page 80.
2. For the period under discussion, there were no more than about 200,000 Chinese Christians, with this maximum reached around 1700, when the Chinese population is estimated to have been about 150 million (Standaert 2001, 380–386).

in the mathematical sciences knew about *xi xue* 西學 'western learning' whatever their attitude towards it might have been.

Two factors contributed to shaping Jesuit science in China: on the one hand, the importance that the Society of Jesus gave to mathematics (in the usual sense of this term in early modern Europe)—what we might call the offer; on the other hand, the renewal of interest in *shi xue* 實學 'solid learning' among late Ming scholars—what we might call the demand. Accordingly, I shall first outline the place of mathematics in Jesuit education and briefly describe the state of mathematics in China around 1600. I will then go on to discuss translations of works on the mathematical sciences during the first decades of the mission, and recount how participation in the Calendar Reform of 1629 and integration in the Chinese civil service in 1644 shaped the Jesuits' practice and teaching of mathematics. Chinese responses to western learning entailed competing propositions for structuring the discipline. Mei Wending's 梅文鼎 integration of Chinese and western mathematics was the most elaborate reconfiguration of the field. His synthesis and the Kangxi emperor's appropriation of western science were instrumental in reshaping the landscape of mathematics in China.

Mathematics in the Jesuit curriculum

Founded in 1540, the Society of Jesus soon started setting up colleges across Europe. The sons of the elites of Catholic countries were educated in them, as were most members of the Society. The latter often trained to be teachers, and for some of them this remained their main occupation. The content and structure of the education provided by the Society were crucial in shaping Jesuit culture, in Europe as well as in China. Having previously studied the *trivium* (grammar, logic, and rhetoric), students entering a Jesuit college would typically begin with further training in rhetoric. This was followed by three years devoted to logic, philosophy, and metaphysics. Early in the Order's history, natural philosophy (or physics) and mathematics were both grouped under philosophy. According to the Aristotelian classification, physics and mathematics addressed two of the ten categories, respectively quality and quantity. Physics provided a qualitative explanation of natural phenomena; it was based on the four-elements theory, according to which all matter was composed of earth, air, fire, and wind, and the earth lay motionless at the centre of concentric crystalline spheres. In the scholastic tradition, mathematics consisted of the four disciplines of the *quadrivium*, namely arithmetic, music, geometry, and astronomy.

However, it was somewhat redefined in the Jesuit curriculum. The Roman College, founded in 1551, set the standards for the Society's educational network. The *Ratio studiorum* (final version 1599), which defined the Jesuit system of teaching, gave a new importance to mathematics (*Ratio studiorum* 1997). Christoph Clavius, architect of the Gregorian Calendar Reform of 1582, had taught mathematics at the Roman College since 1565, and was the first to hold the chair of mathematics there. He was instrumental in establishing it as a subject independent from philosophy and in asserting its status as a science (Baldini 1992; Romano 1999, 133–178). This was the outcome of a debate within the Society that was mainly epistemological. However the new importance of the mathematical arts in sixteenth-century Italy must also have played a role in the inclusion of mathematics in the subjects in which the Jesuits strove to be eminent (Gorman 1999, 172).

While establishing mathematics as an independent discipline in the Jesuit curriculum, Clavius redefined its structure and produced textbooks for its teaching. Following Proclus, he divided mathematics into 'pure' and 'mixed', the former consisting of arithmetic and geometry, the latter comprising six major branches (which were further divided into subordinate disciplines): natural astrology (astronomy), perspective, geodesy, music, practical arithmetic, and mechanics. This structure broadened the scope of mathematics and extended its fields of application (Engelfriet 1998, 30–32; Feldhay 1999, 110). The works authored by Clavius, first and foremost his editions of and commentaries on Euclid's *Elements* and Sacrobosco's *Sphaera* 'Sphere' (a thirteenth-century treatise on astronomy), as well as his textbooks on arithmetic and algebra, formed the basis of mathematical education as he defined it for the Society (Feldhay 1999, 109).

Jesuit education was not uniform: there were local variants in the mathematics taught,[3] and, as with any school curriculum, a number of updates occurred. Thus, after the 1620s, the Ptolemaic system defended and taught by Clavius was gradually replaced by the Tychonic system, in which the sun, while revolving around the earth, was the centre of the orbit of the planets (Baldini 2000, 77). By and large, the tradition Clavius had established was continued in the sense that many teachers produced their own textbooks that were conceived as continuations of his, though departing from them in their approach (Feldhay 1999, 114). Two examples are relevant to the mathematics transmitted to China. First, the number of textbooks entitled *Elements of geometry* produced in the seventeenth century, within and without the Society, was such that the phrase, and even the name of Euclid, came to refer to a genre—that of geometry textbooks—rather than merely to editions of the Greek classic. Second, whereas Clavius' *Algebra* was one of the last representatives of the medieval tradition of cossic algebra, in which the unknown and its powers are denoted by abbreviations of their names (Reich 1994), Viète's new notation, with vowels denoting the unknowns and consonants the given quantities, was introduced into Jesuit teaching in the 1620s

---

3. On Portugal see Leitão (2002); on France see Romano (1999, 183–354; 2006).

(Feldhay 1999, 116–126). Matteo Ricci, the first Jesuit to enter China at the end of the sixteenth century, had studied with Clavius at the Roman College and brought with him Clavius' mathematics, but some of his successors in the China mission would present mathematics as it evolved in Jesuit colleges later in the century.

Mathematics in Late Ming China: the 'Unified lineage of mathematical methods'

It is widely admitted that by 1600, the most significant achievements of the Chinese mathematical tradition had fallen into oblivion. The *Jiu zhang suan shu* 九章算術 'Nine chapters on mathematical procedures' (first century AD), regarded as the founding work of the Chinese mathematical tradition (Chemla and Guo 2004) and included in the *Suan jing shi shu* 算經十書 'Ten mathematical classics' (656), had effectively been lost. Furthermore the sophisticated *tian yuan* 天元 'celestial element' algebra developed in the thirteenth century had been forgotten.[4] The calculating device on which both were based, the counting rods, had fallen into disuse; the abacus had become the universally used calculating device.[5]

By contrast with this picture of decline in mathematics, some historians describe the sixteenth century as a 'second Chinese Renaissance'. In reaction against Wang Yangming's 王陽明 philosophy of the mind, which, around 1500, gave priority to introspection over concern with the outside world, as well as in response to a more and more perceptible political crisis, the last decades of the sixteenth century witnessed a strong renewal of interest in technical learning and statecraft (Cheng 1997, 496–530). The advocates of 'solid learning' emphasized the social role of literati, underlining that scholarship was of value only if it contributed to welfare and social harmony, while being grounded in verifiable evidence. At the same time, the lowering of the cost of printing resulted in a significant broadening of the book market, which facilitated the circulation of knowledge. The renewal in many fields of scholarship is exemplified by such major works as *Ben cao gang mu* 本草綱目 'Compendium of medical material' (1593) by Li Shizhen 李時珍, *Lü lü jing yi* 律呂精義 'Essential meaning of pitchpipes' (1596) by Zhu Zaiyu 朱載堉, and *Tian gong kai wu* 天工開物 'Exploitation of the works of nature' (1637) by Song Yingxing 宋應星. The mathematical treatise *Suan fa tong zong* 算法統宗 'Unified lineage of mathematical methods' (1592) by Cheng Dawei 程大位 can be regarded as belonging to this trend and is representative of

4. General accounts of the history of Chinese mathematics in western languages include Li and Du (1987), Martzloff (1997), Yabuuti (2000).

5. On counting rods see Volkov (1998); Lam and Ang (2004, 43–112). On the abacus see Jami (1998a).

the state of mathematics in China by 1600. It was to remain a bestseller to the end of the imperial age (1911).

In mathematics as in many other fields, late Ming scholars blamed the perceived loss of ancient traditions on their more recent predecessors' indulgence in self-centred and esoteric pursuits. Far from claiming to innovate radically, Cheng Dawei aimed at providing a compilation of earlier treatises that he had spent decades collecting. This is reflected in the book's title:[6] it is not unlikely that Cheng saw himself as the heir of a lineage of scholars versed in mathematics. In his work he gave a bibliography of all earlier works on the subject that he knew of. *Suan* 算 was the usual term to refer to mathematics, *fa* 法 referred to the methods by which each problem was solved; *suan fa* occurred in the title of many of the works known to Cheng Dawei. The 'Unified lineage of mathematical methods' represents a synthesis of a tradition of popular mathematics based on the abacus that can be traced back to Yang Hui 楊輝 (fl. 1261), in the Southern Song dynasty (Lam 1977; Yabuuti 2000, 103–121). This tradition is usually contrasted with the learned tradition of the Song and Yuan dynasty that culminated with celestial element algebra.

Like most of the predecessors known to him, Cheng Dawei referred to a canonical nine-fold classification of mathematics that can be traced back to the 'Nine chapters on mathematical procedures', although the book itself was evidently unavailable to him. In fact, during the late Ming and early Qing period, the phrase *jiu zhang* 九章 'nine chapters' mostly referred to that classification rather than to the classic work itself. But like most if not all authors before and after him, Cheng Dawei failed to fit all the mathematical knowledge at his command into the headings of the nine chapters: his work is divided into seventeen chapters. It opens with a general discussion of some ancient diagrams then thought to represent the origins of mathematics. Chapter 1 contains some general prescriptions for the study of mathematics, a list of the nine chapters, concise glosses of more than seventy terms used thereafter (*yong zi fan li* 用字凡例, 'guide to characters used'), lists of powers of tens and units, tables of addition, subtraction, multiplication, and division for the abacus, and brief explanations of some terms referring to common operations such as the simplification of fractions or the extraction of cube roots. Chapter 2 focuses on abacus calculation; it opens with an illustrated description of the instrument. The following fifteen chapters contain 595 problems presented in the traditional form: question, answer, and method of solution. Chapters 3 to 6 and 8 to 12 take up the headings of the nine chapters in the traditional order, whereas Chapter 7 introduces a particular type of problem, which involves *Fen tian jie ji fa* 分田截積法 'Methods for dividing

---

6. This is the reason why I prefer to translate *tong zong* 統宗 literally as 'unified lineage' rather than to use the most common translation: 'systematic treatise'.

fields by cutting off their areas': they deal with the dimensions and areas of figures obtained by cutting off a part of a known figure. Chapters 13 to 16 contain *Nan ti* 難題 'Difficult problems', often stated as rhymes; these problems are again classified according to the headings of the nine chapters. The last chapter gives *Za fa* 雜法 'Miscellaneous methods'; it includes various diagrams such as magic squares and depictions of hand calculation mnemonics. The chapter closes with a bibliography of 52 titles, from the Song edition (1084) of the 'Ten mathematical classics' to works published in Cheng's lifetime, spanning five centuries (Guo 1993, II 1217–1453). As this brief description of the work suggests, while the 'Nine chapters on mathematical procedures' was not accessible to Cheng, and while abacus calculation underlies the whole of his mathematics, his work belongs to a lineage that can be traced back to the first-century AD classic. In this as well as in other respects, the 'Unified lineage of mathematical methods' can be regarded as representative of mathematics as practised in China at the time of the first Jesuits' arrival.

Teaching and translating: Jesuit mathematics in Ming China

The China mission was part of the Portuguese assistancy of the Society of Jesus: since the end of the fifteenth century, all Asian missions were under the patronage of the Portuguese crown. The port of Macao, founded by the Portuguese in 1557, served as their Eastern base. While their Japanese mission flourished, the Jesuits' attempts to settle in China were unsuccessful until 1582. The first Jesuit residence in China was set up in Zhaoqing 肇慶 (Guangdong province). In establishing contact with local elites, Matteo Ricci used both knowledge and artifacts that he had brought with him from Europe. At the same time, he assessed their knowledge in terms familiar to him:

They have acquired quite a good mastery not only of moral philosophy, but also of astrology [that is, astronomy] and of several mathematical disciplines. However, in the past they have been better versed in arithmetic and in geometry; but they have acquired all this and dealt with it in a confused way.[7] (Ricci and Trigault 1978, 95)

In line with this emphasis on the shortcomings of the Chinese as regards mathematics, Ricci turned himself into a teacher. His relations with the first literati interested in Christianity were modelled on a master–disciple relationship, which can be interpreted both in the context of Jesuit education and of Chinese lineages

---

7. Ils ont non seulement acquis assez bonne connaissance de la philosophie morale, mais encore de l'astrologie et de plusieurs disciplines mathématiques. Toutefois il ont autrefois été plus entendus en l'arithmétique et géométrie; mais aussi ils ont acquis ou traité tout ceci confusément.

of scholarship (Jami 2002a). He described the progress of Qu Rukui 瞿汝夔, one of his first sympathizers and advisors, who eventually converted (Standaert 2001, 419–420):

He started with arithmetic, which in method and ease by far surpasses the Chinese one: for the latter all in all consists in a certain wooden instrument in which round beads, strung on copper wire, are changed here and there, to mark numbers. Although in fact it is sure, it is easily subject to misuse, and reduces a broad science to very little. He then heard Christoph Clavius' sphere and Euclid's elements, only what is contained in Book I; towards the end he learnt to paint almost any kind of figures of dials to mark the hours. He also acquired knowledge of the heights of things through the rules and measures of geometry. And being, as I said, a man of wit and well versed in writing, he reduced all this into commentaries in a very neat and elegant language, which he later showed to mandarins. One would hardly believe what reputation this earned to him and to our fathers, from whom he acknowledged having learned it all. For all that he had been taught delighted the Chinese, so that it seemed that he himself could never learn to his heart's content. For he repeated day and night what he had heard, or adorned the beginnings with figures so beautiful that they were by no means inferior to those of our Europe. He also made several instruments, spheres, astrolabes, dials, magnet boxes, mathematical, and other similar instruments very elegantly and artistically set up.[8] (Ricci and Trigault 1978, 308–309)

Ricci's success is evidenced by his student's capacity to produce both instruments and texts that were fit for circulation among literati. The former points to the inclusion of instrument making in mathematics as the Jesuits taught it in China. The latter brings out the fact that the Jesuits needed Chinese scholars' help in order to write in Chinese. During the first decades of the mission, the translation of mathematical texts was the outcome of teaching. After Ricci settled in Beijing in 1601, he taught mathematics to Xu Guangqi 徐光啓 and Li Zhizao 李之藻, two high officials who converted and took on the role of protectors of the Jesuit mission. They collaborated with Ricci in producing works based on some of Clavius' textbooks (Martzloff 1995).

8. Il commença par l'arithmétique qui en méthode et en facilité surpasse de beaucoup la chinoise: car icelle consiste toute en certain instrument de bois auquel des grains ronds enfilés de fil d'archal sont changés çà et là, pour marquer les nombres. Ce qu'encore que véritablement il soit assuré est sujet à recevoir facilement de l'abus et réduit à peu d'espèces d'une science très ample. Il ouït en après la sphère de Christopher *Clavius* et les éléments d'Euclide, ce que seulement est contenu au premier livre; sur la fin il apprit à peindre quasi toutes sortes de figures de cadrans pour marquer les heures. Il acquit aussi la connaissance des hauteurs des choses par les règles et mesures de la géométrie. Et, pour autant, comme je l'ai dit, qu'il était homme d'esprit et fort expert en l'écriture, il réduisit tout ceci en commentaires d'un langage fort net et élégant, lesquels venant par après à montrer aux mandarins ses amis, à peine pourrait-on croire quelle réputation cela acquit tant à lui qu'à nos Pères, desquels il confessait avoir tout appris. Car tout ce qui lui avait été enseigné ravissait par sa nouveauté tous les Chinois en admiration, de façon qu'il semblait que lui même ne pouvait en aucune sorte se saouler et contenter d'apprendre. Car il répétait jour et nuit ce qu'il avait ouï ou ornait ses commencements de figures si belles qu'ils ne cédaient en rien à ceux de notre Europe. Il fit aussi plusieurs instruments, des sphères, astrolabes, cadrans, boîtes d'aimants, instruments de mathématiques et autres semblables fort élégamment et artistement dressés.

Table 1 Chinese translations and adaptations of Clavius' works

Clavius' work	Date of Chinese work	Title of Chinese work	Authors/ translators
Euclidis elementorum, 1574	1607	Elements of geometry (*Ji he yuan ben* 幾何原本)	Matteo Ricci Xu Guangqi
Astrolabium, 1593	1607	Illustrated explanation of cosmographical patterns (*Hun gai tong xian tu shuo* 渾蓋通憲圖說)	Matteo Ricci Li Zhizao
Geometria practica, 1604	1608	Meaning of measurement methods (*Ce liang fa yi* 測量法義)	Matteo Ricci Xu Guangqi
In sphaeram Ioannis de Sacro Bosco commentarius, 1570	1608	On the structure of heaven and earth (*Qian kun ti yi* 乾坤體義)	Matteo Ricci Li Zhizao
Epitome arithmeticae practicae, 1583	1614	Instructions for calculation in common script (*Tong wen suan zhi* 同文算指)	Matteo Ricci Li Zhizao
In sphaeram Ioannis de Sacro Bosco commentarius, 1570	1614	Meaning of compared [figures] inscribed in circles (*Yuan rong jiao yi* 圓容較義)	Matteo Ricci Li Zhizao
Geometria practica, 1604	1631	Complete meaning of measurement (*Ce liang quan yi* 測量全義)	Giacomo Rho

The relationship between the Chinese works and their Latin sources varies. The 'Meaning of measurement methods', a brief treatise on surveying, completed by Ricci and Xu Guangqi at the same time as their translation of the first six books of Euclid's *Elements* in 'Elements of geometry', is not a direct translation from the *Geometrica practica*; it is probably based on Ricci's lecture notes (Engelfriet 1998, 297). The 'Instructions for calculation in common script' takes up a number of problems found in earlier Chinese works such as Cheng Dawei's 'Unified lineage of mathematical methods' and applies to them written arithmetic and the methods given by Clavius in his *Epitome arithmeticæ practicæ* (Jami 1992; Pan 2006).

Collaboration seems to have followed a pattern common to all translations, religious or secular: the Jesuit gave an oral explanation of the meaning of some original text, which the Chinese scholar then wrote down in classical Chinese. New terms were coined when there was no obvious equivalent for a Latin term in Chinese. For example, terms like definition, axiom, postulate, proposition, proof, had to be created during the translation of Euclid's *Elements*. These new Chinese terms were not explicitly defined before being used: whereas their Latin originals were part of the cultural background of those who studied geometry in Europe, such notions would have been entirely alien to a Chinese reader, and would probably remain somewhat of a mystery unless this reader was taught by someone familiar with them. But the vast majority of Chinese scholars who read the 1607 translation did so without the help of such teaching. It is little surprise, therefore, that while there was much interest in the content of the *Elements*, the Euclidean style on the whole aroused more perplexity than enthusiasm (Martzloff 1980; Engelfriet 1998, 147–154; Jami 1996).

Defining and situating mathematics

Whereas Euclidean geometry was presented as a radical innovation, in arithmetic western learning was introduced as an improvement on the Chinese mathematical tradition. The dichotomy between number and magnitude was made explicit in the structure of mathematics described in Ricci's preface to the 'Elements of geometry':

The school of quantity (*ji he jia* 幾何家) consists of those who concentrate on examining the parts (*fen* 分) and boundaries (*xian* 限) of things. As for the parts, if [things] are cut so that there are a number (*shu* 數) [of them], then they clarify how many (*ji he zhong* 幾何眾) the things are; if [things] are whole so as to have a measure (*du* 度), then they point out how large (*ji he da* 幾何大) the things are. These number and measure may be discussed (*lun* 論) in the abstract, casting off material objects. Then those who [deal with] number form the school of calculators (*suan fa jia* 算法家); those who [deal with] measure form the school of mensurators (*liang fa jia* 量法家). Both [number and measure] may also be opined on with reference to objects. Then those who opine on number, as in the case of harmony produced by sounds properly matched, form the school of specialists of pitchpipes and music (*lü lü yue jia* 律呂樂家); those who opine on measure, in the case of celestial motions and alternate rotations producing time, form the school of astronomers (*tian wen li jia* 天文曆家).[9]

This is a description of the *quadrivium*, which, in Chinese terms, proposes to subsume four well-known technical fields under a broader, hitherto unknown discipline: the 'study of quantity'. *Ji he* 幾何 renders the Latin *quantitas*. The title chosen by Ricci and Xu for their translation was apparently intended to encompass not just geometry, but rather the whole *quadrivium*. The claim here is also that the 'Elements of geometry' provides foundations for a discipline that includes

9. 幾何家者專察物之分限者也其分者若截以為數則顯物幾何眾也若完以為度則指物幾何大也其數與度或脫於物體而空論之則數者立算法家度者立量法家也或二者在物體而借其物議之則議數者如在音相濟為和而立律呂樂家議度者如在動天迭運為時而立天文曆家也 Guo (1993, V 1151–1152, cf. Engelfriet (1998, 139); Hashimoto and Jami (2001, 269–270).

the Chinese tradition of *suan fa* as one of its parts. *Ji he* means 'how much' in classical Chinese: it occurred in every problem of the 'Nine chapters on mathematical procedures'. In the 'Unified lineage of mathematical methods', however, *ruo gan* 若干 (a synonym) is used in the question stated in problems; *ji he* appears in the 'Guide to characters used': it is glossed by 'same as *ruo gan*' (Guo 1993, II 1230). Later in the seventeenth century, *ji he* came to refer to the content of the 'Elements of geometry', that is, to Euclidean geometry.[10]

The dichotomy between the two instances of quantity rendered as *shu* 數 'number' and *du* 度 'magnitude' respectively would have been new to a Chinese reader at the time: *shu* was more evocative of numerology and the study of the *Yi jing* 易經 'Classic of change', than of procedures of *suan fa*. By using this last term to refer to procedures, Ricci and Xu again implied that mathematics as hitherto practised in China was to be embedded into a broader discipline.

No matter how unfamiliar Ricci's distinction between *shu* and *du* might have appeared, the translations based on Clavius' works, made in response to the curiosity of a few Chinese scholars, aroused enduring interest among a wider audience. Moreover, bringing together mathematics, surveying, astronomy, and musical harmony was not foreign to their tradition (Lloyd, Chapter 1.1; Cullen, Chapter 7.1 in this volume): surveying was one of the main themes of mathematical problems; mathematical astronomy and musical harmony were discussed in the same section of quite a few dynastic histories. Also, one finds many examples of scholars known both as mathematicians and astronomers: Zhu Zaiyu, mentioned above as the author of the 'Essential meaning of pitchpipes' (1596, the earliest known discussion of equal temperament), strove to unify musical harmony and astronomy (Needham 1962, 220–228).

The translations mentioned above were part of the Jesuits' larger enterprise of 'apostolate through books': their publications merged into the thriving book market of the late Ming (Standaert 2001, 600–631). Their teachings were first presented as a whole in a compendium edited by Li Zhizao, the *Tian xue chu han* 天學初函 'First collection of heavenly learning' (1626). It was divided into two parts: *li* 理 'Principles' (nine works) and *qi* 器 'Tools' (ten works). The first part opens with a description of the European educational system, entitled *Xi xue fan* 西學凡 'Outline of Western learning' (1621). Like Ricci, its author, Giulio Aleni, had been a student of Clavius at the Roman College. It presents the structure of disciplines that was then most common, mathematics consisting of the *quadrivium* and being one subdivision of philosophy (Standaert 2001, 606). The next six works of the collection discuss mainly ethics and religion. The first part closes with an introduction to world geography, also written by Aleni. Illustrated by several maps, including an elliptical world map, the *Zhi fang wai ji* 職方外紀, 'Areas

---

10. The modern Chinese term for geometry is *ji he xue* 幾何學, literally 'the study of *ji he*'.

outside the concern of the imperial geographer' (1623) describes the earth as part of the universe created by God, and Europe as the ideal realm where Christianity has brought long-lasting peace.

The second part of the 'First collection' contains five of Ricci's six works based on Clavius. It also includes three works by another former student of Clavius, Sabatino de Ursis, dealing respectively with hydraulics, the altazimuth quadrant, and the gnomon. A short treatise entitled *Gou gu yi* 句股義 'The meaning of base-and-altitude', written by Xu Guangqi after he had completed the translation of the *Elements* with Ricci, is also included. This was the first attempt to interpret the traditional approach to right triangles (*gou* 句, base, refers to the shorter side of the right angle, and *gu* 股, altitude, to the longer one) in terms of Euclidean geometry (Engelfriet 1998, 301–313; Engelfriet and Siu 2001, 294–303).

In this compilation, Ricci's treatise on the sphere based on Clavius was substituted by another one, the *Tian wen lüe* 天問略 'Epitome of questions on the heavens'. This is the only work pertaining to 'tools' that does not stem from the student lineage of Clavius: its author, Manuel Dias Jr, never left the Portuguese Assistancy of the Society of Jesus. Due to the importance of navigation in Portugal, the study of the sphere was emphasized in Jesuit colleges there (Leitão 2002). Clavius' treatise was one of a genre; it seems that the 'Epitome of questions on the heavens' was an original composition within that genre rather than a translation of a Latin text. It gave a description of Ptolemy's system; in an appendix, it reported Galileo's invention of the telescope and the observations he had made with it (Leitão 2008). This was in keeping with the Society's policy in Europe, where innovations were usually incorporated into teaching. As a whole, the works on instruments in the 'First collection' were part of the mathematical sciences construed and constructed by Clavius for Jesuit colleges.

For converted officials like Xu Guangqi and Li Zhizao, Jesuit teaching met essential concerns of their own agenda. The 'Principles' and the 'Tools' of the 'First collection' formed a coherent whole: whereas the latter could better the material life of the people, the former could contribute to their moral improvement and therefore to social harmony. Heavenly learning was a fitting response to the concerns of 'solid learning', to which *jing shi* 經世 'statecraft' was central. Xu Guangqi's list of the applications of mathematics is revealing in this respect: astrological prediction (for the state), surveying and water control, music (harmony and instruments), military technology, book-keeping and management for the civil service, civil engineering, mechanical devices, cartography, medicine, and clockwork (Wang 1984, 339–342). All these were fields in which any progress would be socially useful. This list includes not only the topics in which western learning proposed innovations but also, more importantly, some of the main fields that 'solid learning' scholars strove to study. The latter's agenda thus oriented the choice of topics for translation; only the subjects that met their concerns had a significant influence. Mathematics in its broader sense was among those subjects.

Mathematics and calendar reform

The field in which converted officials were most successful in promoting the Jesuits and their learning was astronomy. The calendar had always been of utmost symbolical and political importance in China; issued in the emperor's name, it ensured that human activity followed the cycles of the cosmos. The need for calendar reform had been felt before the Jesuits' arrival (Peterson 1968), and Ricci had recommended the Society to send missionaries versed in this matter. In 1613, Li Zhizao proposed that three Jesuits be commissioned to reform the calendar (Hashimoto 1988, 16–17). This may well have fostered opposition to Christianity (Dudink 2001). In 1629 a new proposal put forward by Xu Guangqi was finally approved. Under his supervision, a special *Li ju* 曆局 'Calendar Office' was created (Hashimoto 1988, 34–39). This meant that officials rather than private literati became the main recipients of European science.

The first output of this newly created office was a series of twenty-two works (a few of which had actually been written before 1629). They were presented to the emperor between 1631 and 1634, and formed the *Chong zhen li shu* 崇禎曆書 'Books on calendrical astronomy of the Chongzhen reign'. The knowledge they contained was very different in content and structure from that of the 'First collection': reference was no longer made to an overarching system of knowledge, nor to the Catholic religion. The Ptolemaic system was discarded in favor of the Tychonic system. Thus institutionalized, western learning had become a technical subject organized according to official astronomers' needs.

Three Jesuits, Johann Schreck, Johann Adam Schall von Bell, and Giacomo Rho, were in charge of the work; in 1633 Li Tianjing 李天經 succeeded Xu Guangqi as supervisor. More than twenty Chinese collaborated in this task. Some of these were converts, as were many Chinese who worked at the *Qin tian jian* 欽天監 'Astronomical Bureau' thereafter. In late Ming officials' eyes, calendar reform was to contribute to the restoration of social order and the dynasty's strength, at a time when the military situation in particular was getting worse. However the result of the work done at the Calendar Office ultimately benefited the newly established Qing dynasty, to which Schall offered his service on the fall of the Ming; the calendar he had calculated was promulgated in 1644. The compendium's title was changed to *Xi yang xin fa li shu* 西洋新法曆書 'Books on calendrical astronomy according to the new Western method' and a few works were added to it. This marked the Jesuits' entry into officialdom at the Astronomical Bureau.

According to Xu Guangqi's classification, which he proposed while the 'Books on calendrical astronomy' were being composed, those books should fall into five categories: *fa yuan* 法原 'fundamentals', *fa shu* 法數 'numbers', *fa suan* 法算 'calculations', *fa qi* 法器 'instruments', and *hui tong* 會通 'intercommunication', or correspondence between Chinese and western units. None of these categories correspond to specifically mathematical subjects as opposed to astronomical ones. Once the works were completed, it was not always specified which of these categories they belonged to; the 'Calculation' category remained empty. 'Fundamentals' include practical geometry and trigonometry; *Bi li gui jie* 比例規解 'Explanation of the proportional compass' is among the 'instruments'; trigonometric tables and Napier's rods are included in 'numbers'; this suggests that the latter aid to calculation was understood as a kind of moveable table (Jami 1998b). Neither Euclidean geometry nor the basics of written calculation were deemed necessary for the purposes of calendar reform. On the other hand Ricci's mathematics had to be supplemented, mainly by trigonometry. On the whole, the 'Books on calendrical astronomy' do not bring out astronomy and mathematics as two separate disciplines.

In the 1644 version of the 'Books on calendrical astronomy', a geometry treatise was added, which was not allotted into any of these categories: the *Ji he yao fa* 幾何要法 'Essential methods of geometry' (1631). It was composed of extracts from the 'Elements of geometry', focusing on constructions and leaving out proofs. The work was the result of collaboration between Aleni and Qu Shigu 瞿式穀, Qu Rukui's son, and a Christian like his father (Jami 1997). To paraphrase Xu Guangqi, a recasting of western knowledge into the 'Chinese mould'[11] had occurred between the translation of Euclid's *Elements* and the calendar reform. At the time of the former, astronomy was a branch of 'the study of quantity'. During the latter, mathematics was conversely subsumed under calendrical astronomy for which it provided a series of tools and methods.

Integrating Chinese and Western mathematics: the work of Mei Wending

Whereas conversion to Catholicism remained a marginal phenomenon in officialdom and literati circles, a number of scholars during the late Ming and early Qing period were interested in the Jesuits' mathematics. While the calendar reform took place in Beijing, it was mostly in the Lower Yangzi region, which had been of foremost economic and cultural importance since the tenth century AD, that some scholars read the Jesuits' works. The most thorough and systematic

---

11. For a discussion of this phrase and its posterity, see Han Qi (2001, 367–373).

of them was Mei Wending 梅文鼎, who is the best known mathematician and astronomer of the period.

Mei's syncretistic approach is suggested by the title of a collection of nine of his works that he put together in 1680: *Zhong xi suan xue tong* 中西算學通 'Integration of Chinese and western mathematics'. Only six of these nine works were eventually printed: this reflects the limits of Chinese literati's interest in the mathematical sciences at the time. Mei, however, argued that they were a key to understanding the world: in his view, *li* 理 'principles', a key concept of Neo-Confucian philosophy, could only be fathomed through *shu* 數 'numbers', and the principles thus uncovered were universally valid. For him numbers encompassed the whole of mathematics, which he divided into *suan shu* 算術 'calculation procedures' and *liang fa* 量法 'measurement methods' (*SKQS* 794, 64); accordingly, he proposed to reorganize the traditional nine chapters into two groups. Unlike the Jesuits and Chinese scholars before him, however, he also argued that calculation had primacy over measurement, as only the former could deal with invisible objects; however, the fashion of Euclidean geometry resulted in the neglect of this primordial field. In his view the great contribution of Western learning to both mathematics and astronomy was that it explained *suo yi ran* 所以然 'why it is so', whereas the Chinese tradition stated only *suo dang ran* 所當然 'what must be so' (Engelfriet 1998, 430–431; Jami 2004, 708 and 719). Acknowledging the excellence of Westerners in measurement methods, Mei proposed alternative proofs for some propositions of the 'Elements of geometry', and went on to explore solids (Martzloff 1981, 260–290). In calculation, however, he emphasized the shortcomings of the Westerners. This did not prevent him from adopting and adapting written calculation: in his lengthy *Bi suan* 筆算 'Brush calculation' he transposed the four basic operations by writing all numbers in place-value notation vertically, with the aim of making the orientation of the layout of calculations consistent with writing in China, as it was in the West.

On the other hand, Mei set out to restore what had been lost of the Chinese mathematical tradition. Thus he proposed a reconstruction of the method of *fang cheng* 方程 'rectangular arrays', equivalent to systems of linear equations in several unknowns. The method had been handed down from the eighth of the 'Nine chapters on mathematical procedures' through works like the 'Unified lineage of mathematical methods', in which problems were classified according to the number of unknowns; he regarded it as the acme of calculation. In his *Fang cheng lun* 方程論 'Discussion of the comparison of arrays' (1672),[12] Mei criticized this classification, and also chastised the authors of the 'Instructions for calculation in common script' for failing to recognize the specificity and powerfulness of the *fang cheng* method. Against both works, from which he took up a number

---

12. Unlike today's historians of mathematics, Mei interpreted *fang cheng* as 'comparison of arrays' (*SKQS* 795, 67; cf. Martzloff 1981, 166–168).

of problems, and corrected several mistakes, he proposed an entirely new classification of problems according to the operations involved in their resolution
rather than to the number of unknowns (Martzloff 1981, 161–231; Jami 2004,
706–714). Further, he clarified how the arrays were to be laid out according to the
way the problem was stated, as regarded both the place where each number was
to be and its *ming* 名 'denomination', that is to say, the sign assigned to it for the
purpose of solving the problem. 'Denominations' had been transmitted from the
'Nine chapters on mathematical procedures', so in this respect Mei was indeed
restoring an ancient method rather than innovating.[13] After explaining the 'comparison of arrays' in all its technicalities, he went on to use it in order to solve a
number of problems that pertained to other 'chapters' of the traditional nine-fold
classification, and to astronomy. By showing that his reconstructed method was
a generic tool that could solve problems traditionally associated with more specific methods, he substantiated the claim that it was the acme of calculation. By
applying it to astronomical problems, he also exemplified why he gave primacy to
calculation over measurement.

In several respects the style of the 'Discussion of the comparison of arrays' is
in rupture with that dominant in mathematical works by Chinese authors of the
time. Indeed, the work contained a series of problems, followed by their solution
and the *fa* 法 'method' used to solve them, which included the array associated to
each problem. However, the author warns us, these problems only occupy 30%
of the work, and play the role of *li* 例 'examples', to illustrate *lun* 論 'discussion',
which occupies 70% of the work. Indeed the examples always follow a general
discussion and in turn each of them is followed by further lengthy discussion, for
the purpose of *ming suan li* 明算理 'clarifying the principles of calculation' (*SKQS*
795, 68). Thus after a general discussion of positive and negative denominations,
one particular problem, borrowed from the 'Unified lineage of mathematical
methods', is rephrased four times; four corresponding arrays are given, in order
to illustrate the rule that the first number given in the problem should be laid out
in the top right place of the array, and should always be assigned a *zheng ming* 正
名 'positive denomination' (*SKQS* 795, 76–78).

Mei's choice of the term *lun* 'discussion' to designate the discursive parts of
his text is significant: whereas it was not a term traditionally used in mathematical texts, he knew at least two precedents. In the 'Unified lineage of mathematics', the method for solving a problem was sometimes followed by a discussion
in the form of a poem, most likely with a mnemonic function. *Lun* also rendered
'proof' in the 1607 translation of the 'Elements of geometry' but it is difficult to
tell whether this was independent of its use for 'discussion'. As mentioned before,

---

13. Following the earlier Chinese tradition, Mei considered signs associated to numbers only in the context of *fang cheng* problems. No concept of negative numbers occurs in his works.

no explanation of the deductive structure of Euclidean proofs was given by the Jesuits; on the other hand, the latter themselves frequently used *lun* in the broader sense of 'discussion' or 'discuss', as Ricci did in the passage of his preface of the 'Elements' quoted above. It is not unlikely, therefore, that Mei Wending saw the portions of the 'Elements' entitled *lun* as discussions that clarified the 'why it is so' of each proposition. The presence of lengthy 'discussions' in his own work can be understood as his appropriation of what he felt was a strong point of the Westerners' mathematical style for writing on a subject anchored in the Chinese tradition. Thus the integration of Western learning was not simply a matter of adding a new field, like Euclidean geometry, or choosing, among the methods proposed by the Jesuits and those found in earlier works, the most relevant one. The craft of writing mathematics itself shows signs of hybridization. In discourse on mathematics Chinese and western were often opposed, but in practice they were combined at every possible level.

The Kangxi emperor's appropriation of mathematics

After he was put in charge of the Astronomical Bureau, Schall successfully cultivated the favour of the young Shunzhi emperor. After the death of the latter in 1661, however, the conflicts around Schall culminated in the Calendar Case (1664–1669). Choosing dates and locations for rituals was part of his tasks as the head of the Astronomical Bureau. Therefore, when it was found out that the time of an imperial prince's funeral had been miscalculated, this mistake was added to the charge of promoting heterodox ideas that had previously been brought against him. This brought about his downfall: he was sentenced to death—a sentence soon commuted to house arrest—and all the missionaries who worked in the provinces were expelled to Macao. In 1669, in the process of assuming personal rule at the end of the regency that had followed the death of his father, the young Kangxi emperor had the case reexamined. Ferdinand Verbiest, who succeeded Schall as the main specialist in the sciences after the latter's death, turned out to be more accurate than his Chinese adversaries in predicting the length of the shadow of a gnomon at noon, and the verdict was reversed (Chu 1997). Following this, Kangxi undertook the study of western science, which he was to continue throughout his reign. Verbiest, who was his first tutor, listed the mathematical sciences which thus 'entered the imperial Court' in the wake of astronomy, each presenting to the Emperor some achievement in the form of one or several technical objects: gnomonics, ballistics, hydragogics, mechanics, optics, catoptrics, perspective, statics, hydrostatics, hydraulics, pneumatics, music, horologic technology, and meteorology (Golvers 1993, 101–129). Thus from the early years of the reign, the two-fold pattern of the Jesuits' role at court was settled. On the one hand they were court savants, who built and maintained various machines and instruments and took part in imperial projects. In line with the late Ming trend of 'solid learning', the emperor regarded most of their skills as tools for statecraft. On the other hand the Jesuits were imperial tutors, who wrote textbooks in both Chinese and Manchu. Kangxi's motivations for studying western science were two fold: genuine curiosity was combined with eagerness to be in a position to control all issues and arbitrate all controversies, and to display his abilities to higher officials. The mathematical sciences within western learning were thus integrated into the body of Confucian learning mastered by the emperor—who emulated the Sages of antiquity (Jami 2002b; 2007).

The Jesuits' tutoring of Kangxi in mathematics is best documented for the 1690s, when it seems to have been at its most intensive. There were two different teams of tutors. Geometry was mostly taught by two French Jesuits, Jean-François Gerbillon and Joachim Bouvet, in Manchu; meanwhile, Antoine Thomas was in charge of calculation and he used the Chinese language, with Tomé Pereira as his interpreter. Both teams of tutors produced textbooks that have been preserved as manuscripts (Jami and Han 2003). Kangxi also had his sons trained in the mathematical sciences; Thomas was their tutor. His most talented pupil was prince Yinzhi 胤祉, Kangxi's third son. In 1702 tutor and student were sent on an expedition to measure the length of a degree of a meridian (Bosmans 1926). This was a preliminary to the general survey of the empire that Kangxi commissioned in 1708. A number of Jesuits took part in it, applying the methods used by the Paris Academicians in their survey of France a few years earlier. The outcome of this was the famous *Huang yu quan lan tu* 皇輿全覽圖 'Complete maps of the Empire' (1718) known in Europe as the 'Kangxi Atlas' (Standaert 2001, 760–763).

The tutoring reflected Jesuit mathematical education at the time in Europe. Thus the geometry treatise that the two Frenchmen composed for the emperor was based on one of the many handbooks produced in Europe under the title 'Elements of geometry' in the seventeenth century. Their choice of *Elemens de geometrie* (1671) by Ignace Gaston Pardies for tutoring the emperor—a choice that Kangxi approved—echoed the success of the work in Europe, where it underwent several editions and reprints up to 1724, and was translated into Latin, Dutch, and English (Ziggelaar 1971, 64–68). This work fitted in with Gerbillon and Bouvet's specific agenda in teaching Kangxi. As they were among the five Jesuits sent to China in 1685 by Louis XIV, they saw themselves as representatives of French science as practised under the auspices of the Paris Académie Royale des Sciences. They were in China not only to contribute to its evangelization but also to further French interests in Asia. The latter entailed gathering data for the Académie (Landry-Deron 2001). In their tutoring, which also included medicine and other aspects of philosophy, they claimed that they wrote 'in the briefest and clearest way that [they] could, removing all there is of complicated terms and of pure chicanery, following the style of the moderns'.[14] Pardies, who had dedicated his geometry text-book to the Paris Academicians, discarded the axiomatic and deductive style that characterized Euclid as edited by Clavius, in favour of shortness and ease. This was an adjustment to the widening audience of Jesuit colleges in Europe; it also reflects the idea, common among seventeenth-century mathematicians, that clarity is an intrinsic quality of mathematics (Jami 1996; 2005, 217–221). Both the Manchu and the Chinese versions of the treatise, which are abridged translations, were written under the emperor's personal supervision: some corrections and comments in his hand are found on two copies of the treatise. Like its European counterpart, this new treatise took up the title of the translation of Euclid's *Elements*: in Chinese it was called *Ji he yuan ben* 幾何原本, like the 1607 translation.

Meanwhile, Antoine Thomas composed two lengthy treatises. Before setting sail for Asia, he had taught mathematics in Coimbra, Portugal. For this purpose he had written a kind of *vademecum*, the *Synopsis mathematica*, a work explicitly designed for candidates to the China mission as well as for novices. The first of his Chinese treatises was called *Suan fa zuan yao zong gang* 算法纂要總綱 'Outline of the essentials of calculation', possibly a translation of the title of his Latin treatise. The structure of the former work followed that of the chapters devoted to arithmetic in the latter (Han and Jami 2003, 150–152). However, while the Latin work only gave one example to illustrate each rule of calculation, the Chinese treatise contained a wealth of problems for each of these rules. Some problems were drawn from the 'Instructions for calculation in common script' by Ricci and Li Zhizao. Others evoked subjects that Kangxi discussed with the Jesuits during the tutoring sessions, such as astronomy or the speed of sound (Jami 2007). Another treatise written by Thomas presented a branch of mathematics never before taught by the Jesuits in China, namely algebra. The term was transcribed as *aerrebala* 阿爾熱巴拉 in the foreword of the treatise; however, it was the title of the treatise, *Jie gen fang suan fa* 借根方算法 'Calculation by borrowed root and powers', that gave its name to the mathematical method described in it. Seventy years after some of the Jesuit colleges started to teach Viète's notation, the Kangxi emperor was still being taught cossic algebra. In Chinese, full names in characters were used rather than abbreviations as in European treatises. Thus, for instance, the equation $x^3 + 44 x^2 + 363 x = 1950048$ appears in the *Jie gen fang suan fa* 借根方算法 'Calculation by borrowed root and powers' as:

一立方　　＋　　四四平方　　＋　　三六三根　　＝＝＝＝　　一九五〇〇四八

1 cube　　+　　44 square　　+　　363 root　　=　　1950048
(Bibliothèque Municipale de Lyon, Manuscript 39–43, V 135)

14. [...] de la maniere la plus brieve et la plus claire qu'il nous a esté possible, en retranchant tout ce qu'il y a de termes embrouillés et de pure chicane, conformément au style des modernes (Archivum Romanum Societatis Iesu, Jap Sin 165, f. 101r).

Thomas's textbook may well be an original composition, but the algebra in it is similar to that found in, among other works, Clavius' *Algebra* (1608). Like Clavius, Thomas included some first degree problems in several unknowns, in which he represented the unknowns by the cyclical characters (*jia* 甲, *yi* 乙, *bing* 丙, *ding* 丁...), in a manner equivalent to that in which one would use letters. Coefficients, on the other hand, were always numerical. Thus, more than three decades after Mei Wending's 'Discussion on the comparison of arrays', a Jesuit produced two treatises that appear as refutations of Mei's criticism of Westerners as incompetent in calculation; moreover one of these treatises contained a possible alternative to the *fang cheng* method as reconstructed by Mei. At the time, algebra was not part of elementary mathematical education in Europe. Thomas had not included it in his Latin mathematical treatise, but he was familiar with symbolic algebra. That he nonetheless taught the emperor cossic algebra may reflect his wish to perpetuate the mathematics taught by Clavius and the Jesuits working in China during the late Ming period. It may also simply be due to the fact that symbolic algebra was regarded as more difficult. In 1713, that is, less than fifteen years after Thomas completed his treatise on cossic algebra, another Jesuit, Jean-François Foucquet, attempted to present symbolic algebra to Kangxi; for this purpose, he set out to write a treatise that he entitled *Aerrebala xin fa* 阿巴拉新法 'New method of algebra'. A section on first-degree problems in several unknowns was completed and explained to the emperor; however the tutoring happened to stop just as Foucquet was starting on second-degree equations, so that the emperor did not have the chance to grasp the meaning of the juxtaposition of two unknowns as a representation of their product. The 'New method of algebra' was rejected, and, given the fact that Kangxi actually arbitrated matters to do with mathematics personally, symbolic algebra did not find its way into Chinese mathematical textbooks until the second half of the nineteenth century (Jami 1986).

The emperor strove to integrate the mathematical sciences into imperial scholarship. In 1713 he created a *Suan xue guan* 算學館 'Office of Mathematics' staffed by Chinese, Manchus, and some Mongols. It was modelled on various offices of the same kind for literary or historical projects and headed by his son Yinzhi. The staff of this office compiled a three-part compendium, the *Yu zhi lü li yuan yuan* 御製律曆淵源 'Origins of musical harmony and calendrical astronomy, imperially composed', which was printed at the beginning of the Yongzheng reign (1723–1735). Western learning was dominant in the astronomical part, the *Li xiang kao cheng* 曆象考成 'Thorough investigation of calendrical astronomy' (42 chapters). It was expounded in a separate appendix in the *Lü lü zheng yi* 律呂正義 'Exact meaning of pitchpipes' (5 chapters). It was interspersed with Chinese learning in the mathematical part, entitled *Shu li jing yun* 數理精蘊 'Essence of numbers and their principles' (53 chapters), which set the standard for the study of the subject. The association of the three fields of astronomy, mathematics, and music points to the influence of the *quadrivium*, all the more so as mathematics was constructed on the dual foundations of geometry and calculation. However, the link between calendrical astronomy and the pitchpipes was a traditional one: both were about measuring and setting norms for the cosmos. Since number, that is mathematics, was used in both, putting the three disciplines together would not seem strange to Chinese readers. The rationale put forward to justify it was borrowed from the Classics, the origin of all this learning being said to be the same as that of the *Yi jing* 易經 'Classic of change' (Kawahara 1995).

Most of the content of the 'Essence of numbers and their principles' can be traced back to the Jesuits' tutoring of the 1690s. Some chapters, however, resulted from Chinese scholars' work inspired by the nine chapters tradition. The 'Essence of numbers and their principles' is divided into two parts of very unequal length, followed by some tables. The first five chapters are devoted to *li gang ming ti* 立綱 明體 'Establishing the structure to clarify the substance'. After a discussion of the foundations of mathematics, which roots it into Chinese antiquity, three chapters are devoted to the 'Elements of geometry', a revised version of Gerbillon and Bouvet's textbook. This part closes on a chapter on the *Suan fa yuan ben* 算法原 本 'Elements of calculation', a revised version of one of the textbooks produced in the 1690s, probably authored by Thomas and mostly based on books VII and VIII of Euclid's *Elements*. Thus, while imperial mathematics was asserted to have its origins in ancient China, its foundations stemmed from Western learning, and more precisely from the early modern European appropriation of the Euclidean tradition. The second part, comprising forty chapters, is on *fen tiao zhi yong* 分 條致用 'dividing items to convey their use'. It is divided into five sections: *shou* 首 'initial', *xian* 線 'line', *mian* 面 'area', *ti* 體 'solid', and *mo* 末 'final'. The content is presented in the traditional form, that is, as a sequence of problems and solutions. After basic instruction on the four operations and fractions has been given in the 'beginning section', the three middle sections organize problems according to their dimension. A great part of the material in these first four sections can be traced back to Thomas's 'Outlines of the essentials of calculation', while some material was drawn from Chinese authors as well. Six of the ten chapters in the end section, devoted to cossic algebra, are derived from his 'Calculation by root and powers', with slightly modified vocabulary and notations; three chapters are devoted to a general presentation of the notation and of the techniques for solving equations; the three next chapters give problems that fall respectively in the 'line', 'area', and 'volume' categories. After cossic algebra, there follows a chapter of *Nan ti* 難題 'Difficult problems'; this chapter is one among several clues that suggest that Cheng Dawei's 'Unified lineage of mathematical methods', among other Chinese works, were used to compile the 'Essence of the principles of numbers'. The last three chapters are devoted to the principles of logarithms and to the proportional compass (Guo 1993, III 1143–1235). Logarithms and trigonometric tables were appended. Imperial mathematics, which encompassed most of the knowledge available at the time, integrated an updated version of western learning devised for Kangxi and some revived branches of Chinese learning.

The compilers of 'Essence of numbers and their principles' had at their disposal at least two methods for dealing with problems equivalent to systems of linear equations in several unknowns: Mei Wending's 'comparison of arrays', and Thomas's notation using cyclical characters. Unlike in the case of right triangles, for which they included both the traditional *gou gu* 句股 'base-and-altitude' methods and the techniques of western geometry, they retained only Mei Wending's method, which they presented as an independent chapter of the 'line section'. In the chapter on 'line' problems solved by 'calculation by root and powers', on the other hand, only one root, denoted as usual by *gen* 根, is used. Thus, in the eyes of the compilers, none of the methods proposed by the Jesuits for solving linear problems in several unknowns measured up to the ancient Chinese method as reconstructed by Mei. This can hardly have been the result of a bias in favour of traditional Chinese mathematics on their part: altogether only three chapters of the imperial compendium are titled after the names of the 'Nine chapters'.

In bibliographies compiled during the two centuries that followed its composition, the 'Essence of numbers and their principles' was attributed to Kangxi. The list of editors of the 'Origins of musical harmony and calendrical astronomy, imperially composed', published in 1724, comprises forty-seven names, including Yinzhi and one of his brothers. There is ample evidence that the emperor kept a close eye on the compilation's progress, discussing details such as the layout of numerical tables with Yinzhi (Jami 2002b, 40–41). The compendium was later used for the study of mathematics in imperial institutions (*SKQS* 600, 445). Thus, officials, if not all scholars, were to model their study of mathematics on that of the emperor.

Western learning without the Jesuits

The Rites Controversy, in which the Jesuit policy of accommodation to Chinese customs such as the ritual honouring of ancestors was over-ruled by Rome, brought about a change of imperial policy towards Catholic missionaries. The court Jesuits seem to have lost imperial trust after the visit of a papal legate to Beijing in 1706, bringing the news that Chinese converts must abandon all 'idolatrous' practices. In 1732, all missionaries working in the provinces were expelled to Macao; however, the Beijing Jesuits were allowed to remain and to practise their religion. They continued to be employed as official astronomers and as cartographers, engineers, architects, and artists. Western learning at court remained in the service of imperial magnificence and of control of the expanding Qing territory (Standaert 2001, 358–363, 823–835). During the Qianlong reign (1736–1795), lengthy sequels to the 'Thorough investigation of calendrical astronomy' and to the 'Exact meaning of pitchpipes' were published. By contrast, the 'Essence of numbers and their principles' does not seem to have been regarded as in need of supplementing. Although it never competed with the 'Unified lineage of mathematical methods' for popular readership, the imperial compendium represented the basis of scholarly culture in mathematics.

Eighteenth-century scholars indeed appropriated mathematics and astronomy, but not quite in the way that Kangxi had tried to foster. Instead of becoming an end in itself or a tool for other technical fields, the discipline was integrated into the main intellectual trend of China at the time, *kao zheng xue* 考證學 'evidential scholarship' (Elman 1984, 79–89; Tian 2005, 134–145). The aim was the restoration of the original text of ancient classics, the meaning of which, it was argued, had been distorted, especially by Song dynasty (960–1279) commentators. Scholars who followed this trend developed sophisticated methods in philological disciplines. Mathematics and astronomy were a tool for that purpose: ancient records of astronomical events were used to date documents and events. But they were also an object of study; thus Dai Zhen 戴震, who is regarded as the greatest philologist of the time, reconstructed the text of the 'Nine chapters on mathematical procedures'.

The turn towards ancient texts in the mathematical sciences went together with the development of the idea *xi xue zhong yuan* 西學中源 'western learning originated in China'. While at first he argued for the unity of mathematics East and West, Mei Wending eventually turned to investigating this idea in detail, encouraged by Kangxi (Chu 1994, 184–217; Han 1997). The advantage for the emperor was obvious: if the calendar was based on foreign knowledge, then he could be challenged for applying Barbarian knowledge to regulate the rites that lay at the heart of Chinese civilization. If on the other hand that knowledge had originated in China, he became the personification of the Confucian monarch who retrieved ancient learning for the empire's benefit, which was quite an achievement for a Manchu ruler. For Chinese scholars on the other hand, the Chinese origin of western knowledge neutralized any claim of superiority of the latter. The idea could have some heuristic value as was the case in the field of algebra: identification with calculation by borrowed roots and powers as introduced by Thomas eventually proved instrumental in the rediscovery of thirteenth-century celestial element algebra (Han 2003, 80–81). At the turn of the nineteenth century there were debates over the respective merits of the two methods (Tian 2005, 250–271).

Thus western learning, represented both by late Ming Jesuits' translations and by the 'Essence of numbers and their principles', became an entity opposed to

Chinese learning. Even as eighteenth-century scholars distinguished between these two types of learning, and could side with one against the other, none of them simply ignored western learning; the latter, while keeping its identity, had been appropriated.

Conclusion

Studies of the Jesuits' transmission of mathematics from Europe to China have long focused on Euclid's *Elements of geometry*, arguably to the detriment of other branches of mathematics; this is no doubt a consequence of the role of this work as a supposed embodiment of the essence of either 'western mathematics' or mathematics *tout court*. The story of 'Euclid in China' has been told in terms of European categories, as one of a radical innovation that had universal validity; 'the Chinese understanding' (or misunderstanding) of this innovation supposedly revealed general features of 'Chinese thought'. Writings on geometry by Chinese authors of the seventeenth century have been assessed according to their conformity to the Euclidean model (Martzloff 1980). This fitted in a historiography that modelled Sino-European contacts as (European) 'action' and (Chinese) 'reaction' (Gernet 1982).

Further contextualization of the introduction of Euclidean geometry (Engelfriet 1998; Jami 1996), as well as inclusion of other branches of mathematics into the narrative, have yielded a different picture, one of complex interaction rather than of action and reaction. In introducing European written arithmetic, for example, a synthesis was proposed from the onset between what the Jesuits brought in and what was found in Chinese mathematical works of the time. Looking at the Chinese category *suan* 算, which by 1600 by and large denoted the whole of mathematics, one can trace the restructuring of the field during the hundred and twenty years that followed. The Jesuits first used *suan* as referring to arithmetic, and proposed to embed the Chinese tradition into a broader field, for which their geometry provided a foundation. However, as some Chinese scholars' subsequent interpretations of *suan*, eventually taken up by the Jesuits of the Kangxi court themselves, were broader: a more general category, best rendered by the term 'calculation', was thus constructed, within which a number of competing methods were proposed. In parallel, the term *shu* 數 'number', used by the Jesuits to denote only one of the two instances of quantity, came to name the broader field that encompassed geometry and calculation. Mathematics thus gained a status within scholarship as defined in neo-Confucian philosophy: it was a tool to access *li* 理 'principles' which was the ultimate goal of all learning. Thus the cross-cultural transmission and reception of mathematics entailed its reconstruction at several levels: its methods, branches, the structure of texts, but also the discipline as a whole vis-à-vis other scholarly pursuits, were reshaped by the process of their integration into a different landscape. This conclusion brings out mathematics as a flexible and dynamic system of knowledge and practice, rather than as an immutable body of truths.

Table 2: Names and dates

Song dynasty (960–1279)
Yuan dynasty (1279–1368)
Ming dynasty (1368–1644)
Qing dynasty (1644–1911)
Chongzhen reign (1628–1644)
Shunzhi reign (1644–1661)
Kangxi reign (1662–1722)
Yongzheng reign (1723–1735)
Qianlong reign (1736–1795)
Jesuit mission (1582–1773)

Cheng Dawei 程大位 (1533–1606)
Dai Zhen 戴震 (1724–1777)
Li Shizhen 李時珍 (1518–1593)
Li Tianjing 李天經 (1579–1659)
Li Zhizao 李之藻 (1565–1630)
Mei Wending 梅文鼎 (1633–1721)
Qu Rukui 瞿汝夔 (1549–1611)
Qu Shigu 瞿式穀 (b. 1593)
Song Yingxing 宋應星 (1582– after 1665)
Wang Yangming 王陽明 (1472–1529)
Xu Guangqi 徐光啓 (1562–1633)
Yang Hui 楊輝 (fl 1261)
Yinzhi 胤祉 (1677–1732)
Zhu Zaiyu 朱載堉 (1536–1611)

Giulio Aleni (1582–1649)
Joachim Bouvet (1656–1730)
Christoph Clavius (1538–1612)
Manuel Dias Jr (1574–1659)
Jean-François Foucquet (1665–1741)
Jean-François Gerbillon (1654–1707)
Ignace Gaston Pardies (1636–1673)
Giacomo Rho (1592–1638)
Matteo Ricci (1552–1610)
Johann Adam Schall von Bell (1592–1666)
Johann Schreck (1576–1630)
Antoine Thomas (1644–1709)
Sabatino de Ursis (1575–1620)
Ferdinand Verbiest (1623–1688)

For biographies of the Jesuits who went to China see: http://ricci.rt.usfca.edu/biography/index.aspx

Bibliography

Baldini, Ugo, *Legem impone subactis. Studi su filosofia e scienza dei Gesuiti in Italia, 1540-1632*, Bulzoni Editore, 1992.

Baldini, Ugo, 'The Portuguese Assistancy of the Society and scientific activities in its Asian mission until 1640', in Luis Saraiva ed., in *Història das ciências matematicas: Portugal e o Oriente*, 49–104, Fundação Oriente, 2000.

Bosmans, Henri, 'L'œuvre scientifique d'Antoine Thomas de Namur, S.J. (1644–1709)', *Annales de la société scientifique de Bruxelles*, 44, 46 (1924, 1926), 169–208, 154–181.

Brockey, Liam, *Journey to the East: the Jesuit mission to China, 1579-1724*, Harvard University Press, 2007.

Chemla, Karine, and Shuchun Guo, *Les neuf chapitres: le classique mathématique de la Chine ancienne et ses commentaires*, Dunod, 2004.

Cheng, Anne, *Histoire de la pensée chinoise*, Editions du Seuil, 1997.

Chu, Ping-yi, 'Technical knowledge, cultural practices and social boundaries: Wan-nan scholars and the recasting of Jesuit astronomy, 1600–1800', PhD Dissertation, UCLA, 1994.

Chu, Ping-yi, 'Scientific dispute in the imperial court: the 1664 calendar case', *Chinese Science*, 14 (1997), 7–34.

Cullen, Christopher, *Astronomy and mathematics in ancient China: the Zhoubi suanjing*, Cambridge University Press, 1996.

Cullen, Christopher, *The Suan shu shu 'Writings on reckoning': a draft translation of a Chinese mathematical collection of the second century BC, with explanatory commentary*, Needham Research Institute, 2004.

Dudink, Adrianus, 'Opposition to the introduction of Western science and the Nanjing persecution (1616–1617)', in Catherine Jami, Peter M Engelfriet, and Gregory Blue (eds), *Statecraft and intellectual renewal: the cross-cultural synthesis of Xu Guangqi (1562–1633)*, Brill, 2001, 191–224.

Elman, Benjamin A, *From philosophy to philology: intellectual and social aspects of change in late imperial China*, Harvard University, 1984.

Engelfriet, Peter M, *Euclid in China: The genesis of the first Chinese translation of Euclid's Elements, books I–VI (Jihe yuanben, Beijing, 1607) and its reception up to 1723*, Brill, 1998.

Engelfriet, Peter M, and Man-Keung Siu, 'Xu Guangqi's attempt to integrate Western and Chinese mathematics', in Catherine Jami, Peter M Engelfriet, and Gregory Blue (eds), *Statecraft and intellectual renewal: the cross-cultural synthesis of Xu Guangqi (1562–1633)*, Brill, 2001, 279–310.

Feldhay, Rivka, 'The cultural field of Jesuit science', in John W O'Malley, Gauvin Alexander Bailey, Steven J Harris, and T Frank Kennedy (eds), *The Jesuits: cultures, sciences and the arts 1540–1773*, University of Toronto Press, 1999, 107–130.

Ge Rongjin 葛榮晉 (ed), *Zhongguo shi xue si xiang shi* 中國實學思想史 'History of "solid learning" thought in China', Shoudu shifan daxue chubanshe, 1994, 3 vols.

Gernet, Jacques, *Chine et christianisme: action et réaction*, Gallimard, 1982.

Golvers, Noël, *The Astronomia europaea of Ferdinand Verbiest, S.J. (Dillingen, 1687): text, translation, notes and commentaries*, Steyler Verlag, 1993.

Gorman, Michael John, 'From "the eyes of all" to "usefull Quarries in philosophy and good literature": consuming Jesuit science 1600–1665', in John W O'Malley, Gauvin Alexander Bailey, Steven J Harris, and T Frank Kennedy (eds), *The Jesuits: cultures, sciences and the arts 1540–1773*, University of Toronto Press, 1999, 170–189.

Guo, Shuchun 郭書春 (ed), *Zhong guo ke xue ji shu dian ji tong hui. Shu xue juan* 中國科學技術典籍通彙。 數學卷 'Comprehensive archive Chinese science and technology: mathematics', 5 vols, Henan jiaoyu chubanshe, 1993.

Han, Qi, "Patronage scientifique et carrière politique: Li Guangdi entre Kangxi et Mei Wending", *Etudes chinoises*, 16 (1997), 7–37.

Han, Qi, 'Astronomy, Chinese and Western: the influence of Xu Guangqi's views in the early and mid-Qing', in Catherine Jami, Peter M Engelfriet, and Gregory Blue (eds), *Statecraft and intellectual renewal: the cross-cultural synthesis of Xu Guangqi (1562–1633)*, Brill, 2001, 360–379.

Han, Qi, 'L'enseignement des sciences mathématiques sous le règne de Kangxi (1662–1722) et son contexte social', in Christine Nguyen–Tri and Catherine Despeux (eds), *Education et instruction en Chine. II. Les formations spécialisées*, Centre d'études chinoises and Peeters, 2003, 69–88.

Han, Qi 韓琦, and Catherine Jami 詹嘉玲, 'Kang xi shi dai xi fang shu xue zai gong ting de chuan bo—yi An Duo he "Suan fa zuan yao zong gang" de bian zuan wei li 康熙時代西方數學在宮廷的傳播—以安多和《算法纂要總綱》的編纂為例' 'The transmission of Western mathematics at the Kangxi court—the case of Antoine Thomas' "Outline of the essentials of calculation"', Zi ran ke xue shi yan jiu 自然科學史研究 22 (2003), 145–156.

Hashimoto, Keizo, *Hsü Kuang-ch'i and astronomical reform: The process of the Chinese acceptance of Western astronomy, 1629-1635,* Kansai University Press, 1988.

Hashimoto, Keizo, and Jami, Catherine, 'From the *Elements* to calendar reform: Xu Guangqi's shaping of mathematics and astronomy', in Catherine Jami, Peter M Engelfriet, and Gregory Blue (eds), *Statecraft and intellectual renewal: the cross-cultural synthesis of Xu Guangqi (1562-1633)*, Brill, 2001, 263–279.

Jami, Catherine, 'Jean-François Foucquet et la modernisation de la science en Chine: la "Nouvelle méthode d'algèbre"', Master's Thesis, University of Paris 7, 1986.

Jami, Catherine, 'Rencontre entre arithmétiques chinoise et occidentale au XVIIᵉ siècle', in Paul Benoit, Karine Chemla, and Jim Ritter (eds), *Histoire de fractions, fractions d'histoire*, Birkhaüser, 1992, 351–373.

Jami, Catherine, 'History of mathematics in Mei Wending's (1633–1721) work', *Historia Scientiarum* 53 (1994), 157–172.

Jami, Catherine, 'From Clavius to Pardies: the geometry transmitted to China by Jesuits (1607–1723)', in Federico Masini (ed), *Western humanistic culture presented to China by Jesuit missionaries (17th-18th centuries)*, Institutum Historicum S.I., 1996, 175–199.

Jami, Catherine, 'Giulio Aleni's contribution to geometry in China: the *Jihe yaofa*', in Tiziana Lippiello and Roman Malek (eds), '*Scholar from the West': Giulio Aleni S.J. and the dialogue between Christianity and China*, Steyler Verlag, 1997, 553–572.

Jami, Catherine, 'Abacus (Eastern)', in Robert Bud and Deborah Jean Warner (eds), *Instruments of science: a historical encyclopedia*, Science Museum; National Museum of American History, Smithsonian Institution, 1998a, 3–5.

Jami, Catherine, 'Mathematical knowledge in the *Chongzhen lishu*', in Roman Malek (ed), *Western learning and christianity in China: The contribution of Johann Adam Schall von Bell, S.J. (1592-1666)*, Sankt Augustin: China-Zentrum and Monumenta Serica Institute, 1998b, 661–674.

Jami, Catherine, 'Teachers of mathematics in China: the Jesuits and their textbooks (1580–1723)', *Archives internationales d'histoire des sciences*, 52 (2002a), 159–175.

Jami, Catherine, 'Western learning and imperial control: the Kangxi Emperor's (r. 1662–1722) Performance', *Late Imperial China*, 23 (2002b), 28–49.

Jami, Catherine, 'Légitimité dynastique et reconstruction des sciences en chine au XVIIᵉ siècle: Mei Wending (1633–1721)', *Annales* 59 (2004), 701–727.

Jami, Catherine, 'For whose greater glory? Jesuit strategies and science during the Kangxi reign', in Xiaoxin Wu (ed), *Encounters and dialogues: changing perspectives on Chinese-Western exchanges from the sixteenth to the eighteenth centuries*, Steyler Verlag, 2005, 211–226.

Jami, Catherine, 'Western learning and imperial scholarship: the Kangxi emperor's study', *East Asian Science, Technology and Medicine*, 27 (2007), 144–170.

Jami, Catherine, and Qi Han, 'The reconstruction of imperial mathematics in China during the Kangxi reign (1662–1722)', *Early science and medicine,* 8 (2003), 88–110.

Kawahara Hideki 川原秀城, 'Ritsureki engen to Kato Rakusho 律曆淵源と河圖洛書' 'The "Origins of musical harmony and calendrical astronomy" and the "Hetu" and "Luoshu"' *Chūgoku kenkyū shūkan* 中国研究週刊 16 (1995), 1319–1410.

Lam, Lay Yong, *A critical study of the Yang Hui suan fa: a thirteenth-century Chinese mathematical treatise,* Singapore University Press, 1977.

Lam, Lay Yong, and Tian Se Ang, *Fleeting footsteps: tracing the conception of arithmetic and algebra in ancient China,* World Scientific, 2004.

Landry-Deron, Isabelle, 'Les mathématiciens envoyés en Chine par Louis XIV en 1685', *Archive for the History of Exact Sciences,* 55 (2001), 423–463.

Leitão, Henrique, 'Jesuit mathematical practice in Portugal, 1540–1759', in Mordechai Feingold (ed), *The new science and Jesuit science: seventeenth century perspectives,* Kluwer, 2002, 229–247.

Leitão, Henrique, 'The contents and context of Manuel Dias' *Tianwenlüe*', in Luis Saraiva and Catherine Jami (eds), *History of mathematical sciences: Portugal and East Asia III. The Jesuits, the Padroado and East Asian science,* World Scientific Publishing, 2008, 99–121.

Li, Yan, and Shiran Du, *Chinese mathematics: a concise history,* translated John N Crossley and Anthony W–C Lun, Clarendon Press, 1987.

Martzloff, Jean-Claude, 'La compréhension chinoise des méthodes démonstratives euclidiennes au XVIIᵉ siècle et au début du XVIIIᵉ', in *Actes du IIe Colloque international de sinologie: les rapports entre l'Europe et la Chine au temps des Lumières. Chantilly, 16–18 septembre 1977,* Les Belles Lettres, 1980.

Martzloff, Jean–Claude, *Recherches sur l'œuvre mathématique de Mei Wending (1633–1721),* Collège de France, 1981.

Martzloff, Jean-Claude, 'Clavius traduit en chinois', in Luce Giard (ed), *Les jésuites à la Renaissance: Système éducatif et production du savoir,* PUF, 1995, 309–332.

Martzloff, Jean-Claude, *A history of Chinese mathematics,* Springer, 1997.

Needham, Joseph, *Science and civilisation in China. 4: Physics and Physical technology, part 1: Physics,* Cambridge University Press, 1962.

Pan, Yining 潘亦寧, 'Zhong xi shu xue hui tong de chang shi: yi *Tong wen suan zhi* (1614 nian) de bian zuan wei li 中西數學會通的嘗試：以"同文算指' (1614年) 的編纂為例.' 'Attempts at integrating Chinese and Western mathematics: a case study of the "instructions for calculation in common script"', *Zi ran ke xue shi yan jiu* 自然科學史研究, 25 (2006), 215–226.

Peterson, Willard, 'Calendar reform prior to the arrival of missionaries at the Ming court', *Ming Studies,* 21 (1968), 45–61.

Peterson, Willard, 'Western natural philosophy published in late Ming China', *Proceedings of the American Philosophical Society,* 117 (1973), 295–322.

Peterson, Willard, 'Learning from heaven: the introduction of Christianity and other Western ideas into late Ming China', in Denis Twitchett and Frederick W Mote (eds), *Cambridge history of China: the Ming dynasty 1368–1644,* Cambridge University Press, 1998, 708–788.

Ratio studiorum: Plan raisonné et institution des études dans la Compagnie de Jésus, Bilingual edition Latin and French. Presented by Adrien Demoustier & Dominique Julia, translated by Léone Albrieux & Dolorès Pralon-Julia, annotated by Marie-Madeleine Compère, Belin, 1997.

Reich, Karen, 'The 'Coss' tradition in algebra', in Ivor Grattan-Guinness (ed), *Companion encyclopedia of the history and philosophy of the mathematical sciences,* Routledge, 1994, 192–199.

Ricci, Matthieu, and Trigault, Nicolas, *Histoire de l'expédition chrétienne au royaume de la Chine 1582–1610,* Desclée de Brouwer, 1978 (annotated reprint; original edition: Lille, 1617).

Romano, Antonella, *La Contre-réforme mathématique: constitution et diffusion d'une culture mathématique jésuite à la Renaissance (1560–1640),* Ecole Française de Rome, 1999.

Romano, Antonella, 'Arpenter la "vigne du Seigneur"? Note sur l'activité scientifique des jésuites dans les provinces extra-européennes (XVIᵉ–XVIIᵉ siècles)', *Archives Internationales d'Histoire des Sciences*, 52 (2002), 73–101.

Romano, Antonella, 'Teaching mathematics in Jesuit schools: programs, course content, and classroom practices', in John W O'Malley, Gauvin Alexander Bailey, Steven J Harris, and T Frank Kennedy (eds), *The Jesuits II: cultures, sciences and the arts 1540–1773*, University of Toronto Press, 2006, 355–370.

SKQS: Ying yin Wen yuan ge Si ku quan shu 景印文淵閣四庫全書 'Facsimile of the "Complete library in four sections" kept at the Wenyuan Pavilion' 1500 vols, Taibei shangwu yinshuguan, 1986.

Standaert, Nicolas, (ed), *Handbook of Christianity in China*, Brill, 2001.

Standaert, Nicolas, 'European astrology in early Qing China: Xue Fengzuo's and Smogulecki's translation of Cardano's Commentaries on Ptolemy's *Tetrabiblos*', *Sino-Western cultural relations journal*, 23 (2001b), 50–79.

Tian, Miao 田淼, *Zhong guo shu xue de xi hua li cheng* 中國數學的西化歷程 'The process of Westernization of Chinese mathematics', Shandong jiaoyu chubanshe, 2005.

Volkov Alexei, 'Counting rods', in Robert Bud and Deborah Jean Warner (eds), *Instruments of science: a historical encyclopedia*, Science Museum; National Museum of American History, Smithsonian Institution, 1998, 155–156.

Wang Zhongmin 王重民 (ed), *Xu Guangqi ji* 徐光啓集 'Collected writings of Xu Guangqi', 2 vols, Shanghai guji chubanshe, 1984.

Wardy, Robert, *Aristotle in China: Language, categories, and translation*, Cambridge University Press, 2000.

Yabuuti, Kiyosi, *Une histoire des mathématiques chinoises*, translated Catherine Jami and Kaoru Baba, Belin, 2000.

Ziggelaar, August, *Le physicien Ignace Gaston Pardies, S.J. (1636–1673)*, Odense Universitetsforlag, 1971.

The internationalization of mathematics in a world of nations, 1800–1960

Karen Hunger Parshall

Mathematics has a history with elements of both the contingent and the transcendent. Over the course of the nineteenth century, as the emergence of nation states increasingly defined a new geopolitical reality in Europe, competition among states manifested itself in the self-conscious adoption of the contingent, cultural standards of those states viewed as the 'strongest'. In the case of mathematics, these self-consciously shared cultural standards centred on educational ideals, the desire to build viable and productive professional communities with effective means of communication, and the growing conviction that personal and national reputation was best established on an international stage (Parshall 1995).

In this context, mathematics also increasingly became a 'language spoken' and an endeavor developed internationally, that is, between and among the mathematicians of different nations.[1] For example, in the late nineteenth and early

---

1. The terminology is important. The word 'international' connotes, as indicated here, something *shared between or among* mathematicians. 'Internationalization', the topic of this chapter, is the process by which a globalized community of mathematicians, which *shares* a set of values or goals, has developed. That process, however, has sometimes involved merely 'transnational' communication, that is, communication *across* national borders, whether or not that communication is understood or appreciated. Transnational communication may ultimately lead to mutually appreciated, *shared* values and goals, but this is not a necessary consequence. The words 'international' and 'transnational' will be used in these respective senses in what follows. For more on the terminology that has developed in the historical literature on the process of the internationalization of science, see Parshall and Rice (2002, 2–4).

twentieth centuries, an Italian style of algebraic geometry with its own very idiosyncratic method of theorem formation and proof—a language of algebraic geometry that essentially only Italians spoke—developed in the context of a newly united Italian nation state seeking to demonstrate its competitiveness in the international mathematical arena and in parallel to the very different German tradition (Brigaglia and Ciliberto 1995). By the mid-twentieth century, however, following the advent of modern algebra with its structural approach to, and organization of, mathematics, algebraic geometers whether in the British Isles, Germany, or Italy, or in the United States, China, or Japan all spoke largely the same, nationally transcendent, mathematical language and tackled important, open problems recognized as such by all (Schappacher 2007).

That mutual recognition had stemmed, among other things, from the internationalization of journals and from the institutionalization of the International Congresses of Mathematicians (ICMs) beginning in 1897 for the direct communication of mathematical results and research agendas. It also manifested itself, at least symbolically, in the awarding of the first Fields Medals in 1936 in recognition of that mathematical work judged 'the best' worldwide. This chapter traces the evolution of mathematics as an international endeavor in the context both of the formation of professional communities in a historically contingent, geopolitical world and of the development of a common sense of research agenda via the evolution of a nationally transcendent mathematical language.

The establishment of national mathematical communities in the nineteenth century

Although scientific communities began to coalesce in the seventeenth century around societies like the Accademia dei Lincei in Rome, the Royal Society in London, and the Académie des Sciences in Paris, the evolution of national mathematical communities, indeed the evolution of national communities regardless of the specialty, was largely a nineteenth-century phenomenon. In mathematics as well as in other academic disciplines, Prussia was in the vanguard in the last half of the nineteenth century, serving as a model for other emergent nation states and ultimately supplanting France as the dominant mathematical nation in western Europe (Grattan-Guinness 2002).

Defeated during the Napoleonic Wars at the beginning of the nineteenth century, Prussia had responded with a major political, socioeconomic, and educational reorganization aimed at safeguarding against a similar humiliation in the future. One of the masterminds behind the educational reforms, Wilhelm von Humboldt, used the new University of Berlin (founded in 1810) as a platform from which to launch a neohumanist educational agenda aimed at 'provid[ing] a model for scholarship as well as an idealistic framework for galvanizing the German people into action' (Pyenson 1983, 6). In particular, the classical languages and mathematics, but also the physical sciences, were emphasized in an institutional context that was unfettered by political or religious concerns, and that fostered teaching and pure research over what were perceived as the more utilitarian concerns of the French. This evolved into the twin ideals of *Lehr- und Lernfreiheit*, the freedom to teach and to learn in a politically and religiously disinterested university environment characterized by the tripartite mission of teaching and the production of both original research and future researchers. Universities in Berlin, Königsberg, and ultimately Leipzig, Erlangen, Göttingen, and elsewhere produced a generation of mathematicians who matured as researchers not only in professorial lecture halls but also in targeted mathematical seminars. The research they generated, moreover, appeared on the pages of specialized journals like Crelle's *Journal für die reine und angewandte Mathematik* (founded in 1826) and later the *Mathematische Annalen* (founded in 1869).

In the last half of the nineteenth century and up to the outbreak of World War I, educational reformers in general and mathematical aspirants in particular from China (Dauben 2002, 270), Italy (Bottazzini 1981), Japan (Sasaki 2002, 236–238), Spain (Ausejo and Hormigón 2002, 51), the United States (Parshall and Rowe 1994), and other countries took their lead from Prussia in crafting broad reforms as well as more specific mathematics curricula that aimed at transplanting to, and naturalizing in, their respective soils the perceived fruits of the Prussian system. One result of this transplantation and naturalization was the consolidation and growth of mathematical research communities in a number of national settings between the closing decades of the nineteenth century and the opening decades of the twentieth.

After its defeat in the Franco–Prussian War of 1870–1871, France, too, moved toward reforms of its educational system. French scientists, in fact, had long been warning that they were falling behind the Germans (Grattan-Guinness 2002, 24–25; Gispert 2002). In the United States, the Civil War that had divided the nation in the years from 1861 to 1865 was followed by a so called Gilded Age that witnessed not only the development of federally funded institutions of higher education—the land-grant universities—for the promotion especially of the practical sciences of agriculture, mining, and engineering, but also the establishment of new, privately endowed universities. The presidents of both of these new kinds of institutions consciously looked across the Atlantic for exemplars on which to model their new educational experiments. In importing the research ethos of the Prussian universities, two of the privately endowed universities, the Johns Hopkins University (founded in 1876) and the University of Chicago (founded in 1892), set the tone for higher educational reform in the United States. In mathematics, this translated into the formation of at least two programmes that enabled research-level mathematical training competitive with—although not yet equal to—that attainable, for example, at Berlin or Göttingen (Parshall 1988; Parshall and Rowe 1994, 367, note 9). At the University of Chicago, in particular, two of the three original members of the mathematics faculty—Oskar Bolza and Heinrich Maschke—were Göttingen-trained, German mathematicians, and they, together with their American colleague E H Moore, directly imported the ideas of mathematicians like Felix Klein on elliptic and hyperelliptic function theory and David Hilbert on the foundations of mathematics to their American students (Parshall and Rowe 1994, 372–401). Those students—independently and in concert with their mentors—embraced and extended the mathematical ideas to which they were exposed.[2] In so doing, they participated in what was an increasingly transatlantic mathematical dialogue on research questions of common interest,[3] although this kind of direct importation of mathematical ideas did not dissuade American mathematical aspirants, especially in the 1880s, 1890s, and in the first decade of the twentieth century, from travelling abroad for postgraduate training (Parshall and Rowe 1994, 189–259 and 439–445).

By the outbreak of World War I, America's older colleges, notably Harvard, Yale, and Princeton, had made the transition from undergraduate colleges to research-oriented universities. Together, these and other institutions of higher education contributed to the formation of an American mathematical research community that coalesced around the New York Mathematical Society at its founding in 1888 and then around its reincarnation in 1894 as the American Mathematical Society.[4] This national community also sustained specialized journals like the *American Journal of Mathematics* (founded in 1878), the *Annals of Mathematics* (founded in 1884), and the *Transactions of the American Mathematical Society* (first published in 1900) that actively fostered the communication of mathematical results (Parshall and Rowe 1994, 427–453).

If the United States provides an illustration of a national mathematical community that formed in the nineteenth century in fairly direct emulation of the

2. Students from Italy—notably, Luigi Bianchi, Gregorio Ricci-Curbastro, and Gino Fano—also went to Germany expressly to work with Felix Klein first at the Technische Hochschule in Munich from 1875 to 1880 and then at Göttingen after 1886.

3. See, for example, Fenster (2007) for an account of the transnational development between A Adrian Albert in the United States and Richard Brauer, Emmy Noether, and Helmut Hasse in Germany of the theory of finite-dimensional algebras over the rationals.

4. The American Mathematical Society (AMS) modeled itself on the London Mathematical Society (LMS), which had formed in 1865 (and which, despite its name, was a national society). The LMS was the first such society but other national societies soon followed; for example, the Société mathématique de France began in 1872 and the Tokyo Mathematical (later Mathematico-Physical) Society started in 1877. The more localized Moscow Mathematical Society actually predated them all; it was founded in 1864. By the early decades of the twentieth century even more countries—like Spain (Ausejo and Hormigón 2002, 53–57), Italy, Japan, and China (see below)—had followed suit. The specialized national mathematical society—like the specialized mathematical journal—came to define national mathematical communities internationally.

Prussian model, England represents a country in which a national mathematical community developed with only occasional glances across the Channel, and those perhaps more at France than at Germany. In 1830, Charles Babbage famously caricatured English science in his *Reflections on the decline of science in England*. For Babbage, that decline had resulted from many factors, not the least of which were the ineffectiveness of the Royal Society and the absence of true cultural and professional inducements for science in England.

As with all caricatures, Babbage's contains elements of truth. His rhetorical salvos—as well as those of others like John F W Herschel and Augustus Bozzi Granville—came just as the new British Association for the Advancement of Science was being founded and the Royal Society of London was entering into a period of reorganization and renewal. If English science had been in decline before 1830, its trajectory had a strongly positive slope by the middle of the nineteenth century as exemplified by John Couch Adams's mathematical prediction—independent of that of the French astronomer, Urbain Leverrier—of the existence of the planet Neptune in 1845–6. As the case of Adams also suggests, if, as Herschel famously averred in 1830, 'in mathematics we have long since drawn the rein, and given over a hopeless race', things were improving on that score as well (Babbage 1830, ix).

Although mathematics had long been published in the British Isles in the context of the journals of general science societies, the decades immediately following mid-century witnessed there as in Germany, France, Italy, Russia, and elsewhere the development of specialized, research oriented journals that helped to distinguish a community of mathematical researchers (Despeaux 2002).[5] Of particular importance in the British context was the *Quarterly Journal of Pure and Applied Mathematics* which began under that title in 1855 but which had resulted from an evolutionary process that had transformed the highly localized, undergraduate-oriented *Cambridge Mathematical Journal* (founded in 1837) into the more self-consciously research-oriented and trans-Britannic *Cambridge and Dublin Mathematical Journal* (in 1845) (Crilly 2004).

In 1855 and under the editorial leadership of James Joseph Sylvester and Norman Ferrers, the *Quarterly Journal* not only followed France's *Journal de mathématiques pures et appliquées* in emulating in name Crelle's *Journal für die reine und angewandte Mathematik* but also specifically articulated an internationalist view (albeit with nationalistic overtones) of the mathematical endeavor. As the editors put it in their 'address to the reader' in the journal's first number, their aim was

---

5. Crelle's *Journal für die reine und angewandte Mathematik* and the *Mathematische Annalen* have already been mentioned. In France, among others, were Liouville's *Journal de mathématiques pures et appliquées* (begun in 1836) and later the *Bulletin de la Société mathématique de France* (started in 1873), while Italy supported the publication of, for example, the *Annali di matematica pura ed applicata* (first published in 1858), and mathematicians in Moscow launched *Matematicheskii Sbornik* in 1866.

to 'communicate a general idea of *all* that is passing in mathematical circles, *both at home and abroad*, that can be of interest to Mathematicians as such' (Parshall 2007, 139; my emphasis). To that end, they actively fostered contributions from other countries, and especially from France, thanks both to the presence of Charles Hermite on the editorial board and to the ongoing efforts particularly of Sylvester (Despeaux 2002, 243–271).[6] In this way, they brought some of the latest foreign mathematical research directly to their fellow countrymen in an effort to keep them abreast of what was being done abroad. It was not, however, just a matter of keeping current; it also involved becoming actively competitive on what was recognized as an increasingly international mathematical stage. The editors held 'that it would be little creditable to English Mathematicians that they should stand aloof from the general movement, or else remain indebted to the courtesy of the editors of foreign Journals, *for the means of taking part in a rapid circulation and interchange of ideas by which the present era is characterised*' (Parshall 2007, 139; my emphasis). No longer would the British Isles be mathematically insular.[7] It was a national participant in what was increasingly viewed as a trans-European, if not yet fully international, mathematical endeavor.[8]

Transnational and international impulses in the closing decades of the nineteenth century

Mathematics, as the views expressed by Sylvester and his editorial team illustrate, came to be seen during the last half of the nineteenth century as a body of knowledge that develops effectively through the communication of ideas across national political borders. Sometimes that communication produces—as in the case of Liouville and various of his contributions to, for example, mechanics, potential theory, and differential geometry—new results inspired by and built on the work of mathematicians in other countries (Lützen 2002, 95–100). Or it serves, as in the case of Cesare Arzelà during the 1886–7 academic year, to provide a rich literature—the works of Eugen Netto, Peter Lejeune Dirichlet, Joseph Serret, Camille Jordan—from which to craft the first course of lectures on Galois theory ever to be given in Italy (Martini 1999). As these examples illustrate, *transnational* communication could lead to an *internationally* shared set of research

6. Other 'national' journals also accepted and encouraged contributions from abroad in an effort at international communication, for example, Liouville's *Journal* (Lützen 2002, 91–93).

7. Although some Russian mathematicians like Pafnuti Chebyshev traveled to western Europe to make scientific contacts, and some mathematicians like J J Sylvester journeyed to Russia, the Russian mathematical community experienced first a kind of linguistic isolation and then also a political isolation relative to the rest of Europe in the nineteenth and well into the twentieth century. This did not, however, prevent the formation there of strong mathematical traditions in number theory at St Petersburg University and in function theory at Moscow University.

8. On the development of mathematical Europe, see Goldstein *et al.* (1996).

goals. Communication could, however, be complicated by the growing spirit of active competition not only between individual, emerging national communities but also between individuals within those nations to establish their reputations. A striking example of this phenomenon was the development in the British Isles and in Germany of two distinct approaches to, and languages for, the theory of invariants.

Although examples of what would come to be known as invariants may be found, like the germs of so much other modern mathematics, in Gauss's *Disquisitiones arithmeticæ* of 1801, invariant theory developed in a largely algebraic context in the British Isles and in a primarily number-theoretic and geometric context in Germany beginning in the 1840s and continuing strongly through the 1880s (Parshall 1989). In both settings, the basic question was the same: given a homogeneous polynomial in n (although in practice usually just two or three) variables with real coefficients, find all expressions in the coefficients (invariants) or in the coefficients and the variables (covariants) that remain unchanged under the action of a linear transformation.

As the simplest example, and this example appeared in the *Disquisitiones*, consider $Q = ax^2 + 2bxy + cy^2$ and a nonsingular linear transformation of the variables x and y which takes x to $mx + ny$ and y to $m'x + n'y$, for m, n, m', and n' real numbers and for $mn' - m'n \neq 0$. Applying this transformation to Q gives $Ax^2 + 2Bxy + Cy^2$, where A, B, and C are obviously expressions in a, b, c, m, n, m', and n'. It is easy to see that the following equation holds: $B^2 - AC = (mn' - m'n)^2 (b^2 - ac)$, that is, the expression $b^2 - ac$ in the coefficients of Q, the discriminant, remains invariant up to a power of the determinant of the linear transformation.

Developing a theory of how to find all such expressions occupied Arthur Cayley, J J Sylvester, George Salmon, and others in the British Isles as well as Otto Hesse, Siegfried Aronhold, Alfred Clebsch, Paul Gordan, and others in Germany. The British employed very concrete calculational techniques to seek explicit Cartesian expressions of the invariants, as in the form $b^2 - ac$ above; the Germans developed a more abstract notation and approach, although they, too, aimed at finding complete systems of covariants for homogeneous polynomials of successive degrees. Each group also worked largely in isolation from the other, with the British publishing primarily in their own journals and the Germans in theirs, until 1868 when Gordan proved the finite basis theorem—namely, for any homogeneous form in two variables, a finite (minimum generating) set of covariants generates them all—and explicitly called attention to a major flaw in the British invariant-theoretic superstructure. The British, and especially Sylvester, then went to work to correct the error and to vindicate their techniques. Nothing less than national mathematical pride was at stake, yet neither side could really understand the work of the other. They had literally been speaking different mathematical languages that had been created in their respective national contexts, yet their confrontation over the finite basis theorem also evidenced the increasingly transnational—if perhaps not yet fully international—nature of mathematics by the last quarter of the nineteenth century (Parshall 1989).

Coincidentally, but symptomatic of the kind of situation that had presented itself in invariant theory, a new type of mathematical publication, the reviewing journal, was launched in Germany in 1868 expressly 'to provide for those, who are not in a position to follow independently every new publication in the extensive field of mathematics', to give them moreover 'a means to gain at least a general overview of the development of the science', and 'to ease the efforts of the scholar in his search for established knowledge'.[9] The *Jahrbuch über die Fortschritte der Mathematik* represented a collaborative effort among mathematicians to survey the international mathematical landscape and to report, in German, on the research findings of mathematicians throughout Europe and eventually in the United States and elsewhere. By the end of the century, the *Jahrbuch* had been joined by two additional reviewing journals—the French *Bulletin des sciences mathématiques et astronomiques* (begun in 1895) and the Dutch *Revue semestrielle des publications mathématiques* (started in 1897)—in the ongoing quest effectively to disseminate mathematical results transnationally (Siegmund-Schultze 1993, 14–20).[10]

These reviewing efforts, moreover, were supplemented by great synthetic undertakings like the *Enzyclopädie der mathematishen Wissenschaften,* begun in 1894 under the direction of Felix Klein, and the French translation and update, the *Encyclopédie des sciences mathématiques,* started in 1904 with Jules Molk as editor. Both of these works aimed, in some sense, to go beyond the reviewing journals by surveying contemporary mathematics and indicating promising lines for future research. In so doing, they had the potential to create shared research agendas across national boundaries.[11]

Transnational impulses also manifested themselves at this time in the form of new, expressly international research journals, although these ventures also had nationalistic or regionalistic overtones. As one case in point, the Norwegian mathematician Sophus Lie encouraged his Swedish friend and fellow mathematician Gösta Mittag-Leffler to found a new journal, *Acta mathematica* (first

---

9. For the quote, see the 'Vorrede' of the *Jahrbuch* as translated in Despeaux (2002, 297–298).

10. In the twentieth century, the *Zentralblatt für Mathematik und ihre Grenzgebiete* (begun in 1931 by the German publishing house of Julius Springer) and the *Mathematical Reviews* (started in 1940 by the American Mathematical Society) represented two rival, national, international reviewing journals. The *Mathematical Reviews* was founded largely in response to the dismissal of the Italian Jewish mathematician, Tullio Levi-Civita, as editor of the *Zentralblatt* and to the *Zentralblatt*'s National Socialist policy of debarring Jewish mathematicians from reviewing the work of German mathematicians (Siegmund-Schultze 2002, 340–341). As the case of these two journals makes clear, even the ostensibly international—or at least transnational—reviewing journal was not immune to broader geopolitical currents.

11. Translations were yet another manifestation of efforts at transnational communication. On, for example, a sustained nineteenth-century French translation effort, see Grattan-Guinness (2002, 39–44).

published in 1882), that was to be international in outlook while highlighting the best of Scandinavian mathematical research (Barrow-Green 2002, 140–148). Similarly, the Italian mathematician Giovan Battista Guccia was instrumental not only in founding the Circolo Matematico di Palermo in 1884, a society that despite its local name soon became Italy's *de facto* national mathematical organization, but also the Circolo's *Rendiconti* (first published in 1887). By the outbreak of World War I, both the Circolo and its *Rendiconti* had succeeded in the agenda Guccia had explicitly articulated, namely, 'to internationalize, to diffuse, and to expand mathematical production of the whole world, making full use of the progress made by modern civilization in international relations' (Brigaglia 2002, 187–188).

The International Congresses of Mathematicians and the impact of World War I

Guccia's efforts in Italy, especially in the 1890s and up to the outbreak of World War I, reflected a widely spreading sense among mathematicians that the time was ripe for fostering greater international contact and cooperation. The German mathematician Georg Cantor was one of the first actively to advocate the idea of mounting an actual international congress of mathematicians. Frustrated by the hostile reception that his work on transfinite set theory had received within the hierarchical and paternalistic German university system, Cantor sought as early as 1890 to create a venue for the presentation of new mathematical ideas that would be free of internal mathematical politics and prejudices. In Cantor's view, an international arena would provide the openness and diversity of perspective that he found so lacking in his parochial national context (Dauben 1979, 162–165). By 1895, he had succeeded through what was effectively an international letter-writing campaign in enlisting the support for his efforts of mathematicians like Charles Hermite, Camille Jordan, Charles Laisant, Émile Lemoine, and Henri Poincaré in France, Felix Klein and Walther von Dyck in Germany, and Alexander Vassiliev in Russia, among others (Lehto 1998, 3).

After much discussion and negotiation, the first International Congress of Mathematicians was held in 1897 in Zürich, in politically neutral Switzerland. In all just over two hundred mathematicians from sixteen countries—among them, Austria-Hungary, Finland, France, Germany, Great Britain, Italy, Russia, Switzerland, and the United States—took part in the congress. In addition to hearing a full and rich program of mathematical lectures, the participants succeeded in formulating a set of objectives for future congresses. These events would aim 'to promote personal relations among mathematicians of different countries', to survey 'the present state of the various parts of mathematics and its

applications and to provide an occasion to treat questions of particular importance', 'to advise the organizers of future Congresses', and 'to deal with questions
related to bibliography, terminology, etc. requiring international cooperation'
(Lehto 1998, 7–11, quotes on 9–10). In light of the emphasis on treating 'questions of particular importance' and on issues like terminology that might require
'international cooperation', those present at the Zürich ICM clearly foresaw a
mathematical world in which researchers, regardless of their nationalities, communicated in ever more common mathematical terms in their pursuit of answers
to questions commonly viewed as 'important'. At the second ICM, held in Paris
in 1900, David Hilbert did much to shape this new, international, mathematical
world order.

In the address he gave on 'Mathematical problems', Hilbert famously charted
the courses of a number of mathematical fields by isolating in them what he viewed
as key unsolved problems. As he explained in his introductory remarks, he aimed
'tentatively as it were, to mention particular definite problems, drawn from the
various branches of mathematics, from the discussion of which an advancement
of science may be expected' (Hilbert 1900, 7). Among these, the first six problems
highlighted what became, owing in no small part both to Hilbert's Paris lecture
and to the publication in 1899 of his *Grundlagen der Geometrie*, an emphasis in
twentieth-century mathematics on an axiomatic, foundational, and ultimately
structural approach (Mehrtens 1990, 108–165; Corry 1996, 137–183). In some
sense, this not only provided a vernacular in which mathematicians, regardless
of their nationality, could communicate, but also delineated specific structures—
groups, rings, fields, algebras, topological spaces, vector spaces, probability
spaces, Hilbert spaces, and so on—for further mathematical development.

The import of Hilbert's address at the Paris ICM was sensed immediately. In
addition to its publication in French translation in the Congress proceedings,
the address was published in German in the *Nachrichten von der königlichen
Gesellschaft der Wissenschaften zu Göttingen* and in the *Archiv der Mathematik
und Physik* as well as in English translation in the *Bulletin of the American
Mathematical Society*. German, French, and English speakers could all participate in the agenda that Hilbert had laid out.[12]

The next three ICMs took place in Heidelberg, Rome, and Cambridge, at four-
year intervals from 1904 to 1912. The number of attendees steadily increased as
did non-European participation. At the Cambridge ICM, in particular, of the five
hundred and seventy-four participants, eighty-two were non-European with two
from Africa, six from Asia, sixty-seven from North America, and seven from
South America (Lehto 1998, 14). It was decided on that occasion that, following

---

12. To date, at least sixteen of Hilbert's twenty-three problems can be considered to have been solved in
whole or in part by mathematicians from the Baltic States, France, Germany, Japan, the former Soviet Union,
and the United States (Yandell 2002).

Mittag-Leffler's invitation, the next congress would be held in Stockholm in 1916. Those plans, however, were scuttled owing to the outbreak of World War I in 1914.

The politics of internationalization in the West during the interwar period

At the war's close in 1918, Mittag-Leffler immediately renewed the invitation to Stockholm; he sensed an urgency to resume the ICMs and to get mathematics back on its international track. The new political realities that prevailed in postwar Europe worked counter to his efforts, however. The French, and especially the noted complex analyst and algebraic geometer Émile Picard, actively opposed any relations with the former Central Powers. Picard's answer to the question '*veut-on, oui ou non, reprendre des relations personnelles avec nos ennemis?*', 'do we want, yes or no, to resume personal relations with our enemies?' was a resounding 'no' (Lehto 1998, 16). While some in the British mathematical community agreed, others like G H Hardy strongly supported the resumption of normal scientific relations. Hardy, a well known pacifist, had done his best even during the war to maintain working relations with his mathematical colleagues despite the political agendas of nations. In 1915, for example, the book *General theory of Dirichlet series* that he co-authored with the Hungarian Marcel Riesz appeared as volume twenty-six in the series of Cambridge Mathematical Tracts and bore the avowal '*auctores hostes idemque amici*', 'the authors, enemies, and all the same friends' (Segal 2002, 363).

As these differing opinions make clear, there was little agreement in the immediate aftermath of the war on how best—or even whether—to proceed with the international initiatives that had begun with such promise some two decades earlier. Still, two initiatives did go forward: plans for an ICM to be held not in Stockholm but in Strasbourg in 1920 and plans for an International Mathematical Union (IMU) to be founded officially at the Strasbourg ICM and to oversee, among other things, the planning of future ICMs. Both of these efforts—international only in name in 1920—were fraught with political difficulties from the start.

First, the former Central Powers were barred from attending the Strasbourg ICM and were ineligible both for membership in the IMU and for participation in future ICMs. In the view of the majority, the Central Powers had 'broken the ordinances of civilization, disregarding all conventions and unbridling the worst passions that the ferocity of war engenders'; in order for them to be readmitted into the international confraternity of mathematicians, moreover, they 'would have to renounce the political methods that had led to the atrocities that had shocked the civilized world' (Lehto 1998, 18). As a result, Germany, in particular, the mathematical trendsetter since the mid-nineteenth century, would not be able to participate.

Second, the selection of Strasbourg as the locale for the ICM had blatantly political overtones, given that Alsace-Lorraine in general and Strasbourg in particular had been returned to French control as a result of the Germans' defeat in the war. As Mittag-Leffler bitterly put it, '*ce congrès est une affaire française qui ne peut nullement annuler le congrès international à Stockholm*', 'this congress is a French affair which can in no way annul the international congress in Stockholm' that he had originally proposed (Lehto 1998, 24).

When mathematicians finally convened in Strasbourg in September 1920, it was indeed, as Mittag-Leffler had predicted, 'a French affair'. The unwaveringly anti-German Picard was elected one of the first Honorary Presidents of the Executive Committee of the IMU as well as the President of the Strasbourg ICM, and he took the occasion of his opening ICM address publicly to uphold the decision to debar mathematicians from the former Central Powers. In his words, '*pardonner à certains crimes, c'est s'en faire le complice*', 'to pardon certain crimes is to become an accomplice in them' (Lehto 1998, 29).

These overtly political sentiments clouded not only the Strasbourg ICM but also efforts to mount the next ICM scheduled for 1924. Mathematicians from the United States and British Isles had begun to push for an end to the exclusionary rules imposed by the IMU, and only efforts by the Canadian mathematician John C Fields to host the 1924 ICM in Toronto ultimately rescued it from complete political entanglement. In some sense, matters were no better in 1928 when the Congress met in Bologna. While some in the IMU continued to insist on exclusion, Salvatore Pincherle (IMU President from 1924 to 1928 and President of the 1928 Congress) and his Italian co-organizers, implemented an open door policy at the Bologna ICM. Although some German mathematicians like Ludwig Bieberbach vociferously opposed German participation on political grounds, David Hilbert rallied his countrymen, who ultimately formed the largest non-Italian national contingent at the ICM (Lehto 1998, 33–46).

This ongoing politicization soon took its toll. By the time the next ICM concluded in Zürich in 1932, the IMU had essentially ceased to exist. The prevailing sentiment among the almost seven hundred mathematicians in attendance in Zürich was that the unabashedly political agenda of the IMU had been detrimental to the international health of the community, and that national politics should thenceforth remain separate from mathematics.

One corrective that followed was the establishment in 1932 of the Fields Medal, the equivalent in mathematics to the Nobel Prize, to be awarded on the occasion of the ICMs to acknowledge outstanding achievements made by mathematicians regardless of nationality. The first of these were given at the Oslo Congress in 1936 to the Finnish mathematician Lars Ahlfors for his work on the theory of Riemann surfaces and to the American Jesse Douglas for his solution of Plateau's problem on minimal surfaces (Monastyrsky 1997, 11). Another corrective that had, in fact, already been at work during the troubled postwar years of the ICMs was the Rockefeller Foundation and its International Education Board (IEB), which expressly sought to encourage international scientific and mathematical development in the interwar period. The Foundation, through the IEB, had, for example, funded the building of both the new Mathematics Institute in Göttingen and the Institut Henri Poincaré in Paris in the late 1920s for the international encouragement and exchange of mathematical research. Unfortunately, the activities of the Göttingen Institute were curtailed from 1933 with the rise of National Socialism in Germany and the subsequent ousting of Jews, not least the Institute's director Richard Courant (see Siegmund-Schulze, Chapter 9.4 in this volume); a little later the Institut Henri Poincaré was fundamentally affected by the outbreak of World War II (Siegmund-Schultze 2001).

Internationalization: West and East

The confused political situation in the interwar period in the West did not prevent international mathematical relations more globally, and especially between West and East.[13] Prior to the nineteenth century, Japan and China were largely closed to Western scientific influences, the most notable exception being the introduction of some Western science by Jesuit missionaries in China in the seventeenth century (Jami, Chapter 1.3 in this volume). Following the Meiji Restoration in 1868, however, Japan looked increasingly to the West for educational, scientific, and cultural models that would help them to compete more effectively in the modern world. The same became true of China after its defeat in the first and second Opium Wars (1839–42 and 1856–60) and in the first Sino-Japanese War in 1895.

In the case of Japan, Westernization was officially mandated, and it was swift. Although the infiltration of Western science—notably mathematics and aspects of naval and military science—had begun after Japan opened some of its ports to

---

13. The interwar period also witnessed international mathematical relations between the northern and southern hemispheres. In particular, soon after he took office in 1933, US President Franklin Delano Roosevelt announced what came to be known as the 'Good Neighbor Policy' between the United States and the countries of Central and South America. In the sciences and mathematics, this translated into support from private foundations like the Rockefeller Foundation and the John Simon Guggenheim Foundation for intellectual exchanges beginning in the 1930s, carrying on through the war and afterward. In 1942, for example, Harvard mathematician George David Birkhoff went on a mathematical 'good neighbor' lecture tour of Latin America (Ortiz 2003) to be followed in 1943 by his former student and then Harvard colleague Marshall Stone (Parshall 2007). These trips resulted in North American study tours for a number of talented Latin American students and in the establishment of ties between mathematical communities in the Americas.

Western concerns in the mid-1850s, it started in earnest after 1868 with the Meiji Emperor's pronouncement that 'intellect and learning shall be sought throughout the world, in order to establish the foundations of the Empire' (Sasaki 2002, 231–235, quote on 235). The implications of this for mathematics were particularly concrete. In 1872, the government decreed that the traditional Japanese style of mathematics, *wasan*, was no longer to be taught and that it was to be replaced by Western mathematics in the school curriculum. Five years later in 1877, the University of Tokyo was established as Japan's first modern university with English-trained Kikuchi Dairoku as its first professor of mathematics (Sasaki 2002, 235–237).[14]

Prussia, not England, soon came to define the standards on which Japan modeled itself, however.[15] As Inoue Kowashi, outspoken supporter not merely of Westernization but actually of Prussianization, expressed it:

[…] of all nations in present-day Europe, only Prussia is similar to us with regard to the circumstances of its unification […] If we want to make men throughout the land more conservative-minded, we should encourage the study of German and thereby allow it, several years hence, to overcome the dominance now enjoyed by the English and French (Sasaki 2002, 238).

Relative to mathematics, this had the fairly immediate impact of introducing to the University of Tokyo (which became the Imperial University in 1886) the twin objectives of high-level teaching and research in the context of both the classroom and the research seminar.

These changes were implemented by German-trained Fujisawa Rikitaro, who, after studying under Karl Weierstrass and Leopold Kronecker at the University of Berlin, earned his doctorate under the tutelage of Theodor Reye and Elwin Bruno Christoffel at the University of Strasbourg in 1886 and returned to Japan to take up a professorship in Tokyo (Sasaki 2002, 239–240). One of Fujisawa's most famous students, the algebraic number theorist Takagi Teiji, became one of the first Japanese mathematicians to star on the international mathematical stage.

Sent by the Japanese government to Germany for advanced mathematical study at Berlin and Göttingen between 1898 and 1901, Takagi learnt and embraced much of the modern language of, and approach to, algebraic number theory as it had been developing particularly in Germany in the hands of Leopold Kronecker, Hilbert, and others. Takagi joined Fujisawa at Tokyo following his

---

14. Here and below, Japanese and Chinese names are rendered in their traditional order, that is, surname followed by given name.

15. Of course, England did serve as the primary model for those mathematical societies that ultimately formed in the various corners of its far-flung empire. For example, in India, the Analytic Club, founded in 1907, became the Indian Mathematical Society in 1910. Mathematical societies independent of the London Mathematical Society did not form in Australia and New Zealand until the second half of the twentieth century, however.

foreign study tour, but found himself isolated from European mathematical developments, particularly after the outbreak of war in Europe in 1914. Even in isolation, however, he worked on algebraic number-theoretic research. The result was the 1920 publication in German in the *Journal of the College of Science* (Imperial University of Tokyo) of his stunning theorem that every (finite) abelian extension over an algebraic number field is a class field over that field and vice versa (Frei 2007, 128).

In an interesting example of internationalization, Takagi's work became known in the West and contributed fundamentally to the rapidly developing area of class field theory. This owed not so much to his lecture on it at the Strasbourg ICM in 1920 (recall that the Germans were banned from this ICM) but rather to his post-Strasbourg visit to Hamburg where the number theorist Carl Ludwig Siegel was then on the faculty. Siegel told Hilbert and Emil Artin about the result; Hilbert urged Takagi to republish it in the *Mathematische Annalen*; Artin used it essentially as the foundation on which to complete class field theory and so to solve Hilbert's ninth problem (Sasaki 2002, 241–242; Frei 2007, 127–128, 142; Yandell 2002, 219–245).

If Takagi was Fujisawa's most renowned student outside of Japan, several others went on to foster research level mathematics in Japan through their professorships at Kyoto Imperial University (founded in 1897) and at Tohoku Imperial University (opened in 1907). At Tohoku, Fujisawa's student, Hayashi Tsuruichi, founded the *Tohoku Mathematics Journal* in 1911. Japan's first international mathematics journal, *Tohoku* published predominantly in European languages and thereby served as a means of communication both of Japanese mathematical research to those outside Japan and of Western mathematical findings to those within the Japanese community (Sasaki 2002, 238–246). Such inter- and transnational contacts and initiatives were curtailed, however, following the expansion of World War II into the Pacific theater in 1941.

As for China, repeated military defeats in the nineteenth century, and especially by recently Westernized Japan in 1895, had led to the realization that not only knowledge of foreign languages but also a firm grounding in Western engineering, science, and mathematics would be critical to the country's future success in a rapidly changing geopolitical world. In addition to establishing new schools, beginning in the 1860s, in which Western mathematical techniques were often taught alongside traditional Chinese methods, active translation initiatives were launched in an effort to bring Western science and mathematics directly to the students. Moreover, as early as the 1870s, Chinese students were encouraged to pursue their studies in science and engineering in Europe, and especially in France, Germany, and England (Dauben 2002, 254–261).

Foreign study for Chinese students increased dramatically after 1900 in the wake of the Chinese Boxer Rebellion (or Boxer Uprising) against Chinese

Christians and against outside financial interests in China. An indemnity of over $300 million was imposed on the Chinese for their aggression, but ultimately at least part of the exorbitant fine was remitted by several Western countries in the form of educational initiatives aimed at benefiting the Chinese and creating international ties (Dauben 2002, 266–267, 274–275; Xu 2002, 288–296). In the case of the United States, Boxer Indemnity Scholarships and Chinese provincial governments brought hundreds of Chinese students to US institutions of higher education by 1930. These students returned to China—with PhDs from Harvard University, Cornell University, the University of Chicago, and elsewhere—to staff research oriented mathematics programs at, for example, Qinghua University in Beijing. Qinghua, in fact, launched China's first graduate program in mathematics in 1930 (Xu 2002, 290–292). With the establishment of graduate programs on the Western model, China soon began to train its own mathematicians at the research level. Many of these, like the differential geometer Chern Shiing-shen were drawn, beginning in the late 1930s, to the Institute for Advanced Study in Princeton (Xu 2002, 296–301).

And this kind of direct personal contact was two way. German mathematicians like Konrad Knopp, Wilhelm Blaschke, and Emanuel Sperner, French mathematicians such as Émile Borel, Paul Painlevé, and Jacques Hadamard, English mathematicians most notably Bertrand Russell, and American mathematicians like Norbert Wiener spent extended periods of time in China beginning in the 1920s. All of this activity had resulted by 1935 in the founding of the Chinese Mathematical Society and in the establishment of its *Journal* in 1936. These efforts were interrupted between 1938 and 1945 owing to the outbreak of the second Sino-Japanese War (1937–45), and much more severely curtailed with the formation of the People's Republic of China in 1949 (Dauben 2002, 277–280).

Internationalization in the aftermath of World War II

With the resumption of peace in 1945, a new geopolitical order emerged. Germany was divided into East and West; the Cold War between the Soviet Union and its allies and the United States and its allies was already brewing; French, German, Italian, and East European Jews had fled across the globe to the United States, to Central and South America, to Palestine, to Australia, and elsewhere (Siegmund-Schulze, Chapter 9.4 in this volume). In mathematics, these refugees represented some of the strongest researchers Europe had produced in the first half of the twentieth century; as emigrés, they directly transplanted their research agendas to their newly adopted soils. The United States, for example, which absorbed relatively large numbers of immigrants, benefited in mathematics from the introduction of research areas—within, for example, applied mathematics, probability, and statistics—that had previously been underdeveloped there. These 'new areas' took root and grew thanks to this wave of immigration (Bers 1988).

As after the end of World War I, mathematicians, and especially those in the United States like Marshall Stone, sought almost immediately to re-establish international mathematical ties. The International Congress of Mathematicians that had been scheduled to take place in Cambridge, Massachusetts, in 1940 had been cancelled. Stone and others in the United States worked not only to plan the first postwar ICM but also to re-establish the International Mathematical Union. They succeeded in both efforts. The ICM took place in Cambridge, Massachusetts in 1950, and an ICM has occurred every four years thereafter. The IMU was officially reborn in 1951 after much behind the scenes political negotiation on the part of Stone and others, which began in earnest in 1946.

Those behind the organization of a new IMU were adamant that it would not be like the IMU formed in the aftermath of World War I. As Stone put it in a speech in 1947, 'in considering American adherence to a Union, it must be borne in mind that we want nothing to do with an arrangement which excludes Germans and Japanese as such' (Lehto 1998, 76). An explicitly inclusive philosophy thus guided both the writing of new statutes for the union and the political strategies employed to see them successfully put in place. The first ten member countries— Austria, Denmark, France, Germany, Great Britain, Greece, Italy, Japan, the Netherlands, and Norway—were soon followed by Australia, Canada, Finland, Peru, and the United States (Lehto 1998, 87). By the end of the 1950s, even as the Cold War escalated, the Soviet Union and other Iron Curtain countries had also joined (Lehto 1998, 122–126). The new IMU thus came to embody Hardy's vision of international mathematical cooperation and collaboration, even though geopolitics have—repeatedly and perhaps inevitably—affected its efforts to support international mathematical colloquia and to coordinate and organize both the ICMs and the Fields Medal selection process.[16]

Conclusion

By 1962 and the occasion of the fourteenth ICM in Stockholm, it was a well established fact that a community of mathematicians existed not just in individual national settings but internationally as well. Indeed, the IMU had played a critical

16. At the first IMU general assembly held in Rome in 1952, the International Commission on the Teaching of Mathematics (ICTM) also became an official commission of the IMU. The ICTM had been founded in 1908, dissolved in 1920, and reorganized independently of the IMU in 1928 (Lehto 1998, 97). It was officially renamed the International Commission on Mathematics Instruction (ICMI) in 1954. Discourse and cooperation relative to the teaching of mathematics was thus an early and successful manifestation of internationalization within the broader mathematical community.

role in organizing that ICM by mobilizing its international constituency in order to secure, for the first time, 'the wide experience and knowledge [...] of experts *from all over the world*' in choosing 'the subjects and speakers for the one-hour addresses' and in appointing 'chairmen of the *international panels* which [...] proposed the half-hour speakers' (Lehto 1998, 158; my emphasis).

The process which had led to this 'breakthrough' (Lehto 1998, 156) had in many ways paralleled the tortured path toward international coexistence, if not always actual transnational cooperation, that the nations of the world had taken from the nineteenth into the twentieth century. Like the world's nations, mathematics was immune to the contingent effects neither of politics, nor of particular national agendas, nor of the personal agendas of individuals. Unlike the world's nations, however, mathematics and its practitioners were naturally united in a common goal, the development—increasingly expressed in a common, transcendent language—of a fundamental body of scientific knowledge.

Bibliography

Ausejo, Elena and Hormigón, Mariano, 'Spanish initiatives to bring mathematics in Spain into the international mathematical arena', in Parshall and Rice (eds), 2002, 45–60.

Babbage, Charles, *Reflections on the decline of science in England and on some of its causes*, London, 1830; reprinted in *The works of Charles Babbage*, vol 7, New York University Press, 1989.

Barrow-Green, June, 'Gösta Mittag-Leffler and the foundation and administration of *Acta Mathematica*', in Parshall and Rice (eds), 2002, 139–164.

Bers, Lipman, 'The migration of European mathematicians to America', in Peter Duren *et al.* (eds), *A century of mathematics in America: Part I*, American Mathematical Society, 1988.

Bottazzini, Umberto, 'Il diciannovesimo secolo in Italia', in Dirk J Struik, *Matematica: un profilo storico*, Il Mulino, 1981, 249–312.

Brigaglia, Aldo, 'The first international mathematical community: the Circolo matematico di Palermo', in Parshall and Rice (eds), 2002, 179–200.

Brigaglia, Aldo and Ciliberto, Ciro, *Italian algebraic geometry between the two World Wars*, Queen's University, 1995.

Corry, Leo, *Modern algebra and the rise of mathematical structures*, Birkhäuser Verlag, 1996.

Crilly, Tony, 'The *Cambridge Mathematical Journal* and its descendants: the linchpin of a research community in the early and mid-Victorian age', *Historia mathematica*, 31 (2004), 455–497.

Dauben, Joseph W, *Georg Cantor: his mathematics and philosophy of the infinite*, Harvard University Press, 1979.

Dauben, Joseph W, 'Internationalizing mathematics east and west: individuals and institutions in the emergence of a modern mathematical community in China', in Parshall and Rice (eds), 2002, 253–285.

Despeaux, Sloan, 'The development of a publication community: nineteenth-century mathematics in British scientific journals', University of Virginia, PhD dissertation, 2002.

Fenster, Della Dumbaugh, 'Research in algebra at the University of Chicago: Leonard Eugene Dickson and A Adrian Albert', in Gray and Parshall (eds), 2007, 179–197.

Frei, Günther, 'Developments in the theory of algebras over number fields: a new foundation for the Hasse norm residue symbol and new approaches to both the Artin reciprocity law and class field theory', in Gray and Parshall (eds), 2007, 117–151.

Gispert, Hélène, 'The effects of war on France's international role in mathematics, 1870–1914', in Parshall and Rice (eds), 2002, 105–121.

Goldstein, Catherine, Gray, Jeremy, and Ritter, Jim, *L'Europe mathématique: histoire, mythes, identités/Mathematical Europe: history, myths, identities*, Éditions de la Maison des sciences de l'homme, 1996.

Grattan-Guinness, Ivor, 'The end of dominance: the diffusion of French mathematics elsewhere, 1820–1870', in Parshall and Rice (eds), 2002, 17–44.

Gray, Jeremy J, and Parshall, Karen Hunger (eds), *Episodes in the history of modern algebra (1800–1950)*, American Mathematical Society and London Mathematical Societies, 2007.

Hilbert, David, 'Mathematical problems: lecture delivered before the International Congress of Mathematicians at Paris in 1900', (trans) Mary Winston Newson, *Bulletin of the American Mathematical Society* 8 (1902), 437–479; reprinted in Felix E Browder (ed), *Mathematical developments arising from Hilbert problems*, American Mathematical Society, 1976, 1–34.

Lehto, Olli, *Mathematics without borders: a history of the International Mathematical Union*, Springer-Verlag, 1998.

Lützen, Jesper, 'International participation in Liouville's *Journal de mathématiques pures et appliquées*', in Parshall and Rice (eds), 2002, 89–104.

Martini, Laura, 'The first lectures in Italy on Galois theory: Bologna, 1886–1887', *Historia Mathematica*, 26 (1999), 201–223.

Mehrtens, Herbert, *Moderne, Sprache, Mathematik*, Suhrkamp Verlag, 1990.

Monastyrsky, Michael, *Modern mathematics in the light of the Fields Medals*, A K Peters, 1997.

Ortiz, Eduardo, 'La politica interamericana de Roosevelt: George D Birkhoff y la inclusión de América Latina en las redes matemáticas internationales', *Saber y Tiempo: Revista de história de la ciencia*, 15 (2003), 53–111; 16 (2003), 21–70.

Parshall, Karen Hunger, 'America's first school of mathematical research: James Joseph Sylvester at the Johns Hopkins University 1876–1883', *Archive for History of Exact Sciences*, 38 (1988), 153–196.

Parshall, Karen Hunger, 'Toward a history of nineteenth-century invariant theory', in David E Rowe and John McCleary (eds), *The history of modern mathematics*, vol 1, Academic Press, 1989, 157–206.

Parshall, Karen Hunger, 'Mathematics in national contexts (1875–1900): an international overview', in *Proceedings of the International Congress of Mathematicians: Zurich*, vol 2, Birkhäuser Verlag, 1995, 1581–1591.

Parshall, Karen Hunger, 'A mathematical "good neighbor": Marshall Stone in Latin America (1943)', *Revista brasileira de história da matemática*, Especial 1 (2007), 19–31.

Parshall, Karen Hunger, and Rice, Adrian C, 'The evolution of an international mathematical research community, 1800–1950: an overview and an agenda', in Parshall and Rice (eds), 2002, 1–15.

Parshall, Karen Hunger, and Rice, Adrian C (eds), *Mathematics unbound: the evolution of an international mathematical community, 1800–1945*, American Mathematical Society and London Mathematical Society, 2002.

Parshall, Karen Hunger, and Rowe, David E, *The emergence of the American mathematical research community 1876–1900: J J Sylvester, Felix Klein, and E H Moore*, American Mathematical Society and London Mathematical Society, 1994.

Pyenson, Lewis, *Neohumanism and the persistence of pure mathematics in Wilhelmian Germany*, American Philosophical Society, 1983.

Sasaki, Chikara, 'The emergence of the Japanese mathematical community in the modern western style, 1855–1945', in Parshall and Rice (eds), 2002, 229–252.

Schappacher, Norbert, 'A historical sketch of B L van der Waerden's work in algebraic geometry: 1926–1946', in Gray and Parshall (eds), 2007, 245–283.

Segal, Sanford L, 'War, refugees, and the creation of an international mathematical community', in Parshall and Rice (eds), 2002, 359–380.

Siegmund-Schultze, Reinhard, *Mathematische Berichterstattung in Deutschland: Die Niedergang des 'Jahrbuchs über die Fortschritte der Mathematik'*, Vandenhoeck and Ruprecht, 1993.

Siegmund-Schultze, Reinhard, *Rockefeller and the internationalization of mathematics between the World Wars*, Birkhäuser Verlag, 2001.

Siegmund-Schultze, Reinhard, 'The effects of Nazi rule on the international participation of German mathematicians: an overview and two case studies', in Parshall and Rice (eds), 2002, 333–351.

Xu Yibao, 'Chinese–US mathematical relations, 1859–1949', in Parshall and Rice (eds), 2002, 287–309.

Yandell, Benjamin H, *The honors class: Hilbert's problems and their solvers*, A K Peters, 2002.

2. Regional

The two cultures of mathematics in ancient Greece

Markus Asper

The notion of 'Greek mathematics' is a key concept among those who teach or learn about the Western tradition and, especially, the history of science.[1] It seems to be the field where that which used to be referred to as 'the Greek miracle' is at its most miraculous. The works of, for example, Euclid or Archimedes appear to be of timeless brilliance, their assumptions, methods, and proofs, even after Hilbert, of almost eternal elegance. Therefore, for a long time, a historical approach that investigated the environment of these astonishing practices was not deemed necessary. Recently, however, a consensus has emerged that Greek mathematics was heterogeneous and that the famous mathematicians are only the tip of an iceberg that must have consisted of several coexisting and partly overlapping fields of mathematical practices (among others, Lloyd 1992, 569). It is my aim here to describe as much of this 'iceberg' as possible, and the relationships between its more prominent parts, mainly during the most crucial time for the formation of the most important Greek mathematical traditions, the fifth to the third centuries BC.

---

1. General introductions to Greek mathematics are provided by Cuomo (2001); Heath (1921); Lloyd (1973, chapters 4–5); Netz (1999a).

Reconstructions: Greek practical mathematics

Let us begin with a basic observation. Whoever looks for the first time at a page from one of the giants of Greek mathematics, say, Euclid, cannot but realize an obvious fact: these theorems and proofs are far removed from practical life and its problems. They are *theoretical*.[2] Counting, weighing, measuring, and in general any empirical methods, have no place in this type of mathematics. Somebody, however, must have performed such practices in daily life, for example, in financial or administrative fields such as banking, engineering, or architecture. Some of these fields demand mathematical operations of considerable complexity, for example, the calculation of interest or the comparison of surface areas. Occasionally, ancient authors mention such mathematical practices in passing (for example, at the end of the fifth century BC Aristophanes' play, *Wasps* 656–662). What is known about these practical forms of Greek mathematics?

Not much, obviously. Of the social elite who alone wrote and read for pleasure, most were less interested in practical mathematics, which was apparently not part of common knowledge. Occasionally, one comes across obvious arithmetical blunders, mostly by historians.[3] On the other hand, in most cases the practitioners themselves left no texts. Therefore, of all the manifold forms of practical mathematics that must have existed, only two are known a little, partly through occasional references by authors interested in other topics, partly through preserved artifacts, and, rarely, through the textual traditions of the practitioners themselves.

Pebble arithmetic was used in order to perform calculations of all kinds.[4] 'Pebbles' (*psēphoi*, an appropriate translation would be 'counters') that symbolized different numbers through different forms and sizes, were moved and arranged on a marked surface—what is sometimes called the 'Western abacus' (see Netz 2002b, 326, 342, who remarks that backgammon may well illustrate the principle). Several of these have been found, and the practitioners themselves are mentioned occasionally.[5] These must have been professionals that one could hire whenever some arithmetical problem had to be solved, not so different from professionals renting out their literacy. However, manipulating pebbles on an abacus can lead to the discovery of general arithmetical knowledge, for example the properties of even and odd, or prime numbers, or abstract rules of how to produce certain classes of numbers, for example, square numbers. I call this knowledge 'general' because it no longer has any immediate application. Here 'theoretical'

---

2. I avoid here the notions of 'pure' and 'applied' mathematics with their evaluative connotations.

3. For example, Herodotus 7.187.2; Thucydides 1.10.4 f.; Polybius 9.19.6 f. See Netz (2002a, 209–213).

4. Netz (2002b) has recently described this practice and its social setting as a 'counter culture' (for the sake of the obvious pun, he translates *psēphos* as 'counter').

5. Netz (2002b, 325) surveys the archaeological evidence (30 abaci). Pebble arithmetic is mentioned, for example, in Aeschylus, *Aga.* 570; Solon in Diogenes Laertius 1.59.

knowledge emerges from a purely practical-professional background. Some pebble arithmetic probably shows up in later Greek 'Neo-Pythagorean' arithmetic, most notably in Nicomachus of Gerasa (probably second century AD) and, slightly later, in Iamblichus of Chalcis (Knorr 1975, 131–169).

Pebble arithmeticians, as a group or as individuals, never made it into the range of subjects one could write about in antiquity, a fate they shared with most professionals that one could hire to perform specialized tasks (physicians being the most notable of the few exceptions). Therefore, nothing is known about the people who did pebble arithmetic in classical Greece, how their profession was structured, and how they transmitted their knowledge. Their body of knowledge, however, was apparently known to at least some Pythagoreans in fifth-century Greece who used it for their own, semi-religious practices.[6] Also, abstract insight into the properties of numbers, as it is typically gained by arranging pebbles (Becker 1936), must have been already widely known at the beginning of the fifth century in Greece.[7] These two cases show how specialized, practical knowledge could become abstract and move beyond the circle of specialists.

The practitioners of this art in ancient Greece, however, were probably only a tiny part of a long and remarkably stable tradition of such arithmetic professionals that originated somewhere in the ancient Near East (but, admittedly, may have changed along the way). It has recently been demonstrated, by characteristic calculation errors, that Old Babylonian scribes of the early second millennium BC *and* their Seleucid descendants must have used essentially the same accounting board to carry out multiplications of large numbers.[8] In the Middle East, the tradition resurfaces with people that are called 'ahl al-gabr' in Arabic sources of the ninth century AD, the 'algebra people'.[9] At least partly, their knowledge about algebraic problems and solutions goes back to Old Babylonian times (Høyrup 1989). It is not too bold an assumption to understand Greek pebble arithmetic as part of the same tradition (see West 1997, 23–24 for eastern influence on Greek financial arithmetic). Recently, a similar claim has been made concerning the Greek way of dealing with fractions that apparently shows Egyptian influence (Fowler 1999, 359).

The second subgroup of mathematical practitioners was concerned with measuring and calculating areas and volumes. Unlike the pebble arithmeticians, they had textual traditions, of which traces are scarce for ancient Greece, but considerable throughout the ancient Near East. These textual traditions, however, were

6. Aristotle, *Phys.* 203 a 13–15; *Metaph.* 1092 b 10–13; Theophrastus, *Metaph.* 6 a 19–22.

7. Epicharmus (early fifth century BC): fr. 23 B 2.1f. ed. Diels-Kranz; see Knorr (1975, 136); compare Burkert (1972, 434–439).

8. See Proust (2002, esp. 302), who maintains that this device was somehow based on the hand, that is, it would have been an advanced form of finger reckoning.

9. Thābit ibn Qurra in Luckey (1941, 95–96). Al-Nadim, *Fihrist* II, in Dodge (1970, 664–665) (ninth and tenth century, respectively).

sub-literary, that is, they never made it into the traditions of Greek mathematical literature (later we will see why). Therefore, most of these texts have been found inscribed on papyri, mostly written in imperial times, extant only from the Greek population in Egypt because of the favorable conditions of preservation there. There is every reason to assume, however, that in antiquity such texts were widespread in the Greek speaking world, both earlier and later. Here is an example from a first-century AD papyrus, now in Vienna:

Concerning stones and things needed to build a house, you will measure the volume according to the rules of the geometer as follows: the stone has 5 feet everywhere. Make 5 x 5! It is 25. That is the area of the surface. Make this 5 times concerning the height. It is 125. The stone will have so many feet and is called a cube. (Greek text in Gerstinger-Vogel 1932, 17)

The papyrus contained thirty-eight such paragraphs in sixteen columns, obviously meant to codify valid methods or, rather, approved procedures in textbook style (for details, see Fowler 1999, 253). Obviously, these methods are what the text calls 'the rules (*hoi logoi*) of the geometer'. Other papyri contain more difficult procedures, as the following example shows:

If there is given a parallelogram such as the one drawn below: how it is necessary, the 13 of the side squared is 169 and the 15 of the side squared is 225. (Subtract) from these the 169. 56 remains. Subtract the 6 of the base from the 10 of the top. 4 remains. Take a fourth of the 56. It is 14. From these (subtract) the 4. 10 remains, a half of which is 5. So great is the base of the right-angled triangle. Squared it is 25 and the 13 squared is 169. Subtract the 25. 144 remains, the *pleura* (= square root) of which is 12. So great is the perpendicular. And subtract the 5 from the 6 of the base. 1 remains. (Take) the one from the 10 of the top. 9 remains. So great is the remainder of the upper base of the right-angled triangle. And the 12 of the perpendicular by the 5 of the base is 60, a half of which is 30. Of so many *arourōn* (= square units) is the right-angled triangle in it. And the 12 by the 1 is 12. Of so many *arourōn* is the triangle in it. And the 12 by the 9 of the base is 108, a half of which is 54. Of so many *arourōn* is the other right-angled triangle. Altogether it is 96 units. And the figure will be such. (Pap. Ayer, col. III, first to second century AD, transl. after Goodspeed 1898, 31)

The diagram is reproduced according to the papyrus (Goodspeed 1898, 30). The algorithm gives the area of an irregularly shaped figure as the sum of triangles and rectangles, the areas of which have to be found first. In order to ensure that the reader understands the actual procedure and, thereby, the abstract method, the paradigmatic numbers (in Greek, mostly letters) are repeated in the diagram from the text. As with the first text, this is also a part of a collection of such paragraphs. More such collections are known (see Asper 2007, 200): for example, a Berlin papyrus (Schubart 1915/16, 161–170) and the better part of two treatises (*Geometrica, Stereometrica*) ascribed to Hero of Alexandria, an engineer active in Rome and Alexandria in the first century AD.

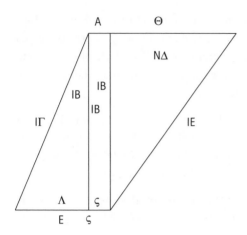

Figure 2.1.1 Diagram in Papyrus Ayer (first to second century AD)

The practical relevance of these procedures is fairly obvious, for example, for practitioners managing construction sites ('how many bricks do I need for a wall with such and such dimensions?') or in surveying ('what is the size of this piece of land?'). Both appeal to commonly shared rules and, thereby, hint at a collective of practitioners whose professional knowledge was codified in such texts.[10]

The rhetorical mood of this codification is clearly one of instruction, of a stylized dialogue between teacher and disciple: strong and frequently iterated imperatives address a second person. More importantly, the method is given as a series of steps, each of which is clearly marked. Often, the end of the procedure is marked as well. That is why these texts remind the modern reader of recipes (Robson 1999, 8). Strangely, the method itself is never explained in general terms, nor is its effectiveness proved. The actual procedure employs paradigmatic numbers that always result in whole numbers (for example, when one has to extract square roots). Obviously, the reader is meant to understand the abstract method by repeatedly dealing with actual, varied cases. The leap, however, from the actual case to the abstract method is never mentioned in these texts. Learning a general method is achieved in these texts by repeatedly performing a procedure, understanding its effectiveness and memorizing the steps by repetition, when one works through the whole text-book. Later, the professional performs his tasks by repeating the method *per analogiam*.

As I have said, these sub-literary Greek texts were written in the first and second centuries AD, mostly in Egypt. The problems they solve are so basic that one can hardly imagine that these methods were not also used much earlier in Greece. They provide, however, a glimpse at a remarkably strong tradition, of which they are probably only a local, rather late branch. Another, older part of

---

10. I understand the peculiar phrase 'how it is necessary' as shorthand for 'how one has to solve this kind of problem according to the experts'. The Greek is *hōs dei*. Goodspeed translates as 'according to the condition of the problem'.

this tradition is much better known, by thousands of texts preserved on clay tablets found in the Near East. Here is an example, problem 17 of a rather substantial textbook (BM 96954), written in Old Babylonian times (first half of the second millennium BC). The text describes a procedure of how to calculate the volume of a 'grain pile' (probably a cone-shaped body):

A triangular grain-pile. The length is 30, the width 10, the height 48. What is the grain? You: multiply 30, the length, by 10, the width. You will see 5 00. Multiply by 48, the height. You will see 4 00 00. Multiply 1 30 by 4 00 00. You will see 6 00 00 00. The grain capacity is 6 00 00 00 *gur*. This is the method. (transl. Robson 1999, 223)

If one leaves aside the differences, mostly the sexagesimal system, one clearly observes the rhetorical features that were so obvious in the Greek texts: a clearly stated practical problem, the intense appeal to the reader, the recipe-like structure, a procedure that operates with paradigmatic numbers, an abstract method that is not mentioned but illustrated by actual procedures. Admittedly, these texts vary in complexity and in their actual textual conventions. The above listed features, however, apply throughout, and not only to Old Babylonian, but also to Egyptian, Hebrew, Coptic, Arabic, and even Latin texts (compare, for example, Gandz 1929-31, 256–258; Høyrup 1996). The tradition illustrated by these occasional glimpses was alive from at least the second millennium BC well into the Middle Ages. It was so stable that some of the younger texts almost appear to be translations of the oldest ones (see the Coptic examples in Fowler 1999, 259). Moreover, some of the Greek texts show the same methods, and sometimes even the same sets of paradigmatic numbers as much older Egyptian or Babylonian ones (Gerstinger and Vogel 1932, 39, 47–50). In this tradition, the textbooks are complemented by lists of coefficients, certain factors, square roots, etc. (the Greek examples are collected in Fowler 1999, 270–276).

Although this knowledge, and the textual conventions that were meant to secure its transmission, originated in the ancient Near East, in time it moved westward and spread over the whole Mediterranean. We have some reasons to believe that people who solved practical problems with these methods were active in fifth-century Greece too (and probably much earlier). This argument relies on reconstruction by analogies: expert knowledge of several kinds came to Greece even before the classical age, especially in technical fields as diverse as architecture, writing, and medicine (to name but a few).[11] I do not see why practices that involved calculation should have been the exception. At least in some fields, 'migrant craftsmen', that is, foreigners seem to have been the transmitters. The argument from analogy seems the more compelling as one would expect numeracy to spread along the lines of literacy, especially when both probably

11. See Burkert (1992, 20–25); West (1997, 23–24, 609–612); for a general introduction to the topic Burkert (2004, 1–15).

took place in the same time and place (and had to be combined in many practices of administration).[12] Early Greek pebble arithmetic, therefore, was quite probably one of the Greek practices that resulted from acculturation with the Middle East in archaic times or even earlier, just like the Greek alphabets (see Netz 2002b, 344) and writing practices more generally.[13]

Back to the texts of these practitioners. There is no notion of definition, proof, or even argument in these texts (and hardly any concept of generality),[14] which has earned them the label 'sub-scientific' (for example, Høyrup 1989). It is important, however, to understand the lack of these features not as a general intellectual 'fault', but to explain them by the social functions of the knowledge concerned: in order to solve important problems, what is needed is not a proof of a general method, let alone of a theorem, but a reliable, accepted procedure that will lead to a reasonable result in every single case. Likewise, it is doubtful whether the notion of an abstract rule was present behind all the actual procedures. It might be a feature of the educational character of these texts that general knowledge is not explicitly stated (*pace* Damerow 2001).

As was the case with the pebble arithmeticians, almost nothing is known about the actual people who were engaged in these practices in Greece. Some assumptions, however, appear to be at least reasonable. First, since this kind of knowledge was of economic importance, it was probably not popular or widespread but rather guarded, perhaps by guild-like social structures. Performing as a practical mathematician in one of these arts was a specialized profession. For some of these people, a Greek name has survived: there was a professional group called *harpēdonaptai* ('rope-stretchers'), obviously surveyors operating with ropes for measuring purposes (Gandz 1929–31). Judging from the stability of the traditions, their group-structures must have been institutionalized somehow, including the education of disciples (maybe in apprenticeship-like relationships). Compared, however, to the complex institutional framework of, for example, the Old Babylonian scribal schools, the migrant craftsmen in Greece must have transmitted their knowledge on a much less institutionalized and, above all, less literate level.

Second, in many realms of professional knowledge, migrant craftsmen had already begun arriving in Greece from the East in the ninth century BC. There existed, for example, Phoenician work shops in seventh-century Athens. The entire vase industry, so prominent especially in Attica, seems to have employed Eastern immigrants with names like Amasis or Lydus (see Burkert 1992, 20–25).

12. See Netz (2002b, 322–324) on the concept of 'numeracy' and its conceptual interdependence with 'literacy' (and even its antecedence to it) in early Greece and the Middle East.

13. Of course, I do not suggest any *direct* contacts between Near Eastern mathematics and Greek *theoretical* mathematics (including astronomy) at any time before second-century Alexandria (see Robson 2005).

14. There are very few exceptions, most notably in the Egyptian Papyrus Rhind (seventeenth century BC, ed. Chace 1927–9). See Høyrup (2002a, 383); Asper (2007, 201 n727).

In archaic times, mathematical practitioners were probably such migrant crafts-men. Later, these traditions certainly became indigenous, but the knowledge retained its structures, even the textual ones.

Third and most importantly, these practitioners, even if occasionally well-paid, must have been of a rather low social level, viewed from the perspective of the well-off polis citizen. Aristotle's judgment of the craftsmen's social status in past and present probably also applied to these experts.[15] In fifth- and fourth-century Greece, most writers were upper-class citizens writing for their peers, which is why we almost never hear about these practitioners. That does not mean, however, that they were a marginal phenomenon. Rather, as I will argue, they provided the background for the emergence of theoretical mathematics. To think of their knowledge as 'sub-scientific' makes sense, as long as one remembers that *our* understanding of what science is has been heavily influenced by Greek *theoretical* mathematics.[16] The 'sub' here should be taken literally: ancient prac-tical mathematical traditions were certainly all-pervasive in ancient Greece, on top of which theoretical mathematics suddenly emerged, like a float on a river's surface—brightly colored and highly visible, but tiny in size.

Greek theoretical mathematics (and its texts)

Compared with practical traditions such as the ones outlined above, Greek theoret-ical mathematics strikes the reader as very different. Most notably, it is almost exclu-sively geometrical. It consists of a body of general propositions proved by deduction from 'axioms', that is, evident assumptions or definitions—hence the designation of this type of mathematics as 'axiomatic-deductive'. Whereas the practitioners' texts collected problems and provided procedures for solving them, the theoreticians' texts collected general statements with proofs. The language and the structure of these texts are highly peculiar, compared both to the practitioners' texts and to Greek prose of the times in general. From a historian's point of view, this form of mathem-atics is no less remarkable: Greek theoretical mathematics suddenly appears at the end of the fifth century BC. Most of the famous mathematical writers (for example, Euclid, Archimedes, and Apollonius) were active in the third century BC. A rather simple theorem in Euclid (*Elements*, I. 15) may provide a suitable introduction:

If two straight lines cut one another, they make the vertical angles equal to one another. For let the straight lines AB, ΓΔ cut one another at the point E; I say that the angle AEΓ is equal to the angle ΔEB, and the angle ΓEB to the angle AEΔ. For, since the straight line AE stands on the straight line ΓΔ, making the angles ΓEA, AEΔ, the angles ΓEA, AEΔ are equal to two right angles. Again, since the straight line ΔE stands on the straight line

---

15. *Politics* III 5, 1278 a 7. 'Craftsmen were either slaves or foreigners.'
16. Maybe one should rather think of them as of a '*science du concret*' (Lévi-Strauss 1966, 1–33).

AB, making the angles AEΔ, ΔEB, the angles AEΔ, ΔEB are equal to two right angles. But the angles ΓEA, AEΔ were also proved equal to two right angles; therefore the angles ΓEA, AEΔ are equal to the angles AEΔ, ΔEB.[17] Let the angle AEΔ be subtracted from each; therefore the remaining angle ΓEA is equal to the remaining angle BEΔ.[18] Similarly it can be proved that the angles ΓEB, ΔEA are also equal. Therefore, if two straight lines cut one another, they make the vertical angles equal to one another. Just what one had to prove. (transl. modified from Heath 1956, I 277–278)

As can be gathered from the text, the reader had to have in front of him a diagram that probably looked like Fig. 2.1.2 (extant in medieval manuscripts of Euclid and probably closely resembling the diagrams illustrating the theorem in the third century BC).[19]

Euclid claims the truth of a general proposition, a theorem (above, given in italics), about what happens when two lines cut each other. First he construes a pseudo-actual case by a diagram, the parts of which are designated by letters. Then he compels his reader to look at the diagram and ask himself which of the already proved or axiomatically accepted truths (both were treated earlier in the first book of the *Elements*) one could use in order to prove the statement. Here, Euclid uses I.13 ('If a straight line set up on a straight line makes angles, it will make either two right angles or angles equal to two right angles.'), and axioms (see notes 10 and 11). From these, already accepted as true, the mathematician can safely deduce the truth of the theorem. The whole proof is implicit, that is, neither does Euclid tell the reader which parts of the axiomatic material or the already proved theorems he uses nor does he ever explain his line of reasoning. At the end of the paragraph, he does facilitate the transition from the pseudo-actual diagram to the general theorem, in exactly the same words that were used in the beginning. The textual unit of theorem, diagram, and proof ends with the explicit and nearly proud assertion that the author has proved what he set out to prove.

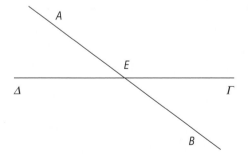

Figure 2.1.2 Diagram illustrating Euclid's *Elements* 1.15 (after Heath 1956, I 277)

---

17. The argument is based on the so-called postulate 4 ('All right angles are equal to one another.') and so-called 'common notion' 1 ('Things which are equal to the same thing are also equal to one another.'). All definitions, postulates, and common notions relevant to the first book of the *Elements* are gathered at the beginning of the book.

18. Presupposes common notion 3 ('If equals are subtracted from equals, the remainders are equal').

19. On diagrams see Saito, Chapter 9.2 in this volume.

Due to its use in schools well into the twentieth century, Euclid's *Elements* are by far the best known text written in this style, but by no means the only one. Here is a proof from the beginning of the treatise *On the sphere and the cylinder* (I.1), written by the famous Archimedes of Syracuse, probably roughly a contemporary of Euclid (Fig. 2.1.3):

If a polygon be circumscribed about a circle, the perimeter of the circumscribed polygon is greater than the perimeter of the circle. For let the present polygon be circumscribed about a circle. I say that the perimeter of the polygon is greater than the perimeter of the circle. For since BA, AL taken together is greater than the arc BL, because they have the same beginning and end, but contain the arc BL, and because similarly LK, KΘ taken together [is greater] than LΘ, and ZH, HΘ taken together [is greater] than ZΘ, and also ΔE, EZ taken together [is greater] than ΔZ, therefore the whole perimeter of the polygon is greater than the perimeter of the circle.[20] (transl. after Heath 1953, 5)

Archimedes proves a fact that is evident to anyone who takes a look at the diagram. His proof utilizes axiomatic material, too (see note 14). Obviously, the two texts share a number of technical and linguistic or, rather, rhetorical features that further illustrate the theoretical character of these mathematical traditions. A third example shows that the theoretical 'culture' betrayed by these texts is almost obligatory for the authors engaged in the field. This is how the astronomer Aristarchus (probably early third century BC), famous for having claimed heliocentricity, talks about the relation between two spheres (the second proposition of his little treatise *On the sizes and distances of the sun and the moon*):

Figure 2.1.3 Diagram illustrating Archimedes, *On the sphere and cylinder* I.1 (after Heiberg, 1972–5, I 13)

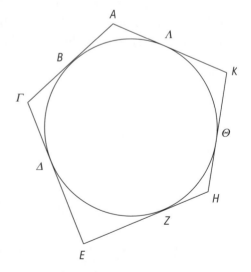

20. The proof is based on the axiomatic 'second assumption' that precedes the first proposition in Archimedes' text.

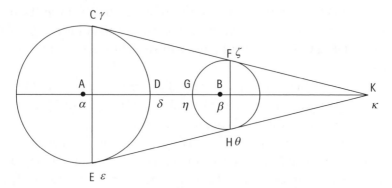

Figure 2.1.4 Diagram illustrating Aristarchus, *On the sizes*, prop. 2 (after Heath 1913, 358)

If a sphere be illuminated by a sphere greater than itself, the illuminated portion of the former sphere will be greater than a hemisphere. For let a sphere the centre of which is B be illuminated by a sphere greater than itself the centre of which is A. I say that the illuminated portion of the sphere the centre of which is B is greater than a hemisphere. For, since two unequal spheres are comprehended by one and the same cone which has its vertex in the direction of a lesser sphere,[21] let the cone comprehending the spheres be (drawn), and let a plane be carried through the axis; this plane will cut the spheres in circles and the cone in a triangle. Let it cut the spheres in the circles CDE, FGH, and the cone in the triangle CEK. It is then manifest that the segment of the sphere towards the circumference FGH, the base of which is the circle about FH as diameter, is the portion illuminated by the segment towards the circumference CDE, the base of which is the circle about CE as diameter and at right angles to the straight line AB; for the circumference FGH is illuminated by the circumference CDE, since CF, EH are the extreme rays. And the centre B of the sphere is within the segment FGH; so that the illuminated portion of the sphere is greater than a hemisphere. (transl. after Heath 1913, 359–361)

Instead of talking about celestial bodies, Aristarchus prefers to 'geometrize' the whole argument and to assume that these are just two given spheres. Illumination is conceptualized as a cone (and illustrated as a triangle). Aristarchus implicitly uses a proposition that has already been proved and accepted by the reader (proposition one, see note 13). Again, both the language and the structure of the argument are completely in line with what Euclid and Archimedes did. This is Greek theoretical mathematics. It seems fair to say that these texts are utterly different from those in which practical mathematicians codified their knowledge. Let us briefly describe the theoretical texts by keeping the characteristic features of the practitioners' textual traditions in mind. They are obviously different in at least three respects:

First, the text's intention is to *prove* a theorem by logical means, which implies that the status of the objects being discussed is general. Therefore, actual numbers

---

21. It had been demonstrated in the first proposition that two unequal spheres are contained by one cone.

or measurements have no place in this type of mathematics. Even the diagram introduces only pseudo-individual forms: the 'two straight lines AB, ΓΔ' or the 'two circles CDE, FGH' are in truth any two straight lines or any two circles, respectively.

Second and accordingly, the mathematical writer is interested in the abstract properties of these general geometrical entities, not in calculating any quantitative properties of real objects or classes of real objects.

Third, the rhetoric of the two textual traditions is completely different. Whereas the recipe-like algorithms of the practical tradition employed strong personal appeals to the reader, the theoretical tradition produced highly *impersonal* texts (see Asper 2007, 125–135). This feature is unique, at least to this extent, in the context of all Greek scientific and technical literature and deserves a closer look. With the exception of exactly one formula that serves to introduce the repeated claim and marks the beginning of the proof ('I say that...' in the three examples above), these texts never introduce an authorial voice nor do they ever address the reader. (The introductory letters of Archimedes and Apollonius are not an exception to this rule: these letters employ a style that is indeed personal, but they are not an integral part of the mathematic texts they introduce. Rather, they have the status of 'paratexts'.) And even this 'I' is not personal in the usual sense, since it merely functions as a marker of the internal structure of the proof and will always show up at exactly the same place. Especially remarkable are the impersonal imperatives that regularly feature in the description of the objects concerned: English has to paraphrase theses imperatives with 'let' which makes them less strong: for example, 'let the straight lines AB, ΓΔ cut one another' (Euclid), 'let the polygon be circumscribed about a circle' (Archimedes), or 'let a plane be carried through the axis' (Aristarchus). In the Greek, these are imperatives in the third person, mostly in the passive voice, and often even in the perfect tense.[22] Hundreds of these strange forms exist in the works of theoretical mathematicians. As one would expect, these forms are hardly extant outside of mathematical language, that is, they are part of an exclusive discourse, of a sociolect. The writers present their objects to the reader as independently given, as something that is just there and can be contemplated objectively (Lachterman 1989, 65–67). They write themselves, their perceptions and their operations out of the picture, as it were,[23] which tends to add an air of timeless truth to what they have to say. The admirable rigor of this discourse is, however, achieved at the cost of explanation and, even more, any context of how the proof was found (famously

---

22. The Greek forms of the translations quoted are: *temnetōsan* (Euclid: 'they shall cut one another!'), *perigegraphthō* (Archimedes: 'it shall be circumscribed!'), and *ekbeblēsthō* (Aristarchus: 'it shall be extended!').

23. The Aristotelian Aristoxenus (third century BC) explained mathematical imperatives in exactly this way (*Harmonica* II 33, pp 42–43 ed. Da Rios).

criticized by Lakatos 1976). This weird way of writing has no parallel in Greek writing and cannot be anything but a rhetorical stylization. The oral discourse of Euclid as he tried to convince a listener of any given proof would have probably contained many more personal markers (like demonstratives, interjections, or personal pronouns).[24] Thus, the main function of the rhetoric of impersonality is to convey objectivity. Generally, this is still (or again) the case in modern 'hard' science (Storer 1967, 79; Rheinberger 2003, 311–315).

As different as the texts of the two traditions may appear at first glance, they also share at least two features. One is their regular use of diagrams, the second is the thorough standardization of their language.[25]

(a) Greek theoretical mathematics always relies on a lettered diagram in order to add to the logical force of the proof by means of visual evidence. The lettered diagram has proved so powerful that it is still used in modern science, in a nearly unaltered form (occasionally, today it is 'numbered' rather than 'lettered'). Neglected for a long time, the diagram in Greek theoretical mathematics has recently been rediscovered as a very important feature (Netz 1999a), crucial to the communicative success of any proof. However, to anyone examining the traditions and texts of practical mathematics, especially the Near Eastern ones, it becomes quite clear that the specific 'theoretical' Greek lettered diagram is somehow related, like a younger member of the same family, to the 'numbered' diagram that we find in the practitioners' texts (see the example taken from the Papyrus Ayer above, Fig. 2.1.1) and, regularly, in Babylonian problem texts.[26] There, the parts of the diagram are, usually, connected to the relevant portions of the text by repeating the paradigmatic measures given in the text. Greek theoretical mathematics uses generalized indices, that is, letters that, in this case, do not signify numbers. The Greek lettered diagram is thus a generalization of the diagrams employed in the textbooks of practical mathematics and therefore closely related to the former (in my opinion, one of several reasons to think that Greek theoretical mathematics must have emerged from a practitioners' background).

(b) Anyone who works through the contents of either tradition will be amazed by how extremely regulated, even standardized, these texts actually are. The practitioners always use the recipe-structure, within a given text the introduction, the appeal to the second person, and the end of the actual

24. The mathematical passages in Aristotle or Plato (minus the general differences between written prose and oral discourse) might be a model of how mathematicians *talked* about their objects.

25. A third, not dealt with here, is the 'discrete' and, for Greek prose, highly unusual status of the major text-units in both traditions (see Asper 2003, 8–9).

26. See, for example, TMS 1, IM 55357, YBC 8633, YBC 4675 (see Damerow 2001, 230, 234, 244, and 280), all from Old Babylonian times.

problem always use exactly the same language. These texts, being standardized, are somewhat remote from oral discourse.[27] Since these structures are very old and easily cross cultures and languages, one could guess that the most effective way to protect this knowledge was to ensure that it was kept *traditional*, as is evident from the standardization of its textual forms. Although Greek theoretical mathematics is comparatively young, it is no less standardized, albeit in a more complex way and on several levels: the lexicon used is small, nearly free of synonyms, and confined by definitions that precede most extant works. Mathematical syntax is even more regulated: the same elements of any given argument show exactly the same form (Netz 1999a, 133–158 has listed more than a hundred such 'formulas'). Even the whole proof always consists of the same parts, always introduced by the same particles.[28] Of these, the famous phrase 'QED' (*hoper edei deixai*, 'just what one had to prove') is still used today.

Therefore, the language of Greek theoretical mathematics strikes one as being far removed from living oral discourse and its common rhetorical strategies, and being just as far removed from other forms of written argument. Already by the fourth century BC, Aristotle's readers did not appreciate mathematical prose aesthetically, because it was too different.[29] To understand why the theoretical tradition produced such unparalleled texts, why it even emerged this way, one has to dive deeply into historical inquiries, all of which greatly benefit from remembering the strong traditions of Greek practical mathematics that must have been constantly present in the environment of the theoreticians.

Theoretical mathematics in Athens: games of distinction

Theoretical mathematics of the kind outlined in the last paragraph existed only in Greece and was, by comparison with the mighty tradition of practical mathematics, clearly a local phenomenon, the distinctive features of which call for historical explanation or, at least, comment. The traditional, that is, ultimately Aristotelian, narrative of how mathematics and philosophy emerged in archaic Greece tells us that it all began in Asia Minor (modern Turkey), in the sixth century BC, with half-mythical characters such as Thales of Miletus (c 600–550 BC)

27. Eleanor Robson, however, mentions that, at least in Old Babylonian mathematics, the syntax is quite natural and the terminology 'local and ad hoc at best' (personal communication).

28. These parts, slightly different depending on whether the proof is a problem or a theorem, and also consistently standardized only in Euclid, were already isolated, described, and named by the ancient tradition, preserved in Proclus (fifth century AD): see Netz (1999b).

29. Aristotle, *Rhet.* 1404 a 12; *Metaph.* 995 a 8–12.

and his Milesian 'school' or, somewhat later, Pythagoras of Samos (c 550–500 BC) and his followers. It is difficult or even impossible to reach firm ground here (radical skepticism in Dicks 1959 and Burkert 1972). Both Thales and Pythagoras are credited with the discovery of geometrical theorems, for example the theorem above quoted from Euclid is ascribed to Thales by ancient tradition. It is disputed, however, whether these ascriptions are to be trusted (rather not, is my guess). Whatever semi-theoretical practices they and their possible successors might have engaged in, it is quite certain that they did not produce texts showing the characteristics discussed above. It was disputed already in antiquity whether they had even produced texts at all. The earliest history of Greek theoretical mathematics that we can lay our hands on begins in the late fifth century BC at Athens, where the center of theoretical mathematics in the Greek world would be located for some 120 years, until other centers emerged in the beginning of the third century BC, most notably at Alexandria in Egypt.

What is known about the persons involved and the contexts in which such a peculiar body of knowledge emerged? As always, too little. Partly because the impersonality of theoretical mathematics prevented authors from telling us anything about themselves (except in occasional introductory letters), partly because the older tradition was obliterated by the star mathematical writers of the third century BC. The sources are thus mostly indirect, scattered, and do not go back further than the fourth century BC: quotations found in Late Antique commentators from the earliest historical account of mathematics, by Eudemus of Rhodes, an Aristotelian scholar (fourth century BC),[30] and occasional remarks in the works of Plato and Aristotle.

After the shadowy prehistory of theoretical mathematics in eastern Greece,[31] the political and economical power of Athens in the second half of the fifth century attracted Greeks from Asia Minor who probably came as political representatives of their Ionian city-states. Two mathematicians from the island of Chios who were active at Athens are still known: Oenopides, who applied geometrical models to astronomical problems and is credited with a couple of methodological achievements, and, more prominently, Hippocrates, who seems to be the founder of the tradition of *Elements*, that is comprehensive axiomatic-deductive treatises in the later style of Euclid. His is the first theoretical *text* that we have in Greek mathematics: a short passage on the quadrature of 'lunules', that is, certain segments of circles, quoted by Simplicius through Eudemus (see Netz 2004). Although disputed in almost all its details, this text shows all the features that struck the reader as peculiar in the above quoted examples, especially the standardization, the

---

30. Proclus in his commentary on the first book of Euclid's *Elements*, Eutocius (sixth century AD) commenting on Archimedes' works, and Simplicius (sixth century AD) on Aristotle's *Physics*. See Zhmud (2002).

31. There are more names, for example Phocus from Samos and Mandrolytus from Priene, but absolutely nothing is known about them.

impersonality, and those strange imperatives. No diagrams have been preserved, but the text obviously relies on several lettered diagrams. Hippocrates is usually dated to 430–420 BC (see Burkert 1972, 314 n 77). Therefore, by this time there must have already existed generic conventions for how to write theoretical mathematics. Furthermore, there must have been a desire to communicate the knowledge to someone, that is, to readers. For Eudemus, writing about a hundred years later, Hippocrates was the founder of the genre that included *Elements* and, apparently, the first 'real' theoretical mathematician in Athens. After him, throughout the fourth century BC, we hear of about twenty names (all in Proclus) and even groups of people in Athens associated with theoretical mathematics, partly in contact with philosophers and astronomers. It is clear, though, that at least some mathematicians were not part of any of these other groups, especially not the Pythagoreans or, later, the followers of Plato. From about 400 BC onwards, at the latest, a mathematical community, however small, must have existed in Athens.

There are reasons to believe that this theoretical knowledge did not suddenly fall from the sky or off the trees, but differentiated itself from the practitioner's knowledge. First, some of the terminology in Euclid betrays a practical origin, for example the term for 'drawing a straight line' (*teinō*, literally 'stretch out'), goes back almost certainly to the aforementioned surveying practices of the 'rope-stretchers'. Similarly, expressions for geometrical entities as angles, certain figures, or the perpendicular go back to craftsmen's traditions.[32] Second, the curious definition of the line in Euclid (Book 1, def. 2) as 'a length without width' makes perfect sense when one realizes that the experts of the practical traditions, when concerned with measuring, always assume a standard width for every line they measure. The theoretician's text is being implicitly, but quite openly polemical here (Høyrup 1996, 61, according to whom the same is true for *Elem*. II.1–10). Third, there is a model for conceptualizing the emergence of theoretical mathematics from practitioners' knowledge. A typical genre of such practitioners' groups and their competitive struggles is the riddle, used by experts to challenge one another. Such riddles are characteristically compound problems, and apply practical methods to improbable problems that already touch upon a theoretical realm (Høyrup 1997, 71–72). According to the traditional histories of Greek mathematics, three problems were at the center of the field's attention from the beginning: how to square the circle, how to trisect an angle, and how to double the cube (Saito 1995). On the one hand, these are *problems to solve*, not theorems to prove, and, therefore, belong to the practical sphere. On the other hand, it is not quite clear in what situation one would be faced with the task of, for example, doubling a cube. Therefore, one model for the transition from practice to

32. Gandz (1929–31, 273–275). More examples in Burkert (1982, 135–136); Høyrup (2002a, 400–405).

theory might indeed be the riddle, which pushed competitive practitioners towards theory.

One can draw another inference from the impersonality and the uniquely coherent terminology of Greek mathematical theory. In other intellectual fields, most notably in medicine and philosophy, one observes the opposite: texts are strongly personal and terminology changes between individuals or, at least, between groups (Netz 1999a, 122 f.).[33] In these cases, both personality and group-related terminology function as instruments in a competitive struggle among the participants in the field, a competition aimed at patients, in the case of the physicians, and at the glory of being right in the philosophers' discussions. As was the case with the authorial 'I', in mathematical texts the reader comes across only one standardized usage of an 'integrative we' (reserved for the formula '…as we will show…'), meant to conjure up a group spirit, so often used in philosophy and medicine. In medicine, personality and polemics in the texts reflect a competitive field, at least partially for economic reasons (Miller 1990, 39). Mathematical practitioners were specialized professionals, paid for their services. The sheer impersonality of the texts, in the case of theoretical mathematics, however, hints instead at a field comparatively free from economic pressures, a field that, for precisely that reason, remained fairly autonomous. The group of theoretical mathematicians in Athens must therefore have been quite homogenous in social terms.

Perhaps it would be adequate to think of theoretical mathematics as some form of *game* rather than something pertaining to a professional occupation, which it has become today, and which practical mathematics has always been. The persons who played this game were certainly at home in the upper circles of Athenian society (evidence collected by Netz 1999a, 279 f.), similarly to Plato and his followers who eagerly absorbed theoretical mathematics. From the majority's perspective, comedians could already make fun of mathematicians in 414 BC.[34] They must have felt like an elitist little group among Athenians. For them, theoretical mathematics was probably a status practice, perhaps enforced by the fact that the most common status practice, that is, politics, became quite dangerous for the old upper class at the end of the fifth century. Mathematics was, as philosophy was to become, a status-conscious way to keep one's head down.[35] There were, however, practitioners around who were, for Athenians from good families, socially unacceptable but who also had some share in mathematical knowledge and its practices. I suggest that many of the odd features of the theorists, such as expressly refusing to mention any practical applications or any useful effect, worked intentionally as distinctive markers, meant to distinguish the precious

33. The last becomes very clear in the writings of the physician Galen (second century AD), especially in his treatise on terminology (*On medical names*, extant only in Arabic).

34. Aristophanes in his *Birds* (v. 1005), targeted at the astronomer Meton.

35. Netz (1999a, 293–294); Fowler (1999, 372). See also Carter (1986, 131–186).

game of distinction[36] from sordid occupations that were carried out by people for hire.[37] Plato himself defines, quite polemically, the difference between practical calculations and theoretical mathematics (*Philebus* 56 D 3–57 A 3, a perspective that is still inherent in the modern opposition of 'pure' and 'applied' sciences). Ancient narratives concerning the emergence of mathematics proper always stress its emancipation from the demands of daily and practical life.[38] Later, this game of distinction effortlessly blended into the Platonic disdain for everything material. (It is difficult to decide whether Platonism adopted theoretical mathematics because it perfectly satisfied the Platonists' desire for immaterial, transcendent truths, or whether Plato and his followers edged theoretical mathematics even further into the ivory tower.) Certainly the game of distinction was already being played before Plato had even dreamed of his forms. A late and, almost certainly, inauthentic anecdote illustrates my point nicely:

Someone who had taken up geometry with Euclid, asked after he had understood the first theorem: 'What is my profit now that I have learned that?' And Euclid called for his servant and said: 'Give him a *triōbolon*, since he must always make a profit out of what he learns'.[39]

The point of theoretical mathematics is precisely that one does *not* gain any material profit from it. The *triōbolon*, here probably synonymous with 'small change', was the day's wage of an unskilled worker in classical Athens, which would bring out the contempt for 'work' on behalf of the mathematicians even better. The anecdote, one of several about Euclid that are all best met with skepticism, probably belongs to a Platonist milieu which began to dominate theoretical mathematics some time after Euclid's lifetime.

For an Athenian gentleman devoted to theoretical mathematics between 420 and 350 BC, however, professional experts of lower social status were not the only group from which it was necessary to demarcate his own pursuits: since the middle of the fifth century, there had been the sophists and, increasingly, the philosophers, both of whom had their own ways for intruding into pure and agoraphobic mathematics. The sophists were professional experts of knowledge and, as such, promised to teach political success, which in a society largely based on public debate depended largely on the use of rhetoric. Some of the sophists, apparently trying to top *all* existing forms of knowledge, tackled the conventions of theoretical mathematics with rather silly objections.[40] Others tried their ingenuity by

---

36. The expression is taken from Bourdieu (1979, 431 (*jeu distinctif*)). Of course, ancient Athenian upper-classes had, just as their modern equivalents, several 'games of distinction' at their disposal, for example, chariot races.

37. What Netz (2002b) has termed the 'marginalization of the numerical' in theoretical mathematics, I see as one more of these distinguishing moves targeted against the practitioners whose practices were almost exclusively numerical.

38. For example, Aristotle, *Metaph.* 981 b 20–25; Proclus, *In Eucl.* pp. 25.12–26.9.

39. Stobaeus, *Anthologium* 2.31.114 (pp. 228.25–29 ed Wachsmuth/Hense).

40. Protagoras in Aristotle, *Metaph.* 998 a 2 f.; compare *Anal. post.* 76 b 39–77 a 4.

solving the quadrature of the circle, again in pointedly amateurish fashion.[41] By doing this, the sophists aimed probably not at the mathematicians, but at potential customers to whom they could demonstrate that they 'knew everything', which was the typical sophistic claim (Hippias in Plato, *Hippias maior* 285 B 7–286 A 5). Viewed from this perspective, the mathematical excursions of the sophists indicate that they had found a body of theoretical mathematical knowledge at Athens, against which they tried to set themselves up as experts. There is no indication, however, that the mathematicians even bothered to engage with these dilettante newcomers.

A similar argument can be made about the philosophers in Athens.[42] Followers of Socrates took a great interest in theoretical mathematics: First, the Socratic Bryson (c 365 BC) tried to square the circle (Aristotle, *Anal. post.* 75 b 40–76 a 3). Similar to Antipho, he proceeded from premises that were too general and thus failed. Aristotle calls this attempt 'sophistic', thereby indicating an outsider's approach. Second, and far more importantly, Plato and his circle discovered in mathematical knowledge a paradigm of the epistemological quality they were generally after. Plato's criticisms of mathematical methods show that, again, his is an outsider's interest.[43] Most of the people in the Academy were not mathematicians, but were eager to discuss meta-mathematical questions and to apply the deductive logic of mathematical proofs to dialectics and even to science in general (as Aristotle did in his *Second analytics*). There is no reason to assume that the mathematicians were interested in these generalizations. Proclus has preserved a significant statement of the otherwise unknown mathematician Amphinomus (first half of the fourth century BC) who boldly contended that it is *not* the business of the mathematician to discuss the epistemological foundations of his work.[44] The mathematicians' desire to distinguish themselves from other discourses of knowledge obviously worked to distance themselves from the philosophers, too. The clearer the distinction and the more exclusive the group, the more enjoyable were the games of distinction.

The lack of institutions for theoretical mathematics

Since games require peers, but not necessarily readers, the remarkable form of mathematical *texts* deserves our attention, too. They probably also served a

41. Hippias in Proclus, *In Eucl.* p. 272.7–10; Antipho in Simplicius, *In Arist. Phys.* I 2, p. 54.12–55.11 ed Diels.

42. There is a dubious tradition that Anaxagoras, a friend of Pericles, engaged in theoretical mathematics, in approximately 450 BC (Ps.-Plato, *Amat.* 132 A 5 f.; Proclus, *In Eucl.* p. 65.21–66.1).

43. Plato, *Resp.* 510 C 2–D 2; 533 B 6–C 4.

44. Proclus, *In Eucl.* p. 202.9–12.

certain function within their original contexts of communication. Many of the features of the theoretical tradition have the effect of ensuring the correct understanding of the texts, especially multi-leveled standardization. In a social context where mathematics is regularly explained by a teacher to disciples, texts used in instruction can afford to be less rigorous and less standardized because there is always the option of live dialogue to ensure that the knowledge is transmitted. The texts of the practical tradition were probably always accompanied by personal, oral instruction that filled in the gaps, explained terms, and so on. Unlike these practical texts, the theoretical tradition produced *autonomous* texts, that is, texts that were able to exclude misunderstandings all by themselves, that were able to force readers into a consensus by realizing the mathematical truth. This is the reason for defining crucial terms, standardizing the structure of proofs, and for excluding the context of discovery, every personal trace, and all controversy. Greek mathematical prose in the theoretical tradition is a paradigm of written knowledge transmission, rigorous in a way that still works today for any reader of Euclid or Archimedes. The lettered diagram plays not a small part in this achievement because it transports visual evidence from the author to the reader.[45] For all these reasons, one can happily read these texts alone, without a teacher, and *still* be fairly sure (as much as is possible in written communication, and compared to, say, poetic or historiographic texts) that one understands the argument in the way the authors intended it to be understood. Thus, the theoreticians have created a powerful, very reliable means of purely written communication.

The practitioners imparted their knowledge from generation to generation within a guild-like institutionalized framework. *For theoretical mathematics*, however, *there was no institutional background* in fifth- and fourth-century Greece (of course not, since the point of this socially distinctive game was its being 'useless'). In its infancy, Greek theoretical mathematics lacked institutionalization, which Netz (1999a) has shown convincingly. Plato and Aristotle lament that the city-state has no esteem for and, accordingly, provides no structures for theoretical mathematics.[46] True, Plato makes his guardians learn abstract mathematics—but his point seems to be that in real-life Athens nobody did (*Resp.* 525 B 3–528 E 2). Initially, Greek mathematicians had too few people around to talk to, so they resorted to writing and travel. Mathematicians were forced to write rather than discuss (compare Plato, *Theaetetus* 147 D 3) and, therefore, developed textual forms that could function perfectly in writing alone. In places such as Syracuse or Cyrene, there might not have even been any continuous oral tradition, a scenario quite different from our practical mathematicians whose group-structure ensured that the trade was handed down from generation to generation.

---

45. See, however, Saito, Chapter 9.2 in this volume on 'over-specification'.
46. Plato, *Resp.* 528 B 6–C 8; Aristotle, fr. 74.1 ed Gigon (= Iamblichus, *Comm. math. sci.*).

Standardization of proof-structure and the theoretical lexicon may have helped to increase the probability of successful knowledge-transmission.

The situation was different in fourth-century Athens and in third- and second-century Alexandria but, by then, the genre of mathematical prose had already emerged with its distinctive features. Besides, even in the third and second centuries, letters and travel were typical for theoretical mathematicians outside of Alexandria, as can be glimpsed from the introductory letters of Archimedes, of Diocles (ed Toomer 1976), and of Apollonius. Instead of walking into a classroom and presenting a new theorem to his graduate students, Archimedes in Syracuse sent letters to the other end of the world challenging his friends in Alexandria to find the proofs of the theorems he has just found.[47] Further, the genre of *Elements* with its peculiar linguistic characteristics emerged as an ideal medium of how to store the pertinent knowledge and is still used in this capacity today.

From a modern perspective it is difficult to imagine that (theoretical) mathematics might not have been institutionalized in some way. In classical Greece, institutionalization proper did not begin with a sudden widespread interest in theoretical knowledge, but with practices of political representation. After the followers of Plato and Aristotle had developed a lively interest in theoretical mathematics throughout the latter half of the fourth century BC, Hellenistic dynasts, above all the Ptolemies in Alexandria, the Seleucids at Antioch in Syria and, on a less grand scale, Hiero at Syracuse in Sicily, sponsored theoretical mathematics just as they funded poets and grammarians: as a contemporary form of pan-Hellenic representation. Intellectuals added to the royal splendor.

Paradoxically, we know next to nothing about Euclid. He may or may not have been the one who migrated from the then thriving mathematical scene of Athens to Alexandria, some time between 320 and 280 BC, and with whom mathematical institutionalization began at Alexandria.[48] There, the Ptolemies had also established some center of engineering, not least because they were keenly interested in siege engines, for which the successful construction and use of practical mathematics was of great importance. There, a tradition of teaching and writing practical mathematics continued into Byzantine or even Arabic times. The most important author of this tradition is the aforementioned Hero of Alexandria, who himself bridged both the practical and the theoretical traditions. Furthermore, the Ptolemies assembled mathematically minded astronomers in Alexandria, for instance Conon of Samos (third century BC) who also wrote on conic sections. Apparently there was a nearby observatory. All these persons[49] must have constantly met and debated with one another. In Archimedes' and Apollonius'

---

47. Archimedes, *Sph. cyl.* praef. vol 1, pp. 168.3–5 ed Heiberg 1972; *Lin. spir.* praef. vol 2, pp. 2.2–6; *Meth.* praef. vol 2, pp. 426.4–7.

48. At least, this is what Pappus of Alexandria (fourth century AD) tells us (*Coll.* p. 678 ed Hultsch).

49. More names and affiliations in Asper 2003, 27.

introductory letters, we strongly sense the existence of a small 'scientific community', again with notions of elitist distinction.[50] There were several libraries that served scholarly purposes and the famous Mouseion, an institution that gathered and awarded royal stipends to scholars from a number of disciplines, including grammar and, probably, medicine. It was here, with this concentration of various sorts of mathematicians that a stable tradition of theoretical mathematics emerged that betrays signs of teaching and canonization (editions of mathematical 'classics', commentaries, and collections), with the later works more firmly embedded in Platonist philosophy and curriculum. But even then, theoretical mathematics remains a discourse based on writing and confined to very small, socially elevated circles. They were still not professionals in the modern sense, as the mathematical practitioners had always been. Despite the astonishing prominence of theoretical mathematics in modern times, which invites anachronistic re-projections, in ancient Greece theoretical mathematics must always have been an epiphenomenon, or rather, a marginal differentiation, of strong practical traditions.

Conclusion: the two mathematical cultures of ancient Greece compared

Practical mathematics must have been present in all the previously mentioned times and places, albeit socially invisible. Mostly, its practitioners worked with their long-established methods without ever paying attention to the theorists and their games. On the other hand, upper-class theorists must have aimed at staying clear of modest craftsmen. Occasionally, one can suspect a direct, polemical reference by theoretical mathematics directed against the practitioners. Platonic ideology contributed its share to the dichotomy, which was apparently quite strict at times. Rarely did somebody bridge the two traditions, which must have occurred regularly in the very beginnings of the theoretical tradition. One might understand these respective bodies of knowledge as complementary and, almost, as mutually explanative:

Greek practical mathematics:

- was derivative of older traditions that, ultimately, originated in the ancient Near East;

- solved 'real-life' problems;

- communicated actual procedures in order to convey general methods;

- used written texts (if at all) as secondary means of knowledge storage and instruction;

50. Both ask their addressees to distribute their findings only to those who are deserving: Apollonius, *Con.* II praef. vol 1 pp. 192.5–8 ed Heiberg 1893; Archimedes, *Sph. cyl.* I praef., vol 1 pp. 4.13 f.

- employed 'social' technologies of trust, that is a rhetoric based on insti-tutional authority; for example, the guild's pristine tradition, the special-ist status of its practitioners, and the knowledge's commonly accepted usefulness;

- worked within a stable and highly traditional social—that is, institu-tional—framework.

Greek theoretical mathematics:

- emerged in sixth- to fifth-century Greece, at least partly from a practical background;

- was a theoreticians' game with artistic implications, pointedly removed from 'real life';

- communicated general theorems concerning ideal geometrical entities;

- depended on writing and produced autonomous texts;

- employed epistemological technologies of trust based on evidence and logic;

- was not institutionalized, at least not during its formative stages.

The two fields differ so greatly with respect to their practices, traditions, milieus, functions, methods, and probably also the mindsets of their participants that I could not resist adopting the catchphrase of the 'two cultures'. Thus, when approaching mathematics in ancient Greece, perhaps one should rather think of two mathematical cultures, in many respects neat opposites.[51] To the leading circles of any given ancient Greek community, the practitioners were probably almost socially invisible. As far as sizes of groups and social presence in everyday life are concerned, however, the theorists were never more than an epiphenom-enon. Apparently, the unusual characteristics of theoretical mathematics evolved as markers of differentiation, meant to stress a distance from the social and epi-stemic background that was associated with practical mathematics.

Bibliography

Asper, Markus, 'Mathematik, Milieu, Text. Die frühgriechische(n) Mathematik(en) und ihr Umfeld', *Sudhoffs Archiv*, 87 (2003), 1–31.

51. One might also find a dichotomy similar to the one described in this paper in Greek medicine (ration-alist Hippocratic medicine versus 'magicians'), and perhaps even in historiography ('serious' historiography versus mere 'storytelling'). Compare also Rihll and Tucker (2002, 297–304).

Asper, Markus, *Griechische Wissenschaftstexte. Formen, Funktionen, Differenzierungs-geschichten*, Franz Steiner Verlag, 2007.

Becker, Oskar, 'Die Lehre vom Geraden und Ungeraden im neunten Buch der Euklidischen Elemente,' *Quellen und Studien zur Geschichte der Mathematik*, B 3 (1936), 533–553.

Bourdieu, Pierre, *La distinction. Critique sociale du jugement*, Éditions de Minuit, 1979.

Burkert, Walter, *Lore and science in ancient Pythagoreanism*, Harvard University Press, 1972.

Burkert, Walter, 'Konstruktion und Seinsstruktur: Praxis und Platonismus in der grie-chischen Mathematik', *Abhandlungen der Braunschweiger Wissenschaftlichen Gesellschaft*, 34 (1982), 125–141.

Burkert, Walter, *The Orientalizing revolution. Near Eastern influence on Greek culture in the early Archaic Age*, Harvard University Press, 1992.

Burkert, Walter, *Babylon—Memphis—Persepolis: eastern contexts of Greek culture*, Harvard University Press, 2004.

Carter, L B, *The quiet Athenian*, Oxford University Press, 1986.

Chace, Arnold B (ed), *The Rhind mathematical papyrus. British Museum 10057 and 10058*, 2 vols, Mathematical Association of America, 1927-9.

Cuomo, Serafina, *Ancient mathematics*, Routledge, 2001.

Damerow, Peter, 'Kannten die Babylonier den Satz des Pythagoras? Epistemologische Anmerkungen zur Natur der babylonischen Mathematik', in Jens Høyrup and Peter Damerow (eds), *Changing views on ancient Near Eastern mathematics*, Reimer, 2001, 219–310.

Dicks, D R, 'Thales', *Classical Quarterly*, NS 9 (1959), 294–309.

Dodge, Bayard, *The Fihrist of al-Nadīm: a tenth-century survey of Muslim culture*, Columbia University Press, 1970.

Fowler, David H, 'Further arithmetical tables', *Zeitschrift für Papyrologie und Epigraphik*, 105 (1995), 225–228.

Fowler, David H, *The mathematics of Plato's Academy. A new reconstruction*, Oxford University Press, 2nd ed, 1999.

Gandz, Solomon, 'Die Harpedonapten oder Seilspanner und Seilknüpfer', *Quellen und Studien zur Geschichte der Mathematik*, B I (1929–31), 255–277.

Gerstinger, Hans, and Vogel, Kurt, 'Eine stereometrische Aufgabensammlung im Papyrus Graecus Vindobonensis 19996', *Mitteilungen aus der Papyrussammlung der Nationalbibliothek in Wien*, 1 (1932), 11–76.

Goodspeed, Edgar J, 'The Ayer papyrus: a mathematical fragment', *American Journal of Philology*, 19 (1898), 24–39.

Heath, Sir Thomas L, *Aristarchus of Samos, the ancient Copernicus. A history of Greek astron-omy to Aristarchus together with Aristarchus' treatise on the sizes and distances of the sun and the moon*, Oxford University Press, 1913.

Heath, Sir Thomas L, *A history of Greek mathematics*, 2 vols, Oxford University Press, 1921.

Heath, Sir Thomas L, *The works of Archimedes, including the Method*, Dover Publications, 1953.

Heath, Sir Thomas L, *The thirteen books of Euclid's Elements, translated from the text of Heiberg, with introduction and commentary*, 3 vols, 2nd ed, Dover Publications 1956.

Heiberg, Johan L (ed), *Apollonius, Conica I–IV*, 2 vols, Teubner, 1893.

Heiberg, Johan L (ed), *Archimedis opera omnia cum commentariis Eutocii*, 4 vols, Teubner, 1972-5.

Heiberg, Johan L, and Stamatis E S, (eds), *Euclid, Elementa*, 5 vols, Teubner, 1969–77.

Høyrup, Jens, 'Sub-scientific mathematics: observations on a pre-modern phenomenon', *History of Science*, 27 (1989), 63–87.

Høyrup, Jens, 'The four sides and the area. Oblique light on the prehistory of algebra', in Ronald Calinger (ed), *Vita mathematica*, Mathematical Association of America, 1996, 45–65.

Høyrup, Jens, 'Hero, Ps-Hero, and Near Eastern practical geometry', *Antike Naturwissenschaft und ihre Rezeption*, 7 (1997), 67–93.

Høyrup, Jens, *Lengths, widths, surfaces. A portrait of Old Babylonian algebra and its kin*, Springer, 2002a.

Høyrup, Jens, 'A note on Old Babylonian computational techniques', *Historia Mathematica*, 29 (2002b), 193–198.

Knorr, Wilbur R, *The evolution of the Euclidean elements*, D Reidel Publishing Co, 1975.

Lachterman, David R, *The ethics of geometry. A genealogy of modernity*, Routledge, 1989.

Lakatos, Imre, *Proofs and refutations. The logic of mathematical discovery*, Cambridge University Press, 1976.

Lévi-Strauss, Claude, *The savage mind (La pensée sauvage)*, Weidenfeld & Nicolson, 1966.

Lloyd, G E R, *Greek science after Aristotle*, W W Norton & Co, 1973.

Lloyd, G E R, 'Methods and problems in the history of ancient science', *Isis*, 83 (1992), 564–577.

Luckey, Peter, 'Tabit b. Qurra über den geometrischen Richtigkeitsnachweis der Auflösung der quadratischen Gleichungen,' *Berichte der saechsischen Akademie der Wissenschaften zu Leipzig. Mathematisch-physikalische Klasse*, 93 (1941), 93–114.

Miller, Gordon L, 'Literacy and the Hippocratic art', *Journal of the History of Medicine*, 45 (1990), 11–40.

Netz, Reviel, *The shaping of deduction in Greek mathematics. A study in cognitive history*, Cambridge University Press, 1999a.

Netz, Reviel, 'Proclus' division of the mathematical proposition into parts: how and why was it formulated?', *Classical Quarterly*, 49 (1999b), 282–303.

Netz, Reviel, 'Greek mathematicians: a group picture', in C J Tuplin, and T E Rihll (eds), *Science and Mathematics in Ancient Greek Culture*, Oxford University Press, 2002a, 196–229.

Netz, Reviel, 'Counter culture: towards a history of Greek numeracy', *History of Science*, 40 (2002b), 321–352.

Netz, Reviel, 'Eudemus of Rhodes, Hippocrates of Chios and the earliest form of a Greek mathematical text', *Centaurus*, 46 (2004), 243–286.

Proust, Christine, 'La multiplication Babylonienne: la part non écrite du calcul', *Revue d'histoire des mathématiques*, 6 (2002), 293–303.

Rheinberger, Hans-Jörg, '"Discourses of circumstance". A note on the author in science', in Mario Biagioli and Peter Galison (eds), *Scientific authorship. Credit and intellectual property in science*, Routledge, 2003, 309–323.

Rihll, Tracey E, and Tucker, John V, 'Practice makes perfect: knowledge of materials in classical Athens', in Christopher J Tuplin, and Tracey E Rihll (eds), *Science and mathematics in ancient Greek culture*, Oxford University Press, 2002, 274–305.

Robson, Eleanor, *Mesopotamian mathematics, 2100–1600 BC. Technical constants in bureaucracy and education*, Clarendon Press, 1999.

Robson, Eleanor, 'Influence, ignorance, or indifference? Rethinking the relationship between Babylonian and Greek mathematics', *Bulletin of the British Society for the History of Mathematics*, 4 (2005), 1–17.

Saito, Ken, 'Doubling the cube: a new interpretation of its significance for early Greek geometry', *Historia Mathematica*, 22 (1995), 119–137.

Schubart, Wilhelm, 'Mathematische Aufgaben auf Papyrus', *Amtliche Berichte aus den Königlichen Kunstsammlungen*, 37 (1915–16), Berlin, 161–170.

Storer, Norman W, 'The hard sciences and the soft', *Bulletin of the Medical Librarians' Organization*, 55 (1967), 75–84.

Toomer, Gerald J (ed), *Diocles. On Burning Mirrors. The Arabic translation of the lost Greek original*, Springer, 1976.

West, M L, *The east face of Helicon: West Asiatic elements in Greek poetry and myth*, Oxford University Press, 1997.

Zhmud, Leonid, 'Eudemus' History of Mathematics', in Istvan Bodnár and William W Fortenbaugh (eds), *Eudemus of Rhodes*, Transaction Publishers, 2002, 263–306.

Tracing mathematical networks in seventeenth-century England

Jacqueline Stedall

In 2000, Cambridge University Library purchased a unique and important collection of mathematical papers from the library of the Earls of Macclesfield at Shirburn Castle in Oxfordshire. The Macclesfield Collection, previously inaccessible to scholars since the early nineteenth century, consists of items once owned by William Jones, tutor to a son of the family in the early eighteenth century. It includes not only a number of Newton manuscripts (see Mandelbrote 2002), but also letters and papers that Jones inherited from John Collins, who in the later seventeenth century had himself been an assiduous collector of mathematical writings. One item that has come to light in the collection is of particular interest to us here: a copy made by Collins of an unpublished treatise by his friend Nicolaus Mercator, who had arrived in England from Denmark in 1653 and earned a living as a mathematics tutor and occasional translator.[1] The treatise is entitled 'The doctrine of differences' (CUL Add MS 9795/9/15). Investigation of its contents has revealed that several of its pages are in fact copied directly from papers written many years earlier by Walter Warner, who in his lifetime was best known for his posthumous edition of Thomas Harriot's algebra (Harriot 1631). This discovery was quite unexpected because Warner died in 1643, ten years before Mercator arrived in England, and there has never been reason to think that the two had any mathematical or social connection. The appearance of

---

1. Nicolaus Mercator is not to be confused with Gerard Mercator, inventor of the cylindrical map projection, who lived a century earlier.

identical material in their separate writings therefore raises several intriguing questions: how did Mercator come to have such intimate knowledge of the papers of a man he had never met? And why did their content matter to him?

Investigation of the connections between Warner and Mercator uncovers networks of informal mathematical communication that spanned more than half a century, and which involved not only practitioners but also patrons and interested bystanders. It also demonstrates a preoccupation with a kind of mathematics that had little to do with the major achievements of the seventeenth century: it was not analytic geometry or the new methods of indivisibles that intrigued Warner or Mercator and their various acquaintances, but a method (to be described below) of interpolating tables using constant differences. Their pursuit of this method consumed much time and effort but produced almost nothing in the way of published work, and until now the method has gone almost entirely unnoticed as a topic of seventeenth-century English mathematical discourse.

The method of differences was not an isolated instance of a long-running mathematical problem. In the later part of the chapter we will look briefly at a second example, this time based on algebra. Taken together, these two case studies demonstrate a lively interest in mathematics amongst people of widely differing social backgrounds, an enthusiasm that nourished a thriving mathematical subculture.

All intellectual activity in England in the mid-seventeenth century has to be seen in the context of a volatile and sometimes violent political background. From 1642 until the early 1650s, the entire population of Britain suffered from a series of civil wars in England, Scotland, and Ireland. The causes and changing alliances were complex: in the early years the factions divided around the relative powers of King and Parliament, but there were also profound divisions over matters of doctrine and religious authority between Anglicans of the established church and dissenting Presbyterians. As in any civil war there was widespread social breakdown. Oxford and Cambridge virtually closed down as universities, and Oxford colleges became Royalist garrisons. Elsewhere families and individuals were dislocated, and homes, books, and papers were lost. After the execution of King Charles I in 1649, the country was ruled for nine years by Oliver Cromwell, and for two more years after his death by his son Richard, until Charles II was restored to the throne in 1660. Those who during the 1640s and 1650s privately discussed mathematics and experimental science must have found in such subjects a welcome respite from more intractable and threatening debates.

Harriot and Warner and the method of differences

To understand Warner's mathematical ideas we must begin with those of his longtime friend and colleague Thomas Harriot. Warner was employed as librarian to the Earl of Northumberland, who was also Harriot's patron from the early

1590s until Harriot's death in 1621. Warner was never in Harriot's intellectual league: the only modern scholar to have studied his non-mathematical writings describes him as a 'not too clear-thinking minor philosopher' (Prins 1992, xviii). Nevertheless, Warner lived and worked in close proximity to Harriot for long enough to learn a good deal about his ideas and their potential, and almost all his own mathematical writings can be connected back to Harriot's.

Some of Harriot's most sophisticated and time-consuming calculations, carried out over several years, were to do with problems of navigation on the curved surface of the globe. Late sixteenth-century navigators knew that when they steered on a given compass bearing they needed to correct for the curvature of the earth, and that the nearer they were to one of the earth's poles the greater the correction had to be. Harriot calculated extensive tables of corrections, or 'meridional parts' (see Pepper 1968). The details of his computations are complex, but only one aspect is needed here. Borrowing Jon Pepper's notation, we may denote a particular constant that arises in the calculations by β (its value happens to be 0.99970915409725778). Harriot needed values of β^n for $n = 0.1, 0.2, 0.3, \ldots, 0.99$. To find them, he used a method of interpolation that allowed him to 'subtabulate' these 99 new values between the (known) values of β^0 and β^1. This method became so valuable in this and other contexts that in outline it is described here, using one of Harriot's own, much simpler, examples.

Consider Table 2.2.1, in which the tabulated figures are in column C. The entries in column B are first differences taken in the positive direction ($48 - 3 = 45$ and so on) and those in column A are second differences ($95 - 45 = 50$ and so on). Clearly the numbers in column C have been chosen (or generated) in such a way that the numbers in column A are constant.

Table 2.2.1 A constant difference table

N	C	B	A
0	3		
		45	
1	48		50
		95	
2	143		50
		145	
3	288		50
		195	
4	483		50
		245	
5	728		

Now suppose, keeping the differences in column A constant, that we wish to interpolate four new entries between 3 and 48 in column C. What must the new constant difference in column A be? What changes will we see in column B? And what will the new entries be in column C? Harriot was able to derive general formulae to answer each of these questions, and in this particular case his solution was as shown in Table 2.2.2.

Table 2.2.2 An interpolated version of Table 2.2.1

$\frac{N}{n}$	c		
			b
0	3		a
		5	
$\frac{1}{5}$	8		2
		7	
$\frac{2}{5}$	15		2
		9	
$\frac{3}{5}$	24		2
		11	
$\frac{4}{5}$	35		2
		13	
$\frac{5}{5}$	48		

The numbers in the leftmost column are not part of the main table, but are margin entries, or indices: thus $\frac{3}{5}$ is not a fraction, but shorthand for 'the 3rd entry of a table interpolated to 5 times its original length'. In general, Harriot wrote this index as $\frac{N}{n}$ and gave formulae for calculating its value for any (integer) values of N and n, and for constant first, second, third, or higher differences.[2]

Harriot's tables of powers of β, being essentially exponential, do not produce constant differences, but may be assumed to do so over relatively short intervals, and Harriot used the above method extensively in his calculations of meridional parts. There are also signs that he began to recognize other uses for it. He experimented, for instance, with evaluating polynomial expressions for the first few integers. Table 2.2.3 is his table of values of 'a cube plus three times its root' or, as Harriot wrote it, $C + 3R$ (BL Add MS 6782, f. 246).

The figures in the three columns on the right are first, second, and third differences calculated from the values 4, 14, 36, 76, 140. Those below the stepped line are calculated directly from these values, but those above it appear to have been obtained by extrapolating upwards, always keeping 6 in the final column. In a second table, written alongside this one, Harriot tested the extrapolation for negative values of R as well. His interpolation method would allow him to insert further figures between those listed here (in the manuscript the relevant formulae appear immediately below the tables) and therefore in principle to solve the equation $C + 3R = N$ for any required value of N.

2. For fuller details and transcripts of the relevant manuscripts see Beery and Stedall (2008).

Table 2.2.3 A cube plus three times its root, with first, second, and third differences, for $R = $ 1, 2, 3, 4, 5

		4	−6	6
1.	$1 + 3 = 4$	4	0	6
2.	$8 + 6 = 14$	10	6	6
3.	$27 + 9 = 36$	22	12	6
4.	$64 + 12 = 76$	40	18	6
5.	$125 + 15 = 140$	64	24	6

There is ample evidence in Warner's manuscripts that he experimented with Harriot's constant difference method, both for interpolating tables and for solving simple polynomial equations. Unfortunately there is no clue to the date of his work except from his handwriting. By the late 1620s Warner was over seventy and his handwriting was visibly shaky.[3] His notes on difference methods are still in a relatively firm hand, suggesting that he made them some years earlier, possibly before Harriot's death and with his guidance.

For reference later, four particular extracts from Warner's manuscripts are briefly described here, labelled for convenience as (W1) to (W4).[4]

(W1) (CUL Add MS 9597/9/15, ff. 72–73) is headed 'Problema Arithmeticum ad doctrinam de differentium Progressionibus pertinens', 'An arithmetic problem pertaining to the doctrine of progressions of differences', and gives Table 2.2.4, for which the first differences are a, $a + e$, $a + 2e$, $a + 3e$,..., and the second difference is always e.

In accompanying notes, Warner explains that he denotes the difference between b and c by f, the difference between c and d by h, and the difference between f and h by g. We may rewrite the first two relationships as

$$f = c - b = 10a + 45e$$
$$h = d - c = 10a + 145e$$

from which it is easy to see that

$$g = h - f = 100e$$

It therefore follows that $e = \dfrac{g}{100}$, and that $a = \dfrac{f - 45e}{10}$ Warner gave a rather longer derivation but arrived at these same formulae for e and a. In other words, given just b, c, and d, it is possible to calculate e and a, and hence to fill in all the intermediate values of the table. This is in fact a special case of Harriot's more general interpolation formulae.

(W2) (BL Add MS 4396, ff. 20–29) is a run of ten sheets concerned with the following two problems for a given constant difference table: (1) given any index

---

3. See, for example, his note on the Preface to the *Praxis* in BL Add MS 4395, f. 92.
4. For fuller details on each of them see Beery and Stedall (2008).

Table 2.2.4 An algebraic difference table

$$b = b$$
$$b + 1a$$
$$b + 2a + e$$
$$b + 3a + 3e$$
$$b + 4a + 6e$$
$$b + 5a + 10e$$
$$b + 6a + 15e$$
$$b + 7a + 21e$$
$$b + 8a + 28e$$
$$b + 9a + 36e$$
$$c = b + 10a + 45e$$
$$b + 11a + 55e$$
$$b + 12a + 66e$$
$$b + 13a + 78e$$
$$b + 14a + 91e$$
$$b + 15a + 105e$$
$$b + 16a + 120e$$
$$b + 17a + 136e$$
$$b + 18a + 153e$$
$$b + 19a + 171e$$
$$d = b + 20a + 190e$$

$\frac{N}{n}$ in the margin, find the corresponding entry in the table; (2) conversely, given an entry in the table find its index $\frac{N}{n}$.

The first worked example is to find the entry corresponding to $\frac{N}{n} = \frac{3}{5}$ in Table 2.2.5.

Table 2.2.5 Warner's worked example from (W2)

N	C	B	A
0	3		
		45	
1	48		50
		95	
2	143		50
		145	
3	288		50
		195	
4	483		50
		245	
5	728		

Clearly this problem is identical to the one posed by Harriot shown in Table 2.2.1. The other problems in (W2) are numerical and algebraic variations on the same theme.

(W3) (BL Add MS 4396, f. 19) is concerned with the interpolation of tables of antilogarithms, or (since Warner was working in base 10) values of 10^x. As we will see later, Warner devoted the final years of his life to such calculations. The method is illustrated by two worked examples: the first shows how to interpolate nine new entries at the beginning of the table, between those for 0 and 0.0001, namely, those for 0.00001, 0.00002,..., 0.00009. Over this range the second difference (taken as far as the twelfth decimal place) can be considered constant, and Warner's method is exactly Harriot's interpolation method for constant second differences. Warner's second example, on the reverse of the same sheet, gives a calculation for the end of the table, for the entries to be interpolated between those for 0.9999 and 1, namely those for 0.99991, 0.99992,..., 0.99999. (Antilogarithms need only be tabulated between 0 and 1 because any others are found by scaling by an appropriate power of 10.) This, of course, was the problem that Harriot had also worked on when computing his tables of meridional parts.

(W4) (BL Add MS 4395, ff. 166–166v and 181) consists of just two pages and a postscript, on the subject of solving equations by the method of differences. Warner begins:

There is another way for the solution of equations by differentiall progressions to be considered of. As for squares $aa + 3a = 130 \; // \; a = 10$

He then gives Table 2.2.6, which shows the first few values of $aa + 3a$ together with first and second differences (f. 166).

Table 2.2.6 Values of $aa + 3a$ for $a = 1, 2, 3, 4$

$a = 1$	//	4	– – –	4	– – –	2
$a = 2$	//	10	– – –	6	– – –	2
$a = 3$	//	18	– – –	8	– – –	2
$a = 4$	//	28	– – –	10	– – –	2

&c. usque ad 130 ('etc. as far as 130').

Clearly this is similar in layout to Harriot's table for a cube plus three roots, in Table 2.2.3 above. It is followed in Warner's manuscript by other tables showing the first few values of $aaa + 2aa + 3a$ (with a constant difference of 6 in the final column), and of $aaaa + 1aaa + 3aa + 2a$ (with a constant difference of 24). All the tables are set out in neat rectangular blocks, and Warner gave rules for writing down the crucial first row simply by inspection:

The primes [first terms] of these progressions [columns] are given thus, first of the squares, the first is the summe of the coefficients, the second is the same, the third the caracteristik 2. Of the cubes the first is the summe of the coefficients, the second the same, the third the double of the coefficients of the squares, the fourth the caracteristik 6. [...] the like rules are to be found for the rest [...]

The rules are correct, but Warner gave no hint as to how he had found them.

Suppose now that one wishes to solve the equation $aa + 3a = N$, where N is a positive number. All one has to do is extend the table downwards and hope that N will appear in the column of values on the left. It may happen, of course, that N will fall somewhere *between* the calculated entries. Warner returned to this problem with a note added fifteen pages later (f. 181), and gave the rules for interpolating new entries between each of those already calculated, using the worked example $aa + 6a = N$.

Warner and Pell and analogics

The Earl of Northumberland died in 1632 and from then on Warner was supported by Sir Thomas Aylesbury, one of the few surviving members of Harriot's circle. By then Aylesbury and Warner had published Harriot's *Praxis*, and planned to edit more of Harriot's papers. The project came to the attention of Samuel Hartlib, a scientific 'intelligencer' who noted such mathematical and scientific advances in his *Ephemerides* (see Clucas 1991). Hartlib in turn introduced Aylesbury to his own mathematical protegé, John Pell, then in his late twenties and seeking employment.

From Pell we have an unusual account of a mathematical conversation between him and Aylesbury in January 1638 (BL Add MS 4419, f. 139), a conversation of a kind that is instantly recognizable but rarely written down. The subject under discussion is the equation $aa + ba = bb$, in which the unknown quantity is a. The technical details are less important, however, than the verbal negotiation between Aylesbury [A] and Pell [P].

P. You are farre enough already Sr for a numerous exegesis. A. No. P. Yes, I can exhibit a by that equation for it is no other than Harriot's $aa + da = ff$. A. True. Let us goe on. Let b be equall to $2c$.

Heere I was not willing to stop him and say why doe you take $b = 2c$. But let him goe on. (More he might have sayd $4cc$ for bb, but he changed not that.)

A. $aa + 2ac = bb$

(adde cc to both)

$aa + 2ca + cc = bb + cc$

A. The former is a true square, therefore let the other be so, viz $= xx$, whose roote $x =$ to the roote of $aa + 2ca + cc$ which is $a-c$. P. No Sr $a + c$. A. No? are we mistaken? tis true.

$a+c=x$

$a=x-c$

and heere he left off as having found a.

'Harriot's $aa + da = ff$' is an equation discussed in the *Praxis* (Harriot 1631, 119–121). Earlier in the conversation Aylesbury had also described Harriot's idiosyncratic equals sign (with two short vertical stokes between the horizontals), a sign that had never appeared in print but only in the manuscripts. Thus, through

Aylesbury, Pell acquired an intimate knowledge of Harriot's work, understanding some of it rather better than Warner did (see Malcolm and Stedall 2005, 273–276).

Aylesbury probably thought that Pell would make an ideal assistant for Warner who, though over eighty, was engaged in a new project: the calculation of antilogarithms or 'analogics'. When Warner and Pell met for the first time in November 1939, Warner explained to Pell that he already had a table of 10,000 antilogarithms and that he now wanted to interpolate it to 100,000 entries (BL Add MS 4474, f. 77). As we have already seen, Warner had written out worked examples of such interpolation in (W3).

Pell did not become actively involved until the summer of 1641, by which time Warner must have recognized that without his help the task would never be completed. Pell kept a detailed record of the time he spent on the calculations (BL Add MS 4365, ff. 36–39), from which we can see that he followed exactly the notation and rules outlined by Warner in (W1). Here, for example, is his record of his second hour of work on Friday 25 June 1641:

> I calculated all the 2nd differences ($g = 100e$)
> subtracted their tenths ($10e$)
> bisected their remainders $45e$
> subtracted those halfes from both f and h to find
> the upper and lower $10a$
> Tried them by adding g to the lower $10a$.
> Found and wrote downe a and e in every
> semi-columne of the 500 in a little lesse than an houre.

(The values that Pell calls 'lower' and 'upper' $10a$ are $f - 45e = 10a$ and $h - 45e = 10a + 100e$, respectively. Since $g = 100e$, adding g to 'lower' $10a$ should give him 'upper' $10a$, a useful check.)

Warner had offered Pell the sum of £40 to complete the work, but Aylesbury increased it to £50.[5] Nevertheless, by the middle of August Pell was complaining about his outlay on materials, the project faltered, and Warner asked for all the papers back. Pell's list of items returned on 27 August includes the 'Problema arithmeticum', identified above as (W1).

The financial difficulties must have been resolved because Pell began calculating again on 20 September. By November he had subcontracted some of the work to a Mr Edward Watts and a Mr Turner, to whom he communicated his method 'in writing and instructions viva voce' (see also Croarken, Chapter 4.4 in this volume). As in earlier conversations between Pell and Aylesbury, much of the communication was oral, and once again, Pell's meticulous recording reveals to us a little of the behind-the-scenes discussion. On Thursday 17 November Watts brought him

---

5. This would have been equivalent to about £6000 in 2007 (Officer 2007).

three sheets of completed calculations and one hundred and four sheets prepared for new entries, but Pell immediately saw that something was wrong:

[…] from 10950 to 52350 that is 41400 Analogicalls which at 400 in a sheete makes 103 sheetes $\frac{1}{2}$. So that heere is somewhere halfe a sheete too much. I sought and found it in his 30th sheete, he had written out the numbers of 2 pages into the 2 following pages. And so put all out of Square for from that sheete forward the sheetes began 21500, 21950, etc.

The three sheets completed, however, were correct, more or less:

Wats his 3 sheetes were true save that in one decad he had erred in the first difference and propagated it to the end which he could not finde. I examined 3 sheetes more for him and let him carry it all home again for Turner to examine.

Pell must have decided it was easier to continue alone, because Watts and Turner soon disappear from the record, transient characters of whom we know nothing more.

By July 1642 the calculations were complete, but the outbreak of civil war prevented their publication. The tables are now lost but in the seventeenth century they were known and discussed by those interested in such things. In 1693 John Wallis reported, on information from Pell, on the origins of the project: 'I do not know but that the tables were begun by Harriot; […] which was recently told to me by Pell, who was well known by Warner'.[6] Thus as late as 1693, more than seventy years after his death, Harriot and his achievements continued to be the subject of oral history.

What else might Pell have learned from Warner of Harriot's methods? Many years later, Pell repeatedly claimed that he knew a method of solving equations by tables, though he could never be prevailed upon to describe it. John Collins, with whom Pell lodged for some months during the second half of 1669, did not understand what the method was, but wrote in 1670 that: 'Dr Pell affirmeth he hath for above 30 yeares used to solve Aequations by tables' (Gregory 1939, 142). Indeed, Collins thought that the tables calculated by Warner and Pell were made for precisely that purpose. Describing the tables in 1675 he wrote:

[…] considering the logarithms were already made, and more proper for compound interest and annuity questions for all ratios, I could not conceive but that this table was made properly for algebraical uses in resolving equations; what use it was intended for Dr. Pell is not free to disclose; none of his friends here can render him communicative. (Rigaud 1841, I, 216; see also II, 197, 219)

Collins was almost certainly confused here. The method Pell knew, but for some reason would not explain, was actually the one Warner had experimented with in (W4), relying on extrapolation and interpolation from constant differences.

---

6. *Canonem illum nescio an inchoaverit D. Thomas Harriot; […] Quod mihi nuper inidicavit D. Johannes Pell, qui et Warnero fuerat familiariter notus* (Wallis 1693, 63).

Collins, Mercator, and Warner's papers

Warner died in 1643, and at the end of that year Pell took up a teaching post in Amsterdam. A few months later he feared that Warner's papers would be 'throwen into the fire; If some good body doe not reprieve them for pye-bottoms' (Malcolm and Stedall 2005, 358). Fortunately, however, the papers went to Warner's nephew, Nathaniel Tovey, and were saved from both frying pan and fire. In 1652 Tovey passed them to Herbert Thorndike, a clergyman at Westminster Abbey with an interest in mathematics. Thorndike discussed with Pell the possibility of publishing the tables of antilogarithms, but Pell was reluctant because the section from 29,759 to 39,750 was now missing (BL Add MS 4279, ff. 275–276). Then in 1667 Thorndike loaned some of the papers to John Collins, who was always keen to acquire and peruse new mathematical material. In 1668 Collins wrote to James Gregory in Scotland:

I have some papers of Mr Warner deceased, wherein he proves if parallels be drawn to an asymptote, so as to divide the other into equal parts, the spaces between them, the hyperbola, and asymptote, are in musical [harmonic] progression [...] (Gregory 1939, 45)

The second part of (W4) follows immediately after Warner's treatment of the hyperbola, indeed on the same page, and so if Collins read one he must also have seen the other.

Collins was elected to the Royal Society in 1667, the year after Nicolaus Mercator, so the two would certainly have come to know each other then, if not before. Indeed it was in 1667 or 1668 that Collins engaged Mercator to translate Gerard Kinckhuysen's *Algebra ofte stel-konst* (1661) from Dutch to Latin (see Whiteside 1967–81, II, 280–291). In 1667 Mercator was also writing his own *Logarithmotechnia*, a short treatise offering two new methods for calculating logarithms. The first of his methods is very obscurely explained, but is based on the calculation of differences between known logarithms. The book therefore contains several examples of difference tables and results derived from them (Mercator 1668, 11–14). His second method is based on the calculation of a partial area under a hyperbola. Both of these topics, as we have seen, were touched on in Warner's papers, so it would have been natural for Collins to offer to show Mercator the relevant portions, though we do not know whether he did so before or after Mercator completed the *Logarithmotechnia*.

Mercator took a particular interest in Warner's work on difference methods and, as noted in the introduction to this chapter, his own treatise 'On the doctrine of differences' incorporates several extracts copied verbatim from Warner's papers. Unfortunately, the two sets of manuscripts are now separated in two different libraries and it is impossible to compare them directly. My first inkling of the shared material came from finding (W2) itself folded into the back of Collins' copy of the treatise. There is no name or date on it but Warner's handwriting is unmistakable. The contents have also been copied out in the body of the treatise itself. From a comparison of my notes on Mercator's treatise in Cambridge with those I had made some years earlier on Warner's papers in the British Library, it became obvious that Mercator had in fact copied out (W1) to (W4) in their entirety. 'On the doctrine of differences' opens with a general introduction to solving quadratic and cubic equations, but then turns to the method of solving equations by differences, and Mercator's examples and rules are exact copies of those in (W4). These are followed by a transcript of (W1), the 'Problema arithmeticum', and then by material from (W2), problems and theorems on the method of differences, rearranged into a more logical order, but otherwise unchanged. Finally there is a copy of (W3), on the calculation of antilogarithms by interpolation. The remaining few pages of the treatise are also devoted to further discussion of antilogarithms. Thus almost all of Mercator's treatise is devoted to themes previously treated by Warner, and over half of it is directly copied from his manuscripts. Further, the original of (W1) tucked into the back of Collins' copy strongly suggests that Collins himself was the source of this material.

Further signs of collaboration between Collins and Mercator are to be found in notes written by Collins under the heading 'Improvements of algebra in England' (BL Add MS 4474, ff. 1–4), a subject he was concerned with during 1668 and early 1669 (Collins 1669). One of the improvements hinted at in his notes is a 'Method of Progressional Differences'. Collins did not understand it very well but wrote that it 'seemes to be no other but a Generall method of interpol[at]ing such ranks whose 3d 4th 5th 6th Differences are aequall'. This was all he could say, but right at the end of the notes we find a few examples of difference tables for polynomials, and they are precisely those from (W4) together with the rules for writing down the first row. Perhaps Collins saw that he was more likely to learn about the method from Warner than from Pell. Perhaps it was at this point too that he turned to Mercator for help, and persuaded him to write 'On the doctrine of differences'.

It is clear that the method of differences devised by Harriot around 1610 was still under discussion sixty years later, and still with reference to Harriot's own examples. This is already a remarkable story of transmission through oral report and the circulation of manuscripts. The full picture is actually even more complex. I have not mentioned, for example, the work of Harriot's friend Nathaniel Torporley, who in the 1620s penned his own treatise on the method of differences based on Harriot's findings. This document too has recently come to light in the Macclesfield Collection (CUL Add MS 9597/17/28) and awaits further study by anyone brave enough to grapple with a hundred pages of Torporley's inpenetrable Latin. Nor have I discussed Sir Charles Cavendish, who became acquainted with Warner in the 1630s and copied out not only (W1) and (W2) but many other pages on the method of differences from Harriot's manuscripts.[7] Twenty years later, when Cavendish and Aylesbury were both living in exile in Antwerp,

---

7. Cavendish referred to the method of differences as the 'doctrine of triangular numbers'; see Beery and Stedall (2009).

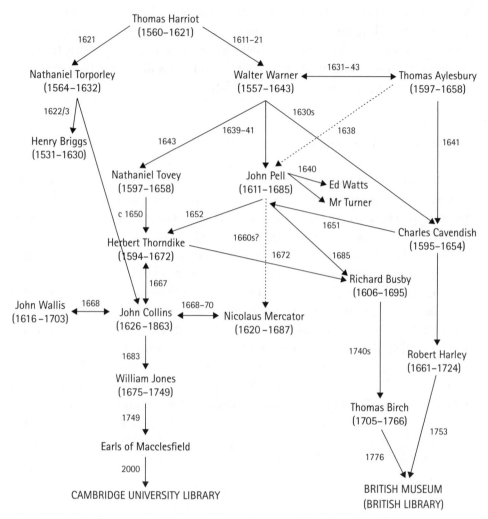

Fig. 2.2.1 The known circulation of Harriot's method of differences

they discussed this material yet again, and Cavendish wrote to Pell asking for an explanation of it (Malcolm and Stedall 2005, 584). His letter is not only evidence of a continued interest in Harriot's method, but also shows that Aylesbury considered Harriot's manuscripts important enough to carry with him into exile.

If one tries to construct some sort of diagram of the many people who knew about or worked on the method of differences between 1610 and 1670, what emerges is not a picture of simple linear transmission, but a tangled web of crossing threads and loose ends.[8] An attempt at such a diagram is shown in Fig. 2.2.1. The threads link individuals who knew and communicated with each other, sometimes by letter but more often through conversations and the passing on of manuscripts. The possibility of reconstructing such networks is very much

8. For further details of the dispersal of Warner's papers see Stedall (2002, 113–116).

a matter of chance. In this case, thanks to the obsessional recording habits of Pell and the willingness of Collins to share everything he knew with anyone who would listen, we have relatively detailed information. At the same time, the evidence remains tantalizingly fragmentary and a new discovery, like Mercator's treatise, can significantly change the existing picture.

Brereton's problem and others

The discussion of the method of differences was not an isolated occurrence of a long running mathematical problem. Here we will look briefly at a second example, this time based on the use of algebra, and in this case too our evidence comes largely from the manuscripts of Pell and Collins.

In 1649 Pell was teaching in Breda in the southern Netherlands, a city that at the time was also the temporary headquarters of the exiled Charles II. The presence of the court brought a number of other Englishmen to Breda, one of whom was Silas Titus, a member of the Presbyterian delegation that was trying to negotiate with the King. Titus took some interest in mathematics, and it is likely that the long friendship between him and Pell was first formed during his months in Breda. Another expatriate was the eighteen-year-old William Brereton, son of a commander in the Parliamentary Army, who had been sent to Breda from Cheshire to complete his education; he too became a lifelong friend to Pell.

A problem brought to Pell by Brereton in 1649, and probably discussed at the time with Titus as well, was the following (BL Add MS 4413, f. 52): to find numbers a, b, and c satisfying the equations:

$$aa + bc = 16$$
$$bb + ac = 17$$
$$cc + ab = 22.$$

The solution in positive integers is easily seen to be $a = 2$, $b = 3$, and $c = 4$, but Pell decided to challenge himself by changing the final equation to:

$$cc + ab = 18.$$

In this form the problem is very much harder, leading to an eighth-degree equation with no easy method of solution.

By 1662 Titus and Pell were both back in England, Titus in a high ranking position as Officer of the King's Bedchamber, and Pell as a clergyman. It seems that in the autumn of 1662 Titus once again brought up Brereton's problem, and Pell tried to reconstruct the solution he thought he had found in Breda, but with little success (BL Add MS 4413, f. 52). Pell now posed it in turn to John Wallis, Savilian Professor of Geometry at Oxford, and the two of them worked on it together over a period of several weeks. In December 1662 Pell left notes on their progress for Titus (BL Add MS 4425, ff. 367, 368), and by April of the following year he and Wallis had successfully solved it, calculating values of a, b, and c to 15 decimal places each (BL Add MS 4425, f. 161). It is not clear whether Wallis and Titus had any direct contact at this time. Wallis was later to describe Titus as 'a very Ingenious Person [...] and very well accomplished in Mathematical and other Learning' (Wallis 1685, 225), but one wonders what he really thought in 1662. Back in 1649, while Pell and Titus had been whiling away their time on algebra, Wallis had been busy deciphering coded letters captured from the Royalists, many of which were intercepted between Breda and London. One of the names that Wallis discovered in them and inserted in red ink into his plain text was that of Silas Titus (Bodleian Library MS e Mus 203, ff. 193, 195, 198, 200). Did Wallis remember this when he and Titus both took up Brereton's problem thirteen years later? Did Titus know or guess that Wallis had once held such dangerous information on him? This is one of the most extreme examples, but by no means the only one, of a common interest in mathematics eventually transcending deep political and religious differences.

Wallis later lodged a copy of the solution with Collins. That copy is now lost, but we know from Pell that Collins labelled it with the words 'Dr Wallis his Resolution of an exercise upon a probleme put by Dr Pell' (BL Add MS 4411, f. 361). Pell possibly resented the implication that he had been unable to solve the problem for himself, and replaced the last three words with 'put by Colonel Titus, who had received it from Dr Pell', which was more or less the form of words used by Wallis when he later published the solution in *A treatise of algebra* (1685, 225–256). Once such a problem was in the hands of Collins it was likely to go further. In 1672 Collins sent it to James Gregory, who referred to it as the 'Breretonian problem' (Rigaud 1841, II, 242). This attribution to Brereton (rather than to Pell or Titus) suggests that information about it came by word of mouth from Pell. It is yet another example of oral mathematical history, in the form of stories and gossip about problems, papers, and people.

In 1677 Robert Hooke noted in his diary that he had borrowed Mr Baker's solution to the problem '$aa + bc = x$. $bb + ac = y$. $cc + ab = z$.' from Collins at the Rainbow, a London coffee house (Hooke 1935, 322). Thomas Baker was a Devon clergyman, a regular correspondent of Collins at this period and later the author of a treatise entitled *The geometrical key* (1684). We thus know of at least eight people who at some time saw or attempted Brereton's problem, and there may well have been others. Wallis's publication of the problem in 1685 was therefore little more than the first public announcement of a problem that had already been passed around for well over thirty years.

We have now seen a significant number of individuals, throughout the seventeenth century and from a wide variety of social backgrounds, who pursued

mathematics at a relatively mundane level out of pleasure or interest. Much of
our information comes from the correspondence of Collins and the manuscript
notes of Pell, but more is to be gleaned from the biographical notes made by
John Aubrey (Bennett, Chapter 4.2 in this volume), from Wallis's *Treatise of
algebra* which took up Collins' theme of promoting 'Improvements of algebra in
England', and from a number of other miscellaneous sources.

From Collins we learn, for example, of 'men of inferior rank that have good skill
in algebra […] One Anderson, a weaver, […] Mr Dary, the tobacco cutter, […]
Wadley, a lighterman' (Rigaud 1841, II, 479–480). Michael Dary, whose occupation
changed frequently and who later became a gauger, published *The general doctrine
of equation* in 1664, a sixteen-page pocket guide to solving equations, and a similarly tiny mathematical *Miscellanies* in 1669, but later turned to Pell for help with a
'soure crabbe' of a biquadratic equation that he could not handle himself (BL Add
MS 4425, f. 57). Robert Anderson took some pleasure in pointing out the errors
in the *Miscellanies,* writing dismissively of Dary's friends that 'I value the snares
of one, the stab of the other, and the envy of the rest, no more than the dirt of my
shoes' (Anderson 1670, 13). Dary and Anderson are men of whom we know a little
because they published a little, but of Wadley the lighterman we know nothing.

From Aubrey we hear of others, amongst them pupils of William Oughtred,
like Charles Scarburgh, who later became Royal physician but who never lost his
interest in mathematics (Aubrey 1898, II, 108, 284; Malcolm and Stedall 2005,
316); or the young Thomas Henshawe, later lawyer and alchemist, who brought
his own candles in order to study with Oughtred in the evenings (Aubrey 1898,
II, 108, 110). Or there was Edward Davenant, who learned mathematics from his
father and taught it in turn to his own ten children, and whose eldest daughter,
Anne Ettrick, Aubrey described as 'a notable Algebrist' (Aubrey 1898, I, 199, 201,
202). In similar vein, Edward Sherburne, in the appendix to his *Sphere of Manilius*
published in 1675, mentioned William Milburne, curate of Brancepeth near
Durham, who around 1640 had made astronomical observations but was also
'very knowing in Arithmetick, particularly in Algebra (having in the year 1628
extracted the roots of the equation $1,000 = aaaaa - aaaa - 4aaa + 3aa + 3a$)'
(Sherburne 1675, 91). Wallis, in a list he wrote in the 1690s of those interested
in algebra, mentioned Scarburgh, and Davenant, but others too, for example,
Adam Martindale, schoolmaster and dissenting clergyman; and the otherwise
unknown George Merry, who went to the trouble of writing two hundred and
thirty-six pages on the factorization of polynomials (Wallis 1693, 233–234; see
also Wallis 1685, 'Additions and emendations', 157–162).

None of these people ever had a theorem named after them or made any significant contribution to mathematical thinking. But nor did they belong to the
class sometimes known as 'mathematical practitioners', who routinely carried
out calculation or mensuration as part of their daily lives and trade (though some

of them, like Dary, may sometimes have done so). These people formed a different group, a kind of mathematical middle class, with enough education and leisure to engage in mathematics for pleasure and intellectual stimulation. Very often, their preferred subject of study was algebra, which must have seemed satisfyingly new and modern, and where even beginners, with some assistance, could make good progress. These loosely connected networks of mathematical enthusiasts came from a broad range of social backgrounds and displayed widely varying abilities. Some, like Cavendish, Aylesbury, or Collins, had only a tenuous grasp on the subject, but nevertheless devoted much of their lives to promoting and supporting it. For the most part, the existence of such people is known to us not through published work but only because discussions or encounters happen to have been recorded in notes, letters, or diaries. This kind of evidence is inevitably rare, fragile, and partial.

Conclusion

Our knowledge of mathematical activity in the past relies upon a combination of circumstances, some of which are very much a matter of chance. First, the mathematics itself must have been recorded in some permanent form. Such a statement seems obvious, but can obscure the fact that most mathematics is not done this way: a mathematical text of any kind almost always arises from, or is meant to be accompanied by, oral discussion and explanation, and such communication is by its nature ephemeral. We can recognize the authenticity of the scraps of conversation recorded by Pell with Aylesbury or Watts, but we also know that such records are extremely rare. Second, the material must survive, having escaped the fate of being 'throwen in the fire'. The records that have come down to us are inevitably incomplete, and sometimes very sparse indeed, and a more comprehensive picture of the mathematics of the past might also be a very different one. Third, the material must be accessible, which for reasons of politics, conservation, or lack of translation it may not be. Finally, the dispersal of material in separate libraries or on different continents means that comparisons and connections are sometimes possible only through the lucky recognition of handwriting or content.

For England in the seventeenth century, the material conditions of existence, survival, and availability have been relatively favourable. What one then reads into the evidence, however, depends on what one is looking for. At one level, the period presents major mathematical stories such as the development and consolidation of the calculus, or the emergence of gravitational theory. But the preoccupations of Warner with antilogarithms, or of Collins with solving equations, show that there were also mathematical concerns of a different kind, still with new and 'modern' ideas, but played out at a more mundane level by men who were followers rather than innovators. Moving yet further into the obscure and little known, we find such disparate figures as Brereton, Titus, Anne Ettrick, Dary, Anderson, Merry, Mercator, and many others, busying themselves for no obvious purpose with mathematical problems of an essentially useless kind. Clearly such people regarded this as a meaningful and worthwhile use of their time. Individually, they are relatively insignificant, but collectively offer us a new perspective on the seventeenth-century mathematical landscape.

Bibliography

Abbreviations:

BL British Library
CUL Cambridge University Library

Anderson, Robert, *Dary's miscellanies examined: and some of his fundamental errors detected*, London, 1670.
Aubrey, John, *Brief lives, chiefly of contemporaries, set down by John Aubrey between the years 1669 and 1696*, ed Andrew Clarke, 2 vols, Oxford, 1898.
Baker, Thomas, *The geometrical key*, London, 1684.
Beery, Janet, and Stedall, Jacqueline, *Thomas Harriot's doctrine of triangular numbers: the 'Magisteria magna'*, European Mathematical Society, 2009.
Clucas, Stephen, 'Samuel Hartlib's *Ephemerides*, 1635–59, and the pursuit of scientific and philosophical manuscripts: the religious e °thos of an intelligencer', *The Seventeenth Century*, 6 (1991), 33–55.
Collins, John, 'An account concerning the resolution of equations in numbers', *Philosophical Transactions of the Royal Society*, 4 (1669), 929–934.
Dary, Michael, *The general doctrine of equation*, London, 1664.
Dary, Michael, *Miscellanies*, London, 1669.
Gregory, James, *James Gregory tercentenary memorial volume*, ed H W Turnbull, London, 1939.
Harriot, Thomas, *Artis analyticae praxis*, London, 1631.
Hartlib, Samuel, 'Ephemerides', The Hartlib Papers, Sheffield University Library.
Hofmann, J E, 'Nicolaus Mercators *Logarithmotechnia* (1668)', *Deutsche Mathematik*, 3 (1938), 446–466.
Hooke, Robert, *The diary of Robert Hooke MS MD FRS 1672-1680*, eds Henry W Robinson and Walter Adams, Wykeham Publications, 1935; reprinted 1968.
Malcolm, Noels, and Stedall, Jacqueline, *John Pell (1611-1685) and his correspondence with Sir Charles Cavendish: the mental world of an early modern mathematician*, Oxford University Press, 2005.
Mandelbrote, Scott, 'Footprints of the lion: at work on the exhibition', *BSHM Newsletter*, 46 (2002), 5–9.
Mercator, Nicolaus, *Logarithmotechnia*, London, 1668.
Mouton, Gabriel, *Observationes diametrorum solis et lunae apparentium*. Lyon, 1670.
Officer, Lawrence H, 'Purchasing power of British pounds from 1264 to 2006', MeasuringWorth.com, 2007.

Pepper, Jon V, 'Harriot's calculation of the meridional parts as logarithmic tangents', *Archive for History of Exact Sciences*, 4 (1968), 359–413.

Prins, Johannes, 'Walter Warner (ca 1557–1643) and his notes on animal organisms', Doctoral thesis, Utrecht, 1992.

Rigaud, Stephen Peter (ed), *Correspondence of scientific men of the seventeenth century*, 2 vols, Oxford 1841; reprinted Hildesheim: Olms, 1965.

Scriba, Cristoph J, 'Mercator's Kinckhuysen translation in the Bodleian library at Oxford', *British Journal for the History of Science*, 2 (1964), 45–58.

Sherburne, Edward, *The sphere of Marcus Manilius made an English poem: with annotations and an astronomical appendix*. London, 1675.

Stedall, Jacqueline, *A discourse concerning algebra: English algebra to 1685*, Oxford University Press, 2002.

Stedall, Jacqueline, *The greate invention of algebra: Thomas Harriot's treatise on equations*, Oxford University Press, 2003.

Wallis, John, *A treatise of algebra both historical and practical*, London, 1685.

Wallis, John, *De algebra tractatus* in Wallis, *Opera omnia*, II 1–482, Oxford, 1693.

Whiteside, D T (ed), *The mathematical papers of Isaac Newton*, 8 vols, Cambridge University Press, 1967–81.

Mathematics and mathematics education in traditional Vietnam

Alexei Volkov

This chapter is devoted to the transmission of mathematical expertise from China to Vietnam. Throughout this chapter China and Vietnam will play the parts of 'centre' and 'periphery'. However, the conventional use of these terms, suggesting that the tradition was established in one specific location (identified as the 'centre') before being transferred to a number of others (the 'periphery'), proves to be somewhat inadequate. This model excludes other possible options, such as interaction between *several* centres, each contributing to the growth of knowledge circulating in the network. Another phenomenon not accounted for in the 'centre-periphery' model is that of 'counter-currents' of scientific expertise transmitted from the presumed 'periphery' to the 'centre'.

Another premiss is invoked by the word 'China' and its derivatives. The modern toponym suggests that the social network within which the older scientific traditions were implemented remained entirely within certain ethnic and cultural boundaries, and that these boundaries, moreover, were identical with the present-day geographical borders. In order to solve or circumvent this problem when speaking about the plethora of mathematical schools that flourished in east and southeast Asia, mainland Chinese scholars coined the term '*Han zi wen hua quan shu xue* 汉字文化圈数学', literally 'mathematics of the sphere of the culture of Chinese characters'. Its conventional English rendering as 'mathematics using Chinese characters' is not only incomplete but also misleading: taken literally, it would imply, for instance, that the Chinese translation of Euclid's *Elements* and other Western mathematical works into Chinese should belong to this category. Moreover, the Chinese term, as well as its English rendering, suggests that the mathematical treatises belonging to this tradition were always written exclusively in (classical) Chinese. However, as we shall see below, in Vietnam a number of mathematical treatises were partly or entirely written in Vietnamese using the local script *Nôm* 喃.

One more way to refer to the mathematical traditions which, as conventionally believed, originated from China, would be to use the term 'Confucian Heritage Culture' adopted by a number of researchers working on mathematics education and its history (for example, Fan *et al.* 2004; Siu 2004). Even though this term avoids the aforementioned difficulties related to the notions of 'centre' and 'periphery', it remains unclear why among the various philosophical and religious teachings that existed in east and southeast Asia, only Confucianism has been chosen to label the scholarly tradition under consideration. The term implies that Confucian culture played a particular role in the development of mathematics, but this is a challenging hypothesis that still awaits confirmation.

These terminological difficulties related to transmission stem from unresolved theoretical problems that deserve discussion but which cannot be treated in the limited scope of this chapter. The conventional model of transmission, in which the relationship between China and Vietnam is pictured as that between a 'centre' and its 'periphery', adopted in this chapter for the sake of conciseness, thus does not necessarily represent the actual historical processes; nevertheless, it can be viewed as a convenient framework for the present discussion.

Transmission of Chinese mathematical expertise:
Korea, Japan, and Vietnam

Standard books in Western languages on the history of Chinese mathematics do not usually pay much attention to the transmission of mathematical expertise from the Chinese 'centre' to the 'cultural peripheries' of Korea, Japan, and, in particular, Vietnam. The Japanese mathematical tradition *Wasan* 和算, historically based on Chinese 'applied' mathematics and on medieval Chinese algebra, is arguably better studied than the other two.[1] During the past three decades, a number of works by Korean authors on traditional Korean mathematics have

---

1. The works on the history of the traditional Japanese mathematics in Japanese language are too numerous to be cited here; I will limit my references to the works published in European languages. They are: the earliest and yet still important monographs written or co-authored by Mikami Yoshio (Mikami 1913; Smith and Mikami 1914), the discussion of the temple mathematical tablets *sangaku* 算額 by Fukagawa and Pedoe (1989), and the monograph of Horiuchi (1994) dealing with the establishment of the most important Japanese mathematical school, that of Seki Takakazu 關孝和 and his disciples.

become available (Kim 1973a; 1973b; 1973c; 1986; Kim and Kim 1978; Cha 2002; Jun 2006), and other publications on Korean mathematics have recently appeared in Japan (Kawahara 1998; 2001), China (Li, Xu, and Feng 1999; Guo 2005), and Taiwan (Horng 2002; 2003; Li 2003; Horng and Li 2007).

It appears that the Chinese mathematical tradition was implemented in Korea and Japan in rather different ways: in Korea, it was the official occupation of professional mathematicians and astronomers, while in Japan it was taught in a network of private schools. We will see that the Vietnamese case is closer to that of Korea: there existed professional mathematicians and astronomers working at the Court, and mathematics was a subject of state examinations. Korean and Japanese mathematicians practised the tradition stemming from the so-called Chinese algebra of the Song dynasty (960–1279 AD) but Vietnamese mathematical treatises more closely resemble treatises on Chinese 'popular mathematics' of the Ming dynasty (1368–1644).

The circumstances of the implementation of the Chinese mathematical tradition in Vietnam remained almost completely unknown for a long time. The first attempt to investigate the extant materials on Vietnamese mathematics was made in the early twentieth century by the outstanding Japanese historian of mathematics Mikami Yoshio 三上義夫. He provided an analysis of the Vietnamese mathematical treatise *Chỉ minh toán pháp* 指明算法 'Guide towards understanding of the methods of calculation',[2] but he did not have access to the extant corpus of Vietnamese mathematical books.[3] In 1938 the Chinese mathematician and historian of science Zhang Yong 章用 visited Hanoi and explored the mathematical books preserved at the École Française d'Extrême-orient. Unfortunately, he died the following year and his findings remained unpublished, except for one paper devoted to the history of the Vietnamese calendar (Zhang 1940). Li Yan 李儼 listed the Vietnamese mathematical treatises found by Zhang Yong (Li 1954), but it remains unclear whether Li Yan had access to the books purchased or (partially) copied by Zhang Yong in Vietnam.[4] In 1991 Han Qi 韓琦 provided a brief introduction to extant Vietnamese astronomical and mathematical texts, on the basis of his study of partial copies of Vietnamese treatises preserved in the Institute for the History of Natural Sciences in Beijing and originating from Zhang Yong's collection (Han 1991). Vietnamese historians themselves have produced only very general descriptions of extant treatises (for example, Ta 1979; Tran and Gros 1993). As far as works in European languages are concerned, the publications of P Huard and M Durand provide a general introduction to the

---

2. According to Mikami (1934), it was purchased in Vietnam in the 1930s and preserved in a private collection in Japan. I was unable to locate copies of this book in Vietnamese or French libraries.

3. I am grateful to Professor Ōhashi Yukio 大橋由紀夫 who drew my attention to Mikami (1934) and kindly sent me a copy of it.

4. Li Yan also mentions papers prepared by Zhang Yong for publication; to my knowledge, they were not published and the whereabouts of the drafts (if any) remain unknown.

sciences in traditional Vietnam but mathematics is treated rather unsatisfactorily (Huard and Durand 1954, 120, 144; Huard and Durand 1963, 538, 540). In his recent monograph on the history of mathematics in China, J-C Martzloff devotes only a dozen lines to Vietnamese mathematics (Martzloff 1997, 110; Volkov 2002, 378 n24). Recently, I published the results of my preliminary investigation of the Vietnamese mathematical treatises preserved in Vietnam and in France (Volkov 2002; 2008a).

All these authors explicitly or implicitly assumed that traditional Vietnamese mathematics was a continuation of the Chinese tradition, but most did not attempt to discuss the details and modalities of transmission. In particular, it remained unclear whether the Vietnamese tradition was the product of long-term 'osmosis' of Chinese mathematical expertise, or whether it resulted from single transmissions of mathematical treatises. Were there attempts by professional practitioners or by the authorities to select the best and the most representative Chinese treatises, or were they selected randomly? Were only governmental institutions involved in the transmission, or did it also take place at grass roots level, through teachers and schools? Where certain treatises were written partly in local Vietnamese script (Nôm), what were the functions of the sections written in Chinese and in Nôm, and why were both languages used rather than only one? What counting instruments were used in Vietnam, how did they differ from their Chinese counterparts, and how did such differences influence computational procedures or the style of mathematical texts? This chapter will offer a preliminary description of the Vietnamese mathematical tradition in order to address at least some of these questions, even if only tentatively.

The extant sources

The available information on extant Vietnamese mathematical treatises can be summarized as follows: the number of the treatises amounts to twenty-two,[5] thirteen of which are written in classical Chinese, while the remaining nine are written partly in Vietnamese using the so-called Nôm 喃 script (discussed further below). One book, the *Toán pháp đại thành* 算法大成 'Great compendium of mathematical methods', is credited to the authorship of the fifteenth-century official Lương Thế Vinh 梁世榮, but his authorship is doubtful and the book well may have been compiled as late as the eighteenth or nineteenth century on the basis of earlier mathematical treatises (see Volkov 2002; 2005; 2006). Another book, the *Cửu chương lập thành tính pháp* 九章立成併法 'Ready-made methods of addition

---

5. The lists published in Volkov (2005; 2008a) contained only nineteen treatises; three more were identified by the author in 2006–2007.

of nine categories' by Phạm Hữu Chung 范有鍾, was block-printed in the early eighteenth century.[6] The dates of nine mathematical treatises are uncertain, but it can be argued that they were compiled after the beginning of the nineteenth century. Three more books were published in 1909. The dates of publication of the remaining eight treatises are still unknown, but on the basis of my preliminary comparison of their contents with those of books with established dates, it seems that they were compiled no earlier than the late eighteenth century, and probably as late as the mid- or late nineteenth century.

All the extant Vietnamese mathematical treatises are written in traditional 'Chinese' format: collections of problems with algorithms for their solution and numerical answers. Fig. 2.3.1, for example, shows two pages from *Toán pháp quyển* 算法卷 (1909) by Đô Đức Tộ 杜德祚, containing four problems. Reading from right to left, the four rightmost vertical lines contain the first three problems (separated by short breaks). The problems are all of the same type and read as follows: 'There are *M* men carrying *B* measuring units of rice each; how much rice do they carry together?' The results are obtained by multiplying *M* and *B*. These lines contain only the conditions and the answers. In the fifth line a four-character title heads a new section: 'Method of establishing the amount of millet in a field', with a brief description of the method written in small characters under the title. The method is then exemplified by a problem: calculate the amount of millet collected from a field of a given area. The amount of grain collected from one *cao* 篙 (unit of area) is supposed to be three 鉢 *bát* so the problem is solved by multiplying the given area by three.

Many problems and methods in the Vietnamese books are similar or identical to those of Chinese treatises that antedate the introduction of European methods in the early seventeenth century. At least nine of the twenty-two treatises contain lengthy quotations from the influential sixteenth-century Chinese treatise *Suan fa tong zong* 算法統宗 'Systematic treatise on methods of computation' by Cheng Dawei 程大位. For instance, *Toán pháp* 算法 'Computational methods' preserved in the Institute for Han-Nôm Studies in Hanoi[7] contains problems on areas of rectilinear and curvilinear figures, and problems devoted to root-extraction and solution of polynomial equations. The computational procedures and geometrical diagrams for the calculation of areas are similar to those found in chapter (*juan* 卷) 3 of the *Suan fa tong zong*,[8] while the problems and methods for polynomials are identical to those found in chapter 6 of the same treatise.[9] Another

---

6. A version of it entitled *Cửu chương lập thành toán pháp* 九章立成算法 with a preface dated 1721 forms an appendix to the manuscript copy of the *Chỉ minh lập thành toán pháp* 指明立成算法 by Phan Huy Khuông 潘輝框 dated 1820.

7. Call number A 3150. The first few pages are missing so the author, date of compilation, and original title are unknown. The provisional title 'Computational methods' was probably given to the manuscript by the copyist(s) or by librarians on the basis of its contents.

8. Compare *Toán pháp*, 24a–28b, with Mei and Li (1990, 246–248, 258–263, 267–272).

9. Compare *Toán pháp*, 40a–60b, with Mei and Li (1990, 454–498).

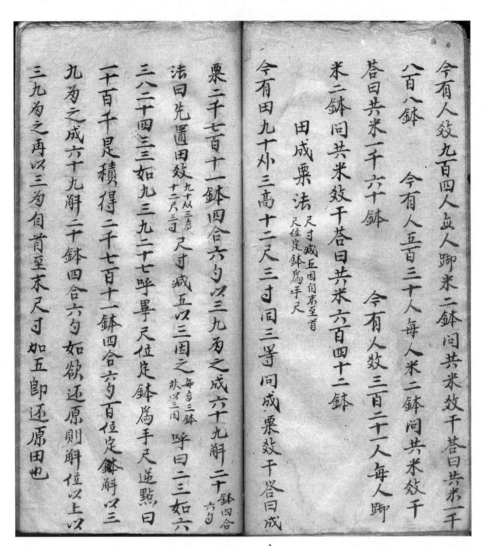

Figure 2.3.1 Pages 21b–22a of *Toán pháp quyển* 算法卷 (1909) by Đô Đức Tộ 杜德祚

example is from the treatise *Thống tông toán pháp* 統宗算法 'Systematic treatise on computational methods'. This manuscript of unknown date was authored by Tạ Hữu Thường 謝有常 and is preserved in the National Library in Hanoi.[10] Its title makes an obvious allusion to the *Suan fa tong zong* 算法統宗. Moreover, certain parts of the Chinese treatise are quoted verbatim (but without references), as, for example, versified rules of calculation for areas of plane figures[11] and the problem of two walkers.[12] Nevertheless, the compiler of the Vietnamese treatise considerably modified certain sections of the Chinese book by adapting the original

10. Call number R 1194. This treatise is not listed in Tran and Gros (1993).
11. Compare *Thống tông toán pháp*, 27–29, with Mei and Li (1990, 226–227).
12. Compare *Thống tông toán pháp*, 207–208, with Mei and Li (1990, 895–896).

problems and solutions to Vietnamese units of measure, inserting a large number of new problems, and providing his own explanations.

In Volkov (2002) I suggested an approach based on the analysis of a number of problems and methods found in the *Toán pháp đại thành* 算法大成, of an unknown date and credited to the state functionary Lương Thế Vinh. More specifically, I compared the following elements found in the Vietnamese treatise with their Chinese (and, in certain cases, Japanese) counterparts: (1) the multiplication table; (2) the hierarchy of 'large numbers' (that is, the terms designating ascending powers of 10 greater than 10^4); (3) rhymed algorithms for computation of areas; (4) methods of remote surveying; and (5) problems on 'numerical divination'[13] and indeterminate analysis.[14] This comparison enables us to identify connections between the Vietnamese treatise and Chinese treatises compiled prior to the seventeenth century. To give one example, a comparison of the problems on numerical divination found in the Vietnamese treatise as well as in the extant version of the *Sun zi suan jing* 孫子算經 (1213),[15] in the Japanese text *Kuchizusami* 口遊 (c 970), in the Chinese treatises *Jiu zhang suan fa bi lei da quan* 九章詳注比類 算法大全 (1450) by Wu Jing 吳敬 and the *Suan fa tong zong* (1592) by Cheng Dawei, showed that one possible reason for the similarity of the methods is that the three latter treatises drew on the same Chinese prototype, presumably a version of the *Sun zi suan jing* of the late first millennium AD, which seems to have contained divinatory calculations different from those found in the versions subsequently printed in China in 1084 and 1213.

The analysis of these elements also showed that the compiler of the Vietnamese treatise adapted Chinese problems and methods to local units of measure, currency, and taxation system, as well as to native animals or plants. However, the analysis did not determine the date of compilation or authorship of the *Toán pháp đại thành*, nor the kind of counting instrument that was supposed to be used for calculations. A full and systematic comparison of extant Vietnamese treatises with their Chinese counterparts still remains indispensable to discerning the history of the transmission of the Chinese mathematical tradition to Vietnam and the independent mathematical development of the latter. The first results of this ongoing project will be reported elsewhere; in this present chapter I will limit myself to discussing the three topics directly related to the transmission of mathematical knowledge to Vietnam: the use of counting instruments, the use of the Vietnamese written language Nôm in mathematical treatises, and the Vietnamese state mathematics examinations.

13. This category includes divination techniques for the sex of an unborn child and prognostications of health for an invalid, using arithmetical operations based on the date of conception and the age of the mother, in the first case, and on the date when the disease was contracted and the age of the invalid, in the second.

14. This category includes a modification of the famous 'hundred fowls for a hundred pence' problem.

15. The *Sun zi suan jing* was compiled at some time between 280 and 473 AD; see Volkov (2002, 384, n39). The earliest extant edition of the treatise (1213 AD) is based on its blockprint edition of 1084.

Counting rods in China and Vietnam

All the algorithms found in Chinese and Vietnamese mathematical treatises were designed for specific counting instruments. In China, the instruments used until the mid-second millennium AD were counting rods, which were then gradually replaced by the abacus;[16] the equivalent instrument used in Vietnam has not been identified. In this section I will argue that counting rods were used in Vietnam until the late seventeenth century and that there is evidence that they may still have been in use as late as the early nineteenth century, concurrently with the abacus.

Counting rods[17] were described for the first time in China around 60 AD in the Lü li zhi 律曆志 chapter of the *Qian Han shu* 前漢書 'History of the early Han dynasty' by Ban Gu 班固.[18] According to his description, the counting rods were round bamboo sticks of six *cun* 寸, that is, 13.8 cm long, and 1 *fen* 分 = 1/10 *cun* = 0.23 cm in diameter.[19] There exist descriptions in even earlier sources of objects that can be interpreted as counting rods, though their meaning is somewhat uncertain. Needham quotes several sources that, according to him, suggest that counting rods were used in China by the late third century BC (Needham 1959, 71); however, some of the sources deserve more detailed investigation.[20] The text usually quoted as evidence of the early existence of counting rods is chapter 27 of the philosophical treatise *Dao de jing* 道德經, which contains the phrase *shan shu bu yong chou ce* 善數不用籌策, literally 'a person good at [operating with] *shu* 數 does not use bamboo tallies and bamboo slips'; this has been conventionally understood to say that 'a skilful calculator does not need to use counting instruments'.[21] On the grounds of this phrase alone Lam Lay-Yong and Ang Tian-Se claimed that 'it can be safely assumed that the invention [of the counting rods] would not be later than the 5th century BC' (Lam and Ang 1992, 22; Needham 1959, 70–71). Modern scholarship, however, dates the compilation of the treatise to the late third century BC (Bolz 1993). Further, the term *shu* 數 has a wide semantic

16. The Chinese abacus, *suan pan* 算盤, is the well-known Chinese instrument constructed as a wooden frame with a number of bars with sliding beads.

17. In China, the most common name of the instrument beginning from the Song dynasty (960–1279) onwards was *suan zi* 算子, yet in earlier texts other terms can also be found; see Volkov (1998; 2001).

18. The chapter, except for the introductory and, probably, concluding remarks of Ban Gu, is traditionally credited to the authorship of Liu Xin 劉歆 (46 BC–23 AD).

19. One *cun* of that time was approximately equal to 2.3 cm, see Wu (1937), 65. Martzloff (1997), 210, miscalculates the diameter of the rods (0.69 cm instead of 0.23 cm).

20. To support the hypothesis of the early origin of the counting rods suggested by Needham, one might add that the term *zhi* 直, 'to represent (literally, "set") [a number with a counting instrument]' later used in calculations with counting rods appears in the mathematical treatise *Suan shu shu* 算數書 completed prior to 186 BC (Cullen 2004) and in the astronomical chapter of the philosophical treatise *Huai nan zi* 淮南子 (c 139 BC) (Volkov 1997, 144). However, neither of these two texts explicitly mentions the counting instrument used, even though the word *suan* 筭 (interchangeable with the character *suan* 算, 'counting rods') appears in the title of the former treatise: it appears in the text (strips 72–73) meaning 'string of cash', as Cullen (2004, 61) suggests.

21. See, for example, Needham (1959, 70): 'Good mathematicians do not use counting-rods'; compare with a rather unconventional English rendering by the translators of Li and Du (1987, 7).

range, including 'fate computation'. The evidence of 'counting-rod numerals' on Chinese coins of the fourth to third centuries BC mentioned by Needham (1959, 70) and Martzloff (1997, 185–186) also appears to be inconclusive (Ch'en 1978, 278–279; Djamouri 1994, 20). The most ancient objects considered to be counting rods were recently excavated from Chinese tombs dated to the second to first centuries BC; some of them are made of bamboo and some of bone (Mei 1983, 58–59; Li and Du 1987, 8; Lam and Ang 1992, 22).

Sizes of counting rods varied considerably,[22] as did the materials from which they were made: historical sources mention bamboo, bone, ivory, iron, and jade. Positive and negative numbers were distinguished by the colours red and black (or white and black),[23] or by different cross-sections, triangular for positive and square for negative.[24] The number of the sides in the cross-sections was explained by the framework of traditional cosmology according to which triangles were associated with odd numbers and the 'positive principle' Yang, while squares were associated with even numbers and the 'negative principle' Yin.[25]

Calculations with counting rods were performed on a flat surface, probably covered with a special cloth.[26] It is not known whether the decimal positions were always marked on its surface as square/rectangular cells or whether they sometimes had only imaginary boundaries.[27] A position on the surface was fixed for units, and positions to its left were then used for the powers 10^1, 10^2, 10^3, ..., while the positions to the right were used for the 10^{-1}, 10^{-2}, ..., respectively. A number n from one to five was represented with n rods set horizontally or vertically (see below), and a number n from six to nine with $(n-5)$ rods set horizontally or vertically plus one rod set orthogonally (thus symbolizing five units); for zero the position was left empty.[28] The orientation of the rods in a given decimal

22. The *Sui shu* 隋書 'History of the Sui dynasty' compiled in the seventh century also mentions 3 *cun* counting rods made of bamboo; the length was therefore 93 mm, 89 mm, or 71 mm, depending on the length of the unit *cun* (Wu 1937, 65).

23. See *Jiu zhang suan shu*, chapter 8, problem 3; the problem mentions the *zheng* 正, literally 'straight', 'upright', numbers, and the *fu* 負, literally, 'borrowed', numbers (the coefficients of simultaneous equations), without indicating any specific medium for representing them. The third century AD commentator Liu Hui 劉徽 mentions red and black rods representing positive and negative numbers, respectively (Guo and Liu 2001, 175). Some sources also mention white and black rods presumably used for the same purpose (Needham 1959, 71).

24. The earliest explicit description is found in the 'Lü li zhi' chapter of the Sui dynasty history *Sui shu* from the seventh century AD, but the underlying idea was already alluded to by Liu Hui. He mentions 'oblique' (*xie* 邪) and 'upright' (*zheng* 正) rods, thus referring to their triangular and square cross-sections, and not to their 'oblique' and 'upright' position on the counting surface, as some modern authors have suggested.

25. Lam and Ang (1992, 21) mention only the rods with square cross-section and do not explain their symbolic significance (suggesting instead that the 'square cross sections prevented them [= rods] from rolling').

26. Such a cloth is mentioned in Liu Hui's commentary on the *Jiu zhang suan shu*, chapter 8, problem 18 (see Guo and Liu 2001, 182). Much later, in Japan, a sheet of paper was used (Horiuchi 1994, 97).

27. Some modern authors believe that operations were performed on a special (wooden?) counting board (Needham 1959, 62–63, 72; Libbrecht 1973, 398). Recently Martzloff argued that there is no evidence in ancient and medieval Chinese sources for the existence of such an instrument (Martzloff 1997, 209).

28. The method of representing digits with the counting rods is described in *Sun zi suan jing* 孫子算 經 'Mathematical treatise of master Sun', third to fifth century AD, and *Xiahou Yang suan jing* 夏侯陽算經 'Mathematical treatise of Xiahou Yang', conventionally dated to the eighth century AD. For translations of

position depended on the power of 10: rods were set vertically for even powers and horizontally for odd powers, whether positive or negative. The conventional explanation for this is that if the rods were all set vertically they might have been easily confused; the number 13, for example, would be represented by (| |||) and therefore could be easily mistaken for 4 (|||||). With alternation of the orientations, 13 is represented by —|||, and it is impossible to make that mistake (Needham 1959, 9; Mei 1983, 59; Chemla 1994, 3 n4). This does not explain, however, why the rods were not always set horizontally. Another way to explain the alternation would be to evoke the symbolic associations according to which 'vertical' and 'horizontal' might have been associated with heaven and earth, and with the cosmic principles Yang and Yin (Kalinowski 1994, 43–44, 57, 59; 1996, 77). However, neither explanation takes into account the fact that horizontal and vertical strokes had been used as early as the Shang dynasty (c 1750–1045 BC) in inscriptions, to designate 1 and 10 respectively (Djamouri 1994, 39). That is, the alternation may have been originally conceived for written numerals, so that representation with counting rods then followed the same conventions.

The exact time that counting rods disappeared in China is unknown. All the surviving Song dynasty (960–1279) mathematical treatises mention them extensively. *Xiang ming suan fa* 詳明算法 'Computational methods explained in detail' (1373) by An Zhizhai 安止齋 and *Suan xue bao jian* 算學寶鑒 'Precious mirror of the learning of computations' (late fifteenth century) by Wang Wensu 王文素 both discuss operations with them. On the other hand, the *Li suan quan shu* 曆算全 書 (seventeenth century) by Mei Wending 梅文鼎 suggests that by the end of the seventeenth century they had fallen out of use, and even professional mathematicians were not certain about their existence in the past.[29] This means that in China the use of counting rods came to an end between the early sixteenth century and the mid-seventeenth century, that is, during the time when the abacus was rapidly becoming the main computational device.[30]

The circumstances of the transmission of Chinese counting rods and related computational methods to Vietnam remain unknown. Huard and Durand (1963, 540) believed that the abacus was introduced to Vietnam from China in the fifteenth century. Han Qi (1991, 6) also mentioned the introduction of the abacus to Vietnam from China, but suggested that Vietnamese use of the instrument was based on the Chinese mathematical treatise *Suan fa tong zong*; he does not specify when the treatise was transmitted to Vietnam. Sometimes the same algorithm had to be written in different ways for counting rods and abacus, so the transition

the relevant passages from the *Sun zi suan jing* and the *Xiahou Yang suan jing*, see Berezkina (1963, 23; 1985, 298); Lam and Ang (1992, 155).

29. See the section 'Gu suan qi kao' 古算器考 in chapter 29 of *Li suan quan shu* 曆算全書 (Mei 1986); for translations, see Vissière (1892); Jami (1994).

30. By the sixteenth century the abacus was widely used in China. Jami (1998) suggests that the two instruments coexisted for some time, being used by two different social groups: the counting rods by professional mathematicians and calendrical astronomers, while the abacus was used in 'popular arithmetic'.

from one to the other may have influenced considerably the style and contents of mathematical treatises.

There is evidence that the counting rods were used in Vietnam as late as the mid-seventeenth century. A description of the rods and their use in Vietnam was written by the Jesuit Giovanni Filippo de Marini, also known as Philippe de Marini, or de Marino (see Dehergne 1973, 72–73), who stayed in Tonkin (in the part of northern Vietnam controlled by the Trịnh 鄭 Lords) in 1647–58, and in Cochinchina (the central part of modern Vietnam, controlled at that time by the Nguyễn 阮 Lords), in 1671–74 (Volkov 2008b). De Marini's *Histoire nouvelle* published in Rome in 1663 contains a description of the work of Vietnamese Court astronomers and mentions their use of counting rods (Marini 1663, 100). De Marini describes the computations as if he actually witnessed them during his stay in Tonkin in 1647–58:

Their Algebra and their way of counting are practiced differently and in another way comparing to those of other Nations, because instead of [written] digits they solely use certain little rods of the length of one palm.[31] They dispose them on the ground sometimes vertically and sometimes horizontally, and by this means they practice all the operations of their Arithmetic, be it Addition, Multiplication, Subtraction, or Division.[32]

De Marini's account suggests that as late as the mid-seventeenth century, Vietnamese mathematicians and astronomers perpetuated methods of computation which were no longer used in China by then.[33] He mentions differing orientations of the counting rods ('[t]hey dispose them on the ground sometimes vertically and sometimes horizontally') but it is unclear whether this was the alternating system used in China.

The last question deserves special attention. Mikami Yoshio found pictures of numbers represented by counting rods accompanying the multiplication table in the Vietnamese treatise *Chỉ minh toán pháp* 指明算法 'Guide towards understanding of methods of calculation' of unknown date; according to him, the book contained no evidence of the use of the abacus (Mikami 1934, 4). Mikami noticed that in this treatise the conventions concerning the orientations of the rods differed from those in China: in the same decimal position the rods could be placed vertically

31. This means that Vietnamese counting rods were longer than the Chinese rods used in the first half of the first millennium AD.

32. 'La ragione de'numeri, e la loro Arithmetica in diuersa maniera si calcula, da quel, che fanno altre nationi, perche a fare presti, e diritti conti, non hanno cifere, mà certi stecchi lunghi vn palmo. Questi dispongono in terra horo per lungo, hora a trauerso in più maniere, e con ciò mettono a fine ogni operatione, sia di sommare, sia di moltiplicare, e sottrare, o diuidere' (Marini 1663, 100). My rendering is based on the French translation 'Leur Algebre & leur façon de nombrer se pratique diuersement & d'vne autre façon que parmi les autres Nations ; parce qu'au lieu de chiffre ils ne se seruent que de certains petits bastons de la longueur d'vne palme. Ils les disposent sur la terre tantost de long & tantost de trauers ; & par ce moyen ils pratiquent toutes les regles de leur Arithmetique, soit l'Addition, la Multiplication, la Soustraction ou la Diuision' (Marini 1666, 182).

33. Marini's description does not mention any kind of special 'counting board' and is thus evidence against the aforementioned theory that computations with counting rods were performed on such a board.

or horizontally, depending on whether the number was odd or even (one vertical rod for 1, two horizontal rods for 2, and so on). The treatise studied by Mikami remains unavailable to me, but three block-printed editions of the *Cửu chương lập thành tính pháp* 九章立成併 'Ready-made methods of addition of nine categories' by Phạm Hữu Chung 范有鍾, published in the early eighteenth century,[34] contain a multiplication table with numbers represented by counting rods in a way that is different from either the Chinese system or the one described by Mikami. In some cases, increasing powers of ten are disposed from left to right, and unconventional configurations of rods are used to represent digits greater than five.[35]

There is evidence for the use of counting rods in Vietnam even as late as the nineteenth century, in a 'model' mathematics examination paper in the treatise entitled *Chỉ minh lập thành toán pháp* 指明立成算法 'Guidance for understanding of the "Ready-made computational methods"', compiled by Phan Huy Khuông 潘輝框 in 1820. Here a 'model student' requests the counting rods in order to solve a problem. The relevant phrase reads: 'I asked for counting rods [and] disposed [them on the counting surface in order] to investigate it [= the problem] 愚請筹而排陳之' (MS A 1240, chapter 4, 30b).

Given that Vietnamese treatises based on the Chinese treatise *Suan fa tong zong* contain pictures of the abacus and describe operations with it,[36] one can conjecture that in the early nineteenth century the two calculating devices were used simultaneously for a time. At present, however, it is impossible to establish exactly how the two traditions coexisted, and whether calculations with both instruments were practised within the same professional community. In the next section the reader will find an example of a computational procedure using, presumably, counting rods.

Use of Nôm in mathematical treatises

Nôm 喃 (or, more precisely, *chữ Nôm*, 'southern script', as distinct from *chữ Nho*, 'script of Confucian scholars') is the original Vietnamese writing system, using both Chinese characters and local modifications (see Nguyen 1990). Knowledge of the 'script of Confucian scholars', using classical Chinese characters, granted

34. Call numbers AB 53, AB 173, and VHb 374 in the library of the Institute of Han-Nôm Studies, Hanoi.
35. One can ask whether the unorthodox configurations resulted from low print quality and the poor mathematical expertise of the editors.
36. See, for example, the anonymous manuscript treatise catalogued in the National Library (Hanoi) under the bogus title *Toán pháp đề cương* 算法提綱 'Presentation of the key points of the computational methods', call number R 1952, unmentioned in Tran and Gros (1993). The first page(s) of the manuscript is (are) missing and *Toán pháp đề cương* is, most probably, the subtitle of its first remaining section. The first part of the manuscript contains a very detailed explanation of operations with the abacus, and a number of diagrams representing configurations of beads on the abacus, see 10a–16b, 19a–24a, 26a–29b, 34b, 35b. The instrument featured in the treatise is the standard Chinese abacus with eleven bars and 2 + 5 beads on each bar.

Vietnamese literati direct access to the immense corpus of Chinese literature, including treatises on science, technology, and medicine. But classical Chinese remained a foreign language, and Nôm was the medium used to write in Vietnamese. To a certain extent the two written languages in scholarly texts can be compared to Latin and vernacular languages in sixteenth-century Europe. Historically, the use of Nôm in official documents and, accordingly, in the state education and examination system was subject to the official language policy of the Court. For example, the reign of the Emperor Minh Mang (1820–41) was marked by a renewal of Confucian learning in Vietnam and a return to classical Chinese in official documents and state examinations.

Since a relatively large number of surviving Vietnamese mathematical treatises include sections written in Nôm, it is worth investigating the role played by these sections, which for local readers were presumably easier to understand than those in classical Chinese. The way Nôm was used in mathematical treatises thus illustrates strategies of 'localization'.

As an example I will examine the way Nôm was used in the treatise *Toán pháp đai thành* 算法大成. The treatise contains 138 mathematical problems, although several problems are incomplete and are known only from the algorithms for their solution or from geometric diagrams. The text is not divided into chapters, but the problems form groups devoted to the same topic or method (see Volkov 2002). Nôm is used only in a few places, but the sections written in it are lengthy. More specifically, it is used in the short rhyming preface that introduces mathematics and is placed before the table of 'large numbers' in the very beginning of the treatise (A 2931, 1b), in commentaries on several problems,[37] and in the concluding part of the manuscript (A 2931, 117a–120b). The distribution of the fragments in Nôm suggest that the main part of the treatise was originally compiled in classical Chinese and that sections and commentaries in Nôm were added later. To confirm this hypothesis more generally, it would be necessary to examine all other mathematical texts containing sections in Nôm.

The remainder of this section will be devoted to a discussion of the Nôm commentaries to problem 18 of the *Toán pháp đai thành*. The problem reads as follows:

Suppose that there is [the amount of] 503 *quan* 貫 7 *bách* 佰 of officials' money to be distributed among 345 military personnel.

Question: how much [one] person [should] obtain?

---

37. In A 2931, in the Institute of Han-Nôm studies, they are: problem 18 and explanations,19a–22b; problem 19 and explanations, 22b–27a; problem 68 and explanations, 62a–62b; problem 69 with a relatively long solution, 62b–66b. Rhyming explanations in Nôm conclude problems 70–85, devoted to the extraction of square roots, 71a–72a. The section on areas of plane figures opens with an introduction in Nôm, 79b, and contains commentaries in Nôm on the solution of problem 94, 81b–82a, and on the method, 82b.

Answer: [The total amount of] money is 302,220 *văn* 文. Each man obtains 876 *văn*, which is 1 *quan* 4 *bách* and 36 *văn*. No remainder left.

(MS A 2931, 19a)

Mathematically, the problem is elementary: *N* units of money are to be distributed among *M* persons, and therefore each person obtains $S = N \div M$ units.[38] However, the problem contains a difficulty: *N* is expressed in *quan* and *bách*, where 1 *quan* = 600 *văn* and 1 *bách* = 60 *văn*.[39] The solution should consist of the following steps: (1) convert *quan* and *bách* to *văn*; (2) divide the number of *văn* by the number of the militaries; (3) express the result in terms of *quan* and *bách*. Let us see how this is implemented in the solution found in the treatise.

The beginning of the computational procedure (or algorithm) is marked with the term *cáo* 告 'report'.[40] This suggests that the data of the problem are to be 'reported' or set out on a counting surface or abacus. It is difficult to decide which of the two instruments was meant to be used, because terms such as 'upper [row]', 上, and 'lower [row]', 下, could refer to positions on the counting surface (if the computations are performed with the counting rods) or to the positions on an abacus (in which case the 'upper row' would be the left part of the abacus, and the 'lower row' the right part). As we have seen, counting rods were mentioned in a mathematical treatise as late as the early nineteenth century, so in what follows I shall present the operations as if they were performed with counting rods. The original disposition of the counting rods is shown in Fig. 2.3.2(a). The number 345 occupies 3 positions. Surprisingly, the total amount of money, 503 *quan* 7 *bách*, is presented on the counting surface occupying six positions as if it were a decimal number 503,700.

The next statement in the text, in classical Chinese, reads as follows:

Take 7 *bách* in position of *bách*. Then multiply 6 and 7, [it will be] 42.

This is followed by a commentary in Nôm which can be rendered as saying:[41]

At this moment 'break' this 7; leave 40; besides, [it] yields 2; no 'injection' [needed].

The first part of the instruction in Chinese 'takes' the 7 *bách*. To convert those 7 *bách* to *văn*, one should multiply 7 x 60 = 420. It is not specified whether the

---

38. One can suggest that the compiler(s) of the problem selected some convenient integer values M and S, and then calculated N (= M x S), and then designed the problem using N and M. The values of M and N/M in this problem, 345 and 876, are both obtained as permutations of consecutive natural numbers: 3, 4, 5 and 6, 7, 8. The reason for choosing 876 and not 678 may have been that 345 x 678 *văn* = 233910 *văn* = 389 *quan* 8 *bách* 30 *văn*, which cannot be expressed solely in terms of *quan* and *bách*.

39. The values for these units can be obtained from external historical sources, but the solution of the problem itself provides the information necessary to reconstruct the values of *quan* and *bách*: if 1 *quan* is equal to *x văn*, and 1 *bách* to *y văn*, we have the simultaneous equations $503x + 7y = 302{,}220$ and $x + 4y + 36 = 876$. The solution is $x = 600$, $y = 60$.

40. To the best of my knowledge, this term is never used in Chinese mathematical treatises, where the terms *shu* 術 'procedure', *fa* 法 'method', or *cao* 草 'record of computations' were used instead.

41. I thank Nguyen To Lan (Hanoi) for her help in translating the Nôm in this section.

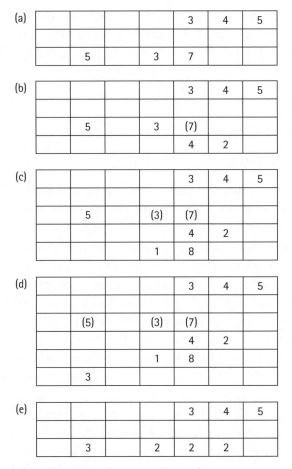

Figure 2.3.2 Division of money using counting rods

product 420 should replace the initial 700 or should be set in another row. For convenience, we will assume temporarily that the product was set in an extra row as shown in Fig. 2.3.2(b). The digit 7 was probably now removed.[42]

The Chinese text goes on:

[Take] 3 *quan*. Then multiply 3 and 6, [it will be] 18.

This states that 3 *quan* (of the total amount of 503 *quan*) should be converted to *văn* (3 *quan* is 3 x 600 = 1800 *văn*). The disposition (before addition) on the counting surface is shown in Fig. 2.3.2(c).

The commentary in Nôm reads:

At this moment 'break' this 3; leave 10; besides, [it] yields 8; 8 then is 'tied-up' [with 4?], 2 returns, outside in second [position?] 'inject' returning 1, [it] becomes 2, no [more?] 'injection' [needed].

42. The result was probably not placed in the row between the 'upper' and 'lower' numbers, since this position was reserved for the quotient of the final division.

This commentary describes the addition of 1800 and 420. One adds 8 and 4 to obtain 12; 2 is set below 8 and 4, and then 1 is added to the 1 of 1800. The meaning of 'injection' in this case is thus the equivalent of 'carrying'. The text goes on in Chinese:

[For] 500 thus multiply 5 times 6, obtain 30.

The commentary in Nôm states:

Then 'break' this 5 to obtain 30.

This instruction states that 500 *quan* (of the total 503 *quan*) should be converted to *văn*, to obtain 500 x 600 = 300,000 *văn*. On our counting surface this will be represented as in Fig. 2.3.2(d).

The next statement mentions the total amount of the money 'as stated in the [question] of the problem'; therefore, at this step the total is obtained by adding the three partial products. The disposition on the counting surface is thus as in Fig. 2.3.2(e). The final stage of the process is the division of 302,220 by 345; this part also contains commentaries in Nôm of a similar kind.

In this particular problem the commentary in Nôm does not add any new information to the text in Chinese; instead the Nôm rephrases the instructions in Chinese in order to render them more understandable. However, it would be incorrect to believe that in this or other mathematical treatises the sections in Nôm were always simply technical comments. For instance, the treatise *Toán điền trừ cửu pháp* 算田除九法 'Methods of computations [of the areas of] fields [using] elimination of nines'[43] contains a number of long sections in Nôm; more specifically, four sections at the beginning of the book explaining elementary arithmetical operations, (MS VHb 50, 1a–3a), as well as two long sections introducing new methods followed by some relevant problems in Chinese (MS VHb 50, 28b–29a, 30b–32b).

Mathematics examinations in traditional Vietnam

State mathematics education and examinations in China were established during the Sui 隋 dynasty (581–618), conducted throughout the Tang 唐 (618–907) and Song 宋 (960–1279) dynasties, and then discontinued for unknown reasons. No Chinese mathematics examination papers have been found so far,

43. This is a provisional translation of the title of the MS VHb 50 preserved at the Institute for Han-Nôm Studies in Hanoi; the date of compilation and the name of the author are unknown. This title is probably not the title of the treatise but of the section on the first page of the surviving copy, since the cover, the table of contents, and probably some of the opening pages are lost.

and reconstructions of instruction and examination procedures in China have been made on the basis of descriptions in other documents.[44] In a number of Vietnamese documents from the eleventh to the late eighteenth centuries there are references to examinations in *toán* 算 'counting'.[45] It is plausible to conjecture that procedures for state mathematics instruction and examination in Vietnam were similar to their Chinese prototypes. No Vietnamese examination papers have been discovered either, but there exists a 'model' mathematics examination paper in a treatise of 1820 by Phan Huy Khuông 潘輝框, who was presumably an instructor in a state or private educational institution. It is in the fourth (and last) chapter of his *Chi minh lập thành toán pháp* 指明立成算法 'Guidance for understanding the "Ready-made computational methods"' (MS A 1240, 30a–32b). This model paper is interesting for two reasons: first, it allows a reconstruction of examination procedures in Vietnam and China; second, it provides important evidence of the reasoning strategies of Vietnamese mathematicians.

The model examination paper consists of a relatively simple problem together with its solution written by a model examinee. The problem is:

[Let us suppose that] now there is money to award [officers], the total amount is 1,000 *cân* 斤 [of silver].[46] [In order] to award [those] officers, the amount of money agreed upon is 5,292 *lượng*. [The superiors] kindly award the aforementioned corpus of 328 officers. [If they] intend to distribute [the silver] equally [in dividing by] the number of people, [then] there would be a remainder of 4 *phan* 8 *li*. [That is, if for those officers] in high [position] and in low [position] [the superiors adopt] the pattern of 'flat rate distribution' (平分) [for all the] ranks, then there is a flaw. The method of 'flat rate distribution' cannot be equally applied [to all ranks], this is already clear. Now, [one] wishes to use this silver to be distributed equally [within one rank], [yet] taking into account the unequal ranks of the aforementioned corpus [of the officers]. There are [officers of] three different ranks: rank *A*, 8 persons, each person receives 7 parts; rank *B*, 20 persons, each person receives 5 parts; rank *C*, 300 persons, each person receives 2 parts.[47]

[If we proceed in this way], then what would be the amounts of the parts received by each person, and the due amount of money for each [of the three groups]?

In other words, three categories of officials, *A*, *B*, and *C* are to be remunerated with an amount S = 5,292 *lượng* 兩. The ratio of the amounts of money to be given

44. Translations of the most important documents relevant to state education during the Tang dynasty can be found in des Rotours (1932); for a discussion of instruction and examinations see Siu and Volkov (1999). A discussion of mathematics instruction during the Song dynasty is in Friedsam (2003).

45. More specifically, in 1077, 1179, 1261, 1363, 1404, 1437, 1472, 1505, 1698, 1711, 1725, 1732, 1747, 1762, 1767, and 1777. In a record from 1762, mathematics examinations were ordered to be held every 15 years (Volkov 2002).

46. The author uses the following measures of weight: 1 *cân* 斤 = 16 *lượng* 兩, 1 *lượng* 兩 = 10 *tien* 錢, 1 *tien* 錢 = 10 *phan* 分, 1 *phan* 分 = 10 *li* 釐.

47. The original denotes the three categories by the cyclical characters 甲, 乙, and 丙 (see Jami in this volume). In the translation and subsequent discussion these have been replaced by the letters *A*, *B*, *C*.

to the individual officials of the three ranks is 7:5:2, and the numbers of officials in each category are $N_A = 8$, $N_B = 20$, and $N_C = 300$, respectively. The questions are: (1) what are the amounts of money to be given to individual officials of the categories A, B, and C? (2) What is the total amount of money allotted to each category?

In modern notation, this is a problem on 'weighted distribution': one has to find values $x_1, x_2, \ldots, x_n$ such that $x_1 + x_2 + \ldots + x_n = S$ and $x_1 : x_2 : \ldots : x_n = k_1 : k_2 : \ldots : k_n$ for given weighting coefficients $k_1, k_2, \ldots, k_n$. Problems of this type are found in the Chinese treatises *Suan shu shu* 算數書 and *Jiu zhang suan shu* 九章算術, where the solution procedure (in modern notation) is as follows:[48] calculate the sum of the coefficients $K = k_1 + k_2 + \ldots + k_n$, and then find the values $x_i = (S \times k_i)/K$. However, the problem in the Vietnamese treatise is modified: for all officials in the same category the weightings are the same, that is, $k_A = k_1 = k_2 = \ldots = k_8 = 7$, $k_B = k_9 = \ldots = k_{28} = 5$, and $k_C = k_{29} = \ldots = k_{328} = 2$; one must therefore find $x_A = x_1 = \ldots = x_8$, $x_B = x_9 = \ldots = x_{28}$, and $x_C = x_{29} = \ldots = x_{328}$ such that $x_A : x_B : x_C = 7:5:2$, and $8x_A + 20x_B + 300x_C = S$. The examinee is also asked to find the total amount allotted to each group as a whole, that is, the values $X_A = 8x_A$, $X_B = 20x_B$, and $X_C = 300x_C$. A problem of this type is found in the *Jiu zhang suan shu* (chapter 3, problem 7), where there are two groups containing three and two persons, respectively, and $k_1 = k_2 = k_3 = 3$, $k_4 = k_5 = 2$, and S = 5 (Guo and Liu 2001, 112; Chemla and Guo 2004, 293). However, the solution in the Chinese treatise simply suggests that the weight coefficients should be written 3, 3, 3, 2, 2 and treats them as individual coefficients without grouping them into 3, 3, 3 and 2, 2.

The structure of the model solution can be summarized as follows. First, the candidate checks that the data given in the problem fits its conditions. At this step he finds that the amount of money, 1000 *cân*, does not give the 5292 *lượng* mentioned in the problem. The discrepancy between the two values leads him to the conclusion that part of the correct amount, 10,708 *lượng*, was taken away. Second, he checks the 'flat-rate distribution'. To do so, he divides the amount of money, 5292 *lượng*, by the number of officials, 328. The remainder, 4 *phan* 8 *li*, is the one mentioned in the condition of the problem, and shows that an unweighted distribution cannot be fairly applied. Third, he describes an algorithm that can be used to solve the problem, different from that in the *Jiu zhang suan shu*.[49]

---

48. Cullen (2004, 44–47); Guo and Liu (2001, 109); Chemla and Guo (2004, 283). The term used for this procedure is 衰分 'distribution [according to] grades', that is, Vietnamese authors used the standard Chinese term from the *Jiu zhang suan shu*.

49. In order to find x_A, x_B, and x_C he calculates the sum $K = k_1 + k_2 + \ldots = k_{328}$. He does this by calculating $N_A \cdot k_A = 56$, $N_B \cdot k_B = 100$, $N_C \cdot k_C = 600$ and adding these to obtain $K = 756$. Then he divides the total amount of money, $S = 5292$ *lượng*, by K to yield the 'constant norm' of 7 *lượng*. The amounts x_A, x_B and x_C to be paid to each officer are then found by multiplying the 'constant norm' by k_A, k_B, k_C, respectively. Perhaps the examinee was not expected to know the algorithm described in the *Jiu zhang suan shu*, and this was why he was not required to reiterate it in the examination. The conjecture that the examinee was familiar only with the *Suan fa tong zong* and not with the *Jiu zhang suan shu* does not explain why the 'classical' method was not used, since the former treatise contains both methods.

Moreover, the examinee suggests an alternative algorithm for the second part of the problem. The solution includes a verification of the answer and a formal ending (see Volkov forthcoming).

Interestingly, the text fits closely to the format of Chinese mathematics examination papers of the Tang dynasty (618–907) (Siu and Volkov 1999): (1) the task is represented by a particular problem; (2) the examinee checks the consistency of the given numerical data, and (3) designs a computational procedure of which (4) the 'structure' is discussed in detail. In this discussion of structure the model Vietnamese examinee used the concept of *lü* 率 'the proportional variable amount'[50] which played a central role in traditional Chinese mathematics as seen in Liu Hui's third-century AD commentary on the *Jiu zhang suan shu*. The Vietnamese examinee thus was expected to produce a text in which the structure and style resembled algorithms of the *Jiu zhang suan shu*, as interpreted with the help of mathematical terms suggested by Liu Hui (Volkov 2004; see also Chemla and Guo 2004, 99–119).

Conclusion

This brief investigation of the surviving documents of Vietnamese mathematics does not provide conclusive answers to the questions posed at the beginning of the chapter. However, it is possible to advance a number of hypotheses concerning the process of transmission, which can be summarized as follows.[51]

It is plausible to conjecture that the transmission of Chinese mathematical expertise to Vietnam may have commenced in the first millennium AD, when Vietnamese candidates started participating in Chinese state examinations. From the eleventh century onwards, when Vietnam was established as an independent political entity, mathematics examinations were perpetuated, unlike their counterparts in China which were discontinued by the late thirteenth century. However, the content of the Vietnamese curriculum remains unknown, because the manuals used for mathematical instruction are lost.

It is also known that the Chinese calendar *Shou shi* 授時 (Vietnamese *Thụ thi*) promulgated in China in 1281 was used in Vietnam as early as 1306 (Hoàng 1982, 142–143). It is therefore possible that mathematics instruction was conducted in two institutional settings, namely, in the network of governmental and private schools preparing candidates for mathematics examinations, and in the

---

50. In modern notation, a number B is a *lü* 率 of another number, B', if one can establish a proportion in which both numbers occupy the same positions: $A : B : \ldots = A' : B' : \ldots$. For discussion of the term, see Chemla and Guo (2004, 135–136, 956–959).

51. These hypotheses are made on the basis of material presented in this chapter and in my earlier publications as listed in the bibliography.

astronomical bureaus focused on computation of the calendar and prediction of eclipses (Volkov 2008b).

After a brief Chinese occupation of Vietnam in 1408–27, the reinstated examination system once again included mathematics; it is plausible that at the time official mathematics textbooks went through radical reform, and were most probably compiled on the basis of treatises brought, once again, to Vietnam from China. However, none of the Vietnamese textbooks has survived, despite the fact that the manuscript *Toán pháp đại thành* 算法大成 is believed to have been authored by the fifteenth-century *literatus* Lương Thế Vinh 梁世榮. Why Lương Thế Vinh, a government officer credited with a certain supernatural bond with the charismatic leader of the country, the Emperor Lê Thánh Tông 黎聖宗, was believed to possess extraordinary mathematical skills still remains to be explored.

The earliest reliably datable mathematical book, the *Cửu chương lập thành tính pháp* 九章立成併法 'Ready-made methods of addition of nine categories' by Phạm Hữu Chung 范有鍾, was block-printed in 1713. The fact that large portions of the treatise are written in Nôm suggests that the Chinese mathematical methods were studied and adapted by Vietnamese mathematicians no later than the early eighteenth century. The fact that the book mentions counting rods suggests that the contents can be dated back to the time when the rods were still in use in China, that is, the late sixteenth to early seventeenth century, or even earlier, thus preceding the adaptation of the influential Chinese treatise *Suan fa tong zong* 算法統宗 (1592) which featured the use of the abacus.

As for the social institutions in which the mathematical tradition was perpetuated, they arguably remained the same as before. On the one hand, de Marini reported that in the mid-seventeenth century, old Chinese methods using counting rods were practised within a narrow circle of governmental officials charged with the duty of updating the calendar. On the other hand, the presence of a large number of applied problems (especially devoted to distribution of money and measuring of surfaces) in the treatises of the eighteenth to twentieth centuries and, in particular, in the model examination paper, suggests that government officials were expected to be capable of solving mathematical problems of certain types. This situation is mirrored in the large number of mathematical treatises produced in the nineteenth and early twentieth centuries; the interesting feature common to these books is that the Western mathematical methods known in China from the seventeenth century onwards were not included, probably intentionally. The Vietnamese compilations were thus styled 'old Chinese' (or perhaps 'old Vietnamese') mathematical treatises. The model examination paper of 1820 offers compelling evidence that mathematical education perpetuated reasoning strategies of the same type as those employed in China of the late first millennium AD. This examination paper, together with abundant explanations in Nôm found in other treatises, suggest that by the nineteenth century the main effort was on the production of explanatory discourses for a rather stale and restricted body of mathematical methods.

The end of the story is well known: being institutionally bounded to the Vietnamese state examination system, the incarnation of the Chinese mathematical tradition ended in a deadly collision with the colonial French education system by the dawn of the twentieth century. A description of this dramatic event is another challenging task awaiting historians.

Bibliography

Berezkina, El'vira I, 'O matematicheskom trude Sun'-tszy. Sun'-tszy: Matematicheskiĭ traktat. Primechaniya k traktatu Sun'-tszy', in *Iz istorii nauki i tekhniki v stranakh Vostoka*, Nauka, 3 (1963), 5–70.

Berezkina, El'vira I, 'Matematicheskii traktat Syakhou Yana', *Istoriko-matematicheskie issledovaniya*, Nauka, 28 (1985), 293–337.

Boltz, William G, '*Lao tzu Tao te ching*', in M Loewe (ed), *Early Chinese texts*, The Society for the Study of Early China, and The Institute of East Asian Studies, University of California, 1993, 269–292.

Cha Jong-chun, 'Reception and appropriation of Chinese mathematics in Joseon dynasty: with special respect to Hushilun', *Sungkyun Journal of East Asian Studies*, 2 (2002), 233–247.

Chemla, Karine, 'Nombres, opération et équations en divers fonctionnements: quelques méthodes de comparaison entre des procédures élaborées dans trois mondes différents', in I Ang and P-E Will (eds), *Nombres, astres, plantes et viscères. Sept essais sur l'histoire des sciences et des techniques en Asie orientale*, Collège de France, 1994, 1–36.

Chemla, Karine, Guo Shucun (présentation, traduction), *Les Neuf chapitres. Le classique mathématique de la Chine ancienne et ses commentaires*, Dunod, 2004.

Ch'en Liang-ts'o [= Chen Liangzuo] 陳良佐, 'Xian Qin shu xue de fa zhan ji qi ying xiang 先秦數學的發展及其影響', *Li shi yu yan yan jiu suo ji kan* 歷史語言研究所季刊, 49 (1978), 263–320.

Cullen, Christopher, *The Suan shu shu 'Writing on reckoning': a translation of a Chinese mathematical collection of the second century BC, with explanatory commentary*, Needham Research Institute, 2004.

Dehergne, Joseph, *Répertoire des jésuites de Chine de 1552 à 1800*, Letouzey & Ané, 1973.

Djamouri, Redouane, 'L'emploi des signes numériques dans les inscriptions Shang', in A Volkov (ed), *Sous les nombres, le monde: Matériaux pour l'histoire culturelle du nombre en Chine ancienne (Extrême-Orient Extrême-Occident*, 16), PUV, 1994, 12–42.

Fan Lianghuo, Wong Ngai-Ying, Cai Jinfa, Li Shiqi (eds), *How Chinese learn mathematics: perspectives from insiders*, World Scientific, 2004.

Friedsam, Manfred, 'L'enseignement des mathématiques sous les Song et les Yuan', in Christine Nguyen Tri et Catherine Despeux, *Éducation et instruction en Chine*, vol 2, Editions Peeters, 2003, 49–68.

Fukagawa Hidetoshi, and Pedoe, Daniel, *Japanese temple geometry problems—Sangaku*, Charles Babbage Research Center, 1989.

Guo Shirong 郭世荣, *Zhong guo shu xue dian ji zai Chao Xian de liu chuan yu ying xiang yan jiu* 中国数学典籍在朝鲜的流传与影响研究, unpublished doctoral dissertation, Beijing: Kexueyuan, 2005.

Guo Shuchun 郭書春, Liu Dun 劉鈍 (eds), *Suan jing shi shu* 算經十書, Taipei: Chiu chang [Jiu zhang], 2001.

Han Qi 韓琦, 'Zhong Yue li shi shang tian wen xue yu shu xue de jiao liu 中越歷史上天文學與數學 的交流', *Zhong guo ke ji shi liao* 中國科技史料, 12 (1991), 3–8.

Hoàng Xuân Hãn, 'Calendrier et calendriers vietnamiens', *Tập san khoa học xã hội*, 9 (1982), 134–144.

Horiuchi, Annick, *Les mathématiques japonaises à l'époque d'Edo (1600–1868) : une étude des travaux de Seki Takakazu (?–1708) et de Takebe Katahiro (1664-1739)*, Vrin, 1994.

Horng Wann-Sheng, 'Sino-Korean transmission of mathematical texts in the 19th century: a case study of Nam Pyong-gil's *Kugo Sulyo Tohae*', *Historia Scientiarum*, 12 (2002), 87–99.

Horng Wann-Sheng 洪萬生, '數學文化的交流與轉化:以韓國數學家南秉吉 (1820–1869)的算學正義為例' 'Transmission and transformation of mathematical cultures: a case study of Korean mathematician Nam Pyung Gil's *Sanhak Chongyi*' 師大學報,人文社會類 '*Communications of the National Normal University, Section of Humanities and Social Sciences*', 48 (2003), 21–38.

Horng Wann-Sheng 洪萬生, Li Chien-Tsung 李建宗, '從東算術士慶善微看十七世紀朝鮮一場數學研討 會' 'Kyong Song-Jin and a mathematics workshop in seventeenth-century Choson Korea', 漢學研究 '*Chinese Studies*', 25 (2007), 313–340.

Huard, Pierre, et Durand, Maurice, *Connaissance du Viet-Nam*, Imprimerie Nationale; Hanoi: École française d'Extrême-Orient, 1954.

Huard, Pierre, et Durand, Maurice, 'La science au Viet Nam', *Bulletin de la Société des études indochinoises*, 38 (1963), 531–558.

Jami, Catherine, 'History of mathematics in Mei Wending's (1633–1721) work', *Historia Scientiarum*, 4 (1994), 159–174.

Jami, Catherine, 'Abacus', in R Bud and D Warner (eds), *Instruments of science: an historical encyclopedia*, Garland, 1998, 3–5.

Jun Yong Hoon, 'Mathematics in context: a case in early Nineteenth-century Korea', *Science in Context*, 19/4 (2006), 475–512.

Kalinowski, Marc, 'La divination par les nombres dans les manuscrits de Dunhuang', in I Ang and P-E Will (eds), *Nombres, astres, plantes et viscères. Sept essais sur l'histoire des sciences et des techniques en Asie orientale. (Mémoires de l'Institut des Hautes Études chinoises*, 35), Collège de France, 1994, 37–88.

Kalinowski, Marc, 'Astrologie calendaire et calcul de position dans la Chine ancienne: les mutations de l'hémérologie sexagésimale entre le IVe et IIe siècle avant notre ère', in K Chemla et M Lackner (eds), *Disposer pour dire, placer pour penser, situer pour agir: pratiques de la position en Chine (Extrême-Orient Extrême-Occident*, 18), PUV, 1996, 71–113.

Kawahara Hideki, 'Tongsan and Chonwonsul—Choson's mathematics in the middle 17th through early 18th century', *Chosen gakko*, 169 (1998), 35–71.

Kawahara Hideki, 'The science of four forms in Joseon dynasty', *Sungkyun Journal of East Asian Studies*, 1 (2001), 283–293.

Kim Yong-Woon, 'Introduction to Korean mathematics history', *Korea Journal,* 13 (1973a), 16–23; 13 (1973b), 26–32; 13 (1973c), 35–39.

Kim Yong-Woon, 'Pan-paradigm and Korean mathematics in the Choson dynasty', *Korea Journal*, 26 (1986), 25–46.

Kim Yong-Woon and Kim Yong-Guk, *Kankoku sugakusi* (A history of Korean mathematics [in Japanese]), Maki Shoten, 1978.

Lam Lay Yong, Ang Tian Se, *Fleeting footsteps. Tracing the conception of arithmetic and algebra in ancient China*, World Scientific, 1992.

Li Chien-Tsung 李建宗, *Chao Xian suan fa jia Qing Shan Zheng* Mo si ji suan fa *chu tan* 朝鮮 算法家慶善微默思集算法初探 'A preliminary investigation of the *Muk sa chip san pŏp* 默思集算 法 by the Korean mathematician Kyong Sŏn Ching', unpublished MA thesis, Taibei: Guoli Taiwan shifan daxue, 2003.

Li Wenlin, Xu Zelin, and Feng Lisheng, 'Mathematical exchanges between China and Korea', *Historia Scientiarum*, 9 (1999), 73–84.

Li Yan, 'Zhang Yong jun xiu zhi Zhong guo suan xue shi yi shi 章用君修治中國算學史遺事' 'The heritage of Mr. Zhang Yong's work on the restoration of the history of Chinese mathematics', in Li Yan, *Zhong suan shi lun cong* 中國算學史論叢 'Collected papers on the history of Chinese mathematics', vol 1, Taibei: Zhengzhong shuju, 1954, 135–146.

Li Yan and Du Shiran, *Chinese mathematics: a concise history*, Clarendon Press, 1987.

Libbrecht, Ulrich, *Chinese mathematics in the thirteenth century: the* Shu-shu chiu-chang *of Ch'in Chiu-shao*, MIT Press, 1973.

Marini, Giovanni Filippo de, *Delle Missioni de padri della compagnia di Giesv Nella Prouincia del Giappone, e particolarmente di quella di Tonkino. Libri cinqve. Del P. Gio: Filippo de Marini Della medisima Compagnia. Alla santita di N.S. Alessandro pp. Settimo*, Rome, 1663.

Marini, Philippe de, *Histoire nouvelle et curieuse des royaumes de Tunquin et de Lao. Contenant une description exacte de leur Origine, Grandeur & Estenduë ; de leur Richesses & leurs Forces ; des Mœurs, et du naturel de leurs Habitants ; de la fertilite de ces contrees, & des rivieres qui les arrosent de tous cotez, & de plusieurs autres circonstances utiles & necessaires pour une plus grande intelligence de la Geographie. Ensemble la Magnificence de la Cour des Roys de Tunquin, & des Ceremonies qu'on observe à leur Enterrements*, Paris, 1666.

Martzloff, Jean-Claude, *A history of Chinese mathematics*, Springer, 1997.

Mei Rongzhao, 'The decimal place-value numeration and the rod and bead arithmetics', in *Ancient China's technology and science*, Foreign Language Press, 1983, 57–65.

Mei Rongzhao 梅榮照 and Li Zhaohua 李兆華 (eds), *Cheng Dawei zhu Suan fa tong zong jiao shi* 程大位著算法統宗校釋 'Systematic treatise on counting methods, by Cheng Dawei, with emendations and explanations', Hefei: Anhui jiaoyu, 1990.

Mei Wending 梅文鼎, *Li suan quan shu* 曆算全書   '*Complete writings on calendrical calculations*, in *Wenyuange Siku quanshu*', vols 794–795, Taibei: Taiwan Shangwu, 1986.

Mikami Yoshio, *Development of mathematics in China and Japan*, Leipzig: Teubner, 1913; reprinted New York: Chelsea, 1974.

Mikami Yoshio 三上義夫, 'Annan-no ichi sansho-ni tsuite 安南の一算書に就て' 'On one mathematical book from Annam', *Gakko sugaku* 學校數學 '*School mathematics*', 14 (1934), 3–11.

Nguyên, Đình-Hoà, 'Graphemic borrowing from Chinese: the case of *chữ nôm*, Vietnam's demotic script', *Bulletin of the Institute of History and Philology, Academia Sinica* [Taiwan], 61 (1990), 383–432.

Needham, Joseph, *Science and civilisation in China*, vol 3 (*Sciences of the Heaven and the Earth*, with the collaboration of Wang Ling), Cambridge University Press, 1959.

des Rotours, Robert, *Le traité des examens, traduits de la Nouvelle histoire des T'ang (chap. 44–45)*, Librairie Ernest Leroux, 1932.

Siu Man-Keung, 'Official curriculum in mathematics in ancient China: how did candidates study for the examination', in Fan *et al.* 2004, 157–185.

Siu, Man-Keung, and Volkov, Alexei, 'Official curriculum in traditional Chinese mathematics: How did candidates pass the examinations?', *Historia Scientiarum*, 9 (1999), 87–99.

Smith, David Eugene, and Yoshio Mikami, *A history of Japanese mathematics*, Chicago, 1914.

Ta Ngọc Liễn, 'Vài nét về toán học ở nước ta thời xưa [Some features of Vietnamese mathematics in pre-modern times]', in *Tìm hiểu khoa học kỹ thuật trong lịch sử Việt Nam* 'The study of science and technology in Vietnamese history', Social Sciences Publishing House, 1979, 289–314.

Tran Nghia et Gros, François, *Catalogue des livres en Han-Nôm*, Nha Xuat Ban Khoa Hoc Xa Hoi, 1993.

Vissière, A, 'Recherches sur l'origine de l'abaque chinois et sur sa dérivation des anciennes fiches à calcul', *Bulletin de Géographie*, 28 (1892), 54–80.

Volkov, Alexei, 'The mathematical work of Zhao Youqin: remote surveying and the computation of π', *Taiwanese Journal for Philosophy and History of Science*, 8 (1997), 129–189.

Volkov, Alexei, 'Counting rods', in R Bud and D J Warner (eds), *Instruments of science: an historical encyclopedia*, Garland, 1998, 155–156.

Volkov, Alexei, 'Le bacchette', in S Petruccioli *et al* (eds), *Storia Della Scienza*, vol. 2, Istituto della Enciclopedia Italiana, 2001, 125–133.

Volkov, Alexei, 'On the origins of the *Toan phap dai thanh* (Great compendium of mathematical methods)', in Yvonne Dold-Samplonius, Joseph W. Dauben, Menso Folkerts, Benno van Dallen (eds), *From China to Paris: 2000 years transmission of mathematical ideas*, Franz Steiner Verlag, 2002, 369–410.

Volkov, Alexei, 'History of ideas or history of textbooks? Mathematics and mathematics education in traditional China and Vietnam', in Horng Wann-Sheng *et al* (eds), *History, culture, and mathematics education in the new technology era*, Proceedings of Asia-Pacific HPM 2004 Conference, May 24–28, 2004, Taichung National Teachers College, 2004, 57–80.

Volkov, Alexei, 'Traditional Vietnamese Mathematics: the case of Lương Thế Vinh (1441–1496?) and his treatise *Toan phap dai thanh* (Great compendium of mathematical methods)', in U Kyi Win (ed), *Traditions of knowledge in southeast Asia*, part 3, Myanmar Historical Commission, 2005, 156–177.

Volkov, Alexei, 'State mathematics education in traditional China and Vietnam: formation of "mathematical hagiography" of Luong The Vinh (1441–1496?)', in Trinh Khac Manh and Phan Van Cac (eds), *Nho Giao o Viet Nam. Confucianism in Vietnam*, Hanoi Social Sciences Publishing House, 2006, 272–309.

Volkov, Alexei, 'Vietnamese mathematics', in Helaine Selin (ed), *Encyclopaedia of the history of science, technology and medicine in non-Western cultures*, Springer-Verlag, 2008a, 1425–1432.

Volkov, Alexei, 'Traditional Vietnamese astronomy in accounts of Jesuit missionaries', in Luis Saraiva and Catherine Jami (eds), *History of mathematical sciences, Portugal and East Asia III: The Jesuits, the Padroado and East Asian Science (1552–1773)*, World Scientific, 2008b, 161–185.

Volkov, Alexei, 'Demonstration in traditional Chinese and Vietnamese mathematics: argumentation for state examinations', in K Chemla (ed), *History and historiography of mathematical proof in ancient traditions*, Cambridge University Press, (forthcoming).

Wu Chengluo, *Zhong guo du liang heng shi* 中國度量衡史 'History of measures, volumes and weights in China', Shangwu yinshuguan, 1937.

Zhang Yong 章用, 'Yue li shuo run kao 越曆朔閏考' 'Sur la concordance des dates néoméniques du calendrier annamite et du calendrier chinois de 1759 à 1886',[52] *Xi nan yan jiu* 西南研究 [*Meridio-occidentale sinense*], 1 (1940), 25–35.

---

52. Zhang Yong's original translation of the Chinese title.

A Balkan trilogy: mathematics in the Balkans before World War I

Snezana Lawrence

E ver since the invention of the term 'the Balkans', the area has attracted the attention of the international community for negative reasons: wars, atrocities, and incessant squabbling about the rightness or wrongness of ethnic or religious origin or denomination. An entirely different and more positive side to the people of the Balkans will be examined here: their efforts to introduce the pursuit of mathematics to the nations to which they belonged.

The development of Balkan mathematics will be examined through the history of three national schools of mathematics. First, we will look at the mathematics of the ruling Ottomans at the end of the eighteenth and the beginning of the nineteenth century, and see how mathematical culture developed under a programme of modernization of the state and its military apparatus. The second focal point will be the mathematics of the Orthodox population of the Ottoman empire. Greeks were the predominant Orthodox *ethnie* within the empire, both in terms of their heritage and cultural influence, and the spread of the Greek diaspora, with merchant communities scattered throughout the empire, gave them an enviable position in terms of their ability to import learning from other countries. Finally, focusing on a particular and relatively small national mathematical culture, that of Serbia, we will examine a personal story that contains many ingredients considered typically Balkan, interwoven with a love of mathematical studies.

A theme such as this poses several difficulties: in defining the geographical region, the historical period we need to consider, and what kind of mathematics may be considered to be Balkan. 'The Balkans', a term that originates in the Turkish word *balkan* 'mountain chain' (Obolensky 1971), was first used by the German geographer Johann August Zeune (1808) in relation to the peninsula that now includes Bosnia, Serbia, Monte-Negro, Albania, Macedonia, Greece, Bulgaria, and Romania (see Fig. 2.4.1). By the end of the century it was used in the context of describing the struggle of some of these nations against the Ottoman empire (Minchin 1886). In the early nineteenth century the national awakening of ethnic groups in the region marked the beginning of European nation-building.[1] This was followed by the establishment of institutions of higher learning for the elite of Balkan societies. This chapter will examine general trends in the development of mathematics by the ruling Ottomans, as well as the parallel evolution of mathematical awareness in the rebelling Orthodox population.

The mathematics of the Ottomans

The Ottoman empire (1299–1922) at the height of its power in the sixteenth and seventeenth centuries stretched from Gibraltar to the Persian Gulf and from Austria to Sudan and Yemen. Relative religious tolerance meant that non-Islamic *ethnies* were allowed to profess their faith.[2] The Ottomans developed schools known in Arabic as *madrasas*, 'places of learning', which were founded throughout the Muslim world from the ninth century (see Brentjes, Chapter 4.1 in this volume). The primary aim of *madrasas* was to teach jurisprudence. It is significant that certain *madrasas* also included teaching on 'rational' sciences such as logic, ethics, Arabic language subjects, and arithmetic, apart from religion and jurisprudence.[3] It is generally believed, however, that individual *madrasas* included mathematical sciences and astronomy only if the *madrasa* professors were themselves expert in such subjects.

Students entered *madrasas* after their basic schooling in the *mektebs* (equivalent to primary schooling), and spent years rising through twelve grades or ranks, corresponding to the ranks of their teachers. After completing his studies, a student might enter the teaching profession or, if he had completed all grades, the

---

1. The Serbian Uprisings and the Greek Revolution happened more than forty years before the unification of either Italy or Germany. The First Serbian Uprising began in 1804, and it was after the Second Uprising that the Serbs gained partial independence from the Ottomans. The Greek War of Independence (1821–31) ended with the Treaty of Constantinople, which proclaimed Greece a free country.

2. Roudometof (1998, 12) describes *ethnie* thus as a pre-modern concept of identity: 'An *ethnie* may have the following characteristics to differing degrees: a collective proper name, a myth of common ancestry, shared historical memories, some elements of common culture (e.g., language, religion), an association with a specific homeland, and a sense of solidarity.' See also Smith (1986, 40).

3. This trend was most widespread under the reign of Kanuni Sultan Suleyman (1495–1566), but became less common after the end of the sixteenth century. See (Sürmen, Kaya, Yayla 2007).

Figure 2.4.1 Alexandre Emile Lapie: map of 'Turkey in Europe', Paris, 1816

hierarchy of *ulema* 'legal scholars' to become a learned man or a religious cleric. This type of education was based on tradition, and the handing on of existing knowledge, rather than the development of new concepts (Zilfi 1983). A major challenge to this system came after the Ottomans lost the Battle of Vienna in 1683. The Ottomans subsequently fought a number of wars with the Russians, until they established the Habsburg–Ottoman–Russian borders in southeast Europe. At the same time they lost large areas of the empire such as Egypt and Algeria to Britain and France. The Ottoman government therefore became increasingly concerned with two major problems: the loss of power and influence on the one hand, and the perceived decline and corruption of the military apparatus on the other. There was thus seen to be an urgent need to modernize the army and its technologies, which in turn led to changes in the learning of mathematics.

At its core, this modernizing process was not motivated by ideas of Western Enlightenment. Rather, the empire sought to create an educational system geared to military needs by introducing Western engineering and scientific learning. Western sciences were therefore imported in relation to the art of war rather than to peace or the pursuit of knowledge itself. One major problem that eventually arose through this process was exclusion: because of its close link with military goals, the new educational programme, at least for the first half-century, excluded de facto the many ethnic groups whose culture was linked to Christianity rather than to Islam.

The French and the Turks had had a long history of cooperation since the sixteenth century, when France received permission to trade in all Ottoman ports, and it was through this longstanding liaison that Western mathematics first came into the Ottoman empire. A Naval Engineering College was founded in Istanbul in 1773 under the guidance of Baron de Tott, a French diplomat of Hungarian origin (de Tott 1785). A Military Engineering College was established in 1795, with a mathematical syllabus almost identical to that of the Naval College, but with the additional subject of fortification. These two institutions were the first in the empire to teach modern mathematics, departing from the traditional Islamic teaching of the *madrasas*, and focusing on the application of the sciences to military and civil engineering. The two colleges later merged and were, in effect, the origin of the Istanbul Technical University.

Among the first group of teachers at the Naval Engineering College was Gelenbevi Ismail, who is credited with introducing logarithms to the empire. Born in the town of Gelenbe in southern Turkey he rose through the ranks of the *madrasas* to become professor at the age of thirty-three. In algebra he worked in the classical Arabic tradition of al-Khwārizmī (c 825), in which problems were presented rhetorically and justified geometrically (Høyrup 1987, 281–329; Berggren 2007). The second generation of teachers initiated a more organized programme of Western mathematics by translating texts into Turkish. Huseyin Rifki, for example, who taught at the Military Engineering School, became a prolific translator of Western works. This was a part of a growing trend among Ottoman scholars to look to the West rather than to the East or to the Arabic mathematical heritage (Sürmen, Kaya and Yayla 2007, 5). Rifki's most important work, in collaboration with Selim Ağa, an English engineer who converted to Islam, was a translation of Bonnycastle's 1789 edition of Euclid's *Elements*. Until this translation appeared in 1825, Euclid was taught from a ninth-century Arabic translation modified in the fifteenth century by Bursali Kadizade-i Rumi, head of the Samarkand *madrasa*. Rumi published shortened and simplified versions of the *Elements* under the title *EsKalü't-te'sis*. His books were studied across the empire and were regarded as the most significant works of Ottoman mathematics, but Rifki nevertheless preferred to translate from an English edition. Perhaps his assistant Selim Ağa persuaded him that this would be more appropriate for modern education than the version of Rumi. The practice of translating into national languages from Western sources rather than from local translations demonstrates how Western mathematics and ideas of education overrode even material that was already known.

Ishak Efendi, a Jewish convert to Islam and a student of Rifki, also became a translator of Western works into Turkish, and created Turkish equivalents for many scientific and mathematical terms. He translated (but never published) Robert Fulton's *Torpedo war and submarine explosions* (1804), but is best known for his four-volume *Mecmua-i Ulum-u Riyaziya* 'Compendium of mathematical sciences', first published in 1831 for the pupils of the Army Engineering School. Another edition was published in Cairo in 1845 for the use of the new military school there. The 'Compendium' was an adapted translation of a number of Western works: although some students could read the originals in French or English, these were not widely available. A substantial work of over two thousand pages, it introduced readers not only to mathematics, but also to physics, astronomy, biology, and chemistry. In mathematics it offered a new approach, heavily influenced by what was perceived as Western materialist science (Özervarli 2007). It emphasized modern mathematics, and claimed that the crucial Islamic notions of *jihād* and *ghāza* were dependent on knowledge of mathematics (Findley 1992, 141). *Jihād* and *ghāza* have a range of meanings, but here referred to the aspiration to live a moral and virtuous life, and the need to spread and defend Islam, the latter being a raison d'etre of the Ottoman state. Although there is evidence that some atheism was included in the learning imported from the West, such religious aspirations were based on the spiritual traditions of Islam (Özervarli 2007).

The first non-military institutions to offer modern science and technology were the *Mekteb-I Maârif-I Adliyye* 'School of Learning', and the *Mekteb-I Ulûm-I Edebiyye* 'School of Literary Sciences', founded in 1839 for training officials. The School of Learning proposed that after learning Arabic, students should take subjects such as French geometry, geography, and history. This marked the beginning of an education that was both secular and non-military

(Somel 2001, 34–35). It coincided with the beginning of a movement known as *Tanzimat* 'Reorganization' (during the period 1839–76), whose primary aim was the secularization of the educational system, but which in view of the rising nationalist movements also promoted inclusiveness amongst all the peoples of the empire.

In terms of educational reform, a compulsory primary system (lasting for four years) was introduced, with a system of further schools, such as *Rushdiya* (four higher grades of primary school, or middle school), *İdadiya* (offering four years of vocational education), and *Sultaniya* (six-year secondary schools, based on the French *lycées*). For the *Rushdiya* there was no mention of mathematics in the official syllabi, but the *İdadiya* and *Sultaniya* had ambitious programmes. The *İdadiya* taught practical mathematics such as bookkeeping, geometry, and algebra, while the *Sultaniya* were supposed to offer descriptive geometry, analytical geometry, perspective, algebra, and trigonometry.

Tanzimat aimed to bring these subjects into general teaching practice, and at the same time initiated a wide ranging reform of teacher education. By the middle of the *Tanzimat* period, however, mathematics was still not widely taught in the teacher training schools. A teacher named Tahsin, for example, who graduated from the School of Civil Service, and who was subsequently appointed administrator in the province of Rumeli (now Macedonia), reported in 1864 that local teachers knew nothing of subjects such as mathematics or geography (Somel 2001, 77).

Thus, during the nineteenth century, Ottoman mathematics was strongly influenced by military preoccupations, and texts were mostly translations of French and English material. Very little original mathematics was done before World War I. A rare exception was the secondary school headteacher, Mehmet Nadir, who obtained some significant results in number theory, particularly in relation to Diophantine equations (İnönü 2006). In 1915 he became the first professor of mathematics at the University for Women (established in 1914 in Istanbul), and was appointed to a chair in number theory at Istanbul University in 1919. The first Ottoman mathematician to be granted a PhD in mathematics was Kerim Erim, at Erlangen in 1919 (İnönü 2006, 234–242). Nor was there much effort to communicate with Western mathematicians: Nadir did so, as did his contemporaries Tevfik Pasha and Salih Zeki, who both published in foreign languages or in foreign journals. The former wrote a *Linear algebra* in English (second edition 1893) for use in teachers' colleges, and the latter published an article on 'Notation algébrique chez les Orientaux', 'Algebraic notation in the East' in the *Journal Asiatique* in 1898. The general lack of communication with the West, other than in importing and translating texts from French and English, has sometimes been seen as having adverse consequences for the development of Ottoman science and mathematics (Rabkin 2003, 196).

The mathematics of Greece

Until the end of the eighteenth century, the influence of Greek culture on the Orthodox community within the Ottoman empire was wide and all-encompassing (Tsourkas 1943). The Greek diaspora spread as far as the Black Sea coast and the Venetian territories, with communities also in cities such as Vienna, Amsterdam, and Budapest. It played an important role not only in commercial development, but also in supporting the intellectual and cultural advancement of Greeks and other Orthodox *ethnies* within the empire. During the nineteenth century, however, various other national influences began to pervade Greek mathematical culture, from France, Germany, and even, for a brief period, the United States (Kastanis 2007).

During the eighteenth century, Greek intellectual and educational life centred around four areas: Bucharest and Jassy (in what is now Romania), Thessaly, the eastern Aegean, and the Ionian islands. The territories around Bucharest and Jassy were run by Greek Phanariots, a community that had grown out of the prominent families who originally lived in the Greek quarter of Istanbul called Phanar. They became the official translators to the Ottomon government during the period when the Ottomans lost northern and eastern territory to the Austrians and Russians in the seventeenth century, and their influence on state apparatus and foreign policy grew substantially from then on. The fact that by the nineteenth century they were sometimes called 'Christian Turks' (Vucinich 1962, 602) indicates their social power as well as their isolation from other Christian communities of the era. The Phanariots, however, fostered a multilingual and metropolitan outlook, and an international network, all of which contributed to the creation of later Greek identity (Dialetis, Gavroglu, and Patiniotis 1977; Sonyel 1991).

The areas around Thessalonika, the eastern Aegean, and the Ionian islands were the most significant in introducing new educational trends into Greek culture at the beginning of the nineteenth century, especially into mathematics. The intellectual prestige of these areas was related to sea trade, which brought with it both relative prosperity and exchange of ideas. The centre of learning in Thessaly was Ampelakia, at the foot of Mount Olympus, where a school was founded in 1749. In the Aegean, the centres were Chios, Kydonies, and Smyrna. Academies were founded in Kydonies and Smyrna in 1800 and 1808, respectively, offering the equivalent of undergraduate studies. The Ionian Islands were under British protection between 1814 and 1864, during which time the Ionian Academy, established in 1824, introduced a Western model of mathematical education.

Some of those who taught mathematics in these institutions were trained abroad with support from Greek individuals or communities outside the empire. For example, Veniamin of Lesvos, who studied at Pisa supported by the Greek community at Livorno and then taught at Kydonies from 1796; Dorotheos Proios who studied in Pisa and Paris and after 1800 taught at Chios (see Kastanis and Kastanis 2006, 523); and Ioannis Carandinos, who studied at the École Polytechnique in Paris and later became Dean of the Ionian Academy. Another person who trained abroad, with support from the brothers Lambros and Simon Maroutsis of Venice, was Evgenios Voulgaris who studied at the universities of Venice and Padua. Rather than returning with modern ideas, however, Voulgaris focused on the re-establishment of ancient teaching in mathematics, believing that classical Greek geometry was the basis for any further progress (Roudemetof 1998, 21). His view was that mathematics was valid in its connection with philosophy rather than with the experimental sciences. Thus the new context of the Enlightenment escaped him, as it did some of his followers (Dialetis, Gavroglu, and Patiniotis 1997). Voulgaris translated many philosophical and some mathematical works into Greek, among them, in 1805, Euclid's *Elements* from Tacquet's 1694 edition. At the time Voulgaris was living at the royal court of Russia, having become a protégé of the empress Catherine. This may be the reason for his having chosen Tacquet's edition, which was available to him in the library there, rather than an existing Greek edition.[4]

The first translations into Greek of modern works on mathematics were made by Spyridon Asanis, a medical doctor who taught mathematics at Ampelakia in Thessaly in the 1790s. His translations drew on works by Nicolas-Luis de Lacaille and Guido Grandi, and two of them were published: *Arithmetic and algebra* (from Lacaille 1741) in Venice in 1797, and *Conic sections* (from Grandi 1744) in Vienna in 1803. Their success encouraged several further translations by others.[5] Iosipos Moisiodax at the Greek Academy in Jassy also drew upon Lacaille. Indeed Lacaille was very popular in both Italy and Austria during the second half of the eighteenth century, having been introduced into their respective educational systems by the Jesuits (Kastanis and Kastanis 2006, 518).

Carandinos in Corfu translated from French all the books he thought necessary to create an undergraduate mathematics department similar to those in western Europe. Between 1823 and 1830, he translated works by Bourdon, Biot, Lagrange, Poisson, Monge, Lacroix, and Legendre (Kastanis 1998, 186; Phili 1998, 303–319). Hence, virtually all the mathematics studied at the Academy was based on the work of French mathematicians. As in other areas under French influence, institutions were modelled on their Parisian counterparts (Mazzotti, Chapter 3.3

4. The *Elements* were first translated into contemporary Greek in 1533 by a German theologian and scholar Simon Grynäus. This edition included Proclus' *Commentary*, given to Grynäus by the then president of Magdalen College, Oxford, John Claymond. Original manuscripts in ancient Greek were by this time no longer available.

5. See Kastanis and Kastanis (2006, 518–520). Algebra was extended by the study and publications of three scholars: Zisis Kavras, who studied in Jena and translated works from German; Dimitrios Govdelas, who studied in Pest and wrote *Stoicheia Algebras* 'Elements of Algebra', based on German sources and published in Halle in 1806; Stefanos Dougas, who studied at Halle, Jena, and Göttingen, and published a four-volume work on arithmetic and algebra, inspired by German tradition, in Vienna in 1816.

in this volume). A Military Academy founded at Napflion in 1829 with the help of the French government was modelled on the École Polytechnique, and the mathematics curriculum was based on books already translated for the Ionian Academy (Kastanis 2003, 136). Later military schools in Greece were also modelled on the École Polytechnique, and there was also a strong French influence in the University of Athens, founded in 1836. The first professor of mathematics there was Constantinos Negris, who had studied in Paris at the Lycée de France and the École Polytechnique (see Kastanis and Kastanis 2006, 531).

Although French mathematics predominated in the Greek academies there were also other influences. Constantinos Koumas, for instance, studied at Vienna from 1804 to 1808 and completed his doctorate at Leipzig. His approach to mathematics has been described as 'Austrian scholastic' (Kastanis and Kastanis 2006, 521) in that his main focus was the work of Jean-Claude Fontaine (see Fontaine 1800; Koumas 1807). He also developed a method of teaching physics experimentally, transforming the Gymnasium of Smyrna (modern day Izmir, Turkey), of which he was a director, into the most famous scientific school for the Greek-speaking community (see Dialetis et al, 1999). The progressive and rationalistic leanings of this school, as well as the teaching of modern mathematics and sciences in the 'Western' manner at the Evangelical School in Smyrna, attracted attention from the Patriarchate. Koumas' school was burnt by an angry mob in 1819, while at the same time Patriarch Grigorios V issued a warning to all students of mathematics: 'cubes and triangles, logarithms and symbolic calculus... bring apathy... jeopardizing our irreproachable faith' (Kastanis and Kastanis 2006, 525; for longer account see Terdimou 2003, 53–62).

Between 1810 and 1820, two mathematics teachers, Stefanous Dougas (educated at Halle, Jena, and Göttingen) and Dimitrios Govdelas (educated at Pest), introduced German-inspired education into the Patriarchic School of Constantinople and the Academy of Jassy. After the appointment of the Bavarian prince Otto Wittelsbach as King of Greece in 1832, Bavarian officials also influenced the Greek educational system. They established a system of secondary education divided into lower and upper Gymnasia. The emphasis was on classical studies, although mathematics was placed as the third most important subject after ancient Greek and Latin. The syllabus in mathematics gave plenty of freedom to the teacher but prescribed the general outline of study and the number of hours of teaching. Mathematics teaching was heavily dominated by Euclidean geometry, and on Diesterweg's principles of teaching geometry by heuristic or discovery learning.[6] The insistence on classicism, and the shortage of mathematics

---

6. Friedrich Adolph Wilhelm Diesterweg (1790–1866) was a German educational thinker whose most famous work, *Wegweiser zur Bildung für deutsche Lehrer* ('A guide to education for German teachers') (1835) set out the principles of teaching based on a theory of development and improvement, heavily coloured by the ideology and philosophy of neo-Classicism (Günther 1993).

teachers, meant that there was a pressing need for suitable textbooks, which were provided through the Bavarian connection. Georgios Gerakis, for example, studied in Germany (with support from the state) and afterwards published textbooks in Greek based on German textbooks: *Elementary geometry and trigonometry* (1842),[7] *Arithmetic and algebra* (1855), and *Plane geometry* and *Stereometry*.[8]

Principles of heuristic teaching similar to those of Diesterweg came to Greece from another source also. During the early 1830s, American missionaries came to Greece, hoping to set up a mathematical programme based on Pestalozzian principles.[9] This was a short lived and fairly unsuccessful episode but one which for a few years exposed Greek mathematical culture to the 'inductive' method as defined by Pestalozzi (Kastanis 2007).

Throughout the nineteenth century Greek mathematical and scientific thinkers were preoccupied with introducing these subjects into educational institutions. In their approach they ranged on the one hand from the view that ancient mathematics was closely linked to ancient Greek thought, and on the other with a concern to model mathematical culture and education on the examples of France (with the École Polytechnique as a model for higher education) and Germany. The influence of Germany blended well with references to ancient Greek thought and mathematics as it was closely linked to ideas of classicism, and geometry was considered to be a de facto embodiment of neo-Classical philosophy. Euclid remained one of the main sources of knowledge at lower levels of education, until the reforms of Greek educational system in the middle of the twentieth century. French influence, however, was most pervasive at the higher levels. The French initiated a long-lasting trend towards analysis but, more than anything else, French mathematical thought and teaching entered Greece through the textbooks and personal experience of the first teachers of mathematics in the schools established after the Greeks gained independence from the Ottomans. The two foremost institutions for the further development of mathematics in Greece, the Ionian Academy and the University of Athens, had the mathematics of the French built into their foundations.

The mathematics of Serbia

The history of Serbian mathematics is interwoven with the colourful lives of several of its most prominent exponents. Indeed, in some respects, their lives epitomize some of the connotations of the term 'Balkan': a bridge between West and

---

7. Based on Snell (1799).
8. Based on Koppe (1836a; 1836b).
9. Johann Heinrich Pestalozzi (1746–1827) was a Swiss educational reformer, whose main contribution to the development of pedagogy was his insistence on a child-centred approach to learning, progressing from the familiar to the new.

East, bohemian passion, and violent confrontation. All these seemingly 'Balkan' traits are exemplified in the life of the mathematically inclined Crown Prince of Serbia, Djordje Karadjordjević, and in his friendship with the most famous of Serbian mathematicians, Mihailo Petrović.

First some background: Serbian mathematics education in the nineteenth century developed first under the Ottomans, and after 1833 under the Austro-Hungarian Empire. The first book on mathematics in Serbian was *Nova serbskaja aritmetika* 'A new Serb arithmetic' (1767) by Vasilije Damjanović, but under-graduate education was established only in 1838, at the Lyceum in Kragujevac. The first mathematics professor there was Atanasije Nikolić, who had studied in Vienna and Pest, and his initial task was to write the first undergraduate text-books in the Serbian language.

Belgrade University grew out of a succession of institutions, the most promin-ent being *Matica Srpska*, literally 'the Serbian Queenbee', founded in 1826 in Pest to promote Serbian culture and science. This institution grew into the Lyceum, and the Lyceum developed into the Superior School. The first trained mathemat-ician to teach at the Lyceum, Dimitrije Nešić, had been educated at Vienna and Karlsruhe Polytechnic, and is credited with defining Serbian terminology for all mathematical concepts and processes known at the time.

At the end of the nineteenth century several Serbian mathematicians stud-ied for doctorates at western universities: Dimitrije Danić at Jena (1885), Bogdan Gavrilović at Budapest (1887), Djordje Petković at Vienna (1893), Petar Vukičević at Berlin (1894), and finally, the most famous Serbian mathematician, Mihailo Petrović, who completed his thesis in Paris in 1894. It is not known why Petrović chose Paris when all his contemporaries had studied in Germany or Austria, but he established important links with the French government during his studies and maintained them later. Thus, although most educational influences in the middle of the century were Austro-Hungarian or German, the most prominent of Serbian mathematicians, who set the future direction of the national mathematical school, introduced French mathematics and French mathematicians to his country.

Petrović, who was from a well-to-do family in Belgrade, completed a degree in natural sciences at the Superior School, sometimes called the Great School, in Belgrade in 1889. He then went on to study at the École Normale, originally a teacher training institution, rather than the École Polytechnique, which earlier in the century had been the preferred place of study for Greek students. The raised prestige of the École Normale at the end of the nineteenth century may have been a deciding factor, but it is not clear whether Petrovic was aware of it. He was awarded his doctorate in 1894 for a thesis entitled *Sur les zéros et les infinis des intégrales des équations différentielles algébriques* 'On zeros and infinities of inte-grals of algebraic differential equations'. The examining commission consisted of Charles Hermite, Charles Émile Picard, and Paul Painlevé.

Petrović dedicated his thesis to Painlevé and Jules Tannery. He had met Tannery during the first year of his studies when he attended courses on differential calculus and differential equations. He had also become good friends with Painlevé, who had come to Paris from Lille in 1891 to become a professor at the Sorbonne, having completed his doctorate in 1887, and they continued the friendship after Petrović returned to Belgrade. Both Petrović and Painlevé later gained friends amongst the political elites of their respective countries.[10] At Petrović's insistence, Painlevé's work on mechanics (Painlevé, 1922) was translated into Serbian by Ivan Arnovljević and published as a textbook in Belgrade in 1828.

Petrović also made friends with Hermite, who had already had another Serbian student, Mijalko Ćirić. Hermite taught Petrović higher algebra, and his son-in-law, Emil Picard, was another of Petrović's examiners. Petrović and Picard became lifelong friends, and Picard drew on work from Petrović's thesis in his *Traité d'analyse* 'Treatise of analysis' (1908) (see Trifunović 1994, 27).

Upon his return to Belgrade in 1894, Petrović was made a professor at the Superior School. At the beginning of 1905, the Superior School was replaced by the University of Belgrade and Petrović was appointed to the Chair in Mathematics, a position he held until his death in 1943. His main area of interest was classical analysis, and he wrote papers on the properties of real and complex functions defined by power series (Petrović 2004, 100). When he first returned to Belgrade, he devoted some time to creating an analogue computer for solving a certain type of analytically unintegrable differential equation (see Petrović 1896). He completed the machine, the Hydro Integrator, for the Universal Exhibition held in Paris in 1900, and was awarded a Bronze medal for it (Petrović 2004).

In the years prior to World War I, he also worked on mathematical phenomenology,[11] with a view to developing a mathematical apparatus that would be able to encompass all facts, and link apparently unconnected phenomena in a mathematical fashion (Petrović 1911). As an outcome of his work on cryptography during the war, he founded the theory of mathematical spectra, which has analogies with the spectral method in chemical analysis. It consists of dispersing unknown quantities from a problem into a numerical spectrum. The quantities are dispersed, separated, and determined in the same way as in spectral analysis in chemistry. This theory was to be applicable to arithmetic, algebra, and infinitesimal calculus, and Petrović taught a course on it at the Faculté des Sciences at the University of Paris in 1928 (Petrović 1919; 1928).

10. In 1906 Painlevé became a Deputy for the fifth arrondissement of Paris, the so-called Latin Quarter. He was later Prime Minister twice, in 1917 and 1925.

11. An approach in the philosophy of mathematics dealing with issues about rational intuition, the place of formal systems in mathematical thinking, and the intuition of essences, the most prominent students of which were Husserl and Gödel. See Tieszen (2005).

In his personal life he was a dedicated fisherman, and passed the examination of Master Fisherman in Serbia in the same year in which he completed his doctorate in Paris. Later he drafted national laws on fishery, and helped define international treaties. He participated in a number of French expeditions to the Arctic and the Antarctic, and to islands in the south Indian Ocean, writing travelogues and a novel based on these journeys.[12] He also had a passion for Gypsy music (see Fig. 2.4.2), and is reported to have written to his mother from Paris after his attendance at a Presidential Ball, forbidding her to tell the neighbours in Belgrade in case they did not believe her and because a local gypsy friend might hear of it, and it could ruin their friendship.

Figure 2.4.2 Petrović (second from the right) played the violin and had his own band, Musical Band-Suz, whose repertoire consisted of traditional Serbian and Gypsy songs. The band recorded more than a thousand pieces for Radio Belgrade between the two World Wars. Reproduced by the kind permission of Aleksandar Petrović, Belgrade.

In 1903 the young crown prince Djordje Karadjordjević returned to Serbia, having spent some years being educated with his younger siblings in Geneva and St Petersburg, and Petrović was asked to teach him mathematics. Petrović was already well connected with the French establishment (Major Levasseur, the first tutor to the prince, had also been chosen through his French connections), as

12. *Kroz polarnu* 'Through the polar field' (1932), *U carstvu pirata* 'In the kingdom of pirates' (1933), *Sa okeanskim ribolovcima* 'With ocean fishermen' (1935), *Na dalekim ostrvima* 'On remote islands' (1936), *Roman o jegulji* 'Novel of the eel' (1940).

well as something of a bohemian. It seems that the Prince and Petrović struck up a friendship immediately, and maintained it through difficult times later. They went fishing together, and also participated in establishing the first fencing club in Serbia, as well as spending time studying mathematics.

The Prince had a reputation for being hot-tempered, and on one occasion attacked his tutor, Major Levasseur, who had to be dispatched back to Paris (Karadjordević 1988). In 1909 a more serious scandal broke when he killed his valet. Although there were moves to cover up this horrible event, the truth came out, and he had to abdicate.[13] Djordje Karadjordjević tried to recant his abdication on a number of occasions, but was unsuccessful. By now Petrović was teaching him mathematics not out of duty, but out of friendship. In the midst of the scandal, the Prince became fascinated by Poincaré's work, and on 3 March 1911 he wrote to Poincaré as follows:

Dear Professor,

I have learned, as an amateur, some elements of the theory of functions, which interests me strongly, and more and more, and I have come to a question on which I have not been able to find precise explanation. Please excuse the liberty I take in addressing myself directly to the Master to clarify the results of modern research on the question.

The question is as follows: What is the least of the limiting values which a polynomial function $F(z)$ may take when the variable z increases indefinitely along the different vectors in its plane?

Begging you to excuse my importunities, I beg you, Monsieur, to accept this expression of my respectful regards.

George.[14] (Trifunović 1992, 66)

The reply arrived on 12 March 1911. The Prince duly thanked the great mathematician, and sent him another letter asking for his signature and a photograph. The only (known) surviving answer comes from a paper by Petrović, which mentions Poincaré's solution to the problem (Petrović 1929).

The letter from the Prince to Poincaré was the high point of his mathematical studies. He continued to study mathematics long after his abdication and into his life as an ordinary citizen. He participated in the Balkan Wars of 1912–13

---

13. In his place came his younger brother, later to become the first King of Yugoslavia (which was founded in 1918), Alexander Karadjordjević. Alexander was assassinated in Marseille in 1934. George survived him and became an ordinary citizen to live and die in Belgrade after the fall of the Kingdom and the establishment of the Federal Republic of Yugoslavia after the end of World War II.

14. 'Monsieur le Professeur, En ayant appris, en amateurs, quelques éléments de la théorie des fonctions qui m'interresse vivement et de plus en plus et en ayant recontré une question sur laquelle je n'ai pu trouver nulle part des renseignements précis. Veuillez bien éxcuser la liberté que je prends en m'adressant directement au Maître pour m'éclaircir sur les résultats acquis par les recherches modernes relative à la question. La question est la suivante:

Quel est la moindre des valeurs limites que puisse avoir une fonction entière F(z) lorsque la variable z augmente indéfiniment suivant différents rayons vecteurs dans son plan.

En vous priant d'éxcuser mes importunités, je Vous prie Monsieur le Professeur d'agréer l'éxpression de ma respectueuse reconnaissance. Georges.'

and World War I, but upon his father's death in 1925 was confined to a mental asylum by his brother, King Alexander. During his years of incarceration, Djordje Karadjordjević became obsessed with Einstein's theory of relativity. Petrović visited him regularly and they spent many hours discussing it. He was eventually released by the German occupying forces in 1941, but in the year of his release, Petrović in turn was imprisoned and deported as a lieutenant colonel of the Yugoslav Royal Army. Djordje Karadjordjević survived another change of regime at the end of World War II, and died an ordinary citizen in Belgrade in 1972.

Serbia progressed rapidly from having little or no mathematical culture at the beginning of the nineteenth century. There were some advantages to this relatively short history. At the International Conference on Mathematics Teaching, in Paris in April 1914, the Serbian delegation was able to report that the introduction of infinitesimal calculus into schools was devoid of problems in their country: modernization did not pose a problem in a place where there was no tradition that could inhibit it:

For the nations that are but at the threshold of civilization in their development, there is no tradition and an idea in general, and especially a new idea, can become very easily an ideal of a new generation. As a consequence, in such circumstances the realization of this ideal is not prevented or delayed by questions of tradition.[15] (*L'enseignement mathématique*, 16 (1914), 332–333)

There was another important factor. Petrović's colourful personality, his friendships in the highest and the lowest echelons of local society, his links with the French establishment, his passion for music and fishing, and his approachable character, all conveyed an image of a bohemian and intellectual elite, at the core of which lay excellence in the study of mathematics. This image was embedded in the national mentality for a further century.

Petrović's work, both in terms of acknowledgement in the international community and his efforts to establish a national school (virtually all mathematical doctorates in Serbia between the two World Wars were done under his supervision),[16] established far-reaching change. This had a long term effect on

15. 'Chez les nations qui ont à peine dans leur développement, passé les premiers seuils de la civilisation, il n'y a pas de tradition et une idée in general et surtout une idée nouvelle, devient très facilement l'idéal meme d'une generation. Par consequent, dans ces circonstances la realisation de cet ideal n'est pas empêchée ou retardée par des questions de tradition.'

16. Petrović's doctoral students were Sima Marković (1904; became a famous Communist, lost his life in Russia under Stalin), Mladen Berić (1912), Tadija Pejović (1923), Radivoj Kašanin (1924; became professor at the University of Belgrade), Jovan Karamata (1926; taught mathematics at the Universities of Belgrade, Göttingen, and Geneva), Miloš Radojčić (1928; professor at the University of Belgrade and the University of Khartoum), Dragoslav Mitrinović (1933; professor of mathematics and founder of mathematical institutes in a number of universities of former Yugoslavia), Danilo Mihnjević (1934), Konstantin Orlov (1934; professor of mathematics at the University of Belgrade), and Dragoljub Marković (1938). These ten produced a further 361 doctoral students during their professional lives.

Serbian, and later Yugoslavian, study of mathematics. In this way the influence of the French school, with Poincaré at its centre, was felt long after his main Serbian student became the founder of the national mathematical school.

Conclusion

The modern idea of a periphery, and in particular of societies, such as those of the Balkans, lagging behind the West which by contrast is seen as progressively onward-moving (Ahiska 2003; Heper 1980), implies a need to catch up with developments at 'the centre' by introducing new technologies, sciences, and cultural innovations. In the case of the Ottoman Empire, the sense of being on the periphery began to emerge at the end of the seventeenth century, which marked the beginning of the empire's decline in military prowess and influence. This was the beginning of the period during which the Ottomans began the process of Westernization, including the adoption of Western mathematics, which entered Ottoman education mainly through the military engineering schools (see Güvenç 1998; Grant 1999; Somel 2001; Ekmeleddin 2003; Gökdogan 2005). In this case the periphery took what it considered useful from the West with a view to regaining military and political prestige, but at the same time filtered out other aspects of imported culture.

In the case of the Greek mathematics of the nineteenth century, the pursuit of mathematics was influenced by the centre to such an extent that the centre often set the agenda for reform in the periphery. This can be seen in the mathematics exported by the French to the Ionian Islands, which at the time were a British protectorate. The mathematics developed at the Ionian Academy in turn impregnated all future developments in Greece after the Wars of Independence in 1821.

Finally, the particular story of Balkan mathematics, described in some detail, focuses on a small national mathematical culture, that of Serbia. In this case, the search for authenticity and individualism was interwoven with the personalities involved, linking the highest and the lowest levels of society. In such a setting mathematics and its narrative became embedded in the national culture, certain elements of which gave rise to an archetypal view of mathematical pursuit linked to a bohemian but also intellectually superior way of life.

With World War I, the intellectual map as well as the political map changed dramatically. First, the centres of political and cultural influence changed irrevocably after the disintegration of the Austro-Hungarian Empire. Second, whilst the choices made by mathematicians and mathematics educators in the nineteenth century were often a matter of inheritance, opportunity, or circumstance, mathematicians of the new era became acutely aware of the seriousness of decisions they had to make in developing their national schools. Through this awareness they began to create their own intellectual landscape, and drew from the influences they considered most appropriate to their national circumstances. The Russian Revolution of 1917 brought another factor into play. Many mathematicians fled from the Russian Revolution towards the west, with a number of them settling in the Balkans. By the end of World War II the division of Europe into Western and Eastern blocs meant further changes to the mathematical cultures of the Balkan societies. One modern historian has likened the developments in the region to the dramatic drawings of M C Escher: weird and frightening, but also containing some wonderful narratives, set in an improbable and often impossible perspective (Glenny 1999, 57). To one degree or another all these elements make up the diverse culture of the Balkans and, as a result, of their mathematics too.

Bibliography

Academie Royale de Serbie, *Notice sur les travaux scientifique de M. Michel Petrovitch*, Paris, 1922.

Ahiska, Meltem, 'Occidentalism: the historical fantasy of the modern', *The South Atlantic Quarterly*, 102 (2003), 351–379.

Arnovljević, Ivan, *Mehanika*, Belgrade, 1828.

Berggren, J Lennart, 'Mathematics in medieval Islam' in Victor Katz (ed), *The mathematics of Egypt, Mesopotamia, China, India, and Islam*, Princeton University Press, 2007, 515–670.

Bonnycastle, John, *The elements of geometry*, London, 1789.

Chambers, Richard L, 'The education of a nineteenth-century Ottoman Alim, Ahmed Cevdet Pasa', *International Journal of Middle East Studies*, 4 (1973), 440–464.

Danić, Dimitrije, *Conforme Abbildung des elliptischen Paraboloids auf die Ebene*, Jena, 1885.

De Tott, Baron, *Memoirs. Russo-Turkish War, 1768-1774. Translation of Mémoires de baron de Tott sur les Turcs et les Tartares*, London, 1785.

Dialetis, D, Gavroglu, C, and Patiniotis, M, 'The sciences in the Greek speaking regions during the seventeenth and eighteenth centuries: the process of appropriation and the dynamics of reception and resistance', in *The Sciences in the European periphery during the Enlightenment*, Kluwer, 1999, 41–72.

Findley, C V, Reviewed works: *Bashoca Ishak Efendi (Turkiye'de Modern Bilimin Oncusu)* by Ekmeleddin Ihsanoglu and *Osmanli Ilmi ve Mesleki Cemiyetleri* by Ekmeleddin Ihsanoglu, *International Journal of Middle East Studies*, 24 (1992), 140–141.

Fontaine, Jean-Claude, *Cours encyclopédique et élémentaire de mathématique et de physique*, Vienna, 1800.

Gavrilović, Bogdan, *Construction of one-valued analytic functions*, Budapest, 1887.

Gelisli, Yucel, 'The development of teacher training in the Ottoman Empire from 1848 to 1918', *South-East Europe Review*, 3 (2005), 131–147.

Glenny, Misha, *The Balkans, 1804-1999*, Granta, 1999.

Gökdogan, Melek Dosay, 'Ottoman mathematical culture in the nineteenth century', *History and Pedagogy of Mathematics Newsletter*, 60 (2005), 15–21.

Grandi, Guido, *Instituzioni delle sezioni coniche*, Firenze, 1744.

Grant, Jonathan, 'Rethinking the Ottoman "decline": military technology diffusion in the Ottoman Empire, fifteenth to eighteenth centuries', *Journal of World History*, 10 (1999), 179–201.

Günther, Karl-Heinz, 'Friedrich Adolph Wilhelm Diesterweg (1790–1866)' *PROSPECTS: The Quarterly Review of Comparative Education*, 23 (1993), 293–302.

Güvenç, B, *History of Turkish education*, Turkish Education Association, 1998.

Heper, Metin, 'Center and periphery in the Ottoman Empire: with special reference to the nineteenth century', *International Political Science Review / Revue internationale de science politique*, 1(1980), 81–105.

Høyrup, Jens, 'The formation of Islamic mathematics: sources and conditions', *Science in Context*, 1 (1987), 281–329.

Ihsanoglu, Ekmeleddin, Chatzis, Kostas, and Nicolaidis, Efthymios (eds), *Multicultural science in the Ottoman Empire*, Turnhout, 2003.

Ihsanoglu, Ekmeleddin, and Al-Hassani, Salim, *The madrasas of the Ottoman Empire*, Foundation for Science, Technology and Civilisation, Manchester, 2004.

İnönü, Erdal, 'Mehmet Nadir: an amateur mathematician in Ottoman Turkey', *Historia Mathematica*, 33 (2006), 234–242.

Karadjordjevic, Djordje, *Istina o mome životu*, Belgrade, 2nd ed, 1988.

Karas, I, *I fysikes kai thetikes epistimes aeon elliniko 18o aiona*, Ekdoseis Gutenberg, 1977.

Kastanis, Andreas, 'The teaching of mathematics in the Greek military academy during the first years of its foundation (1828–1834)', *Historia mathematica*, 30 (2003), 123–139.

Kastanis Iason, and Kastanis Nikos, 'The transmission of mathematics into Greek education, 1800–1840: from individual initiatives to institutionalization', *Paedagogica Historica*, 42 (2006), 515–534.

Kastanis, N, (ed), *Opseis tis neoellinikis mathematikis paideias*, Thessaloniki, 1998.

Kastanis, Nikos, 'American Pestalozzianism in Greek mathematical education 1830–1836' *BSHM Bulletin*, 22 (2007), 120–132.

Kastanis, Nikos, and Lawrence, Snezana, 'Serbian mathematics culture of the nineteenth century', *History and Pedagogy of Mathematics Newsletter*, 59 (2005), 15–19.

Koppe, Carl, *Die Arithmetik, Algebra und allgemeine Grössenlehre für Schulunterricht*, Bädeker, Essen, 1836a.

Koppe, Carl, *Die Planimetrie und Stereometrie für Schulunterricht*, Bädeker, Essen, 1836b.

Koumas, Constantinos, *Seira stoicheiodous ton mathimatikon kai physikon pragmateion ek diaforon syggrafeon syllechtheison*, Vienna, 1807.

Lacaille, Nicolas (Abbé), *Leçons élémentaires de mathématiques*, Paris, 1741.

Lawrence, Snezana, 'Balkan mathematics before the First World War', *BSHM Bulletin*, 4 (2005), 28–36.

Mackridge, Peter, 'The Greek intelligentsia 1780–1830: a Balkan perspective', in Richard Clogg (ed), *Balkan society in the age of Greek independence*, Macmillan, 1981, 63–84.

Minchin, James G C, *Growth of freedom in the Balkan Peninsula*, John Murray, 1886.

Mitrinović, Dragoslav, *Mihailo Petrovic: covek, filozof, matematicar [Mihailo Petrovic: a man, philosopher, mathematician]*, Belgrade, 1968a.

Mitrinovic, Dragoslav, 'Zivot Mihaila Petrovica', *Matematicka biblioteka*, 38 (1968b), 7–32.

Nikolić, Atanasije, *Algebra*, Belgrade, 1839.

Nikolić, Atanasije, *Elementarna geometrija*, Belgrade, 1841.

Obolensky, Dimitri, *The Byzantine Commonwealth: Eastern Europe 500–1453*, Weidenfeld and Nicolson, 1971.

Özervarli, M. Sait, 'Transferring traditional Islamic disciplines into modern social sciences in late Ottoman thought: the attempts of Ziya Gokalp and Mehmed Serafeddin' *The Muslim World*, 97 (2007), 317–330.

Painlevé, Paul, *Les axioms de la Méchanique*, Paris, 1922.

Petković, Djordje, *Abel's theorem proved algebraically and by Riemann's theory of functions*, Vienna, 1893.

Petrović Aleksandar, 'Development of the first hydraulic analog computer', *Archives Internationales d'Histoire des Sciences*, 54 (2004), 97–110.

Petrović, Mihailo, *O asimptotnim vrednostima integrala diferencijalnih jednacina*, Belgrade, 1895.

Petrović, Mihailo, 'Sur l'equation différentielle de Riccati et ses applications chimiques' ['On Riccati's differential equation and its application in chemistry'], *Sitzungberichte der Königlich-Böhmischen gesellschaft der Wissenschaften*, Prague, 1896, 1–25.

Petrović, Mihailo, *Elementi matematičke fenomenologije*, Serbian Royal Academy of Science, 1911.

Petrović, Mihailo, *Les spectres numériques*, Paris, 1919.

Petrović, Mihailo, *Leçons sur les spectres mathématiques*, Paris, 1928.

Petrovic, Mihailo, 'Prilog istoriji jednoga problema teorije funkcija', *Glas*, 134 (1929), 87–90.

Petrović, Mihailo, *Kroz polarnu oblast*, Belgrade, 1932.

Petrović, Mihailo, *U carstvu pirata*, Belgrade, 1933.

Petrović, Mihailo, *Sa okeanskim ribolovcima*, Belgrade, 1935.

Petrović, Mihailo, *Na dalekim ostrvima*, Belgrade, 1936.

Petrović, Mihailo, *Roman o jegulji*, Belgrade, 1940.

Phili, C, 'La reconstruction des mathématiques en Grèce', in N Kastanis (ed), (1998), 303–319.

Rabkin, Y M, 'Attitudes, activities, and achievements: science in the modern Middle East', in Ekmeleddin Ihsanoglu, Kostas Chatzis, and Efthymios Nicolaidis (eds), *Multicultural science in the Ottoman Empire*, Turnhout, 2003, 181–196.

Roudometof, Victor, 'From *rum millet* to Greek nation: enlightenment, secularization, and national identity in Ottoman Balkan society, 1453–1821', *Journal of Modern Greek Studies*, 16 (1998), 11–49.

Smith, Anthony, *The ethnic origins of nations*, Blackwell, 1986.

Snell, Friedrich Wilhelm Daniel, *Leichter Leitfaden der Elementargeometrie und Trigonometrie*, Giessen, 1799.

Somel, S A, *The modernization of public education in the Ottoman Empire (1839–1908)*, Brill, 2001.

Sonyel, Salahi R, 'The protégé system in the Ottoman Empire' *Journal of Islamic Studies*, 2 (1991), 56–66.

Sürmen, Y, Kaya, U, Yayla, H (eds), *Higher education institutions and the accounting education in the second half of the nineteenth century in the Ottoman Empire*, Munich Personal RePEc Archive http://mpra.ub.uni-muenchen.de, 2007.

Tacquet, A, *Elementa euclidea geometriæ planæ ac solidæ*, (books 1–6, 11, 12), Padua, 1694.

Terdimou, M, 'The confrontation of mathematics on behalf of the Eastern Orthodox Church during the Ottoman period', *Multicultural science in the Ottoman Empire*, E Ihsanoglou et al (eds), Brepols, (2003) 53–62.

Terdimou, Maria, 'The confrontation of mathematics on behalf of the Eastern Orthodox Church during the Ottoman period', in Richard Clogg (ed), *Balkan society in the age of Greek independence*, Macmillan, 1981, 53–62.

Tieszen, Richard, *Phenomenology, logic, and the philosophy of mathematics*, Cambridge University Press, 2005.

Toumasis, Charalampos, 'The ethos of Euclidean geometry in Greek secondary education (1836–1985): pressure for change and resistance', *Educational Studies in Mathematics*, 21 (1990), 491–508.

Trifunović, Dragan, 'Prilog istoriji jednoga problema teorije funkcija', *Istorija matematickih I mehanickih nauka*, Belgrade, 1992, 63–70.

Trifunović, Dragan, *Doktorska disertacija Mihaila Petrovica*, Arhimedes, 1994.

Trifunović, Dragan, *Dimitrije Nesic, zora srpske matematike*, Arhimedes, 1996.

Tsourkas, Cléobule, 'Les premiers influences occidentals dans l'Orient Orthodoxe', *Balkania*, 6 (1943), 334–378.

Vucinich, Wayne S, 'The nature of Balkan society under Ottoman Rule' *Slavic Review*, 21 (1962), 597–616.

Vukićević, Petar, *Die Invarianten der linearen homogenen Differential-Gleichungen n-ter Ordnung*, Berlin, 1794.

Zeune, Johann August, *Gea, Versuch einer wissenschaftlichen Erdbeschreibung*, Berlin, 1808.

Zilfi, Madeline C, 'Elite circulation in the Ottoman Empire: great Mollas of the eighteenth century', *Journal of the Economic and Social History of the Orient*, 26 (1983), 318–364.

3. Local

CHAPTER 3.1

Mathematics education in an Old Babylonian scribal school

Eleanor Robson

In the early second millennium BC, southern Iraq—ancient Babylonia—produced large quantities of mathematics, written on small clay tablets in the Akkadian in Sumerian languages using the cuneiform, or wedge-shaped, script. For many decades after its discovery and decipherment in the early twentieth century the study of Old Babylonian (OB) mathematics quite rightly focused on the recovery of knowledge: what was known, and where and when.[1] The 1990s saw a move towards conceptual history: how mathematical language reflected the thought processes behind the techniques. Nevertheless, the corpus of Old Babylonian mathematics was still treated, more or less, as a closed set of disembodied texts: there were few attempts to publish new sources, or to acknowledge that they were recorded on physical objects which could be located in time and space and fruitfully related to other archaeological artefacts. This was in large part due to the academic backgrounds of the small number of core researchers concerned, who were almost exclusively trained in mathematics or the history of science or ideas, and inevitably lacked the technical skills involved in the decipherment of cuneiform tablets or the reconstruction of the archaeological record. Conversely, there

---

1. This chapter is an edited and updated version of Robson (2002). I thank Kai Metzler for permission to reproduce it here.

was a conspicuous lack of cuneiformists willing to do so. It appeared that the great pioneers of cuneiform mathematical studies—Otto Neugebauer, Abraham Sachs, Evert Bruins, and François Thureau-Dangin—had done it all, and there was little left to do but reanalyse their data. The contextual evidence for the material they had published was at best meagre and more commonly non-existent, while the mathematical tablets coming out of more recent excavations were invariably further exemplars of the multiplication tables and lists of weights and measures that Neugebauer had so thoroughly classified in the 1930s and 1940s.

Since the mid-1990s, however, there has been an increasing interest in the material culture of Babylonian scribal schooling, and a growing realization that a wealth of archaeological and artefactual data can be used to counterbalance the traditional sources of evidence, the Sumerian literary narratives about school. This move has gone hand in hand with an increasingly sophisticated approach to textual evidence, which acknowledges that authorial intention was often complex and that literary works in particular cannot be used straightforwardly as a historical source.

This study is situated firmly within that research tradition. It takes as its starting point one single architectural unit and the objects found within it to reconstruct the role of metrology, arithmetic, and mathematics within the curriculum of an individual school. Its aim is not to produce a generalized pedagogical framework for Old Babylonian mathematics, but rather to stress the variety of approaches to mathematics education that existed in the early second millennium BC. Just as modern scholarship is conducted by individuals who are constrained by their environment and education while free to make personal choices about the direction and character of their work, we shall see that ancient education was imparted by people with similar freedoms and constraints.

House F, an eighteenth-century scribal school

House F was excavated in the first months of 1952 by a team of archaeologists from the universities of Chicago and Pennsylvania. It was their third field season in the ancient southern Iraqi city of Nippur and one of their express aims was to find large numbers of cuneiform tablets (McCown and Haines 1967, viii). For this reason they had chosen two sites on the mound known as Tablet Hill, because of the large number of tablets that had been found there in the late nineteenth century. Those previous digs, when the development of recorded stratigraphic archaeology was still in its infancy, had necessarily been little more than hunts for artefacts. The new generation of archaeologists, however, made detailed archaeological records of finds and findspots as a matter of course, so that when they

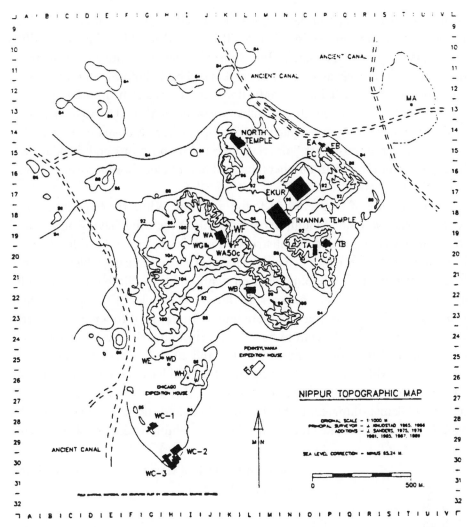

Figure 3.1.1 Topographic map of Nippur. Area TA is south of Inana's temple (Gibson *et al* 2001, fig. 1)

hit upon the large cache of tablets they had been hoping for, the architectural context, stratigraphic location, and physical description of every one of them was noted.[2] They labelled their excavation areas TA and TB (Fig. 3.1.1), deliberately siting TB right next to the pits left by their nineteenth-century predecessors.[3] The tablets, over 1400 of them, were in an unremarkable looking house in the

2. Nevertheless, the 1950s field records still present major problems for researchers: see Zettler (1996, 88–89).

3. Gibson *et al* (2001) give a thorough overview of the excavations at Nippur and their results.

corner of Area TA, one of eight mud-brick dwellings packed into the 20 × 40 m rectangle. Other houses in TA and TB had yielded tablets, but in handfuls or dozens, not in the thousands. The House F tablets were not stored in jars or discarded on the street as some of the others had been, but were part of the very fabric of the house itself, built into the floors and walls and furniture (Fig. 3.1.2). It quickly became apparent that the tablets were not a normal household archive of documents relating to property ownership, debt, and business matters, but comprised in the most part Sumerian literary compositions and standardized lists of signs and words, in numbers that had never before been recovered from a controlled excavation.

The excavators of House F had found a school. While the huge number of school tablets were not enough to confirm this at the time, having been found in secondary context used as construction rubble, the presence of large quantities of unused tablet clay and facilities for soaking and reusing tablets has since been attested in other schooling environments and leaves little room for doubt. The schooling took place, it appears, in the courtyards, loci 192 and 205, where benches and three recycling bins were found. Three small rooms to the northwest, 184, 189, and 191, seem to have been private quarters (a bread oven was discovered in 191, domestic pottery in 184 and 205, and decorative plaques in 191 and 205), while the partially excavated 203 must have served as the entrance hall. The tablets, it seems, were laid down shortly after 1740 BC, the tenth regnal year of Samsu-iluna, king of Babylon and son and successor of Hammurabi. Mud-brick structures like House F needed to be rebuilt or extensively renovated

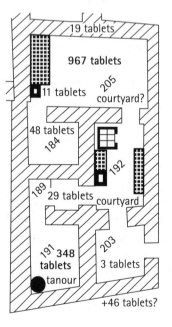

Figure 3.1.2 Plan of House F, level 10 (after Stone 1987, pls. 17–19)

every twenty-five years or so (Gasche and Dekiere 1991) and indeed House F appears to have undergone three or four such remodellings over the course of its life, from the late nineteenth century to about 1721 BC. The use of tablets as building material seems to mark the end of the house's life as a school: when it was later reoccupied the new inhabitants appear to have been engaged in other activities.

After excavation, the finds were distributed between the Oriental Institute in Chicago and the University Museum, Philadelphia. The Iraq Museum in Baghdad also took a share, while sending portions of its allocation to Chicago and Philadelphia on long-term loan for publication purposes. The Chicago loan was returned in the 1980s. The excavation report on TA and TB came out fifteen years after the dig (McCown and Haines 1967) and was later reanalysed in the light of information from about one hundred household archival records from the two sites (Stone 1987). Neither work treated in any detail the school tablets from House F or the rest of the site. Meanwhile, many of those tablets were making their way into critical editions of Sumerian literary and lexical works, sometimes contributing as much as 25 percent of the sources. Only the fragments loaned to Philadelphia were systematically published (Heimerdinger 1979), but even this work consisted solely of sketches of the tablets with no textual commentary, edition, or discussion of the archaeological context. As the tablets were published, it became clear that fragments could be joined to form larger pieces, but that often these joins had to be made virtually, across the three collections. Nevertheless, the preservation in Philadelphia of the original field notebooks containing a complete record of the epigraphic finds as they were excavated, as well as a tablet catalogue drawn up in the 1970s, has made it possible to attempt a reconstruction of the original tablet assemblage. When Baghdad briefly became more accessible to the international community, I started to check those records against the tablets in the Iraq Museum. As this process had only just begun before the 2003 war, there are still some gaps and inconsistencies in the data, which should therefore not be taken as completely accurate. Nevertheless, it is already possible to make some interesting observations and draw some preliminary conclusions about the functioning of the House F school in the mid-eighteenth century BC.

In total, 1425 tablets were recovered from Level 10 of House F, some 9 percent of which are mathematical. All but four of these belong to the tablet typology of elementary schooling, which account altogether for 50 percent of the tablets found in the house. The position, content, style, and purpose of mathematical instruction within the House F elementary curriculum will be discussed in the first two sections below. The bulk of the remaining tablets bear extracts from Sumerian literary compositions (42 percent), which collectively help to shed light on post-elementary mathematical training. This is the topic of the final section.

The remaining 8 percent, household documents and hitherto unidentified fragments, will not come into the discussion.[4]

Metrology in the elementary curriculum

Several types of tablet were used for elementary schooling in Nippur, as classified by a scheme devised by Miguel Civil (e.g., 1995, 2308) to describe lexical lists—standardized lists of signs and words. But, as Niek Veldhuis (1997, 28–39) showed, this tablet typology applies equally to all elementary school exercises, including mathematical ones. It happens that mathematics has survived on just three types of tablet of from House F: the small Type IIIs and the larger Type I and IIs, of which it will be important to distinguish between the flat obverse (Type II/1) and the slightly convex reverse (II/2):

Type I refers to generally large tablets [...], with a full lexical list and a substantial part thereof and nothing else.

   Type II [...] tablets contain divergent material on each of [their] two sides. To the left of the flat side (II/1) there is a carefully written lexical passage extracted from a fuller list, apparently the work of an instructor, while to the right the passage is copied by a student. On the convex side (II/2) of a Type II tablet, there is a multicolumn excerpt from a longer list.

   Type III tablets [...] contain just one column with material extracted from a longer list. (Civil 1995, 2308)

   Because the Type II tablets contain different compositions on the obverse and reverse they can be used to reconstruct the curricular sequence of elementary education in Nippur (Fig. 3.1.4 below). Veldhuis (1997, 40–63) correlated the contents of obverse and reverse on some 1500 Type II tablets from Nippur, working from the hypothesis that they had been written by the same student, who reviewed on the reverse an earlier part of the same composition he was learning on the obverse, or long sections from one he had completed earlier.[5] Veldhuis's results were impressively consistent, enabling him to assign about twenty different compositions to four phases of the elementary curriculum: writing techniques, thematic noun lists, advanced lists, and introductory Sumerian. He discussed the educational function of each phase in turn, showing a steady progression

---

4. Several other assemblages of Old Babylonian school tablets have been identified, all much smaller than the House F corpus (Robson 2002, 329–330). House F falls in the middle of the chronological and geographical spread of these school corpora. Mathematics is present in all of them, in proportions ranging from 5 percent to 16 percent; the 9 percent proportion of mathematics in House F can thus be seen as relatively normal. However, the contents and format of the mathematical tablets in the different assemblages, as far as they can be identified, varies quite remarkably.

5. In some circumstances Old Babylonian women could train to be scribes too: see Lion and Robson (2006); Robson (2007).

from first exposure to the physical form of cuneiform signs, the construction of whole words, and the exploration of the complexities of cuneiform writing, to the use of whole sentences of grammatically correct Sumerian. All phases involved the rote memorization of set texts, mainly of the sort traditionally character- ized in Assyriology as 'lexical texts', namely, lists of cuneiform signs or Sumerian words. But the curriculum also included model legal documents, Sumerian prov- erbs, and—most importantly for our purposes—long sequences of multiplica- tion tables and lists of metrological units. It is impossible, though, on present evidence, to estimate how long the student(s) had been at school, or how old they were, at this or any other point in their educational careers.

Using the same methodology on the 250 or so Type II tablets from House F yields a similarly consistent picture (Table 3.1.1). It differs from Veldhuis' general conclusions primarily in the ordering of the third phase, where mathematical matters are addressed (Robson 2001). Weights and measures were learnt system- atically, by means of a standard series, towards the end of the third phase of the House F curriculum. Multiplication and division facts were memorized imme- diately afterwards. However, metrological matters were first addressed within the second phase, as sequences within the thematic noun list (see below) and later contextualized in the model legal contracts of phase four. Looking at the numbers of tablets attested, it is striking that the series of divisions and multipli- cations is one of the most frequently occurring compositions, while the metro- logical sequence is among the least represented. Why this might be, if it is not simply an accident of preservation, cannot for the moment be determined.

The students' first exposure to metrological notation was in the second phase of elementary education, as sub-sequences within the six-part the- matic noun list. In Division 1, the list of trees and wooden objects, students met the main capacity measures in descending order. Larger capacity measures (c 1,500–18,000 litres) were contextualized as standard sizes of boat, while smaller units (c 0.17–60 litres) were treated later on, in a section of their own (Veldhuis 1997, 157, 163). Weights were treated very briefly in the list of trees and wooden objects, within a five-line section on weighing equipment, but were covered more exhaustively, as stone weights, as a section in the list of stones (Division 4 of the thematic noun list: Veldhuis 1997, 161; Landsberger et al 1970, 60–61). A very few length measures were listed in a section on reed measuring-rods in Division 2 (Landsberger 1959, 191–192), but in general length and area metrol- ogy was not covered, presumably because little of it could be related to the sizes of material objects.

Later in the curricular sequence, the names of some metrological units crop up in the more advanced list now called Proto-Ea, whose function was to list dif- ferent Sumerian readings of single signs. Because the list is ordered by the shapes of the signs, the signs with metrological significance are scattered randomly

Table 3.1.1 The elementary curriculum in House F

	Phase/Composition	Educational function	No. of tablets in House F[6]
	First Phase: writing techniques		*146*
0	Exercises in sign forms	Writing single and combined cuneiform wedges	1
1	Syllable Alphabet B	The proper formation of simple cuneiform signs	70
2	Lists of personal names	The combination of signs into meaningful sense units	82
	Second Phase: thematic noun lists	*Vocabulary acquisition: realia*	*98*
3	Division 1: List of trees and wooden objects		28
4	Division 2: List of reeds, vessels, leather, and metal objects		20
5	Division 3: List of animals and meats		19
6	Division 4: List of stones, plants, fish, birds, and garments		25
7	Division 5: List of geographical names and terms, and stars		6
8	Division 6: List of foodstuffs		7
	Third Phase: advanced lists		*207*
	Nigga }		16
10	Proto-Kagal } order uncertain	Structured by key signs	11
	Proto-Izi }		30
12	Proto-Lu	Thematic vocabulary acquisition: titles and professions	22
13	Proto-Ea	Sumerian readings of signs	17
14	Metrological lists and tables	Weights and measures	15
15	Multiplication and reciprocal tables	Number facts	93
16	Proto-Diri	The readings of compound signs	16
	Fourth Phase: introductory Sumerian		*107*
17	Model contracts	Simple Sumerian prose	54
18	Proverbs	Sumerian literary language	54

6. Of all tablet types, not only Type II. The numbers in the column are not commensurate because of the co-occurrence of different compositions on the Type II tablets.

throughout the 994 entries. For instance the metrological unit *ninda* 'rod' is just one possible reading of the sign GAR:[7]

ni-im₃	*niĝ₂*	The sign GAR can be read as *niĝ*
ĝa₂-ar	*ĝar*	The sign GAR can be read as *ĝar*
in-da	*ninda*	The sign GAR can be read as *ninda*
šu-ku	*šuku*	The sign sequence U GAR can be read as *šuku*
pa-ad	*pad*	The sign sequence U GAR can be read as *pad*
ku-ru-um-ma	*kurum₆*	The sign sequence U GAR can be read as *kurum*.

(After Civil *et al* 1979, 40, ll 208–213)[8]

Immediately after Proto-Ea, it appears, the students of House F moved on to learning the standard metrological series. They had thus already acquired some systematic knowledge of measures in context (in sub-sequences of the thematic noun list), and learnt the contextualized readings of many of the signs for metrological units, before they encountered the system as a whole.

Very little has been studied of the Old Babylonian metrological lists since Neugebauer and Sachs (1945, 4–6) established the organization of the four systems—length, area and volume,[9] weight, and capacity. The standard pedagogical series comprises four sections in the following order and ranges:

Capacity:	1/3 *sila₃* − 1 00 00 *gur*	(5×60^4 *sila₃*)	c 0.3 − 65 million litres
Weight	1/2 *še* − 1 00 *gun*	(3×60^4 *še*)	c 0.05 g − 1,800 kg
Area:	1/3 *sar* − 2 00 00 *bur₃*	(60^4 *sar*)	c 12 m² − 47,000 ha
Length:	1 *šu-si* − 1 00 *danna*	(3×60^4 *šu-si*)	c 17 mm − 650 km (after Friberg 1987–90, 543).

As sources for the metrological history of Babylonia, the series yields nothing more than the approximate sizes of the basic units and the relationships between them. However, when viewed as the product of scribal education it becomes potentially interesting again. Extracts from the series could be written in the form of lists—with each entry containing the standard notation for the measures only, or as tables—where the standard writings were supplemented with their sexagesimal equivalents (Friberg 1987–90, 542–543). For instance, the reverse of

---

7. Metrological units written with more than one sign, such as *ma-na* 'mina' and *šu-si* 'finger' make no appearance in Proto-Ea, and nor do measures such as *ban₂*, *barig*, and *eše₃*, whose units are implicit in the writing of the numerical values.

8. The modern letters ĝ and š represent sounds like 'ng' and 'sh', as for instance in 'shopping'. The subscripts indicate to Assyriologists which sign-forms have been used in the cuneiform.

9. Exactly the same units were used for areas and volumes, volume units being defined as 1 (horizontal) area unit × 1 cubit height (Neugebauer and Sachs 1945, 5).

3N-T 316 contains an extract from the end of the metrological table of lengths (Fig. 3.1.3):[10]

10 *danna* 5	10 leagues = 5 00 00 (rods)
11 *danna* 5 30	11 leagues = 5 30 00 (rods)
12 *danna* 6	12 leagues = 6 00 00 (rods)
13 *danna* 6 30	13 leagues = 6 30 00 (rods)
14 *danna* 7	14 leagues = 7 00 00 (rods)
15 *danna* 7 30	15 leagues = 7 30 00 (rods)
16 *danna* 8	16 leagues = 8 00 00 (rods)
17 *danna* 8 30	17 leagues = 8 30 00 (rods)
18 *danna* 9	18 leagues = 9 00 00 (rods)
19 *danna* 9 30	19 leagues = 9 30 00 (rods)
20 *danna* 10	20 leagues = 10 00 00 (rods)
30 *danna* 15	30 leagues = 15 00 00 (rods)
40 *danna* 20	40 leagues = 20 00 00 (rods)
50 *danna* 25	50 leagues = 25 00 00 (rods)
1 *danna* 30	1 00 leagues = 30 00 00 (rods)

Fifteen tablets with extracts from the standard metrological series survive from House F. Some or all of their contents, tablet type, and compositional format can be determined for twelve of them. Almost all identifiable pieces are Type II/2 tablets. On their obverses are a reciprocal table, sections of Proto-Diri, model contracts, and Sumerian proverbs: metrology thus preceded these topics in the House F curriculum. One Type II/1 table of weights has an extract from Proto-Izi on the reverse: metrology thus followed this composition in the House F curriculum. The other fragments also appear to have come from Type I or Type II tablets; there is no metrology surviving on tablet type III. Of the six tablets whose contents and compositional format are identifiable, all but one are from the start of the sequence, but there is an even split between tabular and list format.

In short, on present evidence little can be said about the standard metrological series within House F, except that its position in the curriculum can be established, and Type II/2 extracts from the beginning of the compositional sequence apparently predominate the meagre extant record. But it is impossible to determine whether the list and tabular formats had distinct pedagogical functions; neither is there much to be deduced from comparative material (primarily because it

---

10. 1 *ninda* or rod = 6m; 1 *danna* or league = 1800 *ninda* = 10.8 km.

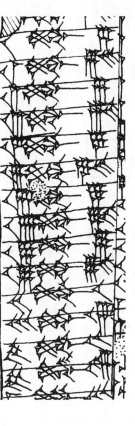

Figure 3.1.3   3N–T 316 = A 30211. Detail of reverse (lines II 3–18), showing large length measures

Figure 3.1.4 3N–T 594 = IM 58573. The obverse of the Type II tablet (left) shows a teacher's copy of the list of reciprocals, with the student's copy to the right erased. The reverse (right) is an extract from the standard metrological list, with capacity measures from 12 to 19 *gur* and 3000 to 360,000 *gur*

is under-published). However, we can do a great deal more with the much more abundant remains from the standard arithmetical series which immediately followed it in the House F curriculum. First, though, we will jump ahead to the end of the elementary curriculum to examine the use of metrology in model contracts.

Towards the end of elementary education in House F students were introduced to whole sentences in Sumerian for the first time, in the form of model legal contracts. The genre as a whole, although apparently a relatively common element in scribal schooling, has not yet been studied in depth. The contracts from House F concern grain and silver loans, inheritance divisions, and sales of slaves and houses. All of them use metrological units in quasi-realistic contexts, as the following example shows:

1/3 *sar* (c 12 m²) of built-up house, next to Dingir-gamil's house

25 *sar* (c 900 m²) of field, the ruin mound of Ahuni, bordering Dingir-gamil's (land)

12 1/2 *sar* (c 450 m²) date orchard of the royal waterway field next to Dingir-gamil's date orchard

[1] large [offering table], 1 wooden pot stand for beer,

[3 wooden spoons] and a quarter of its (i.e., the estate's) equipment.

Apil-ilishu's share.[11] (3N-T 342 = IM 58436, lines 10'–16')

11. 1/3 sar e₂-du₃-a da e₂ diĝir-ga-mi-il / 25 sar a-šag₄ du₆ a-hu-ni uš-a-du diĝir-ga-mil / 12 1/2 sar ĝiš-kiri₆ <a>-šag₄ id₂ «lugal» lugal / zag ĝiš-kiri₆ diĝir-ga-mil / [1 ĝiš-banšur-zag]-gu-la 1 ĝiš-ga-nu-um-kaš / [3 ĝiš-dilim₂] u₃ niĝ₂-gu₂-[un]-a igi-4-ĝal₂-bi / ha-la-ba a-pil₂-i₃-li₃-šu.

In short, some aspects of metrology ran right through the elementary curriculum in House F. However, the focus appears to have been on memorization and contextual use; there is no evidence that the House F students practised metrological conversions or calculations of any kind.

Arithmetic

The standard list of multiplications was described long ago by Neugebauer (1935–7, I 32–67; Neugebauer and Sachs 1945, 19–33) and is very well known. Nevertheless, it is useful to summarize its salient features from an educational standpoint. Systematic differences in content and textual format across tablet types reflect their pedagogical function, while regular omissions from the standard list suggest one or two idiosyncrasies particular to House F.

The series starts with a list of one- and two-place reciprocal pairs, encompassing all the regular integers from 2 to 81. It is followed by multiplication 'tables' for sexagesimally regular head numbers from 50 down to 1 15, with multiplicands 1–20, 30, 40, and 50.[12] Some series also include the squares and inverse squares of each head number. Neugebauer reconstructed the standard sequence on the basis of what he called 'combined multiplication tables', that is, several tables on one tablet—or in curricular terminology long extracts of the standard series on tablet types I and II/2. What he called 'single multiplication tables' turn out to be tablet types II/1 and III.

Neugebauer also identified three main textual formats for multiplications, and four less common variants (Neugebauer and Sachs 1945, 20). We could call Neugebauer's Types A and A' *verbose* formats, in that they repeat the word a-ra$_2$ 'times' in every line of each table (h a-ra$_2$ 1 h, a-ra$_2$ m hm "h times 1 is h, times m is hm"). His Types B, B', B'', C, and C', however, are all *terse*, as a-ra$_2$ 'times' makes at most one appearance in the first line; thereafter the text is entirely numerical (m hm). In fact, it turns out that the formats so far attested in House F are all either Type A or Type C; for that reason they will be referred to simply as Verbose and Terse formats, to prevent the confusing proliferation of Types in the discussion (Fig. 3.1.5). Analogously, the reciprocal tables at the head of the series may be in Verbose format (igi-n-ĝal$_2$-bi 1/n 'Its nth part is 1/n', e.g., Fig. 3.1.4, above) or Terse (n 1/n) (cf. Neugebauer and Sachs 1945, 12).

Of the 97 House F tablets currently known to contain extracts from the standard multiplication sequence, 32 can be identified as Type III, 38 as Type II, and 10 as Type I. Nine fragments may be from Type I or Type II tablets and the

---

12. In the following paragraph, I abbreviate 'head number' as h and 'multiplicand' as m. I have put the word 'tables' in inverted commas because these tablets are not laid out as formal tables with columnar divisions but as lists like the bulk of the rest of elementary school subject matter (see Robson 2003).

Figure 3.1.5 3N–T 261 = UM 55-21-289 (obverse), a verbose Type III multiplication table for 1;40 (left), and 3N–T 608 = UM 55-21-360 (obverse), a terse Type III multiplication table for 3 (right)

typology of the remaining seven is unknown. Eleven of the twenty-five probable Type II/2 tablets have identifiable compositions on their obverses: eight are tables from towards the end of the multiplication series, while there is one model contract, one sequence of Sumerian proverbs, and one composition yet to be distinguished. There are 21 multiplication tables on Type II/1 tablets; apart from the eight multiplication reverses just mentioned, one exemplar each of the thematic noun list (Division 4), Proto-Lu, Proto-Izi, and a metrological list (Fig. 3.1.4) have been identified.

Why were three different types of tablet used to record the multiplication series in House F? Looking first at the 34 tablets which bear just one identifiable multiplication or reciprocal table each, namely Types II/1 and III, attested tables are scattered apparently randomly through the series: there are nine tablets from the first quarter of it, eight from the second, nine from the third, and seven from the last, and there is little difference between the two tablet types. The picture that emerges from the tablets containing longer extracts from the series is very different, however. On both the Type II/2 and the Type I tablets the attested tables are predominantly from the first quarter of the series, namely 18 of the 25 Type II/2 tablets and six of the ten Type Is (69 percent in total). All but one of the remainder are from the second quarter, where it appears that there was a formal section break between the tables for 20 and 18.

This distribution is not peculiar to House F. A simple analysis of the Old Babylonian 'combined multiplication tables' published by Neugebauer before House F was excavated reveals a striking similarity (Neugebauer 1935–7, I 35, II 37; Neugebauer and Sachs 1945, 25–33). Of the 70 tablets he listed, 51 of them (72 percent) apparently begin their sequences of multiplications in the first quarter of the series, six in the second quarter, eight in the third, and five in the last.[13] On the other hand, the number of Neugebauer's 159 'single multiplication tables' (Neugebauer 1935–37, I 34, II 36, Neugebauer and Sachs 1945, 20–23) decreases more or less linearly across the series: 56 are from the first quarter, 44 from the second, 34 from the third, and 25 from the last. While this pattern of attestation does not exactly match the even distribution of tablet types II/1 and III in House F, it is clearly distinct from the heavy skew towards the beginning of the series found in the 'combined' multiplication tables (Tablet Types I and II/2) from House F and elsewhere.

Neugebauer (1935–37, I 62–64) highlighted the strong correlation between tablet type and textual format: some 80–90 percent of his 'combined' multiplication tables (depending on how one defines and counts the tables) are in terse formats and the remainder are verbose. Conversely, about 70–80 percent of the 'single' multiplication tables are verbose and the rest terse. Once again we find similar results in the House F corpus, where formats can be identified: 31 of the 35 Type I and Type II/2 tablets (89 percent), bear tersely formatted tables, while 29 out of the 34 Type II/1 and III tablets (85 percent), are verbose.

In sum, there are two clearly marked distinctions between the 'single' multiplication tables on the one hand and the 'combined' tables on the other. On the one hand, the single tables (on tablet Types II/1 and III) are evenly distributed across the whole series (but with some skew towards the beginning in Neugebauer's sample) and are predominantly verbosely written, while the longer extracts containing sequences of tables are very heavily weighted towards the start of the series and are generally terse. One can also make a further differentiation: it is generally true that the 'single' tables are written in a careful, calligraphic hand with clear line spacing, while the long extracts comprising many tables appear to have been written with little regard for visual appearance: there are generally no line rulings, for instance, and even the columnar divisions are often difficult to make out.

---

13. However, it is difficult to judge from the descriptions given by Neugebauer and Sachs (1945, 25–33) whether the tablets are fragments or not, and therefore whether complete sequences are attested on them.

These three factors combine to suggest a clear pedagogical distinction between the well written, fully worded single tables on the one hand and the hastily scribbled, terse sequences of tables on the other. We have already reviewed Veldhuis's hypothesis that Type II tablets had a dual function: on the obverse (II/1) the student repeatedly copied the teacher's model of an extract (or table) that he was learning for the first time, and then on the reverse (II/2) wrote out a much longer extract from earlier in that same composition, or from one he had already mastered. The evidence from the standard series of multiplication tables presented here not only allows us to confirm that hypothesis but also to draw some further conclusions. First, it appears that Type III tablets were also used in the initial stages of learning an extract, presumably after the student had memorized it well enough to no longer need a model to copy in the Type II/1 pattern. Equally, the Type I tablets appear to have served a similar revision purpose to the Type II/2 tablets, on which students reviewed long stretches of material they were no longer actively working on, or perhaps fitting their most recent achievements into their place in the compositional sequence. Second, and perhaps more interestingly, it seems that while students were given initial exposure to the whole of a composition, by means of short extracts on tablet Types II/1 and III, their revision of that work was much less systematic, starting from the beginning again each time and rarely reaching the end.

This distribution of tablet types across the series is found in other elementary educational compositions too. It is comparable, for instance, to the survival patterns of Old Babylonian tablets from Nippur containing extracts from division one of the thematic noun list, the trees and wooden objects. Dividing the tablet types into their functions of 'first exposure' (Types II/1, and III) and 'revision' (Types I and II/2), we see that there are never more than four 'first exposures' for any one of the lines sampled but more often one or none. Conversely, the 'revision' tablets are very heavily weighted indeed towards the beginning of the composition (taking into account the commonly occurring damage to the corners of tablets which has lowered the number of attestations for the very first two or three lines) (Veldhuis 1997; Robson 2002). In other words, this suggests that although elementary students in Nippur tended to be taught compositions in their entirety, from beginning to end, all revision in the elementary curriculum was slanted towards the opening sections of compositions to the detriment of their middles and closing lines.

Returning to the standard series of multiplications as attested in House F, nine of the 40 known head numbers—namely 48, 44 26 40, 20, 7 12, 7, 5, 3 20, 2 24, and 2 15—do not survive on known tablets. Should we attribute these omissions to the accidents of recovery or to deliberate exclusion from the series? The patterns of attestation make it easier to make definitive statements about the higher head numbers than the lower. The head number 48, for instance, is included in just five of the 71 'combined' tables catalogued by Neugebauer (two of those five are from Nippur), compared to 23 certain omissions. He lists no 'single' tables for 48. Similarly, 2 15 occurs in two out of nine possible 'combined' tables, neither of them from Nippur, and in no 'singles'. It is not surprising, therefore, that the 48 and 2 15 times tables were apparently not taught in House F. The exclusion of 44 26 40, is rather more surprising: given its place near the start of the standard series it is presumably not simply missing by archaeological accident. On the other hand none of Neugebauer's 'combined' tables appear to omit it, while he lists three 'single' tables for 44 26 40. This is a deliberate but idiosyncratic omission then, particular to House F—though perhaps a judicious one; none of the other head numbers are three sexagesimal places long. It is probably best to reserve judgement on the remaining six 'missing' head numbers.

Mathematical imagery in Sumerian literature

As we have seen, the vast majority of mathematical tablets in House F can be assigned to the elementary curriculum on grounds of content and tablet typology, but there are four which cannot be. Three of those tablets bear calculations, while the fourth contains an extra-curricular table. Although the table is difficult to place pedagogically, it is possible to position the calculations within the 'advanced' curriculum, which in House F was dominated by Sumerian literature. First, however, we need to review what is known of the Sumerian literary curriculum in House F.

Over eighty different literary works have survived from the House, attested on around six hundred different tablets. Although we do not have a clear-cut tablet typology from which to deduce a well defined and ordered curriculum, it is possible to at least outline the contents of that curriculum, based on contemporaneous literary catalogues and some basic quantitative methods (Tinney 1999; Robson 2001). First, by simply counting the number of sources for each composition, it becomes clear that there is one 'mainstream' group of twenty-four literary works, each with a mean of eighteen sources, compared to the rest which have on average just three attestations. Second, ten of those twenty-four 'mainstream' works comprise a widely-attested curricular grouping that Steve Tinney (1999, 168–170) has labelled the Decad. The remaining members of that mainstream grouping, which I have called the House F Fourteen (Robson 2001), appear on three of those same catalogues, in a fixed order though not clustered together in a single block like the Decad.[14] The remaining House F literature can be roughly categorized into four groups (Robson 2001): myths, epics and laments

14. Outside House F, the Decad members are found on an average of 41 Nippur tablets each and 35 non-Nippur tablets. For each of the Fourteen there are, on average, 30 Nippur tablets (outside House F) and 10 from beyond Nippur.

(13 works), hymns to kings and deities (11), school narratives, debates, and dialogues (7), and literary letters and related short pieces (25).

At first glance this seems a very eclectic mix of genres and subject matter, but it has recently become clear that they were not randomly selected for use in House F but were chosen with particular purposes in mind. On average the literary works in House F contain ten times as many references to literacy and numeracy as non-pedagogical works of OB Sumerian literature (Robson 2007). Sumerian literary references to mathematical achievement and failure have been collected before, usually in a misguided attempt to use literary works as unproblematic sources of historical evidence about 'Sumerian school' (e.g., Sjöberg 1975; Nemet-Nejat 1993, 5–10). However, once we recognize that those literary works were themselves elements of a scribal curriculum, as for instance in House F, it becomes interesting and important to study them for the messages that they conveyed to the students about mathematics and the scribes' relationships to it.

Mathematical and metrological elements appear in some of the humorous narratives and dialogues about school life (the so-called *eduba* texts, named after the Sumerian word for school). Although we can occasionally verify that particular details in the narratives are in some sense 'true' in that they concur with other evidence, they are highly unlikely to have been straightforward documentary accounts: after all, their intended audience, the scribal students, already knew exactly what school was like. The narratives often make use of very broad humour to get their message across (or at least broad humour is the only type that we, with our unsophisticated understanding of Sumerian, can currently understand). It may be that other elements of humour lay in the contrast between school life as depicted and as experienced by the students; in that case those apparently realistic details would have served simply to add elements of verisimilitude to otherwise highly fictionalized accounts.[15]

In the most famous of these works, often known by its modern title 'Schooldays', the teacher of an incompetent scribal student is invited home for dinner and bribery, in an attempt to make him ease up on the hapless child. The father flatters the stern teacher shamelessly, saying:

My little fellow has opened (wide) his hand, (and) you made wisdom enter there; you showed him all the fine points of the scribal art; you (even) made him see the solutions of mathematical and arithmetical (problems).[16] (*Eduba* composition A 59–61, after Kramer 1963, 239)

---

15. Compare Hogwarts, the boarding school for wizards in training, in the highly popular children's novels and films about Harry Potter. No child reader has ever set foot in an institution anything like Hogwarts, yet it is still recognizably a school. Its fascination and attraction lie in the judicious combination of realism, fantasy, and humour with which the stories are constructed—just as in the Sumerian school narratives. This, of course, is where the similarity ends.

16. lu₂-tur-ĝu₁₀ šu-ni i-ni-in-ba₉-ra₂ kug-zu i-ni-in-kur₉-ra / nam-dub-sar-ra niĝ₂-galam-galam-ma-bi mu-ni-in-pad₃-pad₃-de₃-en / šag₄-dub-ba šid niĝ₂-kas₇ ki-bur₂-bur₂-ra-bi igi mu-un-na-an-sig₉-ga-aš.

An earlier passage in the narrative, however, makes it clear that the teacher had showed him little except the business end of his cane.

In a gentler companion piece, now sometimes called 'Scribal activities', a teacher quizzes a student on what he has learnt, some three months before he is due to leave school. The student lists everything he has mastered so far, much of which can be matched quite closely to the evidence from the archaeologically recovered elementary tablets themselves. (This is hardly surprising, as one aim of the composition must have been to encourage identification with, and emulation of, this paragon of learning.) The standard metrological lists are as closely associated with the model contracts here as they are in the elementary curriculum itself.

In the final reckoning, what I know of the scribal art will not be taken away. So now I am master of the meaning of tablets, of mathematics, of budgeting, of the whole scribal art....

I desire to start writing tablets (professionally): tablets of 1 *gur* of barley all the way to 600 *gur*; tablets of 1 shekel all the way to 20 minas. Also any marriage contracts they may bring; and partnership contracts. I can specify verified weights up to 1 talent, and also deeds for the sale of houses, gardens, slaves, financial guarantees, field hire contracts..., palm growing contracts..., adoption contracts—all those I can draw up.[17] (*Eduba* composition D 27–29, 40–48, after Vanstiphout 1997, 592–3; Friberg 1987–90, 543)

A third piece is often known as 'The dialogue between Girini-isag and Enki-manshum' although it is more of a rumbustious slanging match, in which the advanced student Girini-isag belittles and humiliates his younger colleague Enki-manshum (whose defences are often rather ineffectual):

[19–27](*Girini-isag speaks*): 'You wrote a tablet, but you cannot grasp its meaning. You wrote a letter, but that is the limit for you. Go to divide a plot, and you are not able to divide the plot; go to apportion a field, and you cannot even hold the tape and rod properly; the field pegs you are unable to place; you cannot figure out its shape, so that when wronged men have a quarrel you are not able to bring peace but you allow brother to attack brother. Among the scribes you (alone) are unfit for the clay. What are you fit for? Can anybody tell us?'

[28–32](*Enki-manshum replies*): 'Why should I be good for nothing? When I go to divide a plot, I can divide it; when I go to apportion a field, I can apportion the pieces, so that when wronged men have a quarrel I soothe their hearts and [...]. Brother will be at peace

---

17. nig$_2$-kas$_7$-bi ĝar-ra nam-dub-sar i$_3$-zu-a-ĝu$_{10}$ nu-ub-tum$_3$-da(?) / a-da-al-ta šag$_4$-dub-ba a-ra$_2$ nig$_2$-kas$_7$-še$_3$ ba-e-de$_3$-ĝa$_2$-ĝa$_2$-de$_3$-en / nam-dub-sar-ĝu$_{10}$ nig$_2$-ĝar nig$_2$ nu-u$_3$-ĝar NIĜ$_2$ ba-ba e-ne KA dub sar-re-de$_3$ ga-ĝen / dub 1 še gur-ta zag 600 gur-še$_3$ / dub 1 giĝ$_4$-ta zag kug 10 ma-na-še$_3$ / ki nam-dam-ta(?) lu$_2$ hu-mu-un-DU-[(X)] / nam-tab-ba 1 gu$_2$ kur$_7$ igi hu-mu-da-zaĝ$_3$-[X] / e$_2$ a-šag$_4$ ĝiškiri$_6$ saĝ-geme$_2$-arad$_2$ sa$_{10}$-sa$_{10}$-[(X)] / kug-ta gub-ba-aš a-šag$_4$ nam-apin-la$_2$ [(X)] / ĝiškiri$_6$ ĝiš gub-bu-de$_3$ [...] X / u$_3$ dub dumu-tul$_2$-ta-pad$_3$-da sar-re-[bi] mu-un-[zu].

with brother, their hearts [...].'[18] (Following lines lost). (*Eduba* dialogue 3, Vanstiphout 1997, 589)

Girini-isag's point is that accurate land surveys are needed for legal reasons—inheritance, sales, harvest contracts, for instance. If the surveyor cannot provide his services effectively he will unwittingly cause disputes or prevent them from being settled peacefully.

For the scribal students in House F these passages, and others, helped to define the role of mathematical training within their education. The first extract implies that a truly competent teacher can help even the most hopeless student understand difficult subjects like mathematics. The second outlines what successful students can hope to achieve in the appropriate application of metrological knowledge to legal documents of various kinds, while the last warns of the humiliations of practical incompetence. It is not enough, Girini-isag implies, to have learnt your school exercises well if you are physically incapable of putting them into practice.[19]

Two royal praise poems, widely used in the early stages of the Sumerian literary curriculum (Vanstiphout 1979; Tinney 1999, 162–168), cite mathematical achievement within the repertoire of a good king's accomplishments, bestowed on him by Nisaba, the patron goddess of scribes. Their message to the students is that literacy and numeracy are very desirable skills, valued so highly that even kings boast about acquiring them. The following extract from a linguistically elementary hymn to Lipit-Eshtar of Isin (c 1934–1924 BC) addresses the king as one who is divinely aided in his literacy and divinely endowed with measuring equipment:[20]

Nisaba, woman sparkling with joy, righteous woman, scribe, lady who knows everything: she leads your fingers on the clay, she makes them put beautiful wedges on the tablets, she makes them (the wedges) sparkle with a golden stylus. A 1-rod reed and a measuring rope of lapis lazuli, a yardstick, and a writing board which gives wisdom: Nisaba generously bestowed them on you.[21] (Lipit-Eshtar hymn B 18–24, after Black *et al* 1998–2006, no. 2.5.5.2)

---

18. dub i_3-sar dim_2-ma-aš nu-e-kur_9 / u_3-na-a-dug_4 i_3-sar ki-$šer_{11}$ a-ra-ab-tuku / e_2 ba-e-de_3 ĝen-na e_2 nu-mu-da-ba-e-en / a-$šag_4$ sig_9-ge-de_3 ĝen-na $eš_2$-$gana_2$ gi-1-nindan nu-mu-da-ha-za-an / ^{ĝiš}gag-a ki nu-e-da-du_3-en dim_2-ma nu-e-ni-kur_9 / lu_2 dug_{14}-mu_2-a-ba zi li-ib_2-ib_2-gi_4-gi_4-in / šeš šeš-da $teš_2$-bi bi_2-ib_2-dab_5-be_2-en / lu_2-IM nu-ub-du_7 dub-sar-re-e-ne / a-na-$še_3$ ba-ab-du_7-un me-$še_3$ lu_2 he_2-en-tum_2-mu a-na-aš niĝ$_2$ na-me-$še_3$ la-ba-ab-du_7-un / e_2 ba-e-de_3 ga-ĝen e_2 mu-da-ba-e-en / a-$šag_4$ sig_9-ge-de_3 ga-ĝen ki sig_9-ge-bi mu-zu / lu_2 dug_{14}-mu_2-a-bi $šag_4$-bi ab-huĝ-e zi mu-da-gi_4-gi_4-in / šeš šeš-da $teš_2$-bi bi_2-ib-dug_3-ge-en $šag_4$ mu-da-sed_4-de_3-en.

19. *Eduba* composition A is on eighteen tablets from House F; it is the tenth member of the House F Fourteen. *Eduba* dialogue 3 is on three tablets. No House F sources have yet been identified for *Eduba* composition D but the whole composition is not yet in the public domain.

20. Attested on three tablets from House F (and on tablets from other sources).

21. ^{d}nisaba munus ul-la gun_3-a / munus zid dub-sar nin niĝ$_2$-nam zu / si-zu im-ma si ba-ni-in-sa_2 / $šag_4$ dub-ba-ka gu-$šum_2$ mi-ni-in-sag_9-sag_9 / gi-dub-ba kug-sig_{17}-ka šu mu-ni-in-gun_3 / gi-1-nindan $eš_2$-$gana_2$ za-gin_3 / ĝiš-as_4-lum le-um igi-ĝal$_2$ $šum_2$-mu ^{d}nisaba-ke_4 šu daĝal ma-ra-an-dug_4.

Goddesses carry divine measuring equipment in several myths from House F too. Ninlil receives a rod and rope as a wedding present ('Enlil and Sud', Black *et al* 1998–2006, no. 1.2.2); Inana refuses to relinquish hers until she reaches the penultimate gate of the Underworld ('Inana's Descent', Black *et al* 1998–2006, no. 1.7.1). Male deities, by contrast, are never shown measuring or counting in the House F literature, and tend to use capacity measures as beer mugs (Robson 2007).

A few of the Sumerian literary works use metrological concepts as an essential part of their narrative framework. For instance, a thirty-three-line fictionalized letter from Ishbi-Erra (first king of the Isin dynasty, c 2017–1985 BC) to Ibbi-Suen, last king of Ur (c 2028–2004 BC), describes how he, while still in the latter king's service, has been sent north to buy grain in order to alleviate the famine in the south, but is held back by incursions of nomadic Martu people:[22]

Say to Ibbi-Suen, my lord: this is what Ishbi-Erra, your servant, says:

You ordered me to travel to Isin and Kazallu to purchase grain. With grain reaching the exchange rate of 1 shekel of silver per *gur*, 20 talents of silver have been invested for the purchase of grain.

I heard news that the hostile Martu have entered inside your territories. I entered with 72,000 *gur* of grain—the entire amount of grain—inside Isin. Now I have let the Martu, all of them, penetrate inside the Land, and one by one I have seized all the fortifications therein. Because of the Martu, I am unable to hand over this grain for threshing. They are stronger than me, while I am condemned to sitting around.

Let my lord repair 600 barges of 120 *gur* draught each; 72 solid boats, 20......, 30 bows, [40] rudders (?), 50......and 60 (?) boat doors on the boats (?), may he also......all the boats.[23] (lines 1–16, after Black *et al* 1998–2006, 3.1.17)

The letter reads suspiciously like an OB school mathematics problem: the first paragraph gives the silver–grain exchange rate and the total amount of silver available (72,000 shekels); in the second the silver has been correctly converted into grain. Next that huge capacity measure is divided equally among large boats (cf. the contextualized large capacity measures in the list of trees and wooden objects). As is typical for school mathematical problems, the numbers are conspicuously round and easy to calculate with (Friberg 1987–90, 539). The letter, at one level, is no more than a pretext to show simple mathematics and metrology at work in a quasi-realistic context.

---

22. One attestation from House F; several other sources known.

23. ^{d}i-bi$_2$-^{d}suen lugal-ĝu$_{10}$-ra u$_3$-na-a-dug$_4$ / ^{m}iš-bi-er$_3$-ra arad-zu na-ab-be$_2$-a / kaskal i$_3$-si-in^{ki} ka-zal-lu^{ki}-še$_3$ / še sa$_{10}$-sa$_{10}$-de$_3$ a$_2$-še$_3$ mu-e-da-a-aĝ$_2$ / ganba 1 še gur-ta-am$_3$ še sa$_2$ di / 20 gun$_2$ kug-babbar še sa$_{10}$-sa$_{10}$-de$_3$ ba-an-ĝar / inim mar-tu lu$_2$-kur$_2$-ra šag$_4$ma-da-zu kur$_9$-ra ĝiš bi$_2$-tuku / 72000(ŠAR$_2$×MAN) še gur še du$_3$-[a]-bi$^?$ šag$_4$ i$_3$-si-in^{ki}-na-ke$_4$ ba-an-kur$_9$-re-en / a-da-al-la-bi mar-tu du$_3$-du$_3$-a-bi šag$_4$ kalam-ma-še$_3$ ba-an-kur$_9$-re-en / bad$_3$$^?$ gal-gal dili-dili-bi im-mi-in-dab$_5$-dab$_5$ / mu mar-tu še-ba sag$_3$-ge nu-mu-e-da-šum$_2$-mu / lugal-ĝu$_{10}$ 600 ^{ĝiš}ma$_2$-gur$_8$ 120 gur-ta-am$_3$ he$_2$-em-du$_8$-e / ^{ĝiš}ma$_2$ 72 kalag-ga 20 ^{ĝiš}ZA PI X 30 X X GAN / 50 ^{ĝiš}X 60$^?$ ^{ĝiš}ig ma$_2$ ugu ma$_2$$^?$ ĝa$_2$-ĝa$_2$ / u$_3$ ^{ĝiš}ma$_2$ du$_3$-a-bi he$_2$-X.

A longer composition now known as 'The farmer's instructions'[24] uses school mathematics in a very different way. Ostensibly it is a description of the agricultural year from irrigation to harvest, but it is hardly pastoral in tone. Central to its whole rationale are the standard work obligations by which state institutions of the twenty-first century BC measured out agricultural labour to contract managers and their work gangs (Civil 1994, 75–78, Robson 1999, 138–66). A short extract from the 111-line composition is enough to catch its flavour:

The plough oxen will have back-up oxen. The attachments of ox to ox should be loose. Each plough will have a back-up plough. The assigned task for one plough is 180 *iku* (c.65 ha), but if you build the implement at 144 *iku* (c.2 ha), the work will be pleasantly performed for you. 180 (?) *sila* of grain (c.180 litres) will be spent on each 18 *iku* area (c.6 1/2 ha).

After working one plough's area with a *bardil* plough, and after working the *bardil* plough's area with a *tugsig* plough, till it with the *tuggur* plough. Harrow once, twice, three times. When you flatten the stubborn spots with a heavy maul, the handle of your maul should be securely attached, otherwise it will not perform as needed.[25] (lines 23–34, Black *et al* 1998–2006, no. 5.6.3)

'The farmer's instructions' is reminiscent of a small group of Sumerian literary compositions studied by Niek Veldhuis (2004). He highlighted the intimate lexical and pedagogical relationship between the standard list of fish and birds (division four of the thematic noun list) and two works now known as 'Home of the fish' and 'Nanshe and the birds' (Black *et al* 1998–2007, nos. 5.9.1, 4.14.3). But whereas they provide a literary framework for naming and describing fish and birds, 'The farmer's instructions' sets out to sugar the bitter pill of learning agricultural work rates. It was probably several hundred years behind contemporary scribal practice by the time it was taught in House F, but so was much of the other literature taught there (as can be seen from the regnal dates of the kings referred to in the extracts quoted in this section).

From memorization to calculation

By a great stroke of fortune, one tablet has survived from House F that bears both Sumerian literature and a mathematical calculation. They are on the same sort of tablet as the elementary Type III, which was commonly used to write

---

24. It is the thirteenth member of the House F Fourteen (and well attested elsewhere in Nippur).

25. gud-ĝišapin gud dirig-ga a-ab-tuku-a / gud gud-da dur bi₂-ib-tu-lu-a / ĝišapin-bi ĝišapin-na a-ab-dirig / eš₂-gar₃ ĝišapin 1-e 180 iku-am₃ / za-e 144 iku ĝiš du₃-ba-ab / a₂ʾ šag₄ hul₂-gin₇ a-ra-ab-dim₂-e / 18 iku-ba 3 še gur ba-an-ĝa₂-ĝa₂ / usu ĝišapin 1(DIŠ)-e a-šag₄ ĝišbar-dili-bi u₃-bi₂-ak / ĝišbar-dili-bi ĝišapin-tug₂-saga₁₁ u₃-bi₂-ak tug₂-gur-ra-ab / ĝiš ur₃-ra-ab ĝiš gi₄-a-ab ĝiš peš-bi₂-ib / ki sumur-bi ĝišniĝ₂-gul-ta ĝišgag-dag₂-aš u₃-bi₂-ak / ĝišmud ĝišniĝ₂-gul-zu saĝ-za he₂-ha-za niĝ₂al di-še₃na-du-un.

single-column extracts of up to sixty lines of literary works (and for that reason called Type S in this context (Tinney 1999, 160). The literary extract is from the first lines of a composition now known as 'The advice of a supervisor to a younger scribe', one of the curricular grouping discussed above whose fictionalized setting is the school and whose aim is to instil professional identity and pride into trainee scribes (Black *et al* 1998–2006, 5.1.3). Most of the reverse is taken up with a calculation of regular reciprocal pairs (Robson 2000, no. 2) using a method that Sachs (1947) called The Technique (Fig. 3.1.6):

17 46 40	9
2 40	«2» 22 30
3 22 30	[2]
6 4[5]	
9	6 40
8 53 20	
17 46 40	

Other tablets with similar arrangements of numbers are known, as well as one very damaged tablet of unknown provenance which originally contained twelve

Figure 3.1.6 3N–T 362+366 (reverse) (Robson 2000, fig. 2)

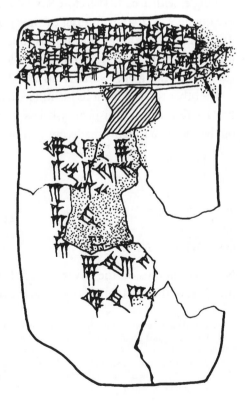

worked examples with instructions.[26] Like much Old Babylonian mathematics, although it first appears to be analogous to modern algebraic operations it can in fact be best understood in terms of very concrete manipulations of lines and areas (Høyrup 1990; 2002). The best preserved of the twelve problems runs as follows:

What is the reciprocal of 2;[13] 20?[27] [You, in your] working: Find the reciprocal of 0;03 20. [You will see] 18. Multiply 18 by 2;10. [You will see] 39. Add 1. [You will see] 40. Take the reciprocal of 40. [You will see] 1 30. Multiply 1 30 by 18. You will see 27. Your reciprocal is 27. [That is the method.][28] (VAT 6505, II 8–16. Neugebauer 1935–37, I 270–273, II pls. 14, 43; Sachs 1947, 226–227)

We can plug the numbers from our House F tablet into this solution. The product of any reciprocal pair is, by definition, 1. We can therefore imagine 17;46 40 as the side of a rectangle whose area is 1 (Fig. 3.1.7a); the task is to find the length of the other side. We can measure off a part of the first side, so that it has a length that is in the standard reciprocal table—in this case 0;06 40, whose reciprocal is 9. We can thus draw another rectangle with lines of these lengths, whose area will also be 1 (Fig. 3.1.7b). This gives us an L-shaped figure. We can fill it in to make a rectangle by multiplying the 9 by 17;40, the part of the original length that we haven't used yet—2 39 (Figure 3.1.7c). Add 1, the area of the 9 by 0;06 40 rectangle. The total area is 2 40. This new large rectangle, 9 by 17;46 40, is 2 40 times bigger than our original rectangle, with area 1. Therefore 9 is 2 40 times bigger than our mystery reciprocal. We divide 9 by 2 40 by finding the inverse of 2 40—0;00 22 30—and multiplying. The reciprocal we wanted to find is thus 9 × 0;00 22 30 = 0;03 22 30 (Fig. 3.1.7d). This is the number in the middle of the calculation. The scribe then checks his result by working backwards from 0;03 22 30 to 17;46 40 again.

The other two calculations identified so far on House F tablets are also attempts to find reciprocals, but conspicuously less successful than the first. The longest, written on the back of a roughly made, approximately square tablet (Fig. 3.1.8), reads:

16	40				
16	40				
16	40				
20	4 37	46	40	9	
50	42 39	[......]			

(3N-T 611 = A 30279)

Nothing remains on the obverse apart from a few apparently random signs. The first part of the calculation is a squaring of the number 16;40, set out in the usual

26. For the most recent discussion, see Robson (2000, 21).

27. I have assigned arbitrary absolute sexagesimal value to the numbers in this problem and those in the following discussion.

28. 2 [13] 20 IGI-[*bu-šu* EN.NAM] / [ZA.E] KID$_x$.TA.[ZU.DE$_3$] / IGI 3 20 DU$_8$.A 18 [*ta-mar*] / 18 *a-na* 2 10 TUM$_2$.A 3[9 *ta-mar*] / 1 DAH.HA 40 [*ta-mar*] / IGI 40 DU$_8$.A 1 30 [*ta-mar*] / 1 30 *a-na* 18 TUM$_2$.A / 27 *ta-mar* 27 IGI-*bu-*[*šu*] / [*ki-a-am ne-pe$_2$-šum*].

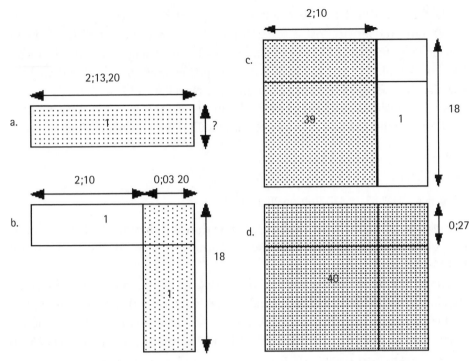

Figure 3.1.7 Finding sexagesimally regular reciprocals using The Technique

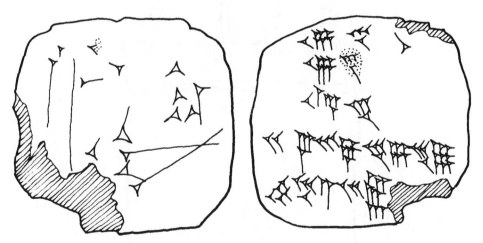

Figure 3.1.8 3N–T 611 = A 30279 (reverse), a student's calculation

way with the multiplicands and product aligned vertically (Robson 1999, 250–252), albeit with an unexplained extra copy of the 16;40. The answer is correctly given as 4 37;46 40, but the 9 written immediately to the right strongly suggests that The Technique was then used to find its reciprocal.

As in our first example, the student has split 4 37;46 40 into 4 37; 40 and 0;06 40. He has appropriately taken the reciprocal of the latter—9—and multiplied it by the former, adding 1 to the result. However, instead of arriving at 41 39 + 1 = 41 40, our student has lost a sexagesimal place and found 41;39 + 1 = 42;39. Unable to go further with his calculation (for the next stage is to find the reciprocal of the number just found, but his is coprime to 60) he has abandoned the exercise there. The correct answer would have been 0;00 12 57 36.[29]

The last calculation of the three is the most pitiful (Fig. 3.1.9). Writing on a Type S tablet like the first example, the student has got no further than:

4 26 40   4;26 40

igi-bi 2 13 20   Its reciprocal is 2;13 20

(3N-T 605 = UM 55-21-357, Robson 2000, no. 1)

The double ruling underneath shows that he thinks he has finished, although he has done nothing more than halve the first number (Fig. 3.1.9). The correct result is 0;13 30.

Two of the three numbers whose reciprocals are to be found come from the standard school sequence of reciprocal pairs to which all other known exemplars of this exercise belong (Robson 1999, 23). The sequence is constructed by successively doubling/halving an initial pair 2 05 and 28 48. Our two are eighth (4 26 40)

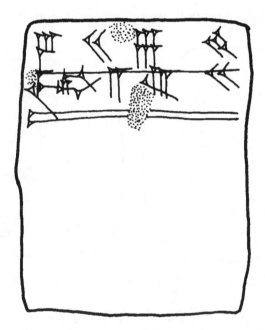

Figure 3.1.9 3N-T 605 = UM 55–21–357 (obverse), an attempted reciprocal calculation

29. I have no explanation for the 20 and 50 written to the left of the calculation; presumably they relate to intermediate steps in the procedure. Compare similarly positioned auxiliary numbers in calculations from Ur, e.g. *UET* 6/2 387 (Robson 1999, 245).

and tenth (17 46 40) respectively. On the other hand, 4 37 46 40 does not, as far as I can ascertain, fit the pattern; presumably it was chosen because, like the other two, it terminates in the string 6 40. One possible interpretation of this commonality is that three students were set similar problems at the same time, using a common method and a common starting point but requiring different numerical solutions. One of three used the method correctly, producing the right answer and checking his results; the second chose the appropriate method but couldn't apply it satisfactorily, while the third had missed the point of the exercise entirely.

Finally, there is just one mathematical tablet from House F which cannot be securely related to other elements of the scribal curriculum. It bears a list of inverse squares, which takes the form n^2-e n ib_2-si_8 'n^2 squares n', where $n = 1$–30. It would perhaps be tendentious to connect it with the squaring exercise on 3N-T 611. It is a well attested table: Neugebauer (1935–37, I 70–71; Neugebauer and Sachs 1945, 33–34) lists 18 other exemplars, 13 of which are in this format; six of those 13 are also from Nippur.

Conclusion

It turns out that a wealth of interesting insights can be gained from mathematical material that has traditionally been dismissed as unimportant and trivial. An awareness of archaeological and social context can illuminate the dullest of texts. Perhaps most importantly, it is now clear that thousands of Old Babylonian multiplication tables and metrological lists survive not because every numerate citizen kept a handy reference collection at home. Conversely—and paradoxically—they are the throw-away by-products of training in an essentially oral, memorized numerate culture passed on by tiny numbers of professionals within a restricted social circle. That is not to say that only the scribes were numerate, but simply that we have very little access to non-scribal numeracy from this time.

Further, the study of imagery in pedagogical literature proves vitally important for understanding the construction of OB scribal identity. Numeracy and literacy, and linear measurement in particular, were considered to be divine gifts, from goddesses not gods, and therefore to some extent gendered female even if most (but not all) human practitioners were male (Robson 2007). Moreover, goddesses such as Nisaba bestowed numeracy and mathematics on humanity for a purpose: to enable kings, and their servants the scribes, to deliver social justice through the equitable distribution and fair management of assets. Although House F contains few examples of mathematical word problems, it is well known that OB mathematics was predominantly concerned with labour management and quantity surveying on the one hand (Robson 1999) and the abstract manipulation of lines and areas on the other (Høyrup 2002). In this light, it becomes clear why this was so: pedagogical mathematical exercises embodied the practices of metrological, numerate justice, a principle that was central to early Mesopotamian royal and scribal self-identity.

However, while these large-scale conclusions about the socio-political context of mathematics apparently hold true for much of early Mesopotamian history, we should be careful not to blithely generalize the particular details of the House F curriculum. Comparative study of contemporary schools has highlighted both the variations large and small between individual corpora of tablets and the virtual impossibility of ascribing those differences to diachronic change, geographical variation or personal choice—although it appears that this last was more pervasive than we might have thought (Robson 2002).

Bibliography

Black, J A, Cunningham, G G, Ebeling, J, Flückiger-Hawker, E, Robson, E, Taylor, J J, and Zólyomi, G G, *The electronic text corpus of Sumerian literature*, <http://etcsl.orinst.ox. ac.uk/>, University of Oxford Oriental Institute, 1998–2006.

Civil, M, *The farmer's instructions. A Sumerian agricultural manual*, Aula Orientalis Supplementa, 5, Editorial AUSA, 1994.

Civil, M, 'Ancient Mesopotamian lexicography', in J Sasson (ed), *Civilizations of the ancient Near East*, Scribner, 1995, 2305–2314.

Civil, M, Green, M W, and Lambert, W G, *Ea A* = nâqu, *Aa A* = nâqu, *with their forerunners and related texts*, Materials for the Sumerian Lexicon, 14, Biblical Institute Press, 1979.

Friberg, J, 'On the big 6-place tables of reciprocals and squares from Seleucid Babylon and Uruk and their Old Babylonian and Sumerian predecessors', *Sumer*, 42 (1983), 81–87.

Friberg, J, 'Mathematik', in D O Edzard (ed), *Reallexikon der Assyriologie und vorderasiatischen Archäologie*, vol 7, Walter de Gruyter, 1987–90, 531–585.

Gasche, H, and Dekiere, L, 'A propos de la durée de vie d'une maison paléo-babylonienne en briques crues', *NABU*, 1991 no. 20.

George, A R, *The Babylonian Gilgamesh epic: introduction, critical edition and cuneiform texts*, Oxford University Press, 2003.

Gibson, McG, Hansen D P, and Zettler, R L, 'Nippur B. Archäologisch', in D O Edzard (ed), *Reallexikon der Assyriologie und vorderasiatischen Archäologie*, vol 9, Walter de Gruyter, 2001, 546–565.

Heimerdinger, J W, *Sumerian literary fragments from Nippur*, Occasional Publications of the Babylonian Fund, 4, University of Pennsylvania Museum, 1979.

Høyrup, J, 'Algebra and naive geometry. An investigation of some basic aspects of Old Babylonian mathematical thought', *Altorientalische Forschungen*, 17 (1990), 27–69, 262–354.

Høyrup, J, *Lengths, widths, surfaces: a portrait of Old Babylonian algebra and its kin*, Sources and Studies in the History of Mathematics and Physical Sciences, Springer, 2002.

Kramer, S N, *The Sumerians: their history, culture, and character*, University of Chicago Press, 1963.

Landsberger, B, *The series HAR-ra* = hubullu, *tablets VIII–XII*, Materials for the Sumerian Lexicon, 7, Biblical Institute Press, 1959.

Landsberger, B, Reiner, E, and Civil, M, *The series HAR-ra* = hubullu, *tablets XVI, XVII, XIX, and related texts*, Materials for the Sumerian Lexicon, 10, Biblical Institute Press, 1970.

Lion, B and Robson, E, 'Quelques textes scolaires paléo-babyloniens rédigés par des femmes', *Journal of Cuneiform Studies*, 57 (2006), 37–53.

McCown, D E, and Haines, R C, *Nippur I: Temple of Enlil, scribal quarter, and soundings*, Oriental Institute Publications, 78, The Oriental Institute of the University of Chicago, 1967.

Nemet-Nejat, K R, *Cuneiform mathematical texts as a reflection of everyday life in Mesopotamia*, American Oriental Series, 75, American Oriental Society, 1993.

Neugebauer, O, *Mathematische Keilschrift-Texte*, I–III, Quellen und Studien zur Geschichte der Mathematik, Astronomie und Physik, A3, Springer, 1935-7.

Neugebauer, O, and Sachs, A, *Mathematical cuneiform texts*, American Oriental Series, 29, American Oriental Society, 1945.

Neugebauer, O, and Sachs, A, 'Mathematical and metrological texts', *Journal of Cuneiform Studies*, 36 (1984), 243–251.

Robson, E, *Mesopotamian mathematics, 2100–1600 BC: technical constants in bureaucracy and education*, Oxford Editions of Cuneiform Texts, 14, Clarendon Press, 1999.

Robson, E, 'Mathematical cuneiform tablets in Philadelphia, I: problems and calculations', *SCIAMVS—Sources and Commentaries in Exact Sciences*, 1 (2000), 11–48.

Robson, E, 'The tablet house: a Nippur scribal school, 1740 BC', *Revue d'Assyriologie*, 95 (2001), 39–67.

Robson, E, 'More than metrology: mathematics education in an Old Babylonian scribal school', in J M Steele and A Imhausen (eds), *Under one sky: mathematics and astronomy in the ancient Near East*, Alter Orient und Altes Testament, 297, Ugarit-Verlag, 2002, 325–365.

Robson, E, 'Tables and tabular formatting in Sumer, Babylonia, and Assyria, 2500–50 BCE', in M Campbell-Kelly, M Croarken, R G Flood, and E Robson (eds), *The history of mathematical tables from Sumer to spreadsheets*, Oxford University Press, 2003, 18–47.

Robson, E, 'Gendered literacy and numeracy in the Sumerian literary corpus', in G Cunningham and J Ebeling (eds), *Analysing literary Sumerian: corpus-based approaches*, Equinox, 2007, 215–249.

Sachs, A, 'Babylonian mathematical texts, I. Reciprocals of regular sexagesimal numbers', *Journal of Cuneiform Studies*, 1 (1947), 219–240.

Sjöberg, Å, 'The Old Babylonian *eduba*', in S Lieberman (ed), *Sumerological studies in honor of Thorkild Jacobsen*, Assyriological Studies, 20, University of Chicago Press, 1975, 159–179.

Stone, E C, *Nippur neighborhoods*, Studies in Ancient Oriental Civilization, 44, The Oriental Institute of the University of Chicago, 1987.

Tinney, S J, 'On the curricular setting of Sumerian literature', *Iraq* 61, (1999), 159–172.

Vanstiphout, H L J, 'How did they learn Sumerian?', *Journal of Cuneiform Studies*, 31 (1979), 118–126.

Vanstiphout, H L J, 'Sumerian canonical compositions. C. Individual focus. 6. School dialogues', in W W Hallo (ed), *The Context of scripture, I: Canonical compositions from the Biblical world*, Brill, 1997, 588–593.

Veldhuis, N, 'Elementary education at Nippur: the lists of trees and wooden objects', unpublished doctoral thesis, University of Groningen, 1997.

Veldhuis, N, *Religion, literature, and scholarship: the Sumerian composition Nanše and the birds*, Brill, 2004.

Zettler, R L, 'Written documents as excavated artifacts and the holistic interpretation of the Mesopotamian archaeological record', in J S Cooper and G M Schwartz (eds), *The study of the ancient Near East in the 21st century: the William Foxwell Albright Centennial Conference*, Eisenbrauns, 1996, 81–101.

The archaeology of mathematics in an ancient Greek city

David Gilman Romano

A modern mapping and research project, the Corinth Computer Project, has been underway in ancient Corinth in Greece, as well as in Philadelphia, for the past twenty years, 1988–2007.[1] By means of highly accurate electronic total station surveys in the field, and advanced digital cartographic techniques in the laboratory, one of the project objectives has been to analyze the ways in which Greek and Roman surveyors worked, both in the city of Corinth and in the

1. Since 1988 a research team from the Mediterranean Section of the University of Pennsylvania Museum of Archaeology and Anthropology has been involved in the study of the planning of the Roman city of Corinth. Known as the Corinth Computer Project, the field work was carried out under the auspices of the Corinth Excavations of the American School of Classical Studies at Athens, Dr Charles K Williams, II, Director (until 1997). I thank Dr Williams for his interest in and support of this project. I would also thank Dr Nancy Bookidis, Associate Director of the excavations, for her assistance during the same years. Since 1997 I thank Dr Guy Sanders, Director, for his assistance. The original objectives of the modern project were to study the nature of the city planning process during the Roman period at Corinth, to gain a more precise idea of the order of accuracy of the Roman surveyor, and to create a highly accurate computer-generated map of the ancient city whereby one could discriminate between and study the successive chronological phases of the city's development. The latest summary of the work is found in Romano (2003). A website has been created to discuss the methodology as well as some of the results of the project, <http://corinthcomputerproject.org>. The final publication of the Corinth Computer Project is planned for a volume in the Corinth Excavation series and will include a text volume, a CD-ROM, and a gazetteer of maps. I thank the more than 140 University of Pennsylvania students who have assisted me in this study, both in the field at Corinth and in the lab in Philadelphia. I am grateful to David Pacifico, Tim Demorest, Nicholas Stapp, and Dan Diffendale for assistance in the production of figures for this chapter.

surrounding landscape. A careful study of their practical techniques at Corinth gives us an opportunity to better understand exactly how the surveyors practiced their skills and, perhaps, some insight into the mathematical principles that their work was based on.

A brief history of Corinth

Corinth was one of the most famous cities of the ancient world. Located adjacent to the isthmus that joined Central Greece to the Peloponnesos, Corinth controlled the land routes across the isthmus as well as those between the adjacent bodies of water, the Corinthian Gulf to the west and the Saronic Gulf to the east. Corinth established harbors on both Lechaion on the Corinthian Gulf and Kenchreai on the Saronic Gulf (Fig. 3.2.1). The city's acropolis, Akrocorinth, is a 573 m high outcropping of rock that rises up only 5 km to the south of the Corinthian Gulf. It was ideally suited to serve as a strategic military location, commanding both land and sea routes. The land immediately to the south of the Corinthian Gulf was a rich agricultural coastal plain that was famous in antiquity for its fertility.

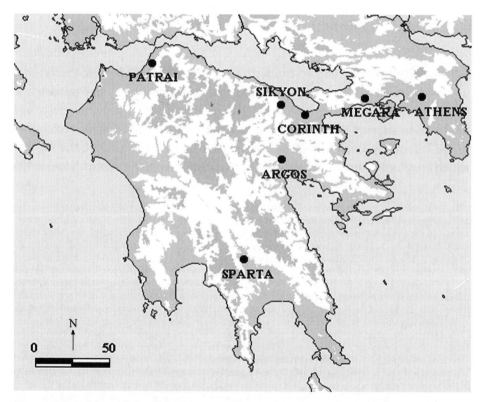

Figure 3.2.1 Map of the Peloponnesos, illustrating the location of Corinth

The Archaic and Classical city of Corinth grew up to the north of Akrocorinth, on a limestone plateau approximately seventy meters above sea level. The earliest 'city' of Corinth was a loose collection of small rural communities in the eighth and seventh centuries (Roebuck 1972; Williams 1982). But it had sent out important colonies to the west, to settlements such as Kerkyra, Apollonia, and Syracuse. Its political history begins with the aristocratic Bacchiadae clan, who were overthrown in around 650 BC and succeeded by the tyrants Cypselus and his son Periander (Salmon 1984, 38–80). It was probably Periander who built the stone roadway between the Corinthian and the Saronic Gulf, the *diolkos*, over which commercial shipments, and perhaps small boats, were transported (Verdelis 1956; Werner 1997). Corinth was an architectural leader among Greek cities: its seventh-century BC Temple of Apollo is among the first to be constructed in Doric style, and is roofed with the first example in the Greek world of terracotta roof tiles.[2] Corinth was also well known for its pottery, traded all over the Mediterranean. Proto-Corinthian style wares, featuring oriental or eastern designs and motifs, were sent out to many of the earliest Greek colonies. In about 720 BC Corinth invented the black-figured technique of vase painting that was copied all over the Greek world.

By the sixth century BC the upper Lechaion Road valley and areas nearby were filled with cults, hero shrines, springs, buildings, and monuments, and a racecourse for athletic foot races. In addition, a series of roadways approached this valley from various directions (Fig. 3.2.2). The successive temples of Apollo dominated the city from Temple Hill and the fountain houses of Glauke and Peirene were close by. By the fifth century BC the city was surrounded by a fortification wall and by the fourth century BC it also included long walls to its harbor at Lechaion on the Corinthian Gulf.[3]

During the Archaic and Classical periods, Corinth often allied itself with Sparta and against Athens, and fought on Sparta's side during the Peloponnesian War of 431–404 BC. During the fourth and third centuries BC a Macedonian garrison occupied Akrocorinth under the control of Ptolemy I, Demetrius Poliorcetes, and Antigonos. When Aratus captured Corinth in 243 BC it joined the Achaean Confederacy of Greek city states. In the second century BC Corinth was a leader of Greek cities' opposition to the coming of Rome (Gruen 1984, 523–527). As a result, in 146 BC Corinth was singled out to be sacked by the Roman consul Lucius Mummius who, according to a literary account, killed all the men and sent all the women and slaves into slavery (Pausanias 7.16.8).

---

2. Coulton (1977, 32–35, 49–50) suggests that the impetus for the large-scale stone Doric architecture at Corinth, and at nearby Isthmia, may have been due to influences coming from Egypt in the second half of the seventh century BC.

3. There is evidence of a defensive wall in the Potters Quarter at Corinth that dates to the seventh century BC (Stillwell 1948, 14–15).

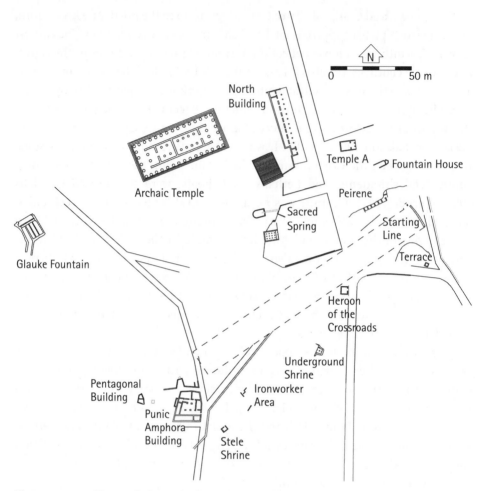

Figure 3.2.2 Plan of Central Corinth, including upper Lechaion Road valley, 450–425 BC

The city was abandoned and deprived of a political identity until 44 BC, when Julius Caesar founded a new Roman colony, *Colonia Laus Iulia Corinthiensis*, on the site of the former Greek city (Walbank 1997). The Roman colony imposed a new urban plan on the city as well as a new rural organization on the land of Corinthia. Later in the first century AD under the Emperor Vespasian a second Roman colony was instituted at Corinth, *Colonia Iulia Flavia Augusta Corinthiensis,* with more new urban and rural planning. Corinth has been continuously occupied through the Late Roman period into the modern day. Following a devastating earthquake in 1858, the modern town of Corinth was moved 8 km to the northeast, to the Gulf of Corinth, where it exists today. A small village remains on the original site. The successive ancient cities of Corinth have been under excavation and study by the Corinth Excavations of the American School of Classical Studies at Athens since 1896.

Part of the work of the Corinth Computer Project has been to undertake a study of the planning of the Roman city and landscape, including a survey of the standing buildings and monuments of all periods in the ancient city. In the course of this work several ancient mathematical questions were encountered, each of which required some concentrated study and analysis to determine the nature of the design process that had been employed by the ancient surveyor. We will discuss two here: the design and construction of a curved starting line to a racecourse in the early fifth century BC, and the planning of field systems in the Roman period.

The curved starting line of a racecourse, c 500–450 BC

During excavations undertaken by the American School of Classical Studies in the area of the Eastern Roman Forum in 1937, successive starting lines of two Greek racecourses were discovered (Morgan 1937, pl. 13, fig. 2). The later starting line, located immediately to the west of the foundations for the Roman Julian Basilica, was excavated completely. It was found to be in the form of a straight line composed of a series of cut rectangular limestone blocks covered in stucco, with a series of small regular rectilinear cuttings in the top surfaces of the blocks. At a lower level a small portion of a second starting line was excavated, of similar composition but at a different orientation.

During the excavations of 1980 by the American School of Classical Studies, a greater portion of the earlier starting line was exposed (Williams and Russell 1981). It became clear that both starting lines were parts of successive rectilinear racecourses, each one *stadion* in length,[4] which extended to the west across the upper Lechaion Road valley (Fig. 3.2.2). The later of the two starting lines dates to after 270 BC, while the earlier dates to 500–450 BC or possibly even the sixth century BC. The most significant difference between the two starting lines is that the later one is straight while the earlier is curved (Fig. 3.2.3). It is between 1.25 and 1.3 m wide and c 12.20 m long, apparently describing the arc of a circle with a radius of c 54 m. The curved shape of this starting line is unique in the Greek world. One of the author's interests in the starting line was to determine how it was planned and constructed in an effort to better understand its design and specifically how its design related to the athletic event that it was intended to serve (Romano 1993).

On the top surface of the earlier starting line a series of painted letters as numbers was discovered, red letters against the black painted surface of the starting

---

4. The *stadion* was a measure of linear distance in ancient Greece always equal to 600 feet, although the absolute measure of the foot could vary from place to place.

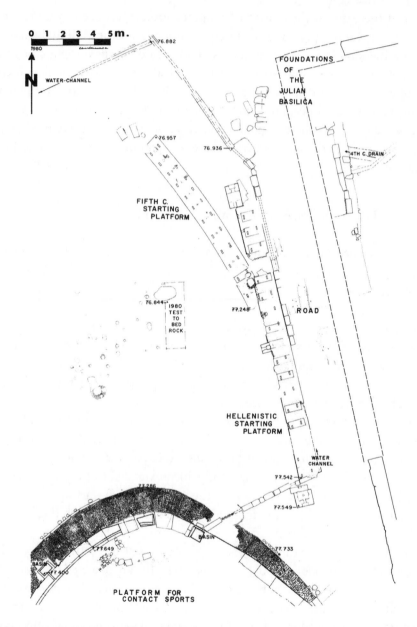

Figure 3.2.3 Corinth, starting lines as preserved to the west of the Julian Basilica, actual-state plan. Courtesy of the Corinth Excavations, American School of Classical Studies at Athens

line. Each of the running lanes was labeled with a number from *alpha* to *pi*, written in the Corinthian epichoric alphabet (Williams and Russell 1981, 2–10), although the first five positions *alpha* to *epsilon* were destroyed by the later starting line. It was thus originally designed for seventeen positions, of which thirteen remain. Each of the starting positions is characterized by grooves cut into the stone blocks for the toes of the front (left) foot and the rear (right) foot of the athlete (Fig. 3.2.4). The distance between the foot positions varies from 0.658 to 0.705 meters (center to center, standard deviation 0.0128 m). The thirteen excavated starting positions are spaced at an average interval of 0.951 m, center to center.[5] The individual starting positions of the excavated curved starting line were found to be almost exactly one degree of a circle, on average 1.019° apart,[6] with a radius of 53.960 m.[7] In modern mathematical terms, it would appear that the architects intended to set out a starting line with starting positions 1° of arc apart from each other.

But this interpretation poses a difficult historical conundrum. How could the Greek architects or surveyors have created such a starting line with the interval so accurately spaced out, on average, at 1° intervals, several centuries before the 360-degree circle became standard mathematical fare?

Until recent years it was common historical methodology to apply 'rational reconstruction' (Kragh 1987, 161) to the remains of ancient structures, whether Neolithic stone circles or Egyptian pyramids, to determine the mathematical methods used in their design. But more recent studies have shown that such work tends to be over-reliant on modern mathematical knowledge and under-sensitive to the concepts and theories of the culture under investigation (e.g., Angell 1976; Rossi 2004). Rather, current scholarship starts from contemporary written mathematics, from the same time and culture as the ancient structure under study, to try and identify planning techniques. It is only if these fail to adequately account for the design features of the structure that previously unattested mathematical knowledge must be adduced. That, however, is a vanishingly rare outcome of carefully researched historical studies.

So, what can be said of the mathematics of circles, as understood in the fifth century BC? No 'strictly mathematical' texts survive in Greek from the fifth or fourth century BC (Cuomo 2001, 5). The best contemporary evidence is from some 2000 km away in Babylonia (southern Iraq), where astronomers had

5. The standard deviation can be computed only for the distances measured between successive positions. The measurements are taken as center to center measurements on the front left groove of each position. The mean of these nine distances is 0.983 m and the standard deviation 0.127 m.

6. The standard deviation can be computed only for those angles measured between successive positions. The mean of these nine angles is 1.038° and their standard deviation is 0.134. I am grateful to Chris Rorres for this explanation.

7. For an earlier discussion of how the circle was created see Rorres and Romano (1997). The archaeological surveyors originally used a 'three points circle' to estimate the radius of the circle, while the 1997 study utilized the 'least-squares circle'.

Figure 3.2.4 Photograph of curved starting line at Corinth, fifth century BC, view toward the south. Courtesy of the Corinth Excavations, American School of Classical Studies at Athens

divided *harrān Sîn* 'the path of the moon' into 360 equal parts by around 460 BC (Neugebauer 1975, 593).[8] These 360 divisions consisted of the twelve zodiacal zones (essentially those that are still in use today), each subdivided into 30. But it would be anachronistic to conclude that they had thereby invented the 360° circle. With hindsight, it is clear that this was certainly the origin of the concept, as it eventually evolved over the following millennium, but no contemporary scholar would have understood it as such. The Babylonian unit of ecliptic measure was the *uš*, which historians of Babylonian astronomy conveniently translate as 'degree'. But the astronomers themselves conceptualized it both as a linear measure—the distance between one celestial body and another—and as a chronological unit—the time between the rising or setting of one celestial body and another (Powell 1990, 467–468). It mapped onto the millennia-old Babylonian ideal of the 360-day year (Englund 1988) and was used only to measure the distances or periods between events on the ecliptic, never to subdivide other circles, whether terrestrial or celestial (Brown 2000).

The narrow conceptual and contextual range of the 360-part circle in fifth-century Babylonia is best exemplified by the writings of Šamaš-iddin of the Šangû-Ninurta family, who lived and worked as an *āšipu* 'healer' in the southern city of Uruk shortly before 410 BC (Oelsner 1993). His family library contained several hundred cuneiform tablets across a wide scholarly spectrum, from mythology to medicine, including astronomy. His mathematical writings on circles use the traditional Babylonian methods, in which the defining component of the circle is its circumference, *kippatum*, from which all other parameters are calculated. These mathematical circles have no radii, and their circumferences are not divided into degree-based arcs (W 23291x (i)–(iii): Friberg *et al* 1990; W 23291 (vii): Friberg 1997).

Šamaš-iddin and his contemporaries worked within small kin-based groups, training sons and nephews in their professional skills and scholarly knowledge while swearing them to secrecy. Curses against those who shared *nişirtī šamê* 'the secrets of the heavens' with *lā mūdû* 'the ignorant', and similar injunctions, were commonly added to Babylonian astronomical writings from the twelfth century BC onwards (Rochberg 2004, 212–219, esp n11). Given these circumstances, it is not surprising that the earliest evidence for the 360-part circle in Greek mathematics is not until the second century BC. A fragmentary astronomical inscription found on Rhodes, dating to about 150–100 BC, states that 'the circle contains 360 degrees' (Neugebauer 1975, 699). A roughly contemporary work by Hypsicles of Alexandria, *On the ascension of the stars*, also divides the zodiacal circle into 360 equal parts. (De Falco and Krause 1966, 47).

---

8. The earliest known attestation is in a diary of astronomical observations taken in Uruk in 462 BC (Sachs and Hunger 1988, 54–55).

So it is chronologically impossible for the surveyors of Corinth to have used the 360-part circle for the construction of their curved starting line in the early fifth century BC. At that time only a few secretive scholars of astronomy, over a thousand miles away, were beginning to divide the ecliptic in this manner—a technique which did not reach the Greek-speaking eastern Mediterranean until 200 BC at the earliest, and which would not be used in non-astronomical contexts for many centuries to come.[9] But the construction of a circular arc does not entail a sophisticated knowledge of angle geometry: all one needs are some pegs, a hammer and a stout rope (Angell 1976). The simplest possible interpretation of the curved starting line, based on the archaeological survey data given above and a foot of 0.27 meters, is that the surveyors marked out two concentric arcs 60 ancient feet (16.2 meters)[10] long with a rope tethered 200 feet (54 meters) away. The athletes had to have enough room between them so that they would not be jostled by their rivals; a distance of 3 1/2 feet (0.95 meters) allowed sufficient space for seventeen competitors.

Thus it can be no more than an extraordinary coincidence that the starting blocks of the earlier starting line at Corinth are placed almost exactly 1° of arc apart on a circular segment. Such coincidences are not unique. For many decades it was widely supposed that the Babylonian mathematical table, Plimpton 322, gave unequivocal evidence for 360° trigonometry in the eighteenth century BC (Neugebauer and Sachs 1945, text A; Calinger 1999, 35–36). Plimpton 322 does indeed list the short sides and hypotenuses of fifteen triangles whose acute angles decrease in approximately 1° increments (mean 0° 55′ 09″, standard deviation 0° 25′ 33″). But that is no more than an accidental outcome of the way in which the table was constructed from mutually reciprocal pairs of sexagesimally regular numbers, drawing on a repertoire of mathematical techniques that are widely attested in early second-millennium Babylonia (Bruins 1949; Robson 2001; 2002).

The question remains as to why the Greek architect and surveyor wanted a curved starting line for the racecourse. What was the nature of the race to be contested and why did it need a curved as opposed to a straight starting line? The

9. Alternative methods for partitioning the circle are attested in Greek mathematics from the mid-third century BC. In a work called *On the sizes and distances of the sun and the moon*, Aristarchus of Samos handles it as four quadrants, each of which can be divided into unit fractions (Heath 1913, 352–3; Neugebauer 1975, 590, 773). According to Theon of Alexandria, a fourth-century AD commentator on Ptolemy's *Almagest*, Hipparchus of Rhodes created a table of chords by successive fractionings of the quadrant, down to 12ths, or 7° 30′ in modern terms, in the early second century BC (Toomer 1973). The origins of these methods are unknown: they may originate with these particular authors, or may have been in common use for some time before. The great paucity of direct evidence for Greek mathematics before about the third century BC means it is virtually impossible to know for certain: (Cuomo 2001, 4–38). But as both scholars were working and thinking in a solely astronomical context, it is unlikely that any hypothetical precursor works were on the mathematics of terrestrial circles.

10. Of which 12.2 meters survive (see above).

answer may be that the Greek architect purposely chose a curved starting line to provide the fairest possible start, as well as an equal distance to be run, by the seventeen athletes towards a distant point. The race was to be multiple lengths of the *stadion*, a distance race in other words. The athletes would have run in lanes from the starting line to a 'break line', so that they would not interfere with each other, at which point they would be equidistant from the middle of the *stadion*. It is likely that from this point onwards the athletes would have run directly toward the turning post at the west end of the racecourse (Fig. 3.2.5) (Romano, Andrianis, and Andrianis, 2006).

Roman centuriation in Corinthia, second century BC–first century AD

The Corinth Computer Project, which primarily studied the planning of the Roman city of Corinth and its territory, also considered the rural agricultural land that surrounded the Caesarian colony of the first century BC and the Flavian colony of the first century AD. The Romans are famous for their systems of rural and urban planning throughout the Empire, and the Roman colonies of Corinth were no exception.

In its simplest form, centuriation was the division of land by the Roman surveyors, *agrimensores*, into large squares or rectangles of land that were then subdivided into smaller regular units (Dilke 1971). The system was applied to urban as well as to rural areas, principally to facilitate the collection of taxes based on known areas of land. The Romans used as their primary unit of measurement the *actus*, or 120 linear Roman feet, approximately 35.4 meters (1 Roman foot = 0.295 m). The word 'centuriation' refers to the fact that the large squares or rectangles of land were typically divided into 100 units or *heredia*. A *heredium* was an area of land equal to 2 *iugera* or 4 square *actus*, in modern units 0.504 hectares.

Much is known about the work of the Roman *agrimensores* from archaeological remains from different areas of Europe and North Africa. There also exist extensive literary accounts of the *agrimensores*, known collectively as the *Corpus Agrimensorum Romanorum*.[11] The *Corpus* is a collection of ancient land surveyors' manuals originally compiled in the fifth century AD, but it includes texts that were composed as early as the first century AD. The surveyors' manuals give very specific information about the training of the *agrimensores*, their methods

---

11. For an early discussion of air photographic evidence used to understand Roman city and land planning see Bradford (1957, 145–216). For a translation of the *Corpus Agrimensorum Romanorum* see Campbell (2000). Campbell's work is critical to the understanding of exactly how the *agrimensores* worked. For a discussion of the modern methods used in the study of Roman centuriation and a summary of the important publications in the area see Campbell (2000, lvii–lviii, nn189, 190).

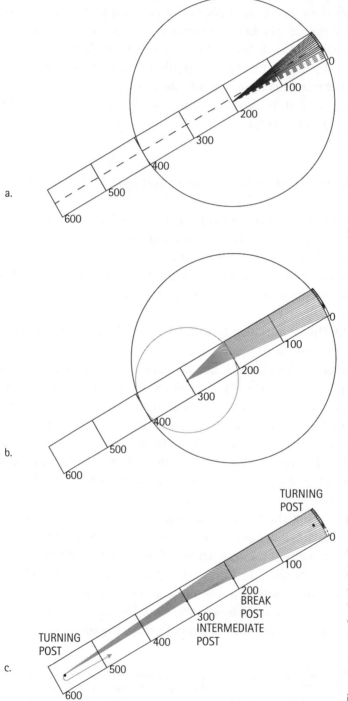

Figure 3.2.5 Corinth drawings illustrating design and use of curved starting line. (a) Circle with radius of 200 feet intersect–ing with curve of starting line and with seventeen starting positions equidistant from centre of circle. (b) Athletes run in lanes to 'break line' to avoid collision, where they have an equidistance to run to a point at 300 feet. (c) Actual nature of race where athletes run to break line at 200 feet and then to turning post at west end of racecourse and return

of work, and the kinds of problems and issues that they encountered during the course of their daily activities. There are also fragments of ancient stone maps from Orange (ancient Aurasio in France) that illustrate the Roman agricultural division of land (Piganiol 1962; Salviat 1977). More commonly, perhaps, maps depicting the work of the *agrimensores* would have been made on perishable materials that have not survived.

The *agrimensores'* principal surveying instrument was the *groma*, a simple device composed of a vertical staff with two horizontal crossbars, connected by a bracket. From each end of the crossbars hung a cord, which was held vertical by a plumb bob (Fig. 3.2.6). Roman surveyors were skilled at using the *groma* together with *decempeda* (sighting rods), to create straight lines and right angles (Schiöler 1994). The *agrimensores* sighted along the *groma* to the survey rods, hammered wooden stakes in the ground at every *actus*, then drew a straight line on the ground connecting the stakes. The surveyors next created a furrow or a shallow ditch to represent the line.[12]

These furrows, or vestiges of them, have been found in various parts of the Roman world. Modern rectilinear fields that are spaced approximately 35 meters apart and have the same orientation as a nearby Roman urban grid may be a part of the Roman land division. In some places vast areas of centuriation have been recognized, for instance in the Po Valley of Italy or in portions of North Africa, where hundreds of square kilometers of centuriated land are known (Bradford 1957, 155–166, 193–207; Mengotti 2002). Elsewhere only fragments of organized land systems have been discovered. The vestiges of orthogonal urban planning are seen in such modern cities as London, Paris, Florence, and Barcelona (Woloch in Grimal 1983, 111–301). The archaeological evidence from Corinth suggests that the urban Caesarian colony was based on four equal quadrants (*centuriae*), each measuring 32 × 15 *actus* = 240 *iugera*, with the forum in the center of the plan and intersecting the *cardo maximus* (Fig. 3.2.7). In Greece, several systems of rural centuriation have been discovered: at Dyme and Patrae in the Peloponnesos, and at Arta in Ambracia (Rizakis 1990; Doukellis and Fouache 1992). Evidence of centuriation has been presented from Nicopolis and Butrint in Epirus (Doukellis 1988; Bescoby 2006).

At least two phases of rural centuriation have been identified from different parts of Corinthia, the region of Corinth, areas of which overlap. The Caesarian system covers a total area of approximately 100 km², while the Flavian system covers about 300 km² (Romano 2003, 293–299). Generally speaking the Caesarian system is found closer to the city of Corinth, and extends only as far as the Longopotamos River to the west and eastwards across the isthmus as far as modern Loutraki

---

12. There were many ways to indicate *limites* (see glossary), including a line of trees or bushes, a stone wall, a pile of stones, a series of wooden stakes, a road, or a stream. See the text of Siculus Flaccus, from the second century AD, 'Categories of land' (Campbell 2000, 103–119).

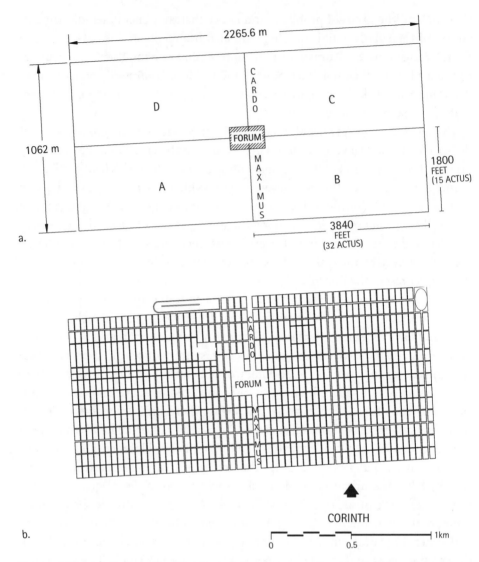

Figure 3.2.6 a) Schematic drawing of the four quadrants of the urban colony, each of which is 32 × 15 *actus,* with centrally located forum and *cardo maximus.* **b)** 'Drawing board' plan of the urban colony of 44 BC

(Fig. 3.2.8). In southern Corinthia it extends as far as Tenea and Kleonai. The Flavian system is wider-ranging, extending into southwest Corinthia as well as along the south coast of the Corinthian Gulf to Sikyon and beyond (Fig. 3.2.9).

The evidence for centuriation associated with the Roman colony of Julius Caesar outside the urban area indicates a division of land into large units of 16 × 24 *actus* (566.4 × 849.6 m) at the same orientation as the city, or approximately 3° west of north. These large units were subdivided into smaller sections, probably typically

Figure 3.2.7 Groma (courtesy of
Mrs O A W Dilke)

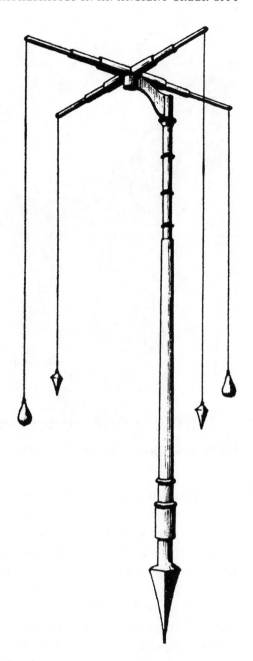

1 × 4 *actus* (35.4 × 141.6 m). Within the area of the defensive long walls, between
the city circuit of Corinth and the Lechaion harbor, there is evidence that parts of
the landscape were laid out and planned prior to the founding of the colony; the
Romans sent out surveyors to begin dividing up the land for sale or rent. During
the century that passed between the destruction of Corinth in 146 BC and colonization in 44 BC, the land that had been under its control became *ager publicus*

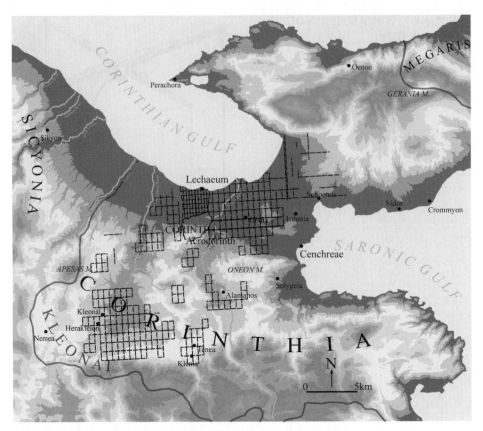

Figure 3.2.8 Extents of Caesarian centuriation in Corinthia

'public land', although several ancient authors suggest that the city of Sikyon had taken over part of the land of Corinth (Livy 27.1; Cicero, *Leg. Agr.* 1.2.5; 2.19.51). The *lex agraria* 'land law' of 111 BC, passed in Rome by the Assembly of Tribes, indicates that some parts of Corinthian territory were measured out for sale, and boundary stones were erected (Crawford 1996, I 139–180; Lintott 1992, 171–285). The archaeological corroboration of this ancient survey is a break through the Greek circuit wall of Corinth by the Romans in order to construct a road in a location that was not previously a Greek gate (Romano 2003, 279–283).

The evidence for centuriation associated with the colony of Vespasian in the 70s AD is more extensive. A series of linking grids borders the south coast of the Corinthian Gulf, with extensions to these grids in southern Corinthia. The grids are composed of 16 × 24 *actus* units, subdivided into smaller units, effectively wrapping around the southern coast of the Corinthian Gulf. Table 3.2.1 lists the specific topographic regions of the Flavian system along the southern coast of the Corinthian Gulf and the orientation of the individual units. The areas, however, represent the total areas of the vestiges of the systems across Corinthia (Fig. 3.2.9).

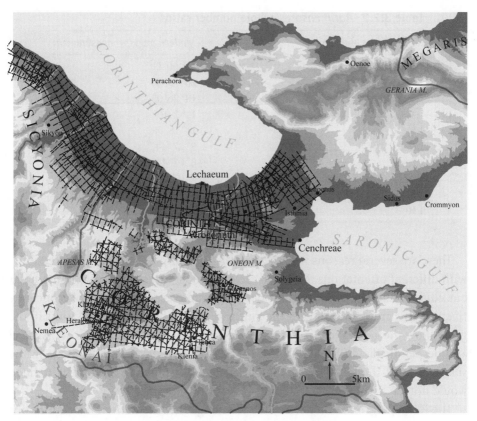

Figure 3.2.9 Extents of Flavian centuriation in Corinthia

Table 3.2.1 Flavian grids south of the Gulf of Corinth

Unit	Geographical region	Orientation	Area in km²
A00	Sikyon Northwest	N20°20′22″E	10
A0	Sikyon North	N34°22′32″E	12
A1	Sikyon	N62°26′52″E	40
A2	Sikyon, coastal region	N48°24′42″E	58
A3	Nemea River area	N34°22′32″E	108
A4	Longopotamos River area	N20°20′22″E	27
A5	Corinth, Lechaion to Kenchreai	N6°18′12″E	142
A6	Corinth to Kenchreai, south corridor	N20°20′22″E	12
A7	Xerias River area	N7°43′58″W	30
A8*	West of Isthmus	N34°22′32″E	11
A9	West of Isthmus	N21°46′8″W	15
A10	East of Isthmus	N35°48′18″W	16
A11*	Area of canal (parallel to canal)	N48°20′02″W	17

* not linked in the same way

Table 3.2.2 *Agrimensores'* whole number ratios

Ratio of sides of triangle	Modern angular measurement
1:1	45° 00′ 00″
1:2	26° 33′ 54″
1:3	18° 26′ 05″
1:4	14° 2′ 10″
1:5	11° 18′ 35″
1:6	9° 27′ 44″
1:7	8° 07′ 48″
1:8	7° 07′ 30″
1:9	6° 20′ 24″
1:10	5° 42′ 38″

There are several reasons why the overall grid comprises separate, linked units with different orientations. First and most obviously, the coastline is not straight but curved, and the land division needed to take account of this. Second, the rivers in this area drain toward the gulf and many of the individual units of the grid follow the general course of the rivers.[13] Third, each unit is roughly parallel and perpendicular to the coastline. This would have been an advantage in the subdivision of the larger units into smaller ones, since the land to be divided would not have irregular shapes and sizes at the coastline.

The Roman *agrimensores* were able to change the orientation of a continuous system of land planning by the creation of right triangles whose long and short sides measured whole numbers of *actus*. They had a number of different ratios to choose from in order to change the orientation of the individual planned unit. Table 3.2.2 shows the possible ratios used to utilize this system of change and the resulting angular values. The angles are measured today in degrees, minutes, and seconds, but these would not have been known to the Roman *agrimensor*.

In the area to the west of Corinth, along the southern coast of the Corinthian Gulf, there is a transition from the A1 system to the A2 system to the A3 system, and so on (Fig. 3.2.10). Each of these transitions creates an angular change to the centuriation grid of 14° 2′ 10″. The *agrimensores* easily achieved that re-orientation, based on a 1:4 triangle, by the simple process illustrated here (Fig. 3.2.11). They would have followed an instruction like the following: 'In the exisiting grid, measure four *actus* from the corner of the section and one *actus* from the outside edge of the grid. Join these two points.'

Related to the linked Flavian centuriation units, A00–A10, along the south coast of the Corinthian Gulf, is a second series of linked units. The secondary units (the so-called B system) are linked to certain of the primary units (of the A system) by

13. Siculus Flaccus, in his discussion of categories of land, discusses questions related to ownership of land bordering rivers and streams (Campbell 2000, 117–119).

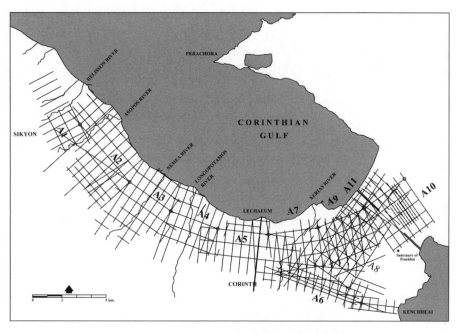

Figure 3.2.10 a) Flavian centuriation 70 AD, showing restored grid of 16 × 24 *actus* units. b) A1, A2 and A3 evidence

triangles of ratio of 1/10 or 5° 42′ 38″. The B grids have been identified in the area of the A2, A3, A4, and A5 grids (Fig. 3.2.12). The linking of the A system with the B system appears to have been an easy way for the surveyors to make smaller adjustments to the orientation of some of the fields. It is not obvious why this was done, and nor is it clear what the chronological relationship was between the two orientations. Were these linked units, the A system and the B system, created at

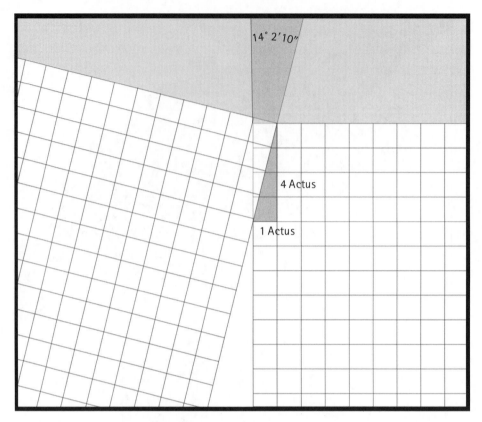

14° 2′10″

4 Actus

1 Actus

Figure 3.2.11 Change of orientation utilizing the ratio of 1 *actus* to 4 *actus*

the same time or at different periods? Based on the location and density of the vestiges of the field lines, as well as the fact that the B series is not known from each unit of the A series, the B series was probably created as a supplement at a later date.

Although the A series and the B series generally do not overlap, the A series does overlap to some degree with the earlier, Caesarian centuriation. Why might this be? It is known that the Emperor Vespasian was interested in recovering non-utilized or under-utilized portions of agricultural land, *subseciva*, around the Roman Empire for the purpose of increasing revenues through taxation (Charlesworth 1936, 1936, 13–19). *Subseciva* could be either land that was outside a centuriated area or land that was within a centuriation, but which had not been formally assigned for one of a number of possible reasons.[14] It would appear that there was available land near Corinth that had been under-utilized, as well as new land in the *territorium* of the city, which was centuriated on the same axis

14. Julius Frontinus, in his consideration of 'land disputes', discusses several types of *subseciva* (Campbell 2000, 5–9).

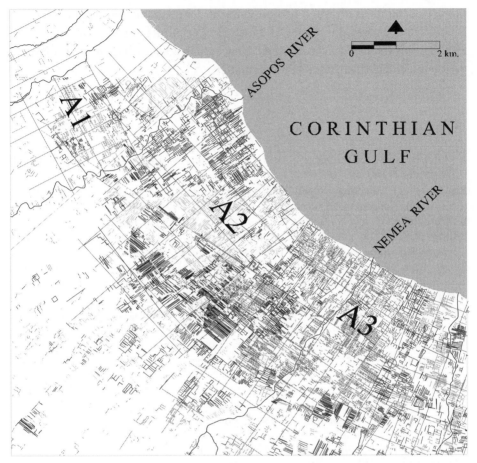

Figure 3.2.12 Relationship of A system to B system in the area of A1, A2, A3

as the under-utilized land in order to be assigned to new colonists by Vespasian (Romano 2000, 102–104).

Conclusion

Modern results of archaeological field work in Corinth have brought to light these two different ancient mathematical applications that are witness to everyday life in the Greek and Roman cities. Both relate to very practical matters: to assure that a Greek athletic foot race has a fair start, and to be able to divide up land in the Roman *territorium* in as judicious and economical a manner as possible. Literary and historical sources relating to ancient mathematics should be supplemented with archaeological evidence whenever possible, especially when the literary accounts are limited or non-existent. Likewise, archaeological research is heavily dependent on literary and historical accounts and the combination of all sources is the best way to proceed. New archaeological discoveries promise to further elucidate the history of mathematics.

Bibliography

Angell, I O, 'Stone circles: megalithic mathematics or Neolithic nonsense?', *Mathematical Gazette*, 60 (1976), 189–193.

Bescoby, D J, 'Detecting Roman land boundaries in aerial photographs using Radon transforms', *Journal of Archaeological Science*, 33 (2006), 735-743.

Bradford, John, *Ancient landscapes: studies in field archaeology*, G Bell and Sons, 1957.

Brown, David, 'The cuneiform conception of celestial space and time', *Cambridge Archaeological Journal*, 10 (2000), 103–122.

Bruins, Evert M, 'On Plimpton 322. Pythagorean numbers in Babylonian mathematics', *Koninklijke Nederlandse Akademie van Wetenschappen. Proceedings,* 52 (1949), 629–632.

Calinger, Ronald, *A contextual history of mathematics: to Euler*, Prentice Hall, 1999.

Campbell, Brian, *The writings of the Roman land surveyors: introduction, text, translation and commentary*, Society for the Promotion of Roman Studies, 2000.

Charlesworth, M P, 'The Flavian dynasty', in S A Cook, F E Adcock, and M P Charlesworth (eds), *Cambridge ancient history*, vol XI/I, Cambridge University Press, 1936, 1–45.

Coulton, J J, *Greek architects at work: problems of structure and design*, Cornell University Press, 1977.

Crawford, Michael, *Roman statutes*, Institute of Classical Studies, University of London, 1996.

Cuomo, Serafina, *Ancient mathematics*, Routledge, 2001.

De Falco, Vittorio, and Krause, Max, *Hypsikles Die Aufgangszeiten der Gestirne*, Vandenhoeck & Ruprecht, 1966.

Dilke, O A W, *The Roman land surveyors: an introduction to the agrimensores*, Barnes and Noble, 1971.

Doukellis, P N, 'Cadastres romains en Grèce: Traces d'un reseau rural à Actia Nicopolis', *Dialogues d'Histoire Ancienne*, 14 (1988), 159–166.

Doukellis, P N, and Fouache, E, 'La centuriation romaine de la plaine d'Arta replace dans le contexte de l'evolution morphologique récente des deltas de l'archatos du Louros', *Bulletin de Correspondance Hellénique*, 116 (1992), 375–382.

Englund, Robert K, 'Administrative time-keeping in ancient Mesopotamia', *Journal of the Economic and Social History of the Orient*, 31 (1988), 121–185.

Friberg, Jöran, '"Seed and reeds" continued: another metro-mathematical topic text from Late Babylonian Uruk', *Baghdader Mitteilungen*, 28 (1997), 251–365.

Friberg, Jöran, Hunger, Hermann, and Al-Rawi, Farouk N H, '"Seed and reeds": a metro-mathematical topic text from Late Babylonian Uruk', *Baghdader Mitteilungen*, 21 (1990), 483–557.

Grimal, Pierre, *Roman cities* (trans and ed G M Woloch), University of Wisconsin Press, 1983.

Gruen, Erich S, *The Hellenistic world and the coming of Rome*, University of California Press, 1984.

Heath, Sir Thomas Little, *Aristarchus of Samos, the ancient Copernicus: a history of Greek astronomy to Aristarchus, together with Aristarchus's treatise on the sizes and distances of the sun and moon*, Clarendon Press, 1913.

Kragh, Helge, *An introduction to the historiography of science*, Cambridge University Press, 1987.

Lewis, M J T, *Surveying instruments of Greece and Rome*, Cambridge University Press, 2001.

Lintott, Andrew, *Judicial reform and land reform in the Roman Republic*, Cambridge University Press, 1991.

Mengotti, C, 'Les centuriations du territoire de Patavium', in M Clavel-Lévêque (ed), *Atlas historique des cadastres d'Europe*, vol 2, Luxembourg: Publications of the Action COST G2, Paysages anciens et structures rurales, 2002, dossier 5 1A.

Morgan, Charles H, II, 'Excavations at Corinth, 1936-1937', *American Journal of Archaeology*, 41 (1937), 549–550.

Neugebauer, Otto, *A history of ancient mathematical astronomy*, 3 vols, Springer, 1975.

Neugebauer, Otto, and Sachs, Abraham, *Mathematical cuneiform texts*, American Oriental Society, 1945.

Oelsner, Joachim, 'Aus dem Leben babylonischer Priester in der 2. Hälfte des 1. Jahrtausends v. Chr. (am Beispiel der Funde aus Uruk)', in J Zabłocka and S Zawadzki (eds), *Šulmu IV: Everyday life in ancient Near East*, Adam Mickiewicz University Press, 1993, 235–242.

Piganiol, André, *Les documents cadastraux de la colonie romaine d'Orange*, Centre Nationale de la Recherche Scientifique, 1962.

Powell, Marvin A, 'Maße und Gewichte', in Dietz O Edzard (ed), *Reallexikon der Assyriologie und vorderasiatischen Archäologie*, vol 7, Walter de Gruyter, 1990, 457–517.

Rizakis, A D, 'Cadastres et espace rural dans le nordouest du Péloponnèse', *Dialogues d'Histoire Ancienne*, 16 (1990), 259–280.

Robson, Eleanor, 'Neither Sherlock Holmes nor Babylon: a reassessment of Plimpton 322', *Historia Mathematica*, 28 (2001), 167–206.

Robson, Eleanor, 'Words and pictures: new light on Plimpton 322', *American Mathematical Monthly*, 109 (2002), 105–120.

Rochberg, Francesca, *The heavenly writing: divination, horoscopy, and astronomy in Mesopotamian culture*, Cambridge University Press, 2004.

Roebuck, Carl W, 'Some aspects of urbanization at Corinth', *Hesperia*, 41 (1972), 96-127.

Romano, David Gilman, *Athletics and mathematics in Archaic Corinth: the origins of the Greek stadion*, American Philosophical Society, 1993.

Romano, David Gilman, 'A curved start for Corinth's fifth century racecourse', in L Haselberger (ed), *Appearance and essence: refinements of Classical architecture: curvature* University of Pennsylvania Museum, 1999, 283–288.

Romano, David Gilman, 'A tale of two cities: Roman colonies at Corinth', in Elizabeth Fentress (ed), *Romanization and the city: creation, transformations and failures, Journal of Roman Archaeology*, 38 (2000), 102–104.

Romano, David Gilman, 'City planning, centuriation, and land division in Roman Corinth: Colonia Laus Iulia Corinthiensis and Colonia Iulia Flavia Augusta Corinthiensis', in C K Williams, II and N Bookidis (eds), *Corinth, results of excavations conducted by the American School of Classical Studies at Athens, Volume XX, The Centenary 1896-1996*, American School of Classical Studies, 2003, 289–301.

Romano, David Gilman, 'Roman surveyors in Corinth', *Proceedings of the American Philosophical Society*, 150/1 (2006), 62–85.

Romano, David Gilman, and Schoenbrun, B C, 'A computerized architectural and topographical survey of Ancient Corinth', *Journal of Field Archaeology*, 29 (1993), 177–190.

Romano, David Gilman, Andrianis, G E, and Andrianis, E G, 'Σχεδιασμος αρχαιου στιβου αγωνων μεγαλων αποστασεων' (The design of an ancient long distance race course), in Αρχαια Ελληνικη τεχνολογια (Ancient Greek technology), Technical Chamber of Greece, 2006, 472–479.

Rorres, Chris, and Romano, David Gilman, 'Finding the center of a circular starting line in an ancient Greek stadium', Society for Industrial and Applied Mathematics Review, 39 (1997), 745–754.

Rossi, Corinna, Architecture and mathematics in ancient Egypt, Cambridge University Press, 2004.

Sachs, Abraham J, and Hunger, Hermann, Astronomical diaries and related texts from Babylonia. Volume 1: Diaries from 652 BC to 262 BC, Österreichischen Akademie der Wissenschaften, 1988.

Salmon, J B, Wealthy Corinth: a history of the city to 338 BC, Clarendon Press, 1984.

Salviat, F, 'Orientation, extension et chronologie des plans cadastraux d'Orange', Revue archéologique de Narbonnaise, 10 (1977), 107–118.

Schiöler, T, 'The Pompeii-groma in new light', Analecta Romana, 22 (1994), 45–60.

Toomer, G J, 'The chord table of Hipparchus and the early history of Greek trigonometry', Centaurus, 18 (1973), 6–28.

Verdelis, N M, 'Der Diolkos am Isthmus von Korinth', Mitteilungen des Deutschen Archäeologischen Instituts, Athens, 71 (1956), 51–59.

Walbank, M E H, 'The foundation and planning of Early Roman Corinth', Journal of Roman Archaeology, 10 (1997), 95–130.

Werner, Walter, 'The largest ship trackway in ancient times: the Diolkos of the Isthmus of Corinth, Greece, and early attempts to build a canal', International Journal of Nautical Archaeology, 26 (1997), 98–119.

Williams, Charles K, 'The early urbanization of Corinth', ASAtene, 60 (1982), 9–20.

Williams, Charles K II and Russell, P, 'Corinth excavations of 1980', Hesperia, 50 (1981), 1–44.

Engineering the Neapolitan state

Massimo Mazzotti

In 1808 a corps of civil engineers was constituted in the southern Italian Kingdom of Naples to provide it with new infrastructures and to implement a plan of economic and technological renewal. The activity of the corps led to controversy, however. In part the debate concerned their mathematical and scientific knowledge: was it sound? And could it legitimate their reshaping of the Neapolitan landscape? Around this debate, two opposite views of mathematics and of its relationship to engineering practices took shape: that of the engineers and that of their critics. While debates over the legitimacy and scope of mathematical methods in this period were not exclusive to Naples, they were experienced there with unparalleled intensity. The analysis of the production and use of mathematical knowledge in this context can therefore be particularly revealing.

The emergence of modern professions in the pre-unitary Italian states has attracted growing interest, and has proved to be an area where the methodological resources of the history of science and those of social theory can be fruitfully combined (Betri and Pastore 1997; Banti 1993; Malatesta 1995; Santoro 1997). Research has highlighted the relationships between changes in curricula, the redefinition of academic disciplines, the creation of new professions, and the emergence of modern administrative and bureaucratic structures (Brambilla 1997). The connection between professionalization and modernization is most evident in the case of the 'modern engineer' (Castellano 1987; Giuntini 1989; Santoro 1989; Zucconi 1992; Bigatti 1995; Foscari 1996; d'Elia 1996; Blanco 2000; di Biasio 2004). Two main issues have emerged so far from the body of literature on engineering around 1800. First, the institutionalization and legitimization of modern engineering was strictly connected to administrative reform. A second and less straightforward issue has to do with the nature of the engineers' knowledge, the new analytic rationality that informed their action, and with its relevance as a cause of social change. It is precisely the relationship between the cognitive-technological level and the socio-historical level to which I shall turn my attention here. How did the scientific and mathematical knowledge of engineers relate to their professionalization and to the modernization of the state?

It is generally assumed that the mathematical knowledge of modern engineers, in the Italian states as elsewhere, was the product of a gradual, quantitative increment of previous forms of mathematical and technical knowledge. In this perspective, the novelty of modern engineering training is described as the presence of *more* and *more advanced* mathematics. The development of a technoscientific culture, and the improved mathematical training of engineers, would thus be the cause of the shift of their practice from art to profession (Foscari 1996, 16). Accordingly, historian Luigi Blanco (2000, 18) has referred to a 'problematic knot', a dilemma that characterizes the current historiographical debate: was the professionalization of engineering primarily an intrinsic effect of scientific and technological innovation, or was it shaped by the intervention of the state? In this chapter I shall address this question by exploring aspects of the process of modernization of the kingdom of Naples in the early nineteenth century, and by focusing on the ways that professional civil engineering emerged and was legitimated (Giannetti 1988; de Mattia and de Negri 1988; di Biasio 1993; Foscari 1996; d'Elia 1996, 1997; Foscari 1997, 2000).

Redistributing power

The implementation of forms of administration that distinguish a modern state from earlier semi-feudal systems can be seen as a process of redistribution of power. In Naples, this process took the form of a long lasting struggle against the feudal system and the juridical and cultural institutions that sustained it, a struggle that reached a turning point in 1806, when this independent kingdom was invaded and occupied by the French army. From 1806 to 1815 Naples and southern Italy were an integral part of Napoleon's imperial system (Davis 2006). The French installed a new government, advised by local reformers and republican sympathizers. The government abolished the feudal system once and for all, and put in place an ambitious project of reform. Essentially, this meant transforming the semi-feudal Neapolitan kingdom into an efficient and centralized administrative monarchy (de Martino 1984; Rao and Villani 1995). The government could count on the collaboration of the reformist intelligentsia and of a significant portion of the local trading and landed middle-classes, to whom the new order could offer unprecedented economic opportunities and new forms of political representation. The new administrative system can be described as vertical, in the sense that it linked each administrative level to the one above and eventually to central government. By contrast, the traditional system can be described as horizontal, because the various administrative and juridical institutions worked almost independently, often following conflicting strategies.

The fate of the entire process of modernization rested upon the successful reshaping of the relation between central government and peripheral administrative institutions. The redistribution of power threatened the interests of elites both in Naples and in the provinces. The traditional semi-feudal system had guaranteed a large degree of autonomy and discretionary power to local elites. It also gave small groups of private investors a monopoly over lucrative financial and commercial enterprises such as public works, the grain trade, and money lending. French reform, by drastically reducing the discretion of provincial institutions, threatened the interests embedded in the administrative status quo (Davis 1981; di Ciommo 1988; Macry 1988). The new centralized and vertical administrative system was designed to allow the government to act quickly and effectively in any portion of the kingdom. Matters that had previously had an exclusively local dimension could now be managed by the central bureaucracy. At the same time, the map of the kingdom was redesigned by dividing it into provinces, districts, and communes, and by ranking towns and ports that hosted public functions according to their newly acquired status. Administrators at all levels acted more and more as civil servants in a hierarchical and meritocratic system, which was a threat to the powerful ancien régime institutions that were still functioning.

Emblematic of the new system of vertical relations was the government representative in each province, the *intendente provinciale* 'provincial superintendent'. Also central to the plan of reform was the creation ex novo of the Ministry of the Interior, which accumulated a large number of functions. All peripheral personnel depended on the Ministry, which was also in charge of controlling the finances of the communes, the administration of prisons, hospitals, hospices, agriculture, trade, manufacture, public education, the production of statistical data and, crucially for our present concerns, public works.

This plan of centralization was by far the most ambitious and radical to be implemented by the French anywhere in the newly occupied territories. As might have been expected, resistance was remarkably strong. Local elites voiced their concerns primarily through the provincial councils, especially after the 1815 restoration of the Bourbon monarchy, when it became clear that the administrative reform would be continued (Spagnoletti 1988).

The reform aimed to shift the decision-making process from the periphery to the centre of the state. This was not only a physical phenomenon: the very discourse of decision-making was shifted to a new cognitive level. French politicians and the Neapolitan intellectuals who advised them did not have first-hand knowledge of local conditions in the provinces. Instead, the new government legitimated its decisions and long-term plans by referring to a new kind of authority: the technical expertise of civil engineers. With this move, traditional interlocutors of the government such as the landed aristocracy, the church, local communities, and the other relevant local institutions, lost their status as decision-makers. They could not speak the mathematical language of the engineers and they could not understand their technical knowledge. Their opinion on local matters could therefore be reasonably dismissed as irrelevant. Thus the future of reform depended essentially upon the authority of new experts: at the core of the modernization programme were the engineer and his highly specialized mathematical knowledge (for a comparison with the French case, see Porter 1995).

The knowledge of the engineer

The mathematical language of the professional engineer and the supposed neutrality of his scientific judgements embodied the hopes and ambitions of the emerging middle class. In the technical reports of the engineers, as well as in the political and literary discourse, a new bourgeois myth was being forged: that of the engineer's exceptional cognitive, technical, and moral stand. He spoke the voice of Reason and fought against the obscure political interests of immoral individuals who threatened the 'common good' and the future of the country.

The *Real Corpo degli Ingegneri di Ponti e Strade* 'Royal Corps of the Engineers of Bridges and Roads' was established by royal decree in 1808. Never before in Naples had engineers been granted such a high social status, comparable to that of professionals in law and medicine (de Mattia and de Negri 1988). The corps was operative early in 1809, and included twenty-three engineers on four levels: Inspectors, Chief engineers, Ordinary engineers, and Assistant engineers. Each Inspector was responsible for a large territorial unit called a *divisione*, and was in charge of technical, financial, and administrative matters for all public works in his area. Chief engineers supervised all projects within their own *dipartimento*, and negotiated with local contractors. Ordinary engineers were responsible for the technical details of specific projects, and were supported by Assistant engineers. A small number of candidate engineers could also be employed as trainees. On a small scale, the corps was an exemplary model of a vertical, centralized, and well-administered institution, based on a meritocratic system of promotion and alien to the myriad particular interests that had shaped social life under the ancien régime.

The Neapolitan liberal intelligentsia rallied to support the modernizing battle of the engineers. Many believed that, due to the particular features of Neapolitan entrepreneurship, modernization guided from above was the only viable option for reforming both the administration and the economy. As for the engineers, their adherence to the values of a moderate political and economic liberalism was never in question, and culminated in the direct involvement of many members of the corps in the 1848 liberal revolution. The periodical *Il progresso delle scienze, delle lettere e delle arti* 'The progress of sciences, letters, and arts', a leading voice of Neapolitan and Italian liberalism, provided constant support for the activities of the corps. One contributor referred to engineering practice before the creation of the corps as greatly imperfect, because traditional civil architects had not been trained in the necessary scientific and practical disciplines. Their shortcomings were cognitive as well as moral: unlike the members of the corps, the writer complained, these architects had been chosen only because they were devoted to powerful ministers; as it might be expected, he wrote, their technical errors had been extremely grave (Rossetti 1835).

In straightforward prose, Carlo Afan de Rivera, director of the corps and its school from 1826 to 1852, offered the same fusion of cognitive and moral virtues, not only in the more rhetorical pieces he wrote for the general public, but also in the technical reports he sent to the ministry. Every technical intervention in the territory was an occasion for lauding the exceptional virtues of the members of the corps, and defending their exclusive prerogatives. The works around Lake Salpi, for instance, included land reclamation, the construction of new channels connecting the lake to the sea, and the creation of a modern fishing industry. Rivera reported that the former feudal owner of the lake lacked the relevant scientific knowledge to manage it, and therefore had been unable to make any profit out of it. With engineers now in charge of it things had changed dramatically: they had transformed a malarial area into a profitable investment for the government, a model site for national industry. The technical activities of the engineers always had a clear economic and political dimension. Rivera himself liked to refer to his activity as *economia politica*, 'political economy': he did not simply design channels and roads but also planned new *colonie* 'villages' the relocation of the population, and the establishment of factories. Rivera criticized Neapolitan entrepreneurs for their conservative investing strategies, and celebrated the engineers' effort to promote the various branches of political economy and hence the public wealth (Rivera 1842, 4, 6).

The intervention of the corps altered the socioeconomic structure of entire regions. Not surprisingly, Rivera had to fight what he regarded as preoccupations and prejudices, and admitted that the greatest obstacles to his plans were moral, rather than physical. He lamented that many discredited his work publicly, guided either by their own *illecito guadagno* 'illicit interest', or by the misguided belief of having lost some legitimate right. A typical example was the opposition of sheep-breeders to his project for agricultural exploitation of the Apulian plan. Rivera judged their traditional breeding system to be a heritage of *barbarie de' tempi andati* 'barbaric times' and the sign of a *vizioso* 'immoral' system of administration (Rivera 1842, 26).

In Rivera's discourse, the superior morality of the engineer was grounded on his understanding of natural and social reality, which legitimated and guided his transforming action. This was not the knowledge produced and transmitted in universities, where architects and civil engineers had traditionally been trained. Modern engineers were trained in an entirely new institution of higher education, the *Scuola di Applicazione di Ponti e Strade* 'School of bridges and roads', based on the model of the *Ecole des Ponts et Chaussées* in Paris, reformed in 1804 (Blanco 1991; Picon 1992). For the first recruits to the corps, in 1809, entry was by direct nomination by the king on the recommendation of the director of the corps; thereafter, suitable candidates were trained at the highly selective *Scuola* where courses started in the autumn of 1811 (Russo 1967). The course lasted three years (in 1818 it was reduced to two; in 1826 it was extended to four), and at the end the top students could enter the corps of engineers at the lower level (the numbers varied: the 1814 class saw three students entering the corps). The curriculum was radically different from that of the university, and mathematics was a central element. Students were selected through an entrance examination designed to test their mathematical knowledge. It included questions of plane and solid geometry, trigonometry, analytic geometry (curves and second degree surfaces), differential and integral calculus, design, French, and Latin. Successful candidates entered the 'first class', a course in which most of the time was devoted to mathematical disciplines (Rossetti 1835, 330). They studied the basics of mechanics, hydraulics, descriptive geometry, perspective, and geodesy. At the end of this cycle, another examination selected those who would enter the 'second class', mainly devoted to practical applications. Here the students studied subjects such as applied chemistry, agronomy, the application of mechanics to constructions and machines, architectonic structures, architectonic machines, and the features of the various kinds of constructions, primarily walls, roads, roofs, dikes, bridges, and suspended metal bridges. At the end of this second cycle, students went through a final one-week examination, on the basis of which they were ranked (Rossetti 1835, 333–335). The level of the entrance examination, and the career prospects, made the *Scuola* an ideal choice for college-trained men of 18 to 20 years old, mostly from the middle and upper levels of the southern bourgeoisie (Foscari 1997, 286–287). From 1812 a new and multifunctional *Scuola Politecnica e Militare* 'Polytechnic and military school' began to train, among others, candidates for the engineering school. In this way, the curriculum of the engineer became entirely independent from the local colleges and universities.

The life of the corps and the *Scuola* became more difficult under the restored Bourbon regime from 1815 to 1860. A powerful conservative block returned to power, representing the local and monopolistic interests that were most threatened by engineering activities. Reducing the autonomy and prerogatives of engineers became a primary objective of conservative political strategy. At first, in 1817, the corps and the *Scuola* were suppressed. The corps was replaced by a *Direzione generale di ponti e strade*, 'General directorate of bridges and roads', which was closely modelled on an ancien régime institution for the control of roads. Personnel were reduced from sixty-nine to fifteen; Inspectors were eliminated; a new figure of contract engineer was created; and the provinces were entitled to supervise some of the works in their own territory. The careers of provincial engineers were influenced by the yearly reports that provincial superintendents sent to the government. The new institution was less structured than the corps, and it was also much less independent of local interests. The juridical status and career prospects of engineers, particularly those on short-term contracts and working under the control of the provincial councils, became uncertain. Even more crucially, public works were now planned on a short-term basis, which made it difficult to modify landscapes (Maiuri 1836).

The aim of these changes was to assimilate engineers under other administrative employees, thus reducing the autonomy and effectiveness of their action in the provinces. The Crown was largely responsible for this weakening of the prerogatives and authority of the corps, and for the closure of its school. Behind these decisions was the political aim of compromising between centralization and the interests of the provincial elites. In the mid 1820s, after the conservative excesses of the early restoration period, the appointment of a new director, Carlo Afan de Rivera, signalled that the balance had again shifted in favour of the centralizing process. Rivera prepared a plan that restored the authority and autonomy of the corps almost to its original form, starting with the name. The plan also included an increase in the number of engineers, limitations on the influence of provincial authorities, and the strengthening of the *Scuola di Applicazione*, which had reopened in 1819. The course was lengthened to four years and the number of teachers was increased. The government accepted the plan almost in its entirety, and it became operative in 1826.

Conservative utopia versus the ideology of progress

In the years that followed, as the debate over modernization intensified, the engineering school increasingly became a target for conservative politicians.

Already in the early 1820s, the director Colonel Francesco de Vito Piscitelli had to defend the special nature of the school, remarking that the distinction between civil engineers and other scientific practitioners was a feature of every civilized nation. He also defended the legislation according to which the school was the only entrance to a career of engineering (de Mattia and de Negri 1988, 464). In 1835, Rivera set up a public demonstration of the didactic effectiveness of the school: a one-week exhibition held at the *Scuola di Applicazione* to celebrate the corps, its many remarkable achievements, and its role in the modernization of the country. The liberal periodical *Il progresso* reported on it enthusiastically, and asserted that the superiority of modern engineering over previous forms of architecture and engineering derived from its being a coordinated collective activity, in contrast to the traditional image of the isolated genius. Modern engineering was described as the practice of a well-structured group of experts, which acted as if guided by a single mind. The exhibition was a success. Projects exhibited by students included plans of ports, prisons, and bridges. Plans of public works recently completed by the corps were also shown, such as those for the metal suspension bridges over the rivers Garigliano and Calore. Students from the school gave lectures on topics such as mechanics, hydraulics, and descriptive geometry. The author of the report in *Il progresso* was confident that the future productions of these alert and educated minds (*sì svegliati e culti ingegni*) would benefit the country greatly, and would keep it apace with the wealthiest European countries. The author touched only cursorily upon the reasons for such an unusual exhibition but the relevant passage is revealing. Rivera had to prove that the scientific training of his engineers was sound, and that the school was not a waste of public money (Rossetti 1835, 329–335). Apparently some influential people were convinced of the contrary, and were putting pressure on the ministry of finance to cut its funding. Leading the critics was a first-rank politician, Giuseppe Ceva Grimaldi, president of a major ancien régime institution and future prime minister.

The activity of Ceva Grimaldi is key to understanding conservative cultural and political strategies. In the 1830s, crucial years for Neapolitan modernization, he repeatedly attacked Rivera on the practices of his engineers and the usefulness of the *Scuola di Applicazione*. He defended the interests of the provincial elites, landowners, and private contractors, and did whatever he could to restrict the autonomy and functioning of the corps. He pressed for a reduction of personnel, for lower wages, and for giving back to the provinces the responsibility for financing and supervising public works (Giannetti 1988, 936–938). By attacking the engineers, conservatives were attacking, although indirectly, the entire centralizing policy of the Crown, and criticizing its absolutist ambitions.

Rivera replied promptly, pointing out the reasons for the socioeconomic developments he considered necessary, and more generally for the need to 'civilize' the provinces. But the technical language in which he described the activities of his engineers, and the large amount of technical data with which he presented Ceva Grimaldi, were now being questioned by conservatives. The modern engineer's 'analytic reasoning' was to Ceva Grimaldi an expression of arid and abstract knowledge, which had little to do with the complexity of practical reality and with the art of economic and political decision-making.

The debate generated various publications, which even in their titles showed two opposing perspectives. Thus Rivera published *Considerazioni su i mezzi da restituire il valore proprio a' doni che ha la natura largamente conceduto al Regno delle Due Sicilie*, 'Remarks on the means to value the gifts that nature has lavished upon the Kingdom of the Two Sicilies' (1832), which emphasized the new possibilities for exploiting the kingdom's natural resources through the application of modern science. In response, Ceva Grimaldi published *Considerazioni sulle opere pubbliche della Sicilia di qua dal faro dai Normanni sino ai giorni nostri*, 'Remarks on public works in continental Sicily from Norman times to the present day' (1839), an historical essay on the art of administration and the Neapolitan tradition of public works since the Middle Ages. Ceva Grimaldi ridiculed the 'prejudice' that civilization in southern Italy had began with the establishment of the Corps of Bridges and Roads, and called the tendency to attribute all administrative evils to the feudal system 'a boring commonplace'. In his writings Rivera referred constantly to such well-known myths of modernity as the inexorable path of progress, the rationality of a liberal economy, and the superior moral status of engineers. In response, Ceva Grimaldi seemed to ground his reasoning on an unmistakably conservative utopia. Behind his sceptical remarks on modernity lay references to a mythical description of the kingdom in the pre-French period. In those days, according to Ceva Grimaldi, all channels were navigable, all roads well kept, the administration functioned perfectly, and the skills of Neapolitan master masons were unsurpassed. He asserted that the Neapolitans knew how to build roads even before the French arrived, and that modifying their routes was a sign of little respect for their ancestors. His descriptions are literary reconstructions of a past that never was, where peasants and artisans lived peacefully around the manor and the church, in a kingdom made up of hundreds of small, highly independent *patrie*, 'fatherlands'. Behind Ceva Grimaldi's utopia, and behind each of his arguments, lay the aim of preserving as much as possible of the current territorial and administrative structure of the country. He claimed to be speaking for the many small *patrie*, which were desperately opposing the 'tyranny' of the capital. The enemy was not the king, of course, but whatever was 'foreign', 'abstract', 'anti-historical'. The new 'industrial feudalism', in Ceva Grimaldi's view, was no less tyrannous than the old one (Ceva Grimaldi 1839, 149, 172–173). In these same years, the mythical image of a happy and timeless Neapolitan countryside was being diffused all over Europe by the extremely successful school of romantic landscapists known as the Posillipo School. What they portrayed, I suggest, was Ceva Grimaldi's conservative utopia: a still and eternal rural society, founded on nature and tradition: a world that did not require any significant change (Causa 1967; Ortolani 1970; Martorelli 1993).

Analytic rationality

The example of the 1835 exhibition makes it clear that in addition to their legislative battle, conservatives also undermined the authority of the engineers by questioning the training on which their power was based. In particular, it was argued that the mathematical knowledge of the engineers did *not* provide them with a superior understanding of physical and social conditions. Ceva Grimaldi led this de-legitimizing campaign, criticizing the kind of mathematical training offered by the engineering school. He argued that the school should be open to everybody, not only those 'few initiates' who passed the entrance examination, and he portrayed its analytic training as an ideological device, and an arbitrary source of 'privilege' (Ceva Grimaldi 1839, 169).

To understand how the conflict over modernization was structured at the cognitive level, we need to look briefly at the mathematical practice of the engineers. The curriculum and teaching of the *Scuola di Applicazione* were shaped by a distinctive set of epistemological and cultural assumptions. Sessions consisted of two hours, one for the teacher's lecture, and the other for the students to engage actively in the learning process. Thus they familiarized themselves with instruments, machines, chemical reactions, or mathematical techniques presented by the teacher. Whenever possible, students were encouraged to handle objects, and to acquire direct experience of the content of the lecture. At the end of every academic year, students took part in a summer campaign, where they assisted the engineers of the corps in their routine operations. Students were to be actively involved in the process of acquiring knowledge, and discover empirical and scientific truths through their own experience. The importance attributed to handling and sensory experience derived from precise beliefs about how the mind works. Teaching was based on a sensationalist concept of knowledge: once exposed to the appropriate sensory inputs and assisted in developing what was supposed to be the most natural way of reasoning, the student would be able to discover *by himself* the relevant scientific truths. But what was 'the natural way of reasoning'? The teachers of the *Scuola* identified it with what they called the 'analytic method'.

In this context, the term 'analysis' was used to refer to different but related kinds of objects and practices. These included problem-solving techniques based on the decomposition and subsequent composition of ideas, the mechanization of thinking according to the combinatorial rules of algebra, the study of economic issues on the basis of individual and collective utility, a philosophy of teaching based on sensation and experience, and the corresponding reorganization of scientific disciplines and curricula. It was a set of cultural resources, styles of thinking, skills, mathematical techniques, relevant problems, and possible solutions that informed the activities of the engineers (Picon 1992). The teachers of the *Scuola* built their experimental and mathematical courses accordingly, so that analytic rationality was incorporated into the training of an engineer from the very beginning.

In its more general sense, analysis was a way of thinking based on the assumption that all complex problems could be broken down into their elementary components, and that these components could be then solved independently thus providing a solution for the original problem. In mathematics, the integration of an equation was taken to be emblematic of this way of proceeding. Mathematical analysis, although a polysemic term, was often used as a general label for the techniques of algebraic analysis and those of calculus, as based on the mechanical manipulation of symbols and in opposition to the intuitive procedures of synthetic geometry. In this sense, analysis was described by its supporters as natural and easy-to-learn. With proper training, any student could master it, because it was simply an extension and formalization of the normal way the human mind works. Once the spirit of the analytic method was grasped, the authors of engineering textbooks claimed, the solution of any relevant mathematical problem would be straightforward, the outcome of a mechanical procedure (Padula 1838, 13). There were no child-prodigies in the engineering school, just enthusiastic students.

It followed that the engineering school taught its students not just more or more advanced mathematics, but a new kind of mathematics altogether. Representative of the analytic approach was a textbook of plane and solid geometry entitled *Raccolta di problemi di geometria risoluti con l'analisi algebrica* 'Geometric problems resolved by algebraic analysis' (1838), by Fortunato Padula, a graduate of the engineering school and one of its future directors. Padula grouped the standard problems of a university course and solved them in a purely analytical way, through the manipulation of general equations that represented the properties of the figures in question. Padula was not simply translating geometrical problems into analytic language in order to make the discovery of the solution easier, and then performing an appropriate final construction, as synthetic mathematicians required. Rather, he did what he had been trained to do: he showed how the solution to each problem could be seen as a particular case of a general structural relationship between families of geometrical entities. The problem-solving process did not begin with inspection of the figure, but with the immediate introduction of the relevant equations, each of them generalized by means of appropriate parameters and variables, and thereafter manipulated without any reference to the initial figure. The particular origin of the problem thus became irrelevant. The necessity for inspecting the figure was eliminated, as was the entire constructive phase. The paradigmatic work, in this respect, was Joseph Lagrange's article (1773) on triangular pyramids. In this article Lagrange completely dispensed with the use of figures, and instead provided a description of the general structural properties of the pyramids. These technical developments accompanied a major epistemological shift in mathematics, as the meaning and heuristic power of algebraic algorithms were now conceived as completely detached from geometrical intuition (Fraser 1989). The Lagrangians—the Neapolitan engineers among them—considered analytic reasoning that detached algebraic reasoning from geometric intuition, as a fully legitimate and autonomous heuristic method.

Standardization

One of the key issues in the debate over modernization was the standardization of weights and measures. Here all the main social, political, and scientific themes that we have touched upon converged in a single, crucial issue. The engineers, Rivera in the first place, campaigned strongly in favour of standardization (Rivera 1840; 1841). Rivera had his engineers adopt a single and all-purpose decimal system as early as 1830. To him standardization was a necessary if the corps was to take full control of the territory of the kingdom. Once again it was centralization versus local autonomy, uniformity versus variety, rational administration versus a world of autarchic communities.

Traditional Neapolitan measuring systems reflected the complex socio-political structure of the country, and the variety of its traditional occupations. So, for instance, land was often measured in terms of the number of days needed to plough it or the number of men needed to harvest it in a day, or else in terms of its monetary value. This meant that metrological units could not be fixed, because working in the plain was different from working in a mountainous region, and the value of land varied from place to place. Further, products that were manufactured by different processes, like wine and oil, were measured in different units, none of which was decimal. Such variety was of no concern to users, who had developed the necessary translating skills. In the personal negotiations that characterized internal trading, buyers and sellers used their discretion to re-shape the rates of exchange between units. To Rivera and the officials of the corps this was chaos, and a sign of bad administration. A uniform plan of development for the country needed a uniform measuring system.

The heated anti-standardization campaign was guided, as might be expected, by Ceva Grimaldi, and has been described by later commentators as rhetorical and historical, and opposed to the rational, technical arguments of Rivera (Giannetti 1988; d'Elia 1997). The problem with this account of the controversy is that it implies that the contemporary scientific establishment was united in defending Rivera's arguments when, in fact, it was quite the contrary. In other words, the two fronts did not map onto humanistic and scientific culture, respectively, but rather onto two different kinds of humanistic and scientific culture. Let us consider Ceva Grimaldi's essay against the reform of weights and measures (1838), one of the most representative texts of the conservative literature. Interestingly enough, this essay contains a long scientific introduction signed by the leading Neapolitan mathematician of the time, Vincenzo Flauti, professor at the University of Naples and secretary of the Royal Academy of Sciences. In opposition to all basic beliefs of the reformist intelligentsia, Flauti noted that mathematical abstractions are beautiful intellectual productions, but they are useless in addressing questions related to public welfare. Flauti openly attacked Rivera and his *banda* 'gang' of engineers for trying to standardize the many systems of weights and measures in the kingdom. The engineers were, he claimed, attempting to provide an abstract solution to a concrete problem, and in so doing they threatened to destroy the inveterate customs of the people. In the past, the people did not need any expert to tell them how to measure and weigh, as they were confident of their own ancient customs. To modify these customs, Flauti concluded, would amount to modifying human nature itself. Flauti agreed that the decimal system was suitable for calculations and for certain scientific purposes, but argued that it was much less so for everyday transactions (Ceva Grimaldi 1838). He believed that decisions regarding weights and measures should be taken by the users themselves, while only economists and historians should advise the government on these matters. This was certainly not the terrain for professional mathematicians, let alone engineer-mathematicians.

Rivera held precisely the opposite view. He thought that a corporation of scientist-artisans should plan and direct public works, and he ridiculed the snobbish behaviour of Neapolitan academic mathematicians, who thought they would degrade themselves if they got close to the factories, and therefore neglected the application of the sciences to the arts (Rivera 1832, II 461). What Rivera aimed for was precisely the application of science to the administration of the state (Rivera 1823, 38).

Another mathematics

On what grounds did Flauti criticize the mathematical practice of engineers? Around 1800 Naples had seen the emergence and rapid institutionalization of a mathematical methodology that was different from the analytic approach.

Following the innovative teaching of Nicola Fergola, a group of young and talented mathematicians had developed a lively school of synthetic mathematics (Loria 1892; Amodeo 1905; Mazzotti 1998). This meant that their approach to mathematics was essentially geometrical. Their practice and techniques were constructed around the basic notion of intellectual intuition as the foundation of all mathematics and the source of its certainty. They emphasized the visual dimension of geometrical knowledge—its distinctive perspicuity—against faith in the unlimited power of analysis and the focus on the mechanical manipulation of algorithms, which had characterized late eighteenth-century French mathematics and which was central to early nineteenth-century engineering training in Naples. The Neapolitan synthetic school denied that there could be something like a universal problem solving method, and insisted on the irreducible specificity of different kinds of mathematical problems and techniques. Crucially, their problem-solving techniques were to be used exclusively in the realm of pure mathematics: they insisted upon the limited scope of mathematical reasoning, which they believed could not guide human action in empirical matters. These beliefs led Fergola and his students to shift their attention from applications to foundational issues, and to prefer the study of classical geometrical problems to that of eighteenth-century analysis.

To solve a geometrical problem elegantly was, according to the synthetics, a matter of long training and exceptional talent. The necessary procedures could not be mechanized in the way the analytics believed and, importantly, they could not be deployed in other disciplines without losing most of their heuristic power. On these grounds, the synthetics asked for a clear-cut demarcation between the speculative pure mathematician and the engineer. The former might lead mathematical research in universities and academies, while the latter would merely use portions of mathematical knowledge for specific practical purposes (Flauti 1820; 1822).

Synthetic teaching was based on discovering and grooming the natural talent of exceptional students through the study of the ancient paradigmatic examples. The best students formed a kind of inner circle that met often at the professor's house. A typical product of this system was the child prodigy Annibale Giordano who, aged fifteen, had been invited by Fergola to solve a certain geometrical problem, and whose classically elegant solution was published in the journal of the most prestigious Italian scientific society of the time (Giordano 1786). Fergola's synthetic school was an innovative phenomenon with respect to previous academic traditions. However, Fergola and his students tended to conceal their novel scientific and philosophical interests by presenting their work as essentially derived from an ancient tradition in geometrical problem solving, which stretched from the Classical era through Galileo and the geometers of the seventeenth century. This tradition, with its emphasis on Greek and Latin sources, was intended to give the synthetic school an authoritative position in conservative Neapolitan academic culture. It also reinforced the impression, among later historians, that Neapolitan synthetics were survivors of a past age. On the contrary, Fergola was well versed in algebraic techniques, and in integral and differential calculus (Fergola 1788a; 1788b; undated manuscript). Fergola's mathematical project was far from being a mere return to the ancients. Rather, this devout professor was clearly concerned by the recent association of certain mathematical techniques—namely analysis—with philosophical and political projects that in his view threatened the stability of traditional society. His scientific work can be seen as a thorough attempt to investigate the foundations of mathematical reasoning, in order to redefine the range of its meaningful applications.

Working along the same lines, Fergola's former pupil Flauti challenged the engineers in 1839 to solve a number of geometrical problems, with the intention of showing the intrinsic superiority of the synthetic over the analytic method. A reply came from Padula, with an essay significantly dedicated to Rivera. Padula stated once again that algebra and calculus should be applied to solve the socioeconomic problems of the country. Anyone who is interested in modern mathematics, Padula argued, should be concerned with its applications to natural philosophy, constructions, and industrial mechanics, and should abandon the sterile and uninteresting exercises favoured by the synthetic school (Padula 1839, 46). He was referring to the concern of the synthetics with pure mathematics, their development of methods of synthetic and projective geometry (Flauti 1807), their study of the history of mathematics, and their attempt to provide a solid logical foundation to the entire edifice of mathematics, including calculus. Indeed, in some respects the synthetics' programme of research was closer to what we now perceive to be the main European trends of the early nineteenth century rather than the dated Lagrangian programme defended by Padula. The engineers remained ostentatiously uninterested in pure mathematics and synthetic geometry until the mid-century (Besana and Galluzzi 1980).

Having emerged together in years of deep political and cultural struggle, the two Neapolitan schools developed goals and practices that were in most respects opposite. They encouraged different skills and techniques, and were differently receptive to novelties coming from abroad. I would not like to argue here that there was a necessary connection between certain political and religious attitudes and certain problem-solving methods. Rather, my interpretation is historical. In early nineteenth-century Naples different cultural resources, including mathematical knowledge, were being mobilized and reshaped to express and support different socio-political projects for the future of the country. On the one hand, engineers and their supporters among the southern bourgeoisie reshaped analysis as an emblem and instrument of modernization. On the other hand, devout and conservative mathematicians like Fergola and Flauti considered the synthetic approach, and the institutionalization of the new subdiscipline of 'pure mathematics', as the most appropriate response to what they perceived as a broad cultural and moral crisis of European civilization.

The analytic rationality of the engineers, and their mathematical and technical knowledge, were openly challenged by university-trained mathematicians until the late 1830s, when the conservative opposition lost ground in the broader political and social arena. Ultimately, the mathematical knowledge of the engineers became legitimated as enough relevant groups within Neapolitan society began to share the orientation of men like Rivera. The emergence and stabilization of analysis, in other words, should not be seen as a pre-existing historical cause for the process of modernization, but rather as an expression of this process, and the emblem of its success.

Conclusions

In this chapter I have explored some aspects of the relationship between mathematical knowledge and its carriers. The Neapolitan case offers an effective illustration of how mathematical knowledge can be shaped, mobilized, and deployed to support the goals of particular collectives. It therefore provides evidence for the socially constructed nature of mathematical knowledge, and more specifically for the contingent character of the relationship between mathematics and engineering. The boundaries between these two sets of practices and the form of their interaction are the product of socio-historical conditions, and as such are constantly open to negotiation and redefinition.

The chapter has also investigated the way in which the institutionalization of new forms of mathematical reasoning in the first half of the nineteenth century was related to the creation of professional elites and to the formation of the modern state. My reconstruction of the Neapolitan case suggests that these should be understood as different aspects of an essentially unitary process of social and cognitive change. The emergence and stabilization of new forms of mathematical reasoning does not appear to have been the cause of the emergence of new and more effective engineering practices. Rather, the establishment of the new mathematics was a condition for the legitimization of engineering activity—and the socio-political vision that shaped it.

In the light of these considerations, the historiographical dilemma mentioned at the beginning of this chapter should be rejected as misleading. One does not have to choose between two competing historical explanations for the professionalization of engineering—either the appearance of a new body of knowledge or the intervention of the state. My interpretation of the Neapolitan case offers further evidence that science and technology do not evolve by virtue of some inner logic independent from the intentions and purposes of those who learn, teach, use, and change them. In particular, engineering practices seem to be the negotiated outcome of specific cultural, political, and economic interactions rather than the result of a straightforward application of scientific knowledge to the solution of practical problems (Alder 1997; Kranakis 1997). Thus, the analytic rationality and specific mathematical practice of Neapolitan engineers were shaped by, and sustained, their theoretical and practical orientations (for the French case, see Picon 1992; Brian 1994). The institutionalization of what they called 'analysis' was part of a broader mobilization and transformation of cognitive resources in support of social reform. In other words, the battle for modernization had to be won not just *with* but also *in* the textbooks of mathematics.

Bibliography

Afan de Rivera, C, *Considerazioni sul progetto di prosciugare il Lago Fucino e di congiungere il mare Tirreno all'Adriatico per mezzo di un canale navigabile*, Naples: Tipografia reale, 1823.

Afan de Rivera, C, *Considerazioni sui mezzi da restituire il valore proprio a' doni che ha la natura largamente conceduto al Regno delle Due Sicilie*, 2 vols, Naples: Fibreno, 1832.

Afan de Rivera, C, *Della restituzione del nostro sistema di misure, pesi e monete alla sua antica perfezione*, Fibreno, 1840.

Afan de Rivera, C, *Tavole di riduzione dei pesi e delle misure delle Due Sicilie in quelli statuiti dalla legge del 6 aprile del 1840*, Fibreno, 1841.

Afan de Rivera, C, *Rapporto del direttore generale di ponti e strade a S.E. il Ministro segretario di stato delle finanze*, Naples, 1842.

Alder, K, *Engineering the revolution: arms and enlightenment in France, 1763–1815*, Princeton University Press, 1997.

Amodeo, F, *Vita matematica napoletana*, 2 vols, Naples: Giannini, 1905; Pontaniana, 1924.

Banti, A, 'Borghesie delle "professioni". Avvocati e medici nell'Europa dell'Ottocento', *Meridiana*, 18 (1993), 13–14.

Besana, L, Galluzzi, M, 'Geometria e latino: due discussioni per due leggi', in G Micheli, *Storia d'Italia. Annali 3: scienza e tecnica nella cultura e nella società dal Rinascimento ad oggi*, Einaudi, 1980, 1287–1306.

Betri, M L, and Pastore, A (eds), *Avvocati, medici, ingegneri. Alle origini delle professioni moderne (secoli XVI–XIX)*, Clueb, 1997.

Bigatti, G, *La provincia delle acque. Ambiente, istituzioni e tecnici in Lombardia tra Sette e Ottocento*, Franco Angeli, 1995.

Blanco, L, *Stato e funzionari nella Francia del Settecento: gli ingénieurs des ponts e chaussées*, Il Mulino, 1991.

Blanco, L (ed), *Amministrazione, formazione e professione: gli ingegneri in Italia tra Sette e Ottocento*, Il Mulino, 2000.

Brambilla, E, 'Università, scuole e professioni in Italia dal primo '700 alla Restaurazione. Dalla "costituzione per ordini" alle borghesie ottocentesche', *Annali dell'Istituto Storico Italo-Germanico in Trento*, 23 (1997), 153–208.

Brian, E, *La mesure de l'Etat. Administrateurs et geometres au XVIIIe siecle*, Albin Michel, 1994.

Castellano, A, 'Il Corpo di Acque e Strade del Regno Italico: la formazione di una burocrazia statale moderna', in *La Lombardia delle riforme*, Electa, 1987, 45–64.

Causa, R, *La scuola di Posillipo*, Fabbri, 1967.

Ceva Grimaldi, G, *Considerazioni sulla riforma de' pesi e delle misure ne' reali dominii di qua dal faro, con note dell'editore, e con una lettera all'autore del prof. Flauti*, [Naples], [1838].

Ceva Grimaldi, G, *Considerazioni sulle opere pubbliche della Sicilia di qua dal faro dai Normanni sino ai giorni nostri*, Flautina, 1839.

Davis, J, *Merchants, monopolists and contractors: a study of economic activity and society in Bourbon Naples, 1815–1860*, Arno Press, 1981.

Davis, J, *Naples and Napoleon. Southern Italy and the European revolutions (1780–1869)*, Oxford University Press, 2006.

de Martino, A, *La nascita delle intendenze: problemi dell'amministrazione periferica nel Regno di Napoli, 1806–1815*, Jovene, 1984.

de Mattia, F, and de Negri, F, 'Il corpo di ponti e strade dal decennio francese alla riforma del 1826', in Massafra 1988, 449–468.

di Biasio, A, *Ingegneri e territorio nel Regno di Napoli. Carlo Afan de Rivera e il Corpo dei Ponti e Strade*, Amministrazione Provinciale, 1993.

di Biasio, A, *Politica e amministrazione del territorio nel Mezzogiorno d'Italia tra Settecento e Ottocento*, Edizioni Scientifiche Italiane, 2004.

di Ciommo, E, 'Elites provinciali e potere borbonico. Note per una ricerca comparata', in Massafra 1988, 965–1038.

d'Elia, C, *Stato padre, stato demiurgo. I lavori pubblici nel Mezzogiorno, 1815–1860*, Edipuglia, 1996.

d'Elia, C, 'La scrittura degli ingegneri. Il Corpo di Ponti e Strade e la formazione di un ceto burocratico a Napoli', in Betri and Pastore 1997, 293–306.

Fergola, N, Undated manuscript, Biblioteca Nazionale di Napoli, Ms.III.C.31–36.

Fergola, N, 'Risoluzione di alcuni problemi ottici', *Atti della Reale Accademia delle Scienze e Belle-Lettere di Napoli*, 1 (1788a), 1–14.

Fergola, N, 'La vera misura delle volte a spira', *Atti della Reale Accademia delle Scienze e Belle-Lettere di Napoli*, 1 (1788b), 65–84 and 119–138.

Flauti, V, *Elementi di geometria descrittiva*, Salvioni, 1807.

Flauti, V, *Tentativo di un progetto di riforma per la pubblica istruzione nel Regno di Napoli*, Naples, 1820.

Flauti, V, *Dissertazioni del metodo in matematiche, della maniera d'ordinare gli elementi di queste scienze, e dell'insegnamento de' medesimi*, Naples, 1822.

Foscari, G, *Dall'arte alla professione. Gli ingegneri napoletani tra Settecento e Ottocento*, Edizioni Scientifiche Italiane, 1996.

Foscari, G, 'Ingegneri e territorio nel Regno di Napoli tra età francese e restaurazione: reclutamento formazione e carriere', in Betri and Pastore 1997, 279–291.

Foscari, G, 'Dalla Scuola al Corpo: l'ingegnere meridionale nell'Ottocento preunitario', in Blanco 2000, 379–395.

Fraser, C, 'The calculus as algebraic analysis: some observations on mathematical analysis in the eighteenth century', *Archive for History of the Exact Sciences*, 39 (1989), 317–335.

Giannetti, A, 'L'ingegnere moderno nell'amministrazione borbonica: la polemica sul corpo di ponti e strade', in Massafra, 1988, 935–944.

Giordano, A, 'Considerazioni sintetiche sopra un celebre problema piano e risoluzione di alquanti problemi affini', *Memorie matematiche e fisiche della società italiana delle scienze, detta dei XL*, 4 (1786), 4–17.

Giuntini, A, 'La formazione didattica e il ruolo nell'amministrazione granducale dell'ingegnere nella Toscana di Leopoldo II', in Z Ciuffoletti, L Rombai (eds), *La Toscana dei Lorena: riforme, territorio, società*, Olschki, 1989, 391–417.

Kranakis, E, *Constructing a bridge: an exploration of engineering culture, design and research in nineteenth-century France and America*, MIT Press, 1997.

Lagrange, J L, 'Solutions analytiques de quelques problèmes sur les pyramides triangulaires' (1773), in J A Serret (ed), *Oeuvres de Lagrange*, Paris, 1869, III 661–692.

Loria, G, *Nicola Fergola e la scuola che lo ebbe a duce*, Tipografia del Regio Istituto Sordomuti, 1892.

Macry, P, 'Le elites urbane: stratificazione e mobilità sociale, le forme del potere locale e la cultura dei ceti emergenti', in Massafra 1988, 799–820.

Maiuri, A, *Delle opere pubbliche nel Regno di Napoli e sugli ingegneri preposti a costruirle*, Naples, 1836.

Malatesta, M (ed), *Society and professions in Italy, 1860–1914*, Cambridge University Press, 1995.

Martorelli, L (ed), *Giacinto Gigante e la scuola di Posillipo*, Electa, 1993.

Massafra, A (ed), *Il mezzogiorno preunitario, Economia, società e istituzioni*, Dedalo, 1988.

Mazzotti, M, 'The Geometers of God: Mathematics and Reaction in the Kingdom of Naples', *Isis*, 89 (1998), 674–701.

Ortolani, S, *Giacinto Gigante e la pittura di paesaggio a Napoli e in Italia dal '600 all'800*, Montanino, 1970.

Padula, F, *Raccolta di problemi di geometria risoluti con l'analisi algebrica*, Fibreno, 1838.

Padula, F, *Risposta al programma destinato a promuovere e a comparare i metodi per l'invenzione geometrica presentato a' matematioci del Regno delle Due Sicilie*, Fibreno, 1839.

Picon, A, *L'invention de l'ingénieur moderne. L'Ecole des Ponts et Chaussées, 1747–1851*, Presses de l'Ecole Nationale des Ponts et Chaussées, 1992.

Porter, T, *Trust in numbers: the pursuit of objectivity in scientific and public life*, Princeton University Press, 1995.

Rao, A M, and Villani, P, *Napoli 1799–1815. Dalla repubblica alla monarchia amministrativa*, Edizioni del Sole, 1995.

R[ossetti], L, 'Sulla scuola di applicazione annessa al Corpo de' Ponti e Strade del Regno di Napoli', *Il progresso delle scienze, delle lettere e delle arti*, 10 (1835), 328–335.

Russo, G, *La scuola di ingegneria di Napoli, 1811–1967*, Istituto Editoriale del Mezzogiorno, 1967.

Santoro, M, 'Professioni, stato, modernità. Storia e teoria sociale', *Annali di storia moderna e contemporanea*, 3 (1997), 383–421.

Santoro, R, 'L'amministrazione dei lavori pubblici nello Stato Pontificio dalla prima restaurazione a Pio IX', *Rassegna degli archivi di stato*, 49 (1989), 45–94.

Spagnoletti, A, 'Centri e periferie nello stato napoletano del primo Ottocento', in Massafra 1988, 379–391.

Zucconi, G, 'Ingegneri d'acque e strade', in G L Fontana, A Lazzarini (eds), *Veneto e Lombardia tra rivoluzione giacobina ed età napoleonica: economia, territorio, istituzioni*, Cariplo, 1992, 400–419.

CHAPTER 3.4

Observatory mathematics in the nineteenth century

David Aubin

'The value of the service of an Assistant to the Observatory', the Astronomer Royal George Biddell Airy wrote in 1861, 'depends very materially on his acquaintance with Observatory Mathematics'.[1] There is a rather strange ring to this expression. One knows, of course, that mathematics has always been used extensively in observatories. Ever since permanent astronomical stations were set up in Europe during the Renaissance, observers have drawn on the most elaborate mathematical tools available to them to correct the observational data they produced and to come up with theoretical predictions to which it could be compared. Up until the nineteenth century, astronomers played a central role in the development of many parts of mathematics. Indeed, together with geometry and arithmetic, astronomy had always been considered as one of the main branches of mathematics.

Still, in what sense can one talk of 'observatory mathematics'? Should one understand the expression as designating the subset of mathematics that was especially relevant to the scientific activities carried out inside observatories? Or is there—has there ever been—a specific character common to all mathematical

---

1. 'Remarks on the neglect, by the Junior Assistants, of the course of education and scientific preparation recommended to them' (4 December 1861). Cambridge University Library, Airy's Papers, RGO 6/43, 235.

tools and concepts used in those places? If so, what sense is there in carving out a portion of mathematics on the basis of it being used in a specific institution?

In this paper I want to introduce the question of place and space in the history of mathematics by looking at the various ways in which mathematics was practised at a specific location—the observatory—in a given period—the nineteenth century. My claim is that this exercise will enrich our understanding of changes undergone by mathematics in that period. As Michel de Certeau (1984) has shown, a focus on place as 'practised space'—that is, space where human practices are deployed—can help the historian identify cultural practices that are common to the users of the same space but that are not necessarily talked about. The issue of place in the history of science was in fact inaugurated by a debate about Tycho Brahe's observatory (Hannaway 1986; Shackelford 1993). Numerous studies have since been devoted to the topic, mapping out various spatial aspects of the laboratory and field sciences (Ophir and Shapin 1991; Livingstone 1995; Kuklick and Kohler 1996; Smith and Agar 1998).

At first sight, the history of mathematics, where disciplinary approaches have been dominant for so long, would seem more immune to spatial approaches than any other part of the history of science. What scientific domain could be less tied to a specific place than mathematics? Mathematicians only need pen and paper. And even those, the Bourbaki mathematician André Weil once wrote, could sometimes be dispensed with.[2] Historians, however, have shown that the universality of mathematics was actually forged in large part in the nineteenth and early twentieth centuries (Parshall and Rice 2002). A few institutional surveys underscore the imprint made on mathematics by particular institutions (the École polytechnique, Göttingen, the Institute for Advanced Study...). Even Weil would at times concur—if only for opportunistic reasons—that institutional histories may be indispensable:

It is unthinkable that anyone would write the history of mathematics in the 20th Century without devoting a large portion of it either to the Institute [of Advanced Study in Princeton] as such, or to the mathematics which have been done here, which comes very much to the same thing.[3]

Similarly, I contend that to study the history of mathematics in the nineteenth century it might be useful to pay special attention to the observatory. It has recently been suggested that a tight focus on observatory techniques can provide new insights about the social organization of science for the nineteenth-century state (Aubin 2002; Boistel 2005; Lamy 2007; Aubin, Bigg, and Sibum

2. 'Let others besiege the offices of the mighty in the hope of getting the expensive apparatus, without which no Nobel prize comes within reach. Pencil and paper is all the mathematician needs; he can even sometimes get along without these' (Weil 1950, 296).

3. André Weil, 'Talk to the Trustees of the Institute for Advanced Study, by Professor André Weil, April 1, 1960', Archives of the Institut des hautes études scientifiques, Bures-sur-Yvette.

forthcoming). Observatory techniques have been defined as the coherent set of physical, methodological, and social techniques rooted in the observatory because they were either developed or extensively used there. Among them, mathematical techniques figure prominently. Whether concerned with astronomy, geodesy, meteorology, physics, or sociology, in their quest for precision observatory scientists were both major consumers and producers of mathematical knowledge and techniques. Most of the founders of the German mathematical renaissance around 1800 had strong ties to observatories (Mehrtens 1981, 414–415). The same is true of other countries. Some observatory scientists, such as Pierre-Simon Laplace or Carl Friedrich Gauss, are even considered among the most outstanding mathematicians of all time. The roster of famous mathematicians who worked in (and often directed) observatories includes Friedrich Wilhelm Bessel, Nikolai Lobatchevski, August Möbius, Adolphe Quetelet, and William Rowan Hamilton. Others, like Augustin-Louis Cauchy, Karl Weierstrass, Henri Poincaré, and David Hilbert, were often passionately interested in celestial mechanics and gravitation theory. So, while mathematical techniques centrally belonged to the arsenal of the observatory, the mathematics developed to serve various observatory sciences equally became prominent areas of mathematics.

The special relation between mathematics and the observatory—or between mathematics and astronomy—is of course in no way a characteristic solely of the nineteenth century. Recall how mathematical analysis has, since Isaac Newton's time, been closely tied with the problems of celestial mechanics. I focus on the nineteenth century because the observatory was, at that time, the place that (as opposed to the Academy of Sciences earlier or the laboratory later) best embodied the intimate link between science, states, and societies in Europe and North America. I do so also because mathematics was then undergoing crucial changes that our look at the observatory will lead us to reinterpret in significant ways.

Indeed, we have a paradoxical view of nineteenth-century mathematics. In historical lectures, Felix Klein said that in earlier times 'independent works of pure mathematics were overshadowed by the powerful creation in which pure and applied mathematics united to answer the demands of the times' (Klein 1979, 2). But in the nineteenth century, mathematics increasingly seemed to be split in two. While the use and application of mathematics went on unabated, pure mathematics—and especially its most abstract and foundational aspects—took centre stage. Mathematics took on larger and larger new territories, providing tools for describing, controlling, and changing the world. In the physical as well as in the social realm, scores of laws expressed in the forms of differential equations were derived by scientists. The number of phenomena that were subjected to precise quantitative measurement increased tremendously. In offices, factories, army barracks, schools, and observatories, people with elementary or advanced mathematical skills multiplied. While the mathematical apprehension of our world progressed, mathematics as an academic discipline became increasingly abstract. Foundational questions started to assume a primary importance for the professional community: 'a revolution [...] characterized by a change in the ontological status of the basic objects of study' (Gray 1992, 226). Pure mathematics was detaching itself from the physical world at the very moment when it seemed more applicable than ever.

Most historians of mathematics have focused on the 'revolution' at the expense of the routine expansion of mathematical territories. Even when they have not, historians have found it difficult to deal with both processes at once. By focusing on the observatory, as a specific place where mathematics was intensely used and produced, I hope to throw new light on those two parallel large-scale processes. When examining mathematical practice in observatories, the major role played by numbers is immediately striking. Numbers are the main mediators between the various parts and functions of the observatory. Faced with an 'avalanche of printed numbers' (Hacking 1990) in their practical work, observatory scientists developed tools and techniques that became prominent factors in both mathematics' move towards abstraction and its increasing appeal as a privileged instrument for understanding nature and society.

There are two aspects to my study. First, I examine the specific spatial arrangement of mathematical work *within* observatories. I want to illuminate mathematical practices at this site, including its social organization. In order to do this, I focus on a social history of numbers, tracing their trajectory from their production with instruments to their insertion in observatory outputs. Second, I consider the observatory as the locus of particular mathematical cultures, which had important effects on the development of the field. I pay particular attention to three domains of mathematics: celestial mechanics, geometry, and statistics. In other words, this paper examines first the place of mathematics in the nineteenth-century observatory and then resituates the observatory in the history of mathematics.[4]

The place of mathematics in the observatory

'Every part of the operations of an observatory is mathematical', Airy wrote in the 1861 memo quoted above. 'Mathematical Mechanics' was involved in the construction of all instruments. 'The action and faults of telescopes and microscopes require for their understanding a knowledge of Mathematical Optics. Every discussion and interpretation of the observations requires Mathematical Astronomy.

---

4. Computing aspects will be slightly downplayed here in order not to overlap too much with Mary Croarken's Chapter 4.4 on human computers.

The higher problems, such as the discovery of the elements of a comet's orbit from observations, require the high Mathematics of Gravitational Astronomy.' In a word, mathematics was omnipresent in the observatory.

Five years earlier, Airy had spelled out the mathematical knowledge he thought was indispensable at each level of the strict hierarchy he had devised for the workings of Greenwich. A first draft was drawn up on 20 November, 1856 and a slightly revised version was adopted on 10 May, 1857.[5] At the bottom of the scale, according to this scheme, were supernumerary computers. In addition to being able to 'write a good hand and good figures' and 'to write well from dictation, to spell correctly and to punctuate fairly', computers were to have rudimentary mathematical knowledge, essentially restricted to arithmetic, including vulgar and decimal fractions, extraction of square roots, use of logarithms, and the use of ±. Next came the Assistant, first grade, who was required to read French and to understand geometry (equivalent to the first four books of Euclid), plane trigonometry, and simple and quadratic equations. Assistants, second grade—like, at the time, Hugh Breen (who had first been hired as a teenager computer in 1839)—needed to read Latin and speak a little French. In mathematics, they ought to understand simple algebraic rules such as the binomial theorem, spherical trigonometry, and differential calculus ('to Taylor's theorem, and applications to small variations of plane and spherical triangles, &c.'), as well as to have some notions in integral calculus. Beyond pure mathematics, they should have elementary knowledge of mechanics and optics and be able to master applications of plane and spherical trigonometry to astronomy. Long-time associates of Airy's had then achieved the higher level of Assistants, third grade. Supposing they conformed to the requirements spelled out by their boss, Edwin Dunkin and James Glaisher would then have understood analytical geometry, conic sections, integrations for surfaces and solids, advanced mechanics, optics, analytical mechanics 'especially in reference to Gravitational Astronomy'. More specifically, they would be conversant in the complete theory of telescopes and microscopes: object glasses, mirrors eyepieces, micrometers, etc. They would be able to apply methods for computing orbits of comets and planets. They should also read ordinary German. Clearly the skills required to work in an observatory were many.

Airy's memo not only sketched a relatively well defined perimeter of the knowledge required for working in an observatory, but also set up a scale of value in mathematical knowledge. While analysis and mixed mathematics (mechanics and optics) clearly stood at the top of his scale, geometry, elementary algebra, the first notions of calculus, and arithmetic especially, lay at the bottom. One may note that contemporary non-university mathematical textbooks reflected such

---

5. Several slightly different copies of this memo are extent in Airy's papers in Cambridge University Library. Above and in the following I quote from RGO 6/43, pp. 170–175.

scales, paying much attention to its lower parts and almost none to the top levels (Rogers 1981). In such textbooks, practical astronomy rather than analysis and rational mechanics often occupied the last and more difficult chapters, thus being set up as the ultimate goal of the 'practical mathematician'.

In addition to establishing a scale of mathematical knowledge, Airy also put mathematics above all the rest in a general scale of knowledge. Recognizing that mathematical skills were not the only ones required from an observatory assistant—he cited foreign languages, 'general or photographic chemistry', 'telegraphic galvanism', and so on—the Astronomer Royal nonetheless insisted on the special value of mathematical knowledge. Routine telescopic observations, 'which a lad acquires in two months, and which a man scarcely improves in many years', Airy added parenthetically, required few mathematical skills. Beyond those, the operations of an observatory required one to expand one's knowledge 'mainly in the mathematical direction'.[6] And this alone allowed one to rise up the hierarchy at the observatory. Astronomers with a more democratic bent similarly concurred that mathematics was what blocked the masses from familiarity with observatory sciences. In his public lectures, the director of the Paris Observatory François Arago (1836) tried to introduce the subject without using advanced mathematics. Like him, Alexander von Humboldt, John Herschel, Auguste Comte, and Laplace were much praised for presenting the public with treatises that circumvented mathematical technicalities.

But, as Yves Gingras (2001) has shown, mathematization also had an important social role as a technology demarcating insiders from outsiders (on this point, see also Schaffer 1994a). Over the course of the nineteenth century, dozens of new observatories were set up all over the world and the number of staff working in major national observatories increased significantly. Conflicts over the best ways to organize collective work inevitably arose. During the French Revolution, observatories in Paris were placed under the direction of the Bureau of Longitudes, a collegial body of astronomers, mathematicians, seamen, and instrument makers specially set up for that purpose. Others opposed the view that collegiality would ensure that national observatories carried out their regular tasks properly. In the early twentieth century, the American astronomer Simon Newcomb drew the following lesson from those discussions: 'The go-as-you-please system works no better in a national observatory than it would in a business institution' (Newcomb 1903, 332). More than a century earlier, in a memoir presented to the Revolutionary Comité de Salut Public (Committee of Public Salvation) in June 1793, the former head of the Paris Observatory Jacques-Dominique Cassini had similarly explained why, as far as the working of an observatory was con-

---

6. Airy, 'Remarks on the neglect' (note 1), RGO 6/43, pp. 235–236.

cerned, he thought it necessary to go against the Republican principle of 'sacred equality':

In astronomy, one distinguishes between the astronomer and the observer: the former is the one who embraces this science as a whole, who knows the facts, the data, and draws results from them. The observer is the one who is more specifically devoted to observation; he only needs to have good eyes, skill, strength and a lot of energy.[7] (Cassini 1810, 207)

At the Paris Observatory as at any national observatory, Cassini went on, a director was needed *pour la même raison que l'on place un pilote dans un vaisseau, un chef dans un bureau* 'for the same reason one places a pilot on a ship, a supervisor in an office' (Cassini 1810, 206–207). There were a whole range of observations that needed to be carried out regularly and without interruption. While young observers could be found with enough zeal to fulfil this task, an experienced astronomer was needed to direct them. His special task would be not only to oversee the work of the observers but also to compile their results in general annual publications, presenting not simply gross observations but nicely reduced ones seamlessly woven into *un narré instructif de l'histoire et des progrès de l'astronomie* 'an instructive narrative of the history and progress of astronomy' (Cassini 1810, 207).

In this context, exceptional mathematical skills were often singled out as those most likely to determine who would make a good observatory director. In the memos quoted above, Airy underscored his opinion that mathematical knowledge was what counted most to head an observatory. As an enticement for studying abstract mathematics, he wrote that the 'acquisition of these attainments would be at least as valuable to the Assistants (particularly if opportunities of quitting the Observatory should occur) as to the Observatory'. Later, especially after the emergence of astrophysics in the second half of the century, other types of skill seemed at least as important as mathematics for rising to the directorship of an observatory. But, as is well known, mathematics never completely lost its prominence as a tool for social selection.

The nineteenth-century observatory was a place where the quantitative spirit was valued most highly. Astronomy in particular was *la science où l'on rencontre de plus fréquentes occasions de faire des calculs longs et compliqués* 'the science where one has most frequently the occasion to carry out long and complicated computations' (Francœur 1830, vii). While Gauss's love of numerical calculation

---

7. On distingue en astronomie l'astronome et l'observateur: le premier est celui qui embrasse l'ensemble de cette science, qui en connaît et approfondit toutes les théories, rassemble et compare les faits, les données, et en tire les résultats. L'observateur est celui qui se livre particulièrement à l'observation; il lui suffit d'avoir de bon yeux, de l'adresse, de la force et beaucoup d'activité.

is legendary (Bourbaki 1994, 153), Airy is also known to have had a special obsession with quantitative results. His son Wilfrid once recalled:

He was never satisfied with leaving a result as a barren mathematical expression. He would reduce it, if possible, to a practical and numerical form, at any cost of labour: and would use any approximations which would conduce to this result, rather than leave the result in an unfruitful condition. He never shirked arithmetical work: the longest and most laborious reductions had no terrors for him, and he was remarkably skilful with the various mathematical expedients for shortening and facilitating arithmetical work of a complex character. This power of handling arithmetic was of great value to him in the Observatory reductions and in the Observatory work generally. (Airy 1896, 7)

The observatory indeed was a true factory of numbers and, as such, it needed competent people able to withstand the 'avalanche'. Simon Schaffer (1988) and Robert Smith (1991) have debated the most proper metaphor to describe the observatory—the factory or the accounting office. What is more significant to us right now is that the production and treatment of numbers on a massive, 'industrial' scale was observatory scientists' main business over most of the nineteenth century. And this implied forms of social organization that put mathematics at the centre of observatory scientists' practices (see also Ashworth 1994; 1998). But which part of mathematics?

A factory of numbers

At the end of the eighteenth century, it appeared clear that an astronomical observatory should be built around its instruments and that the most important of them should generate numbers, accurate numbers. *On ne peut s'occuper de la distribution d'un Observatoire qu'après avoir fixé le nombre, la grandeur, la forme et l'usage des instruments dont on se propose de le meubler.* 'The proper distribution of an observatory can only be addressed after the number, size, shape and use of the instruments intended for it has been fixed' (Cassini 1810, 74). While some of those instruments were portable, others needed to be precisely and firmly positioned in a well-designed, controlled environment. Astronomers in Paris always complained that their observatory, built by Claude Perrault around 1667, was too monumental for this purpose. In a memoir he wrote to make explicit the demands an astronomer wished to place on architects eventually engaged designing a new observatory, Cassini insisted on two special types of telescope, respectively called the transit instrument and the mural quadrant (or circle) (for a detailed architectural memoir on how to design an observatory around 1800, see Borheck 2005).

Both instruments consisted in a combination of telescope and graduated limb, and both were used in conjunction with other instruments (Chapman 1995). Fixed to a wall precisely oriented along the North–South line, the mural quadrant was usually larger and more finely graduated than the transit instrument. On a quadrant, special microscopes on the limb and wire nets, called micrometers, enabled a more precise reading of the graduations. Transit instruments were used together with highly precise astronomical clocks. While the former combination of instruments was used to determine the right ascension of a star or planet above the celestial equator, the second measured the exact time at which it crossed the meridian.

Using mural quadrants and transit instruments, both angular coordinates of a celestial body at a specific moment could therefore be measured to a remarkable degree of precision. In the eighteenth century, graduation had been improved by a factor of 200, from 20 seconds to a tenth of a second, and progress over the nineteenth century was no less spectacular (Frangsmyr *et al* 1990, 6). Achromatic lenses and clock regulators further improved the precision of the measurements made in observatories. In the first half of the nineteenth century, quadrants and transit instruments were combined and significantly transformed by German instrument makers (Chapman 1993), but the telescope's function as observatory scientists' main purveyor of numerical data was unchallenged until the last decades of the century brought the advent of photography, polarimetry, and spectroscopy. At the transit instrument, the astronomer 'listens in silence to the ticking of the clock, and [...] notes exactly the hour, minute, second, and fraction of a second when the star passes each wire' in the telescope (Biot 1810–1, I 55). In the dark, the observer at his eyepiece jotted down a few numbers on a paper slip (Lesté-Lasserre 2004). From then on, those numbers would be copied into large registers, preserved for centuries, averaged, combined with scores of other numbers, and transformed through various computational procedures, tabulated, printed in large folio volumes, distributed across the globe, and eventually picked up by seamen or theoreticians.

While meridian observations, even routine ones, required great manual skill, the level of mathematical sophistication involved at each step of these processes varied greatly. Barely literate teenagers could spend weeks ticking each number in long columns just to make sure that no mistake was made when they were copied from one register to another. But reductions were not trivial computations. In fact it was argued that since the reduction of other people's observations only inspired 'boredom and disgust', it was the observers' task to 'compute' their own observations: 'being the only one to know well the circumstances that go with them, he knows more than anyone else how to choose those that are most trustworthy' (Cassini 1810, 190). This is why any observatory scientist needed to be at least conversant in mathematical techniques.

The mathematical treatment of data served various purposes. Raw data was of little use to the outside community. Reductions made observations at different locations and times comparable with each other. Because a variety of factors affected the interpretation of data, that data was corrected using various mathematical algorithms. Corrections increased the precision of the observation, for example by taking into account flaws in the construction or alignment of the telescope, by compensating for differences in individual perception (the personal equation), or by compensating for aberration (itself dependent on the angle of observation, but also on temperature and atmospheric pressure at the time of measurement). Logarithms were said to be *artifice admirable qui, en abrégeant les calculs, étend pour ainsi dire la vie des astronomes (comme) le télescope avait aggrandi (sic) leur vue* 'admirable artifice that, by shortening computations, extends astronomers' lives (just as) the telescope has increased their sight' (Biot 1803, 26).

In this sense, mathematics was therefore just another instrument in observatory scientists' panoply.[8] As such, mathematics was accordingly taken into account in the spatial arrangement of observatories. At Greenwich, around 1850, it was highly symbolic that, between the transit room, where most transit observations were made, and the east room, previously devoted to Bird's now derelict quadrant and where jotting books and correspondence were stored, stood the computing room—'the grand scene of labour of the whole Observatory':

It is only by exception that the astronomer or his assistants are to be found *using* the instruments, even during the regular hours of observatory work; but they are nearly sure to be found assembled in the Computing Room, busied, at different tables, with their silent and laborious tasks,—the assistants on watch turning an eye now and then to a small time-piece which regulates their task of allowing no celestial object of consequence to pass the meridian unobserved. (Forbes 1850, 449)

Produced in the transit room and stored in the east room, observatory numbers were processed in the computing room located between them. Besides telescopes, there was a wide variety of instruments (thermometers, barometers, magnetometers, polarimeters, and so on), which, alone or in conjunction with clocks, also gave out numbers. In the natural history of numbers churned out by observatories, the operations carried out in the computing room were therefore crucial mediating steps between the instruments producing the numbers and the outside consumers of these numbers.

Prior to the 1830s, the proportion of published numbers with respect to overall production was rather small. When published, numbers often played a part in narratives that underscored difficulties encountered (Terrall 2006). As publication became an indispensable part of the public observatory's mission, the labour that went into preparing such publications was increasingly erased. When Airy endeavoured to make old Greenwich observations public, he coped with the amount of work involved neither by relying on mathematical innovations nor by having recourse to technical advances (contrary to Charles Babbage's hopes, see Schaffer 1994b), but by organizing work hierarchically. A senior wrangler at

---

8. A similar argument is made by Switjink (1987) without making explicit what is owed to observatory culture.

Cambridge, Airy owed much of his professional success to his mathematical talents, and these he attributed in large part to his 'high appreciation of order'. He sometimes went as far as to consider mathematics 'as nothing more than a system of order carried to a considerable extent' (Airy 1896, 6). For reducing the lunar and planetary observations of his predecessors, the Astronomer Royal designed several printed skeleton forms which were used by lower level staff to carry out computations. Mathematical operations involved in data reductions were therefore for the most part reduced to elementary operations. They mainly consisted in carrying out additions and subtractions in decimal and sexagesimal forms, and in using numerical tables. Mathematical tasks were split in two: the execution of computations was rendered as mechanical as possible, while the algorithmic part of the work—deciding on the computations that needed to be done and in what order—remained the Astronomer Royal's responsibility.

In the nineteenth century, observatory mathematics was therefore characterized by the same paradox as the one already mentioned for the whole of mathematics. Obsessively quantitative, it nevertheless put non-numerical practices at the top of its hierarchical scale. Publications streaming out of observatories spread were overfilled with numbers. This type of production was a tremendous boost to the widespread diffusion of mathematical practices not only among physicists and statisticians, but also among craftsmen (such as instrument and clock makers), military officers, and seafarers. Nevertheless, forced to manipulate great quantities of numbers, observatory scientists became famous, won prizes and medals, and were elected to academy seats not because of their computations but for the ingenious ways they devised for avoiding them. 'To the astronomer' belonged the task of 'looking for ways to shorten [computations], since by his constant practice, he is better placed than anyone to perceive the shortcomings of methods and resources to be drawn on to make them more bearable' (Delambre 1810, 100).[9] These methods also played a role, which remains to be studied carefully by historians, in widening the number of mathematically literate people in the nineteenth century. We will now examine in more details the various methods they developed, by focusing on a few famous instances where observatory mathematics had a great impact on the field as a whole.

The observatory in the history of mathematics

Up until the end of the eighteenth century, it went without saying that astronomy had its place in any book on the history of mathematics. Mathematics and astronomy were so close to one other that they were for all purposes united.

9. C'est à l'astronome à chercher les moyens de les abréger, puisque, par un usage continuel, il est plus à portée que personne d'apercevoir les inconvéniens des methods, et les ressources qu'on peut avoir pour le rendre plus supportables.

An astronomer, in order to be skilful, must be a Geometer [that is, a mathematician]; a Geometer, to deal with grand topics, need to have some of the Astronomer's knowledge. [...] The Astronomer in his observatory, the Geometer in his cabinet—this is always the same man who observes and meditates, who applies to the heavens either his senses or his thought. (Bailly 1785, III 208)

As James Pierpont's address to the Saint-Louis International Congress attests, by 1904 the situation had completely changed. As the title of this talk made explicit, 'the history of mathematics in the nineteenth century' could now be written by focusing exclusively on the pure domain (complex variables, algebraic functions, differential equations, groups, infinite aggregates, non-Euclidean geometry, and so on) without even mentioning applications, let alone the observatory sciences.

Meanwhile it seemed that 'mathematics [had] separated from astronomy, geodesy, physics, statistics, etc.', a fact that Klein (1979, 3) attributed to the professionalization and specialization of the sciences that were consequences of the social and cultural upheavals unleashed by the French Revolution. The increasing autonomy of the mathematical field, as well as the growing number of mathematicians earning a living as teachers, had important effects in shaping the evolution of the field towards foundational and structural aspects of mathematics (Mehrtens, Bos, and Schneider 1981). Nevertheless, for most of the nineteenth century the observatory remained one of the central scientific institutions of every nation that wished to be called 'civilized'. It was, as we have seen, a place where mathematics was its workers' daily bread. That observatories went on to play major roles in the development of the physical and mathematical sciences therefore comes as no surprise.

The question, however, is whether observatory mathematics left a specific imprint on nineteenth-century mathematics. In this second section, I would like to suggest that the 'values of precision' (Wise 1995) so dear to observatory culture had in fact everything to do with some of the evolutions of mathematics in that period. In 1846, the alliance between precise observation and precise computation was fully realized when it became possible to predict the presence of a missing planet just by taking into account anomalies in the orbit of its neighbour. Urbain Le Verrier and John C Adams acquired instant universal fame when they computed the orbit of Neptune to explain why Uranus was deviating from the orbit assigned by Newton's gravitational theory. The uncanny fit between theory and observation was a product of the extreme precision that characterized observatory culture. In the second half of the nineteenth century, confidence in the value of Newton's law of gravitation, in observational accuracy, and in the analytical methods brought to perfection by Laplace (the so-called 'French Newton'), was so high that some astronomers actually spent decades of their lives computing numerical tables, developing a single function, or trying to determine the value of a single number such as the solar parallax (Aubin 2006). If a discrepancy was found, instruments, theory, or both were usually blamed. Other times, extreme precision also provided a test for the efficiency of mathematical methods themselves, and even for the soundness of the foundations of mathematics.

In the following I will use an abundant secondary literature to discuss three famous instances where observatory culture seemed to be pushing known mathematics to its limits with considerable impact on its future development: (1) Gauss and non-Euclidean geometry; (2) Quetelet's uses of statistics and his theory of the average man; and (3) Poincaré's solution to the three-body problem. All three episodes have given rise to controversies among historians, and underscore the difficulty of discussing the relationship between mathematical innovations and their social environments. My claim is that by considering each of these contributions as anchored in observatory culture, we may gain insight into how conceptions of space, time, and society are related to the foundations of mathematics.

Geodesy, geometry, and the concept of space

There has been much debate about the exact relationship between Gauss's unpublished anticipations of non-Euclidean geometry and the commission he received in 1820 to carry out the geodetic survey of the state of Hanover. Director of the Göttingen Observatory since 1807, Gauss was a natural choice for this task. For most of the eighteenth and nineteenth centuries, geodesy was closely associated with observatories, since the precise measurement of the earth and that of the heavens were interdependent. Careful astronomical observations of fixed stars were crucial in any geodetic survey, while it had always been important for the purpose of comparing observations to know the exact geodetic position of observatories with respect to one another. In the early nineteenth century, moreover, the skills needed to carry out a geodetic survey were close to those developed in observatories. The lengthy trigonometric computations involved were exactly of the type observatory scientists were well equipped to carry out, intellectually as well as materially. When repeating circles and theodolites were introduced to geodetic practice, the observatory scientist's special skill with the telescope became so indispensable that, even in the turbulent times of the French Revolution, only astronomers could be sent out, at great risk to themselves, to survey the country from Dunkirk to Barcelona (Adler 2002).

Gauss's correspondents, however, thought that the director of the Göttingen observatory could have made better use of his precious time than to spend days and nights crisscrossing the countryside for up to six months a year. His friends' and colleagues' opinions notwithstanding, Gauss seems to have relished this exercise in high numerical and instrumental precision. In September 1823, to link up his triangulation of Hanover with existing ones to the east and the south,

Gauss, together with Christian Ludwig Gerling, measured the angles of a large triangle between Brocken, Hohehagen, and Inselsberg (BHI). Since that time, it has often been said that Gauss undertook the task just to be able to check whether the sum of the angles would add up to 180°, as expected in Euclidean geometry (Miller 1972).

The claim that Gauss made this measurement only to test Euclidean geometry is of course ludicrous. But a detailed examination of his geodetic work concluded that Gauss was bothered enough by the axiom of parallels to bring it up in frequent conversations, sometimes making mention of this large triangle: 'The myth of the BHI triangle as a deliberate test of Euclidean geometry appears a fanciful embroidery upon indubitable fact, encouraged possibly by reports made by Gauss in his inner circle' (Breitenberger 1984, 289). The precision of Gauss's trigonometric surveys was indeed extraordinary (Scholz 2004). In other contemporary surveys (that of Baron von Krayenhoff in the Dutch Counties, for instance) the error in closing triangles was often of the same order of magnitude as the correction that needed to be made to account for the curvature of the earth's surface. In Gauss's survey, however, the closing error was smaller than the latter correction. In this context, Scholz wrote, it was imaginable for Gauss to provide a lower bound for the curvature of physical space, although he never expressed it that way—and for good reason, if we are to follow Gray (2006), since we have no indication that the key concepts of three-dimensional Euclidean geometry were ever truly achieved by Gauss.

If one had no need for non-Euclidean geometry to carry out a precise geodetic survey, nor did one need to be immersed in the tedium of measuring angles in the field to breed doubts about the validity of the parallel postulate, the fact is that to discover—or invent—non-Euclidean geometry one needed to spend much time developing a logically coherent edifice, not checking whether numbers added up. Mathematicians who were versed in observatory techniques knew only too well that absolute precision was not achievable. But they were also acutely aware of whether errors were significant or not. In his geodetic survey (as well as in his magnetic experiments, see Aubin 2005), Gauss used observatory precision technologies to extend the limits of what could be explained mathematically.

Indeed, what may be more significant for the invention of non-Euclidean geometry is the realization that physical and mathematical spaces need not coincide. Neither Girolamo Saccheri, Johann Heinrich Lambert, nor Adrien Marie Legendre, who had tried to show before Gauss that contradicting the parallel postulate led to inconsistencies, ever harboured doubts about the fact that they were working with physical space (Alexander 2006). By contrast, having served as head of the observatory in Kazan, Lobatchevski thought that the nature of physical space could be tested by precisely measuring the angles of a large stellar triangle. An alumnus of the Royal Engineering College in Vienna and a sub-lieutenant in the army engineering corps, Janós Bolyai was certainly familiar with geodetic techniques. When a correspondent of Gauss's, Ferdinand Karl Schweikardt, came up with the basic idea of a geometry where the sum of the angles of a triangle was not equal to 180°, he named it 'astral geometry', because he conjectured that one might be able to observe this departure from Euclidean geometry in triangles drawn in the heavens between stars. It is also highly significant that despite his qualms Gauss expressed his ideas about non-Euclidean geometry quite freely to other observatory directors such as Bessel and Schumacher.

To the scientist working in the observatory and in the field, the difference between physical and mathematical space perhaps went without saying. To illustrate the way in which observatory scientists might be drawn to special ideas about space, let me quote from Emmanuel Liais, the French astronomer who founded the Rio de Janeiro Observatory and extensively surveyed Brazil:

In 1862, I was travelling through the Brazilian campos [...]. Constantly admiring the various but indefinite panoramas in front of me, my thoughts inexorably drifted towards immensity and my attention was caught by our ideas relative to space [*l'espace*]. [...] From the physical point of view, space indeed is another thing than from the point of view of mathematics.[10] (Liais 1882, 6–7)[11]

Liais went on to explain that physical space had many more properties than mathematical space, that even the fact that it could be measured away from the earth was debatable and that mathematical space was a mere abstraction. Experience of space in the Brazilian wilderness or on top of German hills was certainly different to experiencing it in one's armchair. With theodolites, clocks, and numbers, observatory scientists constructed spatial networks. In these networks, observatories were crucial nodes that Bruno Latour (1987) has, for good reason, called 'centres of calculation'. Observatory scientists were thereby reconstructing physical space in a manner that went hand in hand with the reconstruction of the mathematical concept of space.

Quetelet and statistical thinking

From the perspective of the conceptual history of mathematics, the geodetic experience is less significant as an inspiration for non-Euclidean geometry than

10. En 1862, je circulais dans les campos brésiliens (…). En voyant continuellement des tableaux variés mais indéfinis se succéder, ma pensée se reportait invinciblement vers l'immensité, et mon attention se fixait sur nos idées relatives à l'espace. (…) L'espace, en effet, au point de vue physique est autre chose qu'au point de vue mathématique.

11. Although this comment was made long after Gauss's measurement, it is roughly contemporary with Bernhard Riemann's famous *Habilitation* lecture that brought non-Euclidean geometry to a large public (Gray 2005).

as a field where the least-square method was directly and systematically applied, in particular by Gauss (Rondeau Jozeau 1997). If errors in measurement were distributed according to the bell curve, Gauss showed that the most probable value for the 'true measure' was the mean value. In the history of statistics, observatory scientists are quite prominent (Sheynin 1984; Stigler 1986; Porter 1986; Armatte 1995; Desrosières 1998). But pride of place is often given to Quetelet, whose work, it was claimed, 'helped create a climate of awareness [...] that was to lead to truly major advances in statistical methods' (Stigler 1986, 215). With his book *On man* (1835), Quetelet tried to develop statistical methods in order to found sociology. As such, it was a major step in the development of mathematical tools for the social sciences, as well as in the design of general strategies for making mathematics relevant to the social realm. As the founding director of the Brussels Observatory, Quetelet drew extensively from an array of analogies he found in his daily practice. He introduced the central concept of the 'average man' as the formal analogue of the average position of a star deduced from several measurements. The distinction made by Laplace in the study of planetary motion between periodic and secular motion was also mobilized in Quetelet's work on social phenomena.

My claim is that Quetelet's debt to observatory culture is perhaps less deep but much wider than historians have usually been willing to admit. While historians have fallen prey to the temptation of over-interpreting the meaning of his formal analogies, they have neglected to consider the full range of observatory techniques he drew on. In the domain of number manipulation, especially, Quetelet mobilized the whole array of table construction, averaging, corrections, and data standardization. Observation was also organized in ways taken from observatory culture, with standardized instruments distributed across a network of trained observers. In my view, Quetelet therefore had ambitions to understand and perhaps manage the sublunar world (meteors, the weather, plants, animals, and humans) by applying to it the observatory techniques that helped to understand and manage time and space.

The heuristic value of analogies with celestial mechanics first occurred to him at the time of Belgian independence in 1830. He later explained the growing importance such analogies would assume for him:

At a time when passions were vividly excited by the political events, I sought to distract me by establishing analogies between the principles of mechanics and what was happening in front of my eyes. These rapprochements I had made without at first attributing more value than to a spiritual game later came to take the character of truth.[12] (Quetelet 1848, 104)

12. Dans un moment où les passions étaient vivement excitées par les événements politiques, j'avais cherché, pour me distraire, à établir des analogies entre les principes de la mécanique et ce qui se passait sous mes yeux. Ces rapprochements que j'avais faits, sans y attacher d'abord plus de valeur qu'à un jeu de l'esprit, me parurent ensuite prendre le caractère de la vérité.

This retrospective account is corroborated by several other documents, such as the letter Queletet sent to the minister Sylvain Van de Weyer on 22 August 1834:

The most interesting part of my work will be, I think, the theory of population. I was able to import it entirely into the domain of the exact sciences [...]. The great problems of population motion will be as solvable as those concerning the motion of celestial bodies; and what is most remarkable is the astonishing analogy that exists between the formulas that are used for the computations. I think I have partly realized what I have been saying for a long time about the possibility of making a social mechanics, just as we have a celestial mechanics.[13] (quoted in Delmas 2004, 57–58)

In his unpublished thesis, Michel Armatte (1995) also quoted portions from this interesting letter and discussed the way its author was clearly conscious of the analogical transfer of methods from celestial mechanics that he was operating. The question is: what exactly was transferred and how? At the conference organized for Quetelet's bicentennial in 1996, the historian of statistics Stephen M Stigler (1997) opposed the received wisdom according to which it was necessary to insist on Quetelet's astronomical training in order to understand the intellectual sources of his social thinking. Canonical thinking was that Quetelet had sought to repeat in the social sphere what Newton had achieved for the planetary spheres. Stigler thought that this was 'misleading':

The problem, as I see it, is that astronomy, as it was conceived in the 1820s, encompassed a much richer variety of mathematical and empirical problems that can be captured by any simple description; certainly it was much more than Newtonian or Laplacian celestial mechanics. It is quite proper to associate Quetelet with astronomy, but with which part?

The solution offered by Stigler deserves a closer look. According to him, Quetelet was neither the mechanician, nor the physicist, nor even the astronomer of the social, but its 'meteorologist'. It is true that at the Brussels observatory, which for many years lacked proper instruments, Quetelet spent as much—if not more— time working in meteorology and climatology than in either astronomy or the social sciences. He moreover published several books on Belgian meteorology and climatology compared to that of the world. But his scientific practice was intimately linked with the site he was establishing, that is, an observatory. To Quetelet, as far as scientific practice went, the meaningful category was not astronomy,

13. La partie la plus curieuse du travail sera, je crois, la théorie de la population. Je suis parvenu à la *transporter* entièrement dans le domaine des sciences exactes (...). On pourra résoudre les grands problèmes des mouvements de population comme ceux des *mouvements des corps célestes* ; et ce qu'il y a de plus remarquable, c'est l'étonnante *analogie* qui existe entre les formules qui servent à ces calculs. Je crois avoir réalisé en partie ce que j'ai dit depuis longtemps sur la possibilité de faire *une mécanique sociale comme l'on a une mécanique céleste.*

physics, or meteorology, but the observatory sciences. And while the misconception that astronomical practice in the 1820s could be reduced to Laplacian celestial mechanics has been a block to a proper understanding of Quetelet's thinking, there is no doubt that, for someone like Quetelet, underestimating the unity of the observatory sciences and overvaluing disciplinary boundaries would not be of much help either. Like many of his colleagues and correspondents in observatories around the globe, Quetelet was not set on enlarging the dominion of Laplacian determinism. Rather, he was trying to adapt what he perceived as a coherent set of knowledge and techniques that characterized the practice of the observatory sciences to the needs of the world outside the observatory, whether physical or social.

Quetelet's practice in the social sciences is characterized by a strong faith in the quantification of the sciences. One should remember here that the quantification of statistics—that is, the 'science of the state', as it was still understood etymologically—was no trivial business and faced fierce resistance (Quetelet 1830). To him, numbers seemed more objective, less controversial, and less prone to betraying political and ideological a priori opinions than other types of description (on the history of objectivity, see Daston and Galison 1992). But Quetelet could draw on the observatory tradition for material and conceptual techniques to manipulate numbers in large quantities. Tables, equations, averaging, and graphical tools all figure prominently in his social physics, as well as probability theory.

In the 1830s and 1840s, Quetelet's network of collaborators in the physical and in the social sciences, in Belgium and abroad, expanded steadily. Standard instruments were distributed, procedures were shared. By mid-century, it seemed clear that greater coordination was needed. In 1853, Quetelet welcomed two international congresses to Brussels, within two months of one another. The first was devoted to navigation and climate science, under the inspiration of Admiral Matthew Fountain Maury, the head of the US Naval Observatory, while the second founded a series of International Statistical Congresses that is uninterrupted to this day. In both cases, the ideals of the observatory sciences were held in high respect. The aim was to set up vast instrumental networks covering the whole globe and churning out standardized numerical data. Historians have shown the major impact of this vision on the future development of mathematical statistics as well as the social sciences (Armatte 1995; Desrosières 1998).

In this story, the powerful influence of observatories would quickly wane. Quetelet's role in the history of mathematics was therefore not so much to use astronomical analogies at a conceptual level, as it was to adapt the very wide arsenal of tools he had found and developed in the observatory tradition in order to make them pertinent to the sciences of man. In so doing, he mobilized probability theory to an extent rarely done before by physical scientists, leading to important innovations by James Clerk Maxwell and Ludwig Boltzmann, who set out the foundations of statistical physics (Porter 1986). Similarly, this use of statistical and probabilistic tools led to the further development of mathematical statistics (Stigler 1986; Hacking 1990). As the nineteenth century unfolded, it became less and less a characteristic of the observatory to insist on the precise production of numbers, while the mathematical techniques developed for manipulating data were increasingly used outside the observatory. Mathematical statistics was no longer typically associated with the observatory (though some observatory scientists did contribute to it). But, significantly, it was again through the exact quantitative confrontation of mathematics with observations that techniques were developed for standardizing data on an international scale. Numbers extended their empire to society and, by the same token, so did the mathematical techniques for producing and manipulating numbers (Porter 1995).

Poincaré, analysis, and celestial mechanics

International congresses similar to those Quetelet presided over in Brussels— the Congress for establishing a Prime Meridian, in Washington in 1882, the Geodetic International Conference in Rome in 1883, the Solvay Congresses, and so on—loom large in Peter Galison's account of the origins of relativity theory (2003). The close alliance of precision technology (clocks, telegraphs, and theodolites) with numerical precision, in short everything I have associated with the observatory culture of the early nineteenth century, are described as the basis for the material cultures of Albert Einstein and Henri Poincaré. At the beginning of the twentieth century, they had independently developed similar ideas about time and space—although claims in favour of Poincaré's contributions to relativity theory have been greatly exaggerated (Gingras 2007). But a clerk in a Bern patent office could not see things identically to someone sitting on various councils and bureaus. The worldview of a young theoretical physicist in the German cultural sphere was different from that of an established professor of mathematics, physics, and mechanics at the Sorbonne.

What Poincaré's story illustrates well in my opinion is that the extreme precision of observatory science provided incentives to re-examine the inner workings of its mathematical technologies. Poincaré had no intention of revolutionizing physics or mathematics. Instead of questioning Newtonian tenets, he wished to fill the blanks in the picture. In the process, he developed his own philosophy of science, conventionalism. Conventionalism proposes that the statements with which we choose to express the laws of physics, mechanics, and astronomy are used not because they are real but because, due to their simplicity, they are the most convenient we can think of. This was a very different attitude from Einstein's, who thought that new principles were needed to replace old ones.

In Galison's assessment, there thus was a form of 'optimistic modernism' in Poincaré's conventionalism.

Lately, our understanding of Poincaré's work has had to be reconsidered. One reason has been the recent discovery of an error he made. In 1889, he submitted a fundamental essay to a prize competition organized by Gusta Mittag-Leffler on the three-body problem. When Edvard Phragmén started to edit the paper and found the error, Poincaré was devastated. Reworking the argument, he was led to discover 'homoclinic' points, 'the first mathematical description of chaotic motion in a dynamical system' (Barrow-Green 1997, 71).

As opposed to the first draft of Poincaré's prize-winning essay, which 'conveys a sense of optimism about the ultimate resolution of the problem', the tenor of the second draft was 'quite different: the future progress of the problem has lost its air of inevitability' (Barrow-Green 1997, 75). 'Chaos' is of course the second reason why Poincaré's work is now seen in a different light (Aubin and Dahan-Dalmedico 2002). While some scientists and popularizers have hailed chaos as a new scientific revolution—the third of the century after relativity and quantum mechanics—others pointed out that it had first been explored towards the end of the nineteenth or the beginning of the twentieth century (Hirsch 1984; Diacu and Holmes 1996). Most people, however, have agreed on one point—namely, that a new look at many parts of Poincaré's work (his memoirs on curves defined by differential equations, his study of the three-body problem in celestial mechanics, his pioneering work in dynamical systems theory and topology, his contributions to ergodic theory, and so on) played crucial parts in the emergence of chaos theory in the mid-1970s. While it is no doubt true that Poincaré's work foreshadowed concerns, and introduced key concepts and methods used in chaos theory, it is hard to explain why the great burst of activity only took place several decades after his death. This problem has given rise to various attempts to account for this 'nontreatment' (esp. Kellert 1993), but most have eschewed the admittedly arduous task of placing Poincaré among contemporary observatory scientists.

When he submitted his paper in 1889, Poincaré was not directly involved with the observatory.[14] But through his training at the École polytechnique he was fully aware of its scientific culture and trained in the use of theodolites and of the least-square method. Poincaré shared with Cauchy, Le Verrier, and Weierstrass a strong interest in the problem of the stability of the solar system. Further, his main sources very much belonged to the observatory: Hugo Glydén was director of the Stockholm Observatory; Andres Lindstedt had observed at Hamburg and Dorpat; George W Hill worked for the US Nautical Almanac Office.

---

14. Poincaré was nominated as a member the Bureau of Longitudes in 1893, joined the editorial board of the *Bulletin astronomique* published by the Paris Observatory in 1897, and the Paris Observatory Council in 1900.

A source for Poincaré's optimism may be found in observatory culture. Although the social history of celestial mechanics in the nineteenth century remains to be written, there is little doubt that it constitutes one of the most optimistic branches of science at a time when there was particular optimism about science. After the discovery of Neptune, the highpoint of celestial mechanics was perhaps Charles-Eugène Delaunay's publication of his *Moon theory* (2 vols, 1860; 1867). In these books, Delaunay pushed to the extreme the formal analytical expansion of a single function. He spent twenty years of his life developing it to the seventh order (and sometimes even to the ninth order), computing over 1259 terms in the expansion series for the moon's longitude and 1086 for its latitude. Although this extraordinary effort has sometimes been ridiculed, Delaunay's work is emblematic of the tremendous optimism invested both in the precision of the measurements made in the observatory and in the precision of the analytical method.

In the 1860s, however, mathematicians at the university and astronomers in the observatory were already starting to move apart from one another. The rise of astrophysics implied great changes in observatory culture (Le Gars 2007). New instrumentation had given rise to new problems about the physical nature of celestial bodies. To provide answers to these questions, mathematical tools seemed less useful than those taken from physics and chemistry. Similarly, the now fully professionalized mathematical community was shifting its focus (Lützen 2003). Unlike earlier generations of observatory mathematicians, Poincaré was no computer. 'The mathematical style of Poincaré was intensely modern. [...] Few of his results depend on long or difficult computations. He said of himself with a furtive touch of humor [...] that he was poor at arithmetic' (Veblen 1912, viii). Mathematicians were now emphasizing rigour, which led them to reconsider the concept of convergence. Where astronomers had been content with series whose terms decreased rapidly, mathematicians insisted that convergence had to be proved formally (Barrow-Green 1997, 18). For someone like Poincaré, rigour held the key to the elusive proof of the stability of the solar system.

If we follow Galison (2003), we recognize in Poincaré's conventionalism the technical world of diplomats, scientists, and engineers, where international conventions, telegraphy, and maps were used by modern states and businesses to control time and space. My account suggests that it was this same enterprise that required the foundations of mathematics to be opened up and examined anew. But for this task, a new generation of mathematicians, with few ties with the observatory, was coming along: they would focus more on the implications of Poincaré's work in logic, geometry, and philosophy than in old-fashioned celestial mechanics. A product of the mathematical culture of the observatory, Poincaré's homoclinic points did not seem fundamental enough to modern mathematicians, yet too mathematically rigorous to the observatory community. This is probably why very few people at the time were able to understand their significance.

Conclusion

For all the inaccuracies he is known to have perpetrated in his historical work, Eric Temple Bell was drawing attention to an interesting characteristic of nineteenth-century mathematics when he wrote:

Too often for comfort, mathematics in the nineteenth century followed the same formula of glut without digestion as the rest of civilization in that heroic age of expansion at any cost. But according to the abstractionists of 1940, the discarnate spirit of simplicity was then about to descend and bless all mathematics, and the more rococo masterpieces of the nineteenth century were to be preserved only in museums frequented exclusively by historians. (Bell 1992, 410)

Like Bell's abstractionists, historians of mathematics have paid greater attention to the foundational aspects of mathematics than to the bulk of the mathematical work done in the period. By examining the place of mathematics in a specific but significant site, we have been able to grasp the significance of some of the 'rococo masterpieces' of observatory mathematics. Computing astronomical tables, eliminating errors in geodetic surveys, compiling social data, and analytically expanding solutions of differential equations represented massive efforts that led to impressive results. Other sites, like accounting offices, army training grounds, or engineering projects would similarly unveil interesting aspects of the mathematical practice of the period.

In the course of the nineteenth century, precision instruments, mathematical techniques of number manipulation, and social techniques for establishing standardized conventions became ubiquitous. Because of the prominent position occupied by the observatory in the nineteenth-century worldview, it had a special effect on mathematics as a discipline, and many mathematical innovations came out of the work of observatory scientists. But my study has shown that, more than what it directly contributed in terms of mathematical concepts or theories, the importance of observatory mathematics may lie in what it teaches us about transformations in the relationship between mathematics and the world. Or rather, observatory mathematics is an especially good platform from which to look at the way in which mathematics was transformed between 1800 and 1900 so as to become an autonomous logical construct—a construct that was actually made to account for the physical and social worlds that shaped each other.

An anonymous reviewer wrote in 1900 that:

A really good history of mathematics in the nineteenth century has yet to be written; it would probably require the combined labour of an organised body of experts. [...] For the history of modern mathematics is not mainly that of individual discoveries, however brilliant; but that of the systematic investigation of mathematical notions such as 'number', 'continuity', 'function', 'limit' and the like. (GBM 1900, 511)

I hope to have shown that the intense examination of the abstract, foundational, and structural aspects of mathematics that was to characterize the next half-century was a direct consequence of collective efforts made by observatory scientists to construct *both* a world that could be mathematized *and* a mathematics whose basic concepts were precise enough to account for increasingly large chunks of that world.

Bibliography

Airy, George Biddell, *Autobiography* (ed Wilfrid Airy), Cambridge University Press, 1896; repr Kessinger Publishing, 2004.

Alder, Ken, *The measure of all things: the seven-year odyssey that transformed the world*, Little, Brown, 2002.

Alexander, Amir, 'Tragic mathematics: romantic narratives and the refounding of mathematics in the early nineteenth century', *Isis,* 96 (2006), 714–726.

Arago, François, *Leçons d'astronomie professées à l'Observatoire royal*, 2nd ed, Just Rouvier et E Le Bouvier, 1836.

Armatte, Michel, *Histoire du modèle linéaire. Formes et usages en statistique et économétrie jusqu'en 1945*, doctoral thesis, École des hautes études en sciences sociales, Paris, 1995.

Ashworth, William J, 'The calculating eye: Bailly, Herschel, Babbage and the business of astronomy', *British Journal for the History of Science,* 27 (1994), 409–441.

Ashworth, William J, 'John Herschel, George Airy and the roaming eye of the state', *History of Science,* 36 (1998), 151–78.

Aubin, David, 'Orchestrating observatory, laboratory, and field: Jules Janssen, the spectroscope, and travel', *Nuncius,* 17 (2002), 143–62.

Aubin, David, 'The fading star of the Paris Observatory in the nineteenth century: astronomers urban culture of circulation and observation', *Osiris,* 18 (2003), 79–100.

Aubin, David, 'Astronomical precision in the laboratory: the role of observatory techniques in the history of the physical sciences', in Georg Heinrich Borheck (ed), *Grundzüge über die Anlage neuer Sternwarten unter Beziehung auf die Sternwarte der Universität Göttingen*, Universitätsverlag Göttingen, 2005, 31–35.

Aubin, David, 'L'événement astronomique du siècles? Une histoire sociale des passages de Vénus, 1874–1882', *Cahiers François Viète,* 11–12 (2006), 3–14.

Aubin, David, and Amy Dahan Dalmedico, 'Writing the history of dynamical systems and chaos: *longue durée* and revolution, disciplines and cultures', *Historia Mathematica,* 29 (2002), 273–339.

Aubin, David, Bigg, Charlotte, and Sibum, H Otto (eds), *The heavens on earth: observatory techniques in the nineteenth century*, Duke University Press, forthcoming.

Bailly, Jean-Sylvain, *Histoire de l'astronomie moderne depuis la fondation de l'école d'Alexandrie jusqu'à l'époque de MDCCLXXXII*, 2nd ed, 3 vols, de Bure, 1785.

Barrow-Green, June, *Poincaré and the three body problem,* American Mathematical Society and London Mathematical Society, 1997.

Bell, Eric Temple, *The development of mathematics*, Dover, 1992.

Biot, Jean-Baptiste, *Essai sur l'histoire générale des sciences pendant la Révolution française*, Duprat & Fuchs, 1803.

Biot, Jean-Baptiste, *Traité élémentaire d'astronomie physique*, 2nd ed, 3 vols, J Klostermann, 1810–11.

Boistel, Guy (ed), *Observatoires et patrimoine astronomique français* (Cahiers d'histoire et de philosophie des sciences, 54), Société française d'histoire des sciences et des techniques et École normale supérieure, 2005.

Borheck, Georg Heinrich, *Grundsätze über die Anlage neuer Sternwarten mit Beziehung auf die Sternwarte der Universität Göttingen* (ed Klaus Beuermann) Universitätsverlag Göttingen, 2005.

Bourbaki, Nicolas, *Elements of the history of mathematics* (trans John Meldrum), Springer, 1994.

Breitenberger, Ernst, 'Gauss's geodesy and the axiom of parallels', *Archive for History of Exact Sciences*, 29 (1984), 273–289.

Cassini, Jean-Dominique, *Mémoires pour servir à l'histoire des sciences et à celle de l'Observatoire royal de Paris, suivis de la vie de J-D Cassini écrite par lui-même, et des éloges de plusieurs académiciens morts pendant la Révolution*, Bleuet, 1810.

Certeau, Michel de, *The practice of everyday life* (trans Steven Rendall), University of California Press, 1984.

Chapman, Allan, 'The astronomical revolution', in John Fauvel, Raymond Flood, and Robin Wilson (eds), *Möbius and his band: mathematics and astronomy in nineteenth-century Germany*, Oxford University Press, 1993, 32–77.

Chapman, Allan, *Dividing the circle: the development of critical angular measurement in astronomy, 1500–1850*, 2nd ed, John Wiley & Sons, 1995.

Daston, Lorraine J, and Galison, Peter, 'The image of objectivity', *Representations*, 40 (1992), 81–128.

Delambre, Jean-Baptiste, *Rapport historique sur les progrès des sciences mathématiques depuis 1789 et sur leur état actuel*, Paris: imprimerie impériale, 1810.

Delmas, Bernard, 'Pierre-François Verhulst et la loi logistique de la population', *Mathématiques et sciences humaines/Mathematics and Social Sciences*, 42/167 (2004), 27–58.

Desrosières, Alain, *The politics of large numbers: a history of statistical reasoning* (trans Camille Naish), Harvard University Press, 1998.

Diacu, Florin, and Holmes, Philip, *Celestial encounters: the origins of chaos and stability*, Princeton University Press, 1996.

Forbes, James, with Airy, George Biddell, 'National observatories', *Edinburgh Review*, 90 (1850), 299–357; repr *The Eclectic Magazine*, 433–464.

Francœur, Louis-Benjamin, *Astronomie pratique: usage et composition de la 'Connaissance des temps', ouvrage destiné aux astronomes, aux marins et aux ingénieurs*, Bachelier, 1830.

Frangsmyr, Tore, Heilbron, John L, and Rider, Robin E, (eds), *The quantifying spirit in the eighteenth century*, University of California Press, 1990.

Galison, Peter, *Einstein's clocks, Poincaré's maps: empires of time*, W W Norton, 2003.

GBM, 'An illustrated history of mathematics', *Nature*, 61 (1900), 510–511.

Gilain, Christian, 'La théorie qualitative de Poincaré et le problème de l'intégration des équations différentielles', in H Gispert (ed), *La France mathématique. La Société mathématique de France (1872–1914)*, Société française d'histoire des sciences et des techniques/Société mathématique de France, 1991, 215–42.

Gingras, Yves, 'What did mathematics do to physics?' *History of Science* 39, (2001), 383–416.

Gingras, Yves, 'Henri Poincaré: the movie: the unintended consequences of scientific commemorations', *Isis*, 98 (2007), 366–373.

Gray, Jeremy, 'The nineteenth-century revolution in mathematical ontology', in Donald Gillies (ed), *Revolutions in mathematics*, Clarendon Press, 1992, 226–248.

Gray, Jeremy, 'Bernhard Riemann, posthumous thesis "On the hypotheses which lie at the foundation of geometry"', in Ivor Grattan Guinness (ed), *Landmark writings in Western mathematics, 1640–1940*, Elsevier, 2005, 506–520.

Gray, Jeremy, 'Gauss and non-Euclidean geometry', in András Prékopa and Emil Molnár (eds), *Non-Euclidean geometries: János Bolyai memorial volume*, Springer, 2006, 61–80.

Hacking, Ian, *The taming of chance*, Cambridge University Press, 1990.

Hannaway, Owen, 'Laboratory design and the aim of science: Andreas Libavius versus Tycho Brahe', *Isis*, 77 (1986), 585–610.

Hirsch, Morris W, 'The dynamical systems approach to differential equations', *Bulletin of the American Mathematical Society*, 11 (1984), 1–64.

Kellert, Stephen H, *In the wake of chaos: unpredictable order in dynamical systems*, Chicago University Press, 1993.

Klein, Felix, *Development of mathematics in the 19th century*, (trans M Ackerman) Math Sci Press, 1979.

Kuklick, Henrika, and Kohler, Robert E (eds), 'Science in the field', *Osiris*, 11 (1996).

Lamy, Jérôme, *L'Observatoire de Toulouse aux XVIIIe et XIXe siècles: archéologie d'un espace savant*, Presses universitaires de Rennes, 2007.

Latour, Bruno, *Science in action: how to follow scientists and engineers through society*, Harvard University Press, 1987.

Le Gars, Stéphane, 'L'émergence de l'astronomie physique en France (1860–1914): acteurs et pratiques', PhD thesis, Université de Nantes, 2007.

Lesté-Lasserre, Nicolas, 'Le Journal d'observations astronomiques au XVIIIe siècle: entre autocritique raisonnée et affirmation d'un art', mémoire de diplôme d'études approfondies (DEA), École des hautes études en sciences sociales, 2004.

Liais, Emmanuel, *L'Espace céleste ou description de l'univers accompagnée de récits de voyage entrepris pour en compléter l'étude*, 2nd ed, Garnier, 1882 (1st ed 1865).

Livingstone, David N, 'The spaces of knowledge: contributions towards a historical geography of science', *Environment and Planning D: Society and Space*, 13 (1995), 5–34.

Lützen, Jesper, 'The foundation of analysis in the 19th century', in Hans Niels Jahnke (ed) *A history of analysis*, American Mathematical Society, 2003, 155–196.

Mehrtens, Herbert, 'Mathematicians in Germany circa 1800', in H N Jahnke and M Otte (eds), *Epistemological and social problems of the sciences in the early nineteenth century*, Reidel, 1981, 401–420.

Mehrtens, Herbert, Bos, Henk, and Schneider, Ivo (eds), *The social history of nineteenth-century mathematics*, Birkhäuser, 1981.

Miller, Arthur, 'The myth of Gauss' experiment on the Euclidean nature of physical space', *Isis*, 63 (1972), 345–348.

Newcomb, Simon, *The reminiscences of an astronomer*, Boston: Houghton Mifflin, 1903.

Ophir, Adi and Shapin, Steven, 'The place of knowledge: a methodological survey', *Science in Context*, 4 (1991), 3–21.

Parshall, Karen Hunger, and Rice, Adrian C (eds), *Mathematics unbound: the evolution of an international mathematical research community, 1800–1945*, London Mathematical Society and American Mathematical Society, 2002.

Pierpont, James, 'The history of mathematics in the nineteenth century', *Bulletin of the American Mathematical Society*, 11 (1904), 136–159.

Porter, Theodore M, *The rise of statistical thinking 1820–1900*, Princeton University Press, 1986.

Porter, Theodore M, *Trust in numbers: the pursuit of objectivity in science and public life*, Princeton University Press, 1995.

Quetelet, Adolphe, 'Avertissement et observations sur les recherches statistiques insérés dans ce recueil', *Correspondance mathématique et physique*, 6 (1830), 77–80.

Quetelet, Adolphe, *Du Système social et des lois qui le régissent*, Paris: Guillaumin, 1848.

Rogers, Leo, 'A survey of factors affecting the teaching of mathematics outside the universities in Britain in the nineteenth century', in H Mehrtens *et al* (eds), *Social history of nineteenth-century mathematics*, Birkhäuser, 1981, 149–164.

Rondeau Jozeau, M-F, *Géodésie au XIXème siècle: de l'hégémonie française à l'hégémonie allemande. Regards Belges. Compensation et méthode des moindres carrés*, doctoral thesis, Université Paris VII—René Diderot, 1997.

Schaffer, Simon, 'Astronomers mark time: discipline and the personal equation', *Science in Context*, 2 (1988), 115–146.

Schaffer, Simon, 'Machine philosophy', *Osiris*, 9 (1994a), 157–182.

Schaffer, Simon, 'Babbage's intelligence: calculating engines and the factory system', *Critical Inquiry*, 21 (1994b), 203–227.

Scholz, Erhard, 'C F Gauß' Präzisionsmessungen terrestrischer Dreiecke und seine Überlegungen zur empirischen Fundierung der Geometrie in den 1820er Jahren', in Menso Folkerts, Ulf Hashagen and Rudolf Seising (eds), *Form, Zahl, Ordnung. Studien zur Wissenschafts- und Technikgeschichte. Ivo Schneider zum 65. Geburtstag*, Franz Steiner Verlag, 2004, 355–380 [http://arxiv.org/math.HO/0409578].

Shackelford, Jole, 'Tycho Brahe, laboratory design, and the aim of science: reading plans in context', *Isis*, 84 (1993), 211–230.

Sheynin, O B, 'On the history of the statistical method in astronomy', *Archive for History of Exact Science*, 29 (1984), 151–199.

Smith, Crosbie, and Agar, John (eds), *Making space for science: territorial themes in the shaping of knowledge*, MacMillan, 1998.

Smith, Robert W, 'A National Observatory transformed: Greenwich in the nineteenth century', *Journal for the History of Astronomy*, 45 (1991), 5–20.

Stigler, Stephen M, *The history of statistics: the measurement of uncertainty before 1900*, Harvard University Press, 1986.

Stigler, Stephen M, 'Adolphe Quetelet: statistician, scientist, builder of intellectual institutions', in *Actualité et universalité de la pensée scientifique d'Adolphe Quetelet. Actes du colloque organisé à l'occasion du bicentenaire de sa naissance, Palais des académies, 24-25 octobre 1996*, Brussels, 1997, 47–61.

Switjink, Zeno G, 'The objectification of observation: measurement and statistical methods in the nineteenth century', in Lorenz Krüger, Lorraine J Daston, and Michael Heidelberger (eds), *The probabilistic revolution*, 2 vols, MIT Press, 1987, I 261–285.

Terrall, Mary, 'Mathematics in narratives of geodetic expeditions', *Isis*, 96 (2006), 683–699.

Veblen, Oswald, 'Jule Henri Poincaré', *Proceedings of the American Philosophical Society*, 51 (1912), suppl 'Obituary notices of members deceased', iii–ix.

Weil, André, 'The future of mathematics', *The American Mathematical Monthly*, 57 (1950), 295–306.

Wise, M Norton (ed), *The values of precision*, Princeton University Press, 1995.

PEOPLE AND PRACTICES

4. Lives

Patronage of the mathematical sciences in Islamic societies

Sonja Brentjes

Patronage is a term that has no direct equivalent in any of the languages in which mathematical sciences were practised in Islamic societies.[1] If we take one of the ancient Roman meanings of patronage, namely the existence of a clientele relationship between a protector and a protégé, as the starting point for exploring social relationships involving the mathematical sciences in Islamic societies, we will see that practitioners of these sciences could indeed enter relationships with people of higher social status and greater wealth, which provided them with relatively stable positions and various benefits. The complexity of the terminology used to talk about such relationships, and the diversity of forms of exchange involved in them, suggest that there was not one single type of patronage relationship in every Islamic society through the centuries. Indeed, whether all such relationships actually qualify as patronage demands further study.

Patronage and other forms of support for the mathematical sciences constituted part of a much wider phenomenon of social and cultural dependence between men and women of different social status, upbringing, and access to resources. These fields of knowledge shared basic rituals that configured clientele relationships with sectors such as medicine, the arts, administration, and even the

---

1. The names and dates of all the dynasties, scholars, and patrons discussed in this chapter are given in Tables 1 and 2.

Table 4.1.1 Islamic dynasties

Abbasid caliphs (750–1258), capital Baghdad

Umayyad *emirs* and caliphs in al-Andalus (756–1031), capital Cordoba

Samanid *emirs* (819–1005), capital Bukhara

Fatimid caliphs (910–1171), first capital in Mahdiyya, second capital in Cairo

Buyid *emirs* (945–1055), capitals Baghdad and Shiraz, with semi-independent courts in Rayy, Hamadan, Isfahan, Kirman: governed Iran and southern parts of Iraq under the Abbasid caliphs

Khwarazmshahs, capitals Kath, Gurganiye, Urgench (Kunya-Urgench): pre-Islamic title of Afrighids ruling since the early centuries AD, converted to Islam in the early ninth century; destroyed in 995 by their commercial rivals and successors, the Mamunids (995–1017) who were overtaken by the Ghaznavids; the last dynasty with this title existed from 1077 to 1231. They began as Saljuq slave governors; after gaining independence their title was first *emir* and later *shah*; they conquered large parts of the Saljuq empire, were finally destroyed by Mongols

Ghaznavid sultans (975–1187), capital Ghazna

Great Saljuq sultans (1037–1157), capitals Baghdad, Isfahan, Nishapur, govern Iran, Iraq, Anatolia, Syria, and Palestine under the Abbasid caliphs

Zangid *atabegs* and *emirs* (1127–83), capitals Mosul, Aleppo, Damascus, Sinjar, Jazira: recognized Abbasid caliphs as overlords

Ayyubid *maliks* (1169–1260), capital Cairo with semi-independent courts in Damascus, Aleppo, Hama, Hims and other cities in Syria, Palestine and north Iraq: recognized Abbasid caliph as overlord

Mamluk sultans (1250–1516), capital Cairo

Ilkhanid *khans* (1258–1336), capitals Tabriz, Maragha, Sultaniyye: recognized Mongol rulers of China as overlords

Ottoman sultans (1299–1922), capitals Bursa, Edirne, Istanbul

Timurids (1370–1506), capitals Samarkand, Herat, with semi-independent courts in Shiraz, Isfahan, and other cities in Iran

'Ādil Shāhis (1518–1686), capital Bijapur

Mughal shahs (1526–1857), capitals Delhi, Agra, Lahore

---

military. At the same time, there were certain differences within the mathematical sciences and their practitioners' access to patrons and their resources. Studies of these manifold aspects of patronage for the sciences, particularly in the mathematical disciplines, are relatively rare compared to ancient Rome or early modern Catholic and Protestant Europe. In particular, theoretically grounded investigations are lacking (Elger 2007). There is no clarity about the types of patronage and other clientele relationships that existed in different Islamic societies. Only one study has tried to survey the rhetoric, rituals, and, to some extent, the efficacy of patronage relationships and their specific forms, in ninth to eleventh-century Iran and Iraq (Mottahedeh 1980). Being limited to the military and the administration, it pays no attention to the practitioners of the ancient sciences. A second study focuses on patronage in medieval Cairo relating to scholars of Islamic legal theories (*fiqh*), the transmission of Muhammad's tradition (*ḥadīth*), the reading and interpretation of the Qur'an (*qirā'a; tafsīr*), and the fundaments of religion and law (*uṣūl al-dīn; uṣūl al-fiqh*) (Berkey 1992). Berkey's observations and insights agree with most of what I have read in primary sources about patronage relationships for the non-religious sciences, including the mathematical disciplines. As the only available studies of patronage in Islamic societies, these two works gave me guidelines for examining the terminology and social practices of patronage in other Islamic societies, applying their questions whenever possible to the mathematical sciences. In order to enhance the conceptual basis of the paper, I also sought inspiration from studies of patronage in early modern Catholic and Protestant societies, particularly their theoretical and methodological approaches (Kettering 1986; 2002; Asch and Birke 1991; Danneskiold-Samsøe 2004).

In this chapter I use the term 'mathematical sciences' in accordance with its definitions and descriptions in primary sources from Islamic societies. Until the classical system of knowledge was replaced during the nineteenth and early twentieth centuries, the notion of the mathematical sciences, their values, and practices were based on Platonic, Aristotelian, and neo-Pythagorean classifications of philosophy enriched by Stoic and sceptical elements on the one hand and perspectives developed within the religious disciplines of Islam, in particular law and theology, on the other. The mathematical sciences comprised the four main theoretical disciplines of Antiquity, namely number theory, geometry, astronomy, and musical theory (theory of proportions). They also included a variety of applications, such as optics, burning mirrors, mental or written calculation, algebra, magic squares, business arithmetic, surveying, timekeeping, architecture, and water lifting. The twelfth-century scholar Bahā' al-Dīn Kharaqī summarized this view as follows:

The mathematical (sciences) are called the four teaching (disciplines). They are four because their subject matter is quantity. Quantity is either that which is continuous or that which is discrete. The continuous is either in movement or in rest. The moving is *hay'a* (mathematical cosmography) and the non-moving is *handasa* (theoretical geometry). The discrete is either that which has a compound ratio, and this is music, or that which does not have it, and this is the numbers.[2] (Bayhaqī 1996, 173)

Astrology was rarely included explicitly in the classifications of the mathematical sciences. But since it was often seen as the ultimate goal of astronomical theory, and since most practitioners of astronomy also exercised the craft of astrology either as textbook writers or as composers of horoscopes, astrology will be included in the spectrum of disciplines discussed in this chapter.

---

2. Inna l-riyāḍiyyāt tusammā l-ta'līm al-arba'a, wa-innamā kānat arba'a li-anna mawḍū'uhā al-kamiyya, wa-hiya immā an takūna muttaṣila aw munfaṣila, wa-'l-muttaṣila mutaḥarrika aw ghayr mutaḥarrika, wa-'l-mutaḥarrika hiya al-hay'a, wa-ghayr al-mutaḥarikka hiya al-handasa, wa-'l-munfaṣila immā an takūna lahā nisba ta'līfiyya wa-hiya al-mūsīqā, aw lā takūna wa-hiya al-a'dād.

The multiple cultural backgrounds that defined the mathematical sciences and their status explain some of the specifics of patronage for the various disciplines united under this heading. Linked to philosophy, cosmography, *fiqh*, business, administration, and licit magic, the mathematical sciences commanded a rich but contradictory social prestige. Books by Euclid, Archimedes, Ptolemy, and other ancient authors were translated into Arabic, Persian, and occasionally other languages such as Ottoman Turkish. These translations, and works based on them, were studied by scholars involved in mathematical research and by students preparing for religiously sanctioned positions such as that of *muezzin*, the caller for prayer, and *muwaqqit*, the person responsible for determining prayer times, the directions of prayer towards Mecca, and the beginning of the new month. Names of Greek scholars were given as epithets to people seen as experts in their fields. Copies of, and commentaries on, their texts were prepared for individual reading, for classes at schools and colleges, and lavishly illuminated if destined for the library of a collector or wealthy connoisseur.

In the twelfth century, under the impact of Abū Ḥāmid Ghazālī's writings on the relationship between the religious and non-religious sciences, however, theoretical geometry became to be seen by some as a threat to the faith of the believer because of its claim to absolute truth. By contrast, calculation, algebra, surveying, and astronomy were appreciated as useful knowledge for determining inheritance shares and taxes, measuring fields, and predicting eclipses. Excessive practice, however, was also seen as a threat to the mental and psychic balance of the believer (Rebstock 1992, 19–22).

Astronomy, astrology, and music equally found themselves both cherished and challenged. Astrology was heavily disputed and eloquently defended by philosophers, astronomers, theologians, and transmitters of *hadīth*. In the view of some practitioners, their competitors did not know enough philosophy to understand the workings of the universe and the impact of its different spheres on the climate, affairs of the state, and the lives of individuals. Others accused their colleagues of not having mastered the technicalities demanded by these disciplines (Burnett 2002, 206–211). A number of specific issues of theory and reliability of results were also at stake, for instance, whether astrology could deliver forecasts only for universal events in nature and politics or was also applicable to the life of the individual (Burnett 2002, 207). The unreliability of astrological predictions could lead to ridicule and occasionally to imprisonment or even execution. Music, when practiced, not so much when dealt with as a mathematical theory, was regarded by various Sunni and Shi'i legal authorities as compromising the true path of a believer, while Sufis and courtiers appreciated it as support for spirituality and the enhancement of courtly life. As a result, scholars of various legal orientation wrote *fatwas* and other kinds of texts either commending or condemning these fields of mathematical knowledge and their practitioners.

In a sweeping simplification, one can differentiate two major periods for patronage of scientific knowledge, including the mathematical sciences. The first period spans the eighth to the later twelfth century, while the second period stretches over six hundred years to the nineteenth century. The grounds for the earlier period were laid during the two first centuries of the Abbasid dynasty, when the so-called Translation Movement, centred on Baghdad, made Middle Persian, Greek, Syriac, and Sanskrit scientific works accessible, in Syriac and Arabic. Interested readers included physicians, theologians, astrologers, courtiers, princes, and rulers. Scientific enterprises resulting from these translations were first undertaken in the Abbasid capital and other cities of the empire. With the emergence of powerful governors, local rulers, and independent Muslim dynasties such as the Fatimids, the Umayyads, and the Ghaznavids, a richer spectrum of courtly patronage became available. The second period began in the twelfth century, when endowed teaching institutions became a regular feature of scholarly practice in most of the lands from Central Asia to the Atlantic.

Patronage from the eighth to the twelfth centuries

Until the twelfth century, patronage relationships, including those of the mathematical sciences, were mostly located in courts and among wealthy urban groups such as viziers and their families, tutors of princes, and other court officials. The relationships were primarily forged between individuals and between families. Mottahedeh's (1980) study of loyalty and leadership, which also deals with forms of patronage, concentrates on the period of Buyid rule in Iran and Iraq under the formal suzerainty of the Abbasid caliphs. In his view, loyalties were created through three main features: benefits, various forms of formal commitment (oath, vow, guarantee of safe conduct), and gratitude. Benefits (*ni'ma*) given by a ruler to his subjects served to create stable, reciprocal ties. Generosity (*samāḥa*) was expected from the ruler when handing out benefits, but when combined with indulgence (*musāmaḥa*) could diminish the ruler's reputation (Mottahedeh 1980, 90–91). Formal commitments regulated 'duties and obligations that could be enforced without coercion' (Mottahedeh 1980, 43). A personal oath of allegiance (*bay'a*) followed by a payment (*rasm al-bay'a*) bound rulers and (male) members of their families to their soldiers and high-ranking officials. Such personal oaths were also exchanged between leading members of the administration, but were often concealed from the rulers, to maintain the façade of a unilateral top-to-bottom hierarchy of power (Mottahedeh 1980, 70–79).

It is unclear whether such personal oaths were also exchanged between rulers or viziers and practitioners of the mathematical sciences. Such practitioners did benefit, however, from the second type of formal commitment, the vow. The vow was made from man to God and contained the promise to do some good work on the basis of a good intention. Repairs of irrigation channels and bridges, or the building of mosques and later *madrasas*, could be publicly announced in the form of such a vow and contributed, in addition to the actual work of repair, to elevating the symbolic power of the ruler. Practitioners of the mathematical sciences, in particular surveyors and *muhandisūn* (geometers, engineers, architects), were often involved in the repair and replacement work. The recipient of benefits was obliged to show gratitude (*shukr al-niʿma*), since masculine honour was based on acknowledging received benefits, reinforced by the Qurʾan's repeated reminder that the (true) believer is grateful for God's continuous benefits (Mottahedeh 1980, 72–76).

Mottahedeh (1980, 42) argued that patronage during Buyid rule resulted from this triangular net of benefits, formal commitments, and gratitude. A central form of the Buyid patronage system was a relationship called *iṣṭināʿ*, which saw the patron as a parent or protector (*muṣṭanaʿ*) and the client as a child or protégé (*ṣāniʿ*). This relationship was introduced in the middle of the ninth century, together with slavery, by the Abbasid caliphs to rebuild their army, complementing and eventually replacing the older form of clientship (*walāʾ*). In the texts of the tenth and eleventh centuries, *iṣṭināʿ* dominates, while *walāʾ* has disappeared almost completely. Its verb, *iṣṭanaʿa*, often meant that someone's career was fostered. In exchange for this promotion, the patron expected his protégé to serve him in various ways as long as either of the two lived (Mottahedeh 1980, 82–84). Indeed, many of the scholars who worked as astrologers or physicians at the Buyid court, as well as other courts of the period, stayed until they died in courtly service, often through the reign of more than one ruler. It is no exaggeration to claim that many of the most productive and innovative mathematical scholars of this period, such as Thābit b. Qurra, Abu l-Wafāʾ Buzjānī, Abu Sahl Kūhī, Abū Rayḥān Bīrūnī, Ibn al-Haytham, and ʿUmar Khayyām, spent long years in courtly service. This means that the impressive results of nearly five centuries of mathematical research and teaching were produced through the institution of patronage at its major centres—central and local dynastic courts.

The patronage relationships that rulers, viziers, and other courtiers extended to physicians, astrologers, and other scholars were more often discussed in a variety of other terms. A patron extended honour (*ikrām*) and benefits, often in a generous manner, to the men who healed him, cast his horoscope, or observed the stars. The physicians and occasionally also the astrologers are said to have served (*khadama*) their patrons, often until their patron's death. The content of honouring included precious robes, monetary gifts, privileged ranking in seating at court during official occasions, and the inclusion into caliphs', sultans', or viziers' groups of boon companions (*nadīm*). Patrons also shared other, more personal favours with their clients. Ibn abī Uṣaybī'a and Shams al-Dīn Shahrazūrī report, for instance, how Caliph al-Mu'taḍid greatly honoured his astrologer and boon companion Thābit b. Qurra by apologizing for a mistake in etiquette the caliph had committed while walking with Thābit in one of the palace gardens (al-Shahrazūrī 1976, I 4–5; Ibn a. Uṣaybī'a n.d., 295). Thābit was one of the leading translators of Greek mathematical works, himself an excellent mathematician and astronomer, but also a philosopher, physician, and writer on the beliefs of the Sabeans, his religious community. He also took his obligations as a caliph's boon companion very seriously. When al-Mu'taḍid's uncle and official predecessor as caliph, al-Mu'tamid, was imprisoned on the instigation of his brother al-Muwaffaq in the house of Ismā'īl b. Bulbul, the caliph's vizier, Thābit visited the prisoner three times a day at Ibn Bulbul's invitation to keep the caliph company. He entertained him with topics from philosophy, geometry, and astrology. The caliph hungered for Thābit's company and felt pleased in his presence. When freed, al-Mu'tamid described Thābit as the man he valued most after his preferred military slave al-Badr. He made him sit close to him in audiences for the military and civil elite (*al-khāṣṣ*) and the general public (*al-'āmm*) while leaving al-Badr and the vizier to stand (al-Shahrazūrī 1976, I 5–6).[3]

The scholars engaged in patronage relationships were expected to offer expertise in areas such as healing, observing the planets and stars, casting horoscopes, constructing instruments, writing books, making automata, and repairing clocks, water wheels, channels, and other infrastructural components. Some of the best known services of mathematical scholars and instrument makers carried out for courtly patrons are the expeditions for measuring the length of 1° of a meridian, patronized by Caliph al-Ma'mūn in the early ninth century (King 2000). Al-Ma'mūn presented himself as the prince of the Translation Movement. Later he was both praised and cursed for this patronage of the 'foreign' (in particular ancient Greek) sciences (Gutas 1998). His support and encouragement of the geodetic measurements became standard narrative fare in geographical, astronomical, and also some historical writings. Patronage of the sciences was thus transformed into a powerful discursive instrument, either initiating the reader into its heritage and shaping his scientific identity or admonishing him to abstain from this heritage due to its danger to one's religious beliefs.

An important professional service of physicians and astrologers was to accompany their patrons on military campaigns, pilgrimages, and other travels. Abū Ma'shar, a favourite of the caliph al-Muwaffaq, was in his retinue as an astrologer

---

3. Shahrazūrī reports this story for al-Mu'taḍid. But al-Muwaffaq was the latter's father. Al-Muwaffaq usurped the power of Caliph al-Mu'tamid, who was his brother.

when al-Muwaffaq besieged the Zanj, the mutinous slaves of the great estates in Southern Iraq (Ibn al-'Ibrī 1958, 149). Teaching sons of rulers, their relatives, and other courtiers was another service scholars of the mathematical sciences provided for their patrons. Scholars of all disciplines dedicated books and instruments to patrons from the early Abbasid period onwards. It is, of course, not always clear whether the dedication was an expression of gratitude for benefits received or a proposal to enter into a patronage relationship. The vocabulary used in histories of scholars and bibliographies includes translating, writing, composing, or doing a work for a ruler, a vizier or a scholar. Besides texts, instruments too were named after patrons and designed specifically for them (Charette 2006, 133 Tables 2a, 2b). Dedications could take various forms, such as joining a part of the patron's name to the title of the book or including the name and titles of the patron as well as wishes for him in the introduction of the work. Dedications indicate another important service that scholars were supposed to deliver: appealing to God for their patrons' worldly and otherworldly well-being. The belief in such a possibility of intermediation was fundamental to all forms of patronage, including that of endowed teaching institutes through which the donors, in addition to protecting their wealth for their children, aspired to secure their salvation through the continuous recitation of Qur'anic verses and other religious texts. Writing eulogies for rulers was as a rule the obligation of poets and historians, but in dedications mathematical scholars too waxed eloquently about their patron's eminence, goodness, care for his subjects, and religious steadfastness. Occasionally practitioners of the mathematical sciences are even remembered in the biographical literature as writers of panegyric verses.

Other ways of talking about patronage relationships in the mathematical sciences included terms like linking oneself with someone (*ittaṣala bi*), being a *ghulām* (slave/apprentice?), carrying someone away (*ḥamala*), elevating someone or making him great or powerful (*a'ẓama*; *'azzaza*), binding someone by an obligation to someone else or being bound by an obligation to someone (*irtabaṭa*), engaging someone's service (*istakhdama*), summoning someone (*istaḥḍara*), ardently desiring someone (*ishtāqa ilā*), being in someone's favour or good graces (*ḥaẓiya 'inda*), and inviting someone (*da'ā*). Some of these expressions, such as *irtabaṭa*, *istakhdama*, and *da'ā*, were applied to both rulers and viziers, while others, such as *ḥamala*, were used only of rulers. *Irtabaṭa*, *istakhdama*, and *da'ā* describe relationships in which power, although asymmetric, is shared. While scholars looked for patrons, rulers and viziers tried to attract clients to their courts, occasionally going to great lengths to do so. But the scholars did not always accept the prospective patron's offers, invitations, or gifts. The Saljuq sultan Sanjar sent the enormous sum of 1000 dinars to 'Abd al-Raḥmān Khāzinī, a freed slave of Byzantine origin. Khāzinī refused to take the money, saying that he possessed ten dinars, three of which sufficed

Table 4.1.2 scholars and patrons of the mathematical sciences

Abu l-Khayr al-Ḥasan (tenth/eleventh century)

Abu l-Jūd, Khalīl b. Ibrāhīm (fifteenth century)

Abū Maʿshar (c 787–886)

Abu l-Wafāʾ Buzjānī (940–998)

ʿAḍud al-Dawla (r 949–982), Buyid *emir*

Aḥmad b. Khalaf (ninth century)

Akbar (r 1556–1605), Mughal shah

ʿAlam al-Dīn Qayṣar (d 1251)

ʾAlī b. ʿĪsā (ninth century)

al-Ashraf (r 1229–1237), Ayyubid ruler of Damascus

Awrangzīb ʿĀlamgīr (r 1659–1707), Mughal shah

Bīrūnī, Abū Rayḥān (973–1048?)

Fanarī, Shams al-Dīn (d 1435)

Fātiḥ ʿAlī Tippu (1750–1799), second and last sultan of Mysore

Ghazālī, Abū Ḥāmid Muḥammad b. Muḥammad (1058–1111)

Ghāzzān (r 1295–1304), Ilkhanid ruler

Hūlāgū (r 1258–1265), founder of the Ilkhanid dynasty

Ḥasan b. ʿAlī al-Qaymarī (d 1480)

Ibn abī Uṣaybīʿa (1194–1270)

Ibn al-Akfānī (d 1348)

Ibn al-Aʿlam, al-Sharīf (tenth/eleventh century)

Ibn al-ʿAmīd (d 970), Buyid vizier

Ibn al-Fuwatī (1244–1323)

Ibn al-Haytham (d after 1042)

Ibn al-Lubūdī = Najm al-Dīn Yaḥyā b. Muḥammad (1210–c 1265), Ayyubid vizier

Ibn al-Majdī (1365–1447)

Ibn al-Nadīm (d 993)

Ibn Naqqāsh, Nūr al-Dīn (d 1475)

Ibn al-Shāṭir (1306–1375)

Ibn Sīnā (d 1037)

Ibn Wāṣil, Jamāl al-Dīn (1207–1298)

Ismāʿīl b. Bulbul (879–890), Abbasid vizier

al-ʿIzz Abū Naṣr Beg Arslān, Saljuq emir

ʾIzz al-Din al-Ḥasan (fl c 1314)

Karajī, Abū Bakr Muḥammad b. al-Ḥasan (d c 1030)

Kāshī, Ghiyāth al-Dīn (d 1429)

Khafrī, Shams al-Dīn (d 1550)

Kharaqī, Bahāʾ al-Dīn Abū Bakr Muḥammad (twelfth century)

Khayyām, ʿUmar (d 1131)

al-Khāzin, Abū Jaʿfar (tenth century)

Khāzinī, ʿAbd al-Raḥmān (twelfth century)

Kūhī, Abū Sahl (tenth century)

Maḥmūd b. Sebügtegīn (r 998–1030), founder of the Ghaznavid dynasty

Table 4.1.2 *Continued*

---

al-Ma'mūn (r 813–833), Abbasid caliph

al-Māridānī, Jamāl al-Dīn (d 1406)

Maṣʿūd b. Maḥmūd (r 1030–1041), Ghaznavid sultan

Mehmet Fātiḥ (r 1451–1481), Ottoman sultan

Muʾayyad al-Dīn al-ʿUrḍī (d 1266)

Muḥammad ʿĀdil Shāh (r 1626–1660), ʿĀdil Shāh sultan

Muḥammad b. Armaghān (fifteenth century)

Muḥammad b. Khalaf (ninth century)

al-Mustanṣir (r 1226–1242), Abbasid caliph

al-Muʿtaḍid (r 892–902), Abbasid caliph

al-Muʿtamid (r 870–892), Abbasid caliph

al-Muwaffaq (r 875–891), Abbasid regent

Nāṣir Ṣalāḥ al-Dīn Dāʾūd (r 1227–1229), Ayyubid ruler of Damascus

Nāṣir Ṣalāḥ al-Dīn Yūsuf (r 1250–1260), Ayyubid ruler of Damascus

Rāzī, Abū Yūsuf Yaʿqūb b. Muḥammad (tenth century)

Rukn al-Dawla (r 932–976), one of the three founders of the Buyid dynasty

Ṣadr al-Sharīʿa (fourteenth century)

al-Samawʾal, Abū Naṣr (d 1175)

Ṣamṣām al-Dawla (r 989–998), Buyid emir

Sanjar (r 1117–1157), Saljuq sultan

Sebügtegīn (976–997), Samanid governor

Shāh Jahān (r 1628–1659), Mughal shah

Shahrazūrī, Shams al-Dīn (d after 1288)

Shīrāzī, Quṭb al-Dīn (1236–1311)

Sibṭ al-Māridānī, Muḥammad b. Muḥammad (1423–c 1495)

Sijistānī, Abū Sulaymān (c 912–c 985)

Ṣūfī, ʿAbd al-Raḥmān (903–986)

Thābit b. Qurra (d 901)

Timur (r c 1369–1404), founder of the Timurid dynasty

Ṭūsī, Aṣīl al-Dīn (d 1316)

Ṭūsī, Naṣīr al-Dīn (1201–1274)

Ulugh Beg (r 1447–1449), Timurid ruler

---

to finance him for a year, plus a cat. He also rejected the same sizable monetary gift when offered by the wife of one of the Saljuq emirs (Bayhaqī 1996, 181).

The term *ḥamala* 'to carry someone away' reflects the often violent nature of medieval society and the role of scholars as pawns in conflicts between rulers, invaders, rebels, and other claimants for power. When Sebügtegīn and his son Maḥmūd conquered parts of Central Asia and Iran in the late tenth century, they forced numerous scholars, such as the eminent physician and philosopher Ibn Sīnā, to flee westwards into the protection of the Buyid dynasty or else carried them away to Ghazna (in modern Afghanistan). Among those who had to relocate after the final campaign in 1017 were the scientist Bīrūnī and the physician Abu l-Khayr al-Ḥasan (Bayhaqī, 1996, 36; Matvievskaya and Rozenfel'd 1985, II 264). The Ghaznavids were not the only dynasty that forced scholars into their patronage. Various Khwarazmshahs brought scholars to their court in this manner, as did the Mongols in the thirteenth century and Timur and his family in the late fourteenth to fifteenth centuries (Bayhaqī, 1996, 36, 173).

The historical sources paint a complex and rich picture of different forms and formats of patronage. Successful relationships that lasted a lifetime are found, as are those of a more fluid nature. Cases of stability over two or even three successive rulers occurred at several courts during this period. Cases of conflict and failed continuation of patronage were, however, unexceptional. Important factors in the fluidity of courtly patronage were enmity among members of the patron's family and towards his clients as well as military, economic, and political instability. Competition among different factions within a ruling family and among families of administrators, and the growing influence of the Turkish military slaves upon the choice of the next caliph, made patronage an unreliable field of social relations. As a result the figure of the itinerant, often impoverished, scholar emerged. After the Buyid ruler 'Aḍud al-Dawla's death, one of his two teachers of astronomy, al-Sharīf Ibn al-A'lam, fell into poverty when 'Aḍ ud's son, Ṣamṣām al-Dawla, did not take over his father's patronage obligations (Ibn al-'Ibrī 1958, 174). The asymmetry of power and the instability inscribed in it are reflected in the various acts of punishment that a patron could heap on a client, either on his own behalf or on behalf of another client. Extortion, loss of office and property, exile, imprisonment, and execution were widespread means of regulating access to power and funds among the civil elites, in particular at the level of viziers and their immediate entourage. Scholars of the mathematical sciences, even those who were powerful patrons in their own right, also suffered under the arbitrariness of the patronage system. Several were exiled, incarcerated, or killed.

Boundaries that separated patronage from other kinds of relationships, for instance those that existed in the realm of craftsmanship, are often blurred due to a lack of precise information in the sources. The tenth-century bookseller Ibn al-Nadīm, a member of the circle of philosophers and literati surrounding the philosopher Abū Sulaymān Sijistānī and well connected with courtly life under the Buyids, wrote a work about intellectual life in the Abbasid caliphate. He used the term ghulām (plural ghilmān), which primarily meant in the period a military slave, to describe relationships among scholars of the mathematical sciences and instrument makers. In most cases, the owner or master of the ghulām was either an astrologer or an instrument maker. Instrument makers like Aḥmad and Muḥammad, the sons of Khalaf, often started their career as ghilmān and later had ghilmān themselves. Both brothers had been ghilmān of the astrolabe maker

'Alī b. 'Isā and then trained six other men as their *ghilmān,* all of whom became known as instrument makers (Ibn al-Nadīm n.d., 343). Hence it is perhaps reasonable to interpret this relationship as one between a master and an apprentice rather than one of patronage.

Since scholars rarely focused on one discipline alone but were often well educated in several branches of knowledge, neither did patronage focus only on one type of knowledge. Scholars who excelled in the mathematical sciences can be found heading the administration of a hospital founded by their patron or serving as head of courtly protocol, as in the case of Abu l-Wafā' (Kraemer 1993, 182, 216). Abu l-Wafā''s eminence in courtly affairs was such that he also participated, together with elders and representatives of the religious sciences, in political and financial negotiations between his Buyid patron and the city of Baghdad about protecting the populace against the threat from the Byzantine army's invasion of northern Iraq (Kraemer 1993, 100). Another tenth-century mathematical scholar, Abū Ja'far al-Khāzin, acted as ambassador for his patron, the head of the Samanid dynasty in Samarqand, in a war between the Samanids and the Buyids. After he had successfully concluded negotiations between the defeated Samanid general and his Buyid opponent, Abū Ja'far was patronized by the Buyid overlord Rukn al-Dawla, who esteemed him, made his vizier Ibn al-'Amīd employ him, and strongly recommended that the vizier emulate the scholar (Kraemer 1993, 252).

Patronage at courts and endowed teaching institutions from the late twelfth century onwards

The proliferation of richly endowed teaching institutions from the twelfth century changed the framework of patronage of the mathematical sciences. The dynasties that contributed most to this new development were the Saljuqs, the Zangids, the Ayyubids, and the Mamluks. The success of the new system of endowed teaching institutes was such that after the twelfth century many other dynasties also donated funds for building *madrasas,* Sufi *khānqāhs* (lodges), tombs, mosques, houses for teaching the Qur'an and *hadīth,* as well as hospitals. The *waqf* (religious endowment) typically included the costs of new and regularly funded posts as well as stipends for students. The new system extended the scope and complexity of patronage relationships between scholars and courts on the one hand and among scholars themselves on the other, by opening new possibilities for participating in the distribution and redistribution of the donated funds and endowed offices (Berkey 1992, 96). Socially, it was characterized by four major features. First, the donors and appointed administrators of the endowment, the ruling military aristocracy, in particular the rulers themselves and the highest ranking officials of the court hierarchy, the scholars, the head judges of the different legal schools, and other leading religious officials shared responsibilities and opportunities for making appointments to professorships and other posts. Second, professors amassed positions in more than one discipline and more than one endowed institute. Third, it became possible to buy, sell, and outsource professorships and other offices. Fourth, it became feasible to choose one's own successor, preferably from among immediate family members (including minors) or one's own students, which created the phenomenon of hereditary teaching posts (Berkey 1992, 96–97, 102–119, 121–125).

Most of the available teaching posts were within law, *ḥadīth*, Qur'anic studies and languages. But there were also positions for medicine at *madrasas* and mosques in Anatolia, Iraq, Syria, and Egypt; for timekeeping (*'ilm al-mīqāt*) in Mamluk Egypt and Syria as well as in regions of the Ottoman Empire; and for a special mathematical discipline linked to law and dealing with the determination of inheritance shares and legations (*'ilm al-farā'iḍ*) in different regions of the Islamic world (Petry 1981, 65, 428, n90; Berkey 1992, 69; al-Sakhāwī, III 119). In the fifteenth century posts for the mathematical sciences in general were created, when the Ottoman sultan Mehmet Fātiḥ began a tradition of endowing new *madrasas*, while the Mughal shah Akbar initiated a similar practice in some of the teaching institutes. Mathematical sciences were also taught in *madrasas* in north Africa, Iran, and Central Asia. This is verified by colophons and ownership marks in mathematical manuscripts as well as the donation of such books as *waqf* to *madrasa* and mosque libraries. But there is little to no information about whether the teachers held positions specifically linked to the one or the other of the mathematical disciplines in these regions. Occasionally writers of biographical dictionaries report a scholar of the mathematical sciences settling in a *madrasa*, such as al-'Urḍī who lived in the 'Izziya Madrasa in Maragha, northwest Iran, founded by the Saljuq emir al-'Izz Abū Naṣr Beg Arslān (Ibn al-Fuwatī n.d., 387–388). The fluidity of the teaching system in regard to where and what was taught, and its character as a scholarly network, opened wide doors to teachers and students of non-religious disciplines living in *madrasas* and Sufi *khānqāhs*, or simply visiting them for classes.

During the Ayyubid period, the rhetoric of patronage lost some of its previous richness. As main terms there remained service (*khidma*), honour or grace (*ikrām*), and benefit (*ni'ma*). Most of the patronage relationships between a ruler or prince and a physician were described by this terminology (Ibn a. Uṣaybī'a n.d., 584–586, 589–591, 598–601, 635–637 *et passim*). Occasionally a ruler is reported to have esteemed (*iḥtarama*) one of his clients excelling in the mathematical sciences. Although several Ayyubids sponsored astrologers at their courts, the authors of biographical dictionaries seldom applied to them this standard vocabulary of patronage. Writers of historical chronicles and mathematical scholars, however, talked about patronage of the mathematical sciences in such terms. In the mid-thirteenth century, al-'Urḍī described his relationship with the last Ayyubid ruler Nāṣir Ṣalāḥ al-Dīn Yūsuf as one of service. He told his new friends and colleagues in Maragha how he had escaped through his own courage, announcing his professional expertise to the enemy soldiers and his future usefulness to their lord during the bloodbath in which all members of the court died except for himself and two sons of the ruler (Ibn al-'Ibrī 1958, 280). On the day of the attack the scholar had just had a session with the unfortunate Ayyubid prince to cast his horoscope. Al-'Urḍī remained silent, though, about whether he had predicted his patron's imminent demise. Ibn Wāṣil, court historian of the Ayyubid dynasty and well educated in various disciplines including mathematics, characterized his relationship to his father's patron, the Ayyubid prince Nāṣir Ṣalāḥ al-Dīn Dā'ūd, exclusively by one term—service—which he also used of the relationship between his older colleague 'Alam al-Dīn Qayṣar and al-Ashraf, an earlier Ayyubid ruler of Damascus (Ibn Wāṣil 1977, 145–146).

While the Ayyubids continued to exercise their patronage to a substantial degree at court, which for this dynasty was usually the city's fortress, male and some female members of the family put a huge investment into building *madrasas* and other endowed teaching institutions and in maintaining those built by their predecessors in Syria, northern Iraq, and Egypt. This meant that an increasing part of their patronage relationships included endowing such institutions for a particular client, or providing a client with a professorship or another office in an already created institution. Whether practitioners of the mathematical sciences received such dynastic support within the framework of endowed teaching institutions needs to be further explored.

In other regions, the rhetoric of patronage also increasingly privileged the vocabulary of service. Ibn al-Fuwatī was a student of Naṣīr al-Dīn Ṭūsī and head of the library of the Mustanṣiriyya Madrasa in Baghdad, founded by the caliph al-Mustanṣir. He spoke of service when talking about relationships between a variety of scholars, including practitioners of the mathematical sciences, and Ilkhanid rulers, their viziers, and governors of cities (Ibn al-Fuwatī n.d., IV/1, 40, 392, 458, 620; III 103). Ibn al-Fuwatī frequently also used an earlier term that emphasized the character of the patronage as a connection (*ittaṣāl bi*), occasionally adding new ones of a similar vein by speaking of establishing a relationship or link, and being devoted to or depending on (*nasaba, ta'alaqqa*) (Ibn al-Fuwatī n.d., IV/1 40, 132, 149, IV/2 754, II 513, III 513).

The successors of the Ayyubids, the Mamluks, removed patronage of the mathematical sciences from their courts and linked it almost exclusively to endowed teaching institutions. Sultans and high-ranking court officials promoted scholars of the mathematical sciences to professors at prestigious *madrasas*, donated professorships and stipends for students at their *madrasas*, and appointed to the fortress mosque religious functionaries such as *muezzins* who had taken classes in timekeeping, arithmetic, and other mathematical disciplines (Brentjes forthcoming). The rhetoric of patronage under the Mamluks continued to revolve around the same terms as in the Ayyubid period. Clients served, linked themselves with patrons, and patrons honoured or benefited them (al-Sakhāwī, *al-Ḍaw'*, III, 296; VI, 112, 235; VIII, 191; IX, 15; X, 48). New terms are to settle or establish someone finally or firmly (*istaqarra bihi*), to appoint or establish someone (*qarrara hu* or *fī*), and to favour, distinguish, or confer distinction (*ikhtaṣṣa bi*). All reflect the demands and opportunities created by the system of endowed teaching institutions with their secure posts.

The mathematical sciences became part of the general education provided by the endowed teaching institutions, although only some *madrasas* had a professor capable or willing to teach them. Students and teachers of the mathematical sciences were fully integrated into this new framework. They could acquire paid positions and offices as *muezzins*, *muwaqqits*, teachers, preachers, imams, or physicians. Patronage proved useful for acquiring more than one position and keeping them, since competition was strong and interference in the distribution of posts widespread. Offices could also be lost frequently, either due to envy and greed among patrons and scholars or because the office holder was sick or meddled in dynastic policy.

As in the previous period, scholars acted as ambassadors and mediators in moments of war, conflict, or other needs. Naṣīr al-Dīn Ṭūsī, one of the leading philosophers, mathematicians, and astronomers of the thirteenth century, negotiated for his Isma'ili patron the peaceful surrender of the fortress of Alamut in northwestern Iran with the attacking Mongols, who, however, did not honour the negotiated contract. Ṭūsī changed sides and later wrote diplomatic letters for his new Mongol patron Hülägü to the last Ayyubid ruler Nāṣir Ṣalāḥ al-Dīn Yūsuf. His student Quṭb al-Dīn Shīrāzī, a Sufi, physician, philosopher, and gifted practitioner of the mathematical sciences, served the Ilkhanid ruler Ghāzzān as ambassador in Mamluk Cairo. Authors of mathematical treatises, such as Ibn al-Lubūdī, were appointed as viziers or took other administrative positions. Ṭūsī and later his son Aṣīl al-Dīn, in addition to directing the observatory in Maragha, headed the *dīwān* for religious donations of the Ilkhanid dynasty and were thus powerful patrons in their own right who were served by scholars as well as princes.

Dedications of manuscripts and instruments continued during this period. Due to the impact of courtly patronage on the arts of manuscript production, scientific books grew ever more lavish in decoration and design. Several of the most beautifully decorated scientific manuscripts went from one princely library to the next, acquired as gifts, bought for a high price on the market, or taken as booty in a war between neighbours. A few examples are the copy of 'Abd al-Raḥmān Ṣūfī's *Star catalogue* made from a copy owned by Naṣīr al-Dīn Ṭūsī for the library of Ulugh Beg; 'Alī Qushjī's work on mathematical cosmography written for Ulugh Beg in Persian and translated into Arabic for Mehmet Fātiḥ; a partial copy of Bīrūnī's astronomical magnum opus dedicated to the Ghaznavid sultan Maṣʿūd b. Maḥmūd, acquired by a courtier of Shāh Jahān's court in 1649; a copy of Ṭūsī's edition of Euclid's *Elements*, which in 1659 was part of the library of Sultan Muḥammad 'Ādil Shāh in Bijapur and later came into the library of the Mughal ruler Awrangzīb 'Ālamgīr; and a copy of Ghiyāth al-Dīn Kāshī's work on arithmetic dedicated to Ulugh Beg, which was bound for Sultan FātiḥʿAlī Tippu's library (Blochet 1900, 87; Schöler 1990, 172; Loth, 1877, 215–216, 220).

Patronage exercised by scholars, including those in mathematical disciplines, broadened too. Visiting, travelling together, deputizing, finding a position for a former student, marrying one's daughters to one's students, choosing a family member or a former student as one's successor: all were often-used means to secure one's own and one's family's position, influence, and possibly fields of knowledge. Ḥasan b. 'Alī al-Qaymarī was excellent in arithmetic, algebra, timekeeping, the determination of inheritance shares and legations, and prosody, also possessing good knowledge in law and grammar, and was a student of Ibn al-Majdī and Khalīl b. Ibrāhīm Abu l-Jūd from Damietta in Egypt. He received a chair in the field of inheritance shares and legations at the Jawhar al-Ṣafawī Madrasa in al-Ramla after his teacher Abu l-Jūd had talked to its donor (al-Sakhāwī, *al-Ḍawʾ*, III 119).

The vocabulary of courtly patronage found its way into describing relationships between scholars, teachers, and students. Having travelled widely in the East, including China, the son of an educated family from Wasit in Iraq, 'Izz al-Dīn al-Ḥasan turned to Syria where he 'joined the service (*khidma*) of the judge [Muḥammad b. Wāṣil al-Ḥamawī], the judge of Hama. He was well versed in the *Almagest* and mathematics. I [read] with him for some time' (Ibn al-Fuwaṭī n.d., IV/1 101–102).[4] Sons were in the service (*khidma*) of their fathers when studying with them, were launched into the service of princes through serving their fathers, and professors served the books of their intellectual fathers by writing commentaries and glosses on them (Ibn al-Fuwaṭī n.d., IV/3 318, 338; al-Sakhāwī, *al-Ḍawʾ*, X 48). Students formed connections (*ittiṣāl*) with teachers and married their daughters (Ibn al-Fuwaṭī n.d., IV/3 418–419). The obligation embedded in offers of marriage to a professor's student could lead to severe friction if rejected. In the fifteenth century, Muḥammad b. Armaghān, a student of Shams al-Dīn Fanarī and later a prominent scholar, fell out with his teacher's sons because he refused their father's offer to marry their sister. Armaghān justified

---

4. waʾjtamiʿtu bi-khidmat al-qāḍī [Muḥammad ibn Wāṣil al-Ḥamawī] qāḍī Ḥamā wa-huwa ʿārif biʾl-Majisṭī waʾl-riyāḍī [qaraʾ<tu>] ʿalayhi mudda

his decision by pointing to an earlier promise of marriage to another scholar's daughter (Tasköprüzade 1978, 45–46). Occasionally, a holder of a chair preferred to appoint one of his students rather than a relative as his successor. The *muwaqqit* Sibṭ al-Māridānī was the grandson of Jamāl al-Dīn al Māridānī, himself a *muwaqqit* and successful teacher of the mathematical sciences in Cairo who left at least six treatises on timekeeping and astronomical instruments. Sibṭ al-Māridānī, a prolific writer on arithmetic, algebra, timekeeping, astronomical handbooks, and instruments, received his teaching position at the Ibn Ṭūlūn Friday Mosque due to the desire of the previous professor and *muwaqqit* Nūr al-Dīn b. Naqqāsh that he should take the office (al-Sakhāwī, *al-Ḍaw'*, IX 36). The increasing transformation of the positions at endowed teaching institutions into family holdings was not limited to the professorships. Other posts could also be handed down in the family, such as that of the librarian of a *madrasa* or mosque library (al-Sakhāwī, *al-Ḍaw'*, II 154).

Professional identities and remuneration

In the eighth to twelfth centuries, the first period of patronage, courtly patrons paid their clients in two forms, although the information given in the sources regarding them is so irregular and sparse that it is difficult to get a clear picture. The form mentioned most often are gifts such as robes of honour and one-time monetary payments made from the patron's personal treasury (Ṭabarī, 1989, 313; Ibn al-'Ibrī 1958, 131, 137, 182). The second form was regular monthly and yearly payment (*rizq*; *ujra*) allocated mostly from the general treasury of the court or the private treasury of the patron, but also from the *dīwān* for religious donations (King 2000, 211; Ibn a. Uṣaybī'a n.d., 198–200; Halm 1997, 75).

By the twelfth century, regular stipends, paid annually or monthly, seem to have been the more widespread format. It is not always clear whether they came exclusively from the personal treasury of the patron. In addition, other forms of payment were also used, in particular the turning over of an *iqṭā'*, a taxable region, although occasionally already in the first period a village had been given as a gift to a physician or other scholar. The stipends (*jāmakiyya*; *jirāya*) and the *iqṭā'* had previously been standard forms of paying the military. In the twelfth and thirteenth centuries, the Ayyubid dynasty applied such forms to reconcile members of the civil elite of a city with the conquest and destruction wrought by Ayyubid troops, to express their highest appreciation for a local notable or a courtly client, and to pay for services rendered (Eddé 1999, 280). Some Ayyubid court physicians received such remuneration and gratitude but it is unclear whether astrologers or engineers were also recompensed in this manner.

With the proliferation of endowed teaching institutes, regular salaries in conjunction with non-monetary components such as food and lodging for professors, teaching assistants of different kinds, *muwaqqits*, *muezzins,* and other positions linked to these institutes became the norm. The highest salaries often, but not always, went to the professors. The amount depended on the endowment and thus on the wealth and status of the donor. Even among the members of the Mamluk elite in Cairo, the endowments and hence the stipulated salaries varied considerably (Berkey 1992, 77–78). *Muwaqqits* at Mamluk mosques and *madrasas* got substantially less, but a little bit more than the *muezzins*. A *muwaqqit* in a major mosque or *madrasa* in Cairo earned at the highest 60 percent of what a Sufi *shaykh* and 40 percent of what a professor of law received. More often, however, this post was supported by a salary of less than a third of a professor or an *imam* (King 1996, 302). The further bureaucratization of the system under the Ottomans fixed salaries for positions according to their distance from the capital, their status within their city, and the access it could open to positions in the religious and civil administration of the empire. Salaries were paid on a yearly, monthly, or daily basis. The money for the salaries came from profits gained from agricultural production in donated villages, fruit gardens, and shops or were taken from the dynasty's *waqf* treasury, from the *dīwān* for taxes, or from economic enterprises monopolized by a dynasty, such as salt mining. In addition to the salary fixed by the donor or prescribed by the administration, patrons could add substantial monetary and other gifts to honour the person they had appointed and keep him from looking for other lucrative income (Tasköprüzade 1978, 59).

Earning a living through teaching was a lifestyle that was heavily contested over several centuries in a number of Islamic societies. Many remarks and discussions about which kinds of knowledge it was permissible to be remunerated for, and which kinds of offices a devout believer should accept or refuse, fill the pages of manuscripts. Algebra, for instance, could be safely taught for money, while knowledge of *ḥadīth* should be shared for free (Berkey 1992 95–97; Tasköprüzade 1978, 26, 57). Taking the position of a judge or other, less prestigious positions for discharging the law for a salary was seen as an acceptable form of paid employment after long years of learning. The rise of endowed teaching institutions altered the framework of access to remuneration for knowledge, bringing with it new employment opportunities (Berkey 1992, 96). Regularly remunerated positions increased substantially in the centres and spread to the provinces and even villages. Most opportunities were, of course, found in capitals and major administrative centres. But the often fluid forms of power, and the distribution of governance of provinces and smaller cities among male members of a dynasty or members of the military, offered men of the lower ranks their own fields of patronage. In addition, everyone with sufficient wealth could donate an endowed teaching institute, even if it consisted of one position only. Hence families who could afford it were able to create posts for the scholars of their own family and stipulate the kind of disciplines that would be taught. A number of medical *madrasas* were created in this manner, but no case of a scholar of the mathematical sciences is known who founded his own *madrasa* chair specifically dedicated to these disciplines. Most teachers of mathematical disciplines at endowed institutes in Mamluk Cairo, for instance, were appointed by courtly patrons to teach a broad range of religious and other disciplines, or succeeded their former teachers or relatives in their positions. The biographical literature shows clearly that most of them focused on teaching a particular branch of law, the determination of inheritance shares and legates. This focus allowed them to carve out a substantial space for teaching mathematical knowledge covering arithmetic, algebra, practical and theoretical geometry, three astronomical disciplines (timekeeping, planetary theory, compiling astronomical handbooks and ephemeredes), and the construction of scientific instruments. In this sense, the mathematical sciences underwent a process of stabilization and professionalization once they became integrated into the new framework of endowed teaching institutions.

This substantial and important gain of territory, opportunity, and stability is well documented through the thousands of texts introducing these disciplines to many generations of students, extant today in manuscript libraries across the globe. It was accompanied, though, by a focus on elementary content and a submission to the teaching methods and values that governed the religious and philological disciplines. Learning by heart was seen as the highest expression of scholarship and reason for fame. Scholars were praised and admired for the speed and quantity of their memorizing. Studying the texts of one's teachers and of their teachers in turn—that is, chains of texts determined by chains of scholars—became the norm not only for transmitting *ḥadīth*, where it evolved in the early centuries as the only method considered leading to trustworthy knowledge if the transmitters of the chain were deemed reliable, sound and morally worthy, but also in the mathematical sciences. As a result, fewer and fewer texts by scholars of previous generations, centuries, and cultures were directly studied. Editions, paraphrases, commentaries, and super-commentaries of Euclid's *Elements*, for instance, replaced in many classes the study of the *Elements* themselves.

The goal of education in endowed teaching institutes was not to create critical or substantially new knowledge. New ways of looking on taught knowledge and asking questions about it, although lauded as marks of excellence, were considered exceptional. A major way of establishing a scholar's academic credentials, in particular when he was new to town, was to hold a public disputation (*munāẓara*). In such disputations a series of questions was asked to bring out the depth of knowledge of what was taught and discussed in the various disciplines.

Defeat, evaluated by an arbiter who was either a well-established scholar or occasionally a ruler, vizier, or another high-ranking court official, was declared when
a participant was not able to answer these questions satisfactorily. Thus scholarly
excellence did not consist in raising questions beyond the already established
concepts and beliefs but in being broadly and substantively familiar with what
was taught and discussed in scholarly circles. Such broad, encompassing knowledge was worthy of patronage and promotion.

Outcomes of patronage for the mathematical sciences

Outcomes of patronage for the mathematical sciences extant today include
instruments and manuscripts, art objects, and architectural and technological
monuments. Their character as products of patronage can be established through
dedications, frontispieces, colophons, and marks of ownership. Other results of
patronage were expeditions, measurements, observations, and oral performances
as teachers, boon companions, and participants in sessions of serious debate or
conviviality. These survived the centuries through reports and descriptions.
Since it is impossible to describe here all known outcomes of patronage, or to
summarize all the results of patronage relationships, a few examples will have
to suffice.

According to King (2004, II 993–1020) and Charette (2006, 134, n1), over a
thousand astrolabes and several hundred globes, quadrants, sundials, and other
scientific devices made in Islamic societies survive. At least one hundred and
fifty extant astrolabes, some twenty sundials, and about a dozen quadrants go
back to pre-sixteenth-century workshops. Astronomical instruments were part
and parcel of the Translation Movement. The prominent role of instruments
resulted from the political and ideological functions that transformed the act of
translating into a cultural movement sustained for almost two centuries by a fine
net of patronage acts from Abbasid caliphs, courtiers, and practitioners of the
translated knowledge. Gutas (1998, 41–50) convincingly argued that these functions rested on a concept of translation that was part of a pre-Islamic ideology of
kingship from Sasanian Iran.

Closely linked to a new, modified form of this ideology, which now was directed
against the Byzantine Empire, was the organization of geodetic measurements
and astronomical observations between 828 and 833 in Baghdad, Damascus,
Mecca, and the desert of Sinjar near Mosul under the caliph al-Ma'mūn (Gutas
1998, 83–95). The geodetic observations were presented as a means of checking,
comparing, and verifying data found in ancient texts and instruments in use in
different regions of the Abbasid caliphate (King 2000, 215, 217–218, 223–224).
The observation of a lunar eclipse in Mecca is described as a caliphal order to

determine the *qibla* of Baghdad, the direction of prayer towards Mecca (King 2000, 214, 218–219). Other observations most likely served for astrological counselling (Charette 2006, 125). Astrology was probably not merely the practical but also the theoretical context of this astronomical programme, due to the rising impact of Aristotelian natural philosophy (Charette 2006, 135 n18).

Astrologers, instrument-makers, and a judge participated in these measurements and observations—constructing a series of instruments; determining appropriate sites; choosing the team members; supervising their work; witnessing, recording, and communicating the results to the caliphal patron. He in turn formulated successive research questions, informed himself about the reliability of the instruments and evaluated the results (King 2000, 215, 218–220, 223–224). The cultural and scientific results of this first major programme of empirical scientific activities had immediate and long-lasting consequences. They supported the radical shift from Indian and Sasanian models and parameters to Ptolemaic astronomy that took place in the following decades. The series of reports on this programme and its scientific results established astronomical observations and measurements as an important icon of courtly patronage. Elements such as specifically constructing instruments, forming a team of scholars, and inviting witnesses were repeated in later programmes under the Buyids, Saljuqs, and Ilkhanids.

Several princes of the Buyid dynasty either received an excellent education in the mathematical sciences or were presented with high quality writings on theoretical and practical mathematical problems. Berggren (unpublished) has evaluated some aspects of Buyid patronage of the mathematical sciences. He describes a deeply structured network, with patrons in the dynasty and the administration, and clients among the administrators, scholars, and instrument makers. The princely patron of this dynasty was first and foremost ʿAḍud al-Dawla. But several other princes also contributed, among them Rukn al-Dawla—one of the three founders of the dynasty and father to ʿAḍud—and ʿAḍud's sons Sharaf al-Dawla and Ṣamṣām al-Dawla. Rukn al-Dawla, an illiterate mercenary from Daylam on the Caspian Sea, ordered the measurement of the latitude of the city of Rayy and its longitudinal difference from the Abbasid capital. He provided an excellent education for his son by giving him his own vizier, Ibn al-ʿAmīd, as tutor. Ibn al-ʿAmīd collected manuscripts, commissioned a commentary on Book X of Euclid's *Elements* from Abū Yūsuf Yaʿqūb Rāzī, and himself excelled in mechanics. He used the latter for inventing new siege machines and is credited with constructing a mural quadrant used in the astronomical observations requested by Rukn al-Dawla (Sijistānī, *Ṣiwān al-ḥikma*, 321–324; Sayılı 1960, 104). Other Buyid viziers also patronized scholars of the mathematical sciences. Numerous scholars sponsored by Buyid patrons wrote works on geometry, number theory, algebra, and arithmetic in addition to texts on astrology and astronomy.

Works on higher geometry were of two types. One type continued and completed the research of ancient Greek geometers. The two main figures for identification and imitation were Archimedes and Apollonios. The scholars of the Buyid courts wrote on conic sections, loci, polygons, projections of the sphere on the plane, the construction of two mean proportionals, and the trisection of angles and studied the transformation of curves into equations and vice versa (Hogendijk 1981; Berggren 1986, 77–85). The other type began to modify and then to replace the methods and concepts of Greek mathematics with new theorems, procedures, and topics. Works on Menelaus' theorem and new theorems of plane and spherical trigonometry constitute one field of such innovation. Other fields of geometrical research that went beyond the classical heritage, or drew rather on Indian precedent, related to astronomical problems such as interpolation procedures for calculating tables. Approximate solutions were found to practical problems in architecture, surveying, accounting, and determining inheritance shares and legacies.

In number theory, arithmetic, and algebra, Buyid scholars also wrote important books which they dedicated to their patrons. In Karajī's *Fakhrī*, for instance, the author departs from the former understanding of algebra as a set of rules for solving quadratic equations and verifying their solutions, as defined by ninth-century scholars, some of whom had been patronized by Abbasid caliphs. Karajī now presented algebra as a discipline which applied the rules and procedures of arithmetic systematically to unknowns of the type x^n and $1/x^n$ and to polynomials (Rashed 1984/1994). Taken up a century later by the physician and mathematician al-Samaw'al, who wandered through Syria, Kurdistan, Azerbaijan, and other Islamic lands in search of patrons, Karajī's new algebra has had a deep impact on how the field was taught and studied by later generations.

In the period of endowed teaching institutions, patronage of the mathematical sciences took place, as discussed above, at courts, among the civil educated elite, and within the framework of the endowed institutions. The manuscripts, tables, and instruments that were produced in these diverse environments were primarily created within the context of teaching. Luxury specimens were fabricated mainly for princely education, as gifts for members of the ruling family, and as items held and displayed in princely libraries. At times the production of new knowledge, in the form of new solutions to standard problems, variations to extant solutions, and modifications of unsolved problems occupied scholars at court as well as in *madrasas, khānqāhs*, or mosques. Planetary theory and timekeeping are the two prominent fields of such innovative efforts. Debates over modifications to Ptolemaic models, and the relationships between geometrical models, physical properties, and philosophical principles, perhaps began as early as the ninth century. Major scholars patronized by rulers contributed to this debate, among them Ibn al-Haytham, Naṣīr al-Dīn Ṭūsī,

al-'Urḍī, Quṭb al-Dīn Shīrāzī and 'Alī Qushjī. In the second period of patronage, scholars linked to the endowed teaching institutions participated in it
(Roberts and Kennedy 1959; Kennedy and Ghanem 1976; Saliba 1979; 1990;
1993; Langermann 1990; Ragep 1993; 2005). In addition to al-'Urḍī, Shīrāzī,
and Qushjī who also lived in and taught at *madrasas*, Ibn al-Shāṭir, *muwaqqit* and *muezzin* in Damascus, Ṣadr al-Sharī'a in Bukhara, and Shams al-Dīn
Khafrī in Qazvin were important contributors to planetary theory (Kennedy
1957; Saliba 1994; Dallal 1995).

Muwaqqits, and people engaged in similar works and with similar professional
links to mosques and *madrasas*, but named differently, start to appear in the late
thirteenth and early fourteenth centuries, first in Mamluk Egypt and Syria. Later
they also are known from al-Andalus, parts of north Africa, the Yemen, and the
Ottoman Empire. As mentioned earlier, their task consisted in solving the astronomical and mathematical problems connected with determining prayer times,
the direction of prayer towards Mecca, and the beginning of the new month (King
1993; 2004). These tasks had been tackled since the eighth century, often by some of
the most brilliant scholars involved with the mathematical sciences. Independent
of their religious beliefs they contributed to finding exact as well as approximate
numerical and geometrical solutions to these problems. From this perspective it
could well be claimed that a major outcome of courtly patronage for the mathematical sciences in both periods was the development of a rich field of methods for
finding astronomical and mathematical solutions important to religious practices.

When the *muwaqqit* emerged in the late thirteenth and early fourteenth century, such methods became acknowledged as an independent branch of the
mathematical sciences called *'ilm al-mīqāt*. In Charette's view (2006, 129) this
included a broadening of content, as it united spherical astronomy, timekeeping,
astronomical instrumentation, gnomonics, determination of the direction of
prayer, chronology, and the prediction of the beginning of the new month with
the first visibility of the lunar crescent. The *muwaqqits* and the teachers of the
new discipline developed new instruments, calculated multi-entry tables (often
up to 40,000 entries and occasionally even to 415,000 entries), developed sophisticated tools for simplifying the necessary calculations, and created means for
finding solutions valid for all latitudes.[5] The emergence of *'ilm al-mīqāt* also
involved issues of legitimacy within the discursive framework set by al-Ghazālī
in the eleventh century. Ibn al-Akfānī, a physician and *madrasa* professor in
Cairo, declared *'ilm al-mīqāt* to be a discipline obligatory (*wājib*) for a Muslim
(Witkam 1989, 59). Instrument making was mostly seen as socially beneficial
(Charette 2006, 129). This discursive support for the new field stabilized its

5. For the most complete survey of the available corpus of tables, problems, methods, texts and instruments published to date see King (2004; 1975, 83); Charette (2003, 26).

establishment within the teaching system and *'ilm al-mīqāt* indeed became a respectable mathematical discipline studied as part of a general education by numerous, but by no means all, students, including some of the later leading scholars in Mamluk society.

The outcomes of patronage of the mathematical sciences in Islamic societies described here are only a fraction of what was produced. Numerous instruments, more treatises, and wonderfully illuminated copies were produced over the centuries for courtly display in libraries and private settings, for men and women, rulers and their families, viziers, and *emirs*. Many professors, *muwaqqits*, physicians, and other professionals who contributed to the mathematical sciences as teachers, researchers, observers, and instrument makers owed their positions directly or indirectly to military as well as civil patrons. Without patronage, the flourishing mathematical cultures of Islamic societies would have been impossible or at least much poorer.

Conclusion

Sources about patronage of the mathematical sciences in Islamic societies between the eighth and the nineteenth century are rich, but narrow in scope and uncertain in both the meaning and the reliability of their claims. It is possible and useful to collect all dedications of texts, instruments, paintings, and other relevant objects and analyse their rhetoric, focus, and function. Such research will broaden our knowledge about the mathematical fields supported by different kinds of patrons and can elucidate the meaning given to this support.

However, we should not consider dedicated works as the sole outcomes of patronage. They are, without doubt, the central elements in the exchange of benefits, honour, and gratitude that constituted and kept alive the patronage relationship. Patronage, however, is not reducible to the exchange of gifts between patron and client. It also was a relationship of work or, as the medieval authors preferred, of service, which bound the participants to each other in the many ways discussed in the previous sections, work that included studying, researching, observing, interpreting, and writing. Without continuous intellectual work the scholars may have lost their patrons, although in some cases sources indicate that wit and sociability were more valuable commodities than mathematical proficiency. The criticism uttered against the fifteenth-century scholar Kashi, for instance—that he was a bore who had not mastered the refined protocol of Ulugh Beg's court and thus was frowned upon—reminds us that scholars of the mathematical sciences were participants in larger networks of social relationships, expectations, and behaviour. Excellence in the mathematical sciences while important was not sufficient for creating stable and comfortable connections between a client and a patron.

A second substantial field for research about patronage is offered by the rhetoric of patronage and its variations and changes over time and space. Islamic societies differed from each other while sharing a number of elements. The rhetoric of patronage embodies some of these differences and similarities. It reflects different degrees of violence, instabilities in client–patron relationships, the new opportunities provided by the endowed teaching institutes, the spread of patronage forms among the civil elites, and other changes. In this chapter I have tried to trace major differences, similarities, and changes. Focusing on shorter periods of time and smaller territories will help to uncover local and temporal particularities. Such particularities will offer possibilities for understanding why certain, but by no means all or even most, Islamic societies supported the mathematical sciences through patronage.

A third domain for further research is the study of the relative place ascribed to the mathematical sciences in the complex web of courtly patronage. The support given to individual mathematical fields by specific dynasties in relationship to other domains of culture such as medicine, history, law, or the arts and the social loci of these sciences defines their reputation, forms of practices, and various elements of their content. A clearer picture of the specifics of these relations and places will improve our understanding of the substantial changes in productivity, creativity, and focal points in the mathematical sciences in Islamic societies between the eighth and nineteenth centuries.

Bibliography

Asch, Ronald G, and Birke, Adolf M (eds), *Princes, patronage, and the nobility. The court at the beginning of the modern age, c 1450–1650*. Oxford University Press, 1991.

Bayhaqī, Ẓāhir al-Dīn, *Tārīkh ḥukamā' al-Islām* (ed Mamdūḥ Ḥasan Muḥammad), al-Qāhira: Maktabat al-thaqāfa al-dīniyya, 1996/1417.

Berkey, Jonathan, *The transmission of knowledge in Medieval Cairo. A social history of Islamic education*, Princeton University Press, 1992.

Bernards, Monique, and Nawas, John (eds), *Patronate and patronage in early and classical Islam*, Brill, 2005.

Berggren, J Lennart, 'Patronage', unpublished.

Berggren, J Lennart, *Episodes in the mathematics of medieval Islam*, Springer, 1986.

Blochet, E, *Catalogue de la collection de Manuscrits Orientaux de M Ch Schefer*, Ernest Leroux, 1900.

Brentjes, Sonja, 'Shams al-Din al-Sakhawi on *muwaqqits*, *mu'adhdhins*, and the teachers of various astronomical disciplines in Mamluk cities in the fifteenth century', forthcoming.

Burnett, Charles 'The certitude of astrology: the scientific methodology of Al-Qabīṣī and Abū Maʿshar', *Early Science and Medicine* 7/3 (2002), *Special Issue: Certainty, Doubt, Error: Aspects of the Practice of pre- and Early Modern Science in Honour of David A King* (eds Sonja Brentjes, Benno van Dalen, and François Charette), 198–213.

Charette, François, *Mathematical instrumentation in fourteenth-century Egypt and Syria. The illustrated treatise of Najm al-Dīn al-Miṣrī*, Brill, 2003.

Charette, François, 'The locales of Islamic astronomical instrumentation', *History of Science,* 44 (2006), 123–138.

Dallal, Ahmad S, *An Islamic response to Ptolemaic astronomy, Kitab Taʿdil Hayʾat al-Aflak of Sadr al-Shariʿa,* Brill, 1995.

Danneskiold-Samsøe, J F C, *Muses and patrons. Cultures of natural philosophy in seventeenth-century Scandinavia,* Lunds Universitet, 2004.

al-Dhahabī, Shams al-Dīn, *Al-ʿIbar fī khabar man ghabara* (ed Ṣalāḥ al-dīn al-Munjid), al-Juzʾ al-rābiʿ, 1963 [748/1347].

Eddé, Anne-Marie, *La principauté ayyoubide d'Alep* (579–658/1260), Franz Steiner Verlag, 1999.

Elger, Ralph, Review of Bernards and Nawas 2005, *Sehepunkte,* 7 (2007), http://www.sehepunkte.de/2007/07/12053.html.

Gutas, Dimitri, *Greek thought, Arabic culture: the Graeco-Arabic translation movement in Baghdad and early Abbasid society (2nd–4th/8th–10th centuries),* Routledge, 1998.

Halm, Heinz, *The Fatimids and their traditions of learning,* Tauris in association with The Institute of Ismaili Studies, 1997.

Hogendijk, Jan P, 'How trisections of the angle were transmitted from Greek to Islamic geometry', *Historia Mathematica,* 8 (1981), 417–438.

Ibn a Uṣaybīʿa, *ʿUyūn al-anbāʾ fī ṭabaqāt al-aṭibbāʾ* (ed Nizār Riḍā), Dār Maktabat al-Ḥayat, no date.

Ibn al-Fuwatī (642–723 AH), *Talkhīṣ majmaʿ al-ādāb fī muʿjam al-alqāb* (ed Muṣtafā Jawād), no date.

Ibn al-ʿIbrī al-Malaṭī al-maʿrūf bi-Ibn al-ʿIbrī, Ghrīghūrīyūs, *Tārīkh Mukhtaṣar al-Duwal,* Matbaʿa al-Kāthūlīkiyya, 1958.

Ibn al-Nadīm: *Kitāb al-Fihrist li'l-Nadīm* (ed Reza Tajaddod), 2nd ed, Marvi Offset Printing, no date.

Ibn Wāṣil, Jamāl al-Dīn Muḥammad b. Sālim (d 697 AH/1298 AD), *Mufarrij al-kurūb fī akhbār banī ayyūb* (eds Ḥasnīn Muḥammad Rabī and Saʿīd ʿAbd al-Fatāḥ ʿĀshūr), Dār al-Kutub, 1977.

Kennedy, Edward S, 'Late medieval planetary theory', *Isis,* 57 (1957), 365–378.

Kennedy, Edward S and Ghanem, Imad, *The life and work of Ibn al-Shatir, an Arab astronomer of the fourteenth century,* History of Arabic Science Institute, University of Aleppo, 1976.

Kettering, Sharon, *Patrons, brokers and clients in seventeenth-century France,* Oxford University Press, 1986.

Kettering, Sharon, *Patronage in sixteenth and seventeenth-century France,* Cambridge University Press, 2002.

King, David, 'al-Khalīlī's qibla table', *Journal of Near Eastern Studies,* 34 (1975), 81–122.

King, David, *Astronomy in the service of Islam,* Variorum, 1993.

King, David, 'On the role of the muezzin and *muwaqqit* in medieval Islamic societies', in F Jamil Ragep, Sally P Ragep, and Steven J Livesey (eds), *Tradition, transmission, transformation: proceedings of two conferences on premodern science held at the University of Oklahoma,* Brill, 1996, 285–346.

King, David, 'Too many cooks... A new account of the earliest Muslim geodetic measurements', *Suhayl,* 1 (2000), 207–241.

King, David, *In synchrony with the heavens. Studies in astronomical timekeeping and instrumentation in medieval Islamic civilization,* 2 vols, Brill, 2004, 2005.

Kraemer, Joel, *Humanism in the renaissance of Islam. The cultural revival during the Buyid age,* 2nd ed, Brill, 1993.

Langermann, Y Tzvi (ed and trans), *Ibn al-Haytham's On the configuration of the world*, Garland, 1990.

Loth, Otto, *A catalogue of the Arabic manuscripts in the library of the India Office*, London, 1877.

Matvievskaya, G P and B A Rozenfel'd, *Matematiki i astronomy musul'manskogo srednevekov'ya i ich trudy (VIII-XVII vv)*, 3 vols, Nauka, 1985.

Mottahedeh, Roy, *Loyalty and leadership in an early Islamic society*, Princeton University Press, 1980.

Petry, Carl, *The civilian elite of Cairo in the later Middle Ages*, Princeton University Press, 1981.

Ragep, F Jamil, *Nasir al-Din al-Tusi's memoir on astronomy (al-Tadhkira fī 'ilm al-hay'a)*, 2 vols, Springer, 1993.

Ragep, F Jamil, "Alī Qushjī and Regiomontanus: eccentric transformations and Copernican revolutions", *Journal for the History of Astronomy*, 36 (2005), 359–371.

Rashed, Roshdi, *Entre arithmétique et algèbre: Recherches sur l'histoire des mathématiques arabes*, Belles Lettres, 1984. English translation: *The development of Arabic mathematics: between arithmetic and algebra*, Kluwer Academic Publishers, 1994.

Rebstock, Ulrich, *Rechnen im islamischen Orient. Die literarischen Spuren der praktischen Rechenkunst*, Wissenschaftliche Buchgesellschaft, 1992.

Roberts, Victor, and Kennedy, Edward S, 'The planetary theory of Ibn al-Shatir', *Isis*, 50 (1959), 227–235.

al-Sakhāwī, Shams al-Dīn, *al-Ḍaw' al-lāmi' fī ahl al-qarn al-tāsi'*, Bayrūt, no date.

Saliba, George, 'The original source of Quṭb al-Dīn al-Shīrāzī's planetary model', *Journal for the History of Arabic Science*, 3 (1979), 3–18.

Saliba, George, *The astronomical work of Mu'ayyad al-Din al-'Urḍī (d 1266): a thirteenth century reform of Ptolemaic astronomy*, Markaz Dirāsat al-Waḥda al-'Arabiyya, 1990.

Saliba, George, 'Al-Qushjī's reform of the Ptolemaic model for Mercury', *Arabic Science and Philosophy*, 3 (1993), 161–203.

Saliba, George, 'A sixteeenth-century Arabic critique of Ptolemaic astronomy: the work of Shams al-Dīn al-Khafrī', *Journal for the History of Astronomy*, 25 (1994), 15–38.

Sayılı, Aydın, *The observatory in Islam and its place in the general history of the observatory*, Türk Tarih Kurumu, 1960.

Schöler, Georg, unter Mitarbeit von H-C Graf von Bothmer, T Duncker Gökçen, H Jenni, *Arabische Handschriften*, Teil II, Franz Steiner Verlag, 1990.

al-Shahrazūrī, Shamsuddin Muḥammad b. Maḥmood [d after 687 AH/1288 AD], *Nuzhat al-arwah wa rawdhah al-afrah fi tarikh al-hukama wa-al-falasifah*, 2 vols (ed Syed Khurshīd Aḥmed) The Da'iratu'l-Ma'arifu 'l-Osmania, Osmania University, 1976/1396.

Al-Sijistānī, Abū Sulaymān al-Manṭiqī, *Ṣiwān al-ḥikma wa-thalāth rasā'il* (ed 'Abd al-Raḥmān Badawī) Tehrān, 1974.

al-Ṭabarī, *The History of al-Ṭabarī, volume XXX: the 'Abbāsid caliphate in equilibrium; The Caliphates of Mūsā al-Hādī and Hārūn al-Rashīd* (trans and annot C E Bosworth), State University of New York Press, 1989.

Tasköprüzade, *Es-Saqa'iq en-No'manijje* (trans Oskar Rescher), *Gesammelte Werke*, Abteilung IV, Band 1, Biblio Verlag, 1978.

Witkam, J J, *De egyptische arts Ibn al-Akfani (gest 749/1348) en zijn indeling van den wetenschappen*, Ter Lugt Pers, 1989.

CHAPTER 4.2

John Aubrey and the 'Lives of our English mathematical writers'

Kate Bennett

One Sunday evening in February 1680, John Aubrey smoked a pipe of tobacco and dreamed up a new literary project. Stimulated by his collaborative work on a Life of Thomas Hobbes, he resolved to write 'my honoured friend Sir William Petty's life, which will be a fine thing, and which he shall peruse himselfe, and then it shall be left for Posterity herafter, to read (published)'.[1] He started immediately 'to scribble a sheet of paper close' with notes which he intended soon to 'enlarge' into three Lives 'of the worthy and ingeniose Knight Sir William Petty from his cradle; Sir Christopher Wren the like. as also Mr Robert Hooke'.[2] He also recalled that he had 'lodged 5 yeares since a sheet of Minutes of John Dee' with Elias Ashmole.[3] Soon he had drawn up a list of '55 persons' and had 'done 10 of them': these included the Life of Petty, inventor of 'political arithmetic' or political economy, and those of three mathematicians: Edward Davenant, who had been Aubrey's own mathematics tutor, John Pell, and William Oughtred.[4] Thus his biographical collection, which we now call 'Brief lives', had at its core a cluster

---

1. MS Ballard 14, f. 127. All manuscripts cited are in the Bodleian Library, Oxford, unless otherwise stated. For the Hobbes Life, see Hunter 1975, 78–80.

2. MS Ballard 14, f. 126. The 'Brief Life' of Wren was cut out of the manuscript, almost certainly by Anthony Wood, and lost. It was between ff. 28 and 29 of MS Aubrey 6.

3. MS Ballard 14, f. 127.   4. MS Ballard 14, f. 131–131v.

of Lives of English mathematicians.[5] To these Aubrey later added John Wallis, Henry Billingsley, Seth Ward, Sir Jonas Moore, Isaac Barrow, Thomas Harriot, Walter Warner, Francis Line, Henry Coley, Edmond Halley, and many more; as well as some Continental mathematicians: Nicolas Mercator, who had been another of Aubrey's tutors, and Descartes. John Pell contributed a very great deal of information to the 'Lives'—he was the main source for the Lives of Harriot, Briggs, and Warner—and was one of those who read the work in manuscript, although not very attentively.[6]

Aubrey's category of 'English mathematician' thus encompasses a very wide range of people, from astrologers to instrument makers to mechanics. He began to make biographical collections from the early 1650s, as a part of what he hoped would be a wider contribution to the advancement of learning. The Life of Henry Briggs, which includes an account of attempts to carry out Briggs's proposal for a canal to link the Thames and Avon, describes Aubrey's efforts to revive the project in the 1670s. The version found in 'Brief lives' is later than that in his manuscript 'The Wiltshire antiquities' (1671), but in its turn it was superseded by a fuller version for 'The naturall historie of Wiltshire' (1685).[7] Aubrey knew Edward Davenant by 1642, when he was sixteen; he was a close friend of two of Davenant's relations. He attempted, without success, to encourage Davenant to print his work by sending him letters and mathematical books and by introducing him to John Collins FRS, who was hoping to publish a collection of algebraic writings.[8] It seems to have been Aubrey who told Samuel Hartlib in early 1653 that 'Dr Davenant of divinity [. . .] hath made ready several Mathematical Arithmetical Geometrical Astronomical Work's in Latin which he intends shal be publish't after his death'.[9] This death occurred in early 1680; and Aubrey's Life was written two weeks later. Many of these Lives were similarly intended as a contribution to collaborative projects, often ones which Aubrey tried to initiate or revive. An example is the Life of William Oughtred, originally prepared as part of a county history of Surrey in 1673, but abandoned when Aubrey was dropped from the project. In 1680, as we have seen, he wrote another version for 'Brief lives'; later, in 1691, Aubrey also revised what he called his 'Surrey notes'

5. At some stage Aubrey inscribed 'Σχεδιάσματα. Brief Lives' on the cover of Volume 1 (see Clark 1898, I 8; much of the Greek text is now no longer visible). Even after so doing, he never used these titles in his correspondence, continuing to call them 'my minutes of Lives' or 'my Lives'.

6. Aubrey told Anthony Wood that he had 'deposited' his 'Minutes of lives in Dr Pells hands' in autumn 1680 'expecting he would have made additions or amendments, but (poor, disconsolate man!) I recieved it of him without any' (MS Wood F 39, f. 351).

7. MS Aubrey 6, ff. 47v–49; MS Aubrey 3, ff. 22v–23a; 84v; MS Aubrey 1, ff. 49–51. The version of the account of Briggs in the 'Naturall historie' transcribed for the Royal Society in 1690–1 is different again (London, Royal Society, MS 92, 66–69). See Bennett 1993, 114–17. The nineteenth-century editions of Aubrey's Wiltshire manuscripts exclude these biographies.

8. John Collins to Francis Vernon, in Rigaud 1841, I 154–155; MS Aubrey 12, f. 96 is Davenant's letter to Aubrey which shows that in 1667 Aubrey was acting as an intermediary between Davenant and Collins.

9. Hartlib Papers 28/2/49B.

into what is now 'A perambulation of Surrey', with an extended version of the Life of Oughtred.[10] In the previous year, Aubrey had prepared an 'Apparatus for the lives of our English mathematical writers'.[11]

Aubrey shared with Sir William Petty and John Pell a wish to bring about educational reform, as a result of which mathematics and other scientific subjects would be a central subject for study in schools. Aubrey's collections look forward to 'Posterity herafter', in which a more enlightened world might combine the study of mathematics with the pious duty of doing 'right' to those who had in the 'darke time' before the establishment of the Royal Society patronized mathematicians or added to mathematical knowledge. It was towards this idealized future that Aubrey 'religiously collected' the vestigia of the past.[12]

Aubrey's wish to 'doe right' in this way led to his most significant collaboration, that with the antiquary Anthony Wood, whom he sought out in Oxford in 1667. He had been collecting antiquarian and biographical material since the 1640s, and expected that some of it would be published and accredited to him in Wood's writings: the brief biographical parts of Wood's history of the colleges and halls of Oxford University, *Historia et antiquitates universitatis oxonienses* (1674), and the entirely biographical *Athenae oxonienses* (1691), his account of Oxford writers. Aubrey accordingly supplied Wood until the latter's death in 1695 with an extensive range of material on biographical, bibliographical, and antiquarian subjects. As well as a great number of letters, Aubrey sent him books, pamphlets, transcriptions of parish registers, and epitaphs; he copied materials which he also passed to other correspondents; he acted as intermediary between Wood and other informants; he sent him separate manuscript Lives and, most valuable of all, he lent him his manuscripts, including 'Brief lives'. These were in Wood's custody for long periods of time. Aubrey would continue his researches and hastily revise his manuscripts when he occasionally got them back. He regularly sent material to Wood instructing him to add it to the 'Lives'; but Wood usually took no notice, treating the manuscripts as his own: at the end of their collaboration, he even cut out nearly all of the second volume of the 'Lives' and destroyed it. The fact that Aubrey, homeless from 1671, had his own manuscripts in his possession for such a short time, never had all his books and papers together, and relied on an unreliable collaborator is the main cause of the textual confusion that characterizes the 'Lives' manuscripts. Much of the work he sent Wood was not used by him; thus much of Aubrey's work, and of his work on the history of mathematics, remains unprinted.[13]

---

10. The earliest Life of Oughtred is MS Aubrey 4, f. 237v; the Brief Life is MS Aubrey 6 ff. 39–42v; the extended 'Surrey' Life is MS Aubrey 4, ff. 102v–5.

11. MS Aubrey 8, ff. 69–88v.

12. MS Ballard 14, f. 127; MS Wood F 39, f. 229; MS Wood F 45, f. 204; MS Aubrey 6, f. 60.

13. MS Wood F 49, f. 67. In the 1690s Aubrey donated his manuscripts to the Ashmolean Museum, Oxford; they are now held in the Bodleian Library, Oxford.

Aubrey's fine net

Aubrey's chief reason for helping Wood to such an extent was that he found his historical approach so congenial. On one of several occasions on which he found himself defending Wood's historiography, Aubrey said his work mattered because of its inclusiveness of record. Wood, he said, was a 'Candid Historian', who 'made not' himself a 'Judge' of men's 'merites or Abilities' but on the contrary took extraordinary pains to record all his subjects' writings, including, to the scorn of many, details of ephemeral publications such as tracts and unpublished manuscripts.[14] He warmly approved of the way Wood found many strategies to avoid restricting the scope of his publications to Oxford-educated writers: there are plenty of Cambridge-educated, or self-taught, or nonconformist writers, for example, in the *Athenae oxonienses*. However, Aubrey felt Wood did not go far enough. He constantly tried to persuade Wood to add accounts of persons whom Wood did not think proper to include, and details which Wood disdained to notice; and he always wanted Wood to make his entries fuller. Aubrey, as we have seen, expected his papers to be kept safe in Wood's archive for posterity, and the benefit of this for his purposes was that he did not try to make all of his work suitable for print. He was free to include in his manuscripts material that could not be transmitted to print without losing its value, such as the letter in William Oughtred's hand, which Aubrey stitched into his manuscript along- side the mathematician's Life.[15] He was not exclusively interested in the 'Lives of Eminent Men', as a nineteenth-century editor renamed 'Brief lives', although he certainly made them a priority: his interests were wider. Sometimes this led to eccentric judgements: in 1690, eagerly expressing the hope that the autodidact astronomer Thomas Streete would be memorialized by those who talked 'of club- bing towards an Inscription' in Westminster New Chapel where he was buried, Aubrey exclaimed that 'No man living haz deserved so well of Astronomie'.[16] But the reason for Aubrey's championship of lesser mathematical mortals is that in an age when many, like Davenant, who 'being a divine' was 'unwilling to print', were liable to feel conscientious scruples about neglecting their proper religious duties to pursue their studies, Aubrey feared that a focus on major and public achievement would exclude from the notice of posterity many of those who had

14. MS Aubrey 12, f. 8.
15. Oughtred's letter is MS Aubrey 6, ff. 41–2. Aubrey owned Oughtred's annotated copy of Pitiscus 1614, now in the library of Worcester College, Oxford. On the front endpaper, below Oughtred's annotations, Aubrey wrote: 'Sum Johannis Aubrij de Easton-Piers. This was old Mr Oughtreds booke, and the Notes are of his owne handwriting'. It is not known what became of Oughtred's library after his death: it was a 'compleat' mathematical library (see Lloyd 1668, 287). Aubrey does not say how he obtained the Pitiscus, but he was looking for a copy in 1649. In March 1650 Ralph Bathurst found a copy for him at the steep price of 7 shillings, which may have been Oughtred's copy (MS Aubrey 12, ff. 300, 304, 306).
16. MS Aubrey 8, f. 88v.

mething of interest to offer, however fragmentary.[17] He also feared that such squeamishness about printing mathematical works during one's lifetime would result in no publication at all, as family members might not have the skills or inclination to see such recondite works through the press. Aubrey hoped both to record and to stimulate a very wide pattern of participation in the advancement of learning; and this led him to record details of hundreds of people who were not included by Wood in his biographical collections. Aubrey's net was both wider and finer than Wood's, and caught smaller fish.

For example, in 'Idea of education' Aubrey gives details of mathematicians who had taught their female relatives, such as the niece, '6 yeares old goeing in seven', of William Holder, for whom Holder prepared a 'Mathematical Catechism', designed to teach her to add, subtract, multiply, divide, and to understand simple geometrical principles. She is not named, but the likeliest candidate is Sir Christopher Wren's daughter Jane who would have been seven at the time of writing in 1684.[18] Aubrey also transcribed part of Anne Ettrick's mathematical manuscripts, which were prepared as part of her instruction in algebra from her father, Edward Davenant: he describes her as a 'very good Logist'.[19] Aubrey gives a great deal of explicit information about his informants, not disguised as 'an ingenious gentleman of my acquaintance' or a similar formula, but named outright. An example is 'Mr Bayes the Watch-maker', the 'nephew' of Samuel Foster whom Aubrey identifies as his source in the margin of Foster's Life.[20] This example may stand for many. Aubrey looked forward to a new intellectual community in which obscure as well as luminous persons would be recognized for their contribution to the advancement of learning, and in which support networks of family, patronage, intellectual mentoring, and education would become apparent. The 'Lives' were only a part of a wider ambition to transmit their subjects' work to posterity.

One of Aubrey's biographical sources for mathematicians was his 'old cosen' James Whitney (d. 1670), quondam fellow of Brasenose College, Oxford, vicar of the Wiltshire parish of Donhead St Andrew, and 'a great Nomenclator of Oxford men' such as Sir Walter Ralegh. Whitney had an interest in mathematics, and as

17. MS Aubrey 6, f. 43.

18. MS Aubrey 10, f. 36b. Holder, the subject of one of Aubrey's Lives, was a neighbour of the Wrens during this period and had been Wren's mathematical tutor.

19. Worcester College, Oxford, MS 64 (unfoliated).

20. In this case, the detail of the informant allows us to trace an association between Foster and the watch-making trade in Coventry; it also gives evidence of watchmaking in the city earlier than has been previously known. Foster mentions his 'Sister Martha Bayes widdowe living in Coventry' in his will as well as two nephews, 'John Bayes Watchmaker in London' and 'Beniamin Bayes liveing in Coventry' (London Public Record Office, PROB/11/222, f. 42v). Coventry became a major centre for watch-making in the eighteenth century. John was probably the John Bayes of London who became a member of the Clockmakers' Company in 1647 and warden in 1658. A John Bayes, whom I suggest was his father, supplied watches to Charles I and John Pym in 1647 and 1628. Benjamin Bayes was apprenticed to John in 1661 and would also have been living in London (see Britten 1911, 615).

well as donating to Aubrey his copy of Münster's *Rudimenta mathematica*, he also passed on an Oxford tradition of mathematical storytelling.[21] He informed the horrified Aubrey that during the Visitation of Oxford under Edward VI mathematical books were burned 'for Conjuring bookes', and that 'if the Greeke Professor had not accidentally come along, the Greeke Testament had been throwne into the fire for a Conjuring booke too'.[22] He told Aubrey 'by tradition' that Robert Hues, author of *De globis*, was of St Mary Hall, Oxford; and said that Edward Brerewood studied mathematics so intensively because he was too poor to leave his chamber in Brasenose and so was obliged to remain there, in worn-out gown and 'slip-shoes', but profiting 'exceedingly' in 'knowledge'.[23]

This tradition of storytelling, with its gloomy portrayal of neglected mathematical studies in Oxford, is reflected in many places in Aubrey's work. We should be careful not to take it at face value. Mordechai Feingold argues that the passages in the autobiographies of Wallis, Hobbes, and Locke, written in old age, in which these thinkers insist that they were intellectually self-made, alleging that the universities of their undergraduate days were entirely given over to Aristotelianism and were thus devoid of any opportunities to study natural philosophy, should be understood in light 'of the marked tendency of such men towards self-aggrandizement' (Feingold 1997, 359). Biographical chatter, as well as the formal written or orated biography of this period, conveys values rather than fact; and indeed is the particular refuge of those who do not feel their values to prevail. Whitney's mathematical memorializing seems to have consolidated a sense of identification between teller and listener as voices crying in the wilderness.[24] When Aubrey uses such stories in his writing he strikes a melancholy note: many of his lives depict mathematicians as pursuing their interior and private studies in a hostile or uncomprehending world. It was certainly uncomprehending: in his will, Whitney bequeaths 'my Silver Watch which I usually carry about me to know how the day passeth', suggesting that he felt the need to explain the purpose of his watch to those in his rural Wiltshire household, or parish, who did not understand its function or allure.[25] But such hostility as they encountered was normally due to professional rivalry. William Oughtred, publicly accused by his former student of being an 'ignorant mechanick' maker of the slide-rule rather than its inventor, and of neglecting his religious duties for mathematics, retorted furiously that 'the time which over and above those usuall studies I employed upon the Mathematicall sciences, I redeemed night by night from my naturall sleep' (Oughtred 1634, A4v).

---

21. Münster 1551. Aubrey's copy is Ashm. F. 7. Judging by an inscription on the back page, 'Peter Carret is a knaue/so god me saue', it was used as an undergraduate textbook. Bound with it is Whitney's copy of Blagrave 1585.

22. MS Wood F 39, f. 282v.   23. MS Wood F 39, ff. 234, 237, 343v, 347.

24. MS Wood F 39, f. 234.   25. London Public Record Office, PROB/11/333, f. 156

What this dedication meant for Oughtred's family life is suggested by a story his son Ben told Aubrey, 'that his Father did use to lye abed till eleaven or 12 a clock with his Doublet on, ever Since he can remember. Studied late at night went not to bed till 11 a clock, had his tinder-box by him, and on the top of his Bed-staffe, he had his Inkehorne fix't. He slept but little. Sometimes he went not to bed in two or three nights, and would not come downe to meales, till he had found out the Quaesitum' [that which was sought].[26] It could be a lonely business, and its practitioners often equated its satisfactions to those of the devout. Thomas Streete wrote his own epitaph, which Aubrey seems to have been shown by Streete's widow: 'Here lies the Earth of one, that thought some good,/ Although too few him rightly understood:/Above the Starres his heightned Mind did fly,/His hapier Spirit into Eternity.' John Pell similarly defended the value of his mathematical studies by equating them to prayer: Aubrey records that he had 'sayd to me, that he did believe, that he solved some Questions non sine divino auxilio [not without divine help]'.[27] Oughtred asserted that there was 'in all' mathematical 'inventions aliquid divinum [something divine], an infusion beyond human cogitations' (Rigaud 1841, I 35). In 'Idea of education', Aubrey recommends that students should carry in their pockets 'Euclid's Elements: as religiously as a Monke his Breviarie'.[28] A breviary instructs the monk how to mark the passing of the hours. Aubrey implies that geometry is a fundamental discipline governing the conduct and narration of every phase of daily life.

Oughtred's son told Aubrey that his father 'would drawe lines and diagrams on the Dust'. This may have been merely an old-fashioned frugal practice from the Cambridge of Oughtred's undergraduate days, one which saved the expense of paper, but in Aubrey's account, it was a sign of genius: of a 'Witt' which was 'always working'.[29] This reminiscence of Oughtred, a Royalist who suffered for his political allegiances during the English Civil War, shapes him in the likeness of Archimedes, said to have been killed by a Roman soldier during the fall of Syracuse while intent on the geometrical figures he had traced in the dust; oblivious to 'many shameful examples of anger and many of greed'.[30] This tendency to cast mathematicians in the role of 'the English Archimedes', as Oughtred characterized Briggs, testifies to a determination amongst mathematicians and virtuosi to gain the respectability of the Ancients for their Modern

---

26. MS Aubrey 6, f. 39v.   27. MS Aubrey 6, f. 52.

28. MS Aubrey 10, f. 94.

29. MS Aubrey 6, f. 39. A mathematical lectureship established in 1573 at Queens' College, Cambridge, specified that the geometry lectures were to be 'redd [...] with a penn on paper [...] or a sticke or compasse in sand or duste'. See Feingold (1984, 39).

30. MS Aubrey 6, f. 39; Livy, *Ad urbe conditia* XXV.31. Aubrey wrote in the margin of his copy alongside this passage, 'Geometricall figures' (Ashm. D 23, 478), and he made a memorandum of the passage on the back endpaper.

studies (Oughtred 1633, 18). Thomas Sprat, describing the early meetings of those who were to be the founder members of the Society, dramatically claims that they were 'scatt'red by the miserable distractions of that Fatal year' of 1659, when the country seemed on the brink of a second civil war. Sprat has them fearing that the 'continuance of their meetings' at Gresham College 'might have made them run the hazard of the fate of Archimedes: For then the place of their meeting was made a Quarter for Soldiers' (Sprat 1667, 58–59). Sprat uses the innocently-occupied geometrician to represent the Society as constructive, peace-loving, and patriotic. Archimedes was a particularly congenial figurehead because, as John Wilkins argued in his *Mathematicall magick*, unlike those 'ancient Mathematicians' who 'did place all their learning in abstracted speculations', he was willing to apply his mind to 'Mechanicall experiments'. The culture in which Archimedes worked, Wilkins maintains, led him to make a 'superstitious' choice not to 'leave anything in writing', but rather to 'conceale' his learning from the 'apprehension' of the 'vulgar' (Wilkins 1648, 3–4). However, the 'vail' of 'mysticall' language and practices clearly had appeal for Oughtred: there is some evidence that he may have practiced divination.[31]

Sprat's and Wilkins's 'English Archimedes' had no intention of getting killed off; theirs was an improved version, without the mysticism and with more common sense. *Mathematicall magick* is not about magic at all; it offers rational explanation and information, bringing the means of acquiring mathematical knowledge to a wider public. Its title could be that of a twentieth-century textbook. In his Life of Wilkins, Aubrey calls him 'a lustie strong growne, well sett broad shoulderd person. cheerfull, and hospitable', and follows this amiable portrait with the assertion that Wilkins 'was the principall Reviver of Experimentall Philosophy (secundum mentem Domini Baconi) [according to the way of Master Bacon] at Oxford, where he had weekely, an experimentall philosophicall Clubbe', the nucleus of the Royal Society.[32] Aubrey once again makes a positive association between charismatic conviviality and scientific invention in his description of Wren as 'Englands Archimedes [...] a person not only of admirable parts but of a sweet communicative nature'.[33] Biography, concentrating as it does on personal character, proved

---

31. Aubrey observes in the Life that he had seen 'some notes' in Oughtred's 'hand-writing on Cattan's Geomantie'. This was a copy of Cattan's *Geomancie* (1591, 1608), which was among those of Oughtred's books and papers which had been bought by the physician Richard Blackburne. Aubrey says that Oughtred's 'marks and notes' showed that he had 'thoroughly perused' the work (MS Aubrey 6, f. 39; MS Aubrey 4, f. 102v). Aubrey quotes from the work in his 'Remaines of gentilisme', where he defines 'Geomantie' as a form of divination to be performed with a great deal of seriousness and prayers and 'in a very private place; or on the sea shore' (British Library, Lansdowne MS 231, f. 111).

32. MS Aubrey 6, f. 92. See the accounts in Tyacke 1997, 430, 548–550, and Feingold 2005, 167–183. Feingold, who supports Aubrey's version of the origins of the Society, gives an account of scientific and mathematical clubs before the Society's foundation.

33. Wood's transcription from 'Mr Awbrey's collection B. p. 58', that is, the lost 'Liber B'. MS Wood F 39, f. 129v.

especially suitable for the advancement of an understanding of mathematics not merely as a body of knowledge found out by those possessing 'admirable parts', but as a culture. For Aubrey, the well-being of mathematics was dependent on ethical practices and principles, governing the communication of knowledge and skill by the sweet-natured to support, encourage, and give credit to others. Mathematical stories were a medium through which such social values and aspirations might be communicated. In the 'Lives', this understanding is crystallized in anecdotes of personal behaviour and interaction. John Wallis is punished in the 'Lives' for plagiary; those ignorant or drunken souls who threw pocket watches into moats or smashed sundials are rebuked; and 'the old gentleman' William Oughtred is lovingly remembered for his care of his pupils, and for having 'taught all free'.[34] His long-suffering wife, required to support his hospitality on a modest income, is depreciated as 'a penurious woman' who denied her husband candles to light his night-time work.[35] Through 'Brief lives' Aubrey tries to create societies 'secundum mentem Domini Baconi': pursuing links between minor and grand individuals, and between the living and the dead; and seeking to engage the reader's sympathies through a literary style that is itself of a 'cheerfull, and hospitable' nature.

This kind of social breadth, encompassing innkeepers and Wiltshire parsons as well as Hooke and Wren, can be seen across Aubrey's entire career. In his 'Idea of education', Aubrey recommended that young men should follow his example by collecting information in notebooks. 'Semper excerpe [always copy down] in some kind or other', he recommended. 'One may take a Hint from an old Woman, or simple bodie. Contemne no Body: aime still at Trueth. Had I not excerped, these Notions that are here stitch together, and good part whereof I have gott from my learned and deare Friend Dr John Pell, had been utterly lost, and buried in Oblivion.'[36] Aubrey did indeed collect reminiscences from the 'simple bodies' who made up the domestic circle of the mathematicians he wished to memorialize. He suggests that we should regard these as 'a Hint' for further historical research; and certainly as a source of information his testimony was sometimes of mixed quality. This is demonstrated in the case of John Dee, one of the cornerstones of Aubrey's projected 'Lives of our English mathematical writers'.[37] Aubrey visited Mortlake in early 1674 and found Dee's place of burial: he also interviewed an eighty-year-old woman called Goody Faldo, with the intention of including her memories in his contribution to a collaborative Life. Mrs Faldo seems to have

34. Aubrey, like others in the Royal Society, believed that Wallis had plagiarised several members of the Society, in particular William Holder. See Hobbes 1994, II 753–756; Bennett 2001, 216–217.

35. MS Aubrey 6, ff. 39, 42.

36. MS Aubrey 10, f. 114.

37. Aubrey added a note in the Life of John Pell in 'Brief lives' to instruct Wood (who took no notice) that 'I would have the Lives of John Dee, Sir Henry Billingsley, the two Digges, father and sonne, Mr Thomas Hariot, Mr .... Warner, Mr Brigges, and Dr Pells be putt together' (MS Aubrey 6, f. 51. The ellipsis marks, indicating that Warner's first name was not known to him, are Aubrey's).

done laundry and perhaps needlework for Dee's family: she certainly describes his 'black gowne' with 'long sleeves, with slitts but without Buttons and loopes and tufts' in convincing detail. Through her account we are permitted to see Dee as a kind and wise man who used 'a great many stilles', and employed magic to recover hampers of stolen linen; we are also told that children 'dreaded' him because of his reputation as a conjuror. It is a very confused narrative: Elias Ashmole was so discontented with it that he made a special trip to re-interview her; but his account, although clearer, is not remarkably coherent. In fact Aubrey's version seems a very faithful record of the disjointed memories which were all she had to offer him, and he deliberately preserved this incoherence in three texts of the interview, one of which is incorporated into the Life of Dee. Thus his 'digressive and inconsequential treatment' of the interview with 'an old Woman' is not, as Ashmole's editor Josten said of it, an indication of Aubrey's weak psychology: it is an indication of his respect for the terms in which his interviewees transmitted their own memories. In doing so, Aubrey has provided us with a revealing record—or 'Hint'—of how the obscure practices of mathematicians and natural philosophers might appear to their family, neighbours, and domestics (Ashmole 1966, IV 1298–1300, 1332–1335, and 1335 n8; Bennett 1993, 94–104).

This fidelity to the origins of his information is characteristic. Aubrey saw his role, not as a writer, but as an intermediary: as one who was to collect the materials which would allow posterity to write fuller, more accurate history. He was unsure what to call his life of Hobbes and initially wrote 'Supplementum Vitae Thomae Hobbes' [a supplement to the Life of Thomas Hobbes] on the title-page.[38] Wood, puzzled, wrote in the margin 'what need you say Supplimentum pray say The Life of Thomas Hobbs'.[39] But Aubrey resisted this advice, and later called the manuscript a 'supellex'. He also called his astrological work 'Collectio geniturarum' a 'supellex'.[40] This word forms part of Bacon's argument in The advancement of learning that:

'Schollers in Vniversities come too soone, & too vnripe to Logicke & Rhetoricke' which are 'the Rules & Directions, how to set forth & dispose matter; & therfore for mindes emptie & vnfraught with matter, & which haue not gathered that which Cicero calleth Sylua and Supellex, stuffe and varietie, to beginne with those Artes' brings learning into contempt (Bacon 2000, 59).

This Ciceronian term, literally 'household furniture', was by the late-seventeenth century often used by those with scientific interests to mean the equipment or apparatus for an experiment or operation, and sometimes by extension, the literature appropriate for scientific study. In 1684 Sir William Petty prepared a 'Supellex Philosophica', a list of forty scientific 'Instruments requisite to carry on the designs of the' Dublin Philosophical 'Society' (Petty 1927, 29). This document was a proposal, not an inventory; it was one of the lists which Petty, and

38. Aubrey's dissatisfaction with his collaborators came later: see below.
39. MS Aubrey 9, f. 28.   40. MS Ballard 14, f. 125; MS Aubrey 6, f. 12v.

others in his circle including Hooke, were fond of drawing up: Petty included it in his collections towards an ideal school. William Stukeley, antiquary, natural philosopher, and biographer of Newton, described the Exeter library of William Musgrove FRS, the editor of the *Philosophical Transactions*, as a 'good collection of books, coins, and other antiquarian *supellex*' (Chope 1918, 139). Stukeley's phrase implies that something is a 'supellex' when it crosses the boundary of material object or book and is instead, in Petty's words, a thing 'requisite to carry on the designs' of an interest-group.

Aubrey's use of the word in two biographical contexts (for, as there are significant connections between 'Collectio geniturarum' and 'Brief lives', the astrological manuscript can be so considered) may encourage us to link this connotation of the word 'supellex' with the word 'brief' in 'Brief lives': not merely something short, nor merely 'notes towards' a biographical work; but also a condensed instruction to posterity, like a lawyer's brief. These 'briefs' do not simply consist of the utterances made in the biographical texts: as we saw in the case of Davenant, Aubrey's Lives often refer to efforts to preserve the manuscripts, annotated books, instruments, and libraries of mathematicians after their deaths; efforts in which Aubrey was often a prime mover, or even the only mover (Bennett 2001, 213–245).

From this combination of biography and bibliographical collecting derives the peculiar nature of Aubrey's 'Apparatus for the Lives of our English Mathematical Writers', begun 25 March 1690.[41] Aubrey uses the word 'apparatus' in several other contexts to denote materials towards a larger work or a work-in-progress; for example, he uses the word on the title-page of his 'Wiltshire antiquities', which is unfinished and was intended to form part of an ambitious collaborative project; and he told Sir William Petty in 1675 that he was hard at work 'transcribing out of my Description i.e. apparatus for a Description of Wilts all my Naturall observations'.[42] Aubrey began his 'apparatus' with a list of those mathematicians whom he wanted to include, then he wrote their names in the manuscript, leaving room for their biographies to be added later. Some of these, such as Thomas Hobbes, Thomas Allen, Sir Henry Billingsley, John Collins, and William Lilly, are followed by no new material, but are merely marked with the word 'donne'. This is revealing. In the case of Billingsley and Allen, this means that Aubrey had written Lives of these men in 'Brief lives'. In the case of the astrologer William Lilly, 'donne' means that Lilly had written his own manuscript Life, in the hands of Elias Ashmole. John Collins has an entry in the index, directing us to page 27 of the manuscript, which in its turn merely says that the work has been 'donne'.

41. John Britton, who examined the Aubrey papers in the Ashmolean Library before they were transferred to the Bodleian, rebound and repaginated, describes the 'Apparatus' as a separate manuscript: 'These lives occupy sixteen leaves, foolscap folio, written on one side only, and paged by the author. They are stitched together, and fastened inside the cover of part iii. of the "Lives of Eminent Men"' (Britton 1845, 110). This indicates that their current position (MS Aubrey 8, ff. 69–88v) is probably not the position in which they were found before rebinding. They are paginated, rather than foliated (as the Lives are).

42. MS Aubrey 3, f. 1a; British Library Add MS 72850, f. 141.

Only the briefest of biographical notices of Collins survives in Aubrey's letters to Anthony Wood, and nothing is found in 'Brief lives' except for an entry in the index: 'Collins John Mathematician in print hereto annexed'. Yet no printed material relating to Collins survives in any of the remaining 'Lives' manuscripts. These were, however, significantly tampered with after they were deposited in the Ashmolean Museum in the 1690s, and in a volume of papers and pamphlets also donated by Aubrey to the Museum is found Collins's preface for the reissued edition of his *Introduction to merchants-accompts*, which contains his account of his life.[43] Collins probably gave it to Aubrey, whom he knew well: they collaborated in collecting the unpublished papers of mathematicians and natural philosophers such as Davenant, as we have seen, and Briggs. So by calling his text an 'apparatus', Aubrey implies that he is engaged, not in a project of personal authorship, but in the task of constructing a repository of biographical data, of the kind he had supplied to Collins: an equipping of the intellectual world so that it might better undertake the advancement of learning.

The advantages of a mathematical education

We need not take Aubrey entirely at his word. Despite their profoundly collaborative context, Aubrey's 'Lives' are far more than an apparatus. As we have seen, Aubrey intended to write the Lives of his mathematicians from their 'cradle' (or, in the case of Robert Boyle, his Irish nurse's 'pendulous Satchell').[44] The emphasis in 'Brief lives' on early youth, and on education, supports an argument for the 'great advantage' of a mathematical education for the very young. In his life of Hobbes, Aubrey records that Sir Jonas Moore was 'wont much to lament' that Hobbes had not learnt mathematics before the age of forty, when he discovered Euclid.[45] Aubrey devotes a part of his own very inchoate Life to a counter-example, describing his own wasted potential. He tells us how he was 'bred' in the depth of the Wiltshire countryside in an 'Eremiticall Solitude', a boy of great curiosity, and with a 'zeale' to learning, whose 'greatest delight' was 'to be continually with the artificers'; but 'discouraged' in this by his father, a country squire whom he characterizes as educated merely to 'Hawking'. He describes the lack of a sustained period of education, and says that he was obliged to study geometry on horseback and in the privy, with a copy of Oughtred's *Clavis mathematicae* hidden in his pocket for spare moments.[46] Against this he assembles in the 'Lives' the cases of those who were more fortunate: Oughtred, Edward Davenant, Edmond Halley.

43. Ashm. F 4 (45). This preface (Wing 5382A) is not identical with that of the 1674 and 1675 editions, but may be that of 1664; alternatively, the preface may have been issued separately.
44. MS Aubrey 6, f. 16.   45. MS Aubrey 9, ff. 36, 53.
46. MS Aubrey 10, f. 83; MS Aubrey 7, f. 3.

These accounts, dispersed in the 'Lives', are collected together and repeated almost verbatim in the introduction to 'Idea of education':

Without doubt it was a great advantage to the learned Mr W. Oughtred's naturall parts, that his father taught him common Arithmetique perfectly while he was a schoole-boy. The like advantage may be reported of the reverend and learned Edward Davenant D.D. whose father a Merchant of London taught him Arithmetick when a Schoole-boy. The like may be sayd of Sir Chr. Wren; & Mr Edmund Halley R.S.S. <*in margin:* his fathers Prentice taught him Arithmetique & to write at nine yeares old.> and Mr Th. Ax his father taught him the Table of Multiplication at seaven years old.[47]

Halley's father was a 'soap-boiler'; Oughtred's father's duties as a clerk employed by Eton College included keeping financial records; Davenant's was a merchant who had made money fishing pilchards. Aubrey's argument in 'Idea of education' was that, at the age at which boys were usually introduced to Latin grammar, gentlemen should receive the kind of education which was given to those of their social inferiors who were being taught navigation at Christ's Hospital. For 'no Nobleman's son in England [...] can have so good Breeding' as 'the Kings Mathematicall Boyes at Christ-church-Hospital'.[48]

Aubrey's emphasis on 'good Breeding' is deliberate. Along with his loving biographical record of the personal and domestic interactions of mathematicians, it challenges a common prejudice against those who were primarily interested in mathematics (as opposed to those who had studied mathematics as an integral part of their university studies) as unsocial beings. In the leading text of Tudor education, the humanist Roger Ascham famously argued that those who were particularly fond of mathematical studies were solitary by nature, 'unfit to live with others', and 'unapt to serve the world', unskilled in negotiation (Feingold 1997, 363–365). The usual response was to point to the practical benefit of mathematics, as Aubrey does in an early chapter in 'Idea of education' entitled 'Mathematical Prudence'. Here he begins to assemble the grounds of such arguments, by showing the value of a knowledge of mercantile mathematics: lawyers, shopkeepers, and young heirs would all benefit from a capacity to understand interest, calculate rent, and read a balance sheet. Aubrey recommends exercises that schoolboys might perform to acquire these skills, but his researches were not confined to printed works on the subject. He was arguing for a far more ambitious scientific education than the smattering of arithmetic usually thought sufficient to allow a landowner to check his steward's accounts; and the real value of even this level of learning would best be demonstrated if it were firmly grounded in experience. Hence Thomas Axe (as we have seen, an example of one who had benefited by early mathematical teaching) was to be asked for his 'Calculations of Sir William Portman's Old Rents' which Axe had used as the basis of his investment advice

---

47. MS Aubrey 10, f. 9.   48. MS Aubrey 10, f. 7.

to Portman, whose attorney he was.[49] At the outset of the chapter Aubrey quotes Francis Osborne in his *Advice to his son* saying 'that he haz known many men of good Estates, have been undonne meerly for want of skill in a little plaine & common Arithmetick'.[50]

This advice was, Aubrey apparently felt, the weaker for its general nature. As Bacon said, 'knoweledge drawne freshly and in our view out of particulers, knoweth the waie best to particulers againe. And it hath much greater life for practise: when the discourse attendeth vpon the Example, then when the example attenddeth vpon the discourse' (Bacon 2000, 162). Aubrey accordingly substantiates it by adding a biographical 'particuler' in the margin: the name of Colonel Alexander Popham, of Littlecote House, Wiltshire. In a Chancery suit of 1649, Popham told the court that his late brother John had 'wasted his Estate in Hospitalitye', leaving debts of £38,000.[51] In his Brief Life of John's father, Judge Popham, Aubrey says that John 'was a great waster' whose extravagant hospitality stretched as far as inviting three or four lords and their retinue to stay at Littlecote at a time. His equally 'vaine' wife is also said to have spent her substantial marriage portion in inviting 'all the woemen' of the county to heavy-drinking house-parties: 'They both dyed by excesse; and by Luxury, and cosonge by their servants, when he dyed there was I thinke a hundred thousand pound debt': a grievous example of the want of mundane prudence. Further on in 'Mundane Prudence', Aubrey records that 'Alderman Gombleton of London told me, that Colonel Alexander Popham (who had 900*li* [pounds sterling] per annum) was wont to make his complaints of casting-up of long Bills, and Accounts, which terrified him: and many times he lookt only on the foot of the Account'.[52] This statement is entirely substantiated by the Colonel's use of his will to 'make his complaint' of his inheritance of 'many Debts and inconveniences, not of my owne contraction', the discharge of which, he grumbled, had brought upon him 'much travell and disquiet'.[53] The Alderman's anecdote thus forms a counterpart to Aubrey's Life of Judge Popham in 'Brief lives', where the general view of Wiltshiremen that the Pophams were a wild, boorish, and spendthrift lot is retailed. Aubrey's use of such biographical details to support an argument about mathematical learning endorses what Sir William Petty had once told him: that 'the great Logicians of the Schooles are the least persuasive men in the world: the reason is plain: they want practicall, and Prudentiall Mediums'.[54] The best arguments, then, are those

49. MS Aubrey 10, f. 36a. Portman's will is annotated in Axe's hand; he had received it 'for the use' of the executors. Sir William and Axe were both of Orchard, Somerset (London Public Record Office, PROB/11/401, f. 254v). Axe's letters to Aubrey of 1684, the year of the 'Idea', refer to the business Axe was doing for Portman (MS Aubrey 12, f. 13).

50. MS Aubrey 10, f. 27.

51. *Calendar of state papers domestic 1637–8*, 169, 176. Most of the family's landed property survived the financial crisis.

52. MS Aubrey 10, f. 35a.   53. London Public Record Office, PROB 11/334, f. 445v.

54. MS Aubrey 10, f. 51.

which substantiate advice and maxims with 'practicall and Prudentiall' anecdotes drawn directly from personal experience, or second-hand through gossip.
Although Aubrey was aware that his texts were neither unquestionably true nor
exclusively based on well attested experience (but might offer a 'hint' for such
certainty), his collections offer the means of potential forms of persuasion—or
briefs—based on the amassing of experience.

Sir William Petty and the 'Faber fortunae'

The connection between life-writing and mathematics is most forcefully
advanced in the Life of Petty, a man of relatively humble mercantile origins
whose mathematical education brought him spectacular rewards and who was
Mathematical Prudence personified. Petty's Life is 'the first' in Volume I of 'Brief
lives'. This is deliberate: Aubrey draws particular attention to it, and parallels it
in another collection. In the 'Lives' manuscript, Petty's horoscope faces the first
page of the biographical text. A fuller version of this astrological information is
found in 'Collectio geniturarum', where Petty's horoscope and the 1676 judgement on it by the astrologer Charles Snell occupy the equivalent primary place in
the volume immediately after the index.[55] It would seem that, whatever it is that
we are to find out by studying lives, Aubrey believes that Petty's is the example
best worth our investigation.

This is not merely an example of 'gratitude' to a 'singular good friend'
and patron who had supported him through a financial crisis.[56] Petty's leading position signals, once again, a commitment to Baconian historiography.
Aubrey reported to Wood that Petty had 'perused' a draft of his collections
towards a Life of Hobbes, and entirely approved of what he read. But Aubrey's
collaborators, Richard Blackbourne, John Dryden, and John Vaughan, Earl of
Carbery, had reservations, and they did not use all the material that Aubrey had
painstakingly collected. He complained that in 'the compiling' of a work from
a number of sources, they 'agree to leave out all minute things'. Dryden, poet
laureate and historiographer royal, had strong views on biographical genre, and
as a result, Aubrey felt, there would 'be the trueth, but not the whole'. This
displeased Petty, who 'would have [...] all stand'; but Aubrey was obliged to
'submitt to these great Witts'. The printed Life of Hobbes proved to be 'writt
in a high style'. 'Now', Aubrey objected, 'I say the Offices of a Panegyrist, and
Historian, are much different. A Life, is a short Historie: and there minutenes
of a famous person is gratefull. I never yet knew a Witt (unles he were a piece of

55. MS Aubrey 6, f. 12v–15v; MS Aubrey 23, ff. 11v–19v.
56. BL Add MS 72850, ff. 141–142; MS Ballard 14, f. 135.

an Antiquary) write a proper Epitaph, but leave the reader ignorant, what coun-
tryman etc only tickles his eares with Elogies'. So in 'Brief lives' Petty stands
for a principle: that life-writing is, as Bacon defined it in *The advancement of
learning*, a branch of history, not that which 'hath most estimation and glory',
but certainly that which 'excelleth' in 'profit and vse'. Aubrey identified life-
writing with antiquities, the 'Remnants of History' in Bacon's phrase, a matter
of 'industrious [...] diligence and obseruation' (Bacon 2000, 66). Life-writing,
for Aubrey, involved the recording of detailed particulars, however undigni-
fied in content or origin, fragmentary, dry, or even, for the time being, imperfectly sifted and verified. Against this model, Aubrey sets up the figure of the
'great Witt' Dryden, who stands for the view that life-writing is a branch of
rhetoric, not a form to be cluttered with details and trivialities. The role of the
biographer is to conform to classical literary patterns; and the facts should be
manipulated or left out when to record them faithfully would mean sacrificing
the opportunity to create some prized rhetorical effect, such as a Plutarchan
biographical parallel. A page is set aside for Dryden in the 'Lives' manuscript,
among the collection of lives of the English poets and dramatists, honourably
positioned between Ben Jonson and Shakespeare. Yet it is empty, with only the
title and a marginal note announcing that 'he will write it for me himselfe'.[57]
Perhaps this was Dryden's way of declining to have anything more to do with
Aubreyan biography. Petty's, on the other hand, is long and full of 'minute
things'. We may fruitfully consider in what respect, and to what extent, he, too,
had resolved to write it 'himselfe'.

When the 'great Witts' left out many details from Aubrey's Life of Hobbes,
he particularly resented the fact that they would not 'mention his being Page'.[58]
One possible narrative of Hobbes's life, that of the humble boy who had made
good through the exercise of an unusually keen and independent intelligence,
would thus be suppressed. Life-writing of this kind claims to offer the wisdom of
experience, that knowledge 'wherein mans life is most conuersant' (Bacon 2000,
218). Accounts of those who have been able to improve their state are, by implication, most worthy the attention of the Baconian reader. As Bacon did not advocate the study of mathematics, the special connection which Aubrey and some
of his circle made between life-writing and mathematics seems to be in part an
attempt to reconcile mathematics with Baconian principles. In Aubrey's view,
one shared by many in his circle, Petty is the consummate example of one who
'hewed-out his Fortune himselfe' through innovative thinking, skill in modern
as well as ancient languages, and the application of new scientific and technical

57. MS Aubrey 6, f. 108v.
58. In the manuscript of the Life of Hobbes, Aubrey adds notes to those friends among whom the text
would circulate. Against his statement that Hobbes's brother Francis was a glover, Aubrey queries, 'shall I
expresse or conceale this [Glover.] The Philosopher would acknowledge it.' MS Aubrey 9, f. 29v.

knowledge to the advancement of learning.[59] In his 'Idea of education', a work profoundly indebted to Petty's own educational proposal *The advice of W. P. to Mr. Samuel Hartlib* (1647), Aubrey claims of his utopian mathematical school that 'This Institution will teach' boys 'to be Fabri suae Fortunae [makers of their own fortunes], e.g. that noble and ingeniose Knight Sir William Pety my ever honoured Friend, who from a small Stock' gained in Ireland 'by Surveying, etc.' a fortune of 'severall Thousand pounds per annum, honestly and ingeniosly'.[60] The maxim *Faber quisque fortunae suae*, 'each man is maker of his own fortune', derives from Bacon's 'newe and vnwoonted Argumente to teach men how to raise and make theire fortune' which is one of the essential purposes of the educational reform propounded in the *Advancement of learning* (see Bacon 1985, 264; Bacon 2000, 163–179). The belief that although 'Outward Accidents', such as 'Favour, Opportunitie, Death of Others' and the like, 'conduce much to *Fortune*', nevertheless 'the Mould of a Mans *Fortune*, is in his owne hands', forms the theme of Bacon's essay 'Of Fortune', as it does of his 'Faber Fortunae sive Doctrina de ambitu vitae' [The maker of one's fortune or the doctrine of the pursuit of success in life], an essay which the ambitious Samuel Pepys liked to carry in his pocket (Pepys 1971, II 102 and n1). The narrative of the 'Faber fortunae', a distinctively Baconian form, is therefore a Royal Society form also: Evelyn describes another Irish adventurer, the father of Robert Boyle, as 'Faber Fortunae, a person of wonderful Sagacity in Affaires [...] by which he compass'd a vast Estate'.[61]

The narrative of the 'Faber fortunae' is, I suggest, one of the stratagems used by the circle of the Royal Society in attempting to demonstrate the real benefits of their interests and the intellectual enterprise and creativity of their luminaries. It was autodidacticism which was emphasized; what was not appreciated was a narrative of commonplace educational charity and minor patronage. When Aubrey recorded that Seth Ward had been a sizar at Cambridge, a poor student in receipt of financial support in return for performing menial tasks such as waiting at table, he later told Wood to expunge this information from the 'Lives' manuscript 'for the sake of euphony'.[62] Petty's political brilliance made his life history particularly interesting to those who had been persuaded by Bacon's argument that

59. MS Aubrey 6, f. 15.   60. MS Aubrey 10, f. 143.

61. Hunter 1994, 87. This life-model was so identified with practical activity that 'faber fortunae' is also the name of the characteristic late-seventeenth-century list of potential money-making schemes. When Aubrey was in financial difficulties in 1671, Petty provided him with a list of desirable public projects, such as the keeping of statistical records to give 'a true State of the Nation at all times'. Aubrey transcribed these into a manuscript full of ideas of ways of repairing his fortunes, entitled 'Faber Fortunae'. Petty had a brief manuscript among his own papers, also entitled 'Faber Fortunae', with ideas such as a method of maintaining servants 'by shewing rarityes to the curious' (BL Add MS 72891, f. 8v).

62. *Expunge Servitor, euphoniae gratiâ* (MS Aubrey 8, f. 8). Sometimes undergraduates registered with their colleges as 'sizars' in order to escape the highest scale of fees, but were not in fact poor and were not required to do manual work. Even if this was so in his case, Ward very probably did not want his 'sizar' status known once he became a bishop.

'Histories of Liues' are the best 'grounde' for investigating the arts of negotiation and business.

In the 'Life of Petty', Aubrey records that Petty liked to present himself as a faber fortunae. 'He haz told me, that wheras some men have accidentally come into a way of preferment by lying at an Inne, and there contracting an Acquaintance; on the Roade: or as some others have donne; he never had any such like opportunity, but hewed-out his Fortune himselfe.'[63] This is only really true in that Petty did nothing accidentally, particularly not acquiring influential acquaintance; he certainly displayed a skill in networking which gave him an introduction to Hobbes's circle in Paris, and made the most of his college fellowship. However Petty's determination so to represent himself gained him distinction in scientific circles. Aubrey reports admiringly that Petty had said that in choosing St Andrew's Day for their foundation, the Royal Society had chosen the wrong disciple. They should instead have chosen St Thomas, who refused to believe in Jesus's resurrection until he had put his fingers in the Lord's wounded body (John 21: 24–28). This is representing natural philosophy not only as the antithesis of superstition and the unexamined acceptance of received wisdom, but as fundamentally opposed to deference. He represents himself, in Aubrey's words, as one who 'hewed-out' a fortune, as if he were a manual labourer, rather than one who employed charm, wit, and address, like a courtier. Through his quick wit, he deliberately created a reputation for himself. His contemporaries loved the ridiculous anecdote about Petty's having responded to a challenge by suggesting that, as he was very short-sighted, the duelling-weapon should be a carpenter's axe and the venue a dark cellar; it circulated not only in scientific and governmental circles but in the printed jest-books.[64] However, in his extensive plans for his sons' education, he stipulates that they were not only to be taught mathematics, but also to bow gracefully and 'to go to plays, and learn the company' (Fitzmaurice 1895, 303).

Yet the faber fortunae was not an entirely respectable category: in his essay Bacon states that 'Overt, and Apparent vertues bring forth Praise; But there be Secret and Hidden Vertues, that bring forth *Fortune*.' Among these clandestine personal qualities is a freedom from excessive loyalty to persons or to one's country: certainly, Bacon says, there are no more fortunate properties than to have 'a *Little* of the *Fool*; And not *Too Much* of the *Honest*' (Bacon 1985, 122–123). As we shall see, Aubrey's Life of Petty corresponds closely to Petty's own carefully shaped narrative of his rise from obscurity to fortune; but although Aubrey deeply admired Petty, telling Wood in a private letter that his survey of Ireland demonstrated 'the great Elevation' of his intelligence 'which

63. MS Aubrey 6, f. 15.
64. See Evelyn's life of Petty in BL Add MS 4229, f. 56v, and *The complaisant companion, or new jests, witty reparties, bulls, rhodomantado's and pleasant novels*, London, 1674, 24.

like a Meteor moves above the Sphaere of other Mortalls', nevertheless he does record in the Life something which Petty was at pains to conceal.[65] This was the existence of Frances, Petty's illegitimate daughter, who in 1676 had appeared on stage aged eleven in Settle's *Pastor Fido*. She played Dorinda, a very young nymph who ignores the counsel of an older confidante to protect herself from destruction by containing her passionate sexual feelings for Sylvio until she is more mature: she concludes Act II with the innuendo-laden utterance 'Young though I am, I'm Old enough to dye'. This sexualizing of a child actress was intended to titillate; and 'Mrs Petty' left the stage aged thirteen, either because Petty intervened or because she was taken on as someone's mistress. Aubrey thought she was twenty-one in 1680; in fact she was fifteen. She returned to the stage in 1681 and was said to be 'grown a very Woman' in a prologue of 1682. Petty paid her some money to keep her quiet, on condition that she changed her name and stayed away from his family and from London: he did not want a scandal.[66]

Petty began early to shape his own representation. When he commissioned his portrait, he had himself painted holding a skull and with his finger resting on Johann Vesling's *The anatomy of the body*.[67] He employed Isaac Fuller to paint the portrait, choosing an artist who had a special interest in anatomy. The date of the picture is either 1649 or 1650: in 1649 he took his doctorate of physic at Oxford; in 1650 he was appointed deputy to the professor of anatomy. Petty's education is central to his self-presentation and, in turn, to his career.

Petty was publicly accused of getting his money in Ireland by sharp practice: and in later life he took steps 'to shew the World that' he 'was no such horrible Knave, no such Fox or Wolfe as some would make him'.[68] Facing the Life in the 'Brief lives' manuscript is Petty's coat of arms, carefully drawn and coloured by Aubrey. The 'Lives' manuscripts have many coats of arms in them but none so large as this. Aubrey checked it carefully at Petty's funeral, when the house was full of the heraldic decorations expected at an event of the kind, and corrected an error, although he did not get the motto quite right. This was unfortunate, for Petty had given his coat of arms much thought. In his Life of Petty, Evelyn says that 'He Chose for his Coate-Armor (and which he caus'd to be depicted on his Coach) a Mariners Compass, the Needle pointing to the Polar-star, and for his Crest, a Bee-hive; the *Lemma*, (if I remember well) *Operosa et Sedula*. than which nothing could be more apposite'.[69] The motto was not in fact 'operosa et sedula', diligent and zealous, but 'Ut Apes Geometriam': as bees, so geometry. Evelyn's error suggests that he saw Petty's Latin poem on his coat of arms, in which these

65. MS Wood F 39, f. 275.

66. BL Add MS 72850, ff. 110, 159–60v; Summers (1964, 177–178); Highfill (1987, XI 276–277).

67. Now in the National Portrait Gallery, London, NPG 2924.

68. BL Add MS 72853, f. 104.   69. BL Add MS 4229, f. 56v.

words are found.[70] Petty explains that as bees find honey and seek to be useful, so the geometrician seeks knowledge, and thus we should be diligent in the practice of geometry. Petty told Sir Robert Southwell, his intended biographer, that his life would prove that he was 'no musherome nor Upstart, but that my Estate is the Oyle of Flint, and that Ut Apes feci Geometriam'.[71] His reference is to Deuteronomy 32: 9–14, in which Moses instructs the Israelites that Jacob 'is the lot of his inheritance'; and that he 'found him in a desert land, and in the waste howling wilderness; he led him about, he instructed him; he kept him as the apple of his eye [. . .] He made him ride on the high places of the earth that he might eat the increase of the fields; and he made him to suck honey out of the rock, and oil out of the flinty rock'. 'Butter of kine, and milk of sheep', and suchlike riches were provided for the Israelites by their God. A parallel with Moses was also made by Abraham Cowley, in his ode which forms the dedication 'To the Royal Society' of Sprat's *History*. Cowley identified the wandering in the wilderness as the scholastic Aristotelianism in which 'our wandring Praedecessors [. . .] many years did stray' until 'Bacon, like Moses, led us forth at last [. . .] And from the Mountains Top of his Exalted Wit' saw the promised land and 'shew'd us it'.[72] Bacon showed it; but Petty colonized it. Petty, who reminded Southwell that he had shown himself to be 'like Moses upon the Mount', intended his vast wealth to be understood as proof of exceptional virtue; as a sign of his having been born to conquer the 'waste howling wilderness' by subduing Ireland and stripping it of its wealth, not to his having been a vulgar social climber.[73]

Petty's implied distinction between faber fortunae and mere 'musherome' is a delicate one, but for this Modern Moses it hinges on 'instruction': the possession, application, and augmentation of mathematical knowledge. Petty left an alternative list of mottoes, many of which, such as 'ut apes Arithmeticem politicam facere' [to do political arithmetic like the bees], substitute the rational practice of the mathematical sciences and his invention, political economy, for the more usual, if unprofitable, qualities like fidelity or honour.[74] Petty's bland explanation of his coat of arms was designed to protect himself from 'the wildnesse of Imaginacion', but he wittily tempted his biographer to 'pick' an interpretative 'hole' in his 'coate'.[75] For the compass does not only stand, as he says, for the commonplace conceit that the mind is directed by God as the needle is by the stars. It also alludes to a life-narrative which, as Petty tells it, commences with his mathematical education which allowed him, a skilled navigator, to take a job in the navy. From this point, we may extrapolate, he guided his fortunes in a straight

---

70. BL Add MS 72853, f. 103.
71. BL Add MS 72853, f. 105v, printed in Petty (1928, 224–227).
72. Cowley, 'To the Royal Society', verses 5 and 6, in Sprat (1667, B2r–v).
73. BL Add MS 72853, f. 106.   74. BL Add MS 72853, ff. 106v–107.
75. BL Add MS 72853, f. 106v.

path upwards towards the polar star. His motto, 'ut apes geometriam', may suggest that his knowledge of geometry, which permitted him to survey Ireland, allowed him to coordinate a vast number of workers to bring home the honey of Irish revenues; certainly it implies that the bee-hive, studied closely by both natural and political philosophers, demonstrates the fundamental place of geometry in the ideal commonwealth.

Petty's exposition of his coat of arms was made in the context of a series of letters which he sent to Southwell, after he had 'rumaged and Methodized my papers which amount to 53 chests, and are so many monuments of my Labours and Misfortunes' (Petty 1928, 138). In 1686 Petty sent three autobiographical letters to Southwell, and instructed him to 'pick' him 'an Epitaph out of these 3 Letters'.[76] The first alluded to 'a paper shewing what Mony I had at Christmas 1636 which was 1s.[77] how it rise to 4s. 6d. then to 24s. then to 4 li. Then to 70 Next how it fell to 26. then rose to 480 li at my Landing in Ireland Next to 13060 li at finishing the Survey', concluding that 'Perhaps The like hath not been se[en]'.[78] Southwell described the first letter as a 'short hint of Meliorations from the yeare 1636' (Petty 1928, 212). Petty's educational and financial 'Meliorations' amount to a demonstration of political arithmetic in the form of a life-narrative. They do not include the private anecdotes he told Aubrey, about his subsisting a week in Paris on a bag of walnuts, or of his being imprisoned there for debt. Although Petty used Aubrey, Southwell, and Wood to tell his story, he took no risks and placed the 'official biography' where it would not be overlooked: in the preamble to his will of 1685. Petty begins his Life by saying:

That at the full age of fifteene yeares I had obtained the Latine Greeke, and French tongues, the whole Body of Common Arithmetick, the practicall Geometrie and Astronomie conduceing to Navigation, Dyalling etc with the Knowledge of Severall Mecanicall Trades. All which and haveing been att the Universitie of Caen preferred me to the Kings Navye, where att the age of twentie yeares I had gotten up about threescore pounds with as much Mathematicks as any of my Age was knowne to have had. With this Provision Anno 1643. when the Civill Warrs betwixt the King and Parliament grew Hott, I went into the Netherlands and France for three yeares, and haveing vigorously followed my Studyes, especially that of Medecine att Utrech Leyden, Amsterdam and Paris, I Returned to Rumsey where I was borne bringing back with me my Brother Anthonij whome I had bred with about tenn pounds more then I had carried out of England. With this Seventie pounds and my Endeavors within less then four yeares more I obtained my Degree of Dr. of Phisick in Oxford, and forthwith thereupon to be admitted into the Colledge of Phisicians London. and into Severall Clubbs of the Virtuosi. After all which

---

76. BL Add MS 72853, f. 101v.

77. In the currency of Aubrey's day (as up to 1971) there were 12 pence (*d.*) in a shilling (*s.*) and 20 shillings in a pound (*l.* or *li.* or £).

78. BL Add MS 72853, f. 92r–92v.

Expences defrayed, I had left twenty eight pounds, And in the next two yeares being made Fellow of Brason Nose and Anatomie Professor in Oxford and alsoe Reader att Gressham Colledge I advanced my said Stock to about foure Hundred pounds.

This is a remarkably individual statement. It was quite normal to say explicitly that the intention of the will was to prevent conflict over the estate; it was usual to add a brief personal note such as a statement of affection to one's wife; it was a very frequent practice to commend one's soul to God; but it was not in the least usual to give an extensive life-history with the intention of 'justifying in behalfe of my children the manner and meanes of getting and acquiring the estate which I hereby bequeath unto them exhorting them to improve the same'.[79] As we have seen, Petty anticipated a broader readership than his executors; and indeed this life-narrative, offering as it did a new model for the mathematician, proved of great appeal to those within the Royal Society who were interested in writing the Lives of mathematicians and thereby 'justifying' the value of mathematics and implicitly 'exhorting' its improvement. It was transcribed by Abraham Hill, who also wrote Lives of Isaac Barrow and of Seth Ward; and it was also used by Evelyn as the basis for his Life of Petty, 'this Faber Fortunae', which he communicated to William Wotton in a letter of 1703.[80]

Epilogue / conclusion

Because mathematics in the seventeenth century was a recondite intellectual pursuit, without a positive public image and with very few employment opportunities, some of its exponents chose to pursue their studies while living a retired life, or sought encouragement by joining mathematical societies, formal and informal. In print, many appealed to the authority of antiquity, even while refashioning it. Some found support even in private life from the authority of the ancients; some found parallels for their own lives and practices in religious models; some preferred a modest obscurity; but a prominent few, headed by Sir William Petty, took possession of new biographical territory, the 'faber fortunae' narrative. It was a part of their legacy to the eighteenth century. Henry Fielding, in Shamela (1741), his parody of Richardson's wildly successful epistolary novel Pamela, had his charlatan 'editor' claim that in arranging the papers of the gold-digging imposter Shamela, he had 'exactly followed' Euclid's Elements as the 'properest' model 'for Biography', a joke against the Moderns which has

---

79. London Public Record Office, PROB/11/390, f. 166v.

80. BL Sloane MS 2903, ff. 16–18; Add MS 4229, f. 56v. Aubrey had discussed life-writing with Abraham Hill in 1674 when Wood's *Historia et antiquitates universitatis oxonienses* was published, and Hill gave Aubrey a paper of corrigenda to pass on to Wood 'for the amending of them in the next Impression'. See MS Wood F 39, f. 340.

passed without comment from literary critics (Fielding 1980, 318). They might have asked what the portrayal of Shamela, that shameless upstart, a semi-literate servant-maid who succeeds in marrying her employer for his money, has to do with Euclid; and what Euclid has to do with biography. In this chapter we have explored some claims made for mathematics through the medium of biographical narrative which might explain these connections, and which may have provoked Fielding's amused indignation. For *Shamela*, I suggest, testifies to the enduring power of an association between the promotion of practical mathematics, the narrative of the 'faber fortunae', and inflated claims for the centrality of mathematics to the study of ordinary human affairs. 'Euclid', Aubrey warmly asserted, was 'certainly the best booke that ever was writt'. Newton's admiration was more tempered; yet when a friend of Newton, who had been advised by the great man to read Euclid, asked him doubtfully what 'benefit in life' could possibly derive from the reading of his *Elements*, it was the one time in five years that he was seen to be 'merry' (Iliffe 2006, I 283).[81] Aubrey's answer to sceptics like Newton's friend was the conventional one: that mathematics makes one 'tread sure steppes'.[82] Fielding's joke finds its most elaborated form in Laurence Sterne's novel *Tristram Shandy* (1759–69), in the attempts by Uncle Toby to narrate his own life, first by studying geometry, and then by erecting a mathematical model of the fortifications of Namur on his brother's bowling-green: a project of Quixotic perplexity, producing no 'sure steppes' but, on the contrary, as the narrator cautions, 'intricate are the steps! intricate are the mases of this labyrinth!' (Sterne 1983, 73). Neither Sterne's narrators nor Aubrey achieve a perfect structure or a finished narrative; but through, and beyond, the medium of life-writing, Aubrey, and the enormous circle of learned and unlearned persons who contributed information and publications to his project, began to construct a history of mathematics for their times.

Bibliography

Ashmole, Elias, *Elias Ashmole (1617–1692): his autobiographical and historical notes, his correspondence, and other contemporary sources relating to his life and work*, 5 vols, C H Josten (ed), Oxford University Press, 1966.

Bacon, Francis, *The essayes*, Michael Kiernan (ed), Clarendon Press, 1985.

Bacon, Francis, *The advancement of learning*, Michael Kiernan (ed), Clarendon Press, 2000.

Bennett, Kate, 'Notes towards a critical edition of John Aubrey's *Brief Lives*', DPhil dissertation, Oxford, 1993.

Bennett, Kate, 'John Aubrey's collections and the Early Modern museum', *Bodleian Library Record*, 17 (2001), 213–245.

Birch, Thomas, *The history of the Royal Society of London*, 4 vols, London, 1756–57.

81. MS Aubrey 10, f. 41a.   82. MS Aubrey 10, f. 41a.

Blagrave, John *The mathematical jewel*, London, 1585.

Britten, F J, *Old clocks and watches and their makers*, Batsford, 1911.

Britton, John, *Memoir of John Aubrey*, London, 1845.

de Cattan, Christophe, *The geomancie of Maister Christopher Cattan*, London, 1591, 1608.

Chope, R Pearse (ed), *Early tours in Devon and Cornwall*, James G Commin, 1918.

Clark, Andrew, *Brief Lives, chiefly of contemporaries, set down by John Aubrey*, 2 vols, Clarendon Press, 1898.

Fielding, Henry, *Shamela*, Oxford University Press, 1980.

Feingold, Mordechai, *The mathematicians' apprenticeship*, Cambridge University Press, 1984.

Feingold, Mordechai, 'Mathematical sciences and new philosophies', in Nicholas Tyacke (ed), *The history of the University of Oxford*, vol 4, Clarendon Press, 1997.

Feingold, Mordechai, 'The Origins of the Royal Society Revisited' in Margaret Pelling and Scott Mandelbrote (eds), *The practice of reform in health, medicine and science 1500–2000*, Ashgate, 2005.

Fitzmaurice, Edmond, *The life of Sir William Petty*, John Murray, 1895.

Highfill, Philip, *et al*, *A biographical dictionary of actors, actresses, musicians, dancers, managers & other stage personnel in London, 1660–1800*, Southern Illinois University Press, 1987.

Hunter, Michael, *John Aubrey and the realm of learning*, Duckworth, 1975.

Hunter, Michael, *Robert Boyle by himself and his friends*, Pickering, 1994.

Iliffe, Rob, *et al*, *Early biographies of Isaac Newton 1660–1885*, 2 vols, Pickering and Chatto, 2006.

Lloyd, David, *Memories of the lives, actions, sufferings and deaths of those […] that suffered […] for the Protestant Religion*, London, 1668.

Münster, Sebastian, *Rudimenta mathematica*, Basle, 1551.

Hobbes, Thomas, *The correspondence of Thomas Hobbes*, Noel Malcolm (ed), 2 vols, Clarendon Press, 1994.

Osborne, Francis, *Advice to a son; or directions for your better conduct through the various […] encounters of this life*, Oxford, 1656.

Oughtred, William, *The new artificial gauging line or rod*, London, 1633.

Oughtred, William, *To the English gentrie […] The just apologie of Wil. Oughtred, against the slaunderous insimulations of Richard Delamain*, London, 1634.

Pepys, Samuel, *The diary of Samuel Pepys*, R C Latham and W Matthews (eds), 11 vols, Bell and Hyman, 1971.

Petty, William, *The Petty papers: some unpublished writings of Sir William Petty*, Marquis of Lansdowne (ed), 2vols, Constable, 1927.

Petty, William, *The Petty-Southwell correspondence, 16761–1687*, Marquis of Lansdowne (ed), Constable, 1928.

Pitiscus, Bartholomew, *Trigonometry: or, the doctrine of triangles*, London, 1614.

Rigaud, Stephen, *Correspondence of scientific men of the seventeenth century*, 2 vols, Oxford, 1841.

Sprat, Thomas, *The history of the Royal-Society of London*, London, 1667.

Sterne, Laurence, *The life and opinions of Tristram Shandy*, Oxford University Press, 1983.

Summers, Montague, *The Restoration theatre*, Humanities Press, 1964.

Tyacke, Nicholas (ed), *The history of the University of Oxford*, 8 vols, Oxford, 1984–97.

Turnbull, H W, *The correspondence of Isaac Newton*, 7 vols, Cambridge University Press, 1959–77.

Wilkins, John, *Mathematicall magick. Or, the wonders that may be performed by mechanicall geometry*, London, 1648.

Introducing mathematics, building an empire: Russia under Peter I

Irina Gouzévitch and Dmitri Gouzévitch

European mathematics first appeared in Russia at the beginning of the eighteenth century as a result of the modernizing reforms implemented by Peter I, who reigned from 1695 to 1725. Within a quarter of a century Russia was transformed from a country with a traditional and medieval culture to a modern European-style state, with a new capital built on reclaimed lands and a navy dominating the Baltic Sea. In this new empire, mathematics was used in civil and military construction projects and taught in new educational establishments. There were changes in metrology (units of measurement), the monetary system, and the calendar, in the figurative arts (drawing, engraving, and perspective), and in publishing. By 1725 St Petersburg had its own Academy of Sciences, in which mathematics predominated over all other subjects.

For a nation to cross the cognitive distance that separates elementary arithmetic from differential calculus in such a short time is an uncommon phenomenon, and to understand the change we must analyse Peter I's reforms and the reasons for them. The fact that Russia was conducting wars on several frontiers made it essential and urgent to train military experts and engineers. The reforms were also intended to help preserve the territorial and political integrity of Russia, to rebuild its weak economy, and to sustain its international authority.

In the rapid transfer of European knowledge to Russia, the role of Peter I has been likened to a savage visiting a supermarket who, fascinated by the riches on display, shovels everything into his basket without knowing whether he needs it or not.[1] Grotesque as this image may be, it illustrates the controversial character of Peter's reforms. In particular, did he know whether or not he needed mathematics? If so, how would it be used and developed?

Peter I and mathematics: awakening and impulse

The mathematical inheritance that Peter I received from his predecessors was minimal. When he began his reign in 1695 little more mathematics was known in Russia than the four operations of arithmetic and a few elementary geometric constructions. The word 'mathematics' had appeared in Russia in 1524 in its Greek form *mafematik*, in relation to astrology. Mathematicians of the past had indeed been astrologers, almost all foreign, who served the tsars as medical consultants, preachers, or private tutors. To most Russians, mathematics embodied miraculous knowledge, fearsome and impenetrable, transmitted by people thought to be charlatans. Apart from astrology, the applications of mathematics were rudimentary, in land-surveying, primitive mapping, or elementary calculations. Arabic numerals were known but not much used; calculations were written instead with characters from the old Cyrillic alphabet.

The only printed mathematical document published in Moscow before 1703 was a multiplication table in alphabetic notation (1682). A few manuscripts on arithmetic or geometry were compiled and translated from Western works for use at court but only in small numbers (Ustûgov 1974, 81; Simonov 1996, 122–126). They all suffered from the same handicap: the absence of Russian mathematical terminology. Besides, possessing a book on mathematics could be dangerous, leading to accusations of witchcraft.

Peter I was crowned in 1682 at the age of ten but taken away from the court because of dissent in the ruling family. He thereby escaped the traditional curriculum centred mainly on Orthodox theology and Greek philosophy. Rather, his education was based on practical experience and games. Two incidents in particular influenced his later interests. In February 1687, the ambassador Âcov Dolgorukij spoke to him about a marvellous instrument with which one could measure great distances. Peter I ordered one from France, and in May 1688 he received an astrolabe, as well as a pocket compass and drawing instruments. Since no one in his entourage knew how to use the instruments, a Dutchman, Franz Timmerman, was engaged to explain them. Timmerman also taught Peter

---

1. This image was evoked by W Bérélowitch in his seminar at the EHESS in Paris in the mid-1990s.

geometry, fortification, arithmetic, astronomy, and navigation (Bogoslovskij 1940, I 62; Golikov 1837, 73; *Arhiv knâzâ Kurakina* 1890, 70, 83; *Pis'ma i bumagi* 1887, 485; Ustrâlov 1858, II 18–25, 120, 398–399). The last was inspired by the discovery of a *botik*, an old English sailboat, abandoned in the barn of a distant relative, Nikita Romanov. Under instruction from Timmerman the boat was renovated. By 1693 a small fleet had been constructed, comprising three three-masted sailing boats, two large yachts, a galley, and a frigate. The second significant event was that Peter I started frequenting the *Nemeckaâ sloboda* 'German suburb', a small township outside Moscow, inhabited by a heterodox (non-orthodox) population of artisans, traders, soldiers, and technicians from the Western Europe. On these visits he discovered European knowledge and technology, which never ceased to fascinate him. His adult life was punctuated by incessant military campaigns on land and at sea. The first were the two successive sieges of Azov (in 1695 and 1696), a Turkish fortress defending access to the Azov Sea and to the Black Sea. The failure of the first siege and the narrow success of the second one demonstrated to Peter the obsolete state of his war technology and the lack of specialists in fortification, artillery, and shipbuilding. He therefore took reforming initiatives, one of which was the creation of a navy. With such plans in mind he set out incognito on a European voyage, known as the *Velikoe posol'stvo* 'Great Embassy'. In eighteen months, in 1697–98, he crisscrossed Europe and became a hard-working pupil of Western artisans, war specialists, and scientists (Guzevič 2003). He hired foreign specialists and trained his own, purchased technical equipment, and collected books, maps, and drawings. In Europe he was acquiring more, however, than objects, methods, and ready-to-use technologies. He also began to desire mastery of fundamental theories as the key to long-lasting success. In his mind, theory was equivalent to method, and the method was mathematics, in which he saw the quintessence of all sciences. He was dissatisfied, for instance, with the empiricism of Dutch shipbuilders, who were unable to tell him the proportions of their vessels from a drawing, and left for England where, he had heard, naval architecture had been perfected (*Ustav morskoj* 1993, 9). Peter's study in England, at Deptford, from 3 February to 1 May 1698, was directed by Sir Peregrin Osborn, rear-admiral of the British Navy, who introduced him to the proportions used in the English shipbuilding. The theoretical lessons could be practically applied on boats still under construction.[2]

Two other initiatives with their roots in the Great Embassy were essential for the introduction of mathematics in Russia. The first was an accord signed in Amsterdam with the Dutch entrepreneur Jan Tessing, on the publication of books on mathematics and technology in Russian. Even for the tsar, publishing

---

2. Bogoslovskij (1941, II 307, 309); *Žurnal Gosudarâ* (1787), 67; *Ûrnal 206-go* (1867), 5; Ustrâlov (1858, III 602–605).

secular books in his own country was a dissident action. Books were seen as vehicles for 'the divine word' not as a means for individuals to express themselves on a subject of their choice (Pančenko 1984, 172). The publications entrusted to Tessing were not important scientific works, but translations and compilations on subjects of immediate interest: maps, military matters, mathematics, and architecture (*Opisanie izdanij napečatannyh kirillicej* 1958, 321). In this foreign context, between 1699 and 1701, there were printed for the first time in Russian an arithmetic manual, a work on astronomy, and a guide to navigation, integrating elements of geometry and cosmography.[3]

The second initiative arose from contacts that Peter established in England in 1698. There he had met a number of mathematicians and astronomers, for example, John Flamsteed, director of the Greenwich Observatory.[4] Three Scottish mathematicians were hired in Peter's service of the Crown: Henry Farquharson, tutor at Marischal College, Aberdeen, and his two young colleagues, Stephen Gwyn and Robert Grice, students at the Royal Mathematical School at Christ's Hospital of Oxford, from which the British Navy drew personnel. Russian mathematics gained much from this: the Scots developed and taught courses in mathematics in the School of Mathematical Sciences and Navigation, the first Russian technical institution, founded in Moscow in 1699/1701 and in the Naval Academy founded in St Petersburg in 1715. Meanwhile, a Russian of Scottish origin, Jacob Bruce, stayed in England after the Great Embassy to study mathematics under John Colson. He became one of the best educated men in Peter's inner circle and the first Russian Newtonian (Boss 1972).

Mathematical changes under Peter I

It is difficult to overestimate the impact of the Great Embassy on Peter I's reforms: the tsar went to Europe in search of a means to construct a fleet and brought back an aspiration to integrate 'European civilization' into his own country. Most of his subsequent innovations were inspired by his visit to Europe. But for these initiatives to take root it was essential to maintain a regular transfer of scientific and technical knowledge.

Barely two years after Peter's return from Europe, Russia embarked on what was to be a lengthy war with Sweden (the Northern War, 1700–21). The need to engage with an enemy that was militarily and technically much more advanced

---

3. The following books were published by Ilias Kopievskij, Tessing's assistant: *Kratkoe i poleznoe ruko-vedenïe vo aritmetyku* (1699); *Ougotovánïe i tolkovánïe âsnoe i śělw izrwdnoe krásnoóbraznagw poverstánïâ krugóv" nebesnyh"* (1699); *Kniga učašaâ morskogw plavanïâ* (1701). This last work was a Russian version of a treatise by Abraham de Graaf.

4. In Peter's journal of the Great Embassy, Flamsteed is sometimes referred to as *matematik* and sometimes as *astronomic* (*Ûrnal 206-go* 1867, 7, 11–12).

led to the development on an unprecedented scale of construction work (fortresses, arsenals, ports, vessels) and transportation. For this, highly qualified specialists were needed, and training technical experts became a priority. From the beginning of the eighteenth century, groups of technicians were formed in the main branches of engineering, followed or preceded by the establishment of corresponding schools. Besides the School of Mathematical Sciences and Navigation already mentioned, there were also the Gunnery School (1701) and the School of Military Engineering (1709). Foreign specialists were urgently required as directors of projects and as trainers, and they were to introduce mathematics into both realms of activity.

Attracted by the promise of enticing careers, foreign experts brought to Russia their methods, instruments, and units of measurement. As a result, Russia, which had its own traditional systems of measurement,[5] found itself confronted by a wide variety of units (Gouzévitch 2004–05). In the shipyards in Voronež and St Petersburg, four main systems of measurement, English, Dutch, French, and Danish, rivalled each other. Towards the end of Peter's reign the English system prevailed over rival systems in this important industry. The general but incorrect view is that Peter I replaced traditional Russian units with English units. In reality, he hardly modified Russian units at all, but simply did not apply them where Russia did not have experience (Kamenceva 1962; 1975). The shipyards were not the only example of such problems, which also beset building, artillery, and fortification.

An efficient remedy—because it was applicable to most fields involved—was the traditional practice of using a 'module' (originally the length of a log or a beam) as a basic unit of measure. Under Peter I this principle reappeared in shipbuilding, where the unit of measurement was a 'porthole'; in construction, where the unit was a 'window'; and in artillery, where the unit was the diameter (about 5 cm) of a cannonball of specified weight (115 zolotniks, or about 480 g) (Gordin 2003, 788). This loose and almost universal application of modules, and a belief in the possibility of measuring everything, inspired Peter I to use the same principle to govern the structure of Muscovite society by the meritocratic placing of people according to module, or class. This resulted in a famous regulatory administrative chart, a 'table of ranks', established in 1722 and not abolished until 1917.

The ancient Russian monetary system also underwent transformation. Here the need for reform was felt long before military needs made it imperative in the first part of the eighteenth century. The traditional system had a number of defects. On the one hand, it did not include coins of sufficiently varied values. The only coins in circulation were the kopejki 'kopecks', with very little value.

---

5. For example, the archin (aršin, 72 cm) and the sazhen (sažen', 216 cm) for length, the funt (409 gr) and the pud (16.38 kg) for weight.

Other units were used—*altyn* = 3 kopecks, *grivennik* = 10 kopecks, *poltinnik* = 50 kopecks, *rubl'* = 100 kopecks—but these had only a virtual value, for the purposes of calculation. On the other hand, money was based on the decimal principle (1 rubl' = 10 grivenniks = 100 kopecks) which made it easier to calculate with and also easier to mint. But low quality metals and inferior minting practices meant that coins deteriorated easily, with the result that the amount of currency in circulation did not satisfy the needs of the population or the Treasury even in times of peace. Reform was therefore urgently required in order to produce both the necessary quantity and quality. Peter had visited the London Mint, run by Isaac Newton, in 1698. He did not change the traditional structure but now gave each unit its own coin (Ivanov 2001).

The calendar also underwent change. In 1700 Peter decreed that the New Year should be celebrated on 1 January instead of 1 September. This marked the beginning not only of a new century but of a new chronology. Years were no longer counted from the creation of the world, but from the birth of Christ, as elsewhere in Europe. Peter I maintained the Julian calendar then in use in England and Sweden rather than adopting the Gregorian calendar used in other European countries, and for the next two centuries Russia remained ten (later on twelve, and today thirteen) days behind the Gregorian calendar.

A substantial change of attitude can also be noticed in the case of astrological calendars. During this period astrology came to be regarded as a science like any other, its image less demonized. Peter I had never been much interested in it, but in 1721 he ordered that his horoscope should be mathematically produced, and it was probably done by Bruce. This questioning of the stars produced a character very like to his own, but this may perhaps be attributed less to the science of astrology than to his long-lasting friendship with his 'mathematician'.

Drawing and printing

Nowadays, graphic techniques are essential for map-making and projections, but no such techniques were known in Russia in the first half of the eighteenth century (Gouzévitch 2006). Icon painting was a traditional art, taught through apprenticeships in painters' workshops. Subject to strict religious canons, it was based on a principle of reverse perspective, which highlighted the symbolic importance of the subject rather than representing relationships of place and size. It was therefore of little relevance in technology and science. The only deviations from this style that were allowed were in decorative art objects for the court, produced in the Armoury of the Moscow Kremlin by foreign artists. Here European-style perspective was found, but its use was limited and it did not spread to other domains.

Early in the eighteenth century two European techniques, life drawing and technical drawing, provided architects, engineers, cartographers, and artists with new tools. Both first appeared in engraving, which at the time was used for illustrating technical books, for producing maps and plans, and for recording military feats and political achievements. Drawing thus became, along with mathematics, a compulsory element in the curriculum of the technical schools.[6] As early as 1705, the Navigation School worked with a private engraving and printing workshop headed by V Kipriânov to publish teaching materials (maps, calendars, logarithmic tables, handbooks) drawn up by its Scottish teachers (Bruce, Farquharson, Gwyn, and Grice). From 1715, drawing and technical drawing were included in the curriculum of the newly opened Naval academy, and in 1716, they were introduced along with medicine and Latin into the Surgery School of the St Petersburg military hospital. In 1721, they began to be taught at Feofan Prokopovič's Karpovskaâ School, whose purpose was to educate children from all backgrounds to fulfil various functions of public use.

Printing was essential to the spread of such techniques and of mathematics more generally. A rapid expansion of secular printing was a significant feature of this period. In principle publications were aimed at the general public, but for at least the first fifteen years they were directed at professionals involved in the reforms. Technical teachers and students, in particular, needed specialized knowledge in a systematic and understandable form. To achieve this, a number of problems had to be overcome, above all the taboo on secular books in Old Slavic. This required the development of a new language better adapted to secular writing, with its own alphabet, numerical expressions, lexicon, and style. Further, books in European languages had to be selected and translated. Specialists were required for all these aspects of publication.

Reforms in publishing were launched by a royal decree on 1 January 1708, which imposed the civil alphabet. From then on all secular literature had to be published in this alphabet, which resembled Old Slavic in much the same way that Roman script resembles Gothic. Thus two objectives were achieved at once: Old Slavic was relegated to its traditional stronghold of religious literature, and secular writing was relieved of its constraint. The adoption of Arabic numerals was an easier task because it had started earlier: the numerals had made their first appearance in Russian writing on the art of troop formation

---

6. The Russian word *risunok* 'drawing' itself indicates foreign influence: it was borrowed from Polish around 1705 (Fasmer 1987, 485–486) and replaced the former *znamenit'*, which meant the art of painting flags (Gouzévitch, 1990, I 16; II 14–15; Evsina 1975, 15; Vladimirskij-Budanov 1874, 36–37, 141–142).

(Wallhausen 1615; Val'hauzen 1647–49),[7] and then in Russian coinage (1655).[8] The new civil alphabet gave full legitimacy to Arabic numerals in secular literature and marginalized traditional literal numeration along with Old Slavic. At the same time, important changes were also made to printing techniques. The former polygraphic system was completely reformed by Dutch printers invited to Russia specifically for this purpose. A heterogeneous group of translators and writers, working in all crafts and coming from all social classes, also assisted in the urgent publication of texts in all the main branches of knowledge (Gouzévitch 2003, 2006).

To assess the effect of these measures, some figures on mathematical publications between 1708 and 1724 are given in Table 1.[9] The flow of publications dealing with mathematics, science, and technology illustrates Peter I's concerns and interests at different stages of his reign. Table 1 shows that out of a total of 1303 works published, there were 179 (in rows 1–9) related to mathematics and technology. They ranked only behind 682 administrative publications (14–15) and 363 issues of periodicals (16). They outnumber the 110 works of an educational, social, political nature (10–13). Works on mathematics, engineering, and astronomy (including calendars) were issued at an almost steady annual rate for sixteen years. For the first four years the focus was on artillery and fortification, but later on attention turned to the navy. From 1713 to 1723 publications relating to the navy (navigation, astronomy, geodesy, geography) were issued frequently, with peaks of six or seven publications in 1714, 1716, and 1719. The change of emphasis coincided with the transfer from land to sea of military operations against Sweden.

To add a qualitative dimension to this analysis, let us focus on some particular domains. It is usually held that Petrine publications were issued on pragmatic grounds and were intended for immediate application. This is based on the fact that technical and military publications, in particular on fortification and artillery, were issued as soon as the civil alphabet was implemented. However, this statement loses weight when closer attention is paid to some of the technical works amongst the early publications. Fortification, for example, is represented by five European authors. Charles Sturm's *Arhìtektura voìnskaâ*, 'Military architecture' (1709) explained existing systems of fortification. Two works by the Austrian author Ernst Borgsdorf, *Poběždaûšaâ krěpost'*, 'The triumphant fortress' (1708), and *Pověrennye voìnskìe pravila kako neprìâtelskìe krěpostì*

7. Wallhausen's [Val'hauzen's] treatise was published by two printing shops: the text was printed in Moscow whereas the title page and plates with Arabic numerals were engraved in Holland. Arabic numerals appeared in the text only when they referred to a plate. In other words, the use of illustrations engraved abroad encouraged Russian editors to use Arabic numerals. The numerals also infiltrated due to the incorporation into Russia of Ukraine, where they were already in use. Finally, most works on mathematics were of Western origin, and translators generally kept the original numeration.

8. Actually on a coin minted for use in Ukraine (Kamenceva 1975, 185).

9. The classification corresponds to our modern perception but would have been less distinct at the time, when drawing was classed with geometry and engineering with mathematics, which is why these subjects are grouped together in the table.

Table 4.3.1 Mathematical publications from 1708 to 1724

	Subject	Number of books
1	Mathematics, engineering	11
2	Geography, astronomy, geodesy	12
3	Calendars	38
4	Navigation, ship construction	35
5	Hydrotechnology, medical sciences*	7
6	Fortification	14
7	Artillery	5
8	Military	15
9	Architecture, gardening	5
10	Education, lexicons	15
11	History, philosophy, literature, bibliography	34
12	Heraldry, ceremonial, uniforms, pyrotechnics, festivities, theatre, etc.	48
13	Diplomacy, international relations	19
14	Civil, military and administrative law	62
15	Decrees	620
16	Periodicals	363
		1303

* The few publications in medical sciences are on thermal and mineral water springs and so they are grouped here with hydrotechnology.

silou̇ bratì, 'Military rules for taking enemy fortresses by force' (1709), taught methods of attack and defence.[10] The treatises by Georg Rimpler, *Rìmplerova manìra o stroenìì krěpostei*, 'Rimpler's method of building fortresses' (1708) and by François Blondel, *Novaâ manera, ukrěpleniû gorodov"*, 'A new method of fortifying towns' (1711) both presented particular systems of fortification. Lastly, Minno Coehȯrn's *Novoe krěpostnoe stroenìe na mokrom" ìlì nìśkom" gorìśontě* 'A new fortification for wet and low-lying lands' (1709) summed up, on the one hand, the Dutch experience of constructing fortified towns, and, on the other, a new way of fortification based on a close analysis of European methods of attack and defence. Laskovskij, a historian of Russian fortification, has argued that Coehorn's treatise could not be used as a guide to engineering unless it was

10. The interest of these rather unoriginal works was that they had been written in Russia by one of the foreign experts who contributed to the conquest of Azov (Sturm 1702; Borgsdorf 1708).

understood in the context of the author's arguments against the French military engineering school. Such knowledge could hardly be expected from Petrine engineers who had only just acquired the basics of science (Pekarskij 1862, II 219). Coehorn's manual was, in short, too theoretical for any practical application. There is a similar example in a different domain, that of artillery. The four works published between 1708 and 1711 provide an overview on the artillery of the time. Three treatises, by Ernest Braun (1709), Timofej Brinck (1710), and Ioann Buhner (1711) dealt with practical artillery. An earlier publication, of a slightly different kind, was the anonymous *Razsuždenïe ω metanïi bombov"* 'Reasoning on the launching of bombs and the firing of cannons' (1708), translated by Bruce and published in the form of twenty-one engravings, with comments but without text (*Opisanie izdanij graždanskoj pečati* 1955, 80–82; Mandryka 1960). Uniquely, it presented the theory of parabolas that had been known in Europe since the early seventeenth century. In Russia however, the theory would not be used until a century later (Mandryka, 1960). In other words, Bruce's translation published in Moscow in 1708 was probably as obscure to Russian artillerymen as Coehorn's treatise was to builders of fortifications.

The impracticality of some of the early publications resulted neither from ignorance nor from blind actions. Peter I had already led fourteen sieges and taken part in the (re-)construction of forty-seven fortified towns (Gouzévitch 1990, I 25, II 33). From this point of view it is more sensible to question the nature of Peter I's utilitarianism. According to Robert (1973) something 'useful' is that 'of which the use is or *can be* profitable ... to society, which satisfies a need'. Peter I's gift was perhaps one of apprehending what might be 'profitable to society', not only then but in the future.

Unlike publications on fortification and artillery, which were concentrated into the first four years, works on mathematics, mechanics, and astronomy (61 items in all) multiplied noticeably after 1714. Over a period of ten years between the Great Embassy and the reform of the alphabet, there had been only five Russian publications of this kind, among which were two arithmetics: one by Ilias Kopievskij (1699), mentioned earlier, and one by Leontij Magnickij (1703). Among the forty-eight pages of the former, only thirteen are devoted to the operations of arithmetic, while the remaining thirty-five pages contain a collection of classical maxims in Latin and in Russian and a selection of Aesop's fables. Clearly Kopievskij considered mathematics as being essential knowledge in the same way as logic and ethics (as taught through maxims and fables) (Robson, Chapter 3.1 in this volume). All of this was based on Latin writings; hence the bilingual texts (Okenfuss 1998). Magnickij's *Ariθmétïka* (1703), a compilation in Old Slavic of Western works, was used in the training of three generations of Russian engineers, artillerymen, and navigators. Besides these two texts, *Tablicy logariθmωv"* 'Tables of logarithms' (1703), also primarily in

Russian, were produced by Magnickij, Farquharson, and Gwyn on the basis of the similar tables of the Dutchman Adrian Vlacq (1681).

The eleven works on mathematics and mechanics issued after 1708 in the new alphabet were of a totally different type. First was the famous *Geometrìa slavenskì śemlemĕrìe* (1708), a Russian version of Anthon Ernst Burckhardt von Purkenstein's *Ertzherzogliche Handgriffe desz Zirckels und Lineals* (1686). Translated by Bruce and Pause and re-edited four times, this was the first book in Russian devoted to geometry and its practical applications (Pekarskij 1862, II 198, 210; Danilevskij 1954, 72–88; *Opisanie izdanij graždanskoj pečati* 1955, 67–69, 74–75, 85–86). The book was also a means of testing out new presentations, which turned it into a working tool for technicians instead of a luxury object. Thus the second edition of the *Geometrìa*, issued in November 1708 just eight months after the first edition, had lost the beautiful binding and golden edges, the thick characters and wide margins, and the views of Hungarian fortresses. The edition of February 1709 issued under the title *Prìemy cìrkulâ ì lineikì* 'Methods with compass and ruler' inherited its drawings and small print from the previous version. Its size was that of a 'pocket' guide (15 x 9 cm instead of 19.5 x 15.5 cm in March 1708). Meanwhile a hundred pages were added with further drawings and two new chapters. The first of these was on the transformation of plane figures; the second on construction of sundials. The first was attributed to Jacob Bruce and the second one to Peter I himself (Fel' 1952, 151–152).

The study of mechanics in Russia was inaugurated by Magnickij in his *Arĩθmétïka* (1703), aimed at students of the Navigation School. Contrary to what its title suggests, it taught not only arithmetic but also geometry, navigation, and some elementary algebra. Some years later, in 1722, Grigorij Skornâkov-Pisarev published *Nauka statičeskaâ ili mehanïka*, 'The science of statics, or mechanics'. At the time the author was manager of the building works for the Ladoga canal northeast of St Petersburg (1718–23) and president of the Naval Academy (1719–22). From his knowledge of teaching at the Academy he knew the need for such a treatise, and his experience as a civil engineer ensured its practical character.

The gap of nineteen years between one text and the other seems strange, especially compared to the dynamism of publication in other areas of mathematics. In fact a translation of Sturm's *Mathesis juvenalis* (1702–05) had been undertaken by Vinius in 1709 but had failed: Peter I had discarded it as being unreadable. Bruce took up the same task but could do no better, and consequently no translation was ever published. In mechanics the linguistic problems seemed insoluble. The first full text on mechanics was therefore not a translation but a work written in Russian. An important aspect of this story is the personal interest of the tsar and his companions in mechanics, at least in the didactics of the subject.

The importance of these works is better understood in the context of contemporary printing in Russia. The number of printers could be counted on both hands, and no more than two of them were able to publish technical and scientific works. Further, Peter I faced resistance from pious printers who feared God more than their earthly ruler; he was openly called the Antichrist by the Orthodox clergy. The editorial vicissitudes of Christiaan Huygens' *Kniga mìrozrĕnìâ* 'Kosmotheôros' (1717) offers a good example of such a resistance. Mikhail Avramov, director of the St Petersburg printing office, was in charge of printing it. The process took seven years and two drafts: thirty copies were produced in 1717 and 1200 in 1724. In 1741, Avramov wrote about the first draft to the Empress Elizabeth:

Once his majesty had left [for Europe, in 1716], I examined the book in question, entirely despicable and impious, and it made my heart pound and horrified my spirits, and I bowed down before the image of the Virgin Mary, shaken by sobs [...], being scared of printing and of not printing, but thanks to Christ's grace, I relied on what my heart told me to do: in order to expose [...]these wicked atheists, [...] I printed 30 books instead of 1,200 and sealed them, and I hid them until the monarch came back. (Kirsanov 1996, 30)

Avramov thought that the author and translator (who was Bruce), who were clearly insane and atheist, should be burnt alive without delay. Fortunately Huygens, Bruce, and Peter I himself were now already dead, but clearly Avramov had not come to terms with the fact that the book had eventually been printed, by transferring it to Moscow printers.

There were also, of course, linguistic problems. Early translations of Western treatises were often unreadable because of complex and obscure language filled with newly invented Germanisms and Russianisms. Even the simplest words could be hard to translate. Several manuscripts bear corrections made by Peter I himself, who tried to clarify the language, simplify the grammar, and get rid of archaic Slavonisms. Essential words were taken from other languages, with the result that most European languages, whether dead or alive, provided vocabulary for the Russians. Groups of borrowings (around 11,000 words between 1690 and 1725) corresponded to the countries used for the main reference works (Biržakova 1972, 83–84). Thus German provided administrative and military terms; French supplied vocabulary for fortification workers, artillerymen, hydraulic engineers, and diplomats; Italian words entered architecture and navigation; Dutch and English provided maritime vocabulary; Swedish and Danish inspired the lexicon of regulations; for mathematics, most of the words were borrowed from Latin. Modern literary Russian was less the product of natural evolution than of a deliberate effort to appropriate technical and scientific knowledge, a process directed to some extent by the tsar himself.

The Petrine school of mechanics and mathematics

A number of intellectuals, without being scholars in the Western sense of the word, were members of Peter I's entourage and shared his wish to develop the sciences. Their joint efforts, aimed at systemizing empirical and theoretical knowledge, gave birth to a phenomenon we can call the Petrine school of mechanics and mathematics. Since this group's activity was as a catalyst for the later founding of the Academy, let us try to get a better understanding of its nature. First of all, was it a 'school'? For a group of individuals to be called a 'school', three elements generally need to appear together: an original and distinctive output production bearing some distinctive features; a transmission of collective knowledge to following generations; and a programme of activity shared and followed by all the members. The Petrine group seems to fulfil these three conditions. It was an intellectual group that produced formalized knowledge in its texts, passed it on to new generations, and carried out common activities according to an agreed programme. Its protagonists can be classified into three categories depending on their training and the activities in which they were involved.

The first category consists of the pioneers. Besides Peter I himself, it included his companions from his tour to Europe (Bruce, Skornâkov-Pisarev, Šafirov, Korčmin) and others, like Prokopovič, who had also had European training. As the ideologist of the Petrine group, Prokopovič offered a philosophical argument advocating autonomy for science with regard to theology, though for a clergyman this was considered heretical (Ničik 1977, 124). The second category included those who had been trained in Russia, like Magnickij or Vinius. The third category consisted of the Scotsmen Farquharson, Gwyn, and Grice. They too had been recruited during the Great Embassy and their contribution was as important as that of Vinius or Bruce in setting up a technical teaching programme.

The next generation of the group, disciples of the pioneers, were hired by the technical schools. In larger number than their predecessors, they followed a greater variety of professional career paths, commonly punctuated by training courses abroad. Some of them, like Andrej Nartov, later became driving forces of the Russian school of mechanics and mathematics. Others joined for a brief period, for instance Konon Zotov, Boris Volkov, and Vasilij Suvorov, who were technical translators and who, after training abroad, taught in engineering schools. Some others, without directly joining the school, organized their professional activity according to the methodological principles elaborated within it: Abram Hannibal, godson of Peter I; Burkhard Christoph von Münich, a military engineer from Oldenburg; or Ivan Kirillov, a Russian cartographer. This new generation also included former students of the Naval Academy, a specific group of 'Petrine geodesians' (Vasil'ev 1959, 42; Zagorskij, 1969, 44).

At this stage, the Petrine school began to grow offshoots. One was the *učenaâ družina* 'group of learned friends', founded in the late 1720s around Prokopovič.[11] Among its members whose work affected Russian culture we can mention Gavriil Bužinskij and Feofan Krolik, learned priests and experienced translators; Vasilij Tatišev, a future historian; Antioch Kantemir, future poet and translator; A Čerkasskij and N Trubeckoj. Teachers, writers or translators of specialized handbooks, practising engineers, and scientific instrument manufacturers, were some of the many people produced by the activity of the original group. The transmission of knowledge from one generation to the other was thus guaranteed, and its mechanism, the training system, became firmly rooted in Russia.

Paradoxically, Peter I cannot be called the scientific leader of the school bearing his name. A few articles dealing with questions in mechanics (especially clock making) comprise the short list of his scientific output. Two mathematicians, Farquharson and Bruce, could have claimed the title if each had not found a niche of his own. It would be more appropriate to describe the whole group as a polycentric training scheme animated, encouraged, and spurred on by Peter I, who acted as its true driving force. This is the reason we can talk about a 'Petrine' school. Its works bear common characteristics which make them easy to recognize. Everything in these imperfect works testifies to the tensions of crossing uncharted territory: from the still unstable handwriting to the occasionally obscure language, from the essentially pragmatic orientation to the unusual association of new ideas and traditional representations. These were people struggling to overcome the powerlessness of their existing means of expression.

No document seems to have formalized beforehand the series of energetic measures which, within two decades, flung the quasi-medieval civilization of Russia into modern times. The civil alphabet and Arabic numerals, the network of printing shops and the new printing techniques, the flow of secular books, the accelerated training of technicians and an intellectual elite, and the setting up of the engineering schools all arose from the requirements imposed by a constant state of military emergency. Nevertheless, Peter I's initiatives, despite their unifying and structured nature, can hardly be called an organized scientific plan. These actions of the tsar were situated at another level, which we would rather call 'methodological'. How could it be otherwise? Russia had first to absorb basic knowledge in technology and science, an aim that was achieved within two decades. This in itself was a pioneering achievement but one which never yielded to original scientific research.

At the same time, it became clear that a policy of simply transferring knowledge was not always a panacea. Within fifteen years, translators accomplished

---

11. Traditionally *družina* were troops at the prince's disposal; here they are people sharing common interests and grouping voluntarily (Čistovič 1868, 607–617; Pypin 1902, 343–388; Epifanov 1963).

the task of creating fundamental texts in the principal disciplines and helped to make mathematics the foundation on which professional training was based. However, as the example of mechanics showed, translation had its limits. Reaching the level of European science, let alone competing with Western countries in the production of new knowledge, was still a long-distance target. An alternative idea therefore emerged in Peter I's mind: creating within his own empire a centre capable of producing scientific knowledge and thus accelerating the process of development.

The St Petersburg Academy of Sciences

The project of founding an Academy was developed over some time and was the subject of debates inside Russia as well as discussion abroad (Suhomlinov 1885; Pekarskij 1870–73; Vucinich, 1963; Komkov 1977; Kopelevič 1977; Nevskaâ 1984; Graham 1993, 16–20). Western scholars did everything they could to persuade Peter I to give his initiative an institutional form. Leibniz wrote several letters and notes to Peter I between 1697 and 1716, encouraging him to create scholarly institutions, notably a chemistry laboratory, an astronomic observatory, universities, and a college which would include painters and artisans (Kopelevič 1977, 34). On 22 December 1717, Peter I was elected a member of the Paris Academy (Galitzin 1859; Poludenskij 1865, 675–702; Riabouchinsky 1934, 46–50; Knâžeckaâ 1960, 1964, 1972; Kopelevič 1977, 44). From 1721, the mathematician Christian Wolff actively began recruiting scholars for the future St Petersburg Academy.

Inside Russia, the debates were intense. Was it reasonable to create a centre for the development of advanced sciences in a country where the number of schools could be counted on one's fingers? Would it not be more logical first to establish schools capable of producing suitable candidates for the Academy? These arguments never dissuaded Peter I. He responded with a fable about a peasant who, knowing he was dying, insisted on erecting a mill in the hope that it would encourage his children to provide it with water (Vucinich 1963, 78; Kopelevič 1999). The Academy took shape at the end of 1724, but Peter I died without taking part in the inauguration of his 'mill'.

In order to assess the achievement of the first generation of mathematical academicians, we must briefly recall two aspects of the institution which, more than anything else, assured the accelerated development of the sciences in Russia: payment and status. In Russia the pensions and bonuses paid to academicians in other countries were immediately subsumed into a fixed salary system, including rises, which assured academicians of a comfortable life. In return, they could be required to act as experts and advisors to solve problems in monetary production, analysis of mineral ores, technical work, ballistic and mechanical experiments, assessment of works of art, or the teaching of sciences. These diverse functions corresponded to two motivations of the organizers of the Academy: the benefit Russia could expect from the training of its young people and the glory the State could gain from it. Scholars were thus employees of the State working under its orders, civil servants of the Crown.

In the absence of native candidates, invited scholars were proposed to bring their own students to Russia. The scholars who were given priority were not necessarily the most notable ones, but the most promising. Certain countries were held to have reached a degree of generally acknowledged perfection in certain sciences: thus astronomers were sought from France; chemists and builders from Holland and England; specialists in human sciences from Germany; mathematicians from Switzerland.

Among the first seventeen scholars who arrived in St Petersburg between June and December 1725, the majority represented the sciences, equally divided between the natural sciences and the human sciences. The pioneers were Jacob Hermann, Friedrich Christoph Mayer, Nicolas and Daniel Bernoulli, Joseph-Nicolas Delisle, Christian Goldbach and Georg Bernhard Bülfinger. The young Leonhard Euler, a student of Johann I Bernoulli, joined them the following year. Their average age was thirty-three, ranging from nineteen (Euler) to fifty-nine (Leutmann). Some stayed only a year or two, others spent the rest of their lives in Russia, and others, like Euler, would come and go.

What was their collective contribution to the development of mathematics in Russia? When we talk about mathematics in this context, we mean both pure mathematics and sciences influenced by mathematics: astronomy, geodesy, and theoretical mechanics, as well as engineering sciences like ballistics, hydraulics, or map-making. In fact most of the scholars excelled in each of the subjects mentioned.

From the beginning, the Academy held lectures twice a week. In 1726, more than forty presentations dealt with various aspects of mathematics. In 1735 an attempt was made to group these presentations into a special mathematics meeting, but this practice did not last. Of one hundred and thirteen records of scholarly sessions that have been kept until today, eighty deal with mathematics and mechanics. In the early decades, research focused on differential and integral calculus, the theory of numbers, the foundations of mechanics, and the analysis of the infinitely small. To begin with, importance was given to fields of research inspired by Leibniz. Later research was stimulated by Euler.

The scale of the output is demonstrated by some statistics on the work of Daniel Bernoulli and Euler. The former, who spent seven years in Russia, published around fifty works on mathematics and mechanics. Bernoulli's work in hydrodynamics, mechanics, and acoustics extended mathematical methods to the natural sciences. This later became a dominant practice at the Academy. Euler's links with the Academy lasted fifty-six years, thirty-one of which were spent in Russia. In his first Russian decade alone (1726–35) he published one thousand eight hundred pages, a quarter of which dealt with higher mathematics, and more than a third with mechanics. Over the same period, he laid the foundation of the theory of functions of a complex variable and began his work on the calculation of variations, using imaginary numbers for the calculation of integrals. Other subjects of research included ballistics, vessel stability, geodesy, and practical astronomy. Half of the mathematical works he ever produced were issued in the journals of the Academy and their publication did not slow down for one hundred years (1729–1830). Euler left about ten students, who themselves became notable mathematicians, including Fuss, Rumovskij, Kraft, Kotel'nikov, and Golovin. This group is sometimes regarded as St Petersburg's first school of mathematics.

We may also add a few words on Joseph-Nicolas Delisle. On his arrival, the French astronomer joined a project to create a general map of the Empire based on astronomical data. To him we owe the 'Delisle projection' used in map making until the middle of the nineteenth century. He was also behind the training of geodesians, recruited from former students of the Naval Academy. These scientific schools, one founded by Euler, who was Swiss, and the other by Delisle, who was French, have been regarded as the glory of Russian mathematics.

The legacy of Petrine mathematics

In conclusion, we return to the essential character of Petrine reforms. Their controversial character has been underlined in many of the examples mentioned above. We may therefore ask how useful these initiatives were, which could have either anticipated or interfered with the country's natural evolution. They had in common the fact that they were all able to overturn the paradigms of the time. Calendrical change modified man's relationship to historical time; currency reform influenced daily calculations; the civil alphabet and Arabic numerals determined the way measurements were written. It is hard to deny the generally disturbing effect these changes must have had on ordinary people. Further, at least three of the reforms—in chronology, alphabet, and numeration—challenged spheres traditionally controlled by religion. They therefore overturned, both symbolically and practically, traditional ways of living and governing.

At the same time, there is another question, which at first might sound paradoxical. In the story just presented, is the term 'reform' always appropriate? In its broadest sense, 'reform' usually means some kind of 'transformation'. But in most cases the changes did not abolish or destroy an existing system. Instead, they took place in domains void of any native tradition (the military fleet, artillery, the new capital city, technical teaching, graphic and figurative arts, secular books, and so on). The old systems (including Russian measurements; Orthodox chronology; Old Slavic with its alphabet and its literal numeration; icon painting with its reverse perspective; holy texts and religious printings—and the list does not stop here) were simply put aside, or confined to a limited domain, relegated to the margin of the new world.

What can be said about mathematics? Did the forced and accelerated introduction of this complex body of foreign knowledge also constitute a disturbing element? It did so in old Muscovite culture, since it clearly challenged the established order. Peter I's aim in bringing a European-inspired modernization process to his country, however, was to challenge this order by opposing it. He did so by devoting his life to building a new entity with the image and the dimensions of Europe: the Russian Empire. Mathematics was at the heart of this process right from the start. Mathematics took root in unexploited domains, free from tradition and religious censorship. According to the new imperial ideology, mathematics fulfilled both the pragmatic aspirations of the empire (profits) and its intellectual ambitions (glory). Peter I clearly formulated this ambition in a speech he gave at a boat christening ceremony, ten years before the St Petersburg Academy of Sciences was created, when he claimed that one day, the sciences would flee England, France, and Germany and would come and settle in Russia for the coming centuries (Nikolaev 1986, 109). Peter I did everything he could to make this hope come true, above all by bringing to life a new mathematical and scientific culture.

Bibliography

Anonymous, *Razsuždenïe wmetanïi bombov" istrělânïi ispušek"*, Moskva, 1708.

Arhiv knâzâ Kurakina, I, SPb, 1890.

Biržakova, Elena, Vojnova, L A, Kutina, L L, *Očerki po istoričeskoj leksikologii russkogo âzyka XVIII veka: azykovye kontakty i zaimstvovaniâ*, Nauka, 1972.

Blondel, François, *Nouvelle Manière de fortifier les places*, Paris, 1683.

Blondel, François, *Novaâ manera, ukrěplenïû gorodov"*, translated by Ivan Zotov, Moskva, 1711.

Bogoslovskij, Mihail, *Petr I*, 2 vols, Moskva: Gossozèkgiz, 1940–41.

Borgsdorf, Èrnst Friderih, *Poběždaûŝaâ krěpost'*, Moskva, 1708; 2nd ed, 1709.

Borgsdorf, Ernst Friderih, *pověrennye voinskie pravila kako nepriâtelskie krěposti siloû brati*, Moskva, 1709; 2nd ed 1709; 3rd ed 1710.

Boss, Valentin, *Newton and Russia: the early influence, 1698–1796*, Harvard University Press, 1972.

Braun, Ernest, *Novissimum fundamentum et praxis artilleriae*, Danzig, 1682.

Braun, Ernest, *Novejšee osnovanie i praktika artìlerii*, Moskva, 1709; 2nd ed, 1710.

Brink, Timofej, *Opisanïe artìlerii*, Moskva, 1710.

Brink, Trolis Nilsen, *Beschrywinge van de Artillerye*, 'sGraven-Hage, 1681.

Buchner, Johann Siegmund, *Theoria et praxis artilleriæ*, 3 vols, Nürnberg, 1685–1706.

Buhner [Buchner], Ioann Zigmunt, *Učenìe i praktìka artìlerìi*, Moskva, 1711.

Čistovič, Illarion, *Feofan Prokopovič i ego vremâ*, SPb, 1868.

Coehorn, Minno, *Neuve vestingbouw op een natte of lage horizont,* Leeuwarden, 1685.

Coehorn, Minno, *Novoe krěpostnoe stroenìe na mokrom" ìlì nìskom" gorìšontě,* Moskva, 1709.

Danilevskij, Viktor, *Russkaâ tehničeskaâ literatura pervoj četverti XVIII veka*, Moskva-Leningrad: Izdatel'stvo Akademii nauk SSSR, 1954.

Epifanov, Petr, '"Učenaâ družina" i prosvetitel'stvo XVIII veka: nekotorye voprosy istorii russkoj obŝestvennoj mysli"', *Voprosy istorii*, 3 (1963), 37–53.

Evsina, Natal'â, *Arhitekturnaâ teoriâ v Rossii XVIII v*, Nauka, 1975.

Fasmer, Max, *Etimologičeskij slovar' russkogo âzyka*, III, Progress, 1987.

Fel', Sergej, 'Petrovskaâ geometriâ', *Trudy Instituta istorii estestvoznaniâ*, 4 (1952), 141–155.

Galitzin, Augustin, 'Pierre Ier, membre de l'Académie des Sciences', *Bulletin du Bibliophile et du bibliothécaire*, 9 (1859), 611–617.

Geometrìa slavenskì śemleměrìe Ìśdadesâ novotìpografskim" tìsnenìem", Moskva, 1708; republished as *Prìemy cìrkulâ ì lìneìkì*, Moskva, 1709.

Golikov, Ivan, *Deânìâ Petra Velikago, mudrago preobrazovatelâ Rossii, sobrannye iz dostovernyh istočnikov i raspoložennye po godam*, vol I, Moskva, 2nd ed, 1837.

Gordin, Michael, 'Measures of all the Russia: metrology and governance in the Russian Empire', *Kritika*, 4 (2003), 783–815.

Gouzévitch, Dmitri, *Razvitie mostostroeniâ v Rossii v XVIII – pervoj polovine XIX veka i problemy sohraneniâ i ispol'zovaniâ tehničeskogo nasledìâ otečestvennyh mostostroitelej*, 3 vols, Leningrad, 1990.

Gouzévitch, Irina, *De la Moscovie à l'Empire russe: le transfert du savoir technique et scientifique et la construction de l'Etat russe*, Palaiseau, 2003.

Gouzévitch, Irina, 'The editorial policy as a mirror of Petrovian reforms: textbooks and their translators in the early 18th century Russia', *Science and Education*, 15 (2006), 841–862.

Gouzévitch, Irina, and Gouzévitch, Dmitri, 'Mesurer au temps de Pierre Ier', *Cahiers de métrologie*, 22–23 (2004–05), 19–41.

Gouzévitch, Irina, and Gouzévitch, Dmitri, 'La gravure de l'époque pétrovienne et l'introduction de la perspective en Russie', in Marianne Cojannot-Le Blanc (ed), *L'artiste et l'œuvre à l'épreuve de la perspective*, École Française de Rome, 2006, 87–111.

Graham, Loren, *Science in Russia and the Soviet Union: a short history*, Cambridge University Press, 1993.

Guzevič [Gouzévitch], Dimitri, and Guzevič, Irina, *Velikoe posol'stvo*, SPb, Feniks, 2003.

Huygens, Christiaan, *Kosmotheôros, sive de terris cœlestibus, earumique ornatu, conjecturæ*, Hagæ–comitum, 1698; 2nd ed, 1699.

Huygens, Christiaan, *Kniga mìrozrěnìâ, ili mnenìe o nebesnozemnyh" globusah", i ih" ukrašenìâh"*, SPb, 1717.

Ivanov, Andrej, *Čislom i meroû*, Moskva: Encyklopediâ sel i dereven', 2001.

Kamenceva, Elena, 'Mery dliny v pervoj polovine XVIII v', *Istoriâ SSSR*, 4 (1962), 127–132.

Kamenceva, Elena, and Ustûgov, Nikolaj, *Russkaâ metrologiâ*, Moskva, Vysšaâ škola, 2nd ed, 1975.

Knâžeckaâ, Ekaterina, 'O prihčinah izbraniâ Petra I členom Parižskoj Akademii nauk', *Izvestiâ Vsesoûznogo geografičeskogo obŝestva*, 92 (1960), 154–158.

Knâžeckaâ, Ekaterina, *Sud'ba odnoj karty*, Moskva: Mysl', 1964.

Knâžeckaâ, Ekaterina, 'Petr I – člen francuzskoj Akademii nauk', *Voprosy istorii*, 12 (1972), 199–203.

Kirsanov, Vladimir, 'Pervyj russkij perevod "Kosmoteorosa" Gûjgensa', *Voprosy istorii estestvoznaniâ i tehniki,* 2 (1996), 27–37.

Kniga učašaâ morskogw plavaniâ, Amsterdam, 1701.

Komkov, Gennadij, *et al*, *Akademiâ nauk SSSR: Kratkij istoričeskij očerk*, I, Nauka, 2nd ed, 1977.

Kopelevič, Ûdif', 'Udalos' li Petru I "postroit' vodânuû mel'nicu, ne podvodâ k nej kanala"?', *Naukovedenie*, 4 (1999), 144–155.

Kopelevič, Ûdif', *Osnovanie Peterburgskoj Akademii nauk*, Leningrad: Nauka, 1977.

Kopievskij, Ilias, *Kratkoe i poleznoe rukovednie vo aritmetyku*, Moskva, 1699.

Kratkoe i poleznoe rukovedenie. vo aritmetyku, ili v" wbučenie i poznanie vsâkogw sčotu, v" sočtenii vsâkih" vešej, Amsterdam, 1699.

Krajnûkov, Viktor, 'Pervyj otečestvennyj korablestroitel'nyj reglament', *Gangut*, 1 (1991), 4–13.

Kugorn [Coehorn], Minno, *Novoe krěpostnoe stroenie na mokrom" ili niskom" gorisontě*, Moskva, 1709.

Larionov, Aleksej, *Istoriâ Instituta inženerov putej soobsěniâ Imperatora Aleksandra I za pervoe stoletie ego sušestvovaniâ: 1810–1910*, SPb, 1910.

Mandryka, Aleksej, 'Knigi iz biblioteki Â VBrûsa', *Voprosy istorii estestvoznaniâ i tehniki*, 10 (1960), 136–138.

Magnickij, Leontij, *Ariθmétïka, sirěč' nauka čislitelnaâ*, Moskva, 1703.

Nevskaâ, Nina, *Peterburgskaâ astronomičeskaâ škola XVIII v.*, Nauka, 1984

Ničik, Valeriâ , *Feofan Prokopovič*, Mysl', 1977.

Nikolaev, Sergej, 'O stilističeskoj pozicii russkih perevodčikov petrovskoj èpohi: K postanovke voprosa', *XVIII vek*, 15 (1986), 109–122.

Okenfuss, Max, 'Inauspicious beginnings: Jan Tessing, Amsterdam, and the origins of Petrovian printing', *Baltic Studies*, 5 (1998), 15–24.

Opisanie izdanij graždanskoj pečati: 1708–ânvar' 1725, Moskva-Leningrad: Izdatel'stvo Akademii nauk SSSR, 1955.

Opisanie izdanij napečatannyh kirillicej: 1689-ânvar' 1725, Moskva-Leningrad: Izdatel'stvo Akademii nauk SSSR, 1958.

Ougotovánie i tolkovánie âsnoe i šělw izrwdnoe krásnoóbraznagw poverstániâ krugóv" nebesnyh" ko oupotreblâniû s" pisano est' na kartině, s" pódvigami planét" sirěč' solnca, měsâca î zvězd" nebesnyh na pólzu i outešénie lûbâšym" astronómiû, Amsterdam, 1699.

Pančenko, Aleksandr, *Russkaâ kul'tura v kanun petrovskih reform*, Nauka, 1984.

Pekarskij, Petr, *Nauka i literatura v Rossii pri Petre Velikom*, 2 vols, SPb, 1862.

Pekarskij, Petr, *Istoriâ Imperatorskoj Akademii Nauk v Peterburge*, 2 vols, SPb, 1870–73.

Pis'ma i bumagi Imperatora Petra Velikago, I, SPb, 1887.

Poludenskij, Mihail, 'Petr Velikij v Pariže', *Russkij arhiv*, 5–6 (1865), 675–702.

Pypin, Aleksandr, *Istoriâ russkoj literatury*, III, SPb, 2nd ed, 1902.

Riabouchinsky, Dimitry, 'Les rapports scientifiques entre la France et la Russie', *Revue Générale des Science pures et appliquées*, 45 (1934), 46–50.

Rimpler, Georg, *Rimplerova manira o stroenii krěpostei*, Moskva, 1708; 2nd ed, 1709.

Robert, Paul, *Dictionnaire alphabétique & analogique de la langue française*, Nouveau Littré, 1973.

Simonov, Rèm, 'Rol' inostrancev v rasprostranenii znanij po matematike, astronomii i astrologii v Rosii v XV–XVII vv', *Drevnââ Rus' i Zapad*, Nasledie, 1996, 122–126.

Skornâkov-Pisarev, Grigorij, *povelěniem" Vsepresvětlěišago, Deržavněišago Imperatora, i Samoderžca Vserossiiskago, petra Pervago, i Otca Otečestviâ: izobretennaâ siâ nauka Statičeskaâ ili mehanïka*, SPb, 1722.

Sturm, Leonhard Christoph, *Architektura militaris*, Nürnberg, 1702.

Sturm, Lenard Hristof, *Arhitektura voinskaâ*, translated A G Golovkin, Moskva, 1709.

Suhomlinov, Mihail, (ed), *Materialy dlâ istorii Imperatorskoj Akademii nauk*, I, SPb, 1885.

Tablicy logariθmωv", i sinusωv", tángensωv", sěkansov", Moskva, 1703, 1715.

Ûrnal 206-go 1867.

Ustav morskoj: o vsem čto kasaetsâ k dobromu upravleniû v bytnosti flota na more, Moskva: Novator, 1763; reprinted 1993.

Ustrâlov, Nikolaj, *Istoriâ carstvavaniâ Petra Velikago*, II–III, SPb, 1858.

Ustûgov, Nikolaj, *Naučnoe nasledie*, Nauka, 1974.

Val'hauzen [Wallhausen], Iogann Âkobi, *Ûčenie i hitrost' ratnogo stroeniâ pehotnyh lûdej*, Moskva, 1647/49.

Vasil'ev, Vladimir, 'Sočinenie A K Nartova "Teatrum Machinarum": k istorii peterburgskoj "Tokarni Petra I"', *Trudy Gosudarstvennogo Èrmitaža*, 3 (1959), 41–92.

Vlacq, Adrian, *Tabulæ sinuum, tangentium et sectantium, et logarithmi, sinuum, tangentium et numerorum ab unitaté ad 10 000*, Amsterdam, 1681.

Vladimirskij-Budanov, Mihail, *Gosudarstvo i narodnoe obrazovanie v Rossii v XVII veke do učreždeniâ ministerstv*, SPb, 1874.

Vucinich, Alexander, *Science in Russian culture: a history to 1860*, Stanford University Press, 1963.

von Wallhausen, Johann Jacob, *Kriegskunst zu Fuß*, Hanaw: Selbstverlag, 1615; Frankfurt, 1620; Leuwarden, 1630; Akademische Druck- u.Verl.-Anst., 1971.

Zagorskij, Fedor, *Andrej Konstantinovič Nartov: 1693–1756*, Nauka, 1969.

Žurnal Gosudarâ Petra I s 1695 po 1709, sočinennyj baronom Gizenom, pol 1-â, SPb, 1787.

Human computers in eighteenth- and nineteenth-century Britain

Mary Croarken

In eighteenth- and nineteenth-century Britain, people who earned their living either by calculating mathematical tables or by undertaking other computational work were referred to as computers or calculators (see Simpson and Weiner 1989). Today we associate these terms with electronic or mechanical devices, but there are strong parallels between the modern use of the word computer and its application to people two hundred years ago. The most common sphere in which human computers worked was astronomy, but they were also employed in a variety of scientific fields on a lesser scale. The work undertaken by the human computers of the eighteenth and nineteenth centuries was characterized by the repetitive use of arithmetical steps, often of considerable sophistication, to produce tables or tools for use by others, in astronomy, navigation, tide prediction, and scientific calculation. The work required patience and attention to detail and computers were usually required to adhere to strict algorithms with little scope for individualism.

The distinction between computers and clerks was sometimes blurred, especially in the banking and insurance industries, but was usually characterized by the complexity of the work. Clerks tended to count (manufactured goods, population, and so on) or add (columns of figures) and computers tended to calculate (positions of stars, tide predictions, and so on) using several arithmetic operations and logarithms. Some computers were teachers or clergymen who supplemented their income by producing mathematical tables. Others worked as full time professionals and had rarely had a university education. Some were identified in published work but most remained anonymous. Only a very few were women. The work was repetitive, sedentary, and often wearisome. They were employed, often on a freelance or ad hoc basis, by public bodies, private individuals, and commercial enterprises.

This chapter will identify and describe some of the human computers known to have been working in Britain during the eighteenth and nineteenth centuries. It will examine the nature of their work along with their pay and conditions where such information is available. Because the majority of computers were employed on astronomically related projects, the work of computers employed at the Nautical Almanac Office and the Royal Greenwich Observatory will be described in some detail to illustrate the nature of work during this period. Later sections will identify other computers and computing groups. Because computers were often seen as just hired help, records outlining their work are difficult to find, and there existed many whose work in the scientific community remains unidentified and unrecorded. This account should not therefore be seen as definitive.

Nautical almanac computers

From 1767 onwards, increasing numbers of British seamen used the astronomical tables published in the annual *Nautical almanac* to determine longitude at sea.[1] The annual preparation of these tables was a huge undertaking, carried out by a group of computers paid from the public purse, under the supervision of Nevil Maskelyne, Astronomer Royal from 1765 to 1811 (Croarken 2003a).

In the *Nautical almanac*, the tables for each calendar month were contained in twelve pages giving the sun's position, the position of the planets, the position of the moon, lunar distances from the sun (see Fig. 4.4.1), and fixed stars at three hourly intervals along with extra data such as the days of the month, festivals and feast days, eclipses of the moons of Jupiter, and so on. While the calculations required to compute each entry of the *Nautical almanac* were not difficult, they were time consuming. Maskelyne devised a series of precepts (computing principles or algorithms) for each calculation. Most required nothing more difficult than addition or subtraction of sexagesimal numbers extracted from one of fourteen other types of tables (for example, logarithm tables, lunar tables, and

---

1. The use of chronometers did not come into widespread use (largely because of cost) until the beginning of the nineteenth century, see Croarken (2003a).

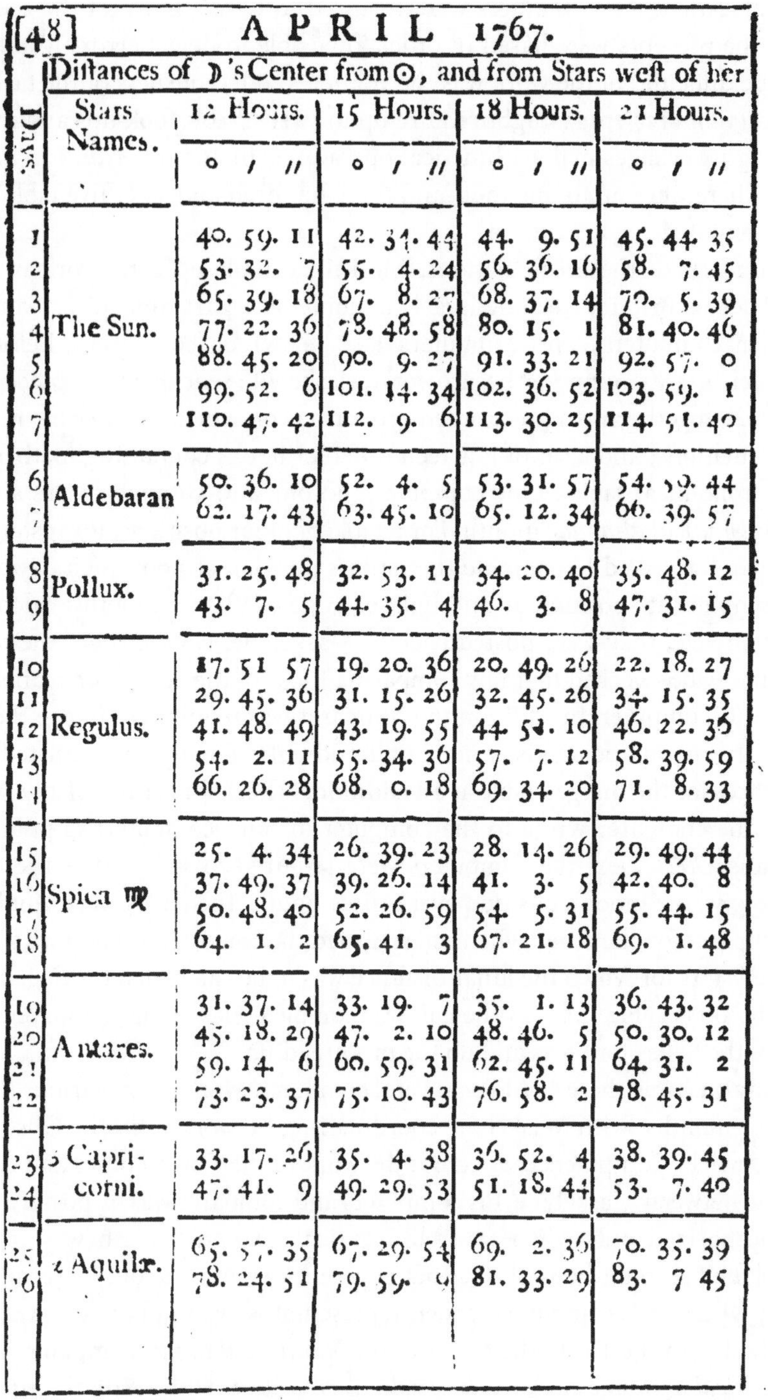

[48]	A P R I L 1767.			
	Diftances of ☽'s Center from ⊙, and from Stars weft of her			
Days	Stars Names.	12 Hours.	15 Hours.	18 Hours.	21 Hours.
		° ′ ″	° ′ ″	° ′ ″	° ′ ″
1		40. 59. 11	42. 34. 44	44. 9. 51	45. 44. 35
2		53. 32. 7	55. 4. 24	56. 36. 16	58. 7. 45
3		65. 39. 18	67. 8. 27	68. 37. 14	70. 5. 39
4	The Sun.	77. 22. 36	78. 48. 58	80. 15. 1	81. 40. 46
5		88. 45. 20	90. 9. 27	91. 33. 21	92. 57. 0
6		99. 52. 6	101. 14. 34	102. 36. 52	103. 59. 1
7		110. 47. 42	112. 9. 6	113. 30. 25	114. 51. 40
6	Aldebaran	50. 36. 10	52. 4. 5	53. 31. 57	54. 59. 44
7		62. 17. 43	63. 45. 10	65. 12. 34	66. 39. 57
8	Pollux.	31. 25. 48	32. 53. 11	34. 20. 40	35. 48. 12
9		43. 7. 5	44. 35. 4	46. 3. 8	47. 31. 15
10		17. 51. 57	19. 20. 36	20. 49. 26	22. 18. 27
11		29. 45. 36	31. 15. 26	32. 45. 26	34. 15. 35
12	Regulus.	41. 48. 49	43. 19. 55	44. 54. 10	46. 22. 35
13		54. 2. 11	55. 34. 36	57. 7. 12	58. 39. 59
14		66. 26. 28	68. 0. 18	69. 34. 20	71. 8. 33
15		25. 4. 34	26. 39. 23	28. 14. 26	29. 49. 44
16	Spica ♍	37. 49. 37	39. 26. 14	41. 3. 5	42. 40. 8
17		50. 48. 40	52. 26. 59	54. 5. 31	55. 44. 15
18		64. 1. 2	65. 41. 3	67. 21. 18	69. 1. 48
19		31. 37. 14	33. 19. 7	35. 1. 13	36. 43. 32
20	Antares.	45. 18. 29	47. 2. 10	48. 46. 5	50. 30. 12
21		59. 14. 6	60. 59. 31	62. 45. 11	64. 31. 2
22		73. 23. 37	75. 10. 43	76. 58. 2	78. 45. 31
23	β Capri-	33. 17. 26	35. 4. 38	36. 52. 4	38. 39. 45
24	corni.	47. 41. 9	49. 29. 53	51. 18. 44	53. 7. 40
25	α Aquilæ.	65. 57. 35	67. 29. 54	69. 2. 36	70. 35. 39
26		78. 24. 51	79. 59. 0	81. 33. 29	83. 7. 45

Figure 4.4.1 Sample page from the 1767 Nautical almanac showing lunar distances. Courtesy of the National Maritime Museum; negative number 7147

solar tables) but the computers were required to have a grasp of the theory on which the precepts were based in order to be able to detect errors in the work. The true difficulty of the work was the volume and tediousness of the task. Each *Nautical almanac* entry might require up to twelve table look-ups and fourteen seven-figure sexagesimal arithmetical operations; in any one *Nautical almanac* month there were up to one thousand three hundred and sixty five table entries to be computed.

To undertake these calculations, Maskelyne built up a network of people spread across Britain, working largely at home on a part time, piece-work basis as either 'computers', 'anti-computers' (see below), or 'comparers'. Before each computer or anti-computer began work, Maskelyne sent them his precepts (see Fig. 4.4.2), together with pre-printed pro formas on which to enter the results of the calculations, and a set of fourteen different books containing mathematical and astronomical tables. One computer and one anti-computer were allocated to each *Nautical almanac* month. For a table of lunar positions, for instance, the computer was asked to compute the moon's position at noon each day and the anti-computer its position at midnight each day. When a month's calculations were complete, they were posted back to Maskelyne. He then sent them to the comparer, who checked them by comparing those of the computer against those of the anti-computer. In addition the comparer was required to merge the tables to give the moon's positions at noon and midnight. Finally he calculated fourth differences for the merged table as an additional check for errors.[2] If an error was found, the comparer wrote to the computer (or anti-computer) to point it out (Hitchins 1792). Next the computers were required to subtabulate the merged table to give the moon's position every three hours. The tables were sent back to the comparer for checking, who then returned the corrected version with a list of the fixed stars for which the lunar distance was to be calculated for each position of the moon (see Fig. 4.4.3). When all was completed it was the comparer's job to prepare the tables for press and do the proofreading.

At any one time there would be several computers working on various *Nautical almanac* months, with Maskelyne coordinating the work from the Observatory in Greenwich. Computers carried out the work in their own homes and communication between Maskelyne, his computers and comparer was by post. Payments were authorized only when a whole *Nautical almanac* month was complete. Maskelyne's account book shows that payments were made only intermittently, usually whenever a computer, or their representative, was in London (Maskelyne 1765–1811a). In the 1770s, the salary of the *Nautical almanac* computers was £75 per *Nautical almanac* year, which gradually rose to £180 in 1810; the equivalent

---

2. First differences are differences between successive entries in the table; second differences are differences between the first differences; and so on.

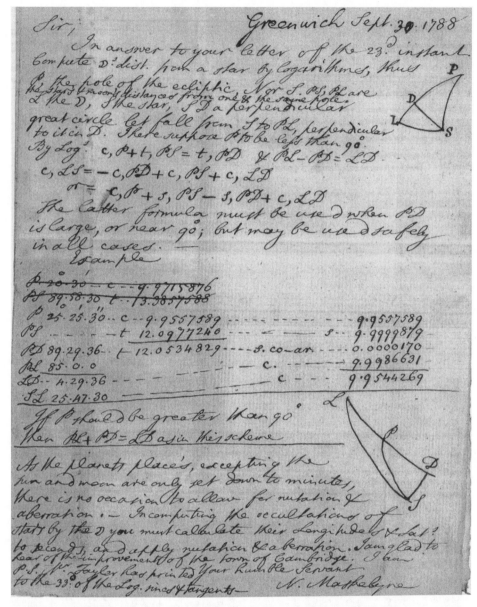

Figure 4.4.2 Letter from Nevil Maskelyne to Joshua Moore 30 September 1788 giving the instructions for computing the moon's distance from a star using logarithms. Courtesy of the Library of Congress Manuscripts Division

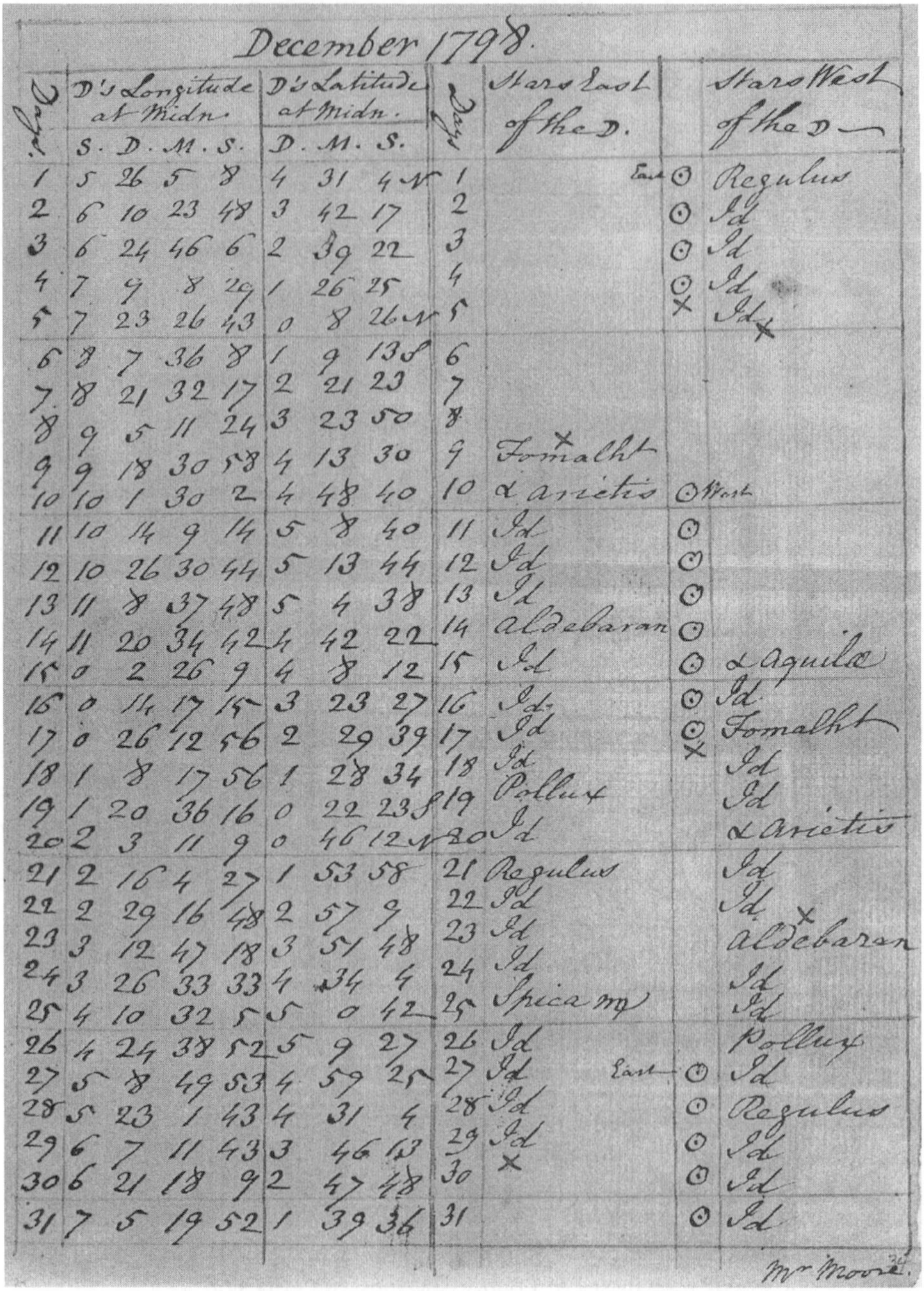

December 1798.

Day	D's Longitude at Midn. S.	D.	M.	S.	D's Latitude at Midn. D.	M.	S.	Day	Stars East of the D.		Stars West of the D
1	5	26	5	8	4	31	4 N	1	East	⊙	Regulus
2	6	10	23	48	3	42	17	2		⊙	Id
3	6	24	46	6	2	39	22	3		⊙	Id
4	7	9	8	29	1	26	25	4		⊙	Id
5	7	23	26	43	0	8	26 N	5		×	Id ×
6	8	7	36	8	1	9	13 S	6			
7	8	21	32	17	2	21	23	7			
8	9	5	11	24	3	23	50	8			
9	9	18	30	58	4	13	30	9	Fomalht ×		
10	10	1	30	2	4	48	40	10	α Arietis	⊙ West	
11	10	14	9	14	5	8	40	11	Id	⊙	
12	10	26	30	44	5	13	44	12	Id	⊙	
13	11	8	37	48	5	4	38	13	Id	⊙	
14	11	20	34	42	4	42	22	14	Aldebaran	⊙	
15	0	2	26	9	4	8	12	15	Id	⊙	α Aquilæ
16	0	14	17	15	3	23	27	16	Id	⊙	Id
17	0	26	12	56	2	29	39	17	Id	⊙ ×	Fomalht
18	1	8	17	56	1	28	34	18	Id		Id
19	1	20	36	16	0	22	23 S	19	Pollux		Id
20	2	3	11	9	0	46	12 N	20	Id		α Arietis
21	2	16	4	27	1	53	58	21	Regulus		Id
22	2	29	16	48	2	57	9	22	Id		Id
23	3	12	47	18	3	51	48	23	Id		Aldebaran ×
24	3	26	33	33	4	34	4	24	Id		Id
25	4	10	32	5	5	0	42	25	Spica ♍		Id
26	4	24	38	52	5	9	27	26	Id		Pollux
27	5	8	49	53	4	59	25	27	Id East	⊙	Id
28	5	23	1	43	4	31	4	28	Id	⊙	Regulus
29	6	7	11	43	3	46	43	29	Id ×	⊙	Id
30	6	21	18	9	2	47	48	30		⊙	Id
31	7	5	19	52	1	39	36	31		⊙	Id

Mr Moore.

Figure 4.4.3 Manuscript tables of the moon's position for December 1798 showing which stars the computer is to use for the lunar distance predictions. Sent by Malachy Hitchins to Joshua Moore. Courtesy of the Library of Congress Manuscripts Division spending power today would be approximately £7,500 and £9,500, respectively.[3] A differential between the computers' pay and the comparers' pay (the computers' work being more time consuming but the comparers' requiring more skill) was not made until 1808, when the comparers' pay exceeded the computers' by £40 per *Nautical almanac* year.[4] The time it took for computers to complete their calculations varied considerably: some took several months to prepare two months' worth of tables while others turned in the same amount of work in less than a calendar month (Maskelyne 1765–1811b).

The computers lived in different parts of England, and included clergymen, teachers, surveyors, ex-Royal Greenwich Observatory assistants, and astronomers on temporary leave from Board of Longitude voyages of exploration. Some computed for the *Nautical almanac* only for a few months, others for forty years or more. For most it was a way of supplementing other sources of income but for a few it was their only employment. Although the work was tedious and required long hours hunched over a desk piled high with books of tables, the computers and comparers of the *Nautical almanac* could choose where and when to do the work and how much to take on at any particular time. They also got to complete the whole range of work needed to compute the tables in contrast to, for example, Gaspard de Prony's table making project in France in the 1790s, where one individual did only one part of a calculation without ever seeing or needing to understand the whole (Grattan-Guinness 2003, 109).

Maskelyne's computing system was designed to make the *Nautical almanac* as accurate as possible, because an incorrect entry in the third or fourth decimal place of some tables could potentially lead to a longitude calculation being out by over thirty miles. The comparer's role was therefore the cornerstone of Maskelyne's methods. Although he and his comparers understood the need for accuracy, his computers were sometimes not so fastidious. In 1770 the comparer Malachy Hitchins found that the work of Joseph Keech and Reuben Robbins (computer and anti-computer respectively) matched too well and deduced that they had copied each other's work instead of computing independently. Keech and Robbins were part of the same London coffee house set and lived near each other. The pair were immediately dismissed and required to pay compensation for the waste of the comparer's time (Board of Longitude 1770). The event taught Maskelyne to ensure that work for the same *Nautical almanac* month was allocated to a computer and anti-computer living in different parts of the country. By the beginning of the nineteenth century, there was a cluster of computers

3. Equivalent modern spending power has been calculated using http://www.MeasuringWorth.com. In 2008, £1 was worth approximately 2 US dollars.

4. The salaries paid to *Almanac* computers and comparers have been gathered from a variety of manuscripts in the Royal Greenwich Observatory archives at Cambridge University Library, namely, RGO 16/55; RGO 14/5 f. 165; RGO 14/6 f. 317; RGO 14/7 ff. 5, 39, 32.

living in Devon and Cornwall close to Malachy Hitchins, the longest serving comparer, but care was taken not to allocate work relating to the same month to near neighbours.

Following Maskelyne's death in 1811, *Nautical almanac* computing continued in much the same way but Maskelyne's successor, John Pond, took no active interest in computational astronomy in general or the *Nautical almanac* in particular. Consequently the accuracy and reputation of the *Nautical almanac* fell to such an extent that the issue was raised in Parliament (Hansard 1818). Steps were taken to restore the accuracy of the publication by appointing the polymath Thomas Young as Superintendent of the *Nautical almanac*, but throughout the 1820s and early 1830s there was considerable clamour from British astronomers to reform both the way the *Nautical almanac* was computed and the tables it contained (Ashworth 1994). Thanks to advances in telescope technology, and the rise of the gentleman astronomer (Chapman 1998), a new rigour was being brought to land-based observation and stellar cataloguing by young mathematicians, astronomers, and business men such as Charles Babbage, John Herschel, Thomas Colby, and Francis Baily. Many of these men had commercial interests in the insurance and banking industries, and their ideas on observation and scientific recording were heavily influenced by their experiences computing life assurance tables, and by the increasing propensity of business culture to record, analyse, and standardize. They perceived a reformed *Nautical almanac* as a vehicle for ensuring standardization of land based astronomy.

The issue of reform of the *Nautical almanac* was part of a wider conflict at the time within the upper echelons of London scientific society (Miller 1983). The conflict centred around the domination of the Royal Society by its President, biologist Joseph Banks, and his followers on the one hand, and young mathematicians and scientists who wanted to push the boundaries of what constituted science on the other. In practical terms this discontent manifested itself in the creation of both the Astronomical Society and the Geological Society, both of which were strongly opposed by Banks. Reform of the *Nautical almanac* should therefore be seen against this background rather than as to do with problems associated with the computation methods themselves (Miller 1983; Ashworth 1994).

In 1830 the Admiralty asked the Astronomical Society (later the Royal Astronomical Society) to report on ways of improving the *Nautical almanac*. The Society set up a committee that included many of the dissatisfied astronomers, as well as mathematicians, naval officers, and others (South 1831, 448). In its 1831 report, the committee called for both radical revision and expansion of the *Nautical almanac*, to make it more useful for astronomers while retaining its usefulness for navigators.

Many of the committee members were as active in accounting and assurance as they were in astronomy, and had strong ideas about the organization of astronomical computing and its parallels with the work of assurance offices (Ashworth 1994). Reflecting the early nineteenth-century drive to mechanize industry (Babbage 1835), reform of the *Nautical almanac* was also seen as a vehicle for centralizing the way the computing was undertaken. The Admiralty accepted the committee's report and all its twenty-six recommendations. Most of the reforms were aimed predominately at land-based astronomers rather than seamen. Examples include: giving the places of the sun and the moon to several places of decimals; giving more details in the tables of the satellites of Jupiter; changing the places of the fixed stars to show time of transit at Greenwich rather than at apparent noon; and extending the list of monthly astronomical phenomena. One reform, replacing apparent time for mean solar time in the *Nautical almanac* tables—a change made possible by the technical improvement of clocks and watches over the previous fifty years—would be of long term benefit to navigators but in the short term would cause confusion at sea for several years to come. At the same time responsibility for the *Nautical almanac* was removed from the Astronomer Royal, John Pond, who had little interest in computational astronomy, and assigned to Francis Beaufort, the Hydrographer of the Navy. The Hydrographic Office remained responsible for the *Nautical almanac* until 1937 when responsibility reverted to the Astronomer Royal. In April 1831 Beaufort, who had been a prominent member of the Astronomical Society committee and influential in preparing the report, appointed William Stratford, another committee member and secretary of the Astronomical Society, as Superintendent of the *Nautical almanac*. Almost immediately Stratford centralized the work in London.

In the late 1820s, just before the reform, the *Nautical almanac* was being prepared by five computers and one comparer, with the work being coordinated by Thomas Young in London using the same techniques that Maskelyne had put into operation sixty years earlier. Three of the computers were from Cornwall (William Dunkin from Truro, Nicholas James from St Hilary, and Richard Martyn from St Mabyn) and had all been recruited to the work by Maskelyne's long-serving comparer Malachy Hitchins. Of the other two, George G Carey was a scientific lecturer and author from Arbroath in Scotland, and Eliza Edwards from Ludlow in Shropshire was the daughter of Mary Edwards, one of the Maskelyne's long-serving computers. The comparer was Thomas Brown, a Church of England vicar from Tideswell in Derbyshire.

In 1831, the computers were told that their services would no longer be required once the preparations for the 1833 *Nautical almanac* were complete, as the work was to be moved to London (Dunkin 1999). Edwin Dunkin, whose father William was one of those computers, remarked that most of the computers were past middle age, fearful of learning new computing methods, and unwilling to leave their homes, and it was also considered that the new office should employ new staff (Dunkin 1999, 45). However, William Dunkin, then in his fifties, did go to London to join the new Nautical Almanac Office. He was the only existing computer to make the move and did so thanks to the patronage of Davies Gilbert, then president of the Royal Society. For Dunkin, however, the move proved far from successful. His son recalled:

I have often heard him express a real regret at the loss of his semi-independent position at Truro, in exchange for the daily sedentary confinement to an office-desk for a stated number of hours in the company of colleagues all junior to himself in age and habits (Dunkin 1999, 45).

Dunkin recognized that computing in a tightly controlled central office was not a profession that was likely to bring advancement to his sons Edwin and Richard, and he tried to educate them for a life in trade. Unfortunately Dunkin died in 1838 and the sons, who quickly needed to become the wage earners of the family, both took up computing positions at the Royal Greenwich Observatory despite their father's wishes. In 1847 Richard transferred to the Nautical Almanac Office; Edwin stayed at the Royal Observatory.

Information about the early organization of the Nautical Almanac Office can be gleaned in part from the incoming and outgoing letter books of the Hydrographic Office. Written communications were exchanged between Stratford, based in the Nautical Almanac Office in Verulam Buildings off the Grays Inn Road, and Beaufort, based in the Hydrographic Office at Somerset House. The buildings were less than a mile apart so much communication would probably also have been face to face. From the records it is difficult to be sure about exactly who was employed when, on what salary, or exactly how the computations were organized. It is likely that Stratford began by hiring approximately seven computers in 1831 but that number had risen to sixteen by 1836. Once the initial computers had been hired and trained, Stratford spent much of his time at his commercial business in the Metropolitan Loan Company and communicated his instructions to his staff by letter. One of the computers, hired at a rate of £100 per annum in 1831,[5] was the twenty-two-year-old Wesley Stoker Barker Woolhouse, recommended to Stratford by a Greenwich schoolmaster (Anon 1838). Woolhouse proved very able and in 1833 was appointed Deputy Superintendent with an increase of £50 to his salary. Stratford himself was earning £500 per annum. Over the next three years Woolhouse supervised many of the computations and wrote several well-regarded technical supplements to the *Nautical almanac*.

In 1836, while the computers were working on calculations for the prediction of the return of Halley's comet (Grier 2005, 49), Stratford requested that

---

5. Approximately £7,000 in spending power today.

the computers work an extra hour per day to catch up with *Nautical almanac* work delayed due to the comet work (Woolhouse 1837). Woolhouse felt that the computers already worked long enough hours, from 9am until 5pm, and refused to comply. Relations deteriorated, accusations were made on both sides, others got involved and the affair continued to be acrimonious even after Woolhouse's resignation in early 1837.[6] Woolhouse later went on to have a successful career as an actuary.

The following years went more smoothly. By the 1840s the size of the Nautical Almanac Office had stabilized at eleven staff plus the Superintendent. The computers were now called 'assistants', a term which reflected change of status rather than a change in the work undertaken. The Superintendent was supported by a Chief Assistant who was responsible for arranging calculation for the *Nautical almanac* as a whole, and undertook independent computations of various parts of the tables to check for accuracy. He selected the stars liable to occultation,[7] and those required for lunar distance calculations (Hind 1878). He was also responsible for seeing the *Nautical almanac* through the press. Under the Chief Assistant were First Class Assistants and Second Class Assistants, whose numbers varied over the period 1840 to 1920.

In 1831, Stratford had begun by hiring computers at £100 per annum. By the 1850s the salaries for second class assistants and junior computers had fallen, as shown in Table 4.4.1. This drop was partly due to the greater availability of clerical labour and partly due to centralization, which meant that tasks could be divided up between computers with varying levels of skill. This was reflected in the pay structure. Not until 1880 was the lowest salary point back near the 1833 level. In the meantime the social status of junior computers had also fallen and they were no longer seen as skilled individuals but as part of the disposable clerical workforce of Victorian London.

Table 4.4.1 Salaries of the *Nautical almanac* assistants 1833–1878[8]

Date	Chief Assistant	First Class Assistant	Second Class Assistant	Junior Computer
1833	£150	£100	£100	£100
1850	£190	£150	£90	£40–£75
1880	£350–£400	£260–£300	£100–£250	disestablished

6. Clippings from *The Times* concerning the affair are held at Cambridge University Library Manuscripts RGO 16/1.

7. That is, stars which are predicted to pass behind another celestial body and therefore out of sight.

8. This data is extracted from Hydrographic Office Incoming letters H798, Hind to Beaufort 27 April 1853; and Royal Greenwich Observatory Archives RGO 16/3 C Packet 1.

Sometimes more than one member of a family was employed in the Nautical Almanac Office, and many individuals spent their entire careers there. Richard Farley, for example, who became Chief Assistant in 1837, was employed for thirty-eight years and his brother for forty-two years. Similarly William Godward, who became Chief Assistant in 1869, was a Nautical Almanac Office employee for forty-three years and his brother John for eighteen years. Even Superintendents tended to stay for long periods.

Although the work was repetitive, it was secure and well paid compared to private observatory assistants and commercial city clerks, but just below the salaries of government clerks in the Board of Trade. This put senior *Nautical almanac* computers in the middle classes of Victorian England, though not of the same social standing as graduates (Chapman 1998, 150; Select Committee 1850, 85). The other traditional employment of the mathematically inclined, schoolteaching, was usually less well remunerated, but wages were very variable and depended to a great extent upon the personality of the teacher and the social class of the pupils.

Very little changed in the way that the Nautical Almanac Office was organized until the early twentieth century when the then Superintendent Phillip Cowell began to employ retired members of staff on a piece-work basis in their own homes, much as Maskelyne had done. The first real change to the computing methods of the Office came in the late 1920s when mechanical calculating machines began to be introduced (Croarken 1990).

Observatory assistants

While the computers and assistants of the *Nautical almanac* spent their days preparing tables of positions of celestial objects as a practical tool for navigators and astronomers, the computers employed at the Royal Greenwich Observatory concentrated on past and current observations (see Aubin in this volume). At the detailed level much of the work was similar in nature—for example, sexagesimal arithmetic and looking up tables—but the context was rather different and, at least in the eighteenth century, the work a little more varied because observing was also part of the job.

Most observations made at the Royal Observatory at Greenwich were carried out as a set of three or five observations over a short time. These observations then had to be 'reduced', that is, averaged, corrected for refraction, parallax, instrument and observer error, and converted from relative positions as observed from earth to absolute positions in the celestial sky. From the appointment of John Flamsteed as First Astronomer Royal in 1675, it had been customary for the Astronomer Royal to hire an assistant to make routine observations, reduce the sets of figures, and carry out whatever other computing work the Astronomer Royal might require. The post thus combined observing and computing. The work was hard, lonely, and tedious: not only did observations need to be taken to a strict timetable, often at night and with the observatory roof open to the elements, but the task of reduction was repetitive and never ending. In addition the assistant was required to live at the Observatory, four miles from London and, at least in the late eighteenth century, was not expected to engage with outside society to any great extent and was forbidden to marry (Croarken 2003b, 289).

Despite the hardship of the post, some assistants stayed at the Observatory for many years but others lasted only a few weeks or months. Before the mid-nineteenth century many Greenwich assistants used the post as a stepping-stone to other, more interesting employment. For example, several assistants left Greenwich to sail as astronomers on the late eighteenth-century voyages of exploration with Cook and others. Others, such as John Hellins and John Brinkley, used the experience to further their own scientific careers, becoming Fellows of the Royal Society and, in Brinkley's case, Astronomer Royal for Ireland.

For others, employment at the Observatory was anything but life affirming. Thomas Evans, assistant at the Royal Observatory from 1796 to 1798, wrote the following:

Nothing can exceed the tediousness and *ennui* of the life the assistant leads in this place, excluded from all society [...] Here forlorn, he spends days, weeks, and months, in the same long wearisome computations, without a friend to shorten the tedious hours, or a soul with whom he can converse. He is also frequently up three or four times in the night [...] (Evans 1810, 333–335)

For another assistant, David Kinnebrooke (1795), not only were the working conditions difficult, but his career was significantly blighted by being dismissed because his observations consistently differed by 800 milliseconds from those of the Astronomer Royal. The case later resulted in the realization by Friedrich Bessel that every observer has different but predictable aural and visual reaction times and that a correction for this should be made on all observations and for all observers (Mollon and Perkins 1996; Shaffer 1988). Bessel's 1821 analysis of what has become known as the personal equation came too late to rescue the career of Kinnebrooke, who had died in Norwich in 1802 (Kinnebrooke 1802).

When George Biddell Airy was appointed Astronomer Royal in 1835, the staff of the Observatory expanded significantly to accommodate extra work that Airy persuaded the Admiralty to pay for, on the backlog of unreduced observations that had built up over the past eighty years (Airy 1896). Airy began both by improving the quality of the assistants (Meadows 1975, 1) and by increasing their number from two in 1835 to six by 1846 with annual salaries ranging from £100 to £400,[9] with additional housing allowances of £70 to £90 (Airy 1846). Many assistants now stayed at Greenwich for much of their working lives. The salaries paid by the Observatory were almost twice those paid by most universities or private observatories and three times those paid to assistants at the Nautical Almanac Office. In addition the terms and conditions at the Royal Observatory were good, with an official five-hour working day, a month's paid holiday per year, and a pension at sixty-five (Chapman 1992). Some assistants started as youths, rising through the ranks to end their careers as senior or sometimes Chief Assistant. The work of the assistants was varied and included much computing alongside observation.

One of Airy's rotas for his assistants is given below.

Day 1 – On duty with the transit circle from 6am until 3am the following day (that is, 21 hours continuously)

Day 2 – 2 to 3 hours computing

Day 3 – Full day's computing followed by night duty with the altazimuth

Day 4 – 2 to 3 hours computing

The 4-day cycle was then repeated. (Meadows 1975, 10)

The transit circle was a telescopic instrument used to observe the time at which a celestial object crossed the meridian. Observations were limited to specific objects, depending on the current pattern of work, but the instrument required long hours of manpower. The altazimuth measured the altitude and horizontal angles of celestial bodies and again had its own observing schedule. The computing part of the work focused on reducing and correcting the observations, or any other work that Airy required. There were other rotas depending on the department and type of additional work being undertaken. The rotas were drawn up by Airy and pinned up above the mantelpiece in the computing room each Monday morning (Dunkin 1862, 26).

Before Airy's appointment as Astronomer Royal in 1835, the Greenwich assistant's life was a lonely one, made up of computing and observing. After Airy's appointment, the staff at the Observatory rapidly expanded and the work became more varied though it still retained a mix of observing and computing. The computers could live with their families within a mile of the Observatory and the terms and conditions improved as the assistants became formally part of the British Civil Service. There was a distinct division, however, between the working conditions of the Greenwich assistants and those of the temporary supernumerary computers who were hired and fired according to need.

In his efforts to reduce the backlog of unpublished observations, Airy began by employing three supernumerary computers specifically to work on the reduction

9. Approximately £6,500 to £25,500 per annum in spending power today.

of planetary observations from 1750 to 1830: James Glaisher, John Hartnup, and a Mr Thomas. Airy placed them in the Octagon Room of the Observatory and set them to work using a strict algorithm. In 1838 he obtained another £2000 to calculate lunar reductions for the same period, and was able to employ first seven and later sixteen supernumerary computers (Airy 1896). The working conditions of the supernumerary computers were not good but, by the standards of the time, they were not dreadful either. The brothers Edwin and Richard Dunkin, aged seventeen and fifteen respectively, whose father William had been an *Nautical almanac* computer (see above), were hired by Airy on 21 August 1838 as temporary supernumerary computers. Edwin was set to work calculating the right ascension and north polar distance for the planet Mercury with a large book of printed skeleton forms and Lindenau's tables of the position of Mercury (1813). His recollections of that day give us an indication of the life of a Greenwich supernumerary computer.

I felt a little nervous at first, and a momentary fear crossed my mind that some time would be required to enable me to comprehend this intricate form, and to fill up the various spaces correctly from the Tables. Though Lindenau's Tables were new to me, I soon found that the Astronomer Royal had so skilfully prepared the skeleton forms, that any intelligent and careful computer could hardly go astray. After very little instruction from Mr. Thomas, the principal computer in charge, I began to make my first entries with a slow tremulous hand, doubting whether what I was doing was correct or not. But after a little quiet study of the example given in the Tables, all nervousness soon vanished; and before 8 p.m., when my day's work was over, some of the older computers complimented me on the successful progress I had made. My brother was employed on a less advanced class of calculations, and he also satisfied Mr. Thomas, the superintendent of the room. We went home tired enough to our lodgings [...] but with light hearts and the happy thought that we had earned our first day's stipends. On looking back [...] I cannot help thinking that it was some what remarkable that these two youths, full of life, and fresh from school, could feel any elation at the prospect of passing eleven hours, day after day, in a situation allowing no time for that physical recreation considered so indispensable for all classes in the present day. (Dunkin 1999, 72)

This was the start of a pattern of work in the Observatory for which Airy is well known and indeed notorious. He is particularly remembered for having brought factory methods of production to the work of the Observatory (Grier 2005). Airy set out each step of a calculation in a logical sequence and used pre-printed forms to guide the computers through the work (see Fig. 4.4.4), just as Maskelyne had done for the *Nautical almanac* computers sixty years before.[10] The work was then carried out by relatively unskilled computers under the supervision of senior

10. Examples of Airy's skeleton forms can be found in the Royal Greenwich Observatory Archives at Cambridge University at RGO 6/42.

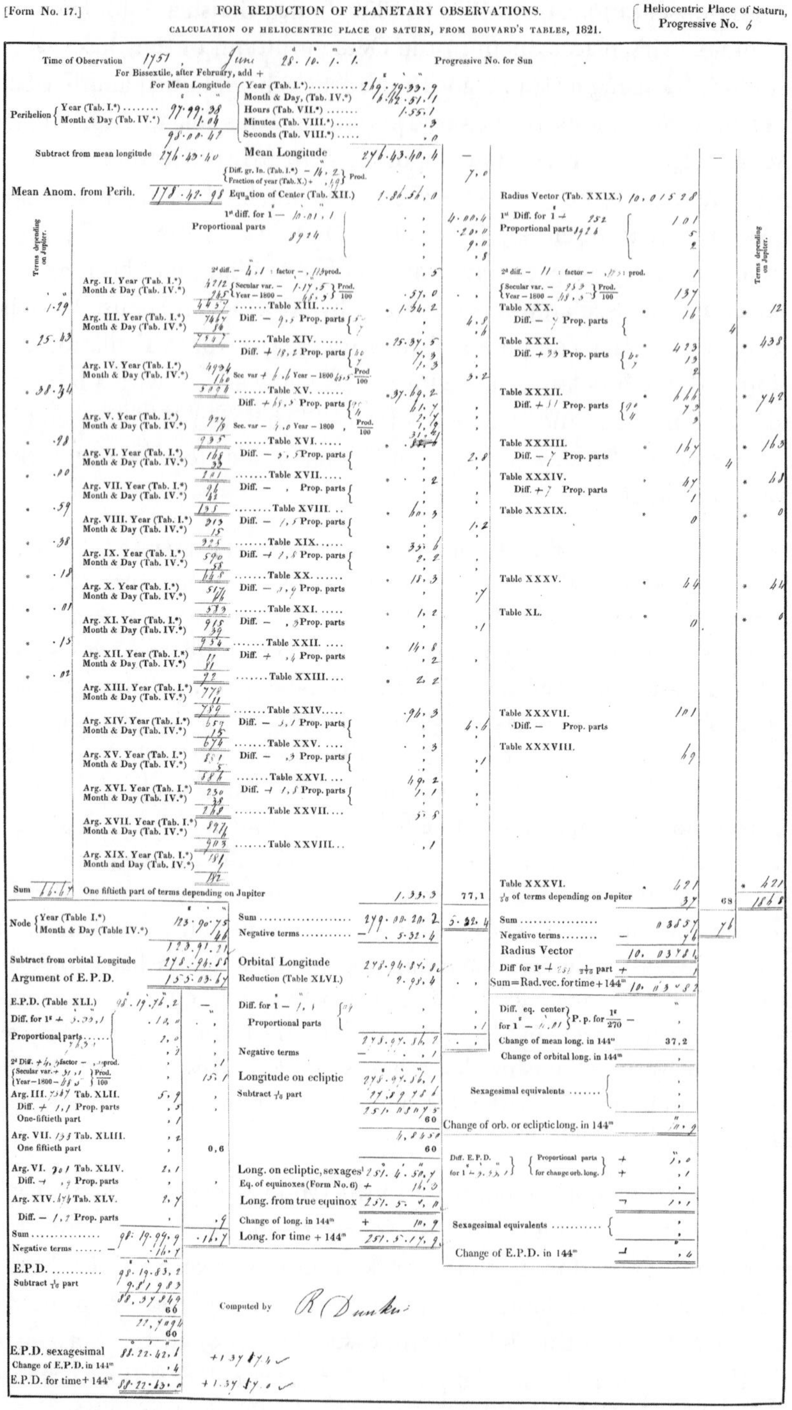

Figure 4.4.4 Skeleton form number 17 from the Royal Greenwich Observatory, for the calculation of the Heliocentric place of Saturn. It was completed by Robert Dunkin in the 1820s from observations made at Greenwich in 1751. Royal Greenwich Observatory Archives, Cambridge University Library RGO 127/53

assistants, and there was no scope for individual computers to adapt or improve on the method (Dunkin 1892, 78; Meadows 1975, 3). Airy used this technique as cheaply as he could by employing boys and young men aged between fourteen and twenty-three rather than men with more experience. The staff of the Observatory was thus divided between senior assistants with considerable astronomical and mathematical ability and junior assistants and computers with much more modest skills.

Simon Newcomb, Director of the US Nautical Almanac from 1877 to 1897 and a computer himself as a young man, summed up the situation of the Greenwich computers:

A third grade [of Greenwich Assistant] is that of computers: ingenious youth, quick at figures, ready to work for a compensation which an American laborer would despise, yet well enough schooled to make simple calculations. Under the new system they needed to understand only four rules of arithmetic: indeed, so far as possible Airy arranged his calculations in such as way that subtraction and division were rarely required. His boys had little more to do than add and multiply. Thus, so far as doing of the work was concerned, he introduced the same sort of improvement that our times have witnessed in great manufacturing establishments, where labor is so organised that unskilled men bring about results that formally demanded a high grade of technical ability. He introduced production of a large scale into astronomy. (Newcomb 1903, 288)

Airy's treatment of the Greenwich computers has had a bad press, with the work being classed by one ex-Greenwich assistant, Edward Walter Maunder, as 'remorseless sweating' (Maunder 1900, 117). However, the truth may be more complex. Dunkin (1999) recalls that when he started work as a computer in 1838 the hours of work were 8am to 8pm with an hour for dinner, but admits that these hours were at the request of the computers themselves, who were paid at an hourly rate of 6 pence to 10 pence an hour and wanted to maximize their income. During the winter, illumination in the computing room was inadequate. By Christmas 1838 Airy noticed that productivity in the second half of the day was poor and that some of the computers' health was suffering. Consequently he offered, and the computers accepted, a fixed rate of pay for a computing day of 8am to 4pm with no midday break.

A further problem was the temporary nature of their employment: computers had no tenure and could be dismissed at a month's notice. In Airy's eyes, however, he was not using boys as disposable labour but was providing training for future careers. Most came from surrounding schools and had to pass an entrance exam that covered arithmetic, use of logarithms, and elementary algebra before being taken on. Airy reports that his computers used their training at Greenwich to find employment as assistants in other observatories, as *Nautical almanac* computers (as in the case of Richard Dunkin), as clerks in the Civil Service, or in London banks (Airy 1872; see also Chapman 1998).

Airy recommended others for particular posts: for example, Charles Todd, the son of a Greenwich tradesman, was employed by Airy in the early 1840s as a computer of lunar reductions before becoming an assistant at the Cambridge University Observatory in 1848 (Meadows 1975, 11). He returned to Greenwich in 1854 to take charge of the telegraphic transmission of time signals before being recommended by Airy to the South Australian Government as a suitable candidate for the post of superintendent of telegraphs. He subsequently had a long and successful career (Symes 1976). There was also the possibility of advancement within the Observatory itself: Edwin Dunkin was promoted from computer to temporary assistant in the new magnetic observatory in 1840,[11] and five years later became an established assistant. Dunkin ended his career as Chief Assistant, the first person without a university education to be appointed to the post.

During the nineteenth century, innovative astronomical research was carried out not only at Greenwich but also at university observatories in Oxford, Cambridge, and Durham, and at public observatories at Liverpool, Edinburgh, and Armagh. Many of the assistants in the public observatories were poorly paid compared to those at Greenwich. There were also a number of private observatories, usually owned or run by men of considerable social and scientific standing (Chapman 1998). Assistants in these observatories held a social position in the household similar to that of a governess, that is, they were intelligent and well educated but were employees just a few steps up from servants (Chapman 1998, 145). Some started out as supernumerary computers at Greenwich: John Hartnup, for example, worked first as a computer at Greenwich, then as an assistant at Lord Wrottesley's observatory in Blackheath, and later as an Assistant Secretary to the Royal Astronomical Society, before being appointed Director of the Liverpool Observatory in 1843.

Tide computers

Another group of human computers were those involved in the preparation of tide tables.[12] Britain's commercial and imperial development depended on reliable tide tables for major world ports, and the sale of accurate tables was a commercially competitive business. Considerable amounts of direct and indirect state funding were channelled into tide research through the British Association for the Advancement of Science, and during the 1830s and 1840s such research

11. The magnetic observatory was designed to observe changes in the force and direction of the earth's magnetism, important to navigation because of advances in compass technology and the increasing use of iron rather than wood for ship building.

12. Tide computers were known as 'calculators' but here we will retain the use of the word 'computer'.

gathered momentum, led by John William Lubbock and William Whewell (Reidy 2007; Cartwright 1999).

By the 1830s the Hydrographic Office employed at least one full time tide computer (Daly 1967). Two employees of the Hydrographic Office, Joseph Foss Dessiou and Daniel Ross, also did tidal computing work at home, paid for by tide table publishers or the British Association. Dessiou collaborated with Lubbock and the Society for the Diffusion of Useful Knowledge to publish tide tables for the Port of London, while Ross, the tide computer for the Hydrographic Office, was able to analyse tidal data from around the world and to contribute to Whewell's research. Thus, unlike the computers at Greenwich and the Nautical Almanac Office, Dessiou and Ross not only carried out calculations but also contributed significantly to the development of tidal theory and computational methods, and their work was reported to the Royal Society (see Lubbock 1833; Whewell 1840; Reidy 2008).

Complementing the computers at the Hydrographic Office were commercial tide table computers like George Holden and Thomas Bywater in Liverpool, and Thomas Gamlen Bunt in Bristol (Reidy 2003; 2008). The importance of accurate tide tables meant that techniques for preparing them were valuable intellectual property, handed down in families from one generation to the next (Reidy 2003). Hence the relationship between Whewell and commercial tide computers was not a straightforward one: Whewell could pay for some of their time to work on theories, but the individuals themselves often depended on the commercial success of their tables to make a living, and were understandably reluctant to reveal their techniques (Reidy 2008).

Freelance table makers

The work of human computers can perhaps most readily be seen in the wide range of mathematical tables and ready reckoners that have been produced since earliest times (Campbell-Kelly *et al* 2003). These include tables of squares, roots, reciprocals, and trigonometric functions, as well as cross-sectional areas of commodities such as wood and steel, and interest calculations. From the mid-eighteenth century to the mid-twentieth century a great many general mathematical tables were published with a wide variation in the accuracy of the computations presented and quality of production. Authors of such tables usually worked freelance and benefited commercially from the sales, which increased with the quality of the tables and the reputation of the table maker.

One such table maker was Michael Taylor (c 1756–1789) who earned his living as a freelance computer for the Board of Longitude. He is best known for two sets of mathematical tables: *Sexagesimal tables* (1780) and *Tables of logarithms* (1792).

His *Sexagesimal tables* were of high quality and were used extensively within astronomical circles and also by surveyors (Glaisher 1873, 40). Glaisher remarked that Taylor's *Tables of logarithms* were a 'near approach to accuracy' with only six uncorrected errors (Glaisher 1872/3, 336), but they never came into general use, possibly because of their size: two thick volumes measuring 32 cm x 28 cm each.

Little is known about Taylor's early life. He came from Cumbria in northwest England, and Wallis (1993) indicates that he was at one time employed as a land surveyor. His life seems to have changed in the mid-1770s, when computing began to take up more and more of his time. It is not clear how he was introduced to Maskelyne, but from 1776 to 1789 he computed for the *Nautical almanac* (Maskelyne 1765–1811a). On average Taylor earned approximately £42 a year from the work, but the income was irregular: some years he received £80, in others nothing at all.

Taylor also undertook various table making and publishing tasks for the Board of Longitude. The first was to compile sexagesimal tables as a practical computing tool for 'Astronomers, Mathematicians, Navigators and Persons in Trade' whose everyday work involved calculations involving time or angle measurements. The worked examples given at the front of the tables (Taylor 1780, iii) make it clear that the majority of users were expected to be astronomers or surveyors. In essence all the worked examples used the tables to simplify division of sexagesimal numbers. The tables facilitated the conversion of English money, weights, and measures to and from sexagesimal numeration, to solve such problems as:

How much superfine cloth can be bought for 15 *l*. 14 *s*. 3 ¼ *d*. of which 43 yards 1 quarter and 2 nails costs 56 *l*. 19 *s*. 8 ½ *d*?

The answer is 11 yards, 3 quarters, 3 nails, and 0.8 inches but it is doubtful that people for whom this kind of calculation was part of daily life would have first converted to sexagesimal, and then reconverted back to imperial measures.

In the preface, Taylor acknowledged Maskelyne's help in designing the precepts (or algorithms) used in compiling the tables and the support of the Board of Longitude, which between June 1778 and March 1781 paid him £450 for his work, significantly more than his *Nautical almanac* income.

As soon as Taylor's *Sexagesimal tables* were available, Maskelyne issued them to other *Nautical almanac* computers, and set Taylor to work on other projects. He was well recognized as a reliable computer, and in 1786, for example, Sir Joseph Banks, then President of the Royal Society, called on him to check the accuracy of parts of the *Nautical almanac* for 1784 and 1785 (Howse 1989, 171).

In spring 1784 Taylor began work on a table of seven-figure logarithms of numbers up to 10,100 and of sines and tangents to every second of the quadrant, designed to be the most extensive and accurate tables then available, again financially supported by the Board of Longitude (Maskelyne 1788–91, f. 44). The origins of the project are not recorded, but it is likely that it developed from a suggestion by Maskelyne, who would have been aware of the lack of such tables in any readily available form. Maskelyne developed the precepts and oversaw the progress of the tables. He advertised the book widely, playing on the reputation of Taylor's *Sexagesimal tables*, and obtained £600 in advance subscriptions with another £490 promised. Every effort was made to ensure that the tables were as accurate as possible. Taylor prepared them to every second of arc by inter-polating from Vlacq's logarithmic sines and tangents (1631) and then abridg-ing to seven figures. With the help of an unknown assistant, he also took great care in correcting the proofs: the first proofs were compared to his manuscript tables; at second proof stage, the second, third, and fourth figures were again compared to the original manuscript, the entries every 36' were compared to Briggs' *Trigonometria britannia* (1633) and every 10' with Vlacq's *Triangulorum logarithmicus* (1631); third proofs were compared with Hutton's *Mathematical tables* (1785) (Maskelyne 1793).

When Taylor died on 24 December 1789, five pages of tables remained calcu-lated but unfinished. Maskelyne completed the work, and wrote the preface and introduction, explaining how the tables had been calculated and how they could be used. They were finally published in 1792 and Maskelyne saw to it that Taylor's son benefited from the sales (£930 after his father's death) (Howse 1989, 178).

Over his lifetime Taylor gained £4050 from his table making work[13] (Maskelyne 1788–91) and secured his son's future. Computing seems to have been his whole life, and family legend has it that he changed his name to Michael Napier Taylor in honour of John Napier, the inventor of logarithms (Dykes 2004). When he named his son he also commemorated Napier's collaborator, Henry Briggs, by calling the boy John Napier Henry Briggs Michael Taylor (Taylor 1792, subscription list).

Another freelance computer was Henry Andrews (1744–1820) who, like Taylor, was a computer for the *Nautical almanac*, but who also published com-mercial almanacs as well as being a schoolmaster, bookseller, and trader of sci-entific instruments. Andrews was born in Lincolnshire and showed an interest in mathematical astronomy from an early age (Anon 1820). He was largely self educated and worked in service, but in his late teenage years he made solar and lunar tables and accurately predicted the 1764 solar eclipse. In the summer of 1766 Andrews moved to Royston, then in Cambridgeshire, where he set up a school in which he taught reading, writing, arithmetic, bookkeeping, mensur-ation, cosmology, astronomy, and the use of globes and maps (Andrews 1767). He remained a schoolmaster until 1805 taking both day and boarding pupils, and also ran a shop selling books and instruments (Croarken 2003c).

---

13. Approximately £385,000 in spending power today.

Andrews is best remembered for contributions during more than half a century to the astronomical data and weather predictions for the annual *Old Moore's almanack*. He prepared the tables for *Old Moore's* from 1766 to 1819, for which he was paid £25 a year. It would seem likely that this drew Andrews to the attention of Maskelyne, who in 1768 was expanding the number of computers on the *Nautical almanac* (Maskelyne 1765–1811b). For the next 47 years Andrews computed an average of three *Nautical almanac* months a year earning approximately £36 a year from the Board of Longitude. The Board also paid him a pension of £250 a year[14] when he retired at the age of seventy-one. Maskelyne's account book reveals that Andrews rarely collected his *Nautical almanac* payments directly but asked Maskelyne to use the money to pay the London booksellers and instrument makers who supplied his shop. By 1782 he was acting as a mortgage lender in Royston and when he died he owned three properties, £600 in cash, and a significant stock of scientific instruments and books.

Taylor and Andrews are just two examples of freelance computers in the late eighteenth century and early nineteenth century. There were many more, as the large number of almanacs, ready reckoners, and mathematical tables published in this period demonstrates. Many freelance computers remain anonymous because they undertook the work on behalf of others, and it is usually impossible to determine whether they were entirely freelance or whether they were employees of the Observatory or Nautical Almanac Office undertaking extra work outside their formal employment. George Biddell Airy, for example, was engaged in numerous projects that required computers or mathematically trained assistants, and probably drew on computers from the Greenwich Observatory who were known to have the requisite skills. The mathematician Charles Babbage published a Life Table in 1826 and acknowledged in the preface that the tables had been computed by unnamed calculators.

Women computers

Almost all of the computers employed in Britain during the eighteenth and nineteenth centuries were men or boys. The only two women known to have made income-generating careers from computing during this period were Mary Edwards and her daughter Eliza (Croarken 2003d). Mary Edwards began computing for the *Nautical almanac* in June 1773 under the name of her husband John (Edwards 1811). John Edwards was a clergyman from Shropshire, earning £30 a year[15] plus a house from his clerical duties, who took in paying pupils to

---

14. Approximately £12,500 per annum in spending power today.
15. Approximately £2,700 in spending power today.

help fund his real passion, which was making telescopes. To supplement the family income further, he became one of Maskelyne's *Nautical almanac* computers, or at least that is what the official records show. On John's death in 1784, however, Mary Edwards officially took over her husband's computing work, which she had been doing all along, but now did on full time basis as her sole source of income.

It is possible that Maskelyne was already aware that Mary had been doing the computing, because he visited the family on more than one occasion. It mattered little to him that Mary was a woman; what mattered was that the job was done and done well. All went well until Maskelyne died in 1811 and Mary Edwards found that the new Astronomer Royal, John Pond, was no longer giving her the same amount of work. Eventually the Board of Longitude ruled that Pond should continue to allocate work to her. Mary Edwards died in 1815 and her daughter Eliza, who had been assisting her over the years, took on the role until 1831, when *Nautical almanac* computing was centralized in London (see above). There was no place in the new Nautical Almanac Office, or the Greenwich Observatory, for women employees, even if Eliza had wanted to make the move to London.

Civil Service rules made the employment of women very difficult until World War I (Zimmeck 1986) and the Nautical Almanac Office did not employ women computers until the 1920s. The same rules applied at the Greenwich Observatory. In 1890, however, William Christie, then Astronomer Royal, was interested in employing some of the highly educated women then beginning to graduate from English universities (Brück 1995). Christie got around the regulations by paying the women as supernumerary computers, as this was a part of the Observatory budget over which he had control, and since the computers were not on the permanent payroll they did not have to conform to Civil Service regulations.

Four women computers were appointed, not to do the work of the supernumerary computers for which they were being paid, but to work in the same way as second assistants undertaking both observing and computing. Most of them stayed only a few years. One of them, Alice Everett, obtained a post at the Astrophysical Observatory at Potsdam in Germany. Another, Annie Russell, married Maunder, the Greenwich assistant who complained that work at the Observatory was 'remorseless sweating' (see above). Russell's job at Greenwich was not routine computing but examining and measuring the daily sunspot photographs taken in the new solar photography department led by Maunder. Russell married Maunder in 1895 and became an independent amateur astronomer, publishing original research and working closely with her husband to found the British Astronomical Association. She continued to be a significant figure in British astronomy and was elected to the Royal Astronomical Society in 1916 (see Ogilvie 2000).

While there is limited evidence of other women in paid employment as computers, we do know that some women did unpaid computing, usually for their husbands. Elizabeth Sabine, for example, wife of the physicist Edward Sabine, was an accomplished woman who translated the works of Humboldt and undertook calculations to support her husband's scientific work. Similarly, Frances Kater assisted her husband Henry with his work (Somerville 2001, 106, 112). Elizabeth and Frances were not employed as computers but were associated with science by having husbands with high social status, who were part of the Royal Society tradition of scientific gentlemen. Were it not for Mary Somerville's mention of them in her memoires, we would be unaware of their work behind the scenes.

The invisibility of women working as computers, or more generally as scientists, is not at all uncommon. Caroline Herschel, for example, was elected an honorary member of the Royal Astronomical Society in 1835 in recognition of her work recording and computing for her more famous astronomer brother William Herschel rather than for her own work in discovering comets (Fara 2004, 146). Caroline and William worked in partnership, with Caroline recording and reducing William's observations, and it was Caroline who carried out any extra computing work required.

In other countries the situation was similar. One of the few eighteenth-century women other than Mary Edwards to be employed as a computer was Nicole-Reine Lepaute who worked for the French astronomer Joseph Lalande. Lepaute met Lalande through her husband, Jean André Lepaute, the Royal clock maker. Their first major collaboration was with Alexis-Claude Clairaut, calculating the predicted return of Halley's comet (1758). In 1759 Lalande was appointed director of the *Connaissance des temps*, an annual almanac similar to the British *Nautical almanac* but smaller in scale and without the lunar distances for longitude calculations which made up such a large part of the *Nautical almanac* computations. He immediately appointed Nicole-Reine as his paid assistant, a post she held for fifteen years. The division of labour between Lelande and Lepaute was simple and reflective of the social norms of the day: Lalande, a respected figure in European astronomical circles, prepared the computing plans and checked the results, while back in the office Lepaute calculated values for the tables that appeared annually in *Connaissance des temps* (Grier 2005).

In her discussion of women in science in the enlightenment, Patricia Fara (2004, 10) makes the point that women of high social standing with no need to earn a living could become involved in scientific pursuits by assisting a male relative, as an adjunct to their domestic responsibilities. During the eighteenth and nineteenth centuries women had much more freedom to become involved with science and mathematics within the home than in the workplace. For this reason the number of women working as paid computers is small. Those working as unpaid assistants to male relatives or in the privacy of their own home, however, are virtually invisible in the historical record.

Amongst those women who needed to earn a living, computing was not usually an option unless they came to the work through husbands or brothers. While many boys had a basic mathematical education, this was not the case for girls who, if from the upper classes, were usually taught domestic skills and/or social accomplishments. The number of women in white collar occupations in Britain began to rise only in the mid-nineteenth century, and then only slowly, until the 1890s saw the introduction of large-scale offices. Later, World War I led to a shortage of male clerks whose positions were filled by women (Anderson 1976; Zimmeck 1986).

Conclusion

Astronomy, navigation, tide tables, and table making were four areas where human computers were employed to great effect during the eighteenth and nineteenth centuries, yet they are seldom remembered. Most were of relatively low social standing without the advantages of a university degree, yet all were intelligent, hard working people who spent much of their life surrounded by mathematical tables and using their mathematical skills to perform repetitive but necessary tasks in order to earn a living.

The *Nautical almanac* computers of the eighteenth century were required to undertake entire computations, to understand the work they were doing and understand checking techniques such as differencing. They worked in their own homes and could control how much they worked in any one day, week, or month. Many combined the work with other occupations, so though it was repetitive and done largely for financial gain rather than intellectual stimulation, it could not be described as exploitative. Even those like Mary Edwards, for whom computing the *Nautical almanac* was a full time job and her only source of income, the timing of the work was within her own control.

Maskelyne's outsourcing of *Nautical almanac* computing work can be compared to other cottage industries such as lace making and glove making, which employed hundreds of women in the towns and villages on the English–Welsh border at that time. Computing too entailed a never ending pile of work, to be done amidst other domestic duties, but it was a much more solitary occupation. Glove and lace makers could congregate and talk as they worked, but computing was not a social task. The computer had to concentrate fully because not only their hands but their minds were employed on the task. In any case there were often no other computers in the same geographical location (Cornwall being a notable exception), so there was no sense of comradeship with a neighbour, or anyone with whom to share the everyday trials and tribulations of the job. It is therefore not surprising that Mary Edwards drew on the help and support of her daughter Eliza.

Although working to prescribed methods, the eighteenth-century computers who worked for the Board of Longitude were both more independent and more skilled than the later, office bound, computers who worked in the Nautical Almanac Office. Computers at the Nautical Almanac Office and Royal Observatory became deskilled through the nineteenth century as their work became more and more centralized and systematized. This was a reflection of the spread of factory methods, as in other industries, as opposed to cottage industry and small work-shops, and to more mechanistic ways of organizing work (Babbage 1835; Berg 1985). Bringing all the Nautical Almanac computers together in a single Office allowed jobs to be divided up so that they could be undertaken by workers with different skills and on different salary scales. Under Maskelyne all the computers had needed the same skills and had earned the same piecework rates.

The same pattern of differentiation is also seen in the Royal Observatory under Airy, who brought in lower-paid young men to do the ad hoc computing, allow-ing better skilled assistants to do more interesting or complex tasks. By contrast, the tide table computers working with Lubbock and Whewell were highly skilled and knowledgeable, and contributed significantly to tidal theory. In addition to institutional computers, there was considerable need for computers to carry out one-off pieces of computing work, or to prepare mathematical tables. Computers did not move in the same scientific and social circles as the mathematicians and astronomers of their day. They were the labourers behind the scenes, creating mathematical tools that others could use. As such they contributed greatly to scientific work in Britain while at the same time earning a living with which to support themselves and their families.

Bibliography

Airy, George Biddell to Francis Beaufort, 23 Nov 1846, Great Britain Hydrographic Office, Incoming Letters A618.

Airy, George Biddell to Hydrographic Office, 15 October 1870, Hydrographic Office Minute Book 17, ff. 421–422.

Airy, George Biddell, *Astronomer Royal's report* 1872, Royal Greenwich Observatory Archives at Cambridge University RGO 17/3.

Airy, George Biddell, *Numerical lunar theory*, London, 1886.

Airy, Wilfred (ed), *Autobiography of Sir George Biddell Airy*, Cambridge University Press, 1896.

Anderson, Gregory, *Victorian clerks*, Manchester University Press, 1976.

Andrews, Henry, advertisement in *Cambridge Chronicle and Journal* January 1767, quoted in Alfred Kingston, *A history of Royston*, London, 1906, 213.

Andrews, Henry, *The royal almanack and meteorological diary for the year of Our Lord 1778*, London, 1778.

Anon 1820, 'Mr Henry Andrews', *Gentleman's Magazine*, xc/1 (1820) 182 and xc/2 (1820) 639–40.

Anon 1838, 'The Newcastle lad and the Admiralty mecaenates', *The Times*, 9 November 1838, page 2.

Ashworth, William, 'The calculating eye: Baily, Herschel, Babbage and the business of astronomy', *British Journal for the History of Science*, 27 (1994), 409–441.

Babbage, Charles, *A comparative view of the various institutions for the assurance of lives*, London, 1826.

Babbage, Charles, *Economy of machinery and manufacture*, London, 1835.

Berg, Maxine, *The age of manufactures: industry, innovation and work in Britain 1700–1820*, Fontana Press, 1985.

Board of Longitude, Minutes 13 January 1770, Cambridge University Library Manuscripts RGO 14/5, f. 188.

Briggs, Henry, *Trigonometrica britannica*, Gouda, 1633.

Brück, Mary, 'Lady computers at Greenwich in the early 1980s', *Quarterly Journal of the Royal Astronomical Society*, 36 (1995), 83–95.

Campbell-Kelly, Martin, Croarken, Mary, Flood, Raymond, and Robson, Eleanor (eds), *The history of mathematical tables: from Sumer to spreadsheets*, Oxford University Press, 2003.

Cartwright, David Edgar, *Tides: a scientific history*, Cambridge University Press, 1999.

Chapman, Allan, 'George Biddell Airy, F.R.S. (1801–1892): a centenary commemoration', *Notes and Records of the Royal Society of London*, 46 (1992), 103–110.

Chapman, Allan, *The Victorian amateur astronomer: independent astronomical research in Britain 1820–1920*, John Wiley and Sons, 1998.

Croarken, Mary, *Early scientific computing in Britain*, Clarendon Press, 1990.

Croarken, Mary, 'Computing the Nautical Almanac in eighteenth-century England', *IEEE Annals of the History of Computing*, 25 (2003a), 48–61.

Croarken, Mary, 'Astronomical labourers: Maskelyne's assistants at the Royal Observatory, Greenwich 1765–1811', *Notes and Records of the Royal Society of London*, 57 (2003b), 285–298.

Croarken, Mary, 'Henry Andrews (1744–1820): an astronomical calculator from Royston', *Herts Past and Present*, 3 (2003c), 21–27.

Croarken, Mary, 'Mary Edwards: computing for a living in eighteenth-century England', *IEEE Annals of the History of Computing*, 25 (2003d), 4, 9–15.

Daly, Archibald, *The Admiralty hydrographic service 1795–1919*, Her Majesty's Stationary Office, 1967.

Dunkin, Edwin, 'A day at the Royal Observatory', *Leisure Hour*, 1862, 22–26, 39–43.

Dunkin, Edwin, 'Sir George Biddell Airy', *The Observatory: a monthly review of Astronomy*, xv/185 (1892), 74–94.

Dunkin, Edwin, Hingley, Peter and Dabiel, Tamsin (eds), *A far off vision: a Cornishman at the Greenwich Observatory: auto-biographical notes by Edwin Dunkin*, Royal Institution of Cornwall, 1999.

Dykes, Sylvia, email to Mary Croarken 27 July 2004.

Edwards, Mary, Petition to the Board of Longitude 5 December 1811, Cambridge University Library Manuscripts RGO 14/11 part 1, f. 140.

Evans, Thomas, quoted in John Evans, *Juvenile Tourist*, London, 1810, 333–335.

Fara, Patricia, *Pandora's breeches: women, science and power in the Enlightenment*, Pimlico, 2004.

Glaisher, James Whitbread Lee, 'On the progress to accuracy of logarithmic tables', *Monthly Notices of the Royal Astronomical Society*, 33 (1872/3), 330–345.

Glaisher, James Whitbread Lee, 'Report of the Committee…on Mathematical Tables', *Report of the British Astronomical Association for the Advancement of Science*, (1873), 1–175.

Grattan-Guinness, Ivor, 'The computation factory: de Prony's project for making tables in the 1790s', in Campbell-Kelly, Martin, Croarken, Mary, Flood, Raymond, and Robson, Eleanor (eds), *The history of mathematical tables: from Sumer to spreadsheets*, Oxford University Press, 2003, 104–121.

Grier, David Alan, *When computers were human*, Princeton University Press, 2005.

Hansard, T C, *Parliamentary Debates*, XXXVII, 6 March 1818, 877–878.

Hind, John to Admiralty 31 December 1878, Cambridge University Library Manuscripts RGO 16/3 Packet C.

Hitchins, Malachy to Joshua Moore 17 November 1792, Library of Congress Manuscripts Division, Papers of Joshua Moore, MMC–2121.

Howse, Derek, *Nevil Maskelyne: the seaman's astronomer*, Cambridge University Press, 1989.

Hutton, Charles, *Mathematical Tables…containing…logarithms…with tables useful in mathematical calculations. To which is prefixed a large…history of the…writings relating to those subjects, etc*, London, 1785.

Kinnebrooke, David Jr to David Kinnebrooke Sr, 24 December 1795, Cambridge University Library Manuscripts RGO 207/1 f. 51.

Kinnebrooke, David Sr to William Kinnebrooke, 28 December 1802, Cambridge University Library Manuscripts RGO 207/1 f. 71.

Lindenau, Bernard August von, *Investigatio nova orbitae a Mercurio circa Solem descriptae. Accedunt tabulae planetae ex elementis recens reporitis, et theoria gravitatis…de Laplace, constructae*, Gotha, 1813.

Lubbock, John William, 'Note on tides', *Philosophical Transactions of the Royal Society of London*, 123 (1833), 19–22.

Maskelyne, Nevil, Accounts with computers and comparers of the Nautical Almanac, (1765–1811a), Cambridge University Library Manuscripts RGO 4/325. Summarized in RGO 16/55 by H P Richards.

Maskelyne, Nevil, Diary of Nautical Almanac Work, (1765–1811b), Cambridge University Library Manuscripts RGO 4/324.

Maskelyne, Nevil, Letter to Joshua Moore, 1 January 1790, Library of Congress Manuscripts MMC–2121, f. 74.

Maskelyne, Nevil, Memorandum Book 3 1788–1791, National Maritime Museum Library Microfilm MRF/184, f. 44, 1790.

Maskelyne, Nevil, 'Preface' dated 1793 in Taylor Michael, *Tables of logarithms of all numbers from 1 to 10100 and of the sines and tangents to every second of arc*, London, 1792.

Maskelyne, Nevil, 'Rules for computing the immersion of a star behind, or the emersion of a star from the moon's limb', n.d. Cambridge University Library Manuscripts RGO 4/216.

Maunder, Edward Walter, *The Royal Observatory Greenwich*, Religious Tract Society, 1900.

Meadows, Arthur Jack, *Greenwich Observatory volume 2: recent history (1836–1975)*, Taylor and Francis, 1975.

Miller, David Philip, 'Between hostile camps: Sir Humphrey Davy's presidency of the Royal Society of London, 1820–1827', *British Journal for the History of Science*, 16 (1983), 1–47.

Mollon, George and Perkins, Adam, 'Errors of judgement at Greenwich in 1796, *Nature*, 380 (1996) 101–102.

Newcomb, Simon, *Reminiscences of an astronomer*, Harper, 1903.

Ogilvie, Marilyn Baily, 'Obligatory amateurs: Annie Maunder (1868–1947) and British women astronomers at the dawn of professional astronomy', *British Journal for the History of Science*, 33 (2000), 67–84.

Reidy, Michael S, 'Masters of tidology: the cultivation of the physical sciences in early Victorian Liverpool', *Transactions of the Historic Society of Lancashire and Cheshire*, 152 (2003), 45–71.

Reidy, Michael S, 'The Tide Predictors', www.airmynyorks.co.uk, nd, accessed 31 March 2007.

Reidy, Michael S, *Tides of history: ocean science and Her Majesty's navy*, University of Chicago Press, 2008.

Schaffer, Simon, 'Astronomers mark time: discipline and the personal equation', *Science in Context*, 2 (1988), 115–145.

Select Committee on Official Salaries, 'Report' 25 July 1850, *Parliamentary Papers, House of Commons, 1837–1850*.

Simpson, J A and Weiner, E S C (eds), *Oxford English Dictionary*, Clarendon Press, 2nd ed, 1989.

Somerville, Mary, *Queen of science: personal recollections of Mary Somerville*, Dorothy McMillan (ed), Canongate, 2001.

South, James 'Report of the committee of the Astronomical Society of London relative to the improvement of the Nautical Almanac', *Memoirs of the Astronomical Society*, 4 (1831), 447–470.

Symes, G W, 'Todd, Sir Charles (1826–1910)', *Australian dictionary of biography*, Melbourne University Press, 1976, 280–283.

Taylor, Michael, *A sexagesimal table, exhibiting, at sight, the result of any proportion, where the terms do not exceed sixty minutes*, Commissioners of Longitude, 1780.

Taylor, Michael, *Tables of logarithms of all numbers from 1 to 10100 and of the sines and tangents to every second of arc*, London, 1792.

Vlacq, Adriaan, *Canon triangulorum logarithmicus; das ist, Kunstliche logaritmische Tafeln der Sinuum, Tangentium und Secantium*, Augsburg, 1631.

Wallis, Ruth, and Wallis, Peter, *Index of British mathematicians part 3 1701–1800*, Newcastle Upon Tyne: Project for Historical Biobibliography, 1993.

Whewell, William, 'Researches on tides – twelfth series', *Philosophical Transactions of the Royal Society*, 130 (1840), 255–272.

Woolhouse, Wesley Stoker Barker, 'Letter from Woolhouse to Lord Commissioners of the Admiralty' 10 April 1837, *The Times*, 9 November 1838.

Zimmeck, M, 'Jobs for the Girls: the expansion of clerical work for women 1850–1914', in Angela V John (ed), *Unequal opportunities*, Basil Blackwell, 1986.

5. Practices

Mixing, building, and feeding: mathematics and technology in ancient Egypt

Corinna Rossi

The number of surviving mathematical sources from ancient Egypt is relatively small, but they all give a consistent picture of mathematics as deeply intertwined with a variety of practical activities. Mathematics, after all, is a constant presence in the daily life of any population, even if in different forms or contexts (Selin 2000; Cuomo 2001; Asper, Chapter 2.1 in this volume). Beside the presence of geometrical patterns in decorative or functional objects (e.g., Robson 2000; Wendrich 2000; Whitley 2001, 77–133; Brezine, Chapter 5.4 in this volume), counting and measuring are naturally embedded in a large number of technological activities.[1]

Our knowledge of ancient Egyptian mathematics relies on a small number of sources, including five papyri, a leather roll, and a pair of wooden tablets, all dating to the Middle Kingdom (2055–1650 BC). The most important is the Rhind papyrus, which contains the most complete and varied list of mathematical problems, followed by the Moscow papyrus, organized in a similar way but narrower in scope (Peet 1923; Chace, Bull, and Manning 1929; Struve 1930). They appear to be school texts, specially conceived to teach mathematics to young

1. I thank Dr Serafina Cuomo for discussing several key issues of this chapter. Thanks also to Professor Mahmoud Ezzamel and to the editors of this volume for their comments and suggestions.

scribes (Ritter 2000, 120), and their contents may be divided into two categories: tables and problems (Imhausen 2002). Table texts are ready-made collections of mathematical data to be consulted while performing calculations (such as the table listing all the results of the number 2 divided by the odd numbers from 3 to 101). Problem texts, instead, show how to solve sample tasks that scribes might be assigned: measuring the area of fields, calculating the volume of granaries, dividing loaves of bread among men, or calculating the slope of a pyramid. Other types of texts, although not strictly mathematical, nevertheless provide important evidence of how mathematics was used: the Reisner papyri, for instance, contain building records and administrative accounts relating to the construction of an unidentified building (Simpson 1963; Rossi and Imhausen, forthcoming).

Present knowledge of ancient Egyptian technology, by contrast, derives from a variety of textual, iconographic, and archaeological sources, including ancient representations on tomb walls, objects, and the remains of manufacturing sites unearthed by archaeologists. To this must be added the invaluable support of modern technology, which provides information on the physical and chemical composition of ancient objects and products (Nicholson and Shaw 2000).

At first sight, the scarcity of ancient Egyptian mathematical sources appears to prevent a detailed reconstruction of the way mathematics was involved in technological processes. However, we can compensate for this lack of direct information in at least two ways: on the one hand by turning to less obvious sources, such as ancient records of building activities or medical recipes, with the aim of extracting from them indirect but valuable information; on the other hand by reconsidering the well-known mathematical texts in search of further clues. The study of ancient Egyptian mathematics has in fact recently taken a new methodological turn that aims to reassess the nature and structure of the ancient documents and to cast a new light on their contents (Imhausen, Chapter 9.1 in this volume). Even if the number of actual sources is limited, the information they can provide has not been exhausted.

Counting, measuring, calculating

A potentially productive area of study concerns the various units of measurements and the different ways they were used. It may be useful at this point to distinguish between three different mathematical actions: counting, measuring, and calculating.

Ancient Egyptians counted in base 10, which superficially makes their system similar to our own. Measuring, however, relied on a number of units of measurement with different characteristics: some closely resemble their modern counterparts, whilst others differ in more or less substantial ways. It is therefore necessary to pay close attention to the way metrological units were used: only by fully appreciating their nature can we reconstruct the ancient mathematical (and/or technological) processes in which they were employed. A clear example of these difficulties is represented by the main linear unit of measurement, the cubit, corresponding to the forearm. In particular, one 'royal' cubit corresponded to 7 palms (c 52.3 cm), whereas 1 'small' cubit was equal to 6 palms (c 45 cm); each palm (c 7.4 cm) was divided, in turn, into 4 fingers (c 1.8 cm). Volumes were generally expressed in cubic cubits; however, as we shall see below in the section on stone, the subunits of this cube with length of 1 cubit appear to have been significantly different from what one might expect: the 'volume' palm, instead of being a cube with a length of 1 palm, was a 'slice' of cubic cubit with a width of 1 palm.

Another important influence on the ancient Egyptian mathematical system was the fundamental role played by commensurability, which derived from the ancient Egyptian practice of performing multiplications and divisions by means of progressive doubling or halving, and/or working with powers of ten. This explains not only the general preference for even numbers, but in particular the specific focus on the numbers belonging to the progression 2, 4, 8, 16, 32, 64 and their reciprocals 1/2, 1/4, 1/8, 1/16, 1/32, 1/64, and on those derived by multiplying these by 10 or powers of 10. A clear example is the subdivision of the capacity unit *heqat*, corresponding to c 4.8 litres: it could either be divided into 10 parts, called *henu* (thus corresponding to about half a litre) or progressively halved into 1/2, 1/4, 1/8, 1/16, 1/32, and 1/64 of a *heqat*. The smallest subdivision was the *ro*, corresponding to 1/320 of a *heqat*; the convenient combination of progressive halving and powers of 10 determined the commensurability of these subdivisions, with 5 *ro* corresponding to 1/64 of *heqat*. A similar example referring to the calculations of the area of fields will be discussed below in the section on food production.

Finally, the way calculations can be performed may be an important source of information, because the mechanisms that regulated them are a faithful mirror of the mental processes that lay behind them. Reconstructing the algorithms involved can provide important leads for further research: first, within ancient Egyptian mathematics, by better understanding its internal mechanisms; and second, in ancient mathematics more generally, by studying the chronological, geographical, and cross-cultural transmission of such mechanisms (Imhausen 2002). Within Egyptology, the close study of calculations may help to clarify the role of mathematics in various areas of knowledge.

Technology is one field that may benefit from such an approach. Here we examine the technologies of extracting and processing stones and metals and those involved in food production. The characteristics of the relevant units of measurement and the ways in which they were employed (or *not* employed) offer a means of exploring and reconstructing ancient technological processes, and also provide hints about the attitudes that lay behind both mathematics and technology.

Understanding metals with the aid of medicines

The mathematical sources relating to the technologies of metal extraction and processing are extremely scanty. There is only one surviving mathematical problem on the general subject of metal objects, namely problem 62 of the Rhind papyrus (henceforth pRhind 62). It is centred on the relationship between the *shaty* 'value', and weight of three different metals. The weight unit *deben* corresponded to 13–14 g in the Old and Middle Kingdoms (2686–1650 BC) (Imhausen 2003, 156; 270–271; Clagett 1999, 169–170).

The task, as stated by the scribe, can be translated as follows: 'A bag containing [equal weights] of gold, silver and lead is bought for 84 *shaty*. What is the amount of each precious metal?' For 1 *deben* of each metal the values in *shaty* (12, 6, and 3 respectively) are given. The procedure starts by adding together the values in *shaty* of 1 *deben* of each metal (12 + 6 + 3 = 21), then moves on to finding how many times one should multiply the result to reach the value in *shaty* of the bag (in other words, it calculates 84:21). The result is 4 *deben* of each metal. At first sight, this little problem provides only rather banal and artificial information: for instance, that the value of gold was twice the value of silver, and four times the value of lead; and that a fixed price for each metal allowed the scribe to shift easily from units of value to units of weight. However, as we shall see below, it may contain indirect information on how to deal with more complex issues. Dealing with metals in practical contexts, in fact, is likely to have entailed more complicated calculations.

The identification of several mining sites in the Egyptian desert has enabled the reconstruction of the political and strategic background of mining expeditions commissioned by several pharaohs (Shaw 1998) and, to a certain extent, of the various extraction methods used by them (Klemm and Klemm 1994; Shaw 1996). Abundant information on how metals were processed and worked comes from ancient representations carved on tomb walls (Wainwright 1944; Chappaz 1983), from the study of the archaeological remains of ancient mining and manufacturing areas (Pusch 1990), and from analysis of the objects produced (Scheel 1989). What is currently known of the structure and composition of ancient Egyptian metals, by contrast, depends almost entirely on modern technology: binocular, metallurgical, and scanning electron microscopes are especially useful for analysing surface and technological details, whilst a variety of spectrographic analyses, including atomic absorption and plasma spectroscopy, are generally used to determine chemical composition (Ogden 2000, 171–172).

These analyses reveal that the Egyptians made substantial use of alloys. In fact, because pure metals are extremely difficult to find and to obtain, ancient metallic objects invariably contain a combination of elements. Small percentages of secondary components are generally regarded as natural impurities, but above a certain threshold the combination of elements is considered to be artificial: copper was voluntarily mixed with arsenic, lead, or tin (the latter combination is generally called bronze), whereas gold might be combined with silver, copper, or occasionally other metals. The reasons may have been practical, as in the case of copper, the strength of which was greatly increased by the addition of other metals, or aesthetic, as in the addition of a significant amount of copper to gold to give a distinctive reddish colour to the final product (Ogden 2000, 164).

The proportions of the various components of alloys may vary. In the case of gold alloys, the quantity of gold ranged from 50% to over 85%, the latter purity rarely achieved before the Late Period (664–332 BC). Whilst it is clear that gold was often deliberately combined with large quantities of silver, it is difficult to distinguish between artificial and natural combinations of gold with small quantities of other metals (Weill 1951; Ogden 1993, 39). In the case of copper, it is easier to provide figures: archaeological scientists consider a percentage up to 2% of iron to be natural, whereas higher figures (up to 20%) suggest artificial additions; the threshold for the natural presence of arsenic is 1%, but it might be artificially added up to 7%. The line dividing the natural from the artificial presence of tin is also 1%, although it might reach 10% in an alloy. Finally, a natural and beneficial presence of up to 2% of lead in copper could be artificially increased to 25% by volume (Ogden 2000, 152–155).

No extant ancient document contains direct information on how alloys were created in practice, in particular on how the various parts were measured and mixed. However, indirect information may be derived by comparing the result of modern analyses with information on how the Egyptians mixed other substances, for instance for pharmacopoeia (Nunn 1996, 140–143). The various quantities to be combined in a single remedy were often expressed in units of capacity, mainly using the small *henu*, corresponding to 1/10 of a *heqat* (c 450 ml). Strangely enough, the use of the even smaller and rather convenient *ro*, corresponding to 1/320 of a *heqat* (c 14 ml) is not unequivocally attested. The commonest method appears to have been combining parts of substances, often expressed by means of fractions. Two examples may be a 'prescription for renewing the skin: honey 1, red natron 1, northern salt 1, ground into a compound and smeared on' (pEdwin *verso* 4, 3–6; Allen 2005, 113) and another involving 'fresh bread cooked in oil and honey; absinthe 1/32; resin of the umbrella pine of Byblos 1/16; valerian 1/8; add it together, cook as one thing. To be drunk for four days' (pEbers, 190; Nunn 1996, 142).

The creation of metal alloys might have been achieved in a similar way by combining parts by volume rather than weights, that is, by establishing ratios between the various components. For instance, iron may have been added to copper in proportions varying from 1:30 (corresponding more or less to 3%) to

1:5 (20%); arsenic may have contributed a proportion up to 1:14 or 1:15 (c 7%), tin may have reached 1:10 (10%), and finally lead may have varied from 1:30 to 1:4 (c 3%–25%).

Combining parts of elements instead of weighing them would have been an easy and effective way to create any desired quantity of alloy. Such a method would not have directly depended on any unit of measurement but, as pRhind 62 reminds us, if the weight was known then the value of the metal would have been easier to establish. In fact, if the value of an alloy corresponded to the sum of the values of its components, the problem stated in pRhind 62 (centred on a bag containing equal quantities of metals) might have also applied to objects made of alloys. This would be in line with the general nature of the Rhind papyrus (and ancient Egyptian mathematical sources in general): that is, it contains sample problems that would have covered the majority of tasks that might have been assigned to a scribe. Even if the data differed, the scribe could search for the most similar problem and use it to carry out his particular task (Clagett 1999, 94). Therefore, the procedure contained in pRhind 62 might have been applied to the general issue of mixtures of metals, either in separate pieces or melted together.

Stone

For stone, in contrast to metals, a range of ancient sources collectively provides substantial information on the way mathematics was involved in its use. It is useful to distinguish between two phases: extraction, quarrying, and processing procedures on the one hand, and final destination on the other—the shape and function eventually adopted by a block in a building, for instance.

In general, only linear and volumetric measurements are found in relation to stone: masses larger than small vessels, statuettes or tools, of course, would have been impossible to weigh. The dimensions of stone blocks were generally expressed in royal cubits and their subunits, palms and fingers. As already mentioned above, papyrus Reisner I, suggests that the cubic cubit was subdivided into 'volume palms' corresponding to 'slices' of cubic cubits 1 palm wide, rather than small cubes with a side-length of 1 palm (Rossi and Imhausen, forthcoming). Such a subdivision would have been useful both in theory for performing calculations and in practice for quarrying trenches or rock-cut chambers. In calculations, it would avoid converting all cubits into palms or fingers in order to shift from large to small units, or vice versa. In practice, volumes with a square base of 1 by 1 cubit and a thickness corresponding to 1 (or more) palm(s) would correspond to parts of the cubic cubits in a straightforward and evident way. This practical approach to subunits is not an isolated case: as we shall see below in the section on food production, it strongly resembles the way in which areas were also subdivided.

Stone quarries provide important information, not only on ancient technology but also on wider issues such as the intentions, strategies, and organization of the quarrying process (Shaw 1998). The marks left by the ancient quarrymen, for instance, help to reconstruct the geometry involved in the process of stone removal, which depended on the hardness of the rock. Soft stones (such as limestone and sandstone) were quarried by digging narrow trenches, generally 3 palms wide, between the outlines of the blocks. As the remains of the quarry north of the Fourth Dynasty pyramid of Khafra at Giza show (2558–2532 BC), larger blocks required larger and proportionally deeper trenches to accommodate a standing or kneeling stonecutter (Arnold 1991, 31). The process for quarrying hard stones (granite, quartzite, and gneiss) was similar, but required more time, energy, and patience.

The unfinished obelisk abandoned in the granite quarry of Aswan during the Eighteenth Dynasty (1550–1295 BC) (Fig. 5.1.1), is the main piece of evidence on the subject. A 10-palm wide trench (c 75 cm) was cut all around the outline of the obelisk; then vertical lines marked in red on the walls of the trench divided the latter into 8-palm wide spaces (c 60 cm), each allocated to a stonecutter. The men squatted in their positions, their tools next to them, and rhythmically pounded the surface of the stone with heavy dolerite balls. The division of the area under their feet into four quadrants shows that they regularly turned around and changed position according to a precise scheme (Arnold 1991, 37). The progress of the excavation was managed by a foreman, who periodically lowered a cubit rod into every working space of the trench, and marked on its wall the position

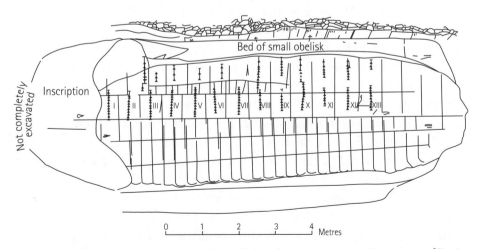

Figure 5.1.1 Control marks on the inner face of the Aswan granite quarry (Clarke and Engelbach 1930, fig 29, reprinted by permission of Oxford University Press)

of the top of the rod with a little red triangle. The progressive lowering of the bottom of the trench was therefore indicated by a sequence of red triangles pointing upwards (Engelbach 1923). Measuring the workmen's progress was an important element of the entire process, which would have allowed the foreman both to keep under control the efficiency of every single workman and to report to higher authorities on the status of the work.

To sum up, quarries tell us that the work was organized on the basis of a simple geometry, easy to measure, apply, and keep under control. This observation becomes even more interesting if we extend the discussion to rock-cut tombs and temples. Quarrying stone always involves the same problems and is carried out with similar (if not identical) techniques and tools, thus implying an intrinsic similarity between stone quarries and, for instance, rock-cut tombs (Owen and Kemp 1994). Moreover, a study of the internal dimensions of rock-cut tombs shows that the starting point of the ancient builders was generally a plan composed of simple measurements, expressed in whole cubits (Rossi 2001a; 2001b). Thus, it appears that every quarrying activity was carried out on the basis of simple geometry, which helped to organize the work and the working space allocated to the stonecutters. In general, workers were expected to remove a known quantity of material per day, which allowed foremen to calculate in advance how many man-days of work were required to complete a certain task (e.g., pReisner I, Section G; Simpson 1963).

Even leaving aside rock-cut temples and tombs and focusing on standing constructions, the dividing line between quarrying and building may have been very thin. For instance, some stone blocks were partially processed at the quarry, to an extent that depended on the hardness of the stone. Dressed blocks of soft stone might be damaged during the transport to the building site, whereas blocks of hard stone would suffer less (Arnold 1991, 52; Aston, Harrell, and Shaw 2000, 15). Equally, the more hard stone that could be removed at the quarry, the lighter the block would be for transport. The partial processing of stones at the quarry implies that the stonecutters were informed of the final dimensions (and probably the function) of the blocks they were quarrying and dressing (Kemp, Rossi, and Harrell, in preparation). Assuming that a final polishing would have been undertaken at the building site, the ancient builders must have known how much stone was to be removed in this last stage, so the dimensions requested of the quarrymen would have included an appropriate margin.

Finally, a direct transition from quarrying to building is represented by the *talatats*, the blocks used to build the stone temples of Amarna during the reign of Akhenaten (1352–1336 BC). This king abandoned the traditional polytheistic religion in favour of a monotheistic cult centred on the sun-disk, and moved the capital from the strong religious centre of Thebes to the newly-founded capital

Akhetaten (modern Amarna) in Middle Egypt. There, in just a few years, he built new temples with an innovative design and palaces and tombs for the royal family, around which a large city grew. The main building material was mud brick, whilst stone was reserved for the most important parts of the main buildings. The rectangular stone blocks used in these cases, called *talatats*, were 1 cubit long, and were already quarried to this shape and dimensions in the mountains around the ancient city. Beside official quarries, such as the one bearing the name of Queen Tiy, mother of Akhenaten, recent studies suggest that commoners might have freely quarried stone blocks from the rock surface of nearby mountains in order to contribute to the construction of local temples (Kemp, Rossi, and Harrell, in preparation). Their standardized dimensions and their relatively small size made the *talatats* easy to manage, both by professional stonecutters and builders and by private citizens, contributing significantly to the speed and efficiency of the building process.

Building with stone relied on the usual set of linear units of measurement, which were often used to calculate volumes as well. Again, no weight is ever recorded, and no traces of structural calculations exist: the ancient builders relied on combining large blocks of stone into more or less stable compositions (Arnold 1991, 109–115), and probably adapted their plans to the available material (Wysocki 1984).

The calculation of volumes to be removed or built played an important role. In the case of excavations, it helped both to calculate in advance how many men would be needed and how long the task would take, and to record how much work had already been carried out. This applied both to rock-cut and standing monuments. Ostracon Strasbourg H.112, for instance, contains a list of work done in the tomb of Khaemweset, son of Ramses III (c 1170 BC): besides the linear dimensions of all chambers, the scribe also recorded the volume of some of them and the total volume excavated, up to the twentieth regnal year of the Pharaoh (Koenig 1997, 9, pls. 44–47; Rossi 2004, 144). Similarly, the Reisner papyri record construction activities on unidentified buildings of stone and mud-brick: the text lists the length, breadth, height, and volume of each room, plus volumes of sand and rubble to be removed (Simpson 1963, 124–126; 1969, 27; Clagett 1999, 261–279).

In the case of standing buildings, calculating in advance the volume to be constructed would have been particularly easy in the case of simple geometrical shapes, such as pyramids. Problem 14 of the Moscow papyrus calculates the volume of a truncated square pyramid (Clagett 1999, 221; Imhausen 2003, 88–89, 330–331). The method is exactly the same as the one we use today, and entails the calculation of the volume of a whole pyramid. Therefore, even if no mathematical problem involving the volume of a whole pyramid has survived, we know that ancient Egyptians could calculate it and also, indirectly, how they did so (Gillings 1972, 189). It may therefore be inferred that the ancient builders were able to calculate in advance the volume of the royal pyramids that they were about to build. Such an estimate might have helped them to establish whether or not a quarry could provide the necessary amount of stone, and how much (for instance, for the outer casing) needed to be imported from elsewhere.

Problems 56–60 of the Rhind papyrus clearly inform us on how the slope of a pyramid (or any sloping surface) might be calculated using the *seked*, that is, the horizontal displacement of a sloping surface for a vertical height of 1 cubit (Clagett 1999, 166–168; Imhausen 2003, 166–168; see Fig. 5.1.2). In other words, ancient Egyptian surveyors would measure or calculate how much the sloping surface had 'moved' from the vertical line at the height of 1 cubit. They basically constructed a right-angled triangle in which the hypotenuse corresponded to the sloping surface (the length of which was irrelevant), the height to 1 cubit, and the horizontal top to the seked. The *seked* of real pyramids was expressed in palms and fingers, and generally varied between 6 and 3 palms (giving an elevation of c 49° 30'–c 67°), with rare exceptions outside these values (Rossi 2004, fig 99). Although no direct evidence has survived, it is possible that the workmen involved in pyramid building might have actually constructed wooden triangles in the shape of the chosen *seked* in order to check their work (Hinkel 1982, figs 19 and 20; Lehner 1997, 220; Rossi 2004, 188–199).

From a geometrical point of view, pyramids were relatively simple to build. Apart from the case of the Bent Pyramid (the earliest 'true' pyramid, dating around 2600 BC, where structural problems forced the builders to drastically reduce the slope twice during construction; Maragioglio and Rinaldi, 1964, 58–62, Stadelmann 1985, 87–92, Rossi 2004, 221–225), the final result must have been identical to the expected design. More complicated buildings, however, might undergo a certain amount of modification during construction. A careful study of ancient sources and finished monuments shows that the ancient builders generally started from simple dimensions, easy to handle and to combine (Arnold and Arnold 1979). In the case of rock-cut tombs, for instance, the corridors were ideally meant to be 30 cubits long, and the length and breadth of

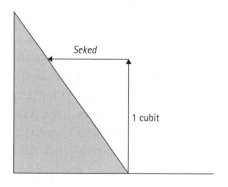

Seked

1 cubit

Figure 5.1.2 The *seked* of pyramids and other sloping surfaces

the various chambers were supposed to correspond to round and simple figures, such as 18 × 16 or 16 × 15 cubits (e.g., Engelbach 1927).

In practice, however, the final dimensions could be different, depending on time constraints or difficulties encountered during the work. This discrepancy between ideal and outcome was not a problem and, in fact, was often completely ignored: papyrus Turin 1885, representing the tomb of Ramses IV (1153–1147 BC), suggests that the completion of the work in a royal tomb was officially 'certified' by an extremely detailed survey of the interior, where the dimensions of each architectural element were recorded with a precision expressed in cubits, palms, and fingers (Carter and Gardiner 1917). The final survey was meant to confirm that the tomb had been completed and that every detail had been taken care of; at this stage, the dimensions used years earlier to start the excavation appear to have been long forgotten (Rossi 2004, 142–147).

One exception to this preliminary (and theoretical) adoption of simple measures can be found in the so-called Building Text, engraved on the walls of the Ptolemaic temples of Edfu (237–142 BC) and Dendera (54–20 BC) (Cauville and Devauchelle 1984; Cauville 1990). Here the dimensions of the various rooms are said to comprise a combination of round figures and a series of fractions, evidently the result of a rather complex planning process (Rossi 2004, 166–173). The degree of correspondence between the text and the finished monument is still unclear but, again, it may be irrelevant: in this case, the builders were evidently proud of these complicated figures, to the point of immortalizing them on the temple walls. Their symbolic meaning might have been more important than their manifestation in the actual building.

In conclusion, building or quarrying a monument was a complex task that required a large number of well-organized workmen, backed by an efficient system that supported their work, their life, and their families. Keeping under control factors ranging from the daily efficiency of each worker to the successful completion of the project required organization based on simple figures that could be easily understood by everyone involved and monitored by those who were in charge of managing and recording the process.

Food production, from field to granary to table

Mathematics was deeply embedded in the seasonal and daily activities of measuring fields, assessing harvests, calculating taxes, storing agricultural products, and producing staple foods. These activities, which all included counting, measuring, and calculating with units of length, area, and capacity, involved the great majority of ancient Egyptians. For many of them it represented their main daily occupation (Samuel 2000, 537).

Agriculture was, and still is, the main source of food in Egypt. In ancient times, emmer wheat and barley played a fundamental role in the Egyptian society, not only as basic elements of the daily diet but also as a measure of wealth (Murray 2000a, 506). Fruit, vegetables, and pulses were also widely consumed (Murray 2000b, 609), whereas fish and meat were probably eaten once or twice a week (Ikram 2000, 669). Cultivation was heavily dependent on the annual Nile floods. In August, water progressively inundated the cultivable land, which was divided into large basins by means of earthen dykes; it slowly receded after about a month and a half, washing away the damaging salts and leaving behind well soaked and effectively fertilized soil (Butzer 1976, 17–18).

It appears that the slightly wetter climate of the Old Kingdom (2686–2160 BC) did not require particular management of the floods, whereas some attention started to be paid to this subject in the Middle Kingdom (2055–1650 BC). In general, however, for most of ancient Egyptian history, human intervention was limited to helping this event to take place in the best possible way. Trenches channelled the flood as far as possible, in case of low floods water would be retained for longer in order to ensure a satisfactory saturation of the soil, and the construction of dykes protected settlements and other installations from particularly high floods (Butzer 1976, 51–56; Murray 2000a, 514–515). In the New Kingdom (1550–1069 BC), the introduction of the *shaduf*, a simple pole and lever device to lift water, improved the agricultural production of small plots but had little influence on large-scale cultivation. A major change took place in the Ptolemaic Period (332–30 BC), with the introduction of the *saqiya*, a 'water-wheel' that ensured a continuous flow of water on a much larger scale (Butzer 1976, 47–50; Venit 1989; Eyre 1994, 63–64). Until then, and even after, the Nile remained the crucial factor in determining successful or unsuccessful agricultural production.

The height of the flood was a potential indicator of the prosperity (or misery) of the coming year. For this reason, the level of the river, carefully measured in cubits, palms, and fingers, appears among the earliest ancient Egyptian records (Clagett 1999, 3). Throughout ancient Egyptian history, a series of 'nilometers' scattered along the river helped the population and the authorities to observe the fluctuations in the water level, to estimate the productivity of the next harvest and, as we shall see below, the ensuing 'taxes' (Murray 2000a, 515). Once the flood water receded, it was necessary to re-establish the rightful boundaries of the agricultural plots (Griffith 1926, 204). The ancient Egyptians must have been very skilled in this practice, since they had to deal with this problem every year: their long-established ability prompted the fifth-century BC Greek historian Herodotus to report that geometry, whose original meaning is simply 'land-measurement', was born in Egypt and from there had been later exported to Greece (*Histories* II, 109).

Beside the basic cubit, fields could be laid out and measured by means of the *khet*, equal to 100 cubits. The commonest unit of area measurement was the *setjat* (also called *aroura*) equal to 1 square *khet* (that is, an area of 100 × 100 cubits), which could be subdivided in two ways: either in progressive halves (1/2, 1/4, and 1/8 of the *setjat*, used in the Pharaonic period, to which 1/16 and 1/32 were added in the Ptolemaic period) or into so-called 'cubit-strips', that is, elongated areas 1 cubit wide and 1 *khet* (100 cubits) long (Clagett 1999, 12–13; Imhausen 2003, 66–67). Several tombs contain representations of men surveying fields by means of ropes, marked at regular intervals by knots (Campbell 1910, 87; Berger 1934; Borchardt 1967; see Fig. 5.1.3). Pairing this information with the fact that the hieroglyphic sign for the number 100 was a coil of rope(s), it may be concluded that ropes 100 cubits long were used to survey the fields (Arnold 1991, 252). It has also been suggested that another hieroglyphic sign ⌁ (*s3*), may represent a measuring rope, 'taking the end loops as handles and the side loops as tags marking the ells [i.e., cubits]' (Reisner 1931, 78).

Long cords made of coarse plant fibres would have inevitably been very thick, and even if the length was subdivided by means of painted marks instead of bulky knots, precision could not be guaranteed (Dorner 1981, 94–95). Approximation might have been acceptable for measuring fields, but the adoption of the same method in other contexts is a matter of debate. It is well known, for instance, that the 'stretching of the cord' was one of the most important ritual steps of the elaborate ceremonies that had been performed since earliest times to mark the foundation of important buildings (Borchardt and Schäfer 1900, 97; Engelbach 1934; Fakhry 1961, 94; Barguet 1962, 31; Redford 1971, 114–115; 1976, pl. 18; Wilkinson 2000, 111–112, 139). In general, cords must have played an important

Figure 5.1.3 Land-surveyors from the Eighteenth Dynasty tomb of Amenhotepsesi at Thebes (Davies 1923, pl X, courtesy of the Egypt Exploration Society)

role in establishing and keeping the alignment of the walls and, to a certain extent, of the corners of buildings (Pendelbury 1951, 6). It is likely, however, that operations requiring a high level of precision (such as establishing right angles) were carried out by technicians outside the ritual ceremony, perhaps using wooden tools or shorter and thinner cords, that ensured a higher degree of accuracy (Rossi 2004, 154–161).

Going back to agriculture and turning to the shape of the fields, beside the most obvious rectangular outline the mathematical sources also include examples of calculation of triangular, trapezoidal, and circular areas as large as fields (respectively, pRhind 51 and pMoscow 4; pRhind 52, reproduced in Fig. 5.1.4; pRhind 50). Whereas the triangular and trapezoidal cases may well have arisen as the result of a regular grid of fields intersecting with an irregular element of the landscape, the third case (a circle with a diameter of about 470 m and a circumference of nearly a kilometre and a half) is likely to have been purely theoretical (Gillings 1972, 139). In fact, the presence of such an unrealistic example might even prompt further comments on the nature of these mathematical papyri: on the occasion of giving practical instructions for one of the commonest tasks that a scribe might be assigned, that is the calculation of the area of fields, the papyri provided in fact a complete list of methods to calculate the areas of the most common geometrical figures, including the circle, under the same practical example. Once more, it appears that the scope of the mathematical papyri was not simply to list specific cases, but also to provide examples and methods that might be applied to wider issues.

The cycle of food production lasted for months: the harvest of cereals took place between February and May, followed by threshing, winnowing, and sieving. The final product was stored still as spikelets, postponing the elaborate process of obtaining the clean grain to the moment immediately prior to consumption (Murray 2000a, 526–527). Before that, it was carefully measured in order to establish the amount that had been produced and to set aside the portion that corresponded to the expected state revenues. The use, in this instance, of the modern word 'taxes' may be problematic, as in New Kingdom Egypt (1550–1069 BC) there seems to be a difference between dues in kind (items counted or items

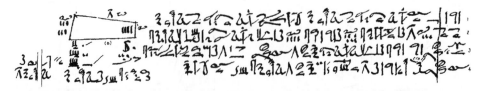

Figure 5.1.4 pRhind problem 52, on the area of a trapezoidal field (Chace, Bull, and Manning 1929, pl 74, reprinted by permission of the Mathematical Association of America)

produced by labour) imposed on officials or on ordinary people on the one hand, and harvest-yield or harvest-tax, dues and tributes or gifts on the other (Gardiner 1941, 60; Katary 1989, 172; Ezzamel 2002a, 20–21). At any rate, it is clear that a fixed proportion of the harvest was taken from the cultivators, transported elsewhere among strict security measures (Gardiner 1941, 23–34, 38–41; Ezzamel 2002a), and used either to supply state institutions or to pay for exchanges of services between them (Menu 1970, 83–91; Katary 1989, 18). Labelling as 'taxations' a number of intra-governmental transactions listed in the ancient sources is a matter of debate, and depends not only on establishing the level of control of the various institutions on the land, but also on whether or not some institutions (such as temples) should be considered part of the state (Katary 1989, 24, 184; cf. Stuchevsky 1974; Janssen 1979).

The amount to be collected by state officials was calculated on the basis of the quality of the fields, which were classified into three groups (Janssen 1975, 143). The assessment, carried out by scribes, depended on a combination of factors including size and position of the fields but also the presence of canals, lakes, wells, and trees (Goedicke 1967, 56, 72). The extent of the land was measured in *setjat* and then multiplied by a factor conventionally called 'measure of corn', which might be 5, 7 1/2, or 10; dues of 1 part in 13 1/3 (that is, 7.5%) were then levied from the resulting figure (see Gardiner 1948, 35). If necessary, the damage inflicted by irregular inundations was taken into account to correct these initial expectations (Ezzamel 2002b, 80; cf. Katary 1989, 214). The crop was surveyed when still standing by an official team including the tax scribe, the clerk of the land, the envoy of the stewards, the 'stretcher of the cord', and the 'holder of the cord' (Ezzamel 2002b, 72–74). Once harvested and threshed, the product was measured again by pouring it into containers of known quantity, thus effectively establishing its volume rather than weight (Murray 2000a, 523, figs 21.9–10). In this respect, the unit of measurement that was used, the *heqat*, may be considered both a unit of volume and a unit of capacity (Chace, Bull, and Manning 1929, 33–34; Clagett 1999, 13). For extremely large volumes, the Egyptians used the double- and the quadruple-*heqat*, the values of which corresponded to twice or four times the value of the *heqat*. The largest unit of volume/capacity was the *khar*, or 'sack', corresponding to 20 *heqat* or 5 quadruple-*heqat* (Clagett 1999, 14–15; Imhausen 2003, 58). The *khar* corresponded to 2/3 of a cubic cubit, thus allowing the scribes to easily shift in their calculations between the volume of a container (expressed in cubic cubits) and the volume of product (expressed in volume/capacity units). Once the former had been calculated, the latter, expressed in *khar*, could immediately be obtained by multiplying the number of cubic cubits by 1 1/2. By first turning the *khar* into quadruple *heqat*, the scribe could have obtained the number of *heqat* stored in the given granary (e.g., pRhind 41–44).

Weighing the grain and establishing how much could be stored in each granary was not only the final act of the long cultivation process that had started a year earlier: it was also the basis for the wealth and well-being of the population in the following year. Large quantities of stored grain ensured a steady supply of basic ingredients. Cooking, baking, and brewing were daily activities, based on traditional practice, and transmitted by means of recipes in which the ingredients were mixed by parts rather than weights, as in pharmacopoeia (Nunn 1996, 140–143; Wilson 2001). Taste, of course, also played an important role in the composition of the mixture, and the Egyptians appear to have been very skilled at manipulating the ingredients to obtain the desired result (e.g., Samuel 2000, 557).

Bread and beer represented a constant on every Egyptian table, from the peasant's to the king's. The two products had a lot in common, as they were made of the same main ingredients (emmer wheat and barley) and were manufactured using similar technologies. The close relationship between brewing and baking is also witnessed by the large number of ancient representations which show these two activities together (Samuel 2000, 596). Bread and beer also shared the use of a peculiar unit of measurement, the *pefsu*, generally translated as 'cooking ratio'. In the case of bread, the *pefsu* corresponded to the ratio between the number of loaves and the quantity of cereal in *heqat* that had been used to make them; in the case of beer, it corresponded to the ratio between jugs of beer and *heqat* of grain. In other words, the *pefsu* measured the strength of the product and, as a consequence, its nutritional contribution and therefore its commercial value. A *pefsu* of 5, for instance, meant that five loaves had been made from one *heqat* of grain, whereas a *pefsu* of 10 meant that one *heqat* had been diluted into ten loaves, and a similar process applied to beer: the lower the *pefsu*, the more valuable were the bread and the beer.

The mathematical sources contain several problems involving the *pefsu* of bread and beer. In the Rhind papyrus the *pefsu* of bread ranges between 5 and 45, whilst that of beer is always 2; in the Moscow papyrus, the *pefsu* of bread is always 20 and that of beer varies between 2 and 6. Some Middle Kingdom administrative texts report *pefsu* values ranging between 60 and 80 for bread and between 1 and 2 for beer (Imhausen 2003, 115–116). By assessing the nutritional value of bread and beer, the *pefsu* played an important role in the ancient barter society, since it established their commercial value and allowed a balanced exchange of products (Clagett 1999, 60). In fact, a number of extant mathematical problems involve not only exchange of loaves of different *pefsu* (pRhind 72–76), but also exchange of bread for beer (pRhind 77–78). In the latter case, a fair exchange could only be performed if the strength of both products was known.

The Rhind and Moscow papyri may not contain the entirety of ancient Egyptian mathematical knowledge, but they certainly contain a good sample of the tasks that might have been assigned to a Middle Kingdom scribe. The predominance of certain types of problems, therefore, very likely reflected the most common events. In the Rhind papyrus, for instance, ten problems deal with *pefsu* of bread and beer (69–78), eight with the calculation of large areas (48–55), six involve various quantities of grain (35–38, 80–81), six concern the calculation of the volume of rectangular and cylindrical granaries (41–46); then eight focus on the distribution of loaves among men (1–6, 63, 65), one on the distribution of fat (66) and one of grain (68), and three on the distribution of food to various animals (82–84). This means that, out of eighty-seven problems contained in this document, thirty deal directly with food production and fourteen with the distribution of food. In the Moscow papyrus, out of twenty-five problems, ten focus on *pefsu* of bread and beer and six on the calculation of areas.

The huge task of re-organizing the land after the annual inundation, the repeated necessity of measuring the fields, the ability to forecast the productivity of the land and to calculate the consequent 'taxes' and the storage of the products, were all activities that required simple, efficient, and flexible mathematical management. Conveniently based on a system of intertwined units of measurement, it allowed scribes and technicians to exchange information easily and ensure an equitable distribution of goods.

Conclusions

The study of the relationship between mathematics and technology in ancient Egypt highlights a number of interesting points and suggests possibilities for further exploration.

Many details of how mathematics was involved in various technological processes remain obscure, but the available material still yields some important information. New studies of ancient mathematical papyri, for instance, can provide new insights into the nature of the ancient sources and extend their realm of influence beyond their obvious field. As we have seen, the extremely practical character of the mathematical papyri might distract our attention from the fact that, aside from providing straightforward examples of precise tasks, they may also indirectly provide solutions for problems that are not specifically mentioned. This may be the case, for instance, for pRhind 62, which directly mentions a bag full of separate pieces of precious metals, and indirectly may imply any mixture of metals. It is possible to push this argument even farther, in a way reversing it, to suggest that in the Egyptian mathematical papyri general issues were normally presented as practical examples. Instead of talking about 'mixtures' of metals, the papyri mention a particular bag with particular contents; instead of indicating how to calculate the areas of various geometrical figures, they list a sequence of field-areas, even including what in practice would be an unrealistic shape for a field—an enormous circle.

The available sources from ancient Egypt are extremely limited both in number and in scope (Ritter 2000, 115–116). Postulating the existence of now lost documents that would cast a completely different light on ancient Egyptian mathematics may be unrealistic, but the possibility that texts of similar nature but listing different topics existed but did not survive should not be dismissed. For instance, pRhind 62 appears to be the only problem dealing with an area of craftsmanship that occupied an important position in the ancient society. Metal working was involved in the production of objects ranging from simple tools to precious jewels, and the technology involved and the organization of the working areas shared several characteristics with glass working and faience production sites.

Modern experimental archaeology and scientific analyses provide important information about technological processes and the chemical and physical composition of the final products. In some cases, such as bread and beer production and the composition of metal alloys, the ancient method has been more or less reconstructed; in other cases, however, such as glass and faience, many important details of the technology employed in their production remain unclear (Nicholson and Henderson 2000; Nicholson and Peltenburg 2000). The lack of mathematical problems referring to these technological areas may be due to several factors; the most obvious is the uneven preservation of the ancient mathematical texts. Another reason may be the nature of the surviving mathematical sources: as already mentioned above, the extant mathematical texts are school exercises meant to teach mathematics to scribes, and it is possible that scribes were not expected to be professionally involved in glass or faience working. If this is true, then one may even infer that mathematics did not play a particularly important role in the technology relating to glass and faience, which instead relied on a long and consolidated practice that did not require specific calculations.

In conclusion, it is vitally important to look at ancient technological processes from an ancient point of view and, in particular, whenever possible, to adopt ancient mathematical language: expressing the slope of pyramids in degrees may help us to visualize their shape, but it tells us nothing about how ancient architects understood the mathematics involved in their work. Similarly, percentages certainly help us to describe the composition of certain mixtures, but only by expressing them as proportions can we visualize the ancient method of making them. Such a flexible and careful approach ensures that modern points of view do not interfere with reconstructing ancient methods, and do not hide the fundamental differences between the two systems.

Bibliography

Allen, James P, *The art of medicine in ancient Egypt*, Metropolitan Museum of Arts and Yale University Press, 2005.

Arnold, Dieter, *Building in Egypt*, Oxford University Press, 1991.

Arnold, Dieter, and Arnold, Dorothea, *Der Tempel Qasr el-Sagha,* Archäologische Veroffentlichungen, 27, von Zabern, 1979.

Aston, Barbara, Harrell, James, and Shaw, Ian, 'Stone', in Paul T Nicholson and Ian Shaw (eds), *Ancient Egyptian materials and technology*, Cambridge University Press, 2000, 5–77.

Barguet, Paul, *Le Temple d'Amon-Rê à Karnak,* Recherches d'Archéologie, de Philologie et d'Historie, 21, Institut Français d'Archéologie Orientale, 1962.

Berger, Suzanne, 'A note on some scenes of land-measurement', *Journal of Egyptian Archaeology*, 20 (1934), 54–56.

Borchardt, Ludwig, 'Statuen von Feldmessern', *Zeitschrift der Ägyptische Sprache*, 42 (1967), 70–72.

Borchardt, Ludwig, and Schäfer, Heinrich, 'Vorläufiger Bericht über die Ausgrabungen bei Abusir im Winter 1899/1900', *Zeitschrift der Ägyptische Sprache*, 38 (1900), 94–103.

Butzer, Karl W, *Early hydraulic civilization in Egypt: a study in cultural ecology*, University of Chicago Press, 1976.

Campbell, Colin, *Two Theban princes, Kha-em-uast and Amen-khepeshf, Menna, a land-steward, and their tombs*, Oliver and Boyd, 1910.

Carter, Howard, and Gardiner, Alan H, 'The tomb of Ramses IV and the Turin plan of a royal tomb', *Journal of Egyptian Archaeology,* 4 (1917), 130–158.

Cauville, Sylvie, 'Les inscriptions dédicatoires du temple d'Hathor à Dendera', *Bulletin de l'Institut Français d'Archéologie Orientale*, 90 (1990), 83–114.

Cauville, Sylvie, and Devauchelle, Didier, 'Les mesures réelles du temple d'Edfou', *Bulletin de l'Institut Français d'Archéologie Orientale*, 84 (1984), 23–34.

Chace, Arnold B, Bull, Ludlow, and Manning, Henry P, *The Rhind mathematical papyrus*, Mathematical Association of America, 1929.

Chappaz, Jean-Luc, 'Working with gold in ancient Egypt', *Aurum*, 14 (1983), 34–40.

Clagett, Marshall, *Ancient Egyptian science: a source book*, vol 3, American Philosophical Society, 1999.

Clarke, Somers, and Engelbach, Reginald, *Ancient Egyptian masonry*, Oxford University Press, 1930.

Cuomo, Serafina, *Ancient mathematics*, Routledge, 2001.

Davies, Norman, *The tombs of two officials of Tuthmosis the Fourth,* Theban Tombs Series, 3, Egypt Exploration Society, 1923.

Dorner, Josef, *Die Absteckung und astronomische Orientierung ägyptischer Pyramiden*, dissertation, Innsbruck, 1981.

Engelbach, Reginald, *The problem of the obelisk*, Fisher Unwin, 1923.

Engelbach, Reginald, 'An architect's project from Thebes', *Annales du Service des Antiquités de l'Égypte*, 27 (1927), 72–76.

Engelbach, Reginald, 'A foundation scene of the Second Dynasty', *Journal of Egyptian Archaeology*, 20 (1934), 183–184.

Eyre, Christopher J, 'The water regimes for orchards and plantations in Pharaonic Egypt', *Journal of Egyptian Archaeology*, 80 (1994), 57–80.

Ezzamel, Mahmoud, 'Accounting working for the state: tax assessment and collection during the New Kingdom, ancient Egypt', *Accounting and Business Research*, 32 (2002a), 17–39.

Ezzamel, Mahmoud, 'Accounting and redistribution: the palace and mortuary cult in the Middle Kingdom, ancient Egypt', *Accounting Historians Journal*, 29/1 (2002b), 61–103.

Fakhry, Ahmed, *The pyramids*, University of Chicago Press, 1961.

Gardiner, Alan H, 'Ramesside texts relating to the taxation and transportation of corn', *Journal of Egyptian Archaeology*, 27 (1941), 19–73.

Gardiner, Alan H, *The Wilbour papyrus*, vol III: Translation, Oxford University Press, 1948.

Gillins, Richard J, *Mathematics at the time of the pharaohs*, MIT Press, 1972.

Griffith, Francis L, 'The teaching of Amenophis the son of Kanakht. Papyrus B.M. 10474', *Journal of Egyptian Archaeology*, 12 (1926), 191–231.

Goedicke, Hans, *Königliche Dokumente aus dem Alten Reich*, Ägyptologische Abhandlungen, 14, Harrassowitz, 1967.

Herodotus, *Histories* (trans Robin Waterfield), Oxford University Press, 1998.

Hinkel, Friederich W, 'Pyramide oder Pyramidenstumpf? (Teil B)', *Zeitschrift für Ägyptische Sprache und Altertumskunde*, 109 (1982), 27–61.

Ikram, Salima, 'Meat processing', in Paul T Nicholson and Ian Shaw (eds), *Ancient Egyptian materials and technology*, Cambridge University Press, 2000, 656–671.

Imhausen, Annette, 'The algorithmic structure of the Egyptian mathematical problem texts', in John M Steele and Annette Imhausen (eds), *Under one sky: Astronomy and mathematics in the ancient Near East*, Ugarit-Verlag, 2002, 147–161.

Imhausen, Annette, *Ägyptische Algorithmen*, Ägyptologische Abhandlungen, 65, Harrassowitz, 2003.

Janssen, Jac J, 'Prolegomena to the study of Egypt's economic history', *Studien zur Altägyptischen Kultur*, 2 (1975), 127–185.

Janssen, Jac J, 'The role of the temple in the Egyptian economy during the New Kingdom', in E Lipinski (ed), *State and temple in the ancient Near East*, vol II, Orientalia Lovaniensia Analecta, 6, Peeters, 1979, 505–515.

Katary, Sally L D, *Land tenure in the Ramesside Period*, Kegan Paul International, 1989.

Kemp, Barry J, Rossi, Corinna, and Harrell, James, 'Quarrying strategies at Amarna', for submission to the *Cambridge Archaeological Journal*.

Klemm, Rosemarie, and Klemm, Dietrich D, 'Chronologischer Abriss der antiken Goldgewinnung in der Ostwüste Ägyptens', *Mitteilungen des Deutschen Archäologischen Institut Kairo*, 50 (1994), 189–222.

Koenig, Yvan, *Les ostraca hiératiques inédits de la Bibliothèque nationale et universitaire de Strasbourg*, Documents de Fouilles de l'Institut Français d'Archéologie Orientale, 33, Institut Français d'Archéologie Orientale, 1997.

Lehner, Mark, *The complete pyramids*, Thames and Hudson, 1997.

Maragioglio, Vito, and Rinaldi, Celeste, *L'architettura delle piramidi memfite*, vol III, Artale, 1964.

Menu, Bernadette, *Le régime juridique des terres et du personel attaché à la terre dans le papyrus Wilbour*, Faculté des lettres et sciences humaines, 1970.

Murray, Mary Ann, 'Cereal production and processing', in Paul T Nicholson and Ian Shaw (eds), *Ancient Egyptian materials and technology*, Cambridge University Press, 2000a, 505–536.

Murray, Mary Ann, 'Fruits, vegetables, pulses and condiments', in Paul T Nicholson and Ian Shaw (eds), *Ancient Egyptian materials and technology*, Cambridge University Press, 2000b, 609–655.

Nicholson, Paul T, and Henderson, Julian, 'Glass' in Paul T Nicholson and Ian Shaw (eds), *Ancient Egyptian materials and technology*, Cambridge University Press, 2000, 195–224.

Nicholson, Paul T, and Peltenburg, Edgar, 'Egyptian faience' in Paul T Nicholson and Ian Shaw (eds), *Ancient Egyptian materials and technology*, Cambridge University Press, 2000, 177–194.

Nicholson, Paul T, and Shaw, Ian (eds), *Ancient Egyptian materials and technology*, Cambridge University Press, 2000.

Nunn, John F, *Ancient Egyptian medicine*, British Museum Press, 1996.

Ogden, Jack, 'Aesthetic and technical considerations regarding the colour and texture of ancient goldwork', in Susan La Niece and Paul T Craddock (eds), *Metal plating and patination: cultural, technical and historical developments*, Butterworth-Heinemann, 1993, 39–49.

Ogden, Jack, 'Metals', in Paul T Nicholson and Ian Shaw (eds), *Ancient Egyptian materials and technology*, Cambridge University Press, 2000, 149–176.

Osman, A, 'Ancient gold mining in Egypt and its geologic aspects', in *Abstracts of the international conference on ancient Egyptian mining and metallurgy and conservation of metallic artefacts (10–12 April 1995)*, Ministry of Culture and Supreme Council of Antiquities, 1995, 8.

Owen, Gwyl, and Kemp, Barry J, 'Craftsmen's work patterns in unfinished tombs at Amarna', *Cambridge Archaeological Journal*, 4/1 (1994), 121–129.

Peet, Eric T, *The Rhind mathematical papyrus*, Liverpool University Press; Hodder and Stoughton, 1923.

Pendlebury, John D S, *The city of Akhenaten*, vol III, Egypt Exploration Society, 1951.

Pusch, Edgar B, 'Metallverarbaitende werkstätten der frühen Ramessidenzeit in Qantir-Piramesse/Nord', *Ägypten und Levante*, 1 (1990), 75–113.

Redford, Donald B, 'The earliest years of Ramses II and the building of the Ramesside court at Luxor', *Journal of Egyptian Archaeology*, 57 (1971), 110–119.

Redford, Donald, *The Akhenaten temple project*, vol I, Aris and Phillips, 1976.

Reisner, George A, *Mycerinus*, Harvard University Press, 1931.

Ritter, James, 'Egyptian mathematics', in Helaine Selin (ed), *Mathematics across cultures. The history of non-western mathematics*, Kluwer Academic Publishers, 2000, 115–136.

Robson, Eleanor, 'The uses of mathematics in ancient Iraq, 6000–600 BC', in Helaine Selin (ed), *Mathematics across cultures. The history of non-western mathematics*, Kluwer Academic Publishers, 2000, 93–113.

Rossi, Corinna, 'Dimensions and slope in the nineteenth and twentieth dynasty royal tombs', *Journal of Egyptian Archaeology*, 87 (2001a), 73–80.

Rossi, Corinna, 'The plan of a royal tomb on O. Cairo 25184', *Göttinger Miszellen*, 184 (2001b), 45–53.

Rossi, Corinna, *Architecture and mathematics in ancient Egypt*, Cambridge University Press, 2004.

Rossi, Corinna and Imhausen, Annette, 'Papyrus Reisner I: architecture and mathematics in the time of Sesostris I', in Salima Ikram and Aidan Dodson (eds), *Festschrift for Barry Kemp*, American University in Cairo Press, forthcoming.

Samuel, Delwen, 'Brewing and baking', in Paul T Nicholson and Ian Shaw (eds), *Ancient Egyptian materials and technology*, Cambridge University Press, 2000, 537–576.

Scheel, Bernd, *Egyptian metalworking and tools*, Shire Egyptology, 13, Shire, 1989.

Selin, Helaine (ed), *Mathematics across cultures. The history of non-western mathematics*, Kluwer Academic Publishers, 2000.

Shaw, C Tim, 'New Kingdom mining technology with reference to Wadi Arabah', in Feisal A Esmael and Zahi Hawass (eds), *Proceedings of the first international conference on ancient*

Egyptian mining and metallurgy and conservation of metallic artefacts (10–12 April 1995), Supreme Council of Antiquities, 1996, 1–14.

Shaw, Ian, 'Exploiting the desert frontier: the logistics and politics of ancient Egyptian mining expeditions', in Bernard Knapp (ed), *Social approaches to an industrial past: the archaeology and anthropology of mining,* Routledge, 1998, 242–258.

Simpson, William Kelly, *Papyrus Reisner I,* Boston Museum of Fine Arts, 1963.

Simpson, William Kelly, *Papyrus Reisner III,* Boston Museum of Fine Arts, 1969.

Stadelmann, Rainer, *Die ägyptischen Pyramiden,* Von Zabern, 1985.

Struve, W W, 'Mathematischer Papyrus des Staatlichen Museums der Schönen Künste in Moskau', *Quellen und Studien zur Geschichte der Mathematik, Astronomie und Physik,* A/1 (1930).

Stuchevsky, I A, 'Data from the Wilbour Papyrus and other administrative documents relating to the taxes levied on the state ("royal") land-cultivators in Egypt of the Ramesside era', English summary, *Vestnik Drevnei Historii,* 1974, 20–21.

Venit, Marjorie S, 'The painted tomb from Wardian and the antiquity of the *saqiya* in Egypt', *Journal of the American Research Center in Egypt,* 26 (1989), 219–222.

Wainwright, Gerald A, 'Rekhmirê's metal-workers', *Man,* 44 (1944), 94–98.

Weill, Adrienne R, 'Une problème de métallurgie archéologique: exams aux rayons X d'un objet égyptien en electrum', *Revue de Metallurgie,* 48 (1951), 97–104.

Wendrich, Willemina, 'Basketry', in Paul T Nicholson and Ian Shaw (eds), *Ancient Egyptian materials and technology,* Cambridge University Press, 2000, 254–267.

Whitley, James, *The archaeology of ancient Greece,* Cambridge University Press, 2001.

Wilkinson, Toby A H, *Royal annals of ancient Egypt: the Palermo Stone and its associated fragments,* Kegan Paul International, 2000.

Wilson, Hilary, *Egyptian food and drink,* Shire Egyptology, 9, Shire, 2001.

Wysocki, Zygmunt, 'The result of research, architectonic studies and of protective work over the northern portico of the middle courtyard in the Hatshepsut temple at Deir el-Bahari', *Mitteilungen des Deutschen Archäologischen Institut Kairo,* 40 (1984), 329–349.

Siyaq: numerical notation and numeracy in the Persianate world

Brian Spooner and William L Hanaway

Siyaq[1] is a system of numerical notation that was introduced during the early Islamic Caliphate, probably under one of the Umayyad caliphs in Damascus between 656 and 751 AD, and was preferred throughout the Persianate regions down to recent times. The basic system is shown in Figs. 5.2.1–5.2.3. It is derived from writing out the words for each of the Arabic decimal numerals. The earliest extant records of *siyaq* are found in financial accounts written for the Abbasid Caliph al-Moqtader in Baghdad in 918–19 AD (Kremer 1887). By this date, each grapheme[2] had already evolved into a cursive shorthand, and was difficult to read for anyone not practised in it, whether or not they were otherwise literate in Arabic. As a system it did not introduce any new principles of numeration, since it is decimal and ciphered, in the sense that there is a distinct grapheme for each numeral. But it offered a new range of advantages and disadvantages, which conditioned its historical role. From the ninth century down into the twentieth, *siyaq* was the primary system for recording quantities, measurements, or

1. This is the most common name, which is pronounced with a long final syllable. Other names include: *khatt-e siyāq, khatt-e raqam, khatt-e roqumi, khatt-e dināri, hesāb-e roqum, hesāb-e dināri.* In Romanizing words from the Arabic script we have followed current Persian usage throughout, in order to avoid using different systems for different languages.

2. We refer to the *siyaq* numerals as graphs (rather than, for instance, *raqam* characters, signs) in order to avoid the associations that come with the other terms.

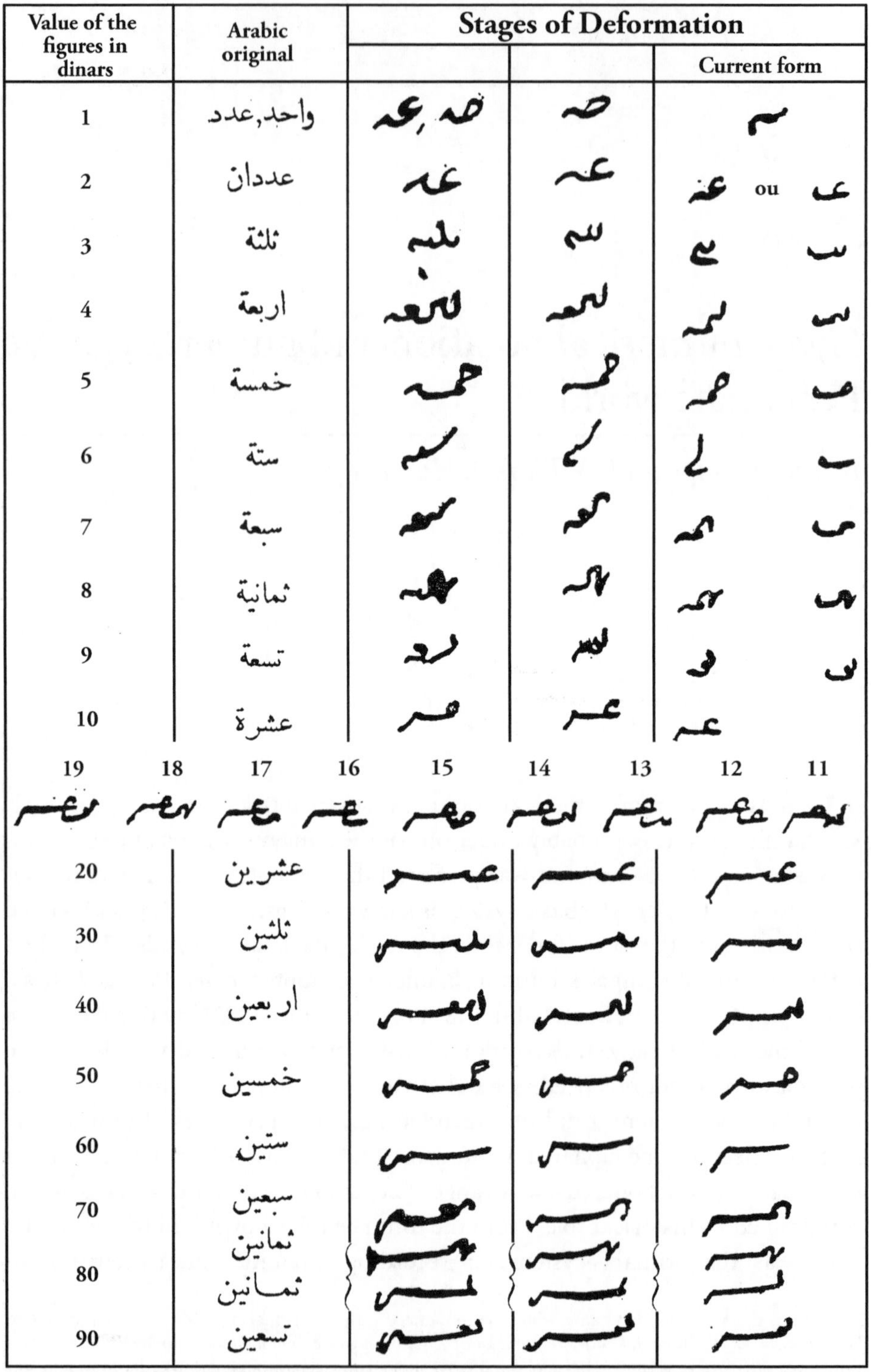

Value of the figures in dinars	Arabic original	Stages of Deformation			
					Current form
1	واحد,عدد				
2	عددان				ou
3	ثلاثة				
4	اربعة				
5	خمسة				
6	ستة				
7	سبعة				
8	ثمانية				
9	تسعة				
10	عشرة				

19	18	17	16	15	14	13	12	11

Value	Arabic original			
20	عشرين			
30	ثلثين			
40	اربعين			
50	خمسين			
60	ستين			
70	سبعين			
80	ثمانين ثمـانين			
90	تسعين			

Figure 5.2.1 *Siyaq* graphs for units, teens, and tens, as used in the central Persianate region, showing how the words for the numbers became progressively deformed (from Kazimzadeh 1915)

Value of the figures in dinars	Arabic original	Stages of Deformation		
				Current form
100	ماة			
200	ماتا			
300	ثلثماة			
400	اربعماة			
500	خمسماة			
600	ستماة			
700	سبعماة			
800	ثمانماة			
900	تسعماة			
1000	الف			
2000	الفى يا الفا			
3000	ثلثة الف			
4000	اربعة الف			
5000	خمسة الف			
6000	ستة الف			
7000	سبعة الف			
8000	ثمانية الف			
9000	تسعة الف			

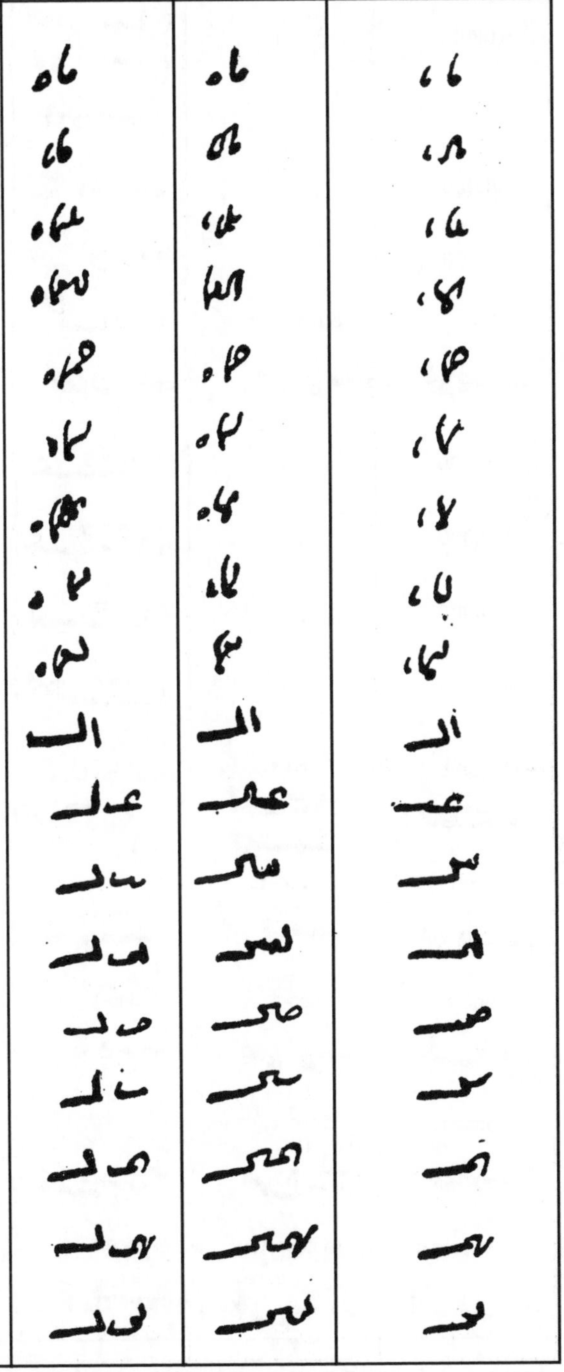

Figure 5.2.2 *Siyaq* graphs for hundreds and thousands, as used in the central Persianate region, showing how the words for the numbers became progressively deformed (from Kazimzadeh 1915)

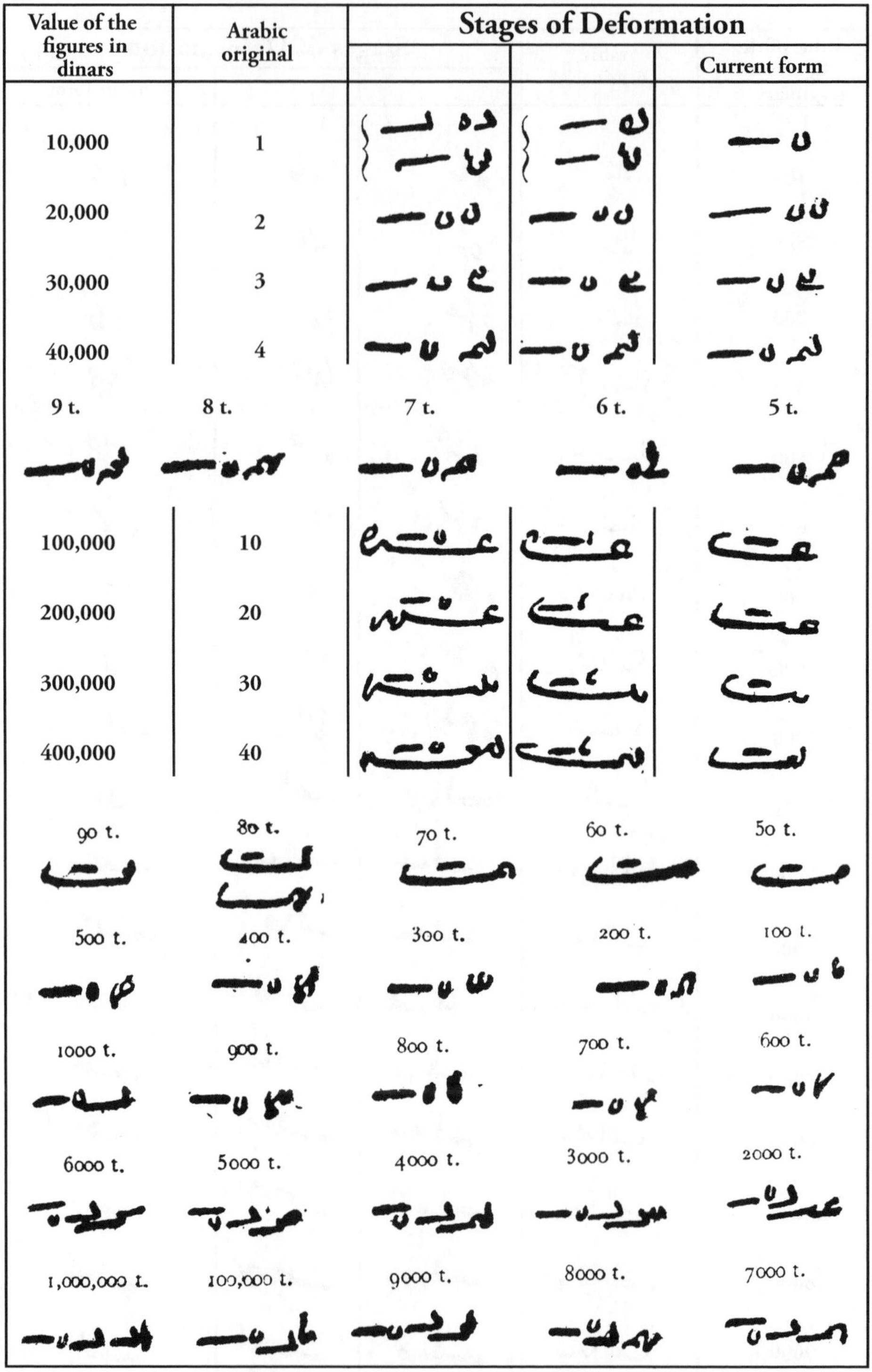

Value of the figures in dinars	Arabic original	Stages of Deformation		Current form
10,000	1			
20,000	2			
30,000	3			
40,000	4			
9 t.	8 t.	7 t.	6 t.	5 t.
100,000	10			
200,000	20			
300,000	30			
400,000	40			
90 t.	80 t.	70 t.	60 t.	50 t.
500 t.	400 t.	300 t.	200 t.	100 t.
1000 t.	900 t.	800 t.	700 t.	600 t.
6000 t.	5000 t.	4000 t.	3000 t.	2000 t.
1,000,000 t.	100,000 t.	9000 t.	8000 t.	7000 t.

Figure 5.2.3 *Siyaq* graphs for larger numbers, as used in the central Persianate region, showing how the words for the numbers became progressively deformed (from Kazimzadeh 1915)

anything involving a quantified amount in certain types of documents through-out the Persianate world.

We use the term 'Persianate' (after Hodgson 1974) for the northern and eastern areas of the Islamic world, where Persian has been the primary language of literacy (except in certain religious contexts) throughout the past millennium. Persian had been the court language of the Iranian Achaemenid (559–330 BC), Parthian (238 BC–224 AD), and Sasanian (224–651 AD) empires, when it was written first in a cuneiform alphabet, later in a form of the Aramaic script. It re-emerged after the Arab conquest of west and central Asia in the court of Samarqand (in modern Uzbekistan) under the local ninth-century Samanid dynasty—in the Arabic script. Persian then spread through the territory we now know as Central Asia, Afghanistan, and Iran (Fig. 5.2.4). In the thirteenth century, under the Mongols, it moved further east along the major trade routes into central China. The Ottomans took it further west through Anatolia into southeastern Europe and the Balkans. Finally, under the Mughals in sixteenth-century India, it became the primary language of administration and belles lettres across most of South Asia. Over this vast area Persian thus became both the common medium of literacy, along with all interaction that was based on literacy, and for a time also the preferred lingua franca for oral communication between people of different language backgrounds (though it surrendered the latter function to Turkic in the later medieval period). Although it began to give way to progressive vernacularization in the peripheral areas as early as the fourteenth century, throughout most of this territory, from roughly the ninth century into the nineteenth and twentieth, Persian remained the primary medium of written communication, irrespective of what was spoken locally, whether for administration, belles lettres, or even trade. In India the British colonial government replaced Persian with English for administration

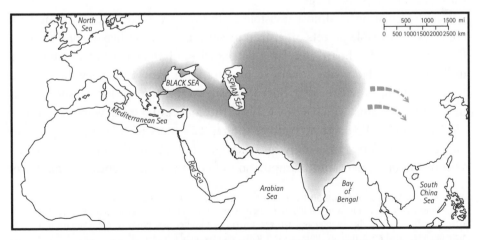

Figure 5.2.4 Persianate area of Eurasia between 850 and 1900 AD, indicating its greatest extent. Map drawn by Kimberly Leaman

in 1837, while elevating Urdu (a vernacular Persian–Hindi creole) to law-court usage. As a result, *siyaq* disappeared from government documents, but continued to be used in matters relating to landholding and other rural accounting.

Not long after *siyaq* was adopted in the early Arab-Islamic administrations in Damascus and Baghdad, new regional political centres in the east (now eastern Iran, northern Afghanistan and the Central Asian republics) began to revive court protocols and genres of literacy from the Sasanian past. They also turned to Persian for administration. They did not, however, return to the Sasanian system of numerical notation.[3] Ironically, it was in the Persianate region, as described above, rather than the Arabic-speaking regions of Syria, Arabia, and Egypt, that *siyaq* continued as established practice down to modern times. The survival of *siyaq* throughout the Persianate world for so long irrespective of other possibilities, and irrespective of changes in neighouring regions, requires explanation, and we shall return to consider it at the end of this article.

Persianate numeracy

Siyaq is one of several complementary components of Persianate numeracy. As was common elsewhere, the various functions we now understand under the single heading of numeracy were learnt and practised independently.[4] In this vast Persianate area from Bosnia to China and India, both in Persian and in the successor Persianate languages of literacy,[5] numbers were written in different ways for different purposes. Only two of these purposes, namely counting or numbering and recording, were written, but they were written differently. Calculating did not entail writing. We will consider each of these functions, in order to clarify the conditions of *siyaq* usage.

First, calculation is almost invisible in the written record. It was done with counters—moveable objects organized in various ways—either in columns on a flat surface, (*takht*, 'board') or, later, on the wires of an abacus, that symbolized decimal place. (No written form of numeration in the Islamic world showed place value until the later adoption of Indian-Arabic numerals.) All money was in the form of coins, which easily fell into the conceptual framework of counters (cf. Netz 2002, 11). Further, some of the region's most popular games, such as backgammon, depend not only on counters but also on the highly developed skill in the movement and manipulation of counters that comes as a matter of

---

3. The Sasanians had used an additive, non-place value system of numerical notation (Nyberg 1964, 173).

4. Compare, for example, Netz (2002) on Greece in the fifth century BC. This article, which came to our attention after most of this chapter was written, pursues a number of arguments from the perspective of ancient Greece that are complementary to the orientation presented here.

5. Such as Ottoman Turkish, which emerged in the fifteenth century, and Urdu, which became dominant in the seventeenth.

course with *takht* or abacus calculation. This non-literate method of calculation satisfied the needs of the majority non-literate population, separating them from the minority literates—a situation that was probably common in most civilizations down to modern times. But the divide may have been deeper and even more historically significant in the Persianate world—for reasons that will become clearer as we proceed.

Second, counting and ranking were, as we would expect, by Persian or vernacular cardinal and ordinal numerals, both oral and (for the literate) written, using the words for the number, and later also the Indian numerals.

Third, the oldest system of numerical notation associated with Arabic and Islamic literacy was the alphabetic ciphered system, which had been inherited from the early versions of the Semitic alphabet. Each Arabic letter carried the same numerical value as its equivalent in Aramaic and Greek. In Arabic-script languages this system was known as *abjad*, after the first four letters of the early Semitic alphabet (*a, b, j/g, d*). It was used to assign simple numbers to successive objects, especially pages, and continues to be used in some of the situations where Roman numerals are still in use in English, such as the pages of a book preface. The same system was also used for chronograms, making possible the construction of words that carry the meaning of significant dates. Though it served these purposes well enough, it was unwieldy for large numbers.

Fourth, the Arabic form of what are known in the West as Arabic numerals arrived in the Islamic world from India not long after the adoption of *siyaq*. They were already used by al-Khwārizmī in early ninth-century Baghdad, but were not adopted generally or applied in common daily use, except sometimes in place of the *abjad* system (Hinz 1950, 6), until the late medieval period. Even then they did not replace *siyaq* but were used alongside it to make falsification more difficult. One reason suggested for the delay in adopting the Indian numeral system is its dependence on 'points' (Bagheri 1998, 301 in reference to Mazandarani 1952, 24)—zero had still not evolved from the Indian *bindu* (a simple dot indicating the absence of a number) to the full-size numeral of modern numeration.

Against this background we can now concentrate on *siyaq*. Literacy was less established among the conquering Arabs than in the ruling class of the Sasanian Empire that they replaced. In Sasanian society professional writers were a privileged class. The status carried over to Islamic chanceries, where the professional writers were called *monshi*, which has generally been translated as scribe. During the early development of Islamic administrative practice, scribes and accountants were recruited mainly from the bureaucracy that had managed the administration of Western Asia for the Sasanians since the early third century, and which probably enjoyed continuity from much earlier. By default these recruits continued the practices of the earlier regime. But there was a gradual process of Arabization—perhaps an explicit policy. Arabization is said to have been launched by the Umayyad Caliph ʽAbduʼl-Malik in 706 AD (Hinz 1950, 3). The use of Arabic in all writing, including administration, spread fast, and it appears to have been assisted, perhaps facilitated, by non-Arabs, especially the literate elite of the Sasanian bureaucracy. Non-Arab scribes, often converts to Islam, practised, rationalized, taught, and promulgated Arabic grammar and the new genres of literacy that came with the language of the Qurʼan. Although the memory of Sasanian practice continued to be a significant influence, at some point, possibly as early as the turn of the eighth century, numerical notation was also Arabized. Since chancery practice in general was in Arabic, it is not surprising that the recording of numbers should also have been Arabized.

The introduction of *siyaq* appears to have preceded the arrival of the Indian numerals: the general Arabizing atmosphere would probably have inhibited the latterʼs adoption. However, it is difficult to assess how strong such ethnic considerations might have been in earlier historical periods. There is circumstantial evidence to suggest that the change was more the result of zeal for the new regime among scribes who were non-Arab converts to Islam (opposed by some of their—perhaps unconverted—peers), than an ethno-political drive on the part of the new rulers, who may not have been fully literate anyway. *Siyaq*, which offered far greater flexibility than other systems available at the time, in particular by easily accommodating the expression of numbers of any magnitude, was both introduced and used by scribes. By contrast, the Indian numerals, when they arrived, were used by scholars in mathematics and science.

How the term *siyaq* came into use, both in its origin and its etymology, is uncertain. Arabic dictionaries list it under the same root as *suq* ʽmarketʼ, but without suggesting the connection. It is later explained as *siyāq al-mostaʽrabin*, meaning the method of those in favour of Arabization, which would fit nicely, but may be a later (thirteenth century) rationalization (ibn Tiqtaqa, in Bagheri 1998, 299). One Iranian scholar argues for tracing it back to a pre-Islamic Persian word meaning ʽnumberʼ (Rahnama 1995).

The formal features of *siyaq*

The Arabization of numerical notation that became known as siyaq took the form of simply writing out the Arabic names of the numerals. It may have been adopted from known models in other languages. For instance, the same procedure was apparently customary in early Greek before the adoption of various acrophonic numeral systems in the seventh century BC (Tod 1911–12). But whether by original design or by simple professional scribal process—there is no textual evidence— the names of the numerals quickly became abbreviated, distorted, stylized into convenient shorthand, producing a form of numerical notation that was uniquely cursive. The transition from the transparently readable words for the numerals, that would have been accessible to anyone who was literate in the Arabic script, to what became known as *siyaq*, which is very difficult to read even for the literate unless they were trained, was helped along by the unique nature of the Arabic script along with the particular sociology of literacy in the Islamic world. The process was no doubt facilitated by the longstanding practices of the scribal profession.

Despite common assumptions about reading in alphabetic scripts, when we read fast we in fact read not letter by letter but by word-shape, ideographically. When we work with numbers, we scan numbers and combinations of numbers similarly. If this is true for us as we read Roman printing, it is more so for readers of Arabic script. In fact before Arabization the Persians were accustomed to what they called *huzvarishn* 'logograms': in writing their pre-Islamic Persian language in a modified form of the Aramaic script they wrote many common words in their Aramaic equivalent and read them as though they had written out the Persian.[6] They were reading ideographically, and could therefore easily take to *siyaq*—writing in a simplified form of one language and reading it straight off in another.

Writing in the Arabic script, compared to other scripts, is not only cursive by default, but cursive to a much greater degree than what we know as cursive in the Latin alphabet. Although alphabetic, it is written not by the individual letter but by the pen-stroke. Each pen-stroke advances the writing through a succession of letters. The number of letters incorporated into each pen-stroke varies according to the nature of the particular letters and the order in which they are to be written. (The letters fall into groups according to their form, each of which connects to preceding and succeeding letters differently.) In practice, therefore, writing in any Arabic-script language draws not simply on the small number of letters in the alphabet (28 for Arabic, 32 for Persian, slightly more for Urdu, and so on), but on a far larger repertoire of pen-strokes, each of which is a particular pattern of letters, usually one to four, but sometimes more. These pen-strokes easily become rushed, condensed, abbreviated, stylized, and distorted—more so than in the case of western handwriting. Since *siyaq* was introduced, handwriting in the Arabic script has moved through a succession of styles, each of which are reflected in the writing of *siyaq*. Therefore, once numeracy became embedded (in the form of *siyaq*) in Arabic literacy, it became subject to all the factors that governed the sociology of literacy in that script down to the time when social changes began to transform and open up literacy in modern times. The relative lateness of the general adoption of printing in this part of the world, especially printing other than lithography, is a consequence of the same sociology.

---

6. Readers of English also do this to some extent: we are more likely to verbalize 'e.g.' as the English phrase 'for example' than as the Latin *exempli gratia*.

Bureaucrats were a small, elite professional community. Recruitment was largely hereditary. They were interested in maintaining, and probably (to further their interests under the new political regime) enhancing, the cultural value of written language (already greatly elevated by the way the Qur'an had come to be understood as the written word of God), reinforcing restrictions on its accessibility and by extension promoting their own status.

Despite its difficulty, and apart from some regional divergence of form between the three main empires of late medieval Islam, the Ottomans, the Safavids, and the Mughals, the practice of *siyaq* continued and remained stable for both administrative and commercial purposes well into the twentieth century. It was used both by accountants and bookkeepers in the financial administration of the Iranian, Ottoman, and Mughal governments, the Uzbek governments of Central Asia, and by merchants, landholders, irrigation officials, and other private individuals from as early as the eighth until well into the twentieth century. It was taught in schools in Iran as part of the standard curriculum until the early 1930s, and in India and Pakistan until later. It was terminated by Romanization in Turkey in 1928. In Iran it ceased to be taught in schools a few years later. But it did not entirely go out of use in rural areas of South Asia until some time in the second half of the twentieth century.

Siyaq differs from other systems of numerical notation in that although it is essentially a ciphered system, the history of its development is akin to the historical transformation of copperplate into bad handwriting. For this reason, it is read ideographically, even more so than may be the case with any handwriting. The basic graphs, written from right to left, are organized in units, tens, hundreds, thousands, and so on. Compound numbers are formed by combining the basic shapes, often with small changes, sometimes with the second component written through or above, as well as at least slightly to the left of the first (Figs. 5.2.5–5.2.6). The higher decimal places are written (and read) first, except that units come before tens (as in the earlier English 'five and twenty'). The value of a number is the sum of the absolute values of the graphs that constitute it.

It is a feature of the Persian language that numbers are always used with counters, or classifiers, in speech as well as in writing. *Siyaq* numbers similarly always implied an amount of something, rather than an abstract number. The *siyaq* graphs used in what is now Iran, which was central to the communication networks throughout the Islamic world, were used for two principal purposes: to record amounts of money, and to record quantities of goods. The unit for recording amounts of money was the *dinar*, multiples of which are:

1.   50 *dinars* = 1 *shahi*

2.   200 *dinars* = 1 `abbasi`

3. 1,000 *dinars* = 1 *qeran* (crown)

4. 10,000 *dinars* = 1 *tuman*.

From 1 to 9,999 all numbers are written in terms of *dinars*, while from 10,000 up they are written in terms of *tumans*. The close conceptual relationship with money should be noted, in view of the reference to coins as counters (see above). The same graphs were used to record quantities of goods measured by weight,

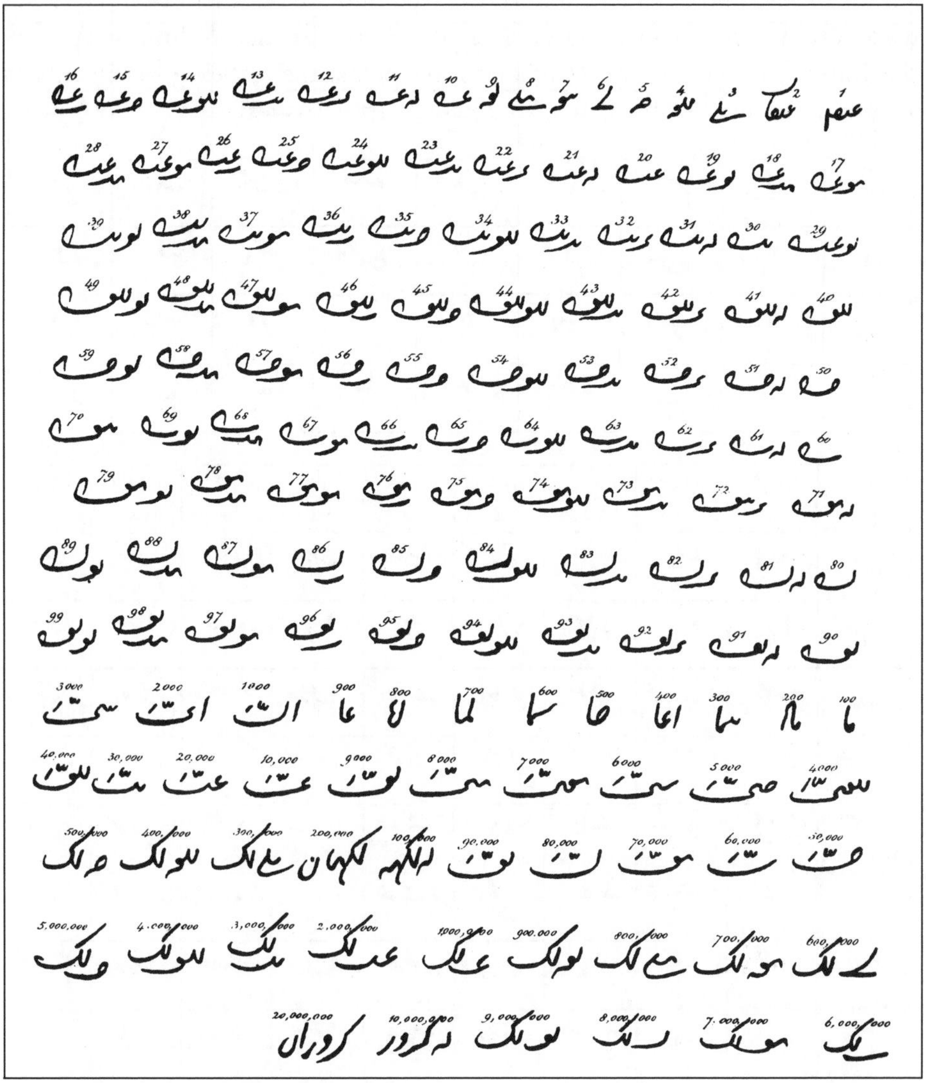

Figure 5.2.5 *Siyaq* graphs from northern India, including examples of combinations of units and tens (Stewart 1825, 25)

sometimes with a determinative figure denoting *man* (a measure of weight that varied in value from city to city) or *kharvar* (donkey-load) appended. For example, one *kharvar* = 100 *man* of Tabriz (the major city of Persian Azarbaijan) or 50 royal *man*. Weights above 99 *man* of Tabriz were recorded in *kharvar*. *Siyaq*

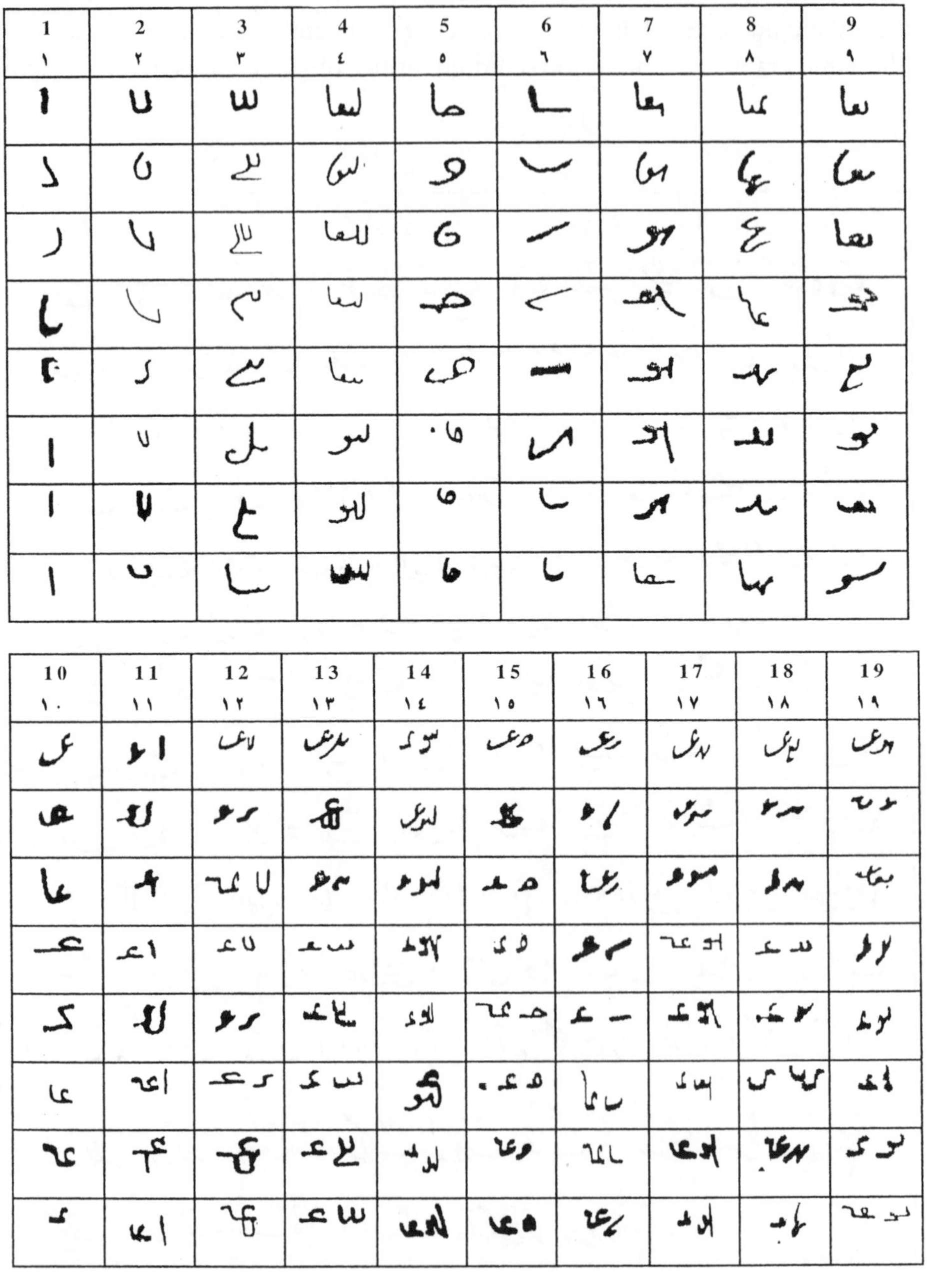

1	2	3	4	5	6	7	8	9
١	٢	٣	٤	٥	٦	٧	٨	٩

10	11	12	13	14	15	16	17	18	19
١٠	١١	١٢	١٣	١٤	١٥	١٦	١٧	١٨	١٩

Figure 5.2.6 Examples of variant *siyaq* graphs from various medieval and early modern Ottoman documents (Öztürk 1996, 66)

was also used to record quantities other than *dinars* and *mans*, such as *mesqal* (a weight equal to 3.64 grams), areas of land, and so forth. Since relatively few persons, even among the numerate, were able to use *siyaq* with ease, it was employed for land records, business accounts, and other information for which privacy or secrecy was at a premium.

In surviving medieval manuals for scribes *siyaq* is generally described in groups of 9 graphs for each decimal place, thus:

1	10	100	1,000	10,000	100,000 etc.
2	20	200	2,000	20,000	200,000
3	30	300	3,000	30,000	300,000
4	40	400	4,000	40,000	400,000
5	50	500	5,000	50,000	500,000
6	60	600	6,000	60,000	600,000
7	70	700	7,000	70,000	700,000
8	80	800	8,000	80,000	800,000
9	90	900	9,000	90,000	900,000

In later Ottoman documents the recording of fractions and decimal numbers was also contrived in the *siyaq* style. Figs. 5.2.1–5.2.3 and 5.2.5–5.2.6 illustrate the actual *siyaq* graphs and show their relationship to the Arabic words for the numerals.[7]

The sociology of *siyaq*

The formal classification of *siyaq* as a ciphered system derived from the words for the numerals obscures its larger historical significance. What distinguishes it from similar systems is the underlying sociology of its introduction and usage. Both the way it was introduced and conditions of its use for some twelve centuries were shaped by particular factors in the organization of society in the Islamic world, that were different from the Christian, Hindu-Buddhist, and Confucian societies to the west and east of it. Instead of being developed by merchants or scientists whose primary interest was in quantity and number, *siyaq* is simply a component of the skills of the Arabizing non-Arab chancery scribe, who wrote mainly for other scribes, in the service of Arab superiors, the state of whose literacy is not known. These scribes operated in small professional communities in

---

7. Further examples of regional variation may be found in Öztürk (1996, 66): various medieval and early modern Ottoman documents; Majma' al-arqām (Mirza Badi' Diwān 1981, 117): Bukhara in Central Asia; Weber (2007, 251): Kashmir; Shakeb (no date, 182): Hyderabad and southern India.

urban centres of government and commerce, making it easier for them to develop all the dimensions of their writing work in ways that came naturally to them but were difficult for outsiders to penetrate. They produced calligraphic documents for public consumption when appropriate, but their writing for other scribes developed into something more like a code. They also communicated routinely with similar scribal communities in other cities (under other governments)— sufficiently to maintain a general standardization of practice over the centuries, but not enough to prevent some regional drift in style—resulting in the stylistic differences that are visible later between examples from the Ottoman, Safavid, and Mughal empires.

For similar reasons *siyaq* remained little known outside these small elite communities in government, land management, and trade. It did not appear in other types of writing, and people outside these professions did not need to know it. It was learnt as part of the scribal apprenticeship, and is introduced in their professional manuals. It was considered part of a restricted professional, rather than a general, education. Western scholars of the region similarly paid little attention to it, except to a limited extent in the specialized fields of economic history and diplomatics.

The history of literacy in general in Persianate civilization contained a paradox. On the one hand there were more people who were able to read the Arabic text of the Qur'an aloud (that is, vocalize the text, irrespective of the ability to comprehend) than who possessed any degree of literacy in other civilizations before the recent spread of mass education. On the other hand, the culture of interactive or functional literacy was confined to a small elite. The ability to read and write *siyaq* was even more restricted. Why should the historical development of numeracy be so different in Persianate civilization from elsewhere? Why, when in its earliest stages Persianate society generated people like al-Khwārizmī, who in their numeracy based on Indian numerals were ahead of other parts of the world, should *siyaq* not have been replaced by these numerals in Persianate society at large centuries before anywhere else?

We now live in a society which depends on universal literacy and numeracy, even though not everyone is fully literate or numerate. The combination of Roman numerical notation and calculation by abacus began to give way to general written numeracy in Indian-Arabic numerals in the West some three hundred years ago. Over the past two centuries this new numeracy has spread throughout the world to become a universal language. It has become the basis of a larger cultural orientation, underlying everything that generates what we recognize as modernity—scientific knowledge, technological capability, and administration. Abacus usage was similarly the basis of a larger cultural orientation. But such non-written calculation did not have the same capability, because it was a face-to-face activity: it was not communicated in writing, and so could not have become the foundation of large organizational frameworks. Since literacy and numeracy are now both taken for granted, there are no social barriers to their acquisition, and they carry no distinctive status.

Why did Persianate societies lag behind in this process? It is interesting to note that one eighteenth-century ruler, Nader Shah, actually gave orders that his registers be written in normal Persian script and that the use of *siyaq* be abandoned (Mohammad Kazem, *Nameh-ye `Alamara-ye Naderi*, I f 11a, in Floor 1998, 95). But after his death in 1747, the succeeding bureaucracies returned to *siyaq* for a further century and a half. The explanation lies in the way Persianate society was organized. It was based in *shari`a*, Islamic law, which specialized in contract. The law was derived from principles that were independent of the government, and it was interpreted and practised by a literate elite of scholars, whose opinions were formally independent of government control. From the early ninth century the caliph was no longer comparable to an emperor, and from 1256 to 1871 there was no caliph to symbolize any centrality of authority. Political centralization was only regional, often no more than local. Long distance trade connected population centres as distant as Morocco and China. The populations of this vast area enjoyed the most open society of the medieval period. Long distance trade, though slow, flourished.[8]

Paradoxically, it may have been as a result of this openness and mobility over such a vast area that Persianate society did not experience the same pressures for socio-political change and the opening up of literacy that eventually ushered in modernity in the West. Literacy had an extremely narrow social accessibility down to the twentieth century, and numeracy remained a narrow specialization within literacy. This accessibility was further narrowed by the fact that, whereas behind writing in general there lay calligraphic models and a cultural awareness of the major Islamic art form of calligraphy, there were no calligraphic models for *siyaq*. Furthermore, the functionality of the abacus and its popularity reduced the pressure for the spread of literate numeracy.

In these conditions scribes were motivated to preserve their largely hereditary social status by increasing their power over the information available to them and maintaining the boundaries on the accessibility to the skills that would unlock it. The drive to preserve status and privilege facilitated the perseverance of *siyaq* and retarded the development of numeracy as well as literacy.

Nothing disturbed the scribes' hold on the spread of numeracy until pressures built up at the end of the nineteenth century as part of the response to colonialism. If Netz (2002, 15) is correct in his discussion of the relationship between numeracy and political organization in classical Greece, we may perhaps argue a similar relationship between numeracy and political history in the Persianate

8. Hodgson's *The venture of Islam* (1974) remains the best entry to the study of this unique social texture.

world: specifically that its political development did not require the counting of citizens, while its economic development required no more than the counting and monetization of their produce on the local level.

Conclusion

Numeracy facilitates larger group awareness. The larger the numbers that can be counted, the larger the entity that can be organized and administered, the larger the empire and the economy, and the more powerful the state. There were no pressures to change the political economy of the Persianate world until the jolt of colonialism. Once the social restrictions on literacy were removed as part of the struggle against colonialism, which began towards the end of the nineteenth century, the restrictive numeracy of *siyaq*, that had served commercial and administrative needs for a millennium, could no longer survive.

However, only a fraction of the documentary sources for this practice have so far been seriously studied by historians. Further study of the uses of number and quantity in Persianate society will in time no doubt improve our understanding of the social conditioning of numeracy.

Annotated bibliography

In addition to full references to the citations in the text the aim of this bibliography is to list the most important and most useful published works on the subject of *siyaq*. It makes no claim to comprehensiveness. Primary sources are not included, except where they are published and available. Some items include bibliographies that lead to primary sources, mostly in manuscript. Probably the richest single source for further bibliography is Mohammad Mehdi Forugh-e Esfahani's *Forughestan*.

Alam, Muzaffar, 'The making of a munshi', *Comparative Studies of South Asia, Africa and the Middle East* 24/2 (2004), 61–72. Discusses *Khulasat al-Siyaq*, written by Indar Sen, probably a Kayastha, in AH 1115 (1703–4 AD). This work is mostly concerned with fiscal management: its three central chapters concern key institutions that dealt with accounting, fiscality, and supplies, that is: the Divān-i A'lā, the Khān-i Sāmān, and the Bakhshi. The conclusion includes examples of arithmetic formulae that would be of use for the *monshi* in his accounting (*siyaq*) practice. The introduction sets out the transition from Hindavi accountancy to Persian in the time of Akbar, and emphasizes the need for the *monshi* class to move with the times. Yet, even more than the *Nigarnāmah* text, this presents a rather narrow conception of the role of the *monshi*. A rather more comprehensive view can be found in the autobiographical materials from the same broad period, insisting as they do on the formation of the moral universe of the *monshi*.

Alparslan, `Ali, 'Khatt', *Encyclopaedia Iranica*, Routledge & Kegan Paul, 1982, 1124–1125.

Amoli, Shams al-Din Mohammad b. Mahmud, *Nafā'es al-fonun fi`arāes al-`Oyun* (ed Mirza Abu al-Hasan Sha`rani), Tehran: Eslāmiya, 1377/1998. See especially vol I, 303–311, *`elm-e estifā*, on accounting and auditing. [In Persian]

Anon, *Mofid nāma*, Cawnpur: Naval Kishore, 1913. Pages 15–16; chapter 19 shows Persian numbers 1–100, 1,000, *lakh* (100,000), and under each of them are given the *siyaq* notation with no explanation or discussion. A book for beginners in Persian. [In Persian]

Bagheri, Mohammad, 'Siyaqat accounting: its origin, history, and principles', *Acta Orientalia* (Budapest), 51 (1998), 297–301. A short introduction to *siyaq* from a modern mathematical standpoint, citing Kremer (1888).

Barker, M A R, *A course in Urdu*, 3 vols, Montreal Institute of Islamic Studies, 1967. See volume I, 356–357.

Elker, Salahaddin, *Divan rakamları*, Ankara: Türk Tarih Kurumu, 1953. A general discussion of the history and use of *siyaq* graphs, in Turkish documents, with useful plates illustrating the figures. [In Turkish]

Farāhāni, Salmān, Bayān al-Saltana, 'Qavā'ed-e dafāter va hesāb', *Farhang-e Iran Zamin*, 23 (1357/1978), 149–177. The author was a Qajar chancery scribe. He presents the theory and practice of writing *siyaq* in accounting. [In Persian]

Fekete, Lajos, *Die Siyaqat-Schrift in der Türkischen Finanzverwaltung*, Akademiai Kiado, 1955. The introduction is very important for understanding the development of *siyaq*. Many plates showing documents, with transcriptions into Latin script.

Floor, Willem, *A fiscal history of Iran in the Safavid and Qajar periods, 1500–1925*, New York: Bibliotheca Persica, 1998. See pages 70, 95, 296. Includes short sections mentioning the use of *siyaq* in administrative documents.

Forbes, William, *A grammar of the Goojratee language*, Bombay: private publication, 1829. One page of *siyaq*, called *hesāb-e roqum*. '[A] short method of writing numbers which is alone used in all Persian writings where sums of money are noted'.

Forugh Esfahāni, Mohammad Mahdi, *Forughestān*, (ed Iraj Afshār), Tehrān: Mirās-e Maktub, 1378/1999. A rich source of information about *siyaq*, plus an extensive bibliography of written and ms sources. [In Persian]

Golchin Ma`āni, Ahmad, 'Resāla dar `elm-e siyāq', *Majalla-ye Dāneshkada-ye Adabiyāt-e Dāneshgāh-e Tehrān*, 12 (1344/1965), 355–623. Describes ms. #7148 by Abu Eshāq Ghiyās al-Din Mohammad Kermāni `Āsheq' (fl 17th century), held in the Āstān-e Qods Library, Meshed. [In Persian]

Gulam Ahmed Munshi, *Farung al-af`al, or dictionary of Persian verbs: amadan nameh*, Bombay: Homee, Sorab & Co, 1925. Includes a brief introduction to the *siyaq* graphs as used in India.

Günday, Dündar, *Arsiv belgelerinde siyakat yazısı Özellikleri va Divan Rakamları*, Ankara: Türk Tarih Kurumu, 1974. Photos of documents reproduced, translated into legible Ottoman script and modern Turkish. Very good tables of the Turkish forms of *siyaq* and *siyaqat*, abbreviations, technical terms, and other material useful for learning to read Ottoman documents. [In Turkish]

Habib, Irfan, *The agrarian system of Mughal India, 1556–1707*, 2nd ed, Oxford University Press, 1999. See pages 470ff for a survey of 'administrative and accountancy manuals'.

Hanaway, William L, and Spooner, Brian, *Reading nasta`liq: Persian and Urdu hands from 1500 to the present*, Costa Messa, CA: Mazda, 1995; 2nd revised ed, 2007. 'An aid to research in historical and other textual materials written in the styles [of Persian] known as *nasta`liq* and *shekasta*' (page 1).

Heywood, C J, 'Siyākat', *Encyclopaedia of Islam*, vol 9, 2nd ed, Brill, 1997, 692–693.

Hinz, Walther, 'Ein Orientalisches Handelsunternehmen im 15. Jahrhundert', *Die Welt des Orients,* 4 (1949), 313–248. Reproduces the ninth section of *Shams al-Siyaq* but with all the original *siyaq* graphs converted to Arabic numerals. Plates 1–2 (#14–15 in the journal) show pages 132a–b of ms. #3986 of the library of Hagia Sophia.

Hinz, Walther, 'Das Rechnungswesen Orientalischer Reichsfinanzämter im Mittelalter', *Der Islam,* 29 (1950), 1–29, 113–141. About bookkeeping. Discusses the various sorts of records that governments and private individuals kept. Useful for the history of *siyaq.*

Hodgson, Marshall G S, *The venture of Islam,* 3 vols, University of Chicago Press, 1974.

Ibn al-Nadim, *Al-Fihrist* (trans [from the Arabic] M R Tajaddod), Tehran: Ebn Sina, 1343/1964.

Ibn Tiqtaqa (Ibn Thiqthaqa), *Fakhrī history* (Arabic text ed W Ahlwardt), Gotha: F A Perthes, 1860.

İnalcik, Halil, *The customs register of Caffa, 1487–1490,* Department of Near Eastern Languages, Harvard University, 1995. 'It is written in a variant of *siyakat (siyaka)* script with some *divani* characteristics. In most cases the figures are rendered in *siyakat* cyphers' (page 3). Interesting in that it shows *siyaq* in use in the fifteenth century.

Jahshiyāri, Mohammad ibn `Abdus, *Ketāb al vozarā va'l-kottāb* (ed M al-Saqqa *et al*), Cairo: Mustafá al-Bābī al-Halabī wa-Awlāduhu, 1938.

Karimi, Asghar, 'Hesāb-i *siyāq*' ya "hesāb-i dināri" ', *Mardom Shenāsi va Farhang-e `Āmma-ye Irān* (AI 2536/1977), 91–100. A brief introduction to *siyaq* and its uses. [In Persian]

Karimi, Asghar, 'Vāhedhā-ye andāza-giri dar il-e bakhtiyāri va hesāb-e siyāq', *Mardom Shenāsi va Farhang-e `Āmma-ye Irān* (1353/1974), 47–57. A description of the units of measurement and the writing of *siyaq* among the Bakhtiyāri. [In Persian]

al-Kāshi, Jamshid ibn Mas`ud, *Miftāh al-hisāb,* Tehran, 1888.

Kazim-zadeh, H, 'Les chiffres siyâq et la comptabilité persane', *Revue du Monde Musulman,* 30 (1915), 1–51. An important, fundamental study of *siyaq* and its use in accounting in Persia.

Kremer, Alfred, Freiherr von, 'Ueber das Einnahmebudget des Abbasiden-Reiches vom Jahre 306 H (918–919)', *Akademie der Wissenschaften, Philosophisch-historische Classe, Denkschriften* (Vienna), 36 (1887), 283–362. Discussion of *siyaq* with examples.

Matin Daftari, Ahmad, 'Asnād-e divāni-ye `ahd-e Qājār', *Rāhnamā-ye Ketāb,* 9 (1345/1966), 31–35. 'These *mostowfis* … kept their account books according to the principles of *siyaq.*' He refers to the *dowra-ye siyāq-nevisi* 'the era of *siyaq*-writing', page 35. A general work on documents of the Qajar period [nineteenth century]. [In Persian]

al-Meyhani, Mohammad ibn `Abd al-Khāleq, *Dastur-e Dabiri* (ed Adnan Sadik Erzi), Ankara: Türk Tarih Kurumu, 1962. The ms is dated 575/1180. The plate following page 99b of the ms shows a very old example of *siyaq.* A manual for beginners entering the chancellery service. [In Persian and Turkish]

Māzandarāni, `Abd-Allah b. Mohammad, *Die Resala-ye falakiyya: ein persicher Leitfaden des stätlichen Rechnungswesens (um 1363)* (ed and trans Walther Hinz), Steiner, 1952.

Menninger, Karl, *Number words and number symbols: a cultural history of numbers,* MIT Press, 1969. Use with caution. He is not correct in his description of how Persian numbers are read and said.

Mirza Badi` Diwān, *Majma` al-arqām (Madzhma` al-arkām* 'Collection of numerals') (Persian text ed A B Vildanova), Nauk, 1981. Facsimile of a ms dated 1212/1798, probably from Bukhara, with translation, commentary in Russian. A general manual of accounting with a section on writing *siyaq,* a glossary, and a table of graphs. [In Persian and Russian]

Mostowfi, `Abd-Allah, *Sharh-e zendagāni-ye man,* Tehran: Zavvār, reprint 1321/1942. Volume II, 335–340 gives a historical account of the development of *siyaq* (no references), and the basic *siyaq* graphs but no useful explanation or description. Use with care. [In Persian]

Navvāb ʿAziz Jang Bahādor Valā, Shams al-Olamā, *Siyāq-e dekkan*, Hyderabad (Deccan): ʿAziz al-Matābeʿ, 1313/1896. A useful introduction to the use of *siyaq* in the Deccan. [In Urdu]

Netz, Reviel, 'Counter culture: towards a history of Greek numeracy', *History of Science*, 40 (2002), 1–27.

Nyberg, Henrik S, *A manual of Pahlavi*, 2 vols, Harrassowitz, 1964. See especially volume I, 173 for a table of Sasanian numerals.

Öztürk, Said, *Osmanlı arsiv belgelerinde siyakat yazısı ve tarihi gelisimi*, Istanbul: OSAV, 1996. A very useful manual with many examples of numbers in Turkish-style *siyaq*. [In Turkish]

Pihan, Antoine Paulin, *Notice sur les divers genres d'écriture ancienne et moderne des Arabes, des Persians et des Turcs*, Paris: L'Imprimerie Impériale, 1856. Short sections on *siyaq*.

Pihan, Antoine Paulin, *Exposé des signes de numération usités chez les peuples orientaux anciens et modernes*, Paris: L'Imprimerie Impériale, 1860. 'Numération arabe', pages 198–214; 'Numération persane', pages 215–26; 'Numération turque', pages 233–37. Includes short sections and discussions of *siyaq*.

Rahnama, Hushang, 'Chand vāzha-ye Irāni; siyāq, bayram, khurvara', *Irānshenāsi*, 7 (1374/1995), 155–157. The author argues for an Iranian origin of the word *siyaq*. [In Persian]

St Clair-Tisdall, W, *Modern Persian conversation grammar*, Heidelberg: Groos, 1923. Includes an excellent table of *siyaq* graphs with their equivalents in western numeration.

Shāhanshāhi (or Shāhanshāhāni) Esfahāni, ʿAbd al-Vahhāb ibn Mohammad Amin Hoseyni, *Bahr al-javāher fi ʿelm al-dafāter*, Tehran, 1306/1888. A general manual of accounting with a section on writing *siyaq*. [In Persian]

Shakeb, Mohammad Ziyāʿ al-Din Ahmad, and Hasan al-Din Ahmad, *Jāmeʿ al-ʿAtiyyāt*, Hyderabad (Deccan): Vala Akademi, no date. Based on the work of Navvāb ʿAziz Jang Bahādor and others, this is a very useful survey of accounting and chancellery practice, focusing on the giving and administration of grants. *Hessa-ye dovvom* (Section two) is devoted to *siyaq* and, among other things, tries to show the difference between the *siyaq* figures used in the Deccan and North India, and those used in Iran and elsewhere. [In Urdu]

Stewart, Charles, *Original Persian letters and other documents, with fac-similes*, London: Kingsbury, Parbury, Allen, & Co, 1825. Includes an excellent table of *siyaq* graphs in their Indian variant.

Storey, A C, *Mathematics*. Vol 2:1-A of *Persian literature: a bio-bibliographical survey*, Brill, 1972, 26. The note mentions some basic texts on ʿelm-e siyaq ʿaccounting'.

Storey, A C, *Arts and crafts*. Vol 2:3-G (a) of *Persian literature: a bio-bibliographical survey*, Brill, 1977, 371–374. Basic bibliography on accounting (ʿelm-e siyaq).

Tod, M N, 'The Greek numeral notation', *The Annual of the British School at Athens*, 18 (1911–2), 98–132.

Vildanova. See Mirzā Badiʿ-Diwān (above).

Weber, Siegfried, *Die persische Verwaltung Kaschmirs (1842–1892)*, 2 vols, Österreichische Akademie der Wissenschaften, 2007, especially vol I, 247–254. Includes tables of *siyaq* graphs as used in Kashmir in the nineteenth century.

Wollaston, Arthur N, *An English-Persian dictionary*, John Murray, 1904 reprint New Delhi: Cosmo, 1978.

Learning arithmetic: textbooks and their users in England 1500–1900

John Denniss

'Numbering', 'reckoning', 'cyphering', and even 'arithmetic' itself are words
that now have an archaic ring to them, but for centuries there have been
textbooks entirely devoted to these subjects. As the Hindu-Arabic numerals
began to pass westwards from Baghdad in the ninth century they were accompanied by treatises that explained how to write and use them. In the Latin West
such treatises became known as 'Algorisms' (after al-Khwārizmī, the author of
the best known of them), and their contents became the essential core of all later
textbooks on arithmetic for many centuries. This chapter will discuss the content
and presentation of Arithmetics published in England from the middle of the
sixteenth century until the end of the nineteenth. In the earlier part of this period
such books were as likely to be used by adults as children. Later, they were written more especially for schoolchildren, and evidence of how they were used in
practice can be discovered from children's manuscripts, a substantial collection
of which is held by the author.

The first Arithmetic published in England was Cuthbert Tunstall's *De arte
supputandi* in 1522. Tunstall, who had studied in Italy, claimed (in his dedication)
to have read all previous Arithmetics. In particular he would have been aware
of some published in the vernacular, but wrote his own work in Latin, perhaps because it was specifically intended for use in the two English universities,

Oxford and Cambridge.[1] It does not seem to have been popular: only one edition was ever published in England, though further editions were printed in Paris. Within twenty years Tunstall's book was followed by two others, both in English: *An introduction for to lerne to rekyn with the pen and with counters*, published anonymously in 1537, and Robert Recorde's *The grounde of artes*, in 1543. Both were to remain in use for many years.

An introduction for to lerne to rekyn was largely a translation from works of the same title in Dutch and French. Allie Wilson Richeson (1947, 49) estimated that the original print run was perhaps of the order of 500. The first edition had a somewhat haphazard appearance, with the later pages elaborating material presented earlier. Changes were introduced in the second edition, in 1539, with less space given to the basic operations of arithmetic and more at the end to the Rule of False Position (see below). After that there were no significant alterations. In all there were eight editions from 1537 to 1629. In the past it has been treated with less than justice: Augustus de Morgan did not mention it in his *Arithmetical books* of 1847, presumably because he never saw a copy. David Eugene Smith, in his *History of mathematics*, gave it only a few lines, claiming that 'it never ranked with *The grounde of artes* either in scholarship or popularity' (Smith 1951, 320), but for almost a century it provided an attractive alternative to *The grounde of artes*. Isaac Newton, for example, owned a copy of *An introduction*, but not *The grounde of artes* (Harrison 1978, 167). The book's lasting popularity can be gauged by the fact that a pared-down version of the original French edition was published as late as 1752.

The first edition of *The grounde of artes* was more than half as long again as *An introduction*. It was a well-constructed work, comprehensive in content, and with extensive explanations given in the form of a dialogue between Master and Scholar. Its homely style must have contributed much to its success. A 'second part' on fractions was added in 1551, taking it to 407 pages. A substantial 'third part', including instruction in calculation of interest, loss and gain, bartering, exchange, a few pages on sports and pastimes, and further material on the rule of three, was added by John Mellis in 1582. The book subsequently went through many more editions, revisions, and extensions. Robert Norton (who had published an English translation of Simon Stevin's seminal treatise on decimals, *De thiende*, in 1608) introduced decimals into *The grounde of artes* in 1615 but Robert Hartwell removed them again in 1618. Hartwell, however, added a short appendix that included the calculation of square and cube roots. In the final edition in 1699, Edward Hatton added a section entitled 'Decimals made easie', but his

---

1. Arithmetics in Latin after Tunstall's were very rare. One was *Elementa arithmeticae numerosae et speciosae* by Edward Wells in 1698. It has the phrase *In usum juventutis academicae* 'For the use of young scholars' on the title page. It is not a commercial arithmetic and focuses mostly on *arithmetica speciosa*, that is, algebra.

attempts to modernize a book that was one hundred and fifty years old were not enough and other textbooks now began to displace it. By then *The grounde of artes* had gone through some forty-seven editions spanning one hundred and fifty-six years.

An introduction and *The grounde of artes* were representative of a genre with a very long history: they reflected the standard pattern of all previous Arithmetics, and most of the topics taught in them were to remain in the curriculum for another three hundred years or more. For that reason we will set the background to the rest of the chapter with a description of their contents, and an explanation of some of the terms used in them.

The contents of early Arithmetics

The chief headings, in more or less the same order in both *An introduction* and *The grounde of artes*, were: numeration, addition, subtraction, multiplication, division (called 'partition' in *An introduction*), reduction, progression, Rule of Three, fractions (in the 1551 and later editions of *The grounde of artes*), Fellowship, the Rule of False Position, Alligation (in *The grounde of artes*), and arithmetic with counters. The first edition of *An introduction* also had sections on duplation (doubling) and mediation (halving) but these were dropped from the second and all subsequent editions.

Both *An introduction* and *The grounde of artes* (until the three final editions), had substantial sections on arithmetic with counters, a topic that continued to be discussed in textbooks until the eighteenth century. The tension between abacist arithmetic, performed with counters manipulated between lines drawn on a board or table, and algorithms worked with numerals on paper or other materials, was a long drawn-out affair (see Chrisomalis, Chapter 6.1 in this volume). Barnard notes that *jettons* 'counters' did not become obsolete in England until the end of the seventeenth century (Barnard 1981, 5, 87). One of the major problems with describing counter arithmetic was that every move on the board

Figure 5.3.1 Recorde's demonstration of 1542 x 365 using counters, from *The grounde of artes*, edition of 1654, page 242

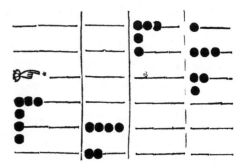

required a new diagram. To multiply 1542 by 365, for example, Recorde needed seven pages, with eight separate illustrations like the one in Fig. 5.3.1, in which the hand indicates the latest move. Recorde devoted further space to showing how, by allocating different values to lines and spaces, the counting board could be adapted to arithmetic with, for example, pounds, shillings, and pence, the currency then in use. Counter arithmetic is, of course, much more easily taught by practical demonstration and oral instruction than by textbook.

Several of the terms used in early Arithmetics will be unfamiliar to modern readers, though all were in use until the late nineteenth century. The 'Rule of Three', or the 'Golden Rule' as it was sometimes called, was taught in all Arithmetics until then, when it was subsumed under the heading of Proportion. The basic version, the Rule of Three Direct, taught how to find a fourth number from three given numbers, in such a way that the ratio of the fourth to the third is the same as that of the second to the first. The rule (multiply the second number by the third and divide by the first) was learned by rote. Another version of such a problem required a different rule, the Rule of Three Inverse. Where more than four numbers were involved, the rule had to be elaborated to the Double (or Compound) Rule of Three Direct, or the Double (or Compound) Rule of Three Inverse. Here is an example of the latter, from *Cocker's arithmetick* (1614), in which five numbers are given and a sixth is to be found:[2]

If a regiment consisting of 939 soldiers, can eat up 351 Quarters of Wheat in 168 Days, how many soldiers will eat up 1464 Quarters in 56 Days at that Rate?

The word 'Alligation' is now obsolete, but it meant the mixing of materials, for example, spices or metals, to produce a blend or alloy at a given price (see Rossi in this volume). Here is an example from *The grounde of artes*:

There are foure sorts of wine of severall prices, one of 6 pence a gallon, another at 8 pence, the third at 11 pence, and a fourth at 15 pence the gallon. Of all these wines I would have a mixture made to the summe of 50 gallons, and so the price of each gallon may be 9 pence. Now demand I: how much must be taken of each sort of Wine?

Recorde devoted more than three pages to this problem. He began by calculating the differences between the actual and required prices and applying the Rule of Three Inverse four times (indicated by the Zs in the diagram in Fig. 5.3.2). Clearly his purpose was to provide a method that could be adapted to any example of this type. A modern solution would be to form two equations in four unknowns, from which we obtain a general set of integer solutions $20 + n$, n, $30 - 4n$, $2n$, but in the sixteenth century algebra was in its infancy and not generally used in Arithmetics.

---

2. Precise page references have not been given for this and other examples because of the large number of editions many of the books went through.

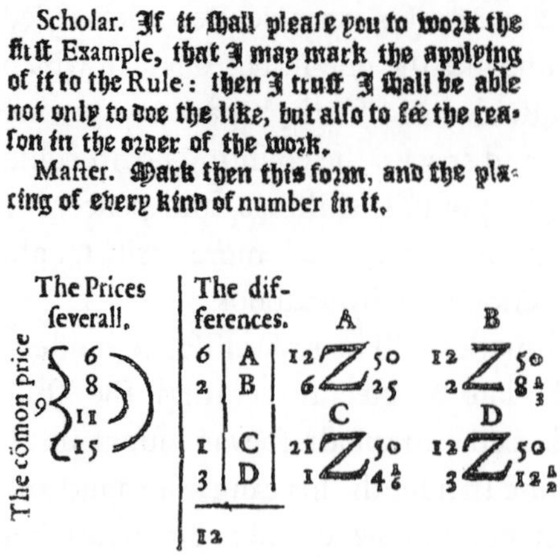

Scholar. Ꝥf it ſhall pleaſe you to wozk the fiꝛſt Example, that Ꝥ map mark the applying of it to the Rule : then Ꝥ truſt Ꝥ ſhall be able not only to doe the like, but alſo to ſée the reaſon in the ozder of the wozk.

Maſter. Mark then this foꝛm, and the placring of every kind of number in it,

Figure 5.3.2 The beginning of Recorde's solution to a problem in alligation, from *The grounde of artes*, edition of 1642, page 355. ('Scholar: It shall please you to work the first Example, that I may mark the applying of it to the Rule: then I trust I shall be able not only to doe the like, but also to see the reason in the order of the work. Master: Mark then this form, and the placing of every kind of number in it.')

'Fellowship' is another term that needs explanation. It arises in situations where several partners invest unequally in an enterprise and want to know how the profit or costs should be apportioned. For example (from Wingate's *Arithmetique* (1630)):

A, B, and C, hold a pasture in common, for which they pay 45,l. [£45] *per annum*; In this pasture A had 24 oxen went 32 dayes, B had 12 there 48 daies, & C had 16 oxen there 24 dayes, now the *question* to be resolved by this *Rule* is, what part each of these tenants ought to pay of the 45,l. rent?

The 'Rule of False Position' begins with an estimated answer and works from that to the true answer. The rule was introduced by Recorde to his Scholar thus:

[the rule]…beareth its name, not for that it teacheth any fraud or falsehood, but for that by false numbers taken at all adventures, it teacheth how to finde those true numbers you seeke for.

Like the Rule of Three, the steps had to be learned by rote, and Recorde offered a verse as a mnemonic (and the equivalent of a modern formula):

Ghesse at this work as hap doth lead,
By chance to truth you may proceed,
And first work by the question,
Although no truth therein be done.
Such falshood is so good a ground,
That truth by it will soon be found.
From many bate too many moe,
From too few take too few also:
With too much joyn too few agayn;
To too few add too many plain;

In cross-wise multiply contrary kinde,
And all truth by falshood for to finde.

One of Recorde's examples concerns a forgetful servant of an affluent master:

There is a servant that hath bought of Velvet and Damask for his master 40 yards, the
Velvet at 20 shillings a yard and the Damask at 12 shillings, and when hee cometh home,
his master demandeth of him, how much he hath bought of each sort; I cannot tell (saith
hee) exactly, but this I know, that I paid for Damask 48 shillings more than I paid for
Velvet: now must you ghesse how many yards of each sort.

In 1363, a law that later became obsolete had prescribed the quality of cloth
different classes could wear: ploughmen and shepherds, for example, could not
wear cloth valued at more than 12 pence a yard (Miles 2006, 280). Even allowing
for inflation the prices in Recorde's question are high, and a sign of social status.
Arithmetics give many examples of goods traded and their relative values, though
one must always allow, as perhaps here, for artificially rounded figures.

All these techniques were taught as rules to be learnt by rote. Pupils were
expected to recognize which type of problem they were faced with, which was the
appropriate rule, and how it was to be applied. The problem with such a system of
teaching is that the rules mount up and any variation in the problem, or even in
the wording of it, can lead to uncertainty. A cautionary note regarding too much
reliance on rules was expressed by Recorde himself:

Yea, but you must prove your self to do some things without my aid, or else you shall not
be able to do any more than you are taught: And that were to learn by rote (as they call
it) then by reason.

Writing more than a century after the publication of *The grounde of artes*, Isaac
Newton expressed similar concerns when he was asked, in 1694, to comment on
the proposed changes to the mathematics curriculum at the mathematical school
at Christ's Hospital:

A Vulgar Mechanick can practice what he has been taught or seen done, but if he is in
error he knows not how to find it out and correct it, and if you put him out of his road, he
is at a stand; Whereas he that is able to reason nimbly and judiciously about figure, force
and motion, is never at rest till he gets over every rub. (Turnbull 1961–77, III 359)

Newton had strong views about the place of arithmetic in the mathematics curriculum; in the same letter about Christ's Hospital curriculum he said:

Arithmetick is set down preposterously in the 12th Article after all the rest of
Mathematicks. For a man may understand and teach Arithmetick without any
skill in Mathematicks, as writing masters usually doe, but without Arithmetick
he can be skilled in noe other parte of Mathematicks, & therefore Arithmetick
ought to have been set down in the very first place as the Foundation of all the rest.
(Turnbull 1961–77, III 357)

Nevertheless, although Newton suggested books that might be suitable for the teaching of algebra, geometry, and trigonometry, he did not suggest one for arithmetic.

The idea that arithmetic should be used as a foundation for mathematics in general, and not just for practical purposes, led to the inclusion of topics of no immediate practical use. *An introduction* had six pages on arithmetical and geometrical progressions, while *The grounde of artes* had no less than thirty-three. The rules for summing such series were clearly stated but the examples were not very realistic. Here, for example, is Recorde's first problem on arithmetic progressions:

A Merchant buyeth 50 pounds of Spices, and agreeth to pay for the first pound 4 pence, for the second 7 pence, for the third 10 pence, for the fourth 13 pence, etc. The question is, how much hee should pay for the last pound, and then how much the 50 pounds cometh to?

For a geometric progression, he gave the traditional nails-in-a-horseshoe problem:

If I sold unto you an horse having 4 shoes, and every shoe 6 nails, with this condition, that you shall pay for the first nayl one ob [a halfpenny], for the second nayl two ob. for the third nayl four ob. and so forth, doubling untill the end of all the nails. Now I ask you, how much would the price of the horse come unto?

After the Scholar has diligently worked his way through to the answer, which was 34,952 pounds, 13 shilling, 7 pence and an ob, the Master remarks mildly: 'That is well done but I think you will buy no horse of the price', to which the Scholar responds: 'No sir, if I be wise'.

Problems of this kind gave practice in calculation, which was the main reason for their inclusion, but they also added variety and even humour to what could be a rather dull diet of computation.

Adult readers and owners

The long and detailed expositions given by Recorde, as compared to the pithier presentations of *An introduction,* raise the question of who the books were written for. Recorde, in his Preface to the Reader in *The grounde of artes*, claimed to be writing especially for those who needed to study alone:

I doubt not but some will like this my Book above any other English Arithmetick hitherto written; and namely, such as lack Instructors, for whose sake I have so plainly set forth the Examples, as no Book that I have seen hath done hitherto: which thing shall be great ease to the rude Readers.

Recorde's reference to 'any other English Arithmetick hitherto written' is ambiguous: would he have regarded Tunstall's work as 'English'? Writing for those who 'lack instructors' was undoubtedly the reason for Recorde's dialogue form: he anticipated the questions a pupil would ask and put them in the mouth of the Scholar. The Master, in turn, explained every step with great care, first giving a clear definition and then illustrating the calculation by means of worked examples. Several later authors adopted the same form.

The readers of Arithmetics were not just children. Indeed there are several examples of otherwise well-educated people who learned arithmetic more or less on their own with the aid of such books. John Wallis, later Savilian Professor of Geometry at Oxford, attended Felsted School in Essex, but describes how he learned arithmetic at home in the holidays from the books of his younger brother:

While I continued a Scholar there, at Christmas 1631 (aged 15), I was, for about a fortnight, at home with my mother at Ashford. I there found that a younger Brother of mine (in Order of a Trade) had for about 3 months been learning (as they call'd it) to *Write and Cypher* or *Cast account* (and he was a good proficient at that time). When I had been there a few days; I was inquisitive to know what it was, they so called. And (to satisfie my curiosity) my Brother did (during the Remainder of my stay there before I returned to School) shew me what he had been Learning in those 3 months. Which was (beside the writing a fair hand) the *Practical* part of *Common Arithmetick* in *Numeration, Addition, Subtraction, Multiplication, Division. The Rule of Three (Direct and Inverse), the Rule of Fellowship (with and without, Time) the Rule of False-Position, Rules of Practice and Reduction of Coins* and some other little things. Which when he had shewed me by steps, in the same method as he had learned them: and I had wrought over all the Examples which he had before done in his book; I found no difficulty to understand it and I was very well pleased with it: and thought it was ten days or a fortnight well spent. This was my first insight into *Mathematicks*; and all the *Teaching* I had. (Scriba 1970, 26–27)

Another future Savilian Professor (of astronomy), Edmund Halley, also learned arithmetic from a man in 'trade'. John Aubrey tells us that Halley's father was a 'Soape-boyler' and that 'At 9 yeares old [in 1665], his father's apprentice taught him to write, and arithmetique' (Aubrey 1992, 120).

The diarist, Samuel Pepys, born in 1633, did not receive any education in mathematics until he was nearly 30, despite attending St Paul's school in London and Trinity Hall, Cambridge. He had to engage a private tutor, a Mr Cooper, who seems to have begun by directing Pepys to get to grips with the multiplication table, which he found particularly hard work (Pepys, *Diary*, 4 July 1662).

As late as 1783, William Taylor wrote, in the Preface to *A complete system of practical arithmetic*:

...there are a great many adult persons, and grown up youth, who through the narrowness of their circumstances, or the neglect of their friends, are forced to endeavour to improve their lost time as well as they can. To such as these the following Treatise will be of great service. (Taylor 1783, vii)

Diversification in the seventeenth and eighteenth centuries

Two new trends emerged during the seventeenth century: one was the publication of ready reckoners, which enabled those with little or no knowledge of arithmetic to find answers to practical problems, and those who did have some knowledge to obtain the answer more rapidly and with less labour. Some authors, like Leonard Digges, had included individual tables in their arithmetical works, but complete books devoted to tables came much later. The other change was the wider range of mathematics taught in Arithmetics.

The simplest ready reckoners were multiplication tables. By the seventeenth century these were being extended: John Darling's *Carpenter's rule made easy* (1658), for example, gives all products up to 100 × 100 and some beyond that. Other tables gave the cost of *n* articles at a given price. Simple interest was also important. An early collection of interest tables was Edward Hatton's *An index to interest* (1711). There were also tables for calculating areas and volumes, to be used by carpenters, surveyors, builders, and merchants. The use of calculating aids such as Napier's bones, the slide-rule, and logarithms was also taught.

One device employed to keep such books pocket-sized and yet cover the range of measurements needed was to adapt the tables for more than one purpose. Darling, for example, who aimed his book particularly at carpenters, gave a 'Table of the square of unequal sided timber' (in effect, a square root table giving the answer in inches) in order to find the volume of a block of wood. If, for example, the dimensions were 8 ½ inches, by 13 ½ inches, by 7 feet, one could consult this table to find the side of a square of the same cross-sectional area (10 ½ inches), then a relatively short 'Table of timber measure' to give the volume as 5.393 cubic feet, an accuracy justified neither by the initial data nor the needs of any timber merchant (see Fig. 5.3.3).

The other notable trend in seventeenth-century Arithmetics was the inclusion of a greater range of material, land measurement being one of the most common. Some authors also began to strive towards a broader based mathematical text, though still keeping a strong arithmetical core. Algebra, in particular, began to gain a foothold. William Leybourn divided his *Arithmetick* of 1657 into four books: vulgar arithmetic (that is, common arithmetic), decimal arithmetic, instrumental arithmetic, and algebraical arithmetic (that is, arithmetic with letters), and managed to do so in 346 small pages. Samuel Jeake's *A compleat body of arithmetick in four books*, first published in 1696 and re-issued in 1701, was a massive volume of 664 folio pages, possibly the most comprehensive work on arithmetic ever written. These and similar works, like Alexander Malcolm's *A new system of arithmetick* (1730) and John Mair's *Arithmetic, rational and practical* (1766), were more like reference books than textbooks.

For general use, something simpler, smaller, and cheaper was needed. Recorde's *Grounde of artes* was gradually replaced by a succession of best-sellers such as Edmund Wingate's *Arithmetique made easie* (1630), James Hodder's *Arithmetick* (1661), and, especially, by Edward Cocker's *Arithmetick* (1694), which went through at least sixty editions over more than one hundred years. Thomas Dilworth, too, was highly successful with *The schoolmaster's assistant* (1743). Like Recorde, Dilworth wrote in dialogue form. His book went through forty-nine editions in England and many more in north America. Many other authors tried their hand but were less popular; Wallis and Wallis (1986) list more than forty Arithmetics published in the first sixty years of the eighteenth century, but most did not go beyond the first edition. The most successful author of all was Francis Walkingame, who established a boarding-school in London. His *The tutor's assistant* (1751) became an immediate best-seller with at least two hundred and forty-six editions over one hundred and thirty years. Print runs were in

60 *A Table of the square*

2	9 1
13	9 2
2	9 2.
14	9 3.
7 ½	
8	7 3
2	7 3.
9	8 0.
2	8 1.
10	8 2.
2	8 3.
11	9 0
2	9 1
12	9 2
2	9 2.
13	9 3
2	10 0
14	10 1
2	10 1.
15	10 2
8 Inch. sq.	
2	8 1
9	8 2
2	8 3
10	8 3
2	9 0
11	9 1
2	9 2
12	9 3
2	10 0

13	10 0.
2	10 1.
14	10 2
2	10 3
15	10 3.
2	11 0.
16	11 1
8 ½	
9	8 3
2	8 3.
10	9 1
2	9 1.
11	9 2.
2	9 3
12	10 0
2	10 1
13	10 2
2	10 2.
14	10 3.
2	11 0
15	11 1
2	11 1.
16	11 2.
2	11 3
17	12 0
9 Inch. sq.	
2	9 1
10	9 2
2	9 2.
11	9 3.

2	10 0.
12	10 1.
2	10 2
13	10 3
2	11 0
14	11 1
2	11 1.
15	11 2.
2	11 3.
16	12 0
2	12 0.
17	12 1.
2	12 2
18	12 3
9 ½	
10	9 3
2	10 0
11	10 0.
2	10 1.
12	10 2.
2	10 3.
13	11 0
2	11 1
14	11 2
2	11 3
15	11 3.
2	12 0.
16	12 1
2	12 2
17	12 2.
2	12 3.

18	13 0
2	13 1
19	13 1.
10 Inc. sq.	
2	10 1
11	10 2.
2	10 3
12	10 3.
2	11 0.
13	11 1.
2	11 2.
14	11 3.
2	12 0
15	12 1
2	12 1.
16	12 2.
2	12 3
17	13 0
2	13 1
18	13 1.
2	13 2
19	13 3.
2	13 3.
20	14 0.
10 ½	
11	10 3
2	10 3.
12	11 1
2	11 1.

A Table of Timber Measure. 89

Leng.	9 *Inch.* *sq.* Feet 1234.	9 ½ Feet 1234.	10 *Inch.* *sq.* Feet 1234.	10 ½ Feet 1234.
1	0.5625	0.6267	0.6944	0.7656
2	1.1250	1.2534	1.3888	1.5312
3	1.6875	1.8802	2.0833	2.2968
4	2.2500	2.5069	2.7777	3.0625
5	2.8125	3.1336	3.4722	3.8281
6	3.3750	3.7604	4.1666	4.5937
7	3.9375	4.3871	4.8611	5.3593
8	4.5000	5.0138	5.5555	6.1250
9	5.0625	5.6406	6.2500	6.8906
10	5.6250	6.2673	6.9444	7.6562
11	6.1875	6.8940	7.6388	8.4218
12	6.7500	7.5208	8.3333	9.1875
13	7.3125	8.1475	9 0277	9.9531
14	7.8750	8.7743	9.7222	10.7187
15	8.4375	9.4010	10.4166	11.4843
16	9.0000	10.0277	11.1111	12.2500
17	9.5625	10.6544	11.8055	13.0156
18	10.1250	11.2812	12.5000	13.7812
19	10.6875	11.9079	13.1944	14.5468
20	11.2500	12.5347	13.8889	15.3125
21	11.8125	13.1614	14.5833	16.0781
22	12.3750	13.7881	15.2777	16.8437
23	12.9375	14.4149	15.9722	17.6093
24	13.5000	15.0416	16.6666	18.3750
25	14.0625	15.6684	17.3611	19.1406
26	14.6250	16.2951	18.0555	19.9062
27	15.1875	16.9218	18.7500	20.6718
28	15.7500	17.5486	19.4444	21.4375
29	16.3125	18.1753	20.1388	22.2031
30	16.8750	18.8020	20.8333	22.9687

Figure 5.3.3 Tables of squares and timber measure, from Darling's *The carpenter's rule made easy*, 1727, pages 60 and 89

the thousands, particularly for nineteenth-century editions. The reason for the book's popularity is not immediately obvious. Its explanations are brief to the point of inadequacy and, unlike Recorde's, are no help to anyone working alone. Here, for example, is Walkingame's description of subtraction:

SUBTRACTION
TEACHETH to take a less Number from a greater, and shews the Remainder, or Difference.

RULE. This being the Reverse of Addition, you must borrow here (if it requires) what you stopped at there, always remembering to pay it to the next.

PROOF. Add the Remainder and less Line together, and if the same as the greater, it is right.

One attraction of Walkingame's book must have been that it was compact and cheap at 1s 6d for the first edition, rising to 2s for the second, and good value in relation to his competitors; another very important factor was that it included copious exercises. Recorde had included virtually no exercises and Cocker very few, but Dilworth had a large number. It is thought that Walkingame modelled his book on Dilworth's, recasting his dialogue into a more straightforward didactic presentation. The use of such exercises is evident in the many children's manuscripts that have survived, particularly from the eighteenth and nineteenth centuries.

Children's manuscripts

Pupils would not normally have had access to printed texts unless they had an individual tutor, but instead created their own manuscript textbooks (see Yeldham 1936; Cline Cohen 1982). Two major collections of children's manuscripts survive, dating from 1684 to 1900. The larger one, of one hundred and ninety-five volumes from one hundred and forty pupils, is in the John Hersee Collection in the Mathematical Association Library in Leicester University. The second, my own, collected over about seventeen years, consists of seventy-three volumes from sixty pupils. The manuscripts come from all over England, from Kent in the southeast to Cumberland in the northwest, and there is also one from Scotland, though none from Wales. Very few of the manuscripts give the child's age. We know the ages of only four of the sixty children represented in the second collection; those four were between ten and fourteen years old. About 15 per cent of the manuscripts in both collections were written by girls, though these appear only after 1809.

Typically each topic in the manuscripts begins with a definition taken verbatim from a textbook. The first worked example is usually the first given in the textbook, and many of the subsequent examples are also from the same book.

The examples are not marked or corrected by the teacher, suggesting they were exercises in copying rather than calculation. The resulting manuscript was meant to be a permanent record of what had been learnt, to be kept and referred to in later life. For this reason they were properly bound, often in leather or vellum.

The earliest manuscript in either collection is from 1684, by Richard Daw. One of its examples is the following (Fig. 5.3.4): 'If 1 hogsett [hogshead] rebate 13 gallons ½ what will 25 Tun ½ rebate?' Daw used reduction and multiplication to bring every quantity to quarts, the rule of three (dividing 1,388,016 by 252) to get the answer in quarts, then division twice more to change quarts to gallons, and gallons to hogsheads. He forgot to add 54 gallons to his final answer, though it is clear he knew he should do so since he worked out the rebated quantity in the right hand calculation.

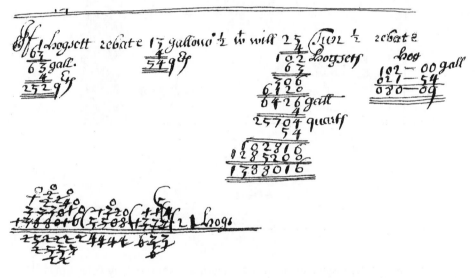

Figure 5.3.4 Example of the 'Rule of Three' from a manuscript by Richard Daw, 1684, f. 93v. ('If 1 hogsett rebate 13 gallons ½ what will 25 Tun rebate')

Children were often allowed, or possibly encouraged, to decorate the pages with elaborate scrolls and sometimes pictures. Figs. 5.3.5 and 5.3.6 show the artwork of G Nicholas in 1832. The problem in verse in Fig. 5.3.5 is:

> As I was beating on the Forest Grounds,
> Up starts a Hare before my two Greyhounds:
> The Dogs, being light of Foot, did fairly run,
> Unto her fifteen Rods, just twenty-one.
> The Distance that she started up before
> Was fourscore sixteen Rods just, and no more:
> Now this I'd have you unto me declare,
> How far they ran before they caught the Hare?

Figure 5.3.5 Problem in verse, illustrated, from a manuscript by G Nicholas, 1832, f. 71r

This comes from a book that Nicholas worked from, *The tutor's guide* by Charles Vyse (edition of 1807, page 67).

Most of the manuscripts (some 70 per cent) were devoted entirely to arithmetic. In every case they appear to be based on textbooks, though it has not always been possible to identify the author. Some contain work taken from more than one author and many used supplementary material. Occasionally the questions are related to individual pupils, for example: 'Barnet Butterfield suppose you are 12 years old what year was you Born in?' (Fig. 5.3.7) Amongst the non-arithmetical material in these books there is often a section on mensuration eliding into practical geometry, surveying, trigonometry, and occasionally a little algebra, its first appearance in these collections being in 1722. There are one or two short excerpts

Figure 5.3.6 Decorative heading from a manuscript by G Nicholas, 1832, f. 67

from Euclid, logarithms rarely, and single instances of spherical trigonometry, fluxions (Newton's calculus), and the arithmetic of infinites.

The manuscripts give an invaluable insight into contemporary teaching methods. Definitions, taken from a textbook, were faithfully copied out by the pupils (presumably from the blackboard). Teachers then set problems taken from textbooks, often drawing on a variety of sources. Answers were helpfully given in the textbooks, in brackets at the end of each question, though teachers must have given some oral explanation of the intermediate steps. Additional help for teachers was provided from the second half of the eighteenth century in separate 'keys', in which many of the problems were shown with their working. The practice of having a pupil's book without answers and a teacher's book with answers was a late nineteenth-century development.

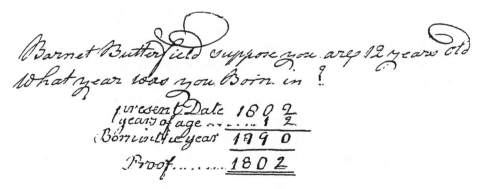

Figure 5.3.7 A personal subtraction, from a manuscript by Barnet Butterfield, 1802, f. 19

The nineteenth century

Consideration of textbooks published in the nineteenth century has to be set against developments in education in that period. Non-governmental agencies had done much to make education more readily available since early in the nineteenth century. The National Society, established in 1811 by the Church of England, opened schools whose aims were primarily religious, but they also taught a wider curriculum, which usually included arithmetic. The Non-Conformist churches soon followed with the establishment of what came to be known as British Schools, run on similar lines. Despite these moves, a Parliamentary Select Committee reported in the 1840s that about one-third of children were receiving no formal education whatever, and of those who did attend school few did so for long. In the 1840s the average length of school attendance was between one and two years (Howson, 1982, 104). As to what these children learned, a Public Commission in 1858 reported that only 69.3 per cent of children attending the 1824 public weekday schools visited were taught arithmetic, and only 33.8 per cent of those attending 3495 private schools (run by individuals).

There was some relevant legislation in the first half of the century. Sir James Graham's Factory Act of 1844, for example, required factory owners to ensure that children in their employ should spend three whole days or six half days in school, though this was somewhat counterbalanced by the lowering of the minimum age of employment from nine to eight in the same Act. But there can be no doubt that the most important development was the 1870 Education Act, which gave every child the right to elementary education and made it the Government's responsibility to ensure its delivery. The expansion of education required a corresponding increase in the number of teachers. The first Teacher Training College was opened in 1839, and by 1860 there were sixty such colleges, though their output was relatively small. Many teachers entered the profession simply as pupil-teachers, supposedly under regular instruction from the headteacher. Untrained child 'monitors' were also employed, until they were gradually phased out in the 1870s by the introduction of 'supplementary' teachers. However, these too were unqualified, the only requirements being that they should be at least eighteen years of age, have been vaccinated against smallpox, and be 'employed during the whole of the school hours in the general instruction of the scholars and in teaching needlework' (such teachers being almost always female) (Horn 1989, 183).[3]

Under such circumstances it is hardly surprising that rote learning and copying from textbooks remained favoured methods of teaching. Many authors

---

3. The relationship between the learning of needlework and the learning of mathematics in the education of girls has been explored in detail by Harris (1997).

continued to model themselves on Walkingame, whose *Tutor's assistant* itself ran on until 1885. John Colenso's *Arithmetic designed for the use of schools*, first published in 1843 and in many further editions, contained more explanatory material than Walkingame, but a much more significant advance was made by John Brook-Smith in his *Arithmetic* of 1860. In the preface to the 1891 edition he wrote:

Every writer on Arithmetic at the present day feels the necessity of explaining the principles on which the rules of the subject are based, but every writer does not as yet feel the necessity of making these explanations strict and complete; or, failing that, of distinctly pointing out their defective character. Difficulties are still avoided or slurred over, and incomplete proofs without one word of remark or warning are used as though they were full and satisfactory. This surely ought not to be.

Brook-Smith was as good as his word; he set out his explanation of Subtraction, for example, very clearly over three pages before giving a set of exercises. Significantly, too, the book ends with twenty-four pages of sample examination papers, from University of London Matriculation, entry to the Royal Military Academy, and so on. The answers to these papers were given at the end of the book, not after each question as had been the custom. This subsequently became common practice.

The examination system for older pupils, particularly the introduction of the School Certificate (for age sixteen) and Higher School Certificate (for age eighteen), developed strongly through the nineteenth century and came to determine the curriculum for secondary schools. The *Arithmetic* published in 1886 by Charles Pendlebury, senior mathematics master at St Paul's School in London, was primarily aimed at older pupils (over eleven) being prepared for examinations. Pendlebury made every effort to keep up with new types of question being set by examiners, and each new edition incorporated examples of questions set for the first time the previous year. The book was immediately successful, going through at least thirty-nine editions and still in print until 1947. Pendlebury's simpler *A shilling arithmetic* (1899), for younger children, was if anything even more popular. Pendlebury's *Arithmetic* finally put paid to the Rule of Three and all its variants, replacing it by a chapter on proportion and its applications.

These works provided little help to teachers of younger children. One obstacle was the adult language in which they were written; children would have had to be able to read proficiently to be able to work from them. Some simple books were published to help with the memorization of the multiplication tables: *Marmaduke Multiply's merry method of making minor mathematicians* (1816) had a picture on each page (Fig. 5.3.8), to be hand-coloured by the pupil, and a short rhyme, such as:

Twice 7 are 14
They're dancing on the Green

Figure 5.3.8 'Twice 7 are 14' from *Marmaduke Multiply's merry method of making minor mathematicians* (1816)

Charming as these were, however, they were not textbooks, nor were they teachers' guides. There was a worrying gap.

The Swiss educational reformer, Johann Heinrich Pestalozzi, active in the early years of the nineteenth century, believed that children should pursue their own interests and learn through activity rather than formal teaching. In England

Horace Grant promoted similar ideas in his two books on arithmetic: *Arithmetic for children*, published sometime before 1835, followed by *Arithmetic for schools and families* in 1841. Introducing the first, he wrote:

Arithmetic is commonly learnt by rote, and is never thoroughly learnt; it is almost always an unpleasant task both to the pupil and the teacher.

A great defect of the existing elementary works on Arithmetic arises from the little variety they offer to the child, either in objects, thought or language.

During the lesson, the teacher should place at hand a small box containing a few counters, pebbles, beans, and small shells. A variety of any kind of small objects, such as wooden cubes, buttons, marbles, nuts, nails, and bits of stick or cork, will also answer the purpose.

He later added to this list a few small weights, coins, and a foot and yard measure, emphasizing his central idea that number teaching must start with the concrete and be applicable to the real world of the child. The rest of his book is a collection of questions and exercises, to be read to the child and to be answered using objects as necessary, the objects to be discarded as the child progresses. In the Introduction to *Arithmetic for schools and families*, Grant says that the book is for the instruction of children between the ages of eight and eleven, but that it is written for teachers (or parents) rather than for the children themselves. It is divided into chapters on the various aspects of arithmetic to be found in any textbook of the period, but explained with more than typical clarity.

Although both of Grant's books went through a number of editions, they did not make the impact in Britain that Warren Colbourn's *An arithmetic on the plan of Pestalozzi* (1821), based on the same principles, made in the US, where new ideas were embraced earlier and more readily (Cline Cohen 1982, 134–138). Shortly before he died in 1891, the teacher and pedagogist, Reverend Robert Herbert Quick, gave a lecture entitled 'The first stage in arithmetic' to the College of Preceptors, an institution concerned with raising the professional standing of teachers, in which he said:

How children should get their first notions of number hardly anybody in this country knows, and, but for the goodly band of ladies who have now begun to study education scientifically, we might add hardly anybody in this country seems to care. (Quick 1896)

Both Recorde and Newton had long ago pleaded for more imaginative teaching of arithmetic, based on understanding rather than on rote learning. At the end of the nineteenth century such ideals had still barely begun to be put into practice.

Conclusion

In the mid-sixteenth century, *An introduction* and Recorde's *The grounde of artes* made arithmetic available for the first time in English. *The grounde of artes* in particular offered material relevant to the everyday and commercial needs of the time, and both books survived well into the seventeenth century. *The grounde of artes*, however, came to have two main drawbacks: it grew too large and yet it contained hardly any exercises. Some Arithmetics attempted to be comprehensive, and became correspondingly larger and more expensive, but in the end those that were pared down became more popular. Space for exercises was made at the expense of the exposition, which was reduced in some cases beyond the point at which it could be helpful.

Arithmetics were used both by students trying to learn alone and by teachers whose own elementary training left them in need of such aids. As the years went by, inertia and the success of existing books meant that Arithmetics tended to perpetuate the content and methods of the past, rather than responding to new methods of education in a changing social context. One would hardly guess from nineteenth-century Arithmetics, for example, that an industrial revolution was under way. The curriculum remained essentially mercantile in nature, and pupils continued to learn through the copying and rote learning of rules and methods. Some of the first hints of permanent change can be seen in the early nineteenth-century writings of Grant, who encouraged children to establish ideas of number and measurement using simple concrete objects or through exploration of their own environment. Major reforms in the teaching of arithmetic in England, however, had to wait until the twentieth century, and it was only then that Arithmetics as a genre finally disappeared.

Bibliography

Dates given for Arithmetics are those of first editions only.

Anonymous, *An introduction for to lerne to rekyn with the pen and with counters*, St Albans, 1537.

Anonymous, *Marmaduke Multiply's merry method of making minor mathematicians*, London, 1816.

Aubrey, John, *Aubrey's brief lives*, Oliver Lawson-Dick (ed), Mandarin, 1992.

Barnard, Francis Pierrepont, *The casting-counter and the counting board: a chapter in the history of numismatics and early arithmetic*, Oxford, 1916; reprinted Fox, 1981.

Bockstaele, P, 'Notes on the first arithmetics printed in Dutch and English', *Isis*, 51 (1960), 315–321.

Brook-Smith, John, *Arithmetic in theory and in practice*, Cambridge, 1860.

Cline Cohen, Patricia, *A calculating people*, University of Chicago Press, 1982.

Cocker, Edward, *Cocker's arithmetick*, London, 1694.

Colbourn, Warren, *First lessons in arithmetic on the plan of Pestalozzi*, Boston, 1821.

Colenso, John Williams, *Arithmetic designed for the use of schools*, London, 1843.

Curtis, Stanley James, *History of education in Great Britain*, University Tutorial Press, 1948.

Darling, John, *The carpenter's rule made easy*, London, 1658.

Digges, Leonard, *A boke named tectonicon*, London, 1556.

Dilworth, Thomas, *The schoolmaster's assistant*, London, 1743.

Grant, Horace, *Arithmetic for young children*, first publication date unknown.

Grant, Horace, *Arithmetic for schools and families*, London, 1841.

Harris, Mary, *Common threads: women, mathematics and work*, Trentham Books, 1997.

Harrison, John, *The library of Isaac Newton*, Cambridge University Press, 1978.

Hatton, Edward, *An index to interest*, London, 1711.

Hodder, James, *Hodder's arithmetick, or, that necessary art made most easie*, London, 1661.

Horn, Pamela, *The Victorian and Edwardian schoolchild*, Alan Sutton, 1989.

Howson, Geoffrey, *A history of mathematics education in England*, Cambridge University Press, 1982.

Jeake, Samuel, *A compleat body of arithmetick in four books*, London, 1696.

Leybourn, William, *Arithmetick*, London, 1657.

Mair, John, *Arithmetic, rational and practical*, Edinburgh, 1766.

Malcolm, Alexander, *A new system of arithmetick*, London, 1730.

Miles, David, *The tribes of Britain*, Phoenix, 2006.

de Morgan, Augustus, *Arithmetical books*, London, 1847.

Pendlebury, Charles, *Arithmetic*, Cambridge, 1886.

Pendlebury, Charles, *A shilling arithmetic*, London, 1899.

Quick, R H, 'The first stage in arithmetic', *Parents' Review*, 7 (1896), 8–17.

Quick, Robert Herbert, 'The first stage in arithmetic', *Parents' Review*, 7 (1896), 8–17.

Recorde, Robert, *The grounde of artes*, London, 1543.

Richeson, Allie Wilson, 'The first arithmetic printed in English', *Isis*, 3 (1947), 47–56.

Scriba, Christoph, 'The autobiography of John Wallis', *Notes and Records of the Royal Society*, 25 (1970), 17–46.

Smith, D E , *History of mathematics*, Boston 1923, reprinted New York: Dover, 1951.

Taylor, William, *A complete system of practical arithmetic*, Birmingham, 1783.

Turnbull, H W, (ed), *The correspondence of Isaac Newton*, 7 vols, Cambridge University Press, 1961–77.

Vyse, Charles, *The tutors's guide*, London, 1770.

Walkingame, Francis, *The tutor's assistant*, London, 1751.

Wallis, P J, and Wallis, R V, *Biobibliography of British mathematics and its applications, 1701–1760*, Epsilon Press, 1986.

Wingate, Edmund, *Arithmetique made easie*, London, 1630.

Yeldham, Florence A, *The teaching of arithmetic through four hundred years*, London 1936.

Algorithms and automation: the production of mathematics and textiles

Carrie Brezine

Mathematics is a diverse field. To the uninitiated, it is a mysterious subject, often presumed to be unintelligible and frighteningly abstract. Many people assume that mathematics is all about numbers, and it is; but it is also about shapes, sets, symmetry, networks, algorithms, and transformations. The specialized definitions and notational systems of mathematics make it almost impossible for a lay person to pick up a recent article on, say, knot theory and make any sense of it. But the mathematics of scholarly publications is not and has never been the only practice of mathematics. Mathematical principles have long been used by artists and engineers, just as anyone dealing with currency practices basic arithmetic to manage income and expenditures. Nor is the academic tradition of western Europe the only method of framing and disseminating mathematical ideas. Indigenous cultures from diverse areas and time periods have developed sophisticated artifacts, games, and methods of record-keeping which show that mathematics has evolved in many different ways (Urton 1997; Gerdes 1999; Selin 2000). The practice of weaving encompasses a multitude of mathematical problems ranging from arithmetical calculation to abstract symmetrical manipulations. Craftspeople who create good fabric are practicing mathematical principles, though they may not communicate them in the way we expect western mathematics to be presented. The sophisticated thought processes behind the production of intricate textiles indicate that even weavers in societies without text have control of abstract mathematical concepts. Framing weaving in terms of mathematics also suggests numerous open problems touching on various modern fields, including combinatorics and geometry.

This chapter broadly describes two methods of textile production and considers their mathematical implications. The two cases are floor loom weaving, as practiced in Europe from medieval times, and weaving on a variable tension (backstrap) loom, the common method in the Andes of South America. In the Andes the machine used to produce cloth is, at first glance, substantially simpler than a western-style loom, yet Andean cloth is some of the most structurally complex in the world. Each technique has its advantages and drawbacks. Both incorporate sophisticated mathematical concepts. Because they are so different, they make good comparative studies for investigating how people solve the many conceptual challenges associated with creating cloth. In order to understand these challenges, and their relationship to mathematics, basic definitions of fabrics will be discussed first.

Definitions and classifications

Words used in speaking about textiles often have different connotations than accorded them in ordinary everyday speech. This article adheres to Irene Emery's terminology as closely as possible (Emery 1994); even among textile scholars the use of terms is not always consistent. *Fabric* is used as 'the generic term for all fibrous constructions' (Emery 1994, xvi). Though the latter part of this chapter focuses on weaving, other structures will be briefly discussed, so that the special characteristics of woven constructions can be better understood. Fabrics are composed of *elements*. In general, the elements are long and flexible. Plant and animal fibers are the most common materials used for fabric construction. Some plants, such as those of the bast family, can yield long unbroken fibers which are strong enough to be stuck together end to end to produce continuous lengths. Silk, unwound from the cocoons of silkworms, is another natural fiber that occurs in very long strands (up to 2,000 m). In contrast, materials such as wool and cotton come in short lengths (approximately 1.5–5.0 cm for cotton, 5.0–30.0 cm for wool, depending on the breed of the animal) (Ross 1983, 89–90, 104–107). To make a continuous thread, short fibers must be twisted together. *Spinning* involves extenuating a mass of fiber to the desired thickness while applying twist. Two twist directions are possible, named S and Z according to the angle the fibers make relative to the axis of the yarn. The twist helps hold the fibers together and endows the thread or yarn with stored energy. Often the energy is neutralized by plying a spun thread with one or more others of its kind, twisting the group together in the direction opposite to that in which they were spun.[1] This process produces a stronger, more stable, and even product and significantly reduces the tendency of the yarn to kink back on itself. Though continuous fibers such as silk and linen do not strictly need to be spun, depending on how they are harvested, it is most common to use them in some spun form.

Once the yarns or threads are prepared,[2] fabrics can be manufactured in numerous ways. An infinite variety of structures is possible, and each structure can be achieved in several ways. The domain of fabric structures has been classified into divisions based on the number and types of elements used (Emery 1994). Fabrics can be made with a single element, a set of equivalent elements, or multiple sets of elements; they make an integral structure by *interworking*. Interworking is a general term which includes specific techniques such as twisting, knotting, looping, and *interlacing* (to be defined later). Knitting is a common example of a single-element construction, so called because it 'is made up of a single continuous element interworked with itself' (Emery 1994, 27). A knitted fabric is composed of a series of loops side by side; on each row, the yarn is drawn up through each loop in turn, creating another row of loops.

A *set* of elements is a group of elements 'all used in a like manner, that is, functionally undifferentiated and trending in the same direction' (Emery 1994, 27). In fabrics made with one set of elements, each element has an equivalent role. The braiding of hair is a common example of this kind of structure: each of three sections interlaces over and under the others in the same pattern. Each strand of the braid performs equivalent motions and they contribute equally to the integrity of the final structure. The elements are fixed at one point, and all tend generally downward from the origin of the braid, though they meander back and forth within it. Assuming all of the elements are of the same grist, any element is isomorphic to any other element in the fabric. This general idea applies to bands made of more than three elements. *Interlacing* is a specific form of interworking in which each element follows a generally linear path (allowing for turns at the edge of a fabric) and passes over or under the elements it crosses (Fig. 5.4.1). Even in structures based on one set of elements the patterns of interlacing need not be restricted to over one, under one; infinite variation is possible (Owen 1995; 2004). This class of structures also includes examples which are not interlaced, but held together through linking, twining, or knotting. Linked and interlaced

---

1. The basic principle of spun energy can be observed in a simple exercise: take a shoelace or a similar short length of string. Hold one end fixed and twist the other end repeatedly in the same direction. When you bring the two ends of the string together, the middle portion will rotate back on itself, creating a twisted cord. It will tend to do this at any point where it is not under tension.

2. The distinction between yarn and thread is not always clear. In general, yarn is less tightly twisted and larger in diameter than thread (Emery 1994, 12–13). In this Chapter I use the terms interchangeably. Regardless of the name applied, it is important to remember that any fabric structure will be heavily influenced by the type of element used to construct it.

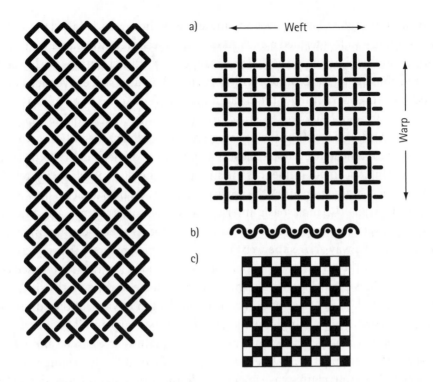

Figure 5.4.1 Ten-strand interlaced braid, a fabric structure made with one set of elements. Each strand plays an equivalent role in the fabric and all move from top to bottom in the process of construction

Figure 5.4.2 a) Diagram of plain weave interlacement
b) Cross-section of plain weave: weft goes over one, under one throughout the cloth
c) Graphic representation of plain weave interlacement. Each square represents one warp–weft intersection; black squares indicate warp on top, white indicate weft on top

fabric structures with one set of elements are direct physical representations of the mathematical definition of braids (Sossinsky 2002, 15–34).

The potential for structural elaboration and variety increases dramatically with each additional set of elements. The simplest class of woven structures is that in which the fabric is made of two sets of elements which interlace perpendicularly to each other (Fig. 5.4.2a). Within each set, the elements are parallel to each other during construction, and do not interwork with others of the same set. The textile is created by holding one set of elements taut, to facilitate threading the other through it. The taut set of threads, by necessity prepared first, is called the *warp*. A single warp thread is called an *end*. Often the warp is longer than it is wide, but this is not a requirement. The second set of elements, which crosses the warp at right angles, passing over and under individual ends, is called the *weft* (or *woof*); a single weft is termed a *pick*. At each intersection of warp and weft, either warp or weft will be uppermost. The arrangement of intersections with weft on top and those with warp on top within each pick and in successive picks determines the woven structure of the cloth. A matrix of squares is often used as visual shorthand for depicting woven structures (Fig. 5.4.2c). Each square represents one warp-weft intersection; the squares are colored black to indicate warp on top, white to indicate weft uppermost (Grunbaum and Shephard 1980, 141). The pattern of interlacement influences the strength of the cloth, its elasticity, texture, and thickness, and determines the apparent pattern on the surface.

Plain weave is the most basic interlacement. In this structure, the first weft pick goes over one warp, under the next, and repeats this over-one-under-one sequence all the way across the web. The next weft pick reverses the sequence, so that it lies over the warps it was previously below, and below those it crossed over on the previous pick (Fig. 5.4.2). This creates a stable, strong cloth with many practical uses. Usually the weft is a continuous length of yarn; at the edge of the textile, it is not cut, but simply turns around the last warp and is re-inserted for the next pick. The appearance and properties of plain weave can be altered dramatically by changing the relative spacing of the warp and weft.[3] Though not very ornamental in itself, plain weave is a perfect foundation for additional decoration such as embroidery. The Bayeux Tapestry is embroidery on a plain weave base, as are richly embellished mantles from the Paracas culture (Paul 2004).

The weft can pass over, or stay under, multiple warp ends. A *float* is a weft thread that passes over more than one warp, or a warp thread that passes over more than one weft. The length and sequence of weft floats can be denoted by numerals indicating the number of warps passed over or under, separated by slashes: 2/2 indicates that the weft passes over two warps and under two, all the way across the cloth. *Twills* are produced by determining a repeating weft interlacement and shifting the pattern one thread to the left or right on each successive pick. They show a diagonal pattern in the cloth, though it may be obscured by the fineness of the thread or the quality of the fiber. The weft can be more prominent on one side of the face than the other; for instance, in a 3/2/2/1 twill the weft floats over five ends for every three it goes under (Fig. 5.4.3). On the face where the weft is less prominent, the warp will be more prominent, and vice versa. Twills and their mathematical classification are discussed by Grunbaum and Shephard (1980). Fabrics in which the diagonal patterns are reflected in one or two directions to form zigzags or diamonds are still considered twills. Floats do not have

---

3. Tapestry is a plain weave structure in which the weft entirely covers the warp. By using different colored wefts in different areas, pictorial designs can be created which appear to break the rectilinear bounds of the fabric grid. At the other extreme, a cloth in which the wefts are entirely hidden by warps is called warp faced.

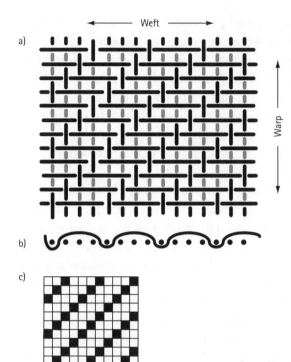

Weft

Warp

a)

b)

c)

Figure 5.4.3 a) Diagram of 3/1 twill interlacement
b) Cross-section of 3/1 twill; weft goes over three, under one
c) Graphic representation of 3/1 twill

to be aligned diagonally; crepes, satins, and lace weaves derive their particular characteristics from varying the length and arrangement of floats. Excellent technical descriptions of weave structures can be found in van der Hoogt (1993) and Strickler (1991).

Compound weaves have more than two sets of elements. The additional sets may be in the warp, the weft, or both. They may be *supplementary*: ornamental only, not essential to the structure of the cloth, or *complementary*: contributing equally to the integrity of the fabric. One example of a compound structure is double cloth, in which two layers of plain weave are woven superimposed on one another. Each layer has its own warp and it own weft. Usually the layers are of different colors and are interchanged at various points in the fabric, so that each surface will have some areas of one color and some of another; a design on one surface of the fabric appears in mirror image with color reversed on the other side of the cloth (Cahlander 1985; Strickler 1991; van der Hoogt 1993, 94–97).

From this brief description, it is apparent that in theory the possible number of distinct fabric structures is huge. Counting the distinct possibilities is not straightforward.[4] How are textile interlacements achieved in practice? There are numerous technical challenges to be overcome. The warp threads must be held

---

4. Grunbaum and Shephard (1980, 149) provide a formula for the number of distinct twills of a given repeat size.

taut and parallel. They must be strong enough to bear tension and to withstand abrasion. Weaving is made much easier if the warps can be divided into two groups and temporarily held apart so that the weft can be inserted between them. Such an opening in the weft is called a *shed*. The most basic warp division is that in which the first warp end and every alternate warp end thereafter belong to one group, and the alternate warp ends belong to the second group. In practice this means that if the first group is raised, the odd numbered warps are raised and the weft goes over every even-numbered end; when the second group is raised, the weft goes over every odd numbered end.

The machines which facilitate weaving are called *looms*. Most are similar in that they allow the warp to be stretched under tension and provide some tool for creating a shed so that the weft can be easily passed from side to side. Looms have been developed and elaborated in different ways in different cultures. The European floor loom,[5] descended from a Chinese invention, is a large, sturdy wooden machine with multiple rotating beams. The Andean variable tension loom can be rolled up and carried from place to place. Though the capabilities of floor looms and variable tension looms overlap somewhat, their comparison points to different ways of thinking about planar embellishment and the structural production of pattern. The creation of cloth involves the application of arithmetic, in the counting and distribution of threads (Urton 1997), but the mathematical concepts at work in fabric go beyond addition and division. The special features of the textile plane—its thickness and directionality—affect the expression of symmetry and geometry. In both the European and Andean traditions, weavers pushed the boundaries of their art to create visually and structurally complex textiles.

The European tradition: floor loom weaving

Floor looms differ in size and complexity. They are designed to hold the warp taut in a horizontal plane and automate the opening of sheds (Fig. 5.4.4). Working from the back of the loom to the front, each loom has a warp beam, to which one end of the warp is attached. This beam usually rotates, so that many meters of warp can be tightly wound on it and released as needed. From the warp beam the warp ends pass over a fixed back beam and up to the harness, which is usually near the middle of the loom. The *harness* is the device that creates the shed. It includes two to thirty-two

---

5. I use the term 'floor loom' to refer to the pre-industrial hand weaving machine and its modern descendants. The boundaries between hand and machine weaving, never clear, are becoming increasingly blurred. Many hand weavers today use computer-controlled looms; some include mechanical assistance for raising and lowering shafts and for passing the shuttle across very wide webs. For the sake of clarity in this chapter, floor looms are assumed to have no computerized components and to work through the input of human power only.

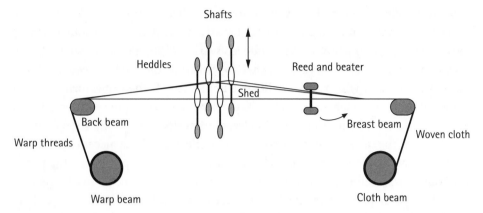

Figure 5.4.4 Schematic of the main features of a floor loom. Not to scale. Arrows indicate the direction of moving parts

shafts.[6] Each shaft holds numerous *heddles*, devices of wire or string with an eye in the middle, at the plane of the warp. Each warp end goes through an eye in exactly one heddle. Each warp end is therefore associated with exactly one shaft. The loom must have some mechanism for raising and lowering the shafts;[7] usually the opening is created by pressing a *treadle* below the loom. When there are more than two shafts, the loom often includes the ability to raise a combination of shafts by depressing only one treadle. Near the front of the loom, each warp end passes through a slot in the *reed*, a device with small equally spaced openings; the reed maintains the distribution of the warp. It is situated in a *beater* which swings back and forth to help push the weft into place. The woven cloth goes over the *breast beam* at the front of the loom, and is wound around the *cloth beam.*

The shafts are numbered from the front of the loom to the back. The order in which the warp ends are put through the heddles is often abbreviated by a sequence of shaft numbers: 1–2–3–4 indicates that the first end is on shaft 1, the second on shaft 2, and so on. When a shaft is raised, all warps which are threaded through heddles belonging to that shaft are raised. Assuming a fixed order for warps, the structure of the cloth is determined by the sequence of shafts through which they are threaded, the combinations determined between shafts, and the order in which the combinations are raised. The number of shafts, and the number of different combinations of shafts (usually determined by the number of available treadles), determines the possible complexity of the pattern that can be woven, including the size of the pattern repeat.

6. In theory, there is no upper bound to the number of shafts. Practical engineering issues tend to limit the number.

7. The shed opening can be created by raising some warps and leaving others fixed, by lowering some and leaving the rest fixed, or by raising some and lowering others (van der Hoogt 1993, 10–11).

Weavers have developed a visual shorthand called a *draft* for recording how to set up a loom to reproduce a particular structure. A draft has four parts: 1) the *threading*, showing on which shaft each warp end should be threaded; 2) the *tie-up*, showing which shafts are attached to which treadles; 3) the *treadling*, indicating the order in which the treadles are depressed; 4) the *drawdown*, a visual representation of the fabric structure. Typically drawdowns are represented on a grid with black and white squares, as already described. Fig. 5.4.5 shows a draft for 2/2 twill. The drawdown is a direct result of the other three pieces of the draft. An examination of the draft will help clarify the relationship between the parts. The warp ends, weft picks, shafts, and treadles are numbered. To insert the first pick, the weaver presses treadle 1. This treadle is attached to shafts 1 and 2, as indicated by the tie-up, so shafts 1 and 2 are raised when treadle 1 is depressed. Therefore any warp end on shaft 1 or shaft 2 will go *over* weft pick 1. This is indicated by the black squares in the drawdown under warp ends 1, 2, 5, 6, and so on. The same relationship between threading, treadling, and tie-up holds for each pick.[8]

A study of drafts helps explain how woven cloth models algebraic relationships between matrices. Note that if there are n shafts and m treadles, the size of the tie-up is n x m. To create matrices representing the tie-up, threading, and treadling, replace each black mark with 1, and the white spaces with 0. The resulting matrices are called binary since they include only 0s and 1s. If the repeat length of the treadling is r and that of the threading is s, the product of these three matrices is an r x s matrix which is equivalent to the drawdown: 1s appear in the positions which are black in the drawdown, and 0s appear elsewhere. A formal proof is given in Hoskins (1983).[9] Using the draft from Fig. 5.4.5, the multiplication of the treadling, tie-up, and threading would look like:

$$
\begin{pmatrix} 1 & 0 & 0 & 0 \\ 0 & 1 & 0 & 0 \\ 0 & 0 & 1 & 0 \\ 0 & 0 & 0 & 1 \end{pmatrix}
\begin{pmatrix} 0 & 0 & 1 & 1 \\ 0 & 1 & 1 & 0 \\ 1 & 1 & 0 & 0 \\ 1 & 0 & 0 & 1 \end{pmatrix}
\begin{pmatrix} 0 & 0 & 0 & 1 \\ 0 & 0 & 1 & 0 \\ 0 & 1 & 0 & 0 \\ 1 & 0 & 0 & 0 \end{pmatrix} =
\begin{pmatrix} 1 & 1 & 0 & 0 \\ 0 & 1 & 1 & 0 \\ 0 & 0 & 1 & 1 \\ 1 & 0 & 0 & 1 \end{pmatrix}
$$

8. The astute reader will notice that in this visual representation, the cloth grows from top to bottom, unlike what the weaver would actually encounter at the loom. Also, the face of the cloth which appears to the weaver will depend on whether the shafts are raised or lowered when treadles are depressed. The reader should be able to convince herself that any given structure can be woven with the other face up by exchanging all white squares for black and black for white in the tie-up. Drafts can be written in any of four orientations— that is, with the tie-up in any of the four corners of the diagram. The relationship between the parts is the same in each case. For a more detailed discussion of drafting, see van der Hoogt (1993).

9. This is a simple example since all the matrices are square and one is the identity matrix. A more general multiplication order is treadling * Transpose(tie-up) * threading. Fuller treatment is found in Brezine (1993).

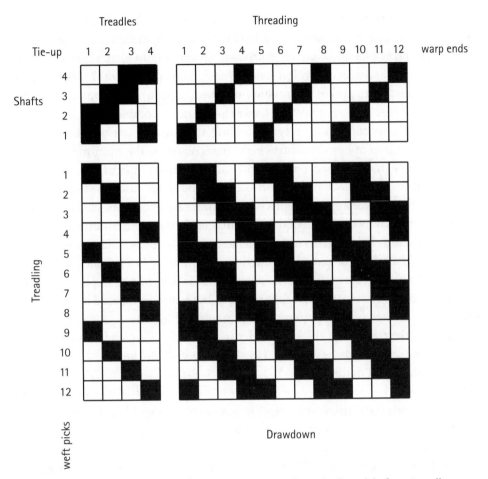

Figure 5.4.5 Draft for a 2/2 twill, woven on four shafts with four treadles

This mathematical modeling of loom interlacement prompts one to ask whether there might be mathematical solutions to weaving problems. Some common questions include: what is the smallest number of shafts and treadles needed to weave a given structure? Given two structures, how can a weaver determine if they can both be woven with the same threading and tie-up? How many different structures can be woven on the same threading and tie-up? (The treadling can easily be changed during weaving, so this is a reasonable question.) How can one determine if plain weave is possible for a given threading? How can the weaver calculate the maximum float length? How can one ensure that the longest warp floats and weft floats are the same length? Can the treadling or threading be rearranged to be easier to remember, without changing the pattern of the cloth? In answering any of these questions, one must remember that not all drawdowns produce viable fabric. Some arrangements of black and white squares represent interlacements that separate into two or more layers (Clapham 1980).

Though the relationship between the structure of cloth and binary matrices is intriguing, one need not assume that weavers in historic times would have expressed principles of loom interlacement in our mathematical terms. What concepts must a weaver control in order to design and create cloth? Though there may be more practical skill than theory involved in passing the shuttle and beating the weft, the setting up of the loom can be quite complex, and in fact was often done by specialists. Each type of fabric structure has its own constraints, and the invention of new structures is far from obvious. The machine itself does not guarantee that cloth woven on it will have structural integrity.

The arithmetic of cloth production is finicky, but straightforward. The weaver must calculate how many warp ends per unit of measurement the cloth is to have, and multiply that by the width of the cloth to prepare the correct number of warps. If the fabric is to have a pattern that is balanced at the edges, adjustment must be made for the number of threads in the repeat. Symmetry is a significant attribute of most cloth structures. The choice of symmetry is determined in part by culture (Washburn and Crowe 1988; Washburn 1999), but is also influenced by the tools at hand. In weaving complex patterns, a common treadling choice is 'tromp as writ': depressing the treadles in the same order in which the warp ends were threaded through the shafts. If the warp order is 1–2–3–4–5–6–7–8–7–6–5–6–7–8, the treadles would be used in the order 1–2–3–4–5–6–7–8–7–6–5–6–7–8. (The number of shafts must be the same as the number of treadles.) Tromp-as-writ produces a diagonal line across the cloth, which helps the weaver remember her place in the treadling sequence after a break from work; it also has the effect of producing bilateral symmetry around the diagonal. If the threading has mirror symmetry, the treadling will too, and the cloth will have p4m symmetry.[10] This patterning is extremely common in floor loom cloth. The traditional North American tradition of coverlet weaving shows an overwhelming preference for such symmetry (Davison 1953). Interestingly, all non-hexagonal symmetries are structurally possible on only four shafts (Brezine 2004). Implementation of these symmetries depends on an understanding of how to mirror, shift, and flip threading orders and treadling orders, what effect the tie-up has on the interlacement, and the interaction between threading and treadling. Effectively, weavers use a practical application of matrix transformations to achieve the interlacements in their cloth.

Certain weave structures depend on numerical relationships for their unique properties. Among these is satin. Originally developed to show off the high sheen of silk, an ideal satin has very long floats and the appearance of a perfectly even surface texture. If n is the number of shafts, the weft floats under $n - 1$ warp ends for a warp-faced satin.[11] The single warps which it goes under are distributed as

---

10. That is, it will have 4-fold rotational symmetry and horizontal and vertical reflective symmetry.

11. Both warp- and weft-faced satins are possible; often a warp-faced satin is woven face down, so that fewer shafts must be lifted.

evenly as possible through the available number of shafts: the goal is to avoid any obvious diagonals, patterns, or irregularities which would mar the surface. In practice, this means that the number of the single shaft raised for each pick shifts by a number relatively prime to the number of shafts (Grunbaum and Shephard 1980, 15; van der Hoogt 1993, 23, 69).

In addition to all the issues of innovating within known rules, there is the ever-present challenge of creating new and better cloth. The number of possible variations on threading, tie-up, and treadling is huge; the variations increase dramatically with each added shaft and treadle. However, not all possibilities are good ones. It is quite possible to come up with combinations that produce two or more layers of loosely interlaced threads instead of one stable fabric. There is a mathematical algorithm based on the matrix representation of a fabric for determining whether a given structure will fall apart (Clapham 1980); historically, weavers had to depend not on algorithms but on their understanding of different weave structures and the relationships between threading, tie-up, and treadling. Given the striking variety of successful cloth that has been produced, one must conclude that weavers' conception of the rules governing geometric interlacement are entirely as successful as the algebraic algorithm given by Clapham, though framed in different terms.

Certain aspects of the western system of cloth production invite comparison to western mathematical concepts. Our traditional geometry is based upon ideas of infinite planes and never-ending lines. Being a part of the real world, lengths of cloth are finite; but in construction they are more often than not treated as infinite in the sense that no particular attention is given to the beginning and ending of the warp. Since the advent of tailored clothing the assumption has been that cloth will be cut and shaped once it is off the loom. The focus is on making the fabric of consistent quality; the ends themselves can be left to ravel. The floor loom provides enormous mechanical advantages of speed and replicability. Because the machine constrains the spacing of the warp and the order in which the threads are raised, and maintains consistent tension throughout the length of the warp, it is relatively easy to duplicate a particular piece of cloth. In patterning, repeats are considered correct only if they exactly duplicate previous motifs. The only equivalence is equality. The downside to the automation of pattern is that once the threading and tie-up are fixed, the structural variations possible are drastically reduced. This is not important if producing yardage in quantity in a plain structure. However, it can be a limitation if elaborate or pictorial designs are desired, or if one wishes to change the pattern along the length of the cloth. The desire to expand the patterning possibilities of loom weaving led to the invention of the Jaquard loom, patented in 1804. This loom used punched cards to individually select which warps to raise and lower. The chains of punched cards were a direct precursor to the earliest computer punchcards (Essinger 2004).

Concerns with replicability led to various systems for writing down the information needed to produce a particular piece of cloth. Early versions of weaving drafts were closely guarded and can be very hard to read. In present times the format is fairly standardized, and computer software makes it possible to quickly view the effects of changes in the threading, tie-up, or treadling before setting up the loom. Before computers simplified the process of creating a visual representation of cloth structure, it was more common to experiment directly at the loom. However, it is not at all impossible that even long ago new structures were devised or inspired by doodling on paper, tweaking existing drafts which had been previously recorded. Perhaps the separate medium of writing played a significant role in the creation of cloth—not only for working out possible new threadings, but in calculating the number of warp ends needed, the total yards of thread required, or making adjustments in the draft to balance the pattern.

Weaving in the Andes: the variable tension loom

Andean civilizations never developed text as we know it. There are no direct correlates to our system of symbols marked onto a flat surface representing language. The communication device of the Andes, at least from the time of the Inkas (approximately 1400–1532 AD), was the *khipu*, an arrangement of colored and knotted cords (Ascher 1981; Urton 2003). Though still undeciphered, the *khipu* were by colonial accounts essential bureaucratic tools of the empire. They may have encoded narrative information; it is well attested that they hold complex numerical data, and that separate *khipu* objects can reference each other (Urton and Brezine 2005). We do not usually consider cloth, or string, to be a medium with explicit textual communicative capacity. However, reviewing the outstanding textiles from ancient Peru and the astonishing variety of complex structures used by ancient Andean weavers, it is clear that a great deal of time and effort were devoted to exploring textile variations and possibilities. It has been suggested that textiles were perhaps the main medium in which Andean cultures met and grappled with theoretical problems (Frame 1986; Lechtman 1993; Urton 1997; Franquemont 2004). Geometry, symmetry, reciprocity, energy, hierarchy, and structure are all explored in Andean weaving.

The Andean method of production can be deduced from existing textiles, and from modern ethnographic weaving studies.[12] The following is a general

---

12. Extrapolating ethnographic observation to make assumptions about ancient practices is always problematic. I am not being chronologically specific in this chapter. A good overview of the development of warp-faced weaving in the Andes is given by Rowe (1977); detailed treatment of the traces left by various weaving techniques can be found in numerous sources including Rodman and Cassman (1995).

description; contemporary details of practice may vary by region or village, and not all ancient cloth adheres to exactly the same production methods. I use the present tense, as these methods are in use today in many areas of the Andes. There is no evidence that Andean cultures ever invented or used a floor loom as described above. Instead, their weaving equipment is the most basic: two stout sticks to which the warp is attached and a third stick which maintains a separation between two sets of warp ends (Fig. 5.4.6). Typically, the warp beam furthest from the weaver is tied to a pole or lashed to stakes firmly stuck in the ground (Fig. 5.4.7). The closer beam may also be held fixed to stakes, but is just as commonly attached to the weaver herself by a belt which goes around the hips. Depending on the height at which the far beam is held, the plane of the cloth may be horizontal or slanted upwards to almost vertical. The unwoven portion of the cloth is not usually wound around the far warp beam; instead, the entire length of the warp is stretched before the weaver. To bring the fell of the cloth within range, the completed part of the weaving is wound around the closer warp beam. There is no distinction between the two beams; in fact, often the weaving progresses from both ends and meets somewhere in the middle, a clear contradiction to the strictly unidirectional sequence of floor loom weaving.

To set up a floor loom, each warp must have at least one cut end, to allow it to be threaded through a heddle and the reed. The very beginning and very end of loom-woven cloth has free hanging warp threads. Andean cloth is much more economical of yarns. The warps are prepared by winding in a continuous figure eight. This produces a length of ordered yarns with loops at both ends. The loops are attached to the warp beams of the loom, and the weaving begins exactly at the sticks, so that there are no dangling loops. The warp selvedges mimic the weft selvedges: the warp reaches the edge of the cloth, turns, and reenters the weave, creating a cloth that is finished on all four sides. As the unwoven portion of the warp gets shorter and shorter, it becomes more and more difficult to insert the weft. A *termination area*, where the last several picks were inserted, is often identifiable by a looser weave, a change in structure, or a discontinuity in pattern. However, it is possible to weave four-selvedge cloth without an obvious termination area, and in the best Andean examples it is impossible to tell where the weaver finished.

The third stick mentioned above maintains one division in the warp yarns. Typically every other thread goes over this *shed stick*. The alternate warp ends go through yarn heddles in front of the shed stick, simple loops applied to the warp after it is attached to the beams. The heddles may be lashed to a *heddle rod* to maintain spacing and make lifting easier. Lifting the heddle rod raises every yarn which goes through a heddle loop. In practice, lifting the heddles can be a time-consuming and laborious process, especially when the warp yarns are of sticky wool and set very close together. Bringing the shed stick forward to

a)

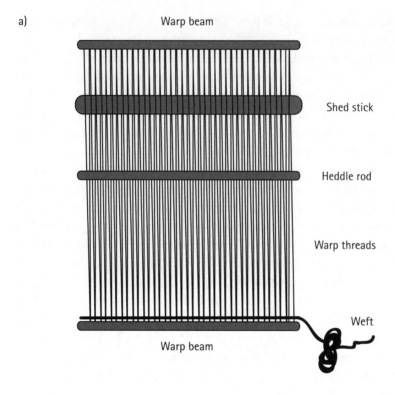

b)

c)

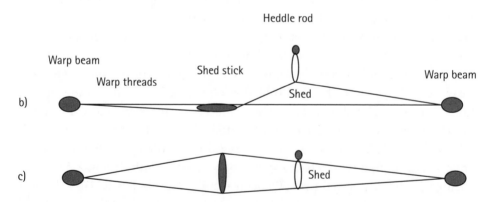

Figure 5.4.6 a) Schematic of a variable tension loom
b) Side view of the variable tension loom, showing heddle rod raised
c) Side view showing shed stick on edge to create the opposite shed

the heddles pushes the heddled yarns down, and raises the yarns which go over the shed stick. This creates the opposite shed opening. The most common arrangement of heddles and shed stick creates a 1–1 interlacement, the familiar plain weave. Multiple sets of heddles are certainly possible and are not

Fig. 5.4.7 Weaver from Accha Alta, Peru, 2001. Her loom is lashed to two stakes at one end and the other end is held around her waist by a belt. She is in the process of selecting threads for a pattern. The complex colored skirt borders can be seen on her skirt and on that of the woman to the right (photo by author)

unknown. For instance, several villages on the east side of the Urubamba valley in Southern Peru weave complex twill skirt borders, in which the twill patterns are automated by eight or more sets of string heddles. However, most Andean textiles, even the most structurally complex, were and are woven with only two opposing sheds.

In the Andean tradition, the machinery for facilitating complex patterns lies not in the elaborateness of the machinery, but in the weaver's head and fingers. Many structures are possible, but complementary warp weave will be described here as a typical example; typical both because it is widespread and because it reaches far back in time. Complementary warp weave also embodies several crucial Andean aesthetic principles: the structure *is* the pattern—neither can be altered without altering the other; the two faces are the same, but with colors reversed; and because of the characteristics of the weave, designs often have no clear figure/ground, but instead favor interlocking shapes which to our eyes may seem almost like optical illusions.

Complementary elements in a fabric are those which 'have the same direction in a fabric and are co-equal in the fabric structure'. (Emery 1994, 150). *Complementary warp weave* is based on a warp of two colors, light (L) and dark (D).

Each light yarn has a dark partner. The colors alternate across the pattern area: L-D-L-D-L-D-L-D. When the threads are arranged into plain weave, on one shed, all the light threads will be up; on the next shed, all the dark threads will be up (Fig. 5.4.8). When the warps are pushed together, so that no weft shows between them, repeating the plain weave sequence results in a series of horizontal light and dark stripes, each stripe one pick high. To create designs, the weaver individually selects which light and dark threads she wants to appear on top for each weft pick. For instance, if the light colored shed is up, but dark threads are required for the design, she will pick up the dark threads in the correct position with her fingers, dropping their light colored partners to the bottom of the shed. Thus for every place a light warp appears on the front of the fabric, a dark one appears on the back (Cason and Cahlander 1976; Rowe 1977; Franquemont 1991). The design is the same on both faces, except that the colors on one side are the reverse of those on the other. Both light and dark threads contribute to the structure of the cloth. Removal of either set would destroy the integrity of the fabric. The designs made in this technique are not merely ornamental, but reflect the deep structure of the fabric (Lechtman 1993).

Though motifs are often repeated without variation through the length of the cloth, the pick-up process offers opportunities for improvisation on every weft. The basic design unit is a band motif that occupies a relatively small number of L-D thread pairs (approximately 6–30; Franquemont 1991; Franquemont and Franquemont 2004). Wider and more complex designs entailing a greater number of threads are built from known band patterns by applying the symmetrical principles of translation, reflection, glide reflection, and/or rotation. The replication of motifs in various orientations often creates spaces in the design that are used for additional improvisation. For instance, one vertical zig zag replicated by reflection creates a series of bounded diamonds; each diamond area is a space where a new motif can be created (Franquemont and Franquemont 2004, 199–201). Because the patterns are not constrained by any previously determined loom set up, the motif in each diamond can be different. Manipulation of motifs is only one level of a hierarchical system of nested symmetry operations in making a woman's shawl. Each shawl is woven in two pieces of equal size which are sewn together so that the shawl has a seam down the center, parallel to the pattern bands (Franquemont and Franquemont 2004, 184–187). One of the two pieces is rotated 180 degrees before the two are seamed, so that termination areas, if visible, occur at different sides of the cloth. Each half of the shawl follows interior principles for placement of the patterned areas and plain weave; each patterned area is framed by symmetrical color stripes, and the pattern bands have their own internal symmetries. Variations in the size, position, and coloring of these elements at any level can denote styles specific to a region or village.

a)

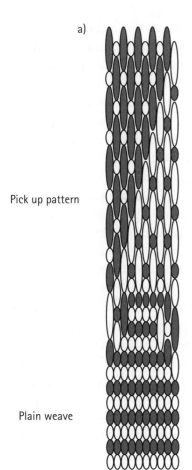

Pick up pattern

Plain weave

b)

Figure 5.4.8 a) Diagram of
complementary warp-faced weave with
a typical Andean pattern. The weft
is not visible because the warps are
packed together

   b) Cross-section showing the equal
but opposite paths taken by one light-
dark warp pair. The dots represent
wefts in this diagram

Unlike a floor loom weaver, an Andean weaver is intimately engaged with the
threads creating the pattern at every pick. Because there is no automation, the
design can change throughout the length of the cloth. It is not uncommon for
one motif to be transformed into another. Most motifs can be expanded or contracted to be woven over more or fewer warps than usual. A textile from Huilloc
in my personal collection has two bands of a common hooked design side by side.
One band of hooks is created over ten pairs of warps, ten light and ten dark, and
the adjoining band takes up eight pairs. Because they are woven over a different
number of warps, they also take a different number of picks for the completion of
each motif. The end result is that the motifs, while they begin near the bottom of
the textile neatly aligned, are soon offset by half the length of a motif, and move
gradually back into alignment along the length of the textile. The repeat length
for such a combination is very long, even though the repeat on each individual band is relatively short. On a floor loom such a long repeat would require a

great number of shafts and treadles. One might think that such a long repeat would also make it difficult for a weaver to recreate by hand. Add to the simple hooks the large zig zags and multiple animal forms that occur on the same textile, and it is difficult to imagine how such complex designs can be created without sophisticated equipment or remembered without written notation.

The answer is that the Andean weaver depends on algorithms rather than on automation. The requirements of the current weft pick are in many cases determined fairly easily by looking at the previous pick. Instead of remembering a series of instructions such as two dark, two light, six dark; three dark, two light, five dark; four dark, two light, four dark…an Andean weaver may have a set of mental instructions for creating a diagonal line by shifting the light threads over by one on each successive pick. This general principle allows her to create diagonals of any width or length, over any number of warp threads. The diagonal line is of course a very simple example, but a similar principle holds for more complex designs. An algorithmic approach simplifies weaving the same design at different scales side by side; the weaver is prompted at each step by what is already woven. Ethnographic observation suggests that Andean weavers remember not every pick in a complete repeat, but the critical picks which form turning points in the design, and a series of transformations (Franquemont and Franquemont 2004, 197). Altering the sequence of transformations in the algorithm can create any desired symmetrical operation on the motif. Weavers' facility in creating the same design in different orientations can be seen in the fact that common motifs are often recreated not just at different scales but in different structures: designs from warp-faced weaving are also seen in weft-faced weaving, requiring a mental shift of 90 degrees. Motifs from weaving may show up on knitted items, involving not just a change in the orientation of the design but the ability to translate the requirements of one structure to those of another. Though complete textiles include many layers of complexity, the total number of decisions required to warp a shawl may be as small as twelve (Franquemont and Franquemont 187).

Complementary warp-faced weave can be replicated on a floor loom, and the patterns can be automated, but most designs would require more numerous shafts and treadles than are commonly available, particularly when multiple patterns are combined in the same cloth. Improvisation of large motifs is impossible on a floor loom without significant manual intervention. Despite the simplicity of their equipment, ancient Andean weavers experimented with every known weave. The flexibility of their simple loom allowed great patterning potential even in structures that are typically 'shaft hungry'. For instance, automated double cloth woven on a floor loom requires four shafts for each independently changing area of color. A simple checkerboard design requires eight shafts; designs of any complexity can easily use sixteen or more. On an Andean loom, just as in complementary warp weave, a double weave pattern is selected pick by pick (Cahlander 1985). This allows the creation of finely detailed shapes including smooth diagonals, and the inclusion of multiple designs across and along the cloth. Double cloth can be extended in theory to three, four, or more layers. This is rarely done on a floor loom, but examples are found from ancient Peru (Cahlander 1985).

Gauzes are an example of a class of textile structures which are difficult to create on a floor loom. In gauze weave, warps cross each other and cross back. Common crossing patterns are one warp crossing its nearest neighbor, two warps crossing the two nearest neighbors, and one warp crossing its next neighbor but one (O'Neale 1948; d'Harcourt 1974; Emery 1994, 180–186). Simple one-over-one crossings can be duplicated on a floor loom with some modifications to the heddling, but complex crossings are practically impossible to achieve.[13] The warp ends on a floor loom are restrained both by their arrangement in the reed and the order of heddles. A variable tension loom has no such restrictions: there is no reed, the heddles are flexible, and the crossings are achieved through finger manipulation of the threads. Gauzes have distinct properties unlike those of any other structure. The crossing of the threads compresses the fabric in the weft direction, but also lends elasticity. Different crossing patterns result in different rates of compression, and different amounts of stretch. The warp crossings hold the wefts apart so that they cannot be packed tightly; the visual effect is of open spaces in the cloth. These openings may be small or large, depending on the structure chosen, the thread used, and the spacing of the yarns. Regardless, most gauzes have a lacy, transparent quality in which pattern is the result of differing weave densities. The most famous Andean gauzes are woven with very fine single cotton yarns with lots of extra energy, adding even more potential buoyancy to the cloth.

Because the peoples of the Andes left no texts that have yet been decoded, we do not know if they defined for themselves a category of intellectual endeavor comparable to our mathematics. Based on the numerical relationships within and between *khipu* (Ascher 1981; Urton 2003; Urton and Brezine 2005), it is clear that they had control over the principles of arithmetic in a base-ten number system. The sophistication of their architecture, terracing, irrigation systems, and textiles suggests that they also had highly developed ideas of geometry, symmetry, and engineering. Because we have no mathematical treatises to help us understand how they approached problems, we can only turn to the artifacts they left and try to deduce what patterns of thought could have led to such complex constructions. It should be clear that the nested hierarchical

---

13. Becker has replicated ancient Chinese gauzes, some of which have the same structure as some Peruvian gauzes. The loom used for these samples is significantly more complex than the floor loom described here; see Becker (1987) for details.

structures of Andean cloth are not the result of felicitous invention. Rather, they echo cultural principles and ways of perceiving the world (Franquemont 2004; Franquemont and Franquemont 2004; Washburn 1999). Algorithms for creating woven pattern allow a degree of improvisation (Franquemont 2004) while ensuring that the structure maintains its integrity. Designs are built up pick by pick and thread by thread. Weavers thus work in a finite plane composed of discrete elements, rather than on a continuous unbounded Euclidean surface. It is tempting to suppose that Andean mathematics before the Spanish conquest lay largely in the fields we would consider to belong to discrete mathematics. Certainly there is no shortage of combinatorial puzzles in Andean textiles. A stunning example is given by the symmetrical designs on Paracas mantles: they are so complex that they defy traditional group theory classifications. Inspired by Peruvian fabrics, Grunbaum pictures 125 distinct planar symmetry schemes and 79 'ribbon' patterns (Grunbaum 2004). The addition of multiple colors to each motif, and the rotation of colors between motifs, adds significant complexity to the cloth as a whole (Grunbaum 2004, 57–60; Paul 2004). Color rotations in modern skirt borders are reminiscent of map coloring problems. All of these examples indicate that Andean weavers do practice mathematics, in the form of sets of principles and transformations that, when properly practiced, produce astounding cloth.

Textile puzzles for future mathematical study

The above descriptions have focused on the classification of different woven structures and two different methods for producing them. Most methods of producing woven textiles, regardless of geographic region or historic period, will have some similarity to one of the two techniques described. There are many additional complexities that have not been covered here. For instance, it is possible for a warp thread to go through more than one heddle and so to belong to two different shafts; this is the basic principle of the drawloom, which was used to produce elaborate pictorial textiles (Becker 1987, Crowfoot *et al* 2001, 23). Two sets of shafts greatly increase the patterning possibilities, and the difficulty of calculating and enumerating possible structures (Hoskins 1983). Andean weavers sometimes use a similar concept: in dual lease weaving, a forked stick holds one warp division fixed, while a set of heddles creates a second, different but non-exclusive division of the warp (Franquemont 1991).

Color is the first thing a viewer notices about a piece of cloth, but interestingly, it is often excluded from classification and analysis. The mathematical issues associated with the use of color are considerable. There is a vast family of so-called 'color and weave effects', in which the apparent pattern on the cloth is a result of the alternation of colors in warp and weft, and does not reflect the actual structure (Hoskins 1983, 308; van der Hoogt 1993, 14, 20–21). It would be interesting to enumerate the different two- or three-color possibilities for a given set of structures. Is it the case that any two-color design can be produced through some color-and-weave effect? What if restrictions are placed on float length? One might also explore the realm of complementary weaves with three or more colors—that is, the class of complementary weaves with more than two sets of warps.

Complex cloths such as the gauzes have not been classified mathematically, in part because of the difficulty of standardizing a diagrammatic representation of warp crossings. This is an area where mathematical investigation could be amply repaid by the discovery of many new structures that have not been utilized in the history of textile production. Each gauze structure has a different density; it would be valuable to specify the relationship between the structure and the elasticity of the cloth. The quality of a gauze is also greatly influenced by the characteristics of the yarn used; the impact of yarn energy on finished cloth is an area that has not been fully studied (Collingwood 1987; van der Hoogt 1993, 19). One of the most fruitful collaborations between mathematicians and weavers could be the mathematical specification of structures with certain characteristics. For example, what kinds of structures will produce fabric that puckers, stretches, or skews? How can one predict from a weaving draft the final texture of the fabric?

Mathematics can offer an inspiring perspective on the study and creation of textiles. Being unfettered by assumptions of what can and cannot be practically achieved on existing equipment, mathematical classifications can offer new fields for exploration. Grunbaum and Shephard's work on isonemal fabrics (1988) is a case in point. Triaxial weaving, in which three sets of elements cross each other at angles of 120 degrees, is not often attempted outside of basketry. Three way isonemal fabrics offer wide scope for experimentation—assuming one can come up with a loom on which to weave them!

Conclusion

Within the general principles of interlacement, the possible number of distinct structures is vast. When pattern is created through structure, rather than applied by dyeing, printing, or embroidery, the woven surface acts as a very particular kind of canvas. It is a plane with thickness and directionality, given by the orientation of warp and weft (Grunbaum 2004). Structural elaboration is subject to the constraints of machinery, in the case of the floor loom, or to algorithms of pattern construction and elaboration, in the case of hand-manipulated weaves created on a body-tensioned loom. There are clearly different cultural priorities in each case. European looms were used to increase efficiency, and to turn out long lengths of consistent cloth with repetitive patterns. Continual development and improvement led eventually to power looms and the industrial revolution, which forever changed the economic landscape of western culture. The desire for increasingly complex woven patterns inspired the Jacquard loom, which in turn inspired the first computers. By contrast, Andean cloth was and is created in complete, finished units. It is never cut. The design potential lies not in the physical machinery, but in the mental concepts each weaver brings to the loom.

Each textile bears traces of its technique of production and the cultural context in which it was created. Fabric structures, their relationship to visual pattern, and the choice of symmetrical manipulations are all evidence of craftspeople's understanding of the geometrical principles of interlacement. Whether one creates cloth by inventing an elaborate machine, or by constructing the intellectual framework for complex mental computations, the existence of cloth is evidence of mathematics at work in the tangible world. Creating pattern through interlaced structure is no less astounding than patterning a floor with an elaborate tiling or a hillside with terraces. It reflects the application of abstract principles to physical objects. In the Andean case, surviving textiles are some of the few hints we have about how people conceived of space, number, and symmetry. European textiles are no less telling as evidence of the careful planning and geometric understanding needed to achieve complex patterning within the restrictions set by the floor loom. Those who make cloth are always interested in new structural possibilities, and many mathematicians are intrigued by unfamiliar puzzles inspired by real-world situations. The two fields have much to offer each other.

Bibliography

Ascher, Marcia, and Ascher, Robert, *Mathematics of the Inkas: code of the quipu*, 2nd ed, Dover, 1981.

Becker, John, *Pattern and loom: a practical study of the development of weaving techniques in China, western Asia, and Europe*, Rhodos International Publishers, 1987.

Brezine, Carrie J, 'The mathematical structure of weaving', Reed College, 1993.

Brezine, Carrie J, 'Symmetry on the loom', in Dorothy K Washburn and Donald W Crowe (eds), *Symmetry comes of age*, University of Washington Press, 2004, 65–80.

Cahlander, Adele, *Double-woven treasures from old Peru*, Dos Tejedoras, 1985.

Cason, Marjorie and Cahlander, Adele, *The art of Bolivian highland weaving*, Watson-Guptill, 1976.

Clapham, C R J, 'When a fabric hangs together', *Bulletin of the London Mathematical Society*, 12 (1980), 161–164.

Collingwood, Peter, *The techniques of rug weaving*, Watson-Guptill Publications, 1987.

Crowfoot, Elisabeth, Pritchard, Frances, and Staniland, Kay, *Textiles and clothing 1150–1450*, The Boydell Press, 2001.

Davison, Marguerite Porter, *A handweaver's sourcebook*, Marguerite P Davison, 1953.

Emery, Irene, *The primary structure of fabrics: an illustrated classification*, 2nd ed, The Textile Museum, 1994.

Essinger, James, *Jaquard's web*, Oxford University Press, 2004.

Frame, Mary, 'The visual images of fabric structures in ancient Peruvian art', in Ann Pollard Rowe (ed), *The Junius B Bird conference on pre-Columbian textiles*, The Textile Museum, 1986, 47–80.

Franquemont, Ed, 'Dual-lease weaving: An Andean loom technology', in Margot Blum Schevill, Janet Catherine Berlo, and Edward B Dwyer (eds), *Textile traditions of Mesoamerica and the Andes*, University of Texas Press, 1991, 283–308.

Franquemont, Ed, 'Jazz: an Andean sense of symmetry', in Dorothy K Washburn (ed), *Embedded symmetries, natural and cultural*, University of New Mexico Press, 2004, 81–94.

Franquemont, E M, and Franquemont, C R, '*Tanka, chongo, kutij*: structure of the world through cloth', in Dorothy K. Washburn and Donald W Crowe (eds), *Symmetry comes of age*, University of Washington Press, 2004, 177–213.

Gerdes, Paulus, *Geometry from Africa: mathematical and educational explorations*, The Mathematical Association of America, 1999.

Grünbaum, Branko, 'Periodic ornamentation of the fabric plane: lessons from Peruvian fabrics', in Dorothy K Washburn and Donald W Crowe (eds), *Symmetry comes of age*, University of Washington Press, 2004, 18–64.

Grünbaum, Branko, and Shephard, G C, 'Satins and twills: an introduction to the geometry of fabrics', *Mathematics Magazine*, 53 (1980), 139–161.

Grünbaum, Branko, and Shephard, G C, 'Isonemal fabrics', *The American Mathematical Monthly*, 954 (1988), 5–30.

Harcourt, Raoul d', *Textiles of ancient Peru and their techniques*, University of Washington Press, 1974.

Hoskins, J A, 'Factoring binary matrices: a weaver's approach', in *Lecture notes in mathematics*, vol 952, Springer-Verlag, 1983, 300–326.

Lechtman, Heather, 'Technologies of power: the Andean case', in John S Henderson and Patricia J Netherly (eds), *Configurations of power: holistic anthropology in theory and practice*, Cornell University Press, 1993, 244–280.

Lemonnier, Pierre, 'Introduction', in Pierre Lemonnier (ed), *Technological choices: Transformation in material cultures since the Neolithic*, Routledge, 2002, 1–35.

O'Neale, Lila, and Clark, Bonnie Jean, *Textiles periods in ancient Peru, III: the gauze weaves*, University of California Press, 1948.

Owen, Roderick, *Braids: 250 patterns from Japan, Peru and beyond*, Interweave Press, 1995.

Owen, Roderick, *Making kumihimo: Japanese interlaced braids*, Guild of Master Craftsmen, 2004.

Paul, Anne, 'Symmetry schemes on *paracas necrópolis* textiles', in Dorothy K Washburn (ed), *Embedded symmetries*, University of New Mexico Press, 2004, 59–80.

Rodman, Amy Oakland, and Cassman, Vicki, 'Andean tapestry: structure informs the surface', *Art Journal*, 54/2 (1995), 33–39.

Ross, Mabel, *The essentials of yarn design for handspinners*, Mabel Ross, 1983.

Rowe, Ann Pollard, *Warp-patterned weaves of the Andes*, The Textiles Museum, 1977.

Schneider, Jane, 'The anthropology of cloth', *Annual Review of Anthropology*, 16 (1987), 409–448.

Seidel, A, Liivak, O, Calve, S, Adaska, J, Ji, GD, Yang, ZT, Grubb, D, Zax, DB, and Jelinski, LW, 'Regenerated spider silk: processing, properties, and structure', *Macromolecules* 33/3 (2000), 775–780.

Selin, Helaine (ed), *Mathematics across cultures: the history of non-western mathematics*, Kluwer Academic Publishers, 2000.

Sossinsky, Alexei, *Knots*, Harvard University Press, 2002.

Strickler, Carol (ed), *A weaver's book of 8-shaft patterns: from the friends of handwoven*, Interweave Press, 1991.

Urton, Gary, *The social life of numbers: a Quechua ontology of numbers and philosophy of arithmetic*, University of Texas Press, 1997.

Urton, Gary, *Signs of the Inka khipu: binary coding in the Andean knotted-string records*, University of Texas Press, 2003.

Urton, Gary, and Brezine, Carrie J, 'Khipu accounting in ancient Peru', *Science*, 309 (2005), 1065–1067.

van der Hoogt, Madelyn, *The complete book of drafting for handweavers*, ShuttleCraft Books, 1993.

Washburn, Dorothy, 'Perceptual anthropology: the cultural salience of symmetry', *American Anthropologist*, 101 (1999), 547–562.

Washburn, Dorothy K, and Crowe, Donald W, *Symmetries of culture: theory and practice of plane pattern analysis*, University of Washington Press, 1988.

6. Presentation

The cognitive and cultural foundations of numbers

Stephen Chrisomalis

Every known human society has the capacity to manipulate quantity; in this sense, mathematics is panhuman. The extent to which any individual or group pays special attention to mathematical concepts, however, is cross-culturally variable. The study of the cultural, (pre)historical, and social aspects of numeration and arithmetic—the foundations of mathematics—is preoccupied with the tension between the universal and specific aspects of the subject. While recent scholarship from cognitive science, generative linguistics, and neuro-psychology provides ample evidence that there are some 'hard-wired' aspects of human numerical faculties, developmental psychologists, ethnomathematicians, anthropologists, historians, and historical linguists stress the differences among cultures and historical periods. How, then, can we synthesize and reconcile these disparate literatures? To assume, a priori, either that universal aspects of numeration or local and historically contingent developments are of primary interest is premature and misguided.

To add to the confusion, the history of numeration lingers under a series of pervasive myths. Nineteenth-century unilinear notions of the evolution of society from 'savagery' or 'primitiveness' to 'civilization' continue to haunt the subject. Such notions have persisted despite a paucity of evidence, partly because the data to refute them are spread among multiple disciplines, and partly because there is a sufficient kernel of truth to them to warrant further investigation. A comparative and historical perspective can help debunk the untenable and shore up the sustainable propositions.

Every known human language has some means of expressing quantity, and in virtually all languages this includes two or more words for specific integers (oral only in non-literate societies, oral or written in literate ones). Most societies, perhaps all, possess one or more means of conducting arithmetic mentally or with the aid of various artifacts and techniques that serve as computational technologies. Finally, a smaller number of societies have a set of visual but primarily non-phonetic numeral-symbols that are organized into a *numerical notation system*. Western numerals and Roman numerals are merely two well-known examples among the many distinct systems used over the past 5,500 years.[1] The linkages between number words, computational technologies, and number symbols are complex, and understanding the functions each serves (and does not serve) will help illustrate the range of variability among the cognitive and social systems underlying all mathematics.

Number words and number concepts

Although linguists, particularly European philologists, have been interested in numeral systems for centuries (Pott 1847; Kluge 1937–42; Salzmann 1950), most of this work was largely non-theoretical description of the numeral systems of the world's languages. In the 1960s, both the generative grammar framework pioneered by Noam Chomsky and the research into cross-linguistic universals led by Joseph Greenberg used numerical evidence as support for panhuman linguistic capabilities, leading to a renewed interest among psychologists, philosophers, and linguists in the foundations of numerical systems in the human language faculty.

Linguists are divided about just what the existence of a universal number concept means in terms of human evolution and the existence of an 'innate' number sense. Chomsky (1980, 38–39) believes that natural selection could not have selected for the human number concept, and thus contends that it is qualitatively different from the quantificational abilities of apes and other animals. Mathematics must, therefore, be no more than a by-product of some other evolved ability, such as the language faculty. For Chomsky, the language faculty and the number faculty (and these two faculties alone) share the concept

---

1. I use the term 'western' to refer to the signs 0123456789 instead of 'Arabic' and 'Hindu-Arabic', not to deny that this innovation was borrowed from a Hindu antecedent through an Arabic intermediary, but to avoid confusion with the distinct Indian and Arabic numerical notations used widely to this day. Rendering these latter notations 'invisible' through nomenclature is counterproductive and potentially ethnocentric.

of 'discrete infinity'—the ability to create an infinite number of things from a smaller and finite set of discrete symbols (Chomsky 1988, 167–168). However, while all languages can, in theory, produce an infinite number of sentences or phrases from their existing vocabulary, there is no natural language whose number sequence can be extended indefinitely without the creation of new numeral terms (Greenberg 1978, 253). For instance, English dictionaries normally end the number sequence at 'decillion', and one needs a new number word to form 'one thousand decillion' in a regular fashion. This is in direct contrast with place-value numeral notation systems, in which one may add zeroes to a number ad infinitum. Unless one wishes to argue that the number concept is divorced entirely from the numeral words, this is poor evidence that the number concept derives from the language faculty. Nevertheless, attested languages display considerable regularities in the structure of their numerical systems, which suggests that some underlying principles severely constrain or even determine the range of outcomes (Hurford 1975; Greenberg 1978; Corbett 1978).

A more sophisticated version of this hypothesis is that promoted by the cognitive linguist Heike Wiese (2003; 2007). Wiese argues, based both on her empirical research and on the mathematical philosophy of Frege (1884), Russell (1903), and others, that numeral words 'do not refer to numbers, they *serve as* numbers' (Wiese 2003, 5). The number concept is a byproduct of the language faculty; numerals developed once language did, as a means of labeling objects within the context of counting activities (for example, naming the fingers, or pebbles, or other physical objects), and thus became the numbers themselves. It is common cross-linguistically for the names of numerals, particularly 'five' and 'ten', to be connected etymologically to the fingers or hands (Bengtson 1987). Hurford (1987) argues along similar lines that one possible origin of number words is through counting rhymes and games such as 'eeny-meeny-miney-mo', rather than as an automatic linguistic expression of an underlying concept. This interdependent co-evolution of numeral words and numerical concepts is a reasonable proposition and is congruent with much of the linguistic literature on the subject of number, but as a historical or evolutionary hypothesis remains untested at present.

In contrast to these perspectives, which stress the universality of the number sense but also deny that it evolved specifically under natural selection, Stanislas Dehaene (1997) and Brian Butterworth (1999), working from neurological and psychological foundations, argue that the structure of the brain will tell us a great deal about how humans count and compute. One line of evidence suggesting that this is so is that various animals have been demonstrated to possess quantificational abilities, particularly the ability to distinguish small quantities up to three or four (McComb *et al* 1994; Hauser 2005). Pre-cultural and pre-linguistic infants have been shown to possess such abilities as well, by means of attention studies that examine the attention and gaze of infants at artificially unexpected or counterintuitive numerical situations (Wynn 1992; 1998). Regardless of language, humans can more easily distinguish four from five objects in a group than nineteen from twenty objects, and more easily distinguish eight from twelve objects than nine from eleven—suggesting some sort of intuitive, plausibly hardwired analog numerical representations.

Yet, as Carey (2001) points out, even if this is the case, the number line itself, based on the successor function of discrete integers, may nonetheless be a cultural construction and may not be evolutionarily hardwired. Geary (1995) and Miller and Paredes (1996) have discussed the differences between Chinese and English numerical systems and their cognitive consequences in terms of ease of learning for children, suggesting that even where there are universal aspects of numeration, variability also plays a major role. While the ability to distinguish two lions from three seems relevant from the perspective of natural selection within the evolutionary history of hominids, the ability to do most arithmetic, or to organize numerical systems in terms of a base and its powers, does not. If these are indeed universal phenomena, other explanations are needed.

Culture, number, and cognition

From early in the study of other societies, the absence or relative paucity of numeral words has been regarded as evidence of savagery among the indigenous peoples of the world. Perhaps the most notable statement of this sentiment is 'On the Numerals as Evidence of the Progress of Civilization' by the Scottish surgeon-scholar John Crawfurd, then president of the Ethnological Society (Crawfurd 1863). Crawfurd's position was that numerals were among the last words invented in any language, and that they 'advance with the progress of civilization', and thus that the 'social condition of a people is, therefore, in a good measure, indicated by its numeral system' (Crawfurd 1863, 84). Nothing about the concept of natural selection implies progress. Nevertheless, following the publication of Charles Darwin's *Origin of species* in 1859, Darwin and his colleagues, working in a Victorian context where British imperial rule was nearly unchallenged, often wrote and behaved as if culture evolved in a single line from simple to complex. In this social and intellectual context, which owed much to Enlightenment speculative histories and racialism, numerals could serve as an easily quantifiable surrogate for measuring cultural progress. Darwin's cousin and friend, Sir Francis Galton reported of the Damara of Namibia that:

they certainly use no numeral greater than three. When they wish to express four, they take to their fingers, which are to them as formidable instruments of calculation as a sliding-rule is to an English schoolboy. They puzzle very much after five, because no spare hand remains to grasp and secure the fingers that are required for units. (Galton 1853, 133)

Galton's report need not be taken at face value; he was not fluent in Damara, which in fact has more numerals than he suggests. His unflattering account was used by Conant (1896) and others as evidence for the proposition that small-scale societies were generally numerically incompetent. The notion that the inventory of numeral words in a language is evidence for or against its speakers' degree of civilization remained current throughout much of the twentieth century. It is found prominently in the psychologically-informed ethnology of Lucien Lévy-Bruhl (1966 [1910]), who distinguished 'primitive' numeration based on 'configurations' of small quantities (pair, triad, etc.) from true cardination and the successor principle. Yet the lack of any meaningful definition of 'primitive' and 'civilized' independent of this assertion renders these conclusions invalid; they do no more than assert that societies that lack extensive series of numeral words are representative of an earlier stage of human development.

It nonetheless cannot be denied that there is *some* correlation between the size of the set of numeral words in a language and other aspects of social life. Greenberg (1978), in his study of universals of numeration, pointed out that all the languages with limited sets of numeral words were small-scale societies, and suggested the need to inquire further into the implications of this finding for cultural evolutionary studies. Divale's (1999) study of two samples of sixty-nine and one hundred and thirty-six societies and their numeral-words revealed a reasonably strong correlation between the highest number normally expressible in a language, and the degree to which the speakers of that language relied on the storage of foods (especially grain cereals) to prevent starvation due to climatic instability. He reasons that one potential explanation is that societies need to quantify such foods in order to collect and distribute them (see also Steensberg 1989). The degree to which a society is able to marshal resources to store food on a large scale, in turn, is related to social size and complexity. This correlation is interesting and deserves further study, but is unlikely to be the sole or even the primary determinant of the size of the numerical lexicon.

A recent controversy involves the ethnographic study of the Pirahã by the linguist Daniel Everett, who asserts that this group of Amazonian horticulturalists possess no numeral words whatsoever, as part of a general cultural constraint against referring to objects and concepts outside of immediate experience (Everett 2005). This represents the most forceful and best-documented instance of such an assertion, and comes from a lengthy period of ethnography over some decades. Although the Pirahã have had centuries of contact with Brazilians of Portuguese descent, including trade relations, Everett has never heard them use any numeral-words, although he recognizes that certain grammatical constructions have a 'quantificational smell' (Everett 2005, 625). The Pirahã themselves are acutely aware of this absence, and expressed concern to Everett that their lack of arithmetic was hindering their trade relations with Portuguese-speaking itinerant traders, but nevertheless had limited success in an educational program designed by Everett to teach them quantification. Gordon (2004), based on a short field period among the Pirahã, presents some evidence from psychological testing that seems to suggest that the Pirahã lack of quantificational words extends to the perceptual and cognitive domains as well. Dixon (1980, 107–108), similarly, has argued that there are no true numeral words in some Australian languages. Everett's assertion that a Pirahã cultural constraint inhibits their use of quantification and structures their thinking about the world is contentious but deserving of further study.

Aside from the work of Everett and Gordon, there is some evidence that speakers of languages that have limited sets of numeral words also have specific limitations in numerical cognition. Findings from developmental psychologists such as Piaget (1952) and Vygotsky (1962) provide an independent set of criteria on which numerical cognition can be judged cross-culturally. Lancy (1983) undertook detailed psychological testing of members of various groups in Papua New Guinea and found that monolingual children who spoke local languages and had no formal education had considerable difficulty with tasks considered simple for their age. This work is supported by the massive linguistic research of Lean (1991). Yet as Gay and Cole (1967) note, traditional mathematical practices can in fact have cognitive advantages over those achieved through Western-style education; the Kpelle of Liberia, among whom they worked, could, for instance, more accurately estimate the volume of a pile of rice than Western-educated individuals, although they performed arithmetical calculations more poorly (see also Reed and Lave 1979). We must be cautious before inferring causation from correlation in these cases, and be wary of ethnocentrically projecting Western interests and values onto tribal societies.

The presence or absence of many numeral words in these languages must be conceived in terms of perceived social needs (or lack thereof) rather than as an intellectual failing. Hallpike (1979, 237), who generally follows Lévy-Bruhl in asserting that abstract number concepts are absent from primitive societies, nonetheless cautions that numerical abstraction 'cannot be deduced merely from the existence of a series of verbal numerals, even a series extending to 100 or 1,000 or more'. Hallpike stresses instead that only the presence of the right sort of social problems leads to the cultural evolution of formal-logical reasoning about quantity. There is abundant evidence that when numeral words are desired, speakers of any language are capable of extending their numeral word sequence, either through modeling new words on older ones in their own language, or through borrowing words from other languages. It cannot be ruled out that when the need is no longer present, higher numeral words cease to be used.

Some languages use different sets of numeral words for counting different classes of object, or *numeral classifiers*. Some linguists and ethnographers argue that numeral classifiers represent evidence of 'concrete' counting as opposed to 'abstract' numeration. While in some cases this variability simply amounts to the use of different morphemes at the end of a single set of numeral words, in other cases the numeral words are radically different. Conant (1896) presented a group of numeral systems from the Tsimshian language of northern British Columbia, Canada (Table 6.1.1). Conant held that the use of multiple systems represented linguistic 'primitivity' and suggested a lack of numerical abstraction, and through Lévy-Bruhl and others this idea enjoys some currency in the contemporary study of numerals. The work of the Near Eastern archaeologist Denise Schmandt-Besserat (1984; 1992) on token systems of the prehistoric Middle East relies heavily on the notion that the use of different symbols (lexical or graphic) for the same numerical referent has cognitive implications for the users of such semiotic systems. This theory has been developed more thoroughly by Peter Damerow (1996), who notes that the multiplicity and semantic ambiguity of the numerical notation systems of late fourth-millennium Mesopotamia suggest an incompletely abstract number concept.

Yet the languages that have numeral classifiers include the Maya languages, whose users developed complex astronomy, mathematics, and architecture (Berlin 1968; Macri 2000) and Japanese (Downing 1996), whose speakers can hardly be accused of non-abstract mathematical thought. In fact, numeral classifiers are no more than a taxonomic system akin to (though more specific than) grammatical gender. They may well reflect particular cultural perspectives on the classification of reality, but they do not imply that their speakers thus have no sense that *gy'ap* and *kpal* have an underlying 'tenness' any more than English

Table 6.1.1 Tsimshian numerals with classifiers (Conant 1896, 87)

No.	Counting	Flat objects	Round objects	Men	Long objects	Canoes	Measures
1	gyak	gak	g'erel	k'al	k'awutskan	k'amaet	k'al
2	t'epqat	t'epqat	goupel	t'epqadal	gaopskan	g'alpēeltk	gulbel
3	guant	guant	gutle	gulal	galtskan	galtskantk	guleont
4	tqalpq	tqalpq	tqalpq	tqalpqdal	tqaapskan	tqalpqsk	tqalpqalont
5	kctōnc	kctōnc	kctōnc	kcenecal	k'etoentskan	kctōonsk	kctonsilont
6	k'alt	k'alt	k'alt	k'aldal	k'aoltskan	k'altk	k'aldelont
7	t'epqalt	t'epqalt	t'epqalt	t'epqaldal	t'epqaltskan	t'epqaltk	t'epqaldelont
8	guandalt	yuktalt	yuktalt	yuktleadal	ek'tlaedskan	yuktaltk	yuktaldelont
9	kctemac	kctemac	kctemac	kctemacal	kctemaetskan	kctemack	kctemasilont
10	gy'ap	gy'ap	kpēel	kpal	kpēetskan	gy'apsk	kpeont

speakers are confused between six eggs and a half-dozen. With regard to late fourth-millennium Mesopotamian tokens and numerals, the accountants and scribes who used them were able to manage complex administrative tasks, and it is implausible that they did not recognize that '8 sheep' and '8 bushels of grain' had something in common. There is simply no evidence from existing human languages for Bertrand Russell's assertion that, 'It must have required many ages to discover that a brace of pheasants and a couple of days were both instances of the number 2' (Russell 1919, 3).

Since the 1980s, the development of ethnomathematics has stressed that numerical concepts develop in different cultures in different ways, and that we should not dismiss too readily the achievements of non-western societies (Ascher 1991; Powell and Frankenstein 1997). Ethnomathematics provides a useful antidote to the sometimes aggressive Eurocentrism of earlier decades, and has brought contemporary anthropological insights to the study of mathematics, but balances a fine line between universalism and radical relativism. It is difficult to know what to make of Mimica's claim (1988) that the Iqwaye of Papua New Guinea developed the concept of transfinite numbers on the basis that an informant used the same numeral word for 'one', 'twenty', and 'four hundred' (that is, $x = x^2 = x^3$). Similarly, despite Urton's fascinating assertion (1997) that the Quechua number concept is strikingly different from the western standard, and thus that the ontology of numbers is culturally relative, it is extremely difficult to evaluate such statements in the absence of some criteria for evaluating the universality of the concept in the first place. Crump's (1990) anthropologically and psychologically-informed synthesis remains the best of this work to date, forcing us to recognize both similarities and differences in number concepts. Ethnomathematics highlights the fact that there can be differences in numerical cognition that do not imply necessary distinctions between right/wrong, simple/complex, or primitive/evolved.

Tallying and abacus methods

Alongside the universal or nearly universal employment of numeral words, the use of notched sticks, knotted strings, and other artifacts for recording number is similarly quite widespread cross-culturally. This has been thoroughly demonstrated for African societies (Lagercrantz 1968; 1970; 1973; Zaslavsky 1973), and more sporadically elsewhere in the world. Yet some objects called 'tallies' are structurally complex and are designed to represent completed enumerations; the Inka *khipu* knot records, for instance, record numbers using a decimal system with place-value (Ascher and Ascher 1980; Urton 1997, 2003). Many of the so-called tallies of medieval Europe are simply wooden slabs or blocks on which

Roman numerals have been carved (Baxter 1989). 'Tallies' of this sort are simply numerical notations that happen to be notated on media different than those used for phonetic scripts.

Tallies that notate quantities serially, using one mark for each object, however, are distinct from numerical notation. Thus, a tally for 15 might read IIIII IIIII IIIII. They are visual and representational, but the function of such artifacts is quite distinct: they are serial records of an ongoing enumeration activity rather than a final cardinal count (as in the Roman XV). Each sign, regardless of its shape or the spacing, represents one unit, and even though the signs can be read as a cardinal count, the process of making them is ordinal. They are immediate aids to computation, albeit of fairly limited flexibility. Because they are not intended primarily for permanent record keeping, their archaeological survival is limited, and in fact some tallies may not survive to be discarded. For instance, Herodotus relates an episode in which Darius of Persia tied sixty knots in a thong and then instructed a group of Ionian despots to untie a knot each day while awaiting his return (Herodotus, *Histories* 4.98).

Tallying is evidently of great antiquity, and probably dates back at least to the Upper Paleolithic period (35,000–10,000 BC), when anatomically modern humans notched bones and possibly other perishable materials (Absolon 1957; Marshack 1972). In some cases, as in the Etruscan/Roman numerals, tally-systems gave rise to numerical notation systems (Keyser 1988). Nonetheless, by no means are tallies primitive, sub-optimal, or simply precursors of written numerals. Their functions are completely different; even in contemporary western societies all sorts of repetitive numerations are taken by marking tallies in groups of five, with the fifth crossing out the first four. Even in an age of widespread electronic computation there is no reason to believe that the humble tally, likely tens of thousands of years old, is at risk of disappearing. Nevertheless, tallying systems are primarily suited to serial counts of objects, rather than general arithmetical functions, and thus merchants and administrators generally require additional computational devices or representational systems to aid in arithmetic.

There is substantial ethnographic and historical evidence demonstrating that computational techniques among non-literate or minimally literate groups are abundant and efficient for the tasks for which they are needed. While school arithmetic is generally decimal, the mental arithmetic of non-literate artisans and traders is often based on doubling, halving, and quartering (Petitto 1982; Rosin 1984). Basque-speaking shepherds in contemporary California use a highly effective array of computational techniques including spoken numeral words, mental arithmetic, and a tallying-system of pebbles and notched sticks (Araujo 1975). The Kédang people of Indonesia accomplish complex mensuration and numeration tasks, such as the measuring and evaluation of elephant tusks as part of the ivory trade (Barnes 1982). In various societies of Melanesia, 'body-counting' is used in place of numeral words, naming various parts of the body in sequence as a means of counting (Saxe 1981; Biersack 1982).

The pebble-abacus was the central technique for performing computations in the ancient eastern Mediterranean (Lang 1957; Taisbak 1965; Schärlig 2001). Roman and Greek numerals were often used to notate the column-values on the device, and to record the results of computations performed on permanent media, but otherwise were irrelevant to the practice of arithmetic. While only around thirty classical abaci have survived, Netz (2002, 327), noting that 'The abacus is not an artefact; it is a state of mind', rightly cautions that any flat surface and set of objects can suffice. Nevertheless, following the classical period, there is no substantial archaeological, textual, or artistic evidence for the use of the abacus between the fifth and tenth centuries AD. In the tenth century in Europe, a sort of abacus was revived under the particular influence of Gerbert of Aurillac (later Pope Sylvester II), who was also one of the primary early adopters of Arabic numeration in the West after being exposed to Arabic arithmetic during his travels in Toledo around 970 (Folkerts 2001). Instead of using multiple pebbles or balls in each column, Gerbert's 'abacus' was a grid on which tokens called *apices* were laid out, each one bearing a western numeral from 1–9, or a zero-sign called *tsiphra* (Berggren 2002, 355–357). These tokens were manipulated by moving them from column to column, and its users were known as 'abacists'. Yet Roman numerals predominated for actually writing the results of computations performed, and not until Leonardo of Pisa (Fibonacci) wrote his *Liber abaci* in 1202 did pen-and-paper arithmetic using western numerals begin to spread across western Europe, among the so-called 'algorismists' (Burnett 2006). Yet until the sixteenth century and the advent of printed arithmetics, most merchants and administrators used neither 'Gerbert's abacus' nor Western numerals, but rather computation on boards with unmarked tokens or pebbles, much like the Greco-Roman abacus (Baxter 1989). Throughout the Middle Ages the English technique of choice was the cloth 'Exchequer board', etymologically related to 'checkerboard' (with results written in Roman numerals); the modern British title 'Chancellor of the Exchequer' preserves the linkage between the counting board and commerce (Murray 1978, 169).

Another reliable, inexpensive, and portable computational 'technology' are the fingers. There is widespread evidence for the use of finger-numbering and arithmetic in classical Greece and Rome, including depictions of individuals reckoning with the fingers, *tesserae* 'gaming tokens' showing particular finger configurations along with Roman numerals, and abundant textual references (Alföldi-Rosenbaum 1971; Williams and Williams 1995). Finger-reckoning was the primary arithmetical technique employed in early medieval Europe, and was strongly praised by Bede in his work on calendrical computation (Wallis 1999). Finger-reckoning systems remained in use in Europe and the Middle East throughout the Middle Ages into the early modern period (Saidan 1996). Like any technique (including pen-and-paper arithmetic), finger-reckoning rests on a foundation of memorized arithmetic facts and/or visual representations such as multiplication tables. Chisanbop, an arithmetical technique developed in Korea in the 1940s, uses the fingers to notate and reckon as if they were a quasi-abacus (Lieberthal 1979).

From at least the fourth century BC until the sixteenth century AD, East Asian arithmetical procedures were centered around the *suan zi*, or counting rods, and their written representation, the rod-numerals (see Volkov, Chapter 2.3 in this volume). The counting rods were thin sticks or strips of bamboo, wood, ivory, or bone, and could be manipulated in columns to represent numbers in a place-value, decimal manner, much as the Roman abacus. As with most technologies, the initial reaction to them involved some skepticism—the *Daodejing*, written around 300 BC, asserts that 'Good mathematicians do not use counting-rods' (Needham 1959, 70–71). Yet they were very quickly adopted, and were the foundation of Chinese mathematical practice until the late Ming dynasty. One of the primary advantages of the system was that the physical rods could easily be transformed into written numerals using horizontal and vertical lines to notate the position of the rods. Many Chinese mathematical terms use the radical for 'bamboo', further signifying the linkage between the counting rods and mathematics (Needham 1959, 72). Although physical rods themselves are no longer used, they survive in written form today in a numerical notation system called *an ma*, used in commercial contexts such as bills and invoices (Martzloff 1997, 189).

The *suan pan* or Asian bead-abacus is of relatively recent origin, probably no earlier than the fourteenth century, and not until the seventeenth century did it definitively supplant counting rods. There is little evidence of a competitive environment between users of the *suan zi* and the *suan pan* to parallel the 'abacist–algorismist' debate in Europe or the later debate between users of the counting board versus users of western numerals. Nevertheless, the transition did occur throughout East Asia, where the *suan pan* (called *soroban* in Japanese) is a central part of mathematics education to the present day. No similar transition seems imminent today that would result in the abandonment of the *suan pan*. Although western numerals are ubiquitous in Japan and commonly used in China, Chinese numerals are rarely used for pen-and-paper arithmetic.

This state of affairs is by no means indicative of a hidebound mindset or stubbornness. Stigler (1984) showed that Japanese master abacus users employ a 'mental abacus'—a mental representation of intermediate and prior positions in a computation that greatly enhances the purely material aspects of the technology. A trained abacus user can normally manipulate multi-digit numbers far more rapidly than any reckoner using pen and paper. On 12 November 1946, the American military service newspaper, *Stars and Stripes*, sponsored a competition between Private

Thomas Wood, an American soldier trained in the use of one of the sophisticated electronic calculators available at the time, and Kiyoshi Matsuzaki, an administrator and abacus master (Kojima 1954). Although the competition was surely designed to impress the audience with the superiority of American technical ingenuity, the Japanese competitor won four of the five events. Zhang and Norman (1995) scorn abacus users for using one technique for arithmetic and another for writing results in numerals, as part of their argument that western numerals are uniquely efficient arithmetical tools. The merit of such a position vanishes once it is recognized that pen-and-paper arithmetic with western numerals cannot be demonstrated to possess this putative superiority.

Although their surviving calendrical and divinatory texts do not notate calculations performed (only results), Landa reported in his *Relación de las cosas de Yucatán* that the Maya and related peoples of lowland Mesoamerica computed using a flat board or on the ground (Tozzer 1941, 98). The Guatemalan Maya at Panajachel in the 1930s reckoned using cacao beans or stones in groups of five or twenty, and this may be a survival of earlier Maya practices (Thompson 1941, 42). In sixteenth-century Peru, Don Felipe Guaman Poma de Ayala, the son of a *conquistador* and an Inka princess, depicted a *khipukamayuq* (khipu-administrator) using the traditional *khipu* system of knotted cords along with an abacus-like grid of black and white pebbles or stones (Wassén 1931; Urton 1998, 417–420). Because the *khipu* could not meaningfully have been manipulated for arithmetic, some abacus-like technique would have been needed to administer the expansive and multi-ethnic Inka Empire.

Computational devices like the abacus were so prevalent in pre-modern states that one might reasonably ask why they would be replaced, given that, at the very least, they seem to have been as efficient as pen-and-paper computation. To answer this question, we need to look seriously at the alternative.

The emergence and spread of numerical notation

While many societies possess visual and/or material tallying techniques, only some societies possess numerical notation. Numerical notation systems are visual but primarily non-phonetic structured systems for representing numbers permanently. Typically they do so using a set between three and forty signs, which combine together by means of a numerical base, often but not always that of the language spoken by its inventors. Over 100 structurally distinct numerical notation systems are known to have been used between 3500 BC and the present day (Chrisomalis forthcoming; see also Cajori 1928; Smith and Ginsburg 1937; Menninger 1969; Guitel 1975; Ifrah 1998). Unlike number words, they represent numbers translinguistically, and do not follow the grammar or lexicon of any specific language. Unlike tallies, they represent completed enumerations, and unlike computational technologies, they create permanent records of numerals. They can be used for computation, but historically this function has been rare. The primary typological distinction among them is between additive and positional (place-value) systems, although this is not the only relevant distinction that can be made (Boyer 1944; Chrisomalis 2004).

The earliest attested numerical notation is the proto-cuneiform system used in the ancient Mesopotamian city-state of Uruk in the late fourth millennium BC (Nissen, Damerow, and Englund 1993). In its initial state, proto-cuneiform writing consisted of a large repertory of at least fifteen different systems for numerical representations of different categories of objects, persons, and capacity measures, along with ideograms and pictograms for the various things being enumerated. It served as an administrative system for the urban temple economy of Uruk and other cities throughout Mesopotamia, and well as the Proto-Elamite area to the east, in modern Iran (Potts 1999).

In a series of articles and books, the Near Eastern archaeologist Denise Schmandt-Besserat has suggested that the Uruk numerical notations, and ultimately writing itself, are the end product of a millennia-long history of accounting and administration. Throughout the Neolithic in Mesopotamia, possibly as early as 8000 BC, clay tokens were used as administrative tools much as tally-sticks and knotted cords might, using one-to-one correspondence between the counters and the objects being enumerated, as part of a sophisticated accounting system. Schmandt-Besserat's hypothesis, itself derived from the earlier work of Amiet (1966), is suggestive, but must be read in the context of severe criticisms such as those of Lieberman (1980) and Zimansky (1993). In particular there is little evidence that the specific forms of the clay tokens bear any resemblance to the proto-cuneiform signs.

It would be erroneous, however, to assume that numerical notation developed independently only in Mesopotamia, or that the developmental trajectory that it took there provides a general template for its development elsewhere. In Egypt, the earliest numerals are found on labels for mortuary offerings, in a royal tomb from a cemetery at the city of Abydos, dating to around 3250 BC (Dreyer 1998). In China of the Shang Dynasty, the first written documents (c 1200 BC) are records of royal divinations (Tsien 2004). In the Middle and Late Formative periods in Mesoamerica (c 600 BC–150 AD), virtually all of the earliest Zapotec, Olmec, and Maya inscriptions contain numerals, but their use is strictly in names and dates, never administrative (Houston 2004). With the possible exception of the Egyptian case, there is very minimal likelihood that the development of numerical notation was spurred by diffusion from Mesopotamia or anywhere else. Rather, the development of both writing and numerical notation is correlated with the formation of early states in each region, but the functions for which these representational systems are used are quite distinct.

Postgate, Wang, and Wilkinson (1995) suggest that the reason we have not found evidence of early administrative writing and numerical notation in Egypt, Mesoamerica, and China is that such documents were written on perishable materials that have not survived. They use this line of reasoning to propose that writing emerges everywhere as it did in Mesopotamia: for bookkeeping and accounting-related functions concerning state administration. Yet in the absence of evidence that this is so, such assertions are unjustified. There is simply no reason to expect that the invention of writing and numerical notation must always have the same underlying function everywhere. It is nonetheless true that numerical notation developed independently only in socially complex societies that had considerable need to represent quantities. Nevertheless, in other cases—as in the expansive, densely populated states of West Africa—numerical notation simply never developed, so it cannot be regarded as an absolute necessity.

Because numerical notation is used widely in exchange and administration, its spread and adoption is strongly correlated with imperialism, long-distance trade, and other political and economic processes associated with states. Although most numerals are used for representation rather than computation, the employment of numerical notation systems for astronomy, mathematics, and related scientific practices has also played a central role in their diffusion. Because numerical notation is not tied to any specific language, is not as difficult to learn as, for instance, a writing system, and is a communication technology used in the context of long-distance commercial and scientific exchanges, it diffuses readily in many circumstances. The well-attested spread of Hindu numeration to Europe through Arabic intermediaries has led diffusionist explanations to be widespread in the literature on numerals, often with good reason. I have argued, on the basis of cultural contact and structural similarities that the Greek alphabetic (or Ionian) numerals developed out of the Egyptian hieratic or demotic numerals used in the 6th century BCE in the context of circum-Mediterranean trade relations (Chrisomalis 2003). Yet the Greek inventors of the alphabetic system were highly innovative; their use of the letters of the alphabet, in sequence, as numeral-signs was unparalleled elsewhere, and the uses to which Greek numerals were put differ substantially from those for which Egyptians used them.

On the other hand the notion that most people are uncreative and therefore most mathematical developments made only once, and spread from a single center, is quite incorrect and frequently tinged with racist assumptions about non-European peoples. Seidenberg's (1960; 1962) pronouncements on the diffusion of mathematics, geometry, and all numbers higher than two as part of a Mesopotamian or Indian Neolithic ritual complex, and his insistence that all Maya numeration and mathematics was derived from Babylonia (Seidenberg 1986), are extreme and unsupported by any textual or archaeological evidence. Joseph Needham's remarks on the subject of the priority and diffusion of Chinese mathematics are more tentative and a necessary counterpoint to Eurocentrism, but nonetheless the notion that the Chinese spread place-value to Babylonia (Needham and Wang 1959, 146–150) or that Chinese mathematics influenced Mesoamerica (Needham and Lu 1984) cannot be sustained.

In fact, it is highly probable that place-value numerical notation, or something quite like it, developed at least five times independently: in Middle Bronze Age Mesopotamia (c 2100 BC), in the Warring States period in China (fourth century BC), in lowland Mesoamerica (no later than 100 AD), in India (c 500 AD), and in the Andes (no later than 1300 AD). No two of these regions are less than 3000 km apart, and the development of place-value occurred centuries apart in each. Each development had antecedents in earlier, local notations and computational techniques, and each is distinct in various ways. For instance, the Chinese rod-numerals have a sub-base of 5, like the Roman abacus, and the Andean *khipu* notation lacks a sign for zero. The most parsimonious explanation for these developments is that place-value is more easily conceived than extreme diffusionists allow. This also provides support for the notion that mathematics is a pan-human activity whose foundations do not differ greatly among different societies.

Numerals and computational efficiency

A connection is frequently asserted between the present ubiquity of the Western numerals in worldwide usage and the utility of this system for performing basic arithmetic using pen-and-paper computation. At first glance, this hypothesis is extremely appealing. Many of the references to positional numerals by 'early adopters' explicitly praised positional numeration in comparison with other techniques and representations. The Syrian Christian bishop Severus Sebokht discussed Hindu mathematics in 662 AD, noting 'their clever method of calculation, their computation which surpasses all words, I mean that which is made with nine signs' (Nau 1910, 225–227). In introducing the system more broadly to the Middle East in the ninth century, the mathematician al-Khwārizmī promoted the use of the nine digits plus zero as an alternative to reckoning with the letter-numerals (*hisāb al-abjad*) with its twenty-seven alphabetic signs, or finger-numeration (*hisāb al-ʿuqūd*), discussed above.

Similarly, the western European debates between the abacists, proponents of the use of the medieval abacus with tokens, with numbers written in Roman numerals and algorismists, proponents of pen-and-paper arithmetic with western numerals, reflect contested narratives of efficiency. The algorismists struck hard with many positive evaluations of the western numerals' efficiency, and eventually became predominant among mathematicians (Burnett 2006). Yet conflict over computational techniques continued heatedly among users of the mercantile counting board as late as the seventeenth century. The famous allegorical representation in Gregor Reisch's *Margarita philosophica* depicts Arithmetic, bedecked with western numerals on her gown, looking approvingly upon Boethius using western numerals while Pythagoras toils at his counting board with pebbles (Reisch 1503). Similarly, in his dictionary of 1530, the lexicographer-priest John Palsgrave included the sentence, 'I shall reken it syxe tymes by aulgorisme or you can caste it ones by counters' as a sample sentence for the verb 'to reckon' (Palsgrave 1530, 337).[2]

Nevertheless, we ought not to assume that the proclamations of advocates and early adopters perfectly reflected reality. The debate surrounding the adoption of place-value numeration, both in the Middle East and later in western Europe, pitted traditionalists against innovators and threatened to overwrite—literally— much of the practice of arithmetic, astronomy, mathematics, and accounting as they had been practiced for centuries. These debates were never solely about efficiency, but had significant ideological components. Struik (1968) argued that the prohibition of western numerals by the Guild of Moneychangers of Florence in 1299 was primarily part of the longstanding conflict between the Guelphs and Ghibellines in the mercantile economy of the city. The denigration of western numerals on the basis that they can be too easily altered, another often-heard reason for their prohibition, may have been a product of xenophobia against an Oriental invention. So, too, authors promoting the use of western numerals might do so, not solely on the asserted technical grounds, but because the promotion of a new arithmetical technique was part of broader social trends within late medieval society.

The argument that the western numerals are computationally more efficient than Roman numerals for doing arithmetic is true, and continues to be raised by authors attempting to explain the decline of Roman numerals to their present vestigial use. The limitations of Roman numerals have been invoked as an explanation for the supposed impoverishment of Roman and early medieval accounting and mathematics (Glautier 1972; Murray 1978; Crosby 1997). The difficulty with this proposition is that Roman numerals, to our knowledge, were never used in written arithmetic in anything like the manner in which western numerals are, but of course through the abacus, through finger-computation, and through mental arithmetic.

To a considerable extent, the preference for positional numerals is an artifact of modern western mathematics. Many additive systems of the past survived for millennia, such as the Egyptian hieroglyphic and hieratic numerals which persisted largely unchanged from the pre-Dynastic to the Roman period, suggesting

---

2. Ironically, though not unusually for the time, Palsgrave's dictionary was foliated in Roman numerals.

that they must have been perceived as desirable or useful for many purposes. There is a trend towards positional notation over time, but it is not inexorable and should not be presumed to now be irreversible (Chrisomalis 2004).

In South Asia, the positional numerals ancestral to our own largely replaced the older additive Brahmi system between the sixth and eleventh centuries AD (Salomon 1998). However, in southern regions of the subcontinent, additive numerals continued to thrive alongside the Tamil, Malayalam, and Sinhalese scripts, right up to the colonial period (Guitel 1975, 614–617). The Tamil additive numerals continue to be used today for many purposes. Cultural resistance against the dominant traditions of northern India probably explains the retention of the additive numerals, but users of these notations suffered no evident disadvantage in their ability to undertake arithmetic. Similarly, the additive Chinese numerals would long ago have been abandoned in favour of the Tibetan numerals (a positional, decimal system transmitted from India) if this were the case. Of course, the reverse is true, for perfectly understandable reasons having to do with Chinese political domination in the region and throughout much of Central Asia. No one would consider efficiency for computation as the explanation in this circumstance. Yet the assumption that the western numerals predominate mainly due to their supreme utility, and that we have reached the timeless pinnacle of the history of numeration, remains commonplace in scholarly and popular works (Dehaene 1997, 101; Ifrah 1998, 592).

A more parsimonious explanation for the current worldwide predominance of western numerals is the predominance of all sorts of western institutions, most notably scientific and economic, since the formation of the modern capitalist world-system with western Europe and later America firmly ensconced within the prestigious and powerful core (Wallerstein 1974). This process was accelerated by the early use of western numerals in printed books, in accounting documents, and on money—both the transmission of wealth and the transmission of information were governed by the western numerals. Changes in patterns of trade and intercultural communication correlate frequently with changes to numeral words as well as numerical notation; as the need for commerce and mathematics with larger, more complex societies increases, the numeral word series expands (Crump 1978, Schuhmacher 1975). The highest basic numeral word in European languages was 'thousand' until the thirteenth-century invention of 'million' among Italian bookkeepers, from which it spread to a wide variety of languages, Indo-European and otherwise.

More significant was the fact that pen-and-paper arithmetic with western numerals produced permanent records of calculations performed, permitting errors to be perceived more easily. The regular practice of writing down calculations and their results also provided early modern landowners, merchants, and administrative officials a high degree of information and control over their economic affairs (Swetz 1987). Nevertheless, much bookkeeping and arithmetic continued to be done without the aid of western numerals (Jenkinson 1926). As prominent a figure as William Cecil (Lord Burghley), Lord High Treasurer to Elizabeth I of England, regularly transcribed economic documents from western back into Roman numerals for his own convenience (Stone 1949, 31). Roman numerals were adequate if not optimal for recording results. The length of Roman numeral-phrases, often cited as a defect of the system, is only one of many factors users consider.

Once the western numerals had achieved a critical mass of popularity among the newly emboldened European middle class, it became likely that others operating within the same economic and communication networks would adopt the system. New users adopted western numerals partly because the current users of the system were prestigious and wealthy—in terms of cultural transmission, this is a *prestige bias* (Richerson and Boyd 2005, 124–126). Their popularity allowed new users to transmit information to more individuals and thereby created a feedback system that further increased their popularity—a *frequency dependent bias* (Richerson and Boyd 2005, 120–123). Their utility cannot be conceived simply in terms of a structured system of signs, but also in terms of who and how many people were using them, and for what purposes. The property of frequency dependence is particularly notable in systems such as numerals for which communication is of central importance. It is linked to the 'QWERTY principle' explaining the persistence of sub-optimal but popular phenomena despite the existence of alternatives, and to the predominance of poor but popular recording media and computer operating systems. The 'cost' of not using the popular system is greater than the advantage of using the technically superior one.

Conclusion

At various times and places, individuals and groups may have adopted new numerical notations because of their perceived efficiency for computation. As a general explanation for the diffusion, adoption, and extinction of numerical notation systems, however, this theory is weak in comparison to the host of political and economic factors operant in any given social context. There is minimal evidence for the widespread use of written numerals as a computational technique prior to the development of Arabic numerals in the ninth century AD and their subsequent spread westward to Europe. Modern scholarly evaluations of the efficiency of various numerical systems for computation are interesting but irrelevant to their diffusion and extinction (Detlefsen *et al* 1975; Lambert *et al* 1980; Anderson 1958). When attempting to show these systems' inferiority (as opposed to demonstrating the feasibility of such work), such analyses are perniciously derogatory. Moreover, with the growth of the electronic calculator industry over the past thirty years, pen-and-paper arithmetic may go the way of the slide rule before too many generations have passed.

Because the relations between number words, arithmetical techniques, and numerical notation symbols are complex, the evaluation of the foundations of mathematics in any society is similarly complex. No two societies are alike, and yet the striking linguistic and cross-cultural parallels observed suggest that human thinking about numerals and arithmetic is highly constrained. The debate between universalistic and particularistic numerical systems will surely continue, as will the comparison of the utility of different numerical systems. An awareness of the common core of features they share make the differences among numerical concepts so much more interesting.

Bibliography

Absolon, Karl, 'Dokumente und Beweise der Fähigkeiten des fossilen Menschen zu zählen im mährischen Paläolithikum', *Artibus Asiae*, 20 (1957), 123–150.

Alföldi-Rosenbaum, Elisabeth, 'The finger calculus in antiquity and in the Middle Ages', *Frühmittelalterliche Studien*, 5 (1971), 1–9.

Amiet, Pierre, 'Il y a 5000 ans les Elamites inventaient l'écriture', *Archeologia*, 12 (1966), 20–22.

Anderson, W French, 'Arithmetical procedure in Minoan Linear A and in Minoan-Greek Linear B', *American Journal of Archaeology*, 62 (1958), 363–369.

Araujo, Frank P, 'Counting sheep in Basque', *Anthropological Linguistics*, 17 (1975), 139–145.

Ascher, Marcia, *Ethnomathematics*, Brooks/Cole, 1991.

Ascher, Marcia, and Ascher, Robert, *Code of the quipu: a study in media, mathematics and culture*, University of Michigan Press, 1980.

Barnes, R H, 'Number and number use in Kedang, Indonesia', *Man* (NS), 17 (1982), 1–22.

Baxter, W T, 'Early accounting: the tally and checkerboard', *The Accounting Historians Journal*, 16 (1989), 43–83.

Bengtson, John D, 'Notes on Indo-European "10", "100", and "1000"', *Diachronica*, 4 (1987), 257–262.

Berggren, J Lennart, 'Medieval arithmetic: Arabic texts and European motivations', in John J Contreni and Santa Casciani (eds), *Word, image, number: communication in the Middle Ages*, Edizioni del Galluzzo, 2002, 351–365.

Berlin, Brent, *Tzeltal numeral classifiers*, Mouton, 1968.

Biersack, Aletta, 'The logic of misplaced concreteness: Paiela body counting and the nature of primitive mind', *American Anthropologist*, 84 (1982), 811–829.

Boyer, Carl B, 'Fundamental steps in the development of numeration', *Isis*, 35 (1944), 153–165.

Burnett, Charles, 'The semantics of Indian numerals in Arabic, Greek and Latin', *Journal of Indian Philosophy*, 34 (2006), 15–30.

Butterworth, Brian, *What counts: how every brain is hardwired for math*, Free Press, 1999.

Cajori, Florian, *A history of mathematical notations*, Open Court, 1928.

Carey, Susan, 'Cognitive foundations of arithmetic: evolution and ontogenesis', *Mind and Language,* 16 (2001), 37–55.

Chomsky, Noam, *Rules and representations,* Columbia University Press, 1980.

Chomsky, Noam, *Language and problems of knowledge,* MIT Press, 1988.

Chrisomalis, Stephen, 'The Egyptian origin of the Greek alphabetic numerals', *Antiquity,* 77 (2003), 485–496.

Chrisomalis, Stephen, 'A cognitive typology for numerical notation', *Cambridge Archaeological Journal,* 14 (2004), 37–52.

Chrisomalis, Stephen, *A comparative history of numerical notation,* Cambridge University Press, forthcoming.

Conant, Levi L, *The number concept,* Macmillan, 1896.

Corbett, Greville G, 'Universals in the syntax of cardinal numerals', *Lingua,* 46 (1978), 355–368.

Crawfurd, John, 'On the numerals as evidence of the progress of civilization', *Transactions of the Ethnological Society of London,* 2 (1863), 84–111.

Crosby, Alfred W, *The measure of reality: quantification and Western society, 1250–1600,* Cambridge University Press, 1997.

Crump, Thomas, 'Money and number: the Trojan horse of language', *Man,* 13 (1978), 503–518.

Crump, Thomas, *The anthropology of numbers,* Cambridge University Press, 1990.

Damerow, Peter, *Abstraction and representation: essays on the cultural evolution of thinking,* Kluwer, 1996.

Dehaene, Stanislas, *The number sense,* Oxford University Press, 1997.

Detlefsen, M, Erlandson, D, Clark, H J, and Young, C, 'Computation with Roman numerals', *Archive for History of Exact Sciences,* 15 (1975), 141–148.

Divale, William, 'Climatic instability, food storage, and the development of numerical counting: a cross-cultural study', *Cross-Cultural Research,* 33 (1999), 341–368.

Dixon, R M W, *The languages of Australia,* Cambridge University Press, 1980.

Downing, Pamela, *Numeral classifier systems: the case of Japanese,* John Benjamins, 1996.

Dreyer, Günter, *Umm el-Qaab I,* von Zabern, 1998.

Everett, Daniel L, 'Cultural constraints on grammar and cognition in Pirahã', *Current Anthropology,* 46 (2005), 621–646.

Folkerts, Menso, 'Early texts on Hindu-Arabic calculation', *Science in Context,* 14 (2001), 13–38.

Frege, Gottlob, *Die Grundlagen der Arithmetik,* Koebner, 1884.

Galton, Sir Francis, *Narrative of an explorer in tropical South Africa,* Ward, Lock & Co, 1853.

Gay, John, and Cole, Michael, *The new mathematics and an old culture: a study of learning among the Kpelle of Liberia,* Holt, Rinehart and Winston, 1967.

Geary, David C, 'Reflections of evolution and culture in children's cognition: implications for mathematical development and instruction', *American Psychologist,* 50 (1995), 24–37.

Glautier, M W E, 'A study in the development of accounting in Roman times', *Revue internationale des droits de l'antiquité,* 19 (1972), 311–343.

Gordon, Peter, 'Numerical cognition without words: evidence from Amazonia', *Science,* 306 (2004), 496–499.

Greenberg, Joseph H, 'Generalizations about numeral systems', in J H Greenberg (ed), *Universals of human language,* vol 3, Stanford University Press, 1978, 249–297.

Guitel, Genevieve, *Histoire comparée des numérations écrites,* Flammarion, 1975.

Hallpike, C R, *The foundations of primitive thought,* Clarendon Press, 1979.

Hauser, M D, 'Our chimpanzee mind', *Nature,* 437 (2005), 60–63.

Houston, Stephen D, 'The archaeology of communication technologies', *Annual Review of Anthropology*, 33 (2004), 223–250.

Hurford, James R, *The linguistic theory of numerals*, Cambridge University Press, 1975.

Hurford, James R, *Language and number*, Basil Blackwell, 1987.

Ifrah, Georges, *The universal history of numbers* (trans David Bellos, E F Harding, Sophie Wood, and Ian Monk), John Wiley and Sons, 1998 (original work published 1994).

Jenkinson, Hilary, 'The use of Arabic and Roman numerals in English archives', *The Antiquaries Journal*, 6 (1926), 263–275.

Keyser, Paul, 'The origin of the Latin numerals 1 to 1000', *American Journal of Archaeology*, 92 (1988), 529–546.

Kluge, Theodor, *Die Zahlenbegriffe*, 5 vols, Stieglitz, 1937–1942.

Kojima Takashi, *The Japanese abacus: its use and theory*, Charles E Tuttle, 1954.

Lagercrantz, Sture, 'African tally-strings', *Anthropos*, 63 (1968), 115–128.

Lagercrantz, Sture, 'Tallying by means of lines, stones, and sticks', *Paideuma*, 16 (1970), 52–62.

Lagercrantz, Sture, 'Counting by means of tally sticks or cuts on the body in Africa', *Anthropos*, 68 (1973), 569–588.

Lambert, Joseph B, Ownbey-McLaughlin, Barbara, and McLaughlin, Charles D, 'Maya arithmetic', *American Scientist*, 68 (1980), 249–255.

Lancy, David F, *Cross-cultural studies in cognition and mathematics*, Academic, 1983.

Lang, Mabel, 'Herodotos and the abacus', *Hesperia*, 26 (1957), 271–287.

Lean, G A, *Counting systems of Papua New Guinea*, 17 volumes, Papua New Guinea University of Technology, Department of Mathematics and Statistics, 1991.

Lease, Emory B, 'The number three, mysterious, mystic, magic', *Classical Philology*, 14 (1919), 56–73.

Lévy-Bruhl, Lucien, *How natives think* (trans Lilian Clare), Washington Square, 1966 (original work published 1910).

Lieberman, Stephen J, 'Of clay pebbles, hollow clay balls, and writing: a Sumerian view', *American Journal of Archaeology*, 84 (1980), 339–358.

Lieberthal, Edwin M, *The complete book of fingermath*, McGraw-Hill, 1979.

Macri, Martha J, 'Numeral classifiers and counted nouns in the classic Maya inscriptions', *Written Language and Literacy*, 3 (2000), 13–36.

Marshack, Alexander, *The roots of civilization*, McGraw-Hill, 1972.

Martzloff, Jean-Claude, *A history of Chinese mathematics*, Springer, 1997.

McComb, K, Packer, C, and Pusey, A, 'Roaring and numerical assessment in contests between groups of female lions, Panthera leo', *Animal Behaviour*, 47 (1994), 379–387.

Menninger, Karl, *Number words and number symbols* (trans Paul Broneer), MIT Press, 1969 (original work published 1958).

Miller, Kevin F and Paredes, David R, 'On the shoulders of giants: cultural tools and mathematical development', in Robert J Sternberg and Talia Ben-Zeev (eds), *The nature of mathematical thinking*, Lawrence Erlbaum, 1996, 83–118.

Mimica, Jadran, *Intimations of infinity*, Berg, 1988.

Murray, Alexander, *Reason and society in the Middle Ages*, Clarendon Press, 1978.

Nau, F, 'Notes d'astronomie indienne', *Journal Asiatique*, 10th series, 16 (1910), 209–225.

Needham, Joseph, and Lu Gwei-Djen, *Trans-Pacific echoes and resonances: listening once again*, World Scientific, 1984.

Needham, Joseph, and Wang Ling, *Science and civilization in China, vol 3: mathematics and the sciences*, Cambridge University Press, 1959.

Netz, Reviel, 'Counter culture: towards a history of Greek numeracy', *History of Science*, 40/3 (2002), 319–332.

Nissen, Hans J, Damerow, Peter, and Englund, Robert K, *Archaic bookkeeping* (trans Paul Larsen), University of Chicago Press, 1993 (original work published 1990).

Nuttall, Zelia, *The fundamental principles of Old and New World civilizations,* Papers of the Peabody Museum, 2, Harvard University, 1901.

Palsgrave, John, *Lesclarcissement de la langue francoyse,* London: Richard Pynson, 1530.

Petitto, Andrea L, 'Practical arithmetic and transfer: a study among West African tribesmen', *Journal of Cross-Cultural Psychology,* 13 (1982), 15–28.

Piaget, Jean, *The child's conception of number,* WW Norton, 1952.

Postgate, Nicholas, Wang Tao and Wilkinson, Toby, 'The evidence for early writing: utilitarian or ceremonial?', *Antiquity,* 69 (1995), 459–480.

Pott, A F, *Die quinare und vigesimale Zählmethode,* Halle: Schwetzke und Sohn, 1847.

Potts, Daniel T, *The archaeology of Elam: formation and transformation of an ancient Iranian state,* Cambridge University Press, 1999.

Powell, Arthur B, and Frankenstein, Marilyn (eds), *Ethnomathematics: challenging Eurocentrism in mathematics education,* SUNY Press, 1997.

Reed, H J, and Lave, Jean, 'Arithmetic as a tool for investigating relations between culture and cognition', *American Ethnologist,* 6 (1979), 568–582.

Reisch, Gregor, *Margarita philosophica,* Freiburg, 1503.

Richerson, Peter J, and Boyd, Robert, *Not by genes alone: how culture transformed human evolution,* University of Chicago Press, 2005.

Rosin, R Thomas, 'Gold medallions: the arithmetic calculations of an illiterate', *Anthropology and Education Quarterly,* 15 (1984), 38–50.

Russell, Bertrand, *The principles of mathematics,* Cambridge University Press, 1903.

Russell, Bertrand, *Introduction to mathematical philosophy,* George Allen and Unwin, 1919.

Saidan, Ahmad S, 'Numeration and arithmetic', in Roshdi Rashed (ed), *Encyclopedia of the history of Arabic Science,* vol 2, Routledge, 1996, 331–348.

Salomon, Richard, *Indian epigraphy,* Oxford University Press, 1998.

Salzmann, Zdenek, 'A method for analyzing numerical systems', *Word,* 6 (1950), 78–83.

Saxe, Geoffrey B, 'Body parts as numerals: a developmental analysis of numeration among the Oksapmin in Papua New Guinea', *Child Development,* 52 (1981), 306–316.

Schärlig, Alain, 'Les deux types d'abaques des anciens grecs et leurs jetons quinaires', *Archives des Sciences,* 54/2 (2001), 69–75.

Schmandt-Besserat, Denise, 'Before numerals', *Visible Language,* 18 (1984), 48–60.

Schmandt-Besserat, Denise, *Before writing,* University of Texas Press, 1992.

Schuhmacher, W W, 'On the liquidation of numerals in "primitive" societies', *Dialectical Anthropology,* 1 (1975), 94–95.

Seidenberg, Abraham, 'The diffusion of counting practices', *University of California Publications in Mathematics,* 3/4 (1960), 215–299.

Seidenberg, Abraham, 'The ritual origin of counting', *Archive for the History of Exact Sciences,* 2 (1962), 1–40.

Seidenberg, Abraham, 'The zero in the Mayan numerical notation', in Michael P Closs (ed), *Native American mathematics,* University of Texas Press, 1986, 371–386.

Smith, David E, and Ginsburg, Jekuthiel, *Numbers and numerals,* National Council of Teachers of Mathematics, 1937.

Steensberg, Axel, *Hard grains, irrigation, numerals and script in the rise of civilizations,* Royal Danish Academy of Sciences and Letters, 1989.

Stigler, James W, ' "Mental abacus": the effect of abacus training on Chinese children's mental calculation', *Cognitive Psychology,* 16 (1984), 145–176.

Stone, Lawrence, 'Elizabethan overseas trade', *The Economic History Review* (New Series), 2 (1949), 30–58.

Struik, Dirk J, 'The prohibition of the use of Arabic numerals in Florence', *Archives Internationales d'Histoire des Sciences*, 21 (1968), 291–94.

Swetz, Frank, 'The evolution of mathematics in ancient China', *Mathematics Magazine*, 52 (1979), 10–19.

Swetz, Frank, *Capitalism and arithmetic: the new math of the 15th century*, Open Court, 1987.

Taisbak, C M, 'Roman numerals and the abacus', *Classica et Mediaevialia*, 26 (1965), 147–160.

Thompson, J Eric S, *Maya arithmetic*, Carnegie Contributions to American Anthropology and History 528, 403, Carnegie Institution, 1941.

Tozzer, A M, *Landa's* Relacion de las cosas de Yucatan, Papers, Peabody Museum of American Archaeology and Ethnology, Harvard University, 18, Harvard University Press, 1941.

Tsien Tsuen-Hsuin, *Written on bamboo and silk: the beginnings of Chinese books and inscriptions*, University of Chicago Press, 2nd ed, 2004.

Urton, Gary, *The social life of numbers*, University of Texas Press, 1997.

Urton, Gary, 'From knots to narratives: reconstructing the art of historical record keeping in the Andes from Spanish transcriptions of Inka *khipus*', *Ethnohistory*, 45 (1998), 409–437.

Urton, Gary, *Signs of the Inka khipu: binary coding in the Andean knotted-string records*, University of Texas Press, 2003.

Vygotsky, L S, *Thought and language* (trans Eugenia Hanfmann and Gertrude Vakar), MIT Press, 1962.

Wallerstein, Immanuel, *The modern world-system*, vol 1, Academic Press, 1974.

Wallis, Faith (ed), *The reckoning of time, by the Venerable Bede*, Liverpool University Press, 1999.

Wassén, Henry, 'The ancient Peruvian abacus', in Erland Nordenskiöld (ed), *Comparative ethnographical studies*, vol 9, AMS Press, 1931, 189–205.

Wiese, Heike, *Numbers, language, and the human mind*, Cambridge University Press, 2003.

Wiese, Heike, 'The co-evolution of number concepts and counting words', *Lingua*, 117 (2007), 758–772.

Williams, Burma P, and Williams, Richard S, 'Finger numbers in the Greco-Roman world and the early Middle Ages', *Isis*, 86 (1995), 587–608.

Wynn, Karen, 'Evidence against empiricist accounts of the origins of numerical knowledge', *Mind and Language*, 7 (1992), 315–332.

Wynn, Karen, 'Psychological foundations of number: numerical competence in human infants', *Trends in Cognitive Sciences*, 2 (1998), 296–303.

Zaslavsky, Claudia, *Africa counts: number and pattern in African culture*, Prindle, Webber & Schmidt, 1973.

Zhang Jiajie, and Norman, Donald A, 'A representational analysis of numeration systems', *Cognition*, 57 (1995), 271–295.

Zimansky, Paul, 'Review of Schmandt-Besserat 1992', *Journal of Field Archaeology*, 20 (1993), 513–517.

Sanskrit mathematical verse

Kim Plofker

The word 'mathematics' nowadays inspires an expectation of numbers and formulas visually presented on a written page. The dominance of literacy in modern learning makes it difficult for us to conceive of advanced mathematical knowledge in any non-literate (or even non-symbolic) format. In ancient and medieval Indian cultures, on the other hand, the veneration of spoken Sanskrit produced an ideal of oral learning that embraced technical subjects including the mathematical sciences, as well as narrative literature and belles lettres. The resulting genre of Sanskrit mathematical verse exemplifies in fascinating and sometimes bizarre ways the challenges and advantages of looking beyond the written word and symbol for the means to express mathematics.

Literacy and orality in Sanskrit learning

The earliest surviving texts in a form of Sanskrit, the sacred Vedas of the ancient Indo-Aryan tradition, were evidently composed and codified long before they were first written down. They are traditionally considered to have been 'heard' by the ancient sages as spoken revelation; the literal meaning of *śruti*, a standard Sanskrit name for these holy texts, is 'hearing'. It is their memorization and recitation that constitutes a sacred act, not the writing or reading of them. Indeed, for many centuries the Vedas were transmitted from generation to generation only via oral learning, in forms carefully designed to preserve them verbatim.

Most of the content of the Vedas is in metrical verse, consisting chiefly of hymns and invocations that are to be chanted during rituals. Sanskrit verse meters are based on the division of spoken syllables into two kinds, 'heavy' and 'light'. Heavy syllables are those that contain a long vowel or diphthong, or are followed by more than one consonant; in recitation they are held for a longer duration than light syllables. The combination of prolonged heavy syllables with briefer light ones in a specified sequence is what gives each verse meter its characteristic form, just as different meters in English prosody combine specified numbers of short and long syllables in regular patterns. However, Sanskrit meters are usually more complex than English ones: their standard repeating unit is not the metrical foot of two to four syllables, but the *pāda* 'quarter-verse' which typically ranges from around six syllables in length to twelve or more.

Early Vedic meters were more flexibly structured than most of the metrical patterns in later Sanskrit verse. For instance, the ancient meter called *Gāyatrī* has an eight-syllable *pāda* with an irregular pattern in its first four syllables, and a *Gāyatrī* verse contains three *pādas* rather than the standard four of later prosody. The following *Gāyatrī* verse, taken from a hymn to the god Indra in the Ṛgveda, is shown in roman transliteration with the syllables marked as light (˘) or heavy (′). Reading it aloud a couple of times will illustrate how the rhythmic pattern of the meter helps fix the words in the mind. The accompanying metrical English translation (Griffith 1896) mimics the iambic-like form of the *Gāyatrī* meter, and has a similar effect.

′   ′   ˘ ′ ˘   ′   ˘   ′   ′   ′   ˘ ′ ˘   ′ ˘ ′

asmāṃ avantu te śataṃ asmān sahasram ūtayaḥ |

′   ′   ′ ˘ ˘ ′ ′ ˘   ′

asmān viśva abhiṣṭayaḥ ||

May thine assistance keep us safe, thy hundred and thy thousand aids: May all thy favours strengthen us. (Ṛgveda 4.31.10)

Not shown here are the archaic pitch accents preserved in this and other early Vedic hymns, where certain syllables are distinguished by a high, low, or falling pitch. The tonal patterns of such accents together with the measured rhythm of the verse structure helped preserve the sacred hymns orally in their canonical form through literally hundreds of generations.

The exegetical and didactic Sanskrit literature that grew up around the Vedas from about the early first millennium BC onwards also reflected this emphasis on the spoken word. In particular, the six scholarly disciplines known as the 'limbs of the Veda' were designed to support the proper performance of Vedic ritual and the proper use of its divine language. The six limbs were pronunciation, grammar, etymology, prosody, ritual practice, and astronomy/calendrics. The purpose of the first four limbs was understanding the meaning, utterance, and structure of the sacred verses, while the other two limbs indicated how and when the sacred rites should be performed. This auxiliary learning was summarized in texts likewise designed to be memorized, and consisting either of verses or of brief prose sentences called *sutras* 'string' or 'rule'.

The extant textual sources for the limbs of the Veda include many of the subjects that later split off into a separate genre of mathematical texts in Sanskrit. Most of them appeared in the Vedic limb of astronomy and calendrics, which of course involved detailed calculations to keep track of the ritual calendar. Others formed part of the discipline of ritual practice, particularly the geometric procedures for constructing sacrificial altars. Still others originated in prosody, which classified metrical patterns in quantitative ways that were appropriated by later Indian mathematicians as belonging to the topic called *aṅkapāśa* 'net of numbers', that is, combinations and permutations. Hence from the very beginning of Sanskrit technical and didactic literature, mathematical subjects were woven into the great tradition of oral learning rooted in the sacred texts.

The mathematical verse treatise in Classical Sanskrit

By the last third of the first millennium BC at the latest, Sanskrit had become primarily a learned and liturgical language rather than a cradle speech. Its use was codified by Indian grammarians of that period in the form known to modern linguists as 'Classical Sanskrit', which remained the chief common language of scholarship in the Indian subcontinent until well into the second millennium AD. During the first several centuries of the ascendancy of Classical Sanskrit, the didactic treatise in short prose *sutras* was largely replaced by its metrical verse counterpart as the typical medium for instructional and scholarly works in almost all disciplines.

By the early first millennium AD writing was widely used in India, but the ancient sacred texts were still ritually passed down in oral form, and memorizing rather than reading was still the ideal for studying a learned composition. A scholarly culture developed which relied heavily on written manuscripts to disseminate and preserve its learning, but which at the same time highly valued the traditional aims and techniques of oral transmission. This culture gave rise to the standard genre of instruction in Classical Sanskrit disciplines: namely, the verse treatise or 'base-text' with accompanying prose commentary, a uniquely hybrid monument to orality and literacy combined.

The disciplines nourished on these hybrid prose-verse texts included mathematical astronomy and mathematics in general. Even the most technical subjects and complicated calculations were evidently not considered unsuitable for the traditional format of mnemonic verse, if it was suitably buttressed with commentarial exposition. The verse treatise was ideally concise and densely written, containing as much information as possible in a small collection of memorized rules.

A prose commentary, on the other hand, was generally ample in extent and broad in scope. The comments on an individual verse or group of verses might consist of any or all of the following: a gloss and grammatical analysis of each of the verse's words; an amplified paraphrase of the verse's content; definitions of technical terms; one or more worked examples illustrating the operation of the rule it described; and an explanation, often in the form of an imagined dialogue between teacher and pupil, of the usefulness or mathematical validity of the rule. Less frequently, a commentator might insert a little biographical information about the author of the verse treatise (particularly if the author was the commentator's own teacher or a pedagogical ancestor of his teacher), an excursus on the philosophical meaning of the concepts in the rule, or a comparison with the statements of other mathematical authors treating the same topic.

The following brief selection from an early seventh-century treatise on mathematical astronomy illustrates the terse style that Sanskrit didactic authors generally favored. The mathematical technique discussed is what we now call the Euclidean algorithm for finding the greatest common divisor of two integers. The algorithm involves dividing the larger integer by the smaller, then dividing the smaller by the remainder, then dividing that remainder by the remainder from the second division, and so on, until the process produces a division with remainder zero. The last nonzero remainder in the sequence is the greatest common divisor. When both integers are divided by this divisor, they are reduced to smaller integers which are relatively prime.

This technique was used in dealing with what Indian mathematicians termed the *kuṭṭaka* 'pulverizer', that is, problems in linear indeterminate equations. In astronomy, such problems were linked to the task of finding some time at which a planet would be located at a particular point in its orbit, if its mean rate of orbital motion was known. That is, if a planet is considered to complete an integer number R of revolutions about the earth in an integer number D of days, then we need to find some other integer number d of days in which the planet moving with the same mean motion will complete some integer number r of revolutions, plus a given fraction C of a single revolution. In modern mathematical terms, this involves solving the indeterminate equation

$$\frac{R}{D} = \frac{r + C}{d}$$

for integer r and d, where R, D, and C are given. The first step is to reduce the given numbers R and D (which are usually very large so that the mean motion parameters can be expressed with reasonable precision) so that they are relatively prime. The seventh-century mathematician Bhāskara describes how to do this, in thirty-two carefully crafted syllables in the somewhat free meter called *anuṣṭubh* ('following in praise', because of its close relation to the twenty-four-syllable hymn meter *Gāyatrī* mentioned above):

˘ ˘ ˘ ˘ ˘ ˘ ˘ ˘ ˘ ˘

kṣmādineṣṭagaṇānyonyabhaktaśeṣeṇa bhājitau |
˘ ˘ ˘ ˘ ˘ ˘ ˘ ˘

hārabhājyau dṛḍhau syātāṃ kuṭṭakāraṃ tayor viduḥ ||

When divided by what remains post division of each by each, Days and cycles become reduced, both divisor and dividend. Once reduced, then they pulverize.

Or, less metrically and more literally:

The divisor [D] and dividend [R], divided by the mutual division-remainder of the numbers of days and [revolutions] of the desired [planet], become fixed [or reduced: *dṛḍha*, literally 'fixed', 'solid']. They [mathematicians] should know the pulverizer in the case of those two [the reduced divisor and dividend]. (Mahābhāskarīya 1.41, Kuppanna Sastri 1957, 53)

This highly compressed formula (requiring a whole extra *pāda* in the first English translation to make it even somewhat comprehensible) would probably not have seemed much more informative to the medieval student than to the modern reader. He would doubtless have turned either to a teacher or to a commentary like the one by the ninth-century astronomer Govindasvāmin, part of whose prose exposition on this verse is translated here:

One should divide the days and the revolutions of the planet by the remainder from mutual division of the days and revolutions of the desired planet. Then the two quotients are fixed [*dṛḍha*]. When the divisor and dividend are reduced [*apavartita*, literally 'taken away' but with the technical meaning 'divided by a common measure without remainder, reduced'], the pulverizer is easy to do: therefore the reduction [procedure, *apavartana*] is stated. And it is to be performed when the two [integers] are not [already] reduced [that is, relatively prime]. On account of that, it is said: 'There is no difference perceived [between] the dividend and divisor reduced [*apavartita*] with the quantity one, and [those two] not reduced.'[1] In such [cases], however, when the two are divided by a number [such as] two, etc., their division without remainder is not produced. But [when they are divided] by the number one, no change in [their] form occurs. Hence

---

1. Correcting the published text from *apavartitayonapavartitayośca* to *apavartitayoranapavartitayośca*.

they are [already] fixed: thus it is explained. Those who understand this should know [to apply] the pulverizer when those two have been fixed.

What is the meaning of the word 'pulverizer' [*kuṭṭākāra*]? It is said: 'pulverizing' [*kuṭṭana, kuṭṭā*] is as much as [to say] 'dividing'. Doing thus is an act: the act of pulverizing is the pulverizer. That by which division with remainder is done is the pulverizer. (Kuppanna Sastri 1957, 54)

Sometimes it was felt that a second layer of commentary was required to elucidate the first, as well as to provide additional information about the original base-text. Such a 'supercommentary' was composed on this joint work of Bhāskara and Govindasvāmin by a fifteenth-century mathematician named Parameśvara. His exposition, in addition to glossing technical terms, closes the verse–prose loop by quoting an unattributed set of verses explaining the rationale of the rule in Bhāskara's original verse. (No attempt is made to translate these metrically in the following rendering of the passage.)

'Desired number': number of revolutions of the desired planet. 'They should know the pulverizer in the case of those two': they should know the operation of the pulverizer when those two become thus fixed. When divided by the remainder from mutual division of the divisor and dividend, those two are without remainder. By the word 'reduction' [*apavartana*] is meant here division. If [one asks]: in the case of division of two quantities without remainder by means of the remainder from [their] mutual division, how is [that] demonstrated? It is said [in verse]:

In the case of mutual division of two quantities, when the large is divided by the small, whatever is the remainder, and the small[er] quantity, are to be divided by a number which itself becomes a quantity to be divided. And the [original] large quantity is omitted, because whatever is the great[er] is divided by the small[er] quantity.

And in the same way again, the 'dividend-ness' of the large and small quantities is to be conceived. The remainder is just another dividend to be divided by the last remainder. Because of that, when the divisor and dividend are divided by the remainder from mutual division, they become without remainder: such is the rationale.

It is not clear what percentage of verse base-texts were actually learned by heart by the mathematicians and students who worked with them. A scholar might encounter in his working lifetime at least dozens of treatises in his field containing scores if not hundreds of verses apiece: could he really memorize every verse of each of them? As evidence of such learning, there are many manuscript copies of commentaries in which most of the verses of the original base-text are left out, with only the opening words of each verse included as a place-marker in the commentary. Either the scribes expected their readers to turn to a separate copy of the base-text to find out what the commentator was talking about (which seems absurd), or else they took it for granted that someone who was studying a commentary would have memorized the treatise it expounded.

The prose commentaries, of course, were not meant to be memorized. But as the above examples show, they were so closely entwined with the Sanskrit ideal of learning embodied in the recited word, with so many layers of quotation and exegesis, that they maintained rather than supplanted the genre of mathematical verse. Sanskrit scholarly mathematics did not incorporate significant numbers of purely prose works until late in the second millennium under the influence of Islamic and European texts, when the institutional basis of Sanskrit learning in general was already losing ground to education and scholarship in vernaculars.

Mathematical verse vocabulary and style

Mathematical verse as a didactic and scholarly genre was not without its drawbacks. As a brief glance at the examples of translation in the previous section will show, one of the most intractable problems is the metrical inflexibility of standard technical terms. An English word like 'divisor' or 'dividend' has a fixed pattern of syllabic stresses, and this affects where it can appear in a verse with a given metrical form. Likewise, in Sanskrit every word has its pattern of light and heavy syllables, which may not fit into the particular place in a verse where the author wants to put it.

To reconcile the conflicting demands of meaning and meter a poet has to pad out the verses with mathematically unnecessary text around the crucial technical terms. Consider Coleridge's attempt at versifying Euclid in 'A mathematical problem', where four stanzas of Pindaric ode (much of it wasted, from a didactic viewpoint, on non-technical poetic imagery) are required just to state and prove Proposition I.1 of the *Elements*.[2] Coleridge, of course, was writing a parody rather than a serious didactic treatise, so he could afford to be long-winded. An author desiring in earnest to explain a mathematical topic in verse as concisely as possible needs a more efficient way of expressing mathematical statements metrically.

To get around this problem, Indian authors ingeniously exploited the structure and literary style of Classical Sanskrit. The vocabulary of Sanskrit is both rich in synonyms and heavily polysemic, meaning that individual words generally have multiple meanings. For instance, there are two common Sanskrit words (and various other obscure ones) meaning 'bow' as in an archer's bow: *dhanus* and *cāpa*. A mathematical author can use either word as a technical term to mean 'arc of a circle', depending on whether he needs its first syllable to be short or long.

2. The opening and first stanza are: This is now—this was erst,/ Proposition the first—and Problem the first./ On a given finite Line/ Which must no way incline;/ To describe an equi-/ –lateral Tri-/ –A, N, G, L, E. / Now let A. B./ Be the given line/ Which must no way incline;/ The great/ Mathematician/ Makes this Requisition,/ That we describe an Equi-/ –lateral Tri-/ –angle on it:/ Aid us, Reason—aid us, Wit!

Similarly, any of the Sanskrit words meaning 'bowstring' can refer to the chord of an arc, or else to its sine: the sense can be left ambiguous or can be clarified with a descriptor such as 'complete bowstring' for the chord or 'half-bowstring' for the sine. The versed sine (that is, the difference in length between the cosine of an arc and the radius of the circle) can be indicated by any word meaning 'arrow', because it extends between the bowstring and the bow. Furthermore, the same quantity can also be called the 'backwards' or 'reversed' sine, as it is in English, because the differences between its successive values in a trigonometric table are the same as those between successive values of the sine, but in reverse order. This multiplicity of technical terms for each mathematical concept greatly eases the task of fitting them into a fixed metrical pattern.

Another handy feature of Sanskrit for the mathematical poet is the almost infinite possibility of creating compound words from combinations of other words. English has a limited capacity to form descriptive or possessive compounds from at most two or three words (as in 'goldfish' for 'a fish that is gold' or 'peabrain' for 'one having a brain the size of a pea'). Sanskrit places no restrictions on the number of terms that can be compounded into a single word. It also recognizes copulative compounds, which in English require a somewhat clumsy hyphenation of terms and conjunctions, as in 'black-and-white (photography)', 'yes-or-no (answer)'. Moreover, Sanskrit stacks and nests multiple compounds of any type to form still more elaborate compounds. Thus in the verse by Bhāskara on Euclidean division quoted above, the first word *kṣmādineṣṭagaṇānyonyabhaktaśeṣeṇa* is a compound constructed from six or seven other compounds. For instance, *anyonya* is a copulative compound literally meaning 'the other and the other' or 'each other'. Juxtaposed with the past participle *bhakta*, 'divided', it becomes the descriptive compound *anyonyabhakta*, 'divided by each other'. In this way the complete compound uses the word-string 'day-desired-number-each-other-divided-remainder' to convey the meaning 'the remainder from dividing by each other the days and number [of revolutions] of the desired [planet]'. What the reader may lose in transparency of meaning, the author gains in conciseness of expression.

The author can also take advantage of the grammatical and philosophical sophistication of Classical Sanskrit to state complex ideas in few words. There are standard constructions for transforming any verb x into various words to express related concepts, such as the abstract noun x-ing, the act of doing x, the quality or state of having been x'd or needing to be x'd, and so forth. Thus, for example, the unique English technical term 'dividend' can be expressed in many ways in Sanskrit just by adding a suffix indicating the future passive participle to any verb meaning 'break', 'cut', or 'divide': hence '[the thing that] will be broken' means 'dividend'. Likewise, when the author of the verses quoted by Parameśvara above wants to say that the larger and smaller numbers in the Euclidean algorithm should both be considered as quantities to be divided, he can simply tack on another suffix, to signify the abstract quality of 'dividend-ness' or condition of being a dividend: 'their dividend-ness is to be conceived'. Delicate shades of meaning within nearly synonymous forms may be emphasized, as in the above use of *dṛḍha* 'fixed, reduced' to suggest 'reduced to its final relatively prime form', as opposed to *apavartita* 'reduced' implying 'in the process of being reduced by division by a common measure'. Or they may be elided, as in the collection of similar nouns *kuṭṭākāra*, *kuṭṭana*, *kuṭṭā* to reinforce the sense of the single mathematical technical term 'pulverizer'.

Modern mathematicians may be inclined to deplore all this verbal flexibility as an unjustifiable sacrifice of clarity and precision. Using multiple words for the same mathematical entity seems superfluous, while using the same word to mean different entities, or neglecting to explain exactly which entity is meant, is a serious violation of standard practice. But we have to bear in mind that semantic ambiguity was seen in Classical Sanskrit literary verse as a virtue rather than a defect. Words in a poem might be deliberately chosen to harmonize with one another's multiple meanings, so that a verse could be read in two different senses simultaneously. Mathematical authors were well aware of this literary use of paranomasia, or double meaning, and occasionally employed it themselves as an ornament. For example, the last verse of the famous twelfth-century arithmetic book by Bhāskara Ācārya (no relation to our seventh-century Bhāskara) is an extended pun on the book's name, *Līlāvatī*, a feminine adjective meaning 'beautiful' or 'charming'. So the same verse can be read as praising either the book called *Līlāvatī* with 'ornamented sections', which brings success to those who 'keep it in their throats' (that is, ready for recitation), or a beautiful woman with 'adorned limbs', delighting those who 'clasp her to their necks' (*Līlāvatī* 272, Āpate 1937, 285).

Indian mathematicians thought of semantic ambiguity as appealing and desirable not just for literary purposes but also for pedagogical reasons. A mathematical rule in verse is evidently intended to be not so much a detailed explanation of a mathematical fact as a helpful reminder of it. The more facts that are simultaneously referenced by the same memorized verse, the more useful that verse will be. Commentators sometimes employed considerable ingenuity in finding additional interpretations of verses to link them to additional mathematical formulas.

Furthermore, ambiguous or imprecise rules require more mathematical awareness in their application than do laboriously detailed algorithms that pupils can use as procedural black boxes to produce correct answers without the need for thinking. When Bhāskara describes Euclidean division with the curt statement that 'the divisor and dividend divided by the remainder from dividing by each other become reduced', he does not specifically explain what 'dividing by each other' means. A student trying to apply this rule to find the relatively prime reduction of two numbers has to put in some effort to understand, or at least to recall from previous drills, the steps of the full procedure. We cannot infer conclusively that authors deliberately courted vague or ambiguous forms of expression in order to trip up lazy readers, but we can tell that they certainly declined to cosset them with foolproof formulas.

Verbal number systems for verse texts

Number words such as 'ten' and 'fifty-two' are special and especially stubborn examples of mathematical terminology that are difficult to shoehorn into a given metrical pattern. Very few number words naturally develop useful synonyms like 'dozen' for 'twelve' or 'score' for 'twenty'. Pre-Classical Sanskrit appears to have been no exception in this regard. In ancient verses like the one quoted above from the *Ṛgveda*, the words for 'hundred', 'thousand', and so on are almost all unique and follow a standard linguistic pattern resembling that of their cognates in other Indo-European languages. Alternatives like 'thrice seventy' may sometimes replace the equivalent standard constructions like 'two hundred and ten', but the range of ways to express a particular number remains very limited.

While it may be possible to insert a few number words here and there in a verse hymn dealing mostly with non-mathematical subjects, the task of constructing usable verses becomes impossible when the content is heavily numerical. Indian mathematicians got around this problem as they had finessed the more general problem of mathematical vocabulary, by deliberately creating synonyms for common number words, as well as by finding ingenious ways to represent numerals by letters.

The method employing verbal synonyms is the simplest to use, and seems to have been devised somewhat earlier than the alpha-numeric method. It works by permitting a number word to be replaced by any word signifying any object that is physically or conventionally associated with that number. Thus any word meaning 'hand' or 'eye' can be used to mean 'two', while a word for 'nail' (as in finger- or toenail) indicates 'twenty'. A far from complete sample of these 'concrete numbers' is shown in the following list (see Sarma 2003 for a fuller enumeration):

- zero: void, sky, dot

- one: earth, moon, the deity Viṣṇu

- two: eye, hand, wing, twin

- three: fire (from the three sacred fires of Vedic ritual), quality (from the three qualities of Indian ontology: darkness, passion, and truth), the hero Rāma (in his three incarnations)

- four: ocean (four principal ones are identified), dice-spot (on the four-sided Indian die), age (one of the four ages in an aeon)

- five: sense-perception, arrow (since the deity Kāma, the Indian Eros, uses the five senses as the five arrows with which he pierces his victims)

- six: flavor (sweet, sour, salt, pungent, bitter, or astringent), limb (as the six limbs of the Veda)

- seven: mountain (seven principal ones are identified)

- eight: elephant (one assigned to each of the eight principal directions; see 'ten' below)

- nine: hook (since the Sanskrit numeral nine looks like a hook)

- ten: direction (the four cardinal directions, the four intercardinal directions, and up and down)

- eleven: the deity Śiva (in his elevenfold incarnation as the beings called Rudras)

- twelve: sun (in each of the twelve solar months), sundial gnomon (conventionally twelve digits long)

- fifteen: lunar day (of which there are fifteen in a half-month)

- twenty: nail

- twenty-four: Jain saint (the 24 tirthankars of Jain tradition)

- thirty-two: tooth

The list thins out drastically on reaching the two-digit numbers, many of which have no recognized 'concrete number' synonyms. Larger numbers were expressed by stringing together small number synonyms, starting with the least significant digit, to form copulative compounds interpreted as decimal place-value numerals. Thus the compound 'eye-fire' would mean 32, while the number 3212 could be represented by the compound 'sun-tooth', far easier to fit into a verse meter than its unwieldy equivalent 'three thousand two hundred and twelve'.

This system of number-word synonyms would seem to offer great scope for literary double meanings, but in fact Indian mathematicians appear to have been generally uninterested in its wordplay possibilities (or else the subtlety of their punning has so far eluded the attention of researchers). A student studying a mathematical verse and encountering a statement such as 'When the planetary disk is two-twin-multiplied and arrow-mountain-Rāma-divided...' (MB 1.39, Kuppanna Sastri 1957, 52) is evidently expected just to multiply by 22 and divide by 375 without pondering the existence of cryptic narratives in the number words. (An apparently unique exception is furnished by the sixteenth-century astronomer-mathematician Jñānarāja, who constructed a score or so of verses that can be read either as computational examples, with their numerical parameters expressed in concrete numbers, or as narrative aphorisms, with the number synonyms interpreted in their literal meanings (Knudsen 2008, section 1.3.6, 41–42).)

The evidence of the concrete-number system has significant implications for the history of decimal place-value numerals. Compounding number synonyms to spell out successive digits of multi-digit numbers presupposes a place-value system: otherwise it would not be clear whether a compound like 'two-twin' is intended to mean '22' or '2 plus 2' or '2 times 2' or some other construct. The earliest known use of concrete-number compounds occurs in an astrological text of the mid-third century AD (Pingree 1978, I 506), several centuries before the date of the oldest surviving inscription using decimal place-value numerals. So the invention of a decimal place-value system for written numerals (including, presumably, a zero symbol as a place-holder between non-zero digits) must date back at least that far.

Decimal place-value notation is also implied by the other main Sanskrit system for representing numbers verbally in verse, namely the alphabetic encoding scheme called *kaṭapayādi*. In this system, each of the thirty-three consonants of the Sanskrit alphabet is assigned to one of the ten decimal digits 1,…, 9, 0, as shown in Table 6.2.1. (Note that aspirated consonants such as *dh* and *bh* are single sounds in Sanskrit, although represented with two consonants in roman transliteration). The name *kaṭapayādi*, literally 'beginning with *k*, *ṭ*, *p*, and *y*', is derived from the fact that those four consonants are assigned to the digit 1. Multi-digit numbers can then be represented by any appropriate succession of syllables, including actual Sanskrit words: for instance, 'Rāma' in *kaṭapayādi* notation would signify 52, since *r* stands for 2 and *m* for 5.

Table 6.2.1 The *kaṭapayādi* encoding

k	kh	g	gh	ṅ	c	ch	j	jh	ñ
1	2	3	4	5	6	7	8	9	0
ṭ	ṭh	ḍ	ḍh	ṇ	t	th	d	dh	n
1	2	3	4	5	6	7	8	9	0
p	ph	b	bh	m					
1	2	3	4	5					
y	r	l	v	ś	ṣ	s	h		
1	2	3	4	5	6	7	8		

The *kaṭapayādi* system apparently dates from around or shortly before the middle of the first millennium AD. It was employed primarily by medieval south Indian authors, some of whom made breathtakingly ingenious use of its capacity to represent numbers and words simultaneously. This is illustrated in the so-called 'sentence' tables of south Indian astronomy, in which long lists of numerical data were encoded in *kaṭapayādi* notation as sequences of short sentences whose syllables correspond to successive digits of the table entries. For example, the last entry in the sine table of Mādhava (fourteenth century), representing the sine of 90° in a circle of circumference 21600, is '*devo viśvasthālī bhṛguḥ*', meaning roughly 'The deity Bhṛgu has the all-containing vessel'. The significant consonants of this phrase (that is, all the ones immediately followed by a vowel) are *d*, *v*, *v*, *v*, *th*, *l*, *bh*, *g*, corresponding in the *kaṭapayādi* table to 8, 4, 4, 4, 7, 3, 4, 3. This digit sequence is actually in mixed decimal and sexagesimal notation: it represents the number 3437; 44, 48, or $3437 + \frac{44}{60} + \frac{48}{3600} = 3437.74666\ldots$, the radius of a circle with circumference 21600 = 360° 00′ (taking $\pi = 3.14159$). Storing a quantity of this precision in eight easily-recollected Sanskrit syllables is a triumph of verbal mathematics, and it is hard to imagine any way in which it could be further compressed and still remain mnemonically useful. Even in English with its briefer number words, nine syllables would be required to express this number merely as a sequence of digits with no mnemonic value at all.

The prize for conciseness in verbal number systems, however, has to go to the unique alphabetic encoding apparently devised by the astronomer Āryabhaṭa around 500 AD. It is not known whether Āryabhaṭa was inspired by (or even aware of) the *kaṭapayādi* notational system,[3] but his own system in some ways resembles it. He arranged the thirty-three Sanskrit consonants in a one-to-one mapping with the thirty-three numbers 1, 2, 3,..., 25, 30, 40,..., 100. Then he assigned vowels to the decimal places as far as 10^{17} (*Āryabhaṭīya* 1.2, Shukla 1976, 7). Hence any consonant-vowel combination, or syllable, had a unique numerical meaning, and individual syllables could be strung together in any order to represent larger numbers. For example, in Āryabhaṭa's system the consonant *b* stands for 23, *ph* for 22, *n* for 20, and *c* for 6, while the vowel *u* represents (in this context) the ten-thousands place, *i* the hundreds place, and *a* the ones place. So the parameter value '*buphinaca*' is $23 \times 10^4 + 22 + 10^2 + 20 \times 10^0 + 6 \times 10^0 = 232226$. And unlike a *kaṭapayādi* number, it would mean exactly the same thing if it were written '*phibucana*' or '*canaphibu*' or any other permutation of the same four syllables.

The drawbacks to Āryabhaṭa's scheme are that it is mnemonically useless and phonetically a mess. While a concrete-number digit sequence like

---

3. Sarma 2003, 27, concludes that he was, but the evidence rests on the assertion of a commentator writing many centuries later who may have been projecting his own knowledge onto Āryabhaṭa.

'earth-hand-tooth' is at least somewhat more memorable than a mere list of numeral names, and a *kaṭapayādi* sentence much more so, Āryabhaṭa's nonsense syllables offer no intrinsic aid to memory. Worse, his assignment of every consonant and vowel to a specific numerical value means that he sometimes needs to combine sounds in ridiculously unpronounceable ways. For instance, Āryabhaṭa let the consonant *ch* stand for 7 and *l* for 50, so he could write the number 57 as '*chala*' or '*lacha*'. But if the verse meter allowed him only one syllable for this number, he had to squeeze both consonants onto one vowel as the syllable '*chla*' (*Āryabhaṭīya* 1.11, Shukla 1976, 39). This phonetic monstrosity, not found in nature, is no more pronounceable in Sanskrit than it is in English (although it is at least preferable to the alternative '*lcha*'). Dysphony of this sort is a definite handicap in a number encoding system that is supposed to conduce to verbal memorization and recitation. These disadvantages may explain why Āryabhaṭa's version of alphabetic numerals never became popular even among followers of his own school, all of whom preferred to stick with the concrete-number or *kaṭapayādi* systems instead.

Graphical features and the verse-text ideal

Of course, verbal formulations by themselves do not suffice to handle all the features of an advanced mathematical corpus. Diagrams and complicated numerical computations, to name but two examples, are impossible to manipulate in words alone. The content of Sanskrit treatises makes it clear that mathematicians were intimately familiar with such non-verbal features of their subject. Even their technical vocabulary, with terms like 'bow' for 'arc' and 'bowstring' for 'chord' or 'sine', frequently suggests visual interpretations of mathematical concepts. But visual and graphical representations, however important they might be to mathematical thinking, could not be directly incorporated into a verse text designed for oral recitation. It was therefore necessary to find some way of reconciling the canonical verse treatise with the requirements of written mathematics.

This reconciliation was primarily the task of commentators and scribes. For instance, after copying in a manuscript a number verbally encoded as concrete-number words or *kaṭapayādi* syllables, a scribe would typically write out the numeral corresponding to it. So the concrete-number compound 'earth-hand-tooth', for example, would appear in a manuscript as 'earth-hand-tooth 3221'. Sometimes this practice led to further confusion instead of clarification. Scribes were generally not very familiar with the technical details of the texts they copied, and thus were not always clear about identifying numerical values. Concrete numbers in astronomical texts were an occasional source of misunderstanding, because a scribe might mistake a literal reference to, say, the sun or moon for a concrete number, and throw in a '12' or '1' after it accordingly. A manuscript of the treatise of Jñānarāja mentioned in the previous section shows a similar error in a verse discussing a sundial gnomon: the scribe read the word 'gnomon' as a concrete number and followed it with the equivalent numeral '12', although in fact the number 12 is irrelevant to this verse (Knudsen, 2008, verse 2.8.8, 234).

To complicate the issue further, scribes might misunderstand the intentions not only of the verse-text author but of their own professional predecessors, the copyists of earlier manuscript versions. Sexagesimal parameters displayed in column format in the middle of lines of text were liable to be unwittingly split up into separate numbers, and assigned to unrelated parts of the verse, as illustrated in Fig. 6.2.1. Or a non-numerical character might simply be misread as a number: for instance, the visual similarity between the Sanskrit consonant *r* and the numeral 2, or between the numeral 1 and the vertical stroke used in Sanskrit script as a punctuation mark, not infrequently interfered with a scribe's effort to represent numbers correctly.

(a)

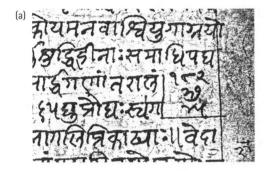

(b)

Figure 6.2.1 Misplaced numerals in a manuscript. The scribe of the top manuscript (a) (Benares 35566, f. 39v) has correctly written the sexagesimal number 182;37,45 as three separate decimal numbers (the integer part and two fractional parts) stacked vertically in a box near the right margin of the text. The scribe of the other manuscript (b) (Bhandarkar Oriental Research Institute 860 of 1887–91, f. 5v) has evidently mis-copied a similar presentation of the same number, splitting its component parts (circled in the reproduction) into two different lines of the text

Commentators were more directly involved than scribes with the technical content of a text, and bridged its verbal–numerical divide in more sophisticated ways. It was the commentary that actually showed how to translate a verse algorithm into a written computation, usually by means of a worked example. The sample problem was frequently stated in a verse, but its solution procedure would be described in prose, beginning with the conventional 'statement' or numeral expression of its parameters. The following verse from the *Līlāvatī*, with the author's own commentary, illustrates the style:

[Verse:] Friend, tell [me] quickly the square of three and one-half, and then the square root of the square, and the cube, and then the root of the cube, if you know fractional squares and cubes.

[Prose:] Statement: $3\frac{1}{2}$. After the integer is multiplied by the denominator, the result is $\frac{7}{2}$. The square of that is $\frac{49}{4}$. Then the root is $\frac{7}{2}$. The cube is $\frac{343}{8}$; its root is $\frac{7}{2}$. (*Līlāvatī* 44, Āpate 1937, 38)

Examples of this sort make it clear that there existed various notational conventions for written computations. Algebra problems, for instance, used a kind of proto-symbolism that represented operations and unknown quantities by the first syllable of their names. In problems of simple proportion, or the 'Rule of Three', where one must find a quantity x in the proportion $a : b = c : x$ in which a, b, and c are known, the known quantities were specified in a standard order $a \mid b \mid c$ (Sarma 2002, 137; see also Denniss Chapter 5.3 in this volume). There are even special Sanskrit names for certain operational layouts, such as 'door-hinge' multiplication (with the digits of multiplier and multiplicand aligned vertically) and 'cow's-urine' multiplication (apparently, zigzagging backwards and downwards through the digits to produce a 'stream' of sub-products to be added up). (Datta and Singh 1962, I 134–149.)

Most of these conventions have to be reconstructed from manuscript content ancillary to the verse base-text. Although verse-treatise authors might briefly allude to particular notational or computational methods, or provide a very vague and general introductory description of them, they did not devote much space to explaining their details. That was evidently considered to be part of written mathematics, subsidiary to the purely verbal content of the treatise itself.

Similarly, in verse treatises diagrams might be mentioned only in passing, leaving their detailed description to commentators. For instance, Āryabhaṭa summed up the geometric procedures for deriving sine values in a single verse as follows:

One should divide up a quarter of the circumference of a circle. And from triangles and quadrilaterals, [find] as many sines of equal arcs as desired. (*Āryabhaṭīya* 2.11, Shukla 1976, 77)

It is the prose commentary of a different author (our seventh-century friend Bhāskara, in fact) that actually undertakes to explain how to use triangles and quadrilaterals inscribed in a quadrant to derive the sines of various arcs (Shukla 1976, 77–83).

But authors could be more painstaking in providing verse descriptions of figures that their users would actually need to draw. For instance, an astronomer was expected to be able to sketch, on the basis of his computations for a

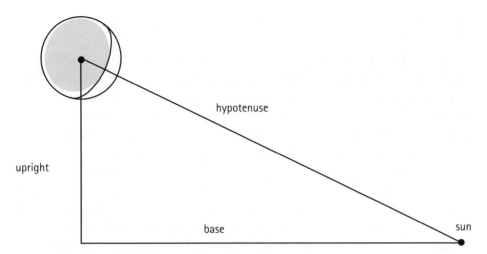

Figure 6.2.2 The diagram of the moon's crescent described by Lalla

given time, the predicted size and orientation of the crescent moon, as shown in Fig. 6.2.2. The 'base' is computed from the difference in azimuth between the moon and the sun, and the 'upright' from their difference in altitude. Those two quantities determine the inclination of the hypotenuse, or the observed degree of tilt in the moon's crescent. Part of the description of this figure in the treatise of the eighth-century astronomer Lalla is translated below:

Some point [marked] on flat ground is considered as the sun, and from it the base is drawn, and the upright, [each] in its own direction […] The hypotenuse [extends] from the tip [of the upright] to the sun-point. The [center of the] disk of the moon is at the juncture of the hypotenuse and the upright. […] (*Śiṣyadhīvṛddhidatantra* 9.15–9.16, Chatterjee 1981, I 142)

As usual, though, a prose commentary is required to fill in all the details, a few of which are given below for comparison with Lalla's more abbreviated description:

[…] Extending the upright perpendicularly, one should make a point at the tip of that. Then, extending the hypotenuse obliquely from the sun-point, one should set it at the center-point at the tip of the upright. One should put the point at the place of junction of the tip of that upright and the tip of the hypotenuse. […] (Chatterjee 1981, I 142)

Like demonstrations and numerical computations, geometric figures clearly had an important place in Classical Sanskrit mathematics, and a recognized technical terminology. But all of them remained officially secondary to the chief vehicle of mathematical learning, the collection of verbal rules in metrical verse.

Conclusion

The modern student of Sanskrit mathematics may be tempted to wonder why the overburdened commentators did not just give up on the verse-text ideal altogether and turn mathematics into a purely literate endeavor, with its true focus on explanations and calculations and drawings. This temptation, of course, reflects the dominance of literacy in modern mathematics, which shapes our expectations of mathematical methodology. It appears that purely literate mathematics of the sort that developed in the West, with its dependence on laborious descriptions of figures and equations, would have seemed to Indian mathematicians simply too, well, prosy. The verbal rule in Sanskrit verse did not lose its charm or the centrality of its truth just because its subject-matter happened to be mathematical. On the contrary, the authors of the verse texts were considered the authorities on the subject, not the commentators who explained their work in prose. As in other Sanskrit disciplines, the title 'teacher' or 'learned one' was bestowed not on commentators or copyists, those drudges in the service of knowledge, but rather on the master of knowledge himself, the speaker of the word.

Bibliography

Āpaṭe, V G, *Līlāvatī*, 2 vols, Puṇe, 1937.

Chatterjee, Bina, *Śiṣyadhīvṛddhidatantra of Lalla*, 2 vols, Indian National Science Academy, 1981.

Datta, B, and Singh, A N, *History of Hindu mathematics: a source book*, 2 vols, Asia Publishing House, 1962.

Griffith, Ralph T H, *The hymns of the Rigveda*, 2nd ed, Kotagiri, 1896.

Knudsen, Toke Lindegaard, 'The Siddhānta-sundara of Jñānarāja', PhD dissertation, Brown University, 2008.

Kuppanna Sastri, T S, *Mahābhāskarīya of Bhāskarācārya with the bhāṣya of Govindasvāmin and the super-commentary Siddhāntadīpikā of Parameśvara*, Madras Government Oriental Manuscripts Library, 1957.

Pingree, David, *The Yavanajātaka of Sphujidhvaja*, 2 vols, Harvard University Press, 1978.

Sarma, K V, 'Word and alphabetic numeral systems in India', in A K Bag, and S R Sarma (eds), *The concept of Śūnya*, Indian National Science Academy, 2003, 37–71.

Sarma, Sreeramula Rajeswara, 'Rule of Three and its variations in India', in Yvonne Dold-Samplonius, *et al* (eds), *From China to Paris: 2000 years transmission of mathematical ideas*, Steiner Verlag, 2002, 133–156.

Shukla, K S, *Āryabhaṭīya of Āryabhaṭa with the commentary of Bhāskara I and Someśvara*, Indian National Science Academy, 1976.

Antiquity, nobility, and utility: picturing the Early Modern mathematical sciences

Volker R Remmert

In Early Modern Europe the term 'mathematical sciences' was used to describe those fields of knowledge that depended on measure, number, and weight, reflecting the much quoted passage from the 'Wisdom of Solomon':[1] 'thou hast ordered all things in measure and number and weight'. The *scientiae mathematicae* 'mathematical sciences' were generally subdivided into those that were *purae* 'pure', dealing with quantity, continuous and discrete, as in geometry and arithmetic, and those that were *mixtae* or *mediae* 'mixed' or 'intermediate', which dealt also with quality, for example, astronomy, geography, optics, music, cosmography, and architecture. The Jesuit Gaspar Schott enumerated more than twenty fields among the *mathematicae mixtae* in his *Cursus mathematicus* 'Course of mathematical sciences' of 1661. The mathematical sciences, then, consisted of various fields of knowledge, often with a strong bent toward practical applications, and only became independent scientific disciplines from the late seventeenth to the early nineteenth century.

In the hierarchy of scientific disciplines of the Middle Ages and up to the late sixteenth century, the mathematical sciences were subordinate to theology, philosophy, and natural philosophy. Even though they then began to challenge the

---

1. *Apocrypha*, Wisdom of Solomon 11:20, King James version.

traditional primacy of philosophy and theology, supremacy did not pass to the mathematical sciences until the seventeenth century. In 1676 the Jesuit Claude-François Milliet de Chales could proudly announce in the letter of dedication in his *Cursus seu mundus mathematicus* 'Course or world of mathematical sciences' that: *plebeiae sunt ceterae disciplinae, mathesis Regia* 'the other disciplines are plebeian, the mathematical sciences are royal'. Modes of explanation informed by the mathematical sciences increasingly dominated the sciences and also infiltrated other parts of society.

The foundations of the mathematical sciences' new status were laid by the work of mathematicians from the mid-sixteenth century onward. In the seventeenth century, efforts to legitimize the mathematical sciences were actively driven forward through various strategies intended to bring them out of seclusion. These strategies usually involved the use of print: mathematical textbooks, books of mathematical entertainments, editions of the classics, inaugural speeches, and so on. In printed media, the processes of legitimization are found not only in the texts themselves, but equally in what Gerard Genette (1987) has called 'paratexts', meaning additional writings such as dedications and prefaces. To these we may also add images, such as illustrations, diagrams, and frontispieces, and other characteristics, such as typography, format, and binding, which convert texts into books. Genette describes paratexts as the *seuils* 'thresholds' of books, a term also adopted by Marc Fumaroli in his discussion of frontispieces (Fumaroli 1994, 325). Paratexts such as dedications and prefaces, as well as frontispieces and illustrations, point directly to the realm of panegyric, public praise, in the world of Early Modern mathematicians.

In the sixteenth and seventeenth centuries the iconographies of the old disciplines of the quadrivium—arithmetic, geometry, astronomy, and music—were fairly standardized, rooted in their representations in liberal arts cycles since the Middle Ages (Scriba 1985). The (female) personification of arithmetic carried a table of numbers, whereas Geometry held a compass and often also surveying instruments (see Fig. 6.3.1). Thus, these disciplines were easy to depict and easy to recognize, as were Astronomy with her heavenly sphere and Music with her lute. With the emergence of more and more subdisciplines during the seventeenth century, the iconographies of the newcomers became increasingly confusing. The mathematician and engineer Johannes Faulhaber, for example, in the frontispiece of the second part of his *Ingenieurs-Schul* (1633), a book on fortification, presented eighteen subdisciplines of the mathematical sciences (Fig. 6.3.1) and helpfully identified each of them to prevent confusion about new disciplines such as logarithmographia, stereometria, algebra, and naval architecture.[2]

---

2. On Faulhaber see Schneider (1993); on his frontispieces see Remmert (2005, 108–111).

Figure 6.3.1 The mathematician and engineer Johannes Faulhaber presented eighteen (female) personifications of the mathematical sciences. As was standard, Arithmetic carried a table of numbers, Geometry held a compass and surveying instruments, Astronomy could be seen with her heavenly sphere and Music held her lute, but Faulhaber also identified new disciplines such as Logarithmographia, Stereometria, and Algebra. (Faulhaber 1633, by permission of Herzog August Bibliothek Wolfenbüttel)

Engraved title pages and frontispieces were an ideal medium for the visual legitimization of the mathematical sciences. In the nexus between the ideas of seventeenth-century scholars (be they mathematicians, mathematical practitioners, experimental scientists, or natural philosophers) and the visual images they used to represent those ideas—the active encoding of ideas into iconographical signs—frontispieces become evidence of scholars' deliberate intentions to make and shape their own images of scientific inquiry. Decoding these visual statements to try to understand these processes offers insights into the self-perception and self-fashioning of their protagonists, the advancement of their cause, and the enhancement of status of their respective disciplines. In the mathematical sciences, frontispieces played specific and significant roles, covering a wide variety of functions and audiences (Remmert 2005; 2006). I shall not be able to consider all of these here. Instead I will use the topics of antiquity, nobility, and utility as pathways into the immensely rich world of imagery representing, praising, and propagating the Early Modern mathematical sciences.

Antiquity, nobility, and utility

During the fifteenth century, three aspects of the mathematical sciences were usually singled out as praiseworthy: their value as preliminary instruction for the study of philosophy, their practical advantage for the community, and their antiquity. During the sixteenth century the arguments in favour of the mathematical sciences became fairly standardized, and drew on a common basis of argumentation (the practical and educational role of the mathematical sciences) and examples (Archimedes' burning mirrors being among the most popular). The educational value of the mathematical sciences, to which their epistemological status was closely related, was usually seen to be in their importance for training the mind and in their recreational potential, but not often in their necessity or value to other disciplines, such as philosophy, medicine, law, or theology (Rose 1975). This situation gradually changed until, in the first half of the seventeenth century, mathematicians, emphasizing the absolute certainty of mathematical knowledge, which had been so hotly debated in the sixteenth-century, declared that the mathematical sciences deserved a new position in the hierarchy of scientific disciplines. In doing so they usually referred to essential characteristics of the mathematical sciences, namely their antiquity, nobility, and utility (Imhausen and Remmert 2006).

Antiquity—mostly Greek antiquity—could, naturally, be referred to in various forms. In the Early Modern mathematical sciences, Archimedes and Euclid featured prominently, both rhetorically and visually. Archimedes was possibly the more prominent as he carried the double symbolism of both pure and mixed

mathematics. We find them both in the engraved title page of Samuel Marolois'
Opera mathematica (1614), with special emphasis on Archimedes at war
(Fig. 6.3.2). While the latter's usefulness to the military was an important aspect,
the other achievements of Archimedes and Euclid, and their undiminished cultural importance, were rather more to the point in emphasizing the antiquity of
the mathematical sciences, as can be seen in a series of frontispieces (Fig. 6.3.3)
used by Oxford University Press for editions of Euclid in 1703, Apollonius in
1710, and Archimedes in 1792 (Fasanelli and Rickey, 2008). The only variations
between the different versions are the geometric drawings in the front left. Their
meaning was explained by a quotation from Vitruvius, which appears below
the picture: 'Aristippus the Socratic philosopher, when he was shipwrecked in
Rhodes, noticed geometrical diagrams drawn on the beach and said to his companions, "We can hope for the best for I see the signs of men"'. Here, the mathematical sciences stood for civilization and, even more, cultural superiority.

The technique of conferring antiquity and nobility through images and frontispieces was firmly rooted in the prevailing political culture, where the elements
they contained were intimately understood. The Danish nobleman and astronomer Tycho Brahe, for example, had a clear understanding of the importance of

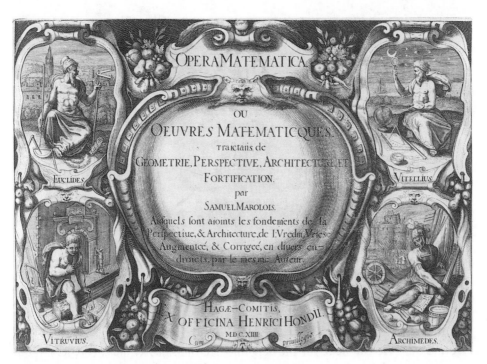

Figure 6.3.2 In the lower right we see Archimedes at war. (Marolois 1614, by
permission of Herzog August Bibliothek Wolfenbüttel)

Aristippus Philosophus Socraticus, naufragio cum ejectus ad Rhodiens. litus animadvertisset Geometrica schemata descripta, exclamavisse ad comites ita dicitur, Bene speremus, Hominum enim vestigia video. *Vitruv. Architect. lib. 6. Præf.*

Figure 6.3.3 'We can hope for the best for I see the signs of men'. (Gregory 1703, by permission of the Museum of the History of Science, Oxford)

artistic legitimization and the glorification of power, and applied this technique to the science of astronomy. His influence on the visual legitimization of astronomy was manifold (Remmert 2005, 125–188; 2007a). In 1574, in his inaugural speech, *De disciplinis mathematicis oratio* 'Oration on the mathematical disciplines', at Copenhagen University he named Timocharis, Hipparchus, Ptolemy, al-Battānī, King Alfonso of Castile, and Copernicus as the most important astronomers of the past, thus citing standard astronomical tradition. He also consistently used these astronomers in the iconographic programmes of his Uraniborg and Stjerneborg institutions on the island of Hven, where he built what amounted to a temple to Urania, the muse of astronomy, full of pictures and sculptures glorifying astronomy. It included material describing the mythological background of astronomy and the evolution of the astronomical tradition, the latter in the form of a canon of authorities beginning with Hipparchus and ending with Brahe himself. Brahe documented his world of images and emblems best of all in his slender volume *Astronomiae instauratae mechanica* 'The restoration of mechanical astronomy' of 1598, which contained numerous engravings of the instruments at Uraniborg and Stjerneborg, and a meticulous description of the institutes themselves.

Brahe's canon of astronomical authorities subsequently influenced a number of frontispieces to Dutch publications, the first of which was the engraved title page of Nicolaus Mulerius' *Tabulae frisicae* 'Frisian tables' (1611) (Fig. 6.3.4). The book was published by Willem Janszoon Blaeu, who had adhered to Brahe's iconographic conventions ever since he had been his assistant in 1595–96. The frontispiece unmistakably conforms to Brahe's canon in its portrayal of Hipparchus, Ptolemy, King Alfonso, Copernicus, and Brahe himself. The astronomical tradition it invokes is that of observational astronomy, as distinct from theoretical astronomy or cosmology: the issue here was the antiquity and nobility of astronomical observation and calculation, not the question of which world system was to be given credence.

Twenty years later, Adriaan Metius' introduction to astronomy, *Primum mobile* 'Primary motion' (1631), appeared with more or less the same engraved title page (Fig. 6.3.5), except that a portrait of Metius is inserted between Brahe and Copernicus. The light of the luminaries of astronomy shines on Metius, thereby enhancing his authority and generating credibility for the book itself. In the years to come, the tradition used by Mulerius and Metius became very popular in Dutch publications. The Copernican Philipp Lansbergen, in the engraved title page of his *Tabulae motuum coelestium* 'Tables of celestial motion' (1632), added only Aristarchus (as the first heliocentric astronomer) and himself to Brahe's original list. With obvious sympathies for the Copernican system, Andreas Cellarius followed the same line in the frontispiece of his impressive *Harmonia macrocosmica* 'Macrocosmic harmony' (1661) portraying Ptolemy, King Alfonso, Lansbergen, Copernicus, Brahe, and Al-Battānī. Finally Jan Luyts

Figure 6.3.4 A reference to Brahe's canon of astronomical authorities: Hipparchus, Ptolemy, King Alfonso, Copernicus, and Brahe himself. (Mulerius 1611, by permission of Herzog August Bibliothek Wolfenbüttel)

Figure 6.3.5 Metius adds himself to Brahe's list of astronomical authorities. (Metius 1631, by permission of Herzog August Bibliothek Wolfenbüttel)

in his *Astronomica institutio* 'Foundations of astronomy' (1692), a Tychonic introduction to astronomy, depicted Hipparchus and Ptolemy, Copernicus and Brahe, Galileo and Hevelius, thus alluding to the earlier iconographic programmes.

These frontispieces demonstrate an intense common awareness of the visual elements involved. The continuity of iconographic elements gave rise to strong local and national conventions, a tradition in itself in the republic of letters. Legitimization and credibility were underpinned by visual references to a tradition stretching back to antiquity, indicating the nobility and value of the astronomical matters treated in the text.

Another approach also became popular in the seventeenth century through Brahe, who introduced the symbolically loaded mythical figures of Atlas and Hercules (Remmert 2007a). Those inspecting his great armillary sphere in the Stjerneborg observatory, for example, found themselves face to face with Atlas, upon whose shoulders the sphere was supported. As can be seen from the frontispiece of Johannes Bayer's *Uranometria* (1603) (Fig. 6.3.6), Atlas and Hercules were the teacher and the disciple of an astronomy that had existed from time immemorial (*Atlanti vetustissimae astronomiae magistro; Herculi vetustissimae astronomiae discipulo* 'Atlas, teacher of the most ancient astronomy; Hercules, disciple of the most ancient astronomy'). Bayer, drawing on Brahe, thus asserts that astronomy is one of the oldest and noblest of all disciplines. But he also identifies Atlas with Ptolemy, who represents the old geocentric astronomy, and shows Hercules, who represents the new Tychonic astronomy, taking over Atlas's task of holding up the world.

To legitimize the new as good and deserving, astronomers of the late sixteenth and the seventeenth centuries could hardly find a better ambassador than Hercules, who signified many positive qualities that could be adopted by propagators of the new astronomy. At the same time, associating the old astronomy with Atlas honoured it as a venerable, indeed royal, part of the tradition. The new astronomy, symbolized by Hercules, is no usurper: presented in benign tandem with Atlas, he becomes the legitimate heir to a regal heritage, and speaks to rulers as an equal.

Antiquity and tradition were fundamental in any argument for the intellectual nobility of the mathematical sciences, and in making them acceptable to the political nobility whose patronage was needed. The Jesuit Gaspar Schott stressed this aspect in the frontispiece of his *Cursus mathematicus* (1661), mentioned above (Fig. 6.3.7). This impressive volume was dedicated to Leopold I, Austrian Emperor since 1658, who can be seen receiving his dedication copy of the *Cursus* from a crowned personification of the mathematical sciences. Schott emphasized that *Mathematica* was the Queen among the sciences, worthy to communicate with the Emperor himself (Remmert 2005, 78–85, 210–217). Thus the nobility of the mathematical sciences could be played out under the eyes of kings and

Figure 6.3.6 Johannes Bayer presents Atlas and Hercules as teacher and disciple of an astronomy that has existed since time immemorial, thus asserting, like Brahe, that it is one of the oldest and noblest of disciplines. Bayer identifies Atlas with Ptolemy, representing the old geocentric astronomy, and Hercules with Brahe, representing the new Tychonic system. (Bayer 1603, by permission of Herzog August Bibliothek Wolfenbüttel)

Figure 6.3.7 The quest for patronage: the nobility of the mathematical sciences could be played out under the eyes of kings and emperors on impressive stages, in vast and beautiful gardens. (Schott 1661, by permission of Herzog August Bibliothek Wolfenbüttel)

emperors on impressive stages, in vast and beautiful gardens—to which I shall return later.

Amongst the Jesuits, visual strategies of legitimization became more and more elaborate. In the frontispiece (Fig. 6.3.8) of Mario Bettini's *Apiaria universae philosophiae mathematicae* 'Beehives of all mathematical philosophy' (1645), for example, the beehives represent ten branches of the mathematical sciences (arithmetic, geometry, cosmography, optics, music, architecture, astronomy, mechanics, hydrology, and fortification), and the bees flying in and out testify to their usefulness (Bennett, Chapter 4.2 in this volume). The Dedication stresses the utility of the mathematical sciences: 'What do you expect from beehives if not honey? The mathematical sciences are the honey of human sciences' (Bettini 1642, I 5).[3] Surveying stands out on the front left and the inevitable burning mirror of Archimedes to the right. Bettini makes use of the nobility of the mathematical sciences too, turning his work into a *hortus mathematicus* 'garden of the mathematical sciences'.

The mathematical sciences were useful both in times of war (Archimedes' burning mirror) and of peace (Remmert 2005, 103–124). The Dutch mathematician Martinus Hortensius in his inaugural speech at the *Athenaeum illustre* 'Illustrious School' in Amsterdam in 1634 succinctly made the point that the mathematical sciences were equivalent to prosperity, because they contributed to navigation, navigation in turn to trade, and trade to the solid and firm prosperity of the country (Imhausen and Remmert 2006, 129). Similar arguments could easily be made for individual branches such as arithmetic. Utility was, of course, central to advertising the mathematical sciences in general and on the book market in particular. Jonas Moores' *New systeme of the mathematicks* (1681) covered, amongst other things, arithmetic, practical geometry, trigonometry, cosmography, and navigation. Its frontispiece (Fig. 6.3.9) deals with the link between these disciplines and trade, thereby stressing the utility of the mathematical sciences in the world at large and, of course, that of the book itself. The framed vignette shows a group of men gathered around a globe occupied with mathematical instruments, while various illustrations represent the mathematical sciences. The sea and maritime commerce can be seen through a window, while Neptune, lord of the seas, and Aeolus, keeper of the winds, watch over the assembly. That Aeolus is smoking a pipe is a tiny but important detail illustrating the importance of maritime trade: tobacco was imported to England from Virginia and re-exported to Amsterdam. Potential buyers were those looking for overviews like this that could provide them with useful basic information on various branches of the mathematical sciences.

---

3. Quid ab Apiariis nisi mella expectes? Humanarum scientiarum mella sunt Mathematicae.

Figure 6.3.8 The beehives represent ten branches of the mathematical sciences (arithmetic, geometry, cosmography, optics, music, architecture, astronomy, mechanics, hydrology, and fortification), and the bees flying in and out testify to their usefulness. Surveying stands out on the front left and the burning mirror of Archimedes to the right. (Bettini 1645, I, by permission of Herzog August Bibliothek Wolfenbüttel)

Figure 6.3.9 Neptune, lord of the seas, and Aeolus, keeper of the winds, watch over the mathematical assembly. That Aeolus is smoking a pipe is a tiny but important detail: tobacco was imported to England from Virginia and re-exported to Amsterdam. (Moore 1681, by permission of Niedersächsische Staats- und Universitätsbibliothek Göttingen)

Casual viewers or readers did not require any special knowledge to understand and decode these advertising frontispieces. Thus they were accessible to a larger audience than those concerned with scientific debates or patronage. Their message was a clear and simple 'Buy this book!' As a result they were often copied or imitated over long periods of time, much like the frontispiece to Mulerius's *Tabulae frisicae*, discussed earlier. Such re-appropriations of frontispieces and the recurrence of their iconographic elements have repeatedly led interpreters into pitfalls. This is not my concern here; the repetition stands as another proof, however, of the power and dissemination of visual strategies in early modern Europe.

Scientific debates: squaring the circle

In 1647 the Jesuit Grégoire de Saint Vincent published his massive *Opus geometricum quadraturae circuli et sectionum coni* 'Geometrical quadratures of the circle and sections of a cone', in which he presented four methods of squaring the circle. He was so proud of this spectacular result that he presented it in the frontispiece (Fig. 6.3.10). On the left we see Archimedes presenting his famous theorem that a circle can be transformed into a right-angled triangle of equal area. From above the sun's rays pass through a square and meet the ground in a circular shape. To this well-known phenomenon of 'Sonnentaler', described by Kepler in his *Optics* (1604), de Saint Vincent adds a quotation from Horace: *Mutat quadrata rotundis* 'He changes square things to round' (*Epistulae* I, I 100). One could hardly devise a clearer visual statement of the fact that the author, in the tradition of Archimedes, had indeed squared the circle. Moreover, in Jesuit imagery the light emanating from the heavens carried the obvious connotation of divine inspiration. And divine inspiration was indispensable if one wished to follow the Habsburg motto seen just above Archimedes' pointer: *plus ultra* 'further beyond', to transcend the limits of traditional knowledge.

The book is dedicated to the House of Habsburg and the frontispiece abounds with Habsburg symbolism. On the lower right two cherubs show two sides of an ancient coin depicting Constantine the Great, founder of the Holy Roman Empire to which the House of Habsburg was heir. One cherub points to Constantine's personal emblem, a sphere on a cubical altar with the motto *beata tranquilitas* 'blessed tranquillity'. This symbolism has led William B Ashworth to suggest that de Saint Vincent 'was attempting to justify the quadrature problem as the mathematical equivalent of the essential imperial problem: good government'. He put his book and discoveries in the context of Habsburg patronage, 'suggesting that the problem of changing a circle into a square is particularly [...] appropriate to the Habsburg House, since the combination of circle and square, orb and

Figure 6.3.10 *Plus ultra* 'further beyond', or transcending the limits of traditional knowledge: 'He changes square things to round' (Horace: *Epistulae* I, I 100). Archimedes, on the left, had not been able to square the circle. (de Saint Vincent 1647, by permission of Herzog August Bibliothek Wolfenbüttel)

altar, has emblematically represented the Empire since the days of Constantine' (Ashworth 1991, 137–167; 149). In the frontispiece (beneath the lion's head) he even renamed the problem as *problema austriacum* 'the Austrian problem'.

De Saint Vincent also borrows the Habsburg motto *plus ultra* and the pillars of Hercules (with the Golden Fleece between), set at the Atlantic end of the Mediterranean to mark the limits of the ancient world, the *ne plus ultra* 'no further beyond'. So it is appropriate that the squaring of the circle is situated beyond the pillars, in the unknown, which not even the great Archimedes, who stands behind the pillars, had been able to reach. This allegory reflects the widespread optimism of seventeenth-century mathematicians, who were convinced that they lived in an age of scientific progress which transcended the boundaries of knowledge of the Ancients.

Among several voices critical of de Saint Vincent's *Opus geometricum*, the booklet *Examen circuli quadraturae* 'Examination of the quadrature of the circle' of his fellow Jesuit Vincent Leotaud (Lyons 1654), is of particular interest because of its engraved title page (Fig. 6.3.11). Leotaud neatly summarizes his critique in the image: the circle cannot be squared, not even by the sheer brutality of a cannon or a sledgehammer (Hofmann 1938). This example nicely illustrates the important function of frontispieces in seventeenth-century scientific debates. While Leotaud's image shows this in its purest form, St. Vincent's frontispiece embeds the scientific message in a rather complex design that also proves his reverence for the Habsburgs.

Patronage among the Jesuits: the example of Francesco Eschinardi

The Jesuits were among the most prolific users of art in the Early Modern period, and in their pictorial worlds various styles and intentions were often combined in arbitrary ways. An intensive culture of frontispieces and engraved title pages began to flourish amongst them in the 1620s. Among Jesuit mathematicians, this interest was first seen in Christoph Scheiner, and later on in Athanasius Kircher, as well as Mario Bettini, Gaspar Schott, and de Saint Vincent, whose frontispieces were carefully designed patronage artefacts, (Remmert 2005, 189–224).

The Italian Jesuit Francesco Eschinardi published widely in the mathematical sciences and contributed regularly to the *Giornale de' letterati* from 1668 to 1675. He had been trained in Jesuit schools since 1637, later read natural philosophy and metaphysics and finally taught the mathematical sciences at the Collegio Romano (Feldhay 1989). In his *Centuria problematum opticorum* 'One hundred optical problems' (1666) and its second part *Centuriae opticae pars altera* 'One hundred optical problems, part two' (1668) he treated topics from optics in

CVRVILINEORVM
AMOENIOR
CONTEMPLATIO.
NECNON
EXAMEN
CIRCVLI QVADRATVRÆ,
A R.P. GREG. A S. VINCENTIO
SOC. IESV PROPOSITÆ.

Figure 6.3.11 The circle cannot be squared, not even by the sheer brutality of a cannon or sledgehammer. (Leotaud 1654, by permission of Herzog August Bibliothek Wolfenbüttel)

textbook fashion, concentrating on microscopes and telescopes, and the theory of mirrors and their applications. The second part was dedicated to Leopold Medici, who from 1657 to 1667 had presided over the Florentine *Accademia del Cimento* 'Academy of Experiment' founded by Ferdinand II and was committed to the experimental programme of Galileo and his school. In his letter of dedication, Eschinardi praised the support Leopold and the Medici had given to the sciences. In particular he emphasized the importance of *'vestro Galileo'* 'your Galileo', who at the same time was the link to optics because he had built a telescope and found the four moons of Jupiter, which he had then dedicated to the Medici as *'sidera medicea'* 'the Medicean stars'.

The frontispiece by François Spierre (Fig. 6.3.12) refers to this connection (Remmert 2005, 218–222). On the upper left we see the crescent moon with the earth-like surface that Galileo had first described. On the opposite side Jupiter and its four moons are clearly visible, observed by a man with a long telescope, possibly Galileo. The personification of optics points to his telescope, using another telescope as a pointer. Her left hand is directed towards the moon as if she is explaining the observation and understanding of celestial phenomena to the girl standing beside her. The girl pays close attention as she holds an over-sized lens in her hands. The Medici coat of arms dominates the scene at front left, where a putto leaning on it has just cut a piece off one of the six *palle* 'balls' and presents it to Optics. This is not an act of vandalism but homage to the generosity of the Medici, because the cut-off piece is itself a freshly polished lens.

What might seem playful to a modern spectator was deeply rooted in the quest for patronage. Accordingly, before publication, a sketch of the frontispiece was sent for approval to Alessandro Segni, a friend of Leopold and secretary of the Accademia del Cimento. Thus when Eschinardi cautiously showed his reverence for Galileo in the final version, he could have been assured of the protection of Leopold. The patronage aspect of a frontispiece would often be prominent, but its other messages were manifold.

The garden of the mathematical sciences

From images and frontispieces it is only a small step to further metaphor in the history of Early Modern mathematical sciences. We have seen that in the frontispiece of Mario Bettini's *Apiaria universae philosophiae mathematicae* (Fig. 6.3.8), bees carry an important message, and one could fill a book with the metaphorical use of bees and beehives in Early Modern natural philosophy and the mathematical sciences. But in the *Apiaria* another, equally powerful metaphor comes up: the *hortus mathematicus*, the 'garden of the mathematical sciences' (Remmert 2004; 2007b).

Figure 6.3.12 Homage to the Medici: the cut–off piece from the Medici coat of arms is a freshly polished lens. (Eschinardi 1668, by permission of Herzog August Bibliothek Wolfenbüttel)

It is more fruitful to take this metaphor seriously, and others too, than to dismiss it as a typical blossom of early modern rhetoric and iconography. Let us take a short glimpse at the *hortus mathematicus* as a field where the theory and practice of gardening and the mathematical sciences interacted. Gaspar Schott in the frontispiece of his *Cursus mathematicus* (Fig. 6.3.7) made good use of the fact that everything needed to create gardens and entertainment in them is covered by the mathematical sciences: practical arithmetic, architecture, perspective, optics, music, and so on. Thus, it was only fitting for the powerful and rich, with whom gardens were fashionable, to support the mathematical sciences.

At the same time, in the 1660s, the English virtuoso, co-founder of the Royal Society and diarist, John Evelyn, pursued the ambitious goal of turning the art of gardening into a science. He spent almost fifty years of his life writing and re-writing a compendium on gardens, his *Elysium britannicum, or the royal gardens*. Of this magnificent encyclopaedia only small parts were published in his lifetime, but the twenty-first century has brought us an annotated printed edition of the whole manuscript (Evelyn 2001). Evelyn drew on the whole range of the mathematical sciences. Accordingly in the chapter 'Of a gardiner, and how he is to be qualified', Evelyn emphasized that knowledge of the mathematical sciences was indispensable for gardening and that 'what *Plato* caused to be inscribed upon the *Architrave* of his Schoole dore, would be set with as much reason over that of our Garden, Αγεωμέτρητος *nemo* [Let no one ignorant of geometry enter].' (Evelyn 2001, 33–34).

In fact, several mathematical textbooks were closely tied to the needs of gardening, in theory and practice. By way of example I mention the *Géométrie pratique* 'Practical geometry' (1702) of Alain Manesson-Mallet, *maître de mathématiques* 'teacher of mathematics' at the court of Louis XIV. Of more than five hundred engravings, many are related to gardening (Fig. 6.3.13). Usually they show castles and gardens in their upper part, so that visually the *Géométrie pratique* is close to architecture and landscape architecture. But Manesson-Mallet uses the engravings not only as a form of reverence to his patron and king, glorified through numerous views of Versailles; he also depicts problems and exercises of everyday garden design. One example concerns a gentleman whose garden is too long and too narrow and who wants to remedy this by exchanging land with his neighbour. Thus he desires the piece QNXY, better proportioned than the original MNOP (Fig. 6.3.14). Manesson-Mallet gives detailed instructions on the basis of Euclid's *Elements*, book VI. The text refers directly to the image: Manesson-Mallet is in operating in the garden itself (Manesson-Mallet 1702, III 248).

From the late sixteenth to the early eighteenth century many authors on gardening were convinced that the mathematical sciences were indispensable in turning the art of gardening into a reliable and calculable enterprise, a science of nature (Remmert 2007b). The mathematical sciences with their many branches

LIV. III. *De la Planimetrie.*   133

PLANCHE XLVIII.

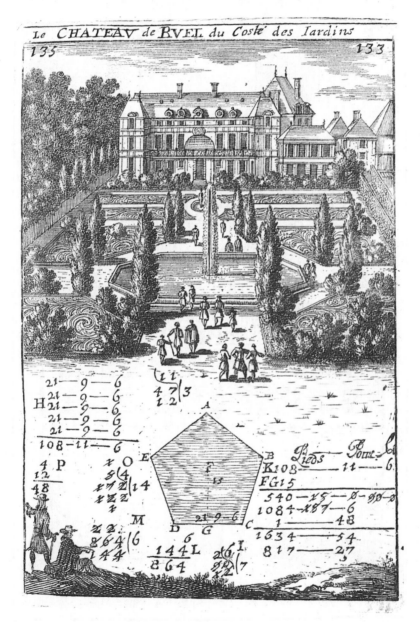

Figure 6.3.13 Mathematics in the garden. (Manesson–Mallet 1702, III 133, by permission of Herzog August Bibliothek Wolfenbüttel)

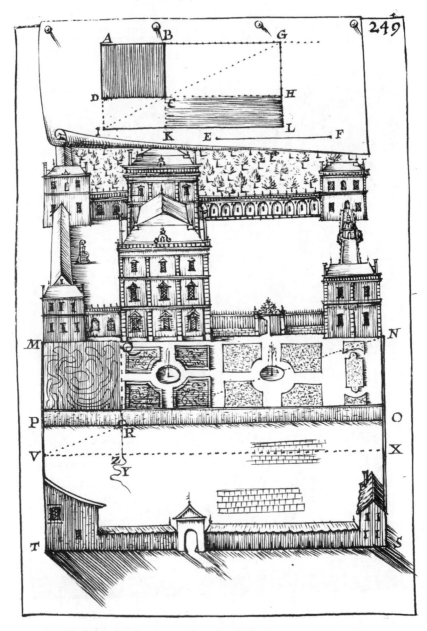

LIV. III. *De la Planimetrie.*   249

PLANCHE CIII.

Figure 6.3.14 Mathematics in the garden: Euclid helps to trade land in order to obtain the piece QNXY, better proportioned than MNOP. (Manesson–Mallet 1702, III 248, by permission of Herzog August Bibliothek Wolfenbüttel)

thus had two important roles to play. On the one hand, gardening as a discipline was transformed and emancipated, for example, from its old rival architecture. On the other hand, gardening reached into the political sphere by offering new possibilities and forms of representation, and not only in the gardens of Versailles, which were, perhaps, the most magnificent example of political representation in seventeenth-century Europe (Mukerji 1997). From this point of view the *hortus mathematicus* was not just one more visual strategy used to legitimize the mathematical sciences, but an essential element of their visual representation.

Conclusion

In the Early Modern period mathematicians, astronomers, and other practitioners of the mathematical sciences attached great importance to the visual strategies for legitimizing the mathematical sciences. From the early seventeenth century the mathematical sciences began to play a leading role in the hierarchy of scientific disciplines, and modes of explanation informed by them increasingly dominated many branches of the sciences and segments of society. In my view, this complex process cannot be understood if we concentrate exclusively on scientific texts, which could be hard to digest then and now. Rather, much of the labour of convincing patrons, readers, and spectators, was invested in paratexts, visual as well as written.

There are considerable gaps in our historical understanding of frontispieces as a means of communication and transmitters of knowledge and opinions. However, there were good reasons for choosing frontispieces as a visual medium because they provided an obvious way of incorporating pictorial methods into the realm of scholarly expression. Visual strategies used by authors and publishers could be based on various models, from the politically loaded symbolism of Atlas and Hercules, to the lines of tradition well known among scholars. Whether frontispieces were suited to a particular audience or a specific discipline depended on whether an appropriate existing iconography could be drawn upon or developed. For astronomy, a huge iconographic repertoire already existed and its range could easily be extended, as, for example, in the Copernican debate and the development of ingenious geo- and heliocentric symbols (Remmert 2003). However, the situation was different for other branches of the mathematical sciences and natural philosophy, where no established iconographic tradition was available (for example, in magnetism). In addition to the availability of suitable forms of iconographic representation, an important issue in designing (or choosing) a frontispiece (aside from who would cover the expense) was its intended audience. The frontispiece as part of a personal patronage scheme had a clearly defined audience, probably extremely small, noble, or wealthy, whereas the frontispiece as advertisement had to be accessible to a larger audience of unknown buyers. But when legitimization was the main concern, patrons, scholars, and laymen were all among the addressees.

Naturally most of the images were designed to persuade more than to prove. And when it came to persuasion, visual material and frontispieces were significant. Whereas readers of the main text would expect proofs, observers of the frontispiece, who might never look at the main text, would not necessarily object to being caught in a persuasive, if unsubstantiated, visual web. The latter were often the main target of various strategies of patronage and legitimization, the success of which was, as it still is, essential for the formation and advancement of ideas and new disciplines.

Bibliography

Ashworth, William B, 'The Habsburg circle', in Bruce T Moran (ed), *Patronage and institutions. Science, technology, and medicine at the European court 1500–1700*, Boydell Press, 1991.

Bayer, Johannes, *Uranometria, omnium asterismorum continens schemata, nova methodo delineata, aeris laminis expressa*, Augsburg, 1603.

Bettini, Mario, *Apiaria universae philosophiae mathematicae*, 3 vols, Bologna, 1645.

Cellarius, Andreas, *Harmonia macrocosmica*, Amsterdam, 1661.

de Chales, Claude-François, *Cursus seu mundus mathematicus*, Lyons, 1674.

Eschinardi, Francesco, *Centuria problematum opticorum*, Rome, 1666.

Eschinardi, Francesco, *Centuriae opticae pars altera*, Rome, 1668.

Evelyn, John, *Elysium britannicum, or the royal gardens*, John E Ingram (ed), University of Pennsylvania Press, 2001.

Fasanelli, Florence D, and Rickey, V Frederick, *Why have a frontispiece? Examples from the Michalowicz Collection at American University*, forthcoming, 2008.

Faulhaber, Johannes, *Ingenieurs-Schul Anderer, Dritter und Vierter theil*, Ulm, 1633.

Feldhay, Rivka and Heyd, Michael, 'The discourse of pious science', *Science in Context*, 3 (1989), 109–142.

Fumaroli, Marc, *L'école du silence. Le sentiment des images au XVIIe siècle*, Flammarion, 1994.

Genette, Gerard, *Seuils*, Éditions du Seuil, 1987.

Gregory, David (ed), *Euclidis quae supersunt omnia*, Oxford, 1703.

Hofmann, Josef Ehrenfried, 'Über die Quadraturen des Artus de Lionne', *National Mathematics Magazine*, 12 (1938), 223–230.

Imhausen, Annette, and Remmert, Volker R, 'The oration on the dignity and the usefulness of the mathematical sciences of Martinus Hortensius (Amsterdam, 1634): text, translation and commentary', *History of Universities*, 21 (2006), 71–150.

Lansbergen, Philipp van, *Tabulae motuum coelestium*, Middelburg, 1632.

Leotaud, Vincent, *Examen circuli quadraturae*, Lyons, 1654.

Luyts, Jan, *Astronomica institutio*, Ultrecht, 1692.

Manesson-Mallet, Alain, *La géométrie pratique*, Paris, 1702.

Marolois, Samuel, *Opera mathematica ou oeuvres mathematiques traictans de géométrie, perspective, architecture et fortification*, The Hague, 1614.

Metius, Adriaan, *Primum mobile astronomice, sciographice, geometrice et hydrographice, nova methodo explicatum*, Amsterdam, 1631.

Moore, Jonas, *A new systeme of the mathematicks*, London, 1681.

Mukerji, Chandra, *Territorial ambitions and the gardens of Versailles*, Cambridge University Press, 1997.

Mulerius, Nicolaus, *Tabulae frisicae*, Alkmaar/Amsterdam, 1611.

Remmert, Volker R, 'In the sign of Galileo: pictorial representation in the seventeenth-century Copernican debate', *Endeavour. A Quarterly Magazine for the History and Philosophy of Science*, 27 (2003), 26–31.

Remmert, Volker R, 'Hortus mathematicus: Über Querverbindungen zwischen Gartentheorie und -praxis und den mathematischen Wissenschaften in der Frühen Neuzeit', *Wolfenbütteler Barock-Nachrichten*, 31 (2004), 3–24.

Remmert, Volker R, *Widmung, Welterklärung und Wissenschaftslegitimierung: Titelbilder und ihre Funktionen in der Wissenschaftlichen Revolution*, Harrassowitz, 2005.

Remmert, Volker R, ' "Docet parva pictura, quod multae scripturae non dicunt." Frontispieces, their functions and their audiences in seventeenth-century mathematical sciences', in Sachiko Kusukawa and Ian Maclean (eds), *Transmitting knowledge: words, images and instruments in Early Modern Europe*, Oxford University Press, 2006.

Remmert, Volker R, 'Visual legitimisation of astronomy in the sixteenth and seventeenth centuries: Atlas, Hercules and Tycho's nose', in *Studies in History and Philosophy of Science*, 38 (2007a), 327–362.

Remmert, Volker R, ' "Of a Gardiner, and how he is to be qualified": John Evelyn, Gartenkultur und mathematische Wissenschaften im 17. Jahrhundert', in Harald Blanke and Berthold Heinecke (eds), *Revolution in Arkadien*, 2007b.

Rose, Paul Lawrence, *The Italian Renaissance of mathematics. Studies on humanists and mathematicians from Petrarch to Galileo*, Droz, 1975.

de Saint Vincent, Grégoire, *Opus geometricum*, Antwerp, 1647.

Schneider, Ivo, *Johannes Faulhaber: 1580–1635. Rechenmeister in einer Welt des Umbruchs*, Birkhäuser, 1993.

Schott, Kaspar, *Cursus mathematicus*, Würzburg, 1661.

Scriba, Christoph J, 'Die mathematischen Wissenschaften im mittelalterlichen Bildungskanon der Sieben Freien Künste', *Acta Historica Leopoldina*, 16 (1985), 25–54.

Writing the ultimate mathematical textbook: Nicolas Bourbaki's *Éléments de mathématique*

Leo Corry

Mathematical textbooks have played a significant role in the history of mathematics. Still, with a few—if important—exceptions, and especially in the twentieth century, mathematical textbooks do not in general convey new results. Rather, they attempt to summarize and present an updated picture of a discipline. Such summaries can hardly be neutral with regard to the body of knowledge they present. Writing a textbook involves much more than simply putting together previously dispersed results. Rather, it requires *selecting* topics and problems, and *organizing* them in a coherent and systematic way, while *favoring* certain techniques, approaches, and nomenclature over others. A mathematical textbook thus privileges certain avenues of research rather than others. Producing a mathematical textbook involves, above all, providing a well-defined structure of the discipline. But this structure is, in general, not forced upon the author in a unique way. The author makes meaningful choices to produce a distinctive image of the discipline.[1] If the textbook turns out to be successful and influential, it will disseminate this image as the preferred one for the discipline in question. Had the

---

1. I will refer to the distinction between 'body' and 'images' of mathematical knowledge (Corry 2001; 2004). Roughly stated, answers to questions directly related to the subject matter of any given discipline constitute the body of knowledge of that discipline, whereas claims and knowledge *about* that discipline pertain to the images of knowledge.

author chosen a different image, or had a book conveying a different disciplinary image been more successful, then the subsequent development of that discipline might have been considerably different. Occasionally a new disciplinary image put forward in a textbook constitutes an innovation no less important than a breakthrough individual result (Grattan-Guinness 2004).

Euclid's *Elements* is, of course, the paradigm example of a textbook compiled from existing knowledge that promoted an enormously influential disciplinary image, definitively shaping mathematics (and more) for millennia. Gauss' *Disquisitiones arithmeticae* is a second prominent example, sometimes compared in importance to the *Elements*, although more clearly circumscribed in its aims (Goldstein *et al* 2007). More recently, Nicolas Bourbaki's *Éléments de mathématique* embodied a unique attempt to play a similarly fundamental role in twentieth-century mathematics, with far-reaching ambitions for its impact on the discipline at large. It comprised a collective undertaking that drew on the efforts of scores of prominent mathematicians and appeared as a multi-volume series published between 1939 and 1998 (with new editions and printings appearing to this very day). Its influence spread throughout the mathematical world and it was instrumental in shaping the course of mathematical research and training for decades.

Bourbaki's extremely austere and idiosyncratic presentation—from which diagrams and external motivations were expressly excluded—became a hallmark of the group's style. The widespread adoption of Bourbaki's approaches to specific questions, concepts, and nomenclature indicates the breadth of its influence. Concepts and theories were presented in a thoroughly axiomatic way and discussed systematically, always going from the more general to the particular and never generalizing a particular result. A noteworthy consequence was that the real numbers could only be introduced well into the treatise, and not before a very heavy machinery of algebra and topology had been prepared in advance.

The Bourbaki phenomenon and the presentation of mathematics embodied in the *Éléments de mathématique* was followed in the mathematics community with a mixture of curiosity, excitement, awe, and, less frequently, criticism or even open disgust. This piece from *Mathematical Reviews* is an inspired description of the difficulties readers faced:

Confronted with the task of appraising a book by Nicolas Bourbaki, this reviewer feels as if he were required to climb the Nordwand of the Eiger. The presentation is austere and monolithic. The route is beset by scores of definitions, many of them apparently unmotivated. Always there are hordes of exercises to be worked painfully. One must be prepared to make constant cross references to the author's many other works. When the way grows treacherous and a nasty fall seems evident, we think of the enormous learning and prestige of the author. One feels that Bourbaki *must* be right, and one can only press onward, clinging to whatever minute rugosities the author provides and hoping to avoid a plunge into the abyss. (Hewitt 1956, 507)

This chapter is devoted to describing the origins and development of the enterprise of writing the *Éléments*, which was often seen, by those who took part in it, as the writing of the ultimate mathematical textbook. The chapter opens with an account of the origins of the group and the first stages of the project. This is followed by a more focused description of the writing of the volumes devoted to algebra and set theory, as well as their relationship to existing textbooks. The following section discusses the centrality of the idea of a mathematical structure for the Bourbakian image of mathematics, and its relationship to the technical contents of the *Éléments*. A final section discusses the conflict that rose in the mid-1950s within the group around the question of whether to adopt the language of categories and functors as a general, unifying language of mathematics.

Bourbaki: a name and a myth

Nicolas Bourbaki is the pseudonym adopted during the 1930s by a group of young French mathematicians who undertook the collective writing of an up-to-date treatise of mathematical analysis adapted to the latest advances and the current needs of the discipline. Among the ten founding members of the group Henri Cartan, Claude Chevalley, Jean Delsarte, Jean Dieudonné, and André Weil—all former students of the *École Normale Supérieure* in the early 1920s—remained the most influential and active within the group for decades. Over the years, many younger mathematicians participated in the group's activities, while the older members were supposed to quit at the age of fifty. All were among the most prominent of their generation, actively pursuing their own research in different specialisms, while the activities of Bourbaki absorbed a part of their time and energies (Chouchan 1995; Mashaal 2006; Beaulieu 2007).

By the early 1930s, the future founders of the group had already launched successful careers and had started publishing important, original research. As was typical in French academic life at the time, their careers started in provincial universities. Weil and Cartan were colleagues at Strasbourg for several years, where they felt increasingly dissatisfied with the way that analysis was traditionally taught in their country and with the existing textbooks written by the old masters (Dieudonné 1970, 136; Weil 1992, 99–100; Beaulieu 1993, 29–30). Edouard Goursat's *Cours d'analyse mathématique* (1903–5) was the most commonly used at the time. Its standards of rigor were unsatisfactory for these representatives of the younger generation. It treated the classical topics of analysis by considering case after case in an extremely detailed fashion, rather than introducing general ideas that could account for many of them simultaneously.

The search for better ways to introduce the basic concepts and theorems of the calculus was a topic of constant conversations between Cartan and Weil. Their predicament also affected their contemporaries teaching at other universities around France and was part of a more general feeling that postwar French mathematics was lagging far behind research in other countries, especially Germany, because of the loss of an entire generation of young mathematicians in the war. This situation provided the central motivation for the deliberations that would lead to the Bourbaki project.

At that time, Cartan and Weil used to meet every fortnight in Paris with their friends Chevalley, Delsarte, Dieudonné, and René de Possel. The framework of the meeting was the 'Séminaire de mathématiques' held from 1933 at the Institut Henri Poincaré under the patronage of Gaston Julia. Visiting mathematicians often participated too, but the 'Séminaire Julia', as it came to be known, was above all a joint production of the proto-Bourbakians. Each academic year, the seminar was devoted to a single general topic in which the participants wished to gain a broader and more systematic knowledge: groups and algebras, Hilbert spaces, topology, and variational calculus. In each meeting, one of the participants was commissioned to prepare a topic for discussion, edit his talk, and then distribute it among the other participants. This approach would later develop into Bourbaki's famous *modus operandi*, described further below.

Over coffee after the meetings of the Séminaire Julia, Weil started to discuss with his friends an ambitious collective initiative to produce the much needed new textbook in analysis. In December 1934, a more clearly delineated plan was stated by Weil, Cartan, Chevalley, Delsarte, Dieudonné, and de Possel: 'to define for 25 years the syllabus for the certificate in differential and integral calculus by writing, collectively, a treatise on analysis. Of course, this treatise will be as modern as possible' (Beaulieu 1993, 28). Following a 'modern' perspective was one of the apparently clear and suggestive ideas that, once the project started to materialize, proved to be in need of a more detailed definition that was not always easily agreed upon. At this meeting several other ideas were suggested concerning the plan of action: subcommittees should be put in charge of the various parts of the treatise; an agreed synopsis should be ready by the summer of 1935; the treatise should be about a thousand pages long; all decisions should be taken by consensus. Even a potential publisher was already in sight: Hermann (whose chief editor, Enrique Freymann was Weil's friend), rather than the leading Gauthier-Villars, where the old masters typically published their treatises.

Under the provisory name of 'Comité de rédaction du traité d'analyse' the group met again in January 1935. This time detailed minutes were taken by Delsarte (2000), who would continue to fulfill this task until 1940. It was decided that the committee would also include Paul Dubreil, Jacques Leray, and Szolem Mandelbrojt. Dubreil and Leray, however, were soon replaced by Charles Ehresmann and Jean Coulomb.

At this second meeting Delsarte and Dubreil presented a list of topics they wanted in the treatise: modern algebra; integral equations with special emphasis on Hilbert space; the theory of partial differential equations with emphasis on more recent developments; and a long section devoted to special functions. Mandelbrojt brought forward a principle that he considered of the utmost importance: whenever a result was intended for discussion in full generality, the general theory needed to prove this result would never be developed in the course of the exposition itself. Rather, all the general, abstract theories would be developed *in advance*. This was in line with the idea of a 'paquet abstrait' that had already been mentioned in the first meeting, and all participants agreed that this principle should be thoroughly pursued. Weil insisted that the treatise should be useful for all possible audiences: researchers, aspiring school teachers, physicists, and 'technicians' of various kinds (Delsarte 2000, 17).

After several preliminary encounters in Paris, the first real Bourbaki working meeting took place in July 1935, at the little town of Besse-en-Chandesse, close to Clermont-Ferrand. It was here that the mythical name was adopted. The expected length of the treatise was now calculated at three thousand two hundred pages and it was planned to be completed within a year. Along with a treatment of the classical themes of analysis, increased attention was given to the basic notions of algebra, topology, and the theory of sets. These now appeared necessary to provide the presentation with the kind of coherence and modern perspective that the group insistently spoke about.

This was the starting point of a long and fascinating endeavor. Its scope, structure, and contents went far beyond the initial plans of the group and their initial assumptions about the amount of work it would require. Except for a break during the war years, over the following decades the group (in its changing membership) continued to organize 'congresses' three times a year at different places around France for a week or two. Minutes of these Bourbaki congresses were circulated among members of the group in the form of an internal bulletin initially called *Journal de Bourbaki* and, from 1940, *La Tribu*.[2] Although *La Tribu* abounds with personal jokes, obscure references, and slangy expressions which sometimes hinder their understanding, they provide a very useful source for the historian researching the development of the Bourbaki project.

At each meeting, individual members were commissioned to produce drafts of the different chapters, which were then subjected to harsh criticism by the other members, and reassigned for revision. Only after several drafts was the final document ready for publication (Cartan in Jackson 1999, 784; Schwartz 2001, 155–163). Each chapter and volume of Bourbaki's treatise was thus the outcome

---

2. For details on the Bourbaki archives and the issues of *La Tribu* quoted here, see Corry (2004, 293 n13; Krömer 2006, 156–158). Direct quotes are taken from volumes in the personal collection of Professor Andrée Ch. Ehresmann, Amiens, and used with her permission. Other issues are quoted indirectly as indicated.

of arduous collective work. The spirit and viewpoint of the person(s) who had written it was hardly recognizable. The personal dynamics at work in the group are a matter of considerable interest and it represents, no doubt, a unique case in the history of science. For many, the most surprising fact about Bourbaki is that it could work at all.

What was initially projected as a modern analysis textbook eventually evolved into a multi-volume treatise entitled *Éléments de mathématique*, each volume of which was meant to contain a comprehensive exposition of a different mathematical subdiscipline. As with any other textbook, the material covered was not meant to be new in itself, but the very organization of the body of mathematical knowledge would certainly embody a novel overall conception of mathematics and, above all, underlying unity would be stressed. The *'paquet abstrait'*, initially conceived as a supporting toolbox of limited scope, gradually took center stage and became the hard core of the treatise, whereas classical topics of courses in analysis were continually delayed and some of them eventually left out of the treatise or relegated to specific sections or to the exercises.[3]

The first chapter of the *Eléments* appeared in 1939. By this time the plan had settled around six basic books: I. Theory of Sets; II. Algebra; III. General Topology; IV. Functions of a Real Variable; V. Topological Vector Spaces; VI. Integration. At a second stage in the 1950s additional chapters were added, including Lie Groups and Lie Algebras; Commutative Algebra; Spectral Theories; Differential and Analytic Manifolds (essentially no more than a summary of results). In its final form the treatise comprised over seven thousand pages, with new chapters continuing to appear until the early 1980s.

In the succeeding decades, Bourbaki's books became classics in many areas of pure mathematics, where the concepts and main problems, nomenclature, and Bourbaki's peculiar style were adopted as standard. The branches upon which Bourbaki exerted the deepest influence were algebra, topology, and functional analysis, becoming the backbone of mathematical curricula and research activity in many places around the world. Notations such as the symbol $\emptyset$ for the empty set, and terms like *injective*, *surjective*, and *bijective*, owe their widespread use to their adoption in the *Éléments de mathématique*. Bourbaki even influenced fields like economics (Weintraub and Mirowski 1994) and, especially in France, anthropology and literature (Aubin 1997).

Yet disciplines like logic, probability, and most fields of applied mathematics, which were beyond Bourbaki's scope, became under-represented in the many places worldwide where Bourbaki's influence was most strongly felt. This was the

---

3. Reviewers of Bourbaki, favorable and critical alike, typically describe the choice of exercises as one of the outstanding features of the collection. In most cases Dieudonné was in charge of this choice (Kaplansky 1953). In fact, for many years Dieudonné was the official scribe of the project and 'every printed word came from his pen' (Senechal 1998, 28).

case for many French and several American universities at various times between 1940 and 1970 (Schwartz 2001, 162–164). Further, group theory and number theory, despite being strong points of some members (notably Weil for number theory) were not treated in the *Éléments*, mainly because they were less amenable to the kind of systematic, comprehensive treatment presented in the collection. As part of an underlying tendency of estrangement from the visual, geometry was completely omitted from the Bourbakian picture of mathematics, except for what could be reduced to linear algebra.

Writing a textbook on modern algebra

As mentioned above, one of the group's declared aims was that their analysis treatise would be 'as modern as possible'. Most likely, the word 'modern' referred in their minds to the current trends in German mathematical research, especially to Bartel van der Waerden's epoch-making *Moderne Algebra* (1930). This book, the most important individual influence behind the entire Bourbaki project, represented the culmination of the deep transformation of algebra that had begun in the last third of the nineteenth century. Before then, algebraic research had mainly focused on theories of polynomial equations and polynomial forms, including algebraic invariants. The ideas implied by the works of Évariste Galois had become increasingly central after their publication by Joseph Liouville (1846). Together with important progress in the theory of fields of algebraic numbers, especially in the hands of Leopold Kronecker and Richard Dedekind, they gave rise to new concepts such as groups, fields, and modules. But this development was only gradually reflected in textbooks.

Towards the end of the 1920s, a growing number of works investigated the properties of the abstractly defined mathematical entities now seen as the focus of algebraic research: groups, fields, ideals, rings, and others. Like many other important textbooks, *Moderne Algebra* arrived at a time when the need was felt for a comprehensive synthesis of what had been achieved since its predecessor, in this case Heinrich Weber's *Lehrbuch der Algebra* (1895). It presented ideas that had been developed by Emmy Noether and Emil Artin—whose courses van der Waerden had attended in Göttingen and Hamburg—and by algebraists such as Ernst Steinitz, whose works he had studied under their guidance (van der Waerden 1975).

Van der Waerden masterfully incorporated many of the important innovations of the early twentieth century into the body of algebraic knowledge. But his book's originality and importance comprises its totally new way of conceiving the discipline. Van der Waerden systematically presented those mathematical branches then related to algebra, deriving all relevant results from a single, unified perspective, and using similar concepts and methods for all of them. The resultant image was based on the realization that a certain family of notions (groups, ideals, rings, fields, and so on) are individual instances of the general idea of an algebraic structure, and that the aim of research in algebra is the full elucidation of those notions. None of them, to be sure, appeared for the first time in this book. Groups had featured in the third edition of Joseph Serret's *Cours d'algèbre supérieure* (1866). Ideals and fields had been introduced by Dedekind in his elaboration of Ernst Edward Kummer's factorization theory of algebraic numbers (1871). But the unified treatment they were accorded in *Moderne Algebra*, the single methodological approach adopted to define and study them, and the compelling new picture of a variety of domains that had formerly been seen as only vaguely related, constituted a striking and original innovation.

One fundamental advance was an implicit redefinition of the conceptual hierarchy underlying the discipline of algebra. Under this image, rational and real numbers no longer had conceptual priority over, say, polynomials. Rather, they were defined as particular cases of abstract algebraic constructs. Thus, for instance, van der Waerden introduced the concept of a field of fractions for integral domains in general, and then obtained the rational numbers as a particular case of this kind of construction, namely as the field of quotients of the ring of integers. His definition of the system of real numbers in purely algebraic terms was based on the concept of a 'real field', recently elaborated by Artin and Otto Schreier, whose seminars van der Waerden had attended in Hamburg.

The task of finding the real and complex roots of an algebraic equation, which was the classical main core of algebra in the previous century, was relegated to a subsidiary role in van der Waerden's book. Three short sections in his chapter on Galois theory dealt with this specific application of the theory, and they assume no previous knowledge of the properties of real numbers. In this way, two central concepts of classical algebra (rational and real numbers) were presented merely as final products of a series of successive algebraic constructs, the 'structure' of which had been gradually elucidated. On the other hand, additional, non-algebraic properties such as continuity and density were not considered at all by van der Waerden as part of his discussion of those systems.

Another of *Moderne Algebra's* important innovations concerns the way in which the advantages of the axiomatic method were exploited in conjunction with all other components of the structural image of algebra. Once one has realized that the basic notions of algebra (groups, rings, fields, and so on) are, in fact, different kinds of algebraic structures, their abstract axiomatic formulation becomes the most appropriate one. The central disciplinary concern of algebra became, in this conception, the systematic study of those different varieties through a common approach, underpinned by the idea of isomorphism. This fundamental recognition is summarized in the *Leitfaden* 'leading threads' in the

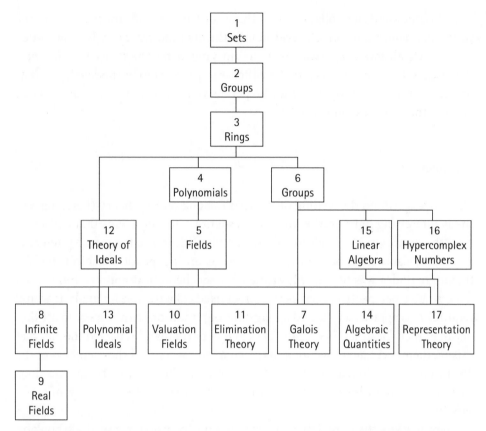

Figure 6.4.1 A translation of the diagram encapsulating the hierarchical relationships between algebraic concepts, presented in the introduction to van der Waerden's
seminal *Moderne Algebra* (1930)

introduction to the book, which pictures the hierarchical, structural interrelation
between the various concepts investigated in it (Fig. 6.4.1).

Obviously, van der Waerden's new image of algebra reflected the current state
of the body of algebraic knowledge. However, that image was *not* a *necessary*
outcome of the body, but rather an independent development of intrinsic value.
Several other contemporary algebra textbooks also contained most of the latest developments in the body of knowledge, but essentially preserved the classical image of algebra. Perhaps the most interesting example is Robert Fricke's
Lehrbuch der Algebra (1924), with the revealing subtitle *Verfasst mit Benutzung
vom Heinrich Webers gleichnamigem Buche* 'based on Heinrich Weber's book of
the same name'.

The main idea embodied in van der Waerden's book—the structural conception of algebra—became highly influential for Bourbaki. Before receiving his
doctorate in 1928, Weil had visited Göttingen, where he came into direct contact

with Noether and her collaborators. This visit left a significant imprint on the young mathematician, which reverberated through the centrality later accorded to modern algebraic approaches as a unifying perspective in the *Éléments*. Bourbaki's volume on *Algebra* (hereafter *A*) is also closely modeled in many respects on *Moderne Algebra*. But the pervasive influence of the book is much broader than that, as I argue below.

Set theory

The process around the writing of Bourbaki's book on Set Theory (hereafter *ST*) sheds interesting light on the kinds of hesitations and problems that accompanied the entire project. Indeed, the initial plan did not envisage a systematic, axiomatic elaboration of the theory of sets as an independent subject. Rather, the original idea was to use only elementary set-theoretical notions, introduced from a naive perspective, such as the direct needs of a treatise on analysis would require. This approach reflected a longstanding tradition with respect to set theory in France (Beaulieu 1994, 246–247), and in particular it reflected the fact that this mathematical field was not a major concern for most of the members of the group. One exception, however, was Chevalley, for whom foundational questions were, especially in his early career, a matter of direct interest (Dieudonné and Tits 1987).

Chevalley was the most active force behind the inclusion of a separate book on set theory as the plan evolved for the contents of the *Éléments*. In 1949 *La Tribu* pointed to the underlying discussions around one of the main questions that had occupied the Bourbaki project from the beginning: the possibility of presenting a self-contained, highly formalized treatment of the entire body of mathematics, with little or no external motivation of the topics treated. Discussions repeatedly arose around the exact way to present many individual topics or theories. This was clearly the case with sets, debates around which continually delayed publication. In the final account, the contents of *ST* were a compromise between the attempt to fully formalize the topic and explore it in detail, as demanded by Chevalley, and the need to produce a relatively easily readable book that would provide a basic language for the treatise while fitting the general reader's interest. Thus, set theory was adopted as a universal language underlying all mathematical domains because of its unifying capabilities (Bourbaki 1968, 9). But this very basic theory was not presented in a truly formalized language, because Bourbaki acknowledged that no mathematician actually works like that: 'his experience and mathematical flair tell him that translation into formal language would be no more than an exercise of patience (though doubtless a very tedious one)' (Bourbaki 1968, 8). Within the entire treatise, only in *ST* does one find explicit statements like this.

The question of the consistency of set theory also arises here. Bourbaki did not attempt to address the question head-on but rather reverted to a strongly empiricist position. In an ironic turn, Bourbaki simply stated that a contradiction was not expected to appear in set theory because none had appeared after so many years of fruitful research (Bourbaki 1968, 13). Yet one of Bourbaki's earlier publications had stated that 'absence of contradiction, in mathematics as a whole or in any given branch of it, thus appears as an empirical fact rather than as a metaphysical principle...We cannot hope to prove that every definition...does not bring about the possibility of a contradiction' (Bourbaki 1949, 3).[4]

A *Fascicule de résultats*, 'Summary of results', on set theory was published as early as 1939. The final volume was published only during the 1950s, comprising the following chapters: 1. Description of Formal Mathematics; 2. Theory of Sets; 3. Ordered Sets, Cardinals, Integers; 4. Structures. This fourth chapter introduced the new concept of *structure*,[5] which was meant to provide a formal notion that supposedly underlies all other mathematical theories described in the remaining parts of the treatise. Briefly put, in order to define this concept Bourbaki considered a finite collection of sets $E_1, E_2, ..., E_n$, and used an inductive procedure, each step of which consists either of taking the Cartesian product $(E \times F)$ of two sets obtained in former steps or of taking their power set $B(E)$. For example, beginning with the sets E, F, G the outcome of one such procedure could be: $B(E)$; $B(E) \times F$; $B(G)$; $B(B(E) \times F)$; $B(B(E) \times F) \times B(G)$ and so forth. Upon such constructs some additional conditions can be imposed to imitate the way in which various known mathematical entities are typically defined. For instance, an internal law of composition on a set A is a function from $A \times A$ into A. Accordingly, given any set A, one can form the scheme $B((A \times A) \times A)$ and then choose from all the subsets of $(A \times A) \times A$ those satisfying certain conditions of a 'functional graph' with domain $A \times A$ and range A. The axiom defining this choice is a special case of what Bourbaki calls an algebraic *structure*. Similarly, Chapter 4 of *ST* showed how the general concept allowed for the definition of other types, such as ordered *structures* or topological *structures*. Finally, the general definition of *structures* led to some further, related concepts such as isomorphism among *structures*, deduction of *structures*, poorer and richer *structures*, equivalent species of *structures*, etc. Chapter 4 was the most idiosyncratic of the volume and of the entire collection, and in an important sense the most problematic one.

4. Imre Lakatos (1978, II 24–42) has called attention to the fact that foundationalist philosophers of mathematics, from Russell onwards, when confronted with serious problems in their attempts to prove the consistency of arithmetic, have not hesitated to revert to empirical considerations as the ultimate justification for it. Although Bourbaki is not mentioned among the profusely documented quotations selected by Lakatos to justify his own claim, it seems that these passages of Bourbaki could easily fit into his argument. See also Israel and Radice (1976, 175–176).

5. Hereafter I write *structures* (italicized) to indicate this specific, Bourbakian technical term, as opposed to the non-formal, general usage of the term.

The *Fascicule de résultats* is strikingly different from the chapters themselves. Whereas the book's stated aim is to show that it is possible to provide a sound, formal basis for mathematics as a whole, the *Fascicule* aims simply to provide the basic lexicon and to explain the non-formal meaning of the terms used. Thus, the opening lines read:

As for the notions and terms introduced below without definitions, the reader may safely take them with their usual meanings. This will not cause any difficulties as far as the remainder of the series is concerned, and renders almost trivial the majority of the propositions. (Bourbaki 1968, 347)

Thus, for example, the painstaking effort invested in Chapters 2 and 3 is here represented by the laconic statement: 'A set consists of *elements* which are capable of possessing certain *properties* and of having relations between themselves or with elements of other sets' (Bourbaki 1968, 347). As for *structures*, the *Fascicule* reduces the whole formal development to a very short, intuitive explanation of the concepts in which the main ideas are explained. The only important related concept which is mentioned is that of isomorphism.

Between the appearance of the *Fascicule* in 1939 and the four chapters in 1954–7 there were many important developments in mathematics, in particular the emergence of category theory. As a consequence, some of the ideas that had perhaps looked very promising in 1939 soon became obsolete. Thus, *ST*, and especially its chapter on *structures*, became one of the least interesting of the entire collection. As a textbook for the discipline, it received little attention and very few of its concepts and notations were widely adopted. As Paul Halmos put it:

It is generally admitted that strict adherence to rigorously correct terminology is likely to end in being pedantic and unreadable. This is especially true of Bourbaki, because their terminology and symbolism are frequently at variance with commonly accepted usage. The amusing fact is that often the 'abuse of language' which they employ as an informal replacement for a technical name is actually conventional usage: weary of trying to remember their own innovation, the authors slip comfortably into the terminology of the rest of the mathematical world. (Halmos 1957, 90)[6]

Even more interesting, the terminology and the concepts introduced in the set theory book, and particularly on the chapter on *structures*, were *hardly used in the other parts of Bourbaki's own book*. And, on the few occasions when it was used, this only made more patent the ad hoc character of the supposedly fundamental part of the treatise. In order to understand this important point in its precise context, it is necessary now to discuss the role of 'structure' in Bourbaki's overall conception of mathematics.

---

6. For a detailed review of Chapters 1 and 2 of *ST*, see Halmos (1955). For an assessment of the technical shortcomings of Bourbaki's system of axioms for the theory of sets, see Mathias (1992).

Two meanings of 'structure'

As work on the treatise developed, an implicit but pervasive idea increasingly came to underlie the overall approach. This was the conception of mathematics as a systematic, elaborate hierarchy of structures: essentially an extension of the idea from van der Waerden's algebra textbook. He had undertaken a unified 'structural' investigation of several concepts that were defined in similar, abstract terms (groups, rings, ideals, modules, fields, hypercomplex systems) while asking similar kinds of questions about them and using similar kinds of tools to investigate them. Now, in Bourbaki's textbooks, algebra, topology, and functional analysis started to appear as individual materializations of one and the same underlying, general idea: the idea of a mathematical structure. Bourbaki attempted to present a unified, comprehensive picture of what they saw as the main core of mathematics, using a standard system of notation, addressing similar questions in the various fields investigated, and using similar conceptual tools and methods across apparently disparate mathematical domains.

In 1950 Dieudonné, under the name of Bourbaki, published an article that came to be identified as the group's manifesto, 'The architecture of mathematics'. Faced with the unprecedented growth and diversification of the discipline, Dieudonné again raised the well-known question of the unity of mathematics. Mathematics was a strongly unified branch of knowledge in spite of appearances, he claimed, and now it was clear that the basis of this unity was the use of the axiomatic method, as the work of David Hilbert had clearly revealed.[7] Mathematics should be seen, Dieudonné added, as a hierarchy of structures at the heart of which lie the so called 'mother structures':

At the center of our universe are found the great types of structures,...they might be called the mother structures...Beyond this first nucleus, appear the structures which might be called multiple structures. They involve two or more of the great mother-structures not in simple juxtaposition (which would not produce anything new) but combined organically by one or more axioms which set up a connection between them...Further along we come finally to the theories properly called particular. In these the elements of the sets under consideration, which in the general structures have remained entirely indeterminate, obtain a more definitely characterized individuality. (Bourbaki 1950, 228–229)

Thus, the idea that van der Waerden had applied successfully and consistently but only implicitly—namely the centrality of the hierarchy of structures—became now explicit and constitutive for Bourbaki. At the same time, an elaborate attempt was made in Chapter 4 of *ST* to present a formal definition of *structure*, which was somehow meant to provide a solid conceptual foundation on which

---

7. Dieudonné frequently described Bourbaki as Hilbert's 'natural heir'. Nevertheless, there were very significant differences between their respective conceptions. See Corry (1997; 2001).

the whole edifice of mathematics could be built. Thus, two different meanings of the term mathematical 'structure' appeared in Bourbakian discourse, which were not always properly distinguished from one another: (1) a non-formal and perhaps even metaphorical meaning, used for example in Dieudonné's manifesto to present the entire science of mathematics as a hierarchy of structures, and implicitly implemented by van der Waerden in his new image of algebra, and (2) a formal technical term, *structure*, appearing in a mathematical theory that was never incorporated into current mathematical research or exposition, and was not even really used by Bourbaki in their own treatise.

As already stated, the main interest of most members of the group was in the various disciplines covered in the treatise but not in *ST* or in its chapter on *structures*. And yet many discussions about the correct way to present those various disciplines were necessarily influenced by the introduction of the basic concepts associated with *structures*. It is remarkable that members of the group tended not to separate the two meanings clearly, thus giving the impression that it was Bourbaki's own formal concept of *structure*, and not the general, structural image of mathematics, that was so central to much of twentieth century mathematics.

Bourbaki's theory of *structures* never received any real attention from working mathematicians, even Bourbaki's members when involved in their own research. When we look at how the concept of *structure* was used in the treatise, all we see is that in the opening chapters of the books on branches such as algebra and topology, some sections were devoted to showing how that branch could, in principle, be formally connected to the general concept of *structure*. This connection, however, was rather feeble, a formal exercise that was forgotten after the first few pages. For instance, while *A* presents vector spaces as a special case of groups, so that all the results proved for groups hold for vector spaces too, this hierarchical relationship is not presented in terms of the concepts defined in *ST*. Likewise, neither commutative groups nor rings are presented as *structures* from which a group can be 'deduced', nor is it proved that **Z**-modules and commutative groups are 'equivalent' *structures*. *Structure*-related concepts do appear in the opening sections of *A*, but the rather artificial use to which they are put and their absence from the rest of the book suggests that this initial usage was an ad hoc recourse to demonstrate the alleged subordination of algebraic concepts to the more general ones introduced within the framework of *structures*. Neither new theorems nor new proofs of known theorems are obtained through the *structural* approach.

As the book advances further into theories of the hierarchy of algebraic structures, the connection with *structures* is scarcely mentioned. Ironically, the need for a stronger unifying framework was indeed felt in later sections. For instance, Chapter 3 discusses three types of algebras defined over a given commutative ring: tensor, symmetric, and exterior algebras. Bourbaki defines each kind and then discusses for each case, 'functorial properties', 'extension of the ring of scalars', 'direct limits', 'free modules', 'direct sums', and so on (Bourbaki 1973, 484–522). Not only would a unified presentation of the three have been more economical and direct but their properties lend themselves naturally to a categorical treatment, a possibility which is not even mentioned. The 'functorial properties' of the algebras are explained through the use of the standard categorical device of 'commutative diagrams', but without mentioning the concepts of functor or category.

The volume on *General Topology* is the most outstanding example of a theory presented through Bourbaki's model of the hierarchy of *structures*, starting from one of the 'mother structures' and descending to a particular *structure*, namely that of the real numbers. And yet, as with A, the hierarchy itself is not introduced in terms of *structure*-related concepts. Thus for instance, topological groups are not characterized as a *structure* from which the *structure* of groups can be 'deduced'. *Structure*-related concepts appear in this book more than anywhere else in the treatise but, instead of reinforcing the purported generality of such concepts, a close inspection of their use immediately reveals their ad hoc character.

The central notion of structure, then, had a double meaning in Bourbaki's mathematical discourse. On the one hand, it suggested a general organizational scheme for the entire discipline, which turned out to be very influential. On the other hand, it comprised a concept that was meant to provide the underlying formal unity but was of no mathematical value whatsoever either within Bourbaki's own treatise or outside it. But Bourbaki's theory of *structures* was only one among several attempts after 1935 to develop a general mathematical theory of structures, and was not even the only such attempt in which members of the group were involved.[8] Thus, in order to understand the full historical and mathematical context of the theory of *structures* and its role within the *Éléments*, we now discuss the conflict created by the rise of its most serious competitor, the theory of categories.

The categorical imperative and its demise

In the early 1940s, Samuel Eilenberg and Saunders Mac Lane, who would both later become involved in Bourbaki, introduced the concepts of category and functor. These concepts and the general perspective they furnished gradually became a widely adopted unifying tool and language for mathematical disciplines, and pursued a structural spirit similar to Bourbaki's. Groundbreaking early instances

---

8. In Corry (2004) I presented a full account of such reflexive theories of structures, their origins and their interrelations.

appear in the works of two younger-generation Bourbaki members, Alexander Grothendieck and Jean-Pierre Serre, who used categories in the early 1950s as basic tools for their own research in homological algebra and algebraic geometry (Krömer 2007, 117–190). Against this background, it is only natural to expect that the categorical approach would easily find its way into Bourbaki's debates as an ideal candidate to support the unifying, structure-oriented perspective that the group had been striving for. Indeed, the idea was discussed at various Bourbaki congresses but in the end it never materialized.[9]

If categorical language were to be adopted by Bourbaki as a unifying language for the *Éléments*, this would entail the reformulation of considerable parts of existing chapters to make them fit the new approach. The chapter on *structures* in *ST* would be a particularly obvious nuisance. As already mentioned, this entire chapter was rather ad hoc and in any case did not represent a main focus of interest for most members of the group. This in itself was a meaningful obstacle to incorporating categories into the treatise; additional obstacles came from diverging views about the intrinsic value of the categorical approach in general. Weil, for one, actively opposed the introduction of categories in any way into the *Éléments*.

Some topics discussed in Bourbaki's book on *Commutative Algebra* were presented in a manner for which the categorical formulation would have been the most natural, but without explicitly doing so (Corry 2004, 327–328). This was also the case with other topics on which Bourbaki had already published by 1950 or would soon publish. During the 1950s *La Tribu* documented recurring attempts to write chapters on homological algebra and categories, and the discussions that ensued. In 1951, Eilenberg was commissioned several times to prepare drafts to be discussed. He had not only created the theory of categories with Mac Lane. In the 1950s he was collaborating on the first two books to systematically use this language to present elaborate mathematical disciplines that had emerged and developed in completely different terms: algebraic topology (Eilenberg and Steenrod 1952) and homological algebra (Cartan and Eilenberg 1956). When it came to Bourbaki, however, he immediately realized the serious difficulties to be expected in the context of the Bourbaki treatise, because it had already introduced *structures*. In an undated, unpublished text possibly written around that time, he said so explicitly:

The method of functors and categories is in some sort of 'competition' with the method of structures as developed at present. Unless this 'competition' is resolved only one of these methods should be presented at the early stage.... The resolution of the 'competition'

9. In Corry (1992) I called attention to the inherent tension between *structures* and categories, and published some illuminating related documents (mainly issues of *La Tribu*), some of which are also included here. More recently, Ralf Krömer (2006) has added significant insights to this important point, using previously unpublished material, some of which I quote below.

is only possible through the definition of the 'structural homomorphism' which would certainly require a serious modification of the present concept of structure. It would certainly complicate further this already complicated concept. Despite my willingness to complicate things I am still unable to produce a general definition that would fit known typical cases.[10]

Over the next few years, the younger-generation Bourbaki members increasingly adopted categorical language for their own research, and repeatedly attempted to introduce it to the *Éléments*. At this time, *structures* had been announced in the *Fascicle* of 1939 but the related chapter in *ST* had not yet been worked out. In principle, there was still room for categories, but, as Eilenberg was quick to see, this would require more than trivial reformulation. *La Tribu* documents heated debates around *structures* and categories throughout the 1950s, which culminated on publication of Grothendieck's famous *Tohoku* article (1957), a milestone in the history of category theory. In it Grothendieck innovatively applied cohomological methods (fully couched in categorical language) to algebraic geometry, thus opening the road for developments that would continue to engage mathematicians for decades. *La Tribu* and the contemporary Serre-Grothendieck correspondence (Colmez and Serre 2001) provide clear evidence that Grothendieck had conceived his famous article as a possible contribution to the Bourbaki treatise. Grothendieck's functorial ideas were well received by most of the group's younger generation, and by Dieudonné, but the continued opposition of others, especially Weil, prevented their adoption in the *Éléments*.

The chapter on *structures* came out in 1957 without the slightest explicit reference to categorical ideas. The incompatibility of the two approaches and the work already invested were the main reasons behind this decision. Cartan wrote that the *structural* point of view should not be abandoned without 'very serious reasons'. Some members of the group, however, notably Grothendieck, were highly dissatisfied. He continued to suggest that a new Chapter 4 of *ST* should replace the old one, 'unusable in all respects' (Krömer 2006, 144).

It is important to delineate more precisely the internal historical context within which this discussion was taking place. By the mid-1950s younger members (Serre, Grothendieck, and others) had started to join the group. Naturally, and partly because of Bourbaki's influence, the mathematical scene was by then very different to that faced by the founding fathers over twenty-five years earlier. At the same time, the age of self-imposed retirement at fifty had arrived for the latter (but was not always strictly adhered to). To the extent that the younger generation members wanted to invest their energies in the Bourbaki project they pursued agendas that differed at various levels from the original one, and also, sometimes,

---

10. Quoted in (Krömer 2006, 142), from an original document in the Eilenberg archive, Columbia University, New York.

from those of each other. Towards the end of the decade, the first six books of the *Éléments* had essentially been completed, covering much of what the group had come to consider as the hardcore of the project. It was time to deal with more advanced and specialized topics, while the younger members wanted a say in the project's overall direction. The possibility of universal participation in each topic and the original view that writing should not be assigned to 'specialists' were both reconsidered. Basic questions about the entire enterprise arose anew, provoking conflicting views and sometimes personal tensions. The debate around the adoption of categories was part of this situation, particularly the opposition between Grothendieck and Weil, two strongly opinionated mathematicians and difficult people to deal with.

Indeed, Weil was a very dominant character whose mathematical prestige and intellectual personality, coupled with his authority as one of the leading forces in the Bourbaki project, bestowed upon him an undisputed, unique position within the group. The retirement of some prominent members over the years has commonly been attributed to conflicts or tensions with Weil. That was certainly the case with de Possel, to whom Evelyn, Weil's wife since 1939, had previously been married. Weil had been the first to suggest that members should retire from active participation at the age of fifty, but ironically, on arriving at that age in 1956 he gave very little sign of wanting to diminish his influence on the project.

Grothendieck, in turn, was a highly unconventional personality even by the standards of this bunch of rather unconventional individuals. He was born in Germany, but escaped during the war to France. He remained an alien citizen, which created obstacles to finding a position in his new country. In 1959, Grothendieck got a research position in the newly created Institut des Hautes Etudes Scientifiques (IHES), where he spent twelve years creating and teaching his revolutionary ideas. In the framework of Bourbaki, he favored the continuation of the generalizing spirit that had permeated the early books, but with more powerful, increasingly abstract, algebraic tools. Not all members, however, approved. Many years later, Armand Borel recalled that Grothendieck's approach was at times 'discouragingly general, but at others rich in ideas and insights', and thus, 'it was rather clear that if we followed that route, we would be bogged down with foundations for many years, with a very uncertain outcome' (Borel 1998, 376).

In Grothendieck's memoirs, a remarkable document called *Récoltes et semailles* 'Reaping and sowing', which initially circulated only within closed circles,[11] he referred to his special status within the group, while pointing to the underlying tension with Weil:

---

11. Two useful websites containing digital editions of *Récoltes et semailles*, and additional material related to Grothendieck, are http://math.jussieu.fr/~leila/grothendieckcircle/index.php and http://kolmogorov.unex.es/~navarro/res.

...until around 1957 I was regarded with certain reservations by more than one member of the Bourbaki group after it had finally co-opted me, I believe, with some reticence....More often than not, I was, moreover, the one most frequently excluded from the Bourbaki congresses, especially during the common readings of the drafts, as I was rather incapable of following the readings and discussions at the pace in which they were conducted. I am possibly not really gifted for collective work. However, the difficulty I had in coping with group-work or the kind of reservation I may have elicited for other reasons from Cartan and others did not once attract sarcastic remarks or rebuffs, or even a shadow of condescension, except once or twice from the part of Weil (evidently a very different case!).[12] (Grothendieck undated I, 142–143)

From Grothendieck's correspondence with Serre in 1956, it is quite evident that both mathematicians disliked Weil's style, although they surely recognized the importance and originality of his ideas for their own concerns (Colmez and Serre 2001, 49–53). Writing retrospectively about this period, Grothendieck put matters in proportion, stressing the positive balance that he attributed to the project and to Weil's role within it (Grothendieck undated, I 46). As it happened, however, Grothendieck quit the group around 1958–59 while some members, such as Serre and Dieudonné, continued to be his close friends and collaborators. In 1970 he completely retired from public scientific life, when he discovered that IHES was partly funded by the military.

Laurent Schwartz, who had directed Grothendieck's dissertation, explained why the latter remained in the group for only a few years: 'he lacked humor and had difficulty accepting Bourbaki's criticism' (Schwartz 2001, 284). There is every reason to accept this explanation, yet there is also clear evidence that the non-adoption of category theory and Weil's attitude towards this question and towards Grothendieck were the main reasons for the latter's decision to quit. An anonymous text (possibly by Serge Lang) was appended to *La Tribu* in the early 1960s under the title *Ad majorem fonctori gloriam*, 'To the greater glory of functors'. It described Grothendieck's departure as a clear indication of a decline in the originally innovative spirit of the Bourbaki enterprise, implying that Weil was to blame:

I have learnt that Grothendieck is no longer a member of Bourbaki. I regret that very much, as I regret the circumstances that led to this decision... [namely,] a systematic opposition, more or less explicit depending on this or that person, against his mathematical point of view, and especially against the use of the latter by Bourbaki....It is

12. Ce fait est d'autant plus remarquable que jusque vers 1957, j'étais considéré avec une certaine réserve par plus d'un membre du groupe Bourbaki, qui avait fini par me coopter, je crois, avec une certaine réticence....J'étais d'ailleurs le plus souvent largué pendant les congrès Bourbaki, surtout pendant les lectures en commun des rédactions, étant bien incapable de suivre lectures et discussions au rythme où elles se poursuivaient. Il est possible que je ne suis pas fait vraiment pour un travail collectif. Toujours est-il que cette difficulté que j'avais à m'insérer dans le travail commun, ou les réserves que j'ai pu susciter pour d'autres raisons encore à Cartan et à d'autres, ne m'ont à aucun moment attiré sarcasme ou rebuffade, ou seulement une ombre de condescendance, à part tout au plus une ou deux fois chez Weil (décidément un cas à part !).

a scandal that Bourbaki not only did not take the lead in the functorial movement, but rather that is not even in its tail....If some of the founding members (e.g., Weil) wish to revert on the decision not to influence the direction that Bourbaki wants to follow, he should say so explicitly....If Bourbaki refuses, not just to join the new movement, but to take the lead in it, then those treatises pursuing the formulation of the elements of mathematics (and not just those dealing with algebraic geometry) will be written by others who will take inspiration not in the spirit of Bourbaki 1960, but in his spirit 1939. That would be a great pity.[13] (Krömer 2006, 152–153)

The consequences of the debate around categories and *structures* continued to be felt for many years, and are manifest in Bourbaki's book on homological algebra, published in 1980 as a chapter 10 of *A*. Categories had become the standard framework for treating homological concepts ever since Cartan and Eilenberg's famous textbook of 1952. In Bourbaki's presentation, however, these concepts are defined within the narrower framework of modules, as using the language of categories here would have gone against the most basic principles that had guided the enterprise since it inception. Thus, whereas Bourbaki's treatment of general topology in the 1940s had embodied a truly innovative approach that many others were to follow, this was hardly the case with homological algebra in the 1980s. This irony is further enhanced by the fact that Bourbaki's own theory of *structures* was not even mentioned in this final volume of the by now truly classic treatise.

Conclusion

The Bourbaki project reached its high-point of success and influence during the 1960s but the impetus that had characterized the project in its early years could not be maintained indefinitely. Dieudonné's catalyzing role could hardly be matched after his retirement. Some new chapters were proposed which never materialized, on topics such as analysis of several complex variables, homotopy theory, spectral theory of operators, and symplectic geometry. Nothing came either of plans to rewrite the first six books. The new books that did appear by

---

13. J'apprends que Grothendieck n'est plus membre de Bourbaki. Je le regrette beaucoup, ainsi que les circonstances qui ont amené cette décision...Ce qui importait, c'est une opposition systématique, plus ou moins explicitée selon les uns ou les autres, contre son point de vue mathématique, ou plutôt son emploi par Bourbaki....C'est un scandale que Bourbaki, non seulement ne soit pas à la tête du mouvement functorial, mais encore n'y soit même pas à la queue....Si certains membres fondateurs (e.g., Weil) désirent revenir sur leur décision de ne pas influencer Bourbaki dnas la direction qu'il désire prendre, qu'ils le disent explicitement....Si Bourbaki refuse, non pas de se mettre dans le nouveau mouvement, mais d'en prendre la tête, alors les traités visant à la redaction des éléments des mathématiques (et pas seulment à ceux de la géométrie algébrique) seront rédigés par d'autres, qui s'inspireront non pas de l'esprit de Bourbaki 1960, mais de son esprit 1939. Ce serait dommage.

1980 included a summary of differential and analytic manifolds, seven chapters on commutative algebra, eight chapters on Lie groups and Lie algebras, and two chapters on spectral theories. In the 1970s the group found itself involved in a legal dispute with its publisher, which absorbed a great amount of energy.

Partly because of the very success and impact of the project, the need for its continued development became much less pressing. The name of Bourbaki also started to elicit negative reactions: for many it represented a style to be avoided, rather than emulated. The backlash was gradually felt by the younger members of the group, which probably affected their own willingness to invest their efforts in the project. Grothendieck, for one, wrote openly about it in his memoirs:

I can recall my astonishment when in 1970 I discovered the extent to which the name itself, Bourbaki, had become unpopular within large circles (theretofore unknown to me) of the mathematical world, which considered it more or less a synonym of elitism, of narrow-minded dogmatism, of a cult of 'canonical' form at the expense of concrete understanding, of hermetism, of castrating anti-spontaneity and so on! (Grothendieck undated I, 49)[14]

Grothendieck also disapproved of the way some of his colleagues (possibly mainly Weil), disparaged interests and approaches that differed from the typical Bourbakian ones:

It was only during the sixties that, as I remember, some of my friends would denigrate mathematicians whose work did not interest them as 'bullshitters'. Since this concerned matters hardly known to me at the time, I tended to accept such appraisals at face value, for I was impressed by such off-hand assurance—until the day when I discovered that such and such 'bullshitters' were persons endowed with deep and original minds who had not had the luck of pleasing my brilliant friend. (Grothendieck undated I, 148)[15]

Of course, one must bear in mind that these memoirs were written from a position of total retirement and deep hostility towards not just individual members of Bourbaki, but the scientific community in general.

Bourbaki's *Éléments de mathématique* became a most influential and widely used classic textbook of twentieth-century mathematics. Generations of students learnt their algebra or topology from the treatise. More than that, it was a

14. Je me rappelle encore de mon étonnement, en 1970, en découvrant à quel point le nom même de Bourbaki était devenu impopulaire dans de larges couches (de moi ignorées jusque là) du monde mathématique, comme synonyme plus ou moins d'élitisme, de dogmatisme étroit, de culte de la forme 'canonique' aux dépens d'une compréhension vivante, d'hermétisme, d'antispontanéité castratrice et j'en passe !

15. C'est au cours des années soixante seulement que je me rappelle tel de mes amis, qualifiant d' 'emmerdeurs' tels mathématiciens dont le travail ne l'intéressait pas. S'agissant de choses dont je ne savais pratiquement rien par ailleurs, j'avais tendance à prendre pour argent comptant de telles appréciations, impressionné par tant d'assurance désinvolte - jusqu'au jour où je découvrais que tel 'emmerdeur' était un esprit original et profond, qui n'avait pas eu l'heur de plaire à mon brillant ami.

highly useful work of reference. Further, the fact that Bourbaki chose to include some disciplines in the treatise while omitting others was itself an influential factor in the way that mathematical careers were built in various places around the world. Some readers may have been aware of the connection between the distinctive mathematical style of the text and the unique collective mechanism that produced it. Most of them surely knew, at least, that the Bourbaki enterprise involved something different from other textbooks authored in the standard way. Very few, of course, knew the details of the internal debates and how they had led to the final product. In all likelihood, no one outside the inner circle was aware of the tension and conflicts surrounding the *structures* versus categories question discussed above. But the truly curious point is that for all of its success and impact, the *Éléments* did not become a textbook of choice for the study of analysis, as originally intended by the founding members. Much less was it used by 'all possible audiences: researchers, aspiring school teachers, physicists, and "technicians" of various kinds' as Weil had initially called for. Goursat's *Cours d'analyse mathématique* was superseded, both in France and elsewhere, by more up-to-date textbooks soon after Bourbaki started its activities and in accordance with their original motivation. Students around the world who took traditional introductory courses in differential and integral calculus went on to study from the many texts that became available over the next decades in a multitude of languages and that followed a multitude of approaches, but never did so with the text that the 'Comité de rédaction du traité d'analyse' had had in mind in their early meetings of 1935.[16]

Bibliography

Aubin, David, 'The withering immortality of Nicolas Bourbaki: a cultural connector at the confluence of mathematics, structuralism, and the Oulipo in France', *Science in Context*, 10 (1997), 297–342.

Beaulieu, Liliane, 'A Parisian café and ten proto-Bourbaki meetings (1934–35)', *Mathematical Intelligencer*, 15 (1993), 27–35.

Beaulieu, Liliane, 'Dispelling the myth: questions and answers about Bourbaki's early work, 1934–1944', in Chihara Sasaki *et al* (eds), *The intersection of history and mathematics*, Birkhäuser, 1994, 241–252.

Beaulieu, Liliane, 'Works about Bourbaki', Archives Henri Poincaré, Nancy: Laboratoire de Philosophie et d'Histoire des Sciences, http://www.univ-nancy2.fr/poincare/. Accessed 26 March 2007.

Borel, Armand, 'Twenty-five years with Nicolas Bourbaki, (1949–1973)', *Notices of the American Mathematical Society*, 45 (1998), 373–380.

Bourbaki, Nicolas, *Éléments de mathématique*, 10 vols, Hermann, 1939.

16. An expanded version of this article, including several additional original documents and a detailed section on the origins of Bourbaki's book on General Topology can be downloaded from http://www.tau.ac.il/~corry/publications/articles/Bourbaki%20-%20OHHM.html.

Bourbaki, Nicolas, 'The foundations of mathematics', *Journal of Symbolic Logic,* 14 (1949), 1–8.

Bourbaki, Nicolas, 'The architecture of mathematics', *American Mathematical Monthly,* 67 (1950), 221–232.

Bourbaki, Nicolas, *Theory of sets,* Hermann, 1968.

Bourbaki, Nicolas, *Commutative algebra,* Hermann, 1972.

Bourbaki, Nicolas, *Algebra,* Hermann, 1973.

Bourbaki, Nicolas, *Homological algebra,* Hermann, 1980.

Cartan, Henri and Eilenberg, Samuel, *Homological algebra,* Princeton University Press, 1956.

Chouchan, Michèle, *Nicolas Bourbaki: faits et légendes,* Edition du choix, 1995.

Colmez, Pierre, and Serre, Jean-Pierre (eds), *Correspondance Grothendieck-Serre,* SMF, 2001.

Corry, Leo, 'Nicolas Bourbaki and the concept of mathematical structure', *Synthèse,* 92 (1992), 315–348.

Corry, Leo, 'The origins of eternal truth in modern mathematics: Hilbert to Bourbaki and beyond', *Science in Context,* 12 (1997), 137–183.

Corry, Leo, 'Mathematical structures from Hilbert to Bourbaki: The evolution of an image of mathematics', in Amy Dahan and Umberto Bottazzini (eds), *Changing images of mathematics in history. From the French revolution to the new millennium,* Harwood Academic Publishers, 2001, 167–186.

Corry, Leo, *Modern algebra and the rise of mathematical structures,* Birkhäuser, 2004.

Delsarte, Jean, 'Compte rendu de la réunion Bourbaki du 14 janvier 1935', *Gazette des Mathématiciens,* 84 (2000), 16–18.

Dieudonné, Jean, 'The work of Nicolas Bourbaki', *American Mathematical Monthly,* 77 (1970), 134–145.

Dieudonné, Jean, 'The difficult birth of mathematical structures. (1840–1940)', in U Mathieu and P Rossi (eds), *Scientific culture in contemporary world,* Scientia, 1979, 7–23.

Dieudonné, Jean, and Tits, John, 'Claude Chevalley (1909–1984)', *Bulletin of the American Mathematical Society,* 17 (1987), 1–7.

Eilenberg, Samuel, and Steenrod, Norman, *Foundations of algebraic topology,* Princeton University Press, 1952.

Fricke, Robert, *Lehrbuch der Algebra—verfasst mit Benutzung vom Heinrich Webers gleichnamigem Buche,* F Vieweg und Sohn, 1924.

Goldstein, Catherine, Schappacher, Norbert, and Schwermer, Joachim (eds), *The shaping of arithmetic after C F Gauss's Disquisitiones Arithmeticae,* Springer, 2007.

Goursat, Edouard, *Cours d'analyse mathématique,* 2 vols, Gauthier-Villars, 1903–5.

Grattan-Guinness, Ivor (ed), *Landmark writings in western mathematics, 1640–1940,* Elsevier Science, 2004.

Grothendieck, Alexander, 'Sur quelques points d'algébre homologique', *Tohoku Mathematical Journal,* 9 (1957), 119–221.

Halmos, Paul, Review of Bourbaki (1939–), Book I, Ch. 1–2 (1954), *Mathematical Reviews* 16 (1955), #454.

Halmos, Paul, 'Nicolas Bourbaki', *Scientific American,* 196 (May 1957), 88–99.

Hewitt, John, Review of Bourbaki (1939–), Book IV (1953–55), *Bulletin of the American Mathematical Society,* 62 (1956), 507–508.

Israel, Giorgio, and Radice, Luca, 'Alcune recenti linee di tendenza della matematica contemporanea', in M Daumas (ed), *Storia della scienza. Vol 2: Le scienze mathematiche e l'astronomia,* Editori Laterza, 1976, 162–201.

Jackson, Allyn, 'Interview with Henri Cartan', *Notices of the American Mathematical Society,* 46 (1999), 782–788.

Kaplansky, Irving, Review of Bourbaki (1939–), Book II, Ch. 6–7, *Mathematical Reviews,* 14 (1953), 237.

Krömer, Ralf, 'La "machine de Grothendieck", se fonde-t-elle seulement sur des vocables métamathématiques? Bourbaki et les catégories au cours des années cinquante', *Revue d'histoire des mathématiques,* 12 (2006), 119–162.

Krömer, Ralf, *Tool and object. A history and philosophy of category theory,* Birkhäuser, 2007.

Lakatos, Imre, *Philosophical papers,* 2 vols, Cambridge University Press, 1978.

Mashaal, Maurice, *Bourbaki: a secret society of mathematicians,* American Mathematical Society, 2006.

Mathias, Adrian, 'The ignorance of Bourbaki', *Mathematical Intelligencer,* 14 (1992), 4–13.

Schwartz, Laurent, *A mathematician grappling with his century,* Birkhäuser, 2001.

Senechal, Marjorie, 'The continuing silence of Bourbaki. An interview with Pierre Cartier', *Mathematical Intelligencer,* 20 (1998), 22–28.

Serret, Joseph, *Cours d'algébre supérieure,* Paris: Gauthier-Villars 1846 (2nd ed 1854, 3rd ed 1866).

Van der Waerden, Bartel L, *Moderne Algebra,* 2 vols, Springer, 1930.

Van der Waerden, Bartel L, 'On the sources of my book *Moderne Algebra*', *Historia Mathematica,* 2 (1975), 31–40.

Weber, Heinrich, *Lehrbuch der Algebra,* F Vieweg und Sohn, 1895.

Weil, André, *The apprenticeship of a mathematician,* Birkhäuser, 1992.

Weintraub, E Roy, and Mirowski, Philip, 'The pure and the applied: Bourbakism comes to mathematical economics', *Science in Context,* 7 (1994), 245–272.

INTERACTIONS AND INTERPRETATIONS

7. Intellectual

People and numbers in early imperial China

Christopher Cullen

For a historian of Chinese mathematics, it is a challenging, and indeed slightly depressing experience to read the first chapter of Thomas Heath's *History of Greek mathematics* (1921, 10–25). For there we find as a natural preliminary to the main topic of the book a careful dissection of the stages of evolution of the reference and content of the various Greek terms that Heath decided to translate as 'mathematics', 'mathematician', and so on. Thus we hear how for Plato *mathemata* simply means any subject of study; we hear how on the other hand for Archytas *toi peri ta mathemata* 'those concerned with *mathemata*' were particularly interested in the speed of the stars, their risings and settings, and about geometry, arithmetic, and sphaeric (astronomy), and also music 'for these *mathemata* seem to be sisters'. Though Heath does not use such language directly, he at least believes in the socially and historically produced *discovery* of the concept of mathematics, even if he would probably not concede that it had been socially and historically *constructed*. The chapter concludes with seven lucid pages on the role of mathematics in Greek education—including a note on Plato's view that elementary mathematical education in Egypt was superior to that in contemporary Greece.

Now the point of this is not to recommend Heath's octogenarian study as an up-to-date reference on such questions. Nor is it to suggest that the historiography of ancient Greek culture should be the reference point for studies of all other places and times. My aim is rather to point out how deficient studies of the history of mathematics in China have generally been in comparison with the level of methodological sophistication that has long been established in the study of the ancient Mediterranean world. Most historical writing on the subject in relation to China simply assumes the category of 'mathematics' and the identity of a 'mathematician' as applicable to China throughout its history, and treats them as unproblematically identical to concepts designated by those names elsewhere—whether in a modern cosmopolitan context, in ancient Greece, or in some supposedly ahistorical world of thought.

This has had some pernicious consequences. Texts have been studied on the assumption that we already have all the concepts needed to understand them. The idea that one might need to think carefully about the historical context of such texts is habitually absent—they are to be understood just in themselves, as 'texts on mathematics', as they simply *are*. Such texts are best explicated (it is thought) by translating them into modern symbolic notation; their statements are assumed not just to be equivalent to equations, but to *be* equations. They must contain proofs—because that is one of the main things that 'mathematicians' produce. And so on.[1]

It would be boring for the reader to read a lengthy critique of the work of particular scholars exemplifying these problems, and it would be ungrateful for me to write in that way about researchers who have laboured for years to elucidate the difficulties of technical writing in Chinese. My aim rather will be to offer a sample of a different way of thinking and writing, partly to see if it works from my own point of view, and partly so that readers can judge for themselves whether it is interesting.

But what am I going to think and write *about*? I have already ruled out the idea that there is a priori a universal ahistorical, cross-cultural 'natural kind' called 'mathematics' that can simply be located and studied once one can penetrate the linguistic barrier to see what it is called in Chinese, and on which one can simply impose all the structures and expectations that a modern person finds in the subject called 'mathematics' in twenty-first-century English.[2] By doing that one effectively insulates oneself from learning anything new. To get out of that impasse, I suggest we might try to use the stratagem put forward by Wittgenstein in Aphorisms 66–67 of his *Philosophical investigations* (1958, 31–32):

66. Consider for example the proceedings that we call 'games'. I mean board-games, card-games, ball-games, Olympic games, and so on. What is common to them all?—Don't

---

1. The move away from this style of scholarship as regards Greece began a generation ago: see the seminal discussion by Unguru (1975).

2. A 'natural kind' in philosophical parlance refers to a grouping of things that is natural rather than artificial. Those who believe that this concept is useful would point to such names as 'elephant' or 'the element potassium' as designating natural kinds, and 'the blue things I thought observed before January 2000' as falling outside the concept. See for instance Quine (1969). There is an illuminating discussion from the point of view of the philosophy of science by Hacking (2000, 103–108).

say: 'There must be something common, or they would not be called "games" '—but look and see whether there is anything common to all.—For if you look at them you will not see something that is common to all, but similarities, relationships, and a whole series of them at that. To repeat: don't think, but look! –

[…]

67. I can think of no better expression to characterize these similarities than 'family resemblances'; for the various resemblances between members of a family: build, features, colour of eyes, gait, temperament, etc. etc. overlap and criss-cross in the same way.—And I shall say: 'games' form a family.

And for instance the kinds of number form a family in the same way. Why do we call something a 'number'? Well, perhaps because it has a direct relationship with several things that have hitherto been called number; and this can be said to give it an indirect relationship to other things we call the same name. And we extend our concept of number as in spinning a thread we twist fibre on fibre. And the strength of the thread does not reside in the fact that some one fibre runs through its whole length, but in the overlapping of many fibres.

In essence, my proposal for identifying what the material for the study of 'Chinese mathematics' should be amounts to reading the final paragraph with the word 'mathematics' substituted for 'number'.

Can we identify an activity in ancient China with a family resemblance to what would nowadays be called 'mathematics'? Or was there a self-conscious and publicly recognized group of people in ancient China with a family resemblance to what would be called nowadays 'mathematicians'? What did these people call themselves? What did they consider their defining skill-set, or their common obsession to be? I suggest that such a question may fruitfully be approached prosopographically, and this is mostly the route that I shall take in what follows. But first let me add a methodological note.

If we are to take the Wittgensteinian route, it will be important not to frustrate that approach at the start by smuggling our modern thinking back into the past through the language we use. For that reason, I shall refrain as far as I can from translating any ancient Chinese term as 'mathematics' in the course of the discussion, but shall resort to transliteration where necessary. I shall not be so scrupulous about rendering the word *shu* 數 as 'number' at times, partly because I believe there is a good case to make for a 'family resemblance' for at least part of the range of uses of this word,[3] but partly too because if I commence the investigation with too many unknowns in the equation the subsequent efforts to find a solution will just take more time and space than can here be spared.

---

3. I say here only 'part of the range of uses' because as we shall see below this word can also refer to regularities much broader than numbers themselves, of the kind exploited in divination. The very varied functions of *shu* 數 in Chinese culture are explored in depth by Ho Peng Yoke (1985).

Beginning the enquiry

A good place to start the investigation might be a famous late imperial biographical compendium, the *Chou ren zhuan* 疇人傳 of Ruan Yuan 阮元 first published in 1799,[4] within whose pages one will find biographies of most of the people who are customarily mentioned in modern books with titles such as *A history of Chinese mathematics*. Inspection of the entries reveals, however, that the people discussed fall into a much broader category than would resemble the modern English group 'mathematicians'. Indeed there is a much closer resemblance to Archytas and his *toi peri ta mathemata*. If we look at the first five names of chapter two, which begins at the start of the imperial age in the late third century BC, we find that the first is that of a generalist high official, Zhang Cang 張蒼, who was famed for calendrical skills and administrative accounting; the second is a great historian, Sima Qian 司馬遷, who also functioned as an astrologer and calendrical expert; and the other three—Luoxia Hong 落下閎, Zhang Shouwang 張壽王, and Geng Shouchang 耿壽昌—were primarily concerned with the study of the heavens.[5] The pattern continues in much of the rest of the book. It seems therefore that from the historical viewpoint adopted by Ruan Yuan, there is no sharp division between what we might today call 'mathematics' and forms of intellectual activity that we might want to distinguish from it—such as astronomy. For Ruan Yuan these things were apparently all of a kind, and all could be subsumed under the same book title. Indeed, had he wished to limit the coverage of his work to 'mathematicians' in the modern sense he might have had a problem: while in modern Chinese there is a term *shu xue jia* 數學家 which can unproblematically be treated as an equivalent to 'mathematician', for much of Chinese history the term *shu xue* 數學 referred to the study of numbers purely in the context of divination (*jia* is a suffix meaning '-ist').

Faced with this situation, perhaps we should try following Wittgenstein's advice, and move from thinking about words to 'just looking'—and by that I mean looking at people, to see which of them in early imperial China seem to have a 'family resemblance' to what a modern English speaker would call a mathematician. Having located some likely candidates (if any exist), we might then ask not what features of their existences seem important to us, but how they saw their place in society and intellectual life, as well as how their contemporaries described them. What did they see themselves as doing, and what did it mean

---

4. I use the two volume reprinted edition, Ruan Yuan (1981). In the title '*Chou ren zhuan*' the final element, '*zhuan*' is a common word that can refer to a biographical account, but '*Chou ren*', which denotes the people who are the subject of those accounts, is decidedly odd. It is in fact an archaic and uncommon term whose early usages suggest that it refers to a hereditary lineage of star-clerks and calendar-makers. See for example *Shi ji* 史記 26, 1258–1259 (see below).

5. Most of these persons are discussed elsewhere in this chapter.

to be doing it well? What were the rewards of success, and how did they relate to wider patterns of career and reputation building? How does all this relate to their use of texts? And how did one acquire and pass on knowledge in relevant fields?

Before we begin this attempt at a prosopography, there are some things that need to be made clear since we are talking about early imperial China, a society that had characteristics very different indeed from (say) the Hellenistic world in which Euclid lived.[6] In China from the second century BC onwards, the Chinese state proved massively successful in co-opting intellectuals to serve it directly as officials as well as inducing them to support it ideologically as teachers and writers. In the period before the forced unification by the Qin 秦 dynasty in 221 BC that crushed the independent and culturally diverse states that had existed until then, the master-disciple lineages that were the basis of the so-called 'hundred schools' of thought could count on no consistent support from government quarters. Under the brief rule of Qin itself, violent and partly effective efforts were made to suppress such lineages and interdict the circulation of their texts, the only exception being the group of advocates of *realpolitik* statecraft favoured by the new rulers. The forces that overthrew Qin and founded the long-lived Han 漢 dynasty initially held no articulate ideological allegiance; the first emperor of Han had begun life as a peasant, and appears to have regarded scholars as social parasites whose help was unnecessary in governing his empire. The Han is conventionally divided into Western (or Former) Han, 206 BC–9 AD, and Eastern (or Later) Han, 23–221 AD. The interregnum (as the Han regarded it) was occupied by the abortive attempt to found a new dynasty by Wang Mang 王莽.

As the decades passed and peace and prosperity came, the government faced problems very different from the straightforward suppression of military opposition, and different policies were accordingly adopted. What has been called the 'Han synthesis' had a number of elements, of which the principal ones were the establishment of the centralized civil service as the preferred career route for literate male gentry (supposedly recruited by merit, though not yet by the open examination system of later dynasties) and the privileging of the Confucian intellectual lineage over all others as the bearer of the state's official ideology. If, then, we try to find out about the forms of intellectual life under the early empire, the people who lived that life at an eminent level are highly likely to have spent at least part of their careers as officials in the state service. Added to that undoubted fact is another influence that will magnify the prominence of literati officials in the historical record: those records that we have from the early imperial period were overwhelmingly officially generated by, and concentrate on a world viewed through the concerns of, central government officials. For the period we are considering,

---

6. An excellent general reference on all aspects of the history of China in the Qin and Han periods is Twitchett, Loewe, and Fairbank (1986).

the vast majority of relevant data are found in three so-called 'standard histories' (*zheng shi* 正史, sometimes called 'dynastic histories'), compiled mainly from official documentary sources, often under more or less direct government patronage. The first is the *Shi ji* 史記 'Records of the historian' by Sima Qian 司馬遷, c 90 BC, (I use the Beijing 1962 edition); the second, which overlaps with it up to about 90 BC is the *Han shu* 漢書 'History of the Western Han dynasty' by Ban Gu 班固, largely complete on his death in 92 AD (Beijing 1962 edition), and the third is the *Hou Han shu* 後漢書 'History of the Eastern Han dynasty' by Fan Ye 范曄, c 450 AD (Beijing 1963 edition). Thus added to the fact that many intellectuals were likely to be in the government service is the fact that those who were not are much less likely to appear prominently in the historical record. Those few indications of relevant activity by those unconnected with government are thus extremely precious.

An excellent example to start with is certainly the first person listed in Ruan Yuan's second chapter: Zhang Cang 張蒼, a former official of the Qin 秦 dynasty who was to serve under four Han 漢 dynasty emperors after the fall of Qin in 206 BC. His career included mainline administrative functions such as a governorship, and like many high officials he was given the title of *hou* 侯 (usually rendered as 'Marquis'), which carried the right to an income from the taxes of a particular district. His biography written half a century after his death by Sima Qian, who had succeeded to several of his functions, tells us of his skills:

He had a clear understanding of the charts, writings, accounts and documents of the empire; he was also good at using *suan*, at harmonics and astronomical systems.[7] (*Shi ji* 96, 2676)

His management work, which involved service as *Ji xiang* 計相 'Chancellor in charge of reckoning' needs little explanation: clearly an empire covering most of the territory of modern China found it essential to have officials capable of managing accounts at a high level and generally controlling the flow of data, particularly numerical data, through the administration.

What, however, was meant by 'using *suan*'? The word *suan* 算, of which 筭 is a more archaic written form, is acting here as a noun referring to what have been called 'counting rods'. At this period, long before the medieval introduction of the abacus in China, all calculation was done using short rods arranged on a flat surface to represent numbers according to conventional patterns. For instance, the numbers one to four might be shown by laying down that number of rods together, but five was shown by a single rod at right angle to the others, so that seven might be two horizontals and one vertical. A place value system operated, so that tens were laid out to the left of units, and so on.[8] *Suan* as a noun can also

---

7. 明習天下圖書計籍；又善用算律曆.

8. For a good discussion of how calculations could be carried out with such rods, see Lay Yong Lam and Tian Se Ang (1992).

refer to what one does with these rods, or to the act of doing it. To avoid begging the question in a discussion whose aim is to seek for activities with a 'family resemblance' to what is nowadays called in English 'mathematics', I have declined to commit myself in advance to a single rendering of *suan* such as 'calculation' or 'mathematics', and have simply transliterated the term throughout in the hope that the reader will thus be able to judge its content from the historical evidence rather than on an a priori basis.[9]

But why the calendrical astronomy and harmonics? To understand this aspect of Zhang Cang's skills, it is essential to realize that an important element in the ideological package of the early imperial synthesis was a cosmology that presented the natural order, the order of human society, and its moral underpinnings as essentially one single system. Why should one submit to the rule of the emperor and his officials? The answer given was not (as in some other times and places) that a divine being had put him in place and commanded obedience, nor was it that a rational evaluation of possible different political systems on the East Asian landmass showed that the imperial system maximized human welfare. Rather the argument was that the imperial system was the *natural* pattern of things. Thus a well-governed empire should be part of an orderly cosmos. Part of the expression of that cosmic order was the annual promulgation on the emperor's behalf of a detailed almanac structured round a luni-solar calendar. At a trivial level, this almanac enabled the emperor's officials, as well as the common people, to plan future activities and to record action taken against an agreed frame of temporal reference. But the most important aspect of such documents was that they made it possible for actions to be taken at the moment that was most cosmically favourable for them to take effect. For ordinary people, the actions in question might be getting married, or concluding a business contract; for the emperor, the vital point might relate to the proper timing of the ritual actions by which he actualized and made effectual the unity of the human and cosmic order.[10]

To use one common example of the latter, it became the custom that the Emperor would, at dawn on the day when winter solstice fell, conduct a solemn ceremony of sacrifice and prayer on behalf of his people. To be effective, this ceremony had to be conducted on the right day. Now in astronomical terms, 'winter solstice' is not a day, but a precise instant of time when the annual cycle of the sun round the ecliptic places it at its maximum distance from the north celestial pole. This instant can occur at any time of day. If it occurs five minutes before midnight on (say) Monday, then Monday is the day when the solstice ceremony must take place. If the solstice falls ten minutes later, so that it is five minutes after the midnight with which Tuesday begins, it must take place on the Tuesday.

9. Another common sense of *suan* is a unit of accounting used in reckoning taxation liability, but this sense is not relevant to the present discussion.

10. Introductory discussions of all the issues in this paragraph are presented by Cullen (1996).

Hence it is essential to have a method that can predict such astronomical events as accurately as possible. This was the job done by the structures of astronomical calculation known as *li* 曆, a word for which I prefer the rendering '[astronomical] system'. The *Zhuan Xu li* 顓頊曆 'Zhuan Xu astronomical system', in which Zhang Cang is said to have been expert, had been in use under the Qin, and continued to be the official system until the great reforms of 104 BC, of which we shall shortly hear more.[11]

As for *lü* 律 'harmonics', this was another expression of the unity of the cosmos, but mediated through the sounds of ritual music, and therefore closely linked to the ordering of time through *li*. From the third century BC one can easily trace the construction of a detailed rationale of musical scales supposedly based on the use of pitch pipes of standard dimensions, and of the attempt to locate the underlying numerical pattern of such dimensions in a broad conception of cosmic order.[12] By the Han, this was a standard part of official cosmology, and, as we shall see, various more or less successful attempts were made to link it closely to the basis of astronomical systems.

This then was Zhang Cang's skill-set, so far as it concerns us in our search for an identity in early imperial China with a Wittgensteinian 'family resemblance' to the modern identity of 'mathematician'.[13] As his official biography puts it, 'He was good at the use of *suan*, at *lü*, and *li*' (*Shi ji* 96, 2676).[14] Despite the relatively full documentation on his life, we have no evidence for how Zhang acquired any of his skills, nor do we hear of him having any disciples to whom he passed them on. No source within three hundred years of his death associates him with a book or indeed any other writing apart from official documents, whether as reader, author, or editor.

So to sum up: in our earliest example of an early imperial elite member skilled in the uses of numbers, we find a bundle of abilities rather than a single discipline. The possession of these skills does not confer on Zhang Cang any kind of named identity apart from his official titles, though two thousand years later it wins him a place in the *Chou ren zhuan*. As later examples will show, although there was some tendency for the abilities given here to be found in association with each other in any given individual, having one of these skills did not necessarily imply having all of them. The one skill that was probably considered essential for any main-line official post was *yong suan* (or just *suan*), as we can see from

---

11. For a more detailed discussion of the political and ideological background of debates on calendrical astronomy in the first half of the Western Han, see Cullen (1993).

12. See for instance Major and Cullen (1993, 106–118).

13. Zhang seems to have had other qualifications besides those we consider here, such as good looks sufficiently striking to have spared him from execution early in his career, and the ability to live on mother's milk obligingly supplied by some of his hundred concubines when he was a toothless centenarian (*Shi ji* 96, 2675, 2682).

14. 善用算律曆.

the certificates of proficiency for Han military officers which say of the person in question, *neng shu kuai ji* 能書會計 'He can write and keep accounts' (Loewe 1967). If that was the basic standard for the military, then given the considerable higher status of the civil official we may assume that numeracy was a basic requirement there too.

The next official figure identifiable as numerically skilled was not as lucky as Zhang Cang in his career. This was Sima Qian 司馬遷, who had to make the hard choice between suicide and castration when he offended the Emperor Wu 武 by supporting a luckless general who had failed in his mission. He chose castration, which he saw as the price of being able to complete his massive history of China—or rather the entire world then known to him—the *Shi ji* 史記 'Records of the historian'. His office as *Tai shi* 太史 'Grand Clerk' put him in charge of state annals, but also gave him responsibility for recording and interpreting celestial omens and for maintaining the astronomical system. His historical work includes major monographs on harmonics and astronomical systems. It was under his term of office that the *Tai chu* 太初 'Grand Inception' astronomical reform took place, but it seems he did not succeed in keeping the reform under his own control. Indeed, the records of the resulting controversy reveal that the government called on the advice of experts without any official position at all, including two persons otherwise almost unknown: Tang Du 唐都, Luoxia Hong 落下閎, and twenty *min jian zhi li zhe* 民間治曆者 'experts on astronomical systems from amongst the people' as officially summoned consultants. Tang Du 'distinguished the divisions of the heavens' while Luoxia Hong 'carried out *suan* (*yun suan* 運算) to revise the astronomical system'. And their approach is said to have been based on *yi lü qi li* 以律起曆 'raising up an astronomical system on the basis of harmonics' (*Han shu* 21a, 975).

Here we have clear evidence that despite the fact that the persons said by our historical sources to be learned in *suan*, *lü*, and *li* are mostly officials of some kind, the skill package already identified in the case of Zhang Cang was also to be found amongst commoners. And although we do not know quite what the relations were between Sima Qian himself and these 'commoner consultants', he says himself that one of them, Tang Du, was his own father's teacher in the subject of *tian guan* 天官 'the celestial offices', which is the title given by Sima Qian to his astrological monograph.[15]

The scribal transmission does not appear to give us any surviving writings on *suan*, *lü*, and *li* attributed to identifiable commoner figures of the Western Han. But the situation has improved considerably with the discovery of the collection of writings on *suan* in the tomb of a medium-rank local official of the early second century BC, bearing the label *Suan shu shu* 筭數書, for which I offer the

---

15. *Shi ji* 130, 3288.

English rendering 'Writings on reckoning'.[16] I have published a fully commented text edition and translation of this work, and have discussed its significance at length in a journal article (Cullen 2004; 2007a). It seems clear from the format of the work that knowledge of *suan* was passed on in a written form similar to that found in contemporary medical knowledge transmission—the short independent section or 'textlet' written on one or more bamboo strips, each of which might itself note material evidently taken from several sources. Within each section one may typically find a statement of a problem, followed by its numerical solution, and finally the prescription of a *shu* 術 'procedure, method' by which the solution may be obtained from the data. Sometimes several different procedures may be given for one type of problem. It is important to grasp that each section of this material is more or less independent of the others; these sections were collected together by whoever compiled the 'Writings', apparently from diverse sources, but there is no trace of an effort to write a single and systematic book on *suan* here. Topics covered range from elementary calculations with fractions to the solution of problems involving what we would nowadays call the Rule of False Position, and finding the volume of solids of various forms.

A fascinating feature of the 'Writings on reckoning' is that a few of the bamboo slips are attributed to one of two people with the common surnames Wang 王 and Yang 楊; two of the attributed sections contain problems on weaving which are so similar in tone and content that it is hard to resist the conclusion that Wang and Yang were in conscious competition with one another—and the fact that one of them makes a mistake in his method of solution that remains uncorrected suggests that the material we have has not passed through much of a filtering process between the authors and the text we have today. But although the 'Writings on reckoning' seems to be good evidence of the existence of individual experts in *suan* at the time of its composition, we have no evidence of their social role or status, nor do we know who they were trying to impress as they competed in showing their skill in *suan*. There is no evidence at this stage of systematic writing on *suan* at book length.

Further evidence from the Western Han underlines *suan* as a skill to be possessed by the successful official. The clear association between the practice of *suan* and the use of the counting rods is illustrated by the fact that someone with the ability to perform *xin ji* 心計 'mental calculation' was considered exceptional enough to qualify for special appointment to a government position at an early age: this is how the specialist in state finance Sang Hongyang 桑弘羊

16. The compound expression *suan shu* 算數 is well attested, and functions as one would expect from the two words from which it is composed, which suggest 'applying *suan* to numbers'—hence 'reckoning'. *Shu* 書 is given here its common sense of 'writing' rather than 'book', since it is clear that the collection of material before us is simply a collection of disparate small units of text rather than a book in any normal sense of the word. For this reason too I refer to the words *Suan shu shu* as a label rather than a title.

first came to official notice in his teens (*Han shu* 24b, 1164; *Shi ji* 30, 1428).[17] For Geng Shouchang 耿壽昌 we have good evidence of his skill in *suan* relating to civil engineering and administration. The *Han shu* tells us that he:

…was good at carrying out *suan* and was able to estimate the profits from works…he was practiced in questions of the estimation of works and the allocation of money.[18] (*Han shu* 24a, 1141)

A few sentences later, Geng is credited with the proposing the establishment of *chang ping cang* 常平倉 'Ever Normal Granaries' to minimize variations in the price of grain. Geng Shouchang was also an observational astronomer whose work on the motions of the moon was discussed by Jia Kui 賈逵 in 92 AD (*Hou Han shu* 2, 3029); the *Han shu* bibliography (see below) also records the presence in the imperial library of two hundred and thirty-two rolls of charts and two rolls of numerical data by him (*Han shu* 30, 1766). Xu Shang 許商, an expert in river works and irrigation, held various high offices under Emperor Cheng (33–7 BC); in the autumn of 29 BC he was called in to advise on flood control, and it was said of him:

The Academician Xu Shang worked on the *Shang shu* [sc. the 'Book of documents', one of the Confucian classics]; being good at *suan*, he was able to plan the use of [resources in public] works.[19] (*Han shu* 29, 1688)

We also have evidence that he wrote on *suan* at some length—as discussed below.

When we reach the transition between the Western and Eastern Han dynasties, we find another figure who ranks in importance with Zhang Cang and Sima Qian, but with the additional feature that he presents us with lengthy reflections on his relations to the disciplines he practices. That figure is Liu Xin 劉歆.[20] As a youth, he was a friend of Wang Mang 王莽, who rose to power through his family connections with the empress. In 9 AD, after having been the effective ruler for some years, Wang caused the boy who held the throne as nominal Han sovereign to go through a ceremony of abdication in his favour, and took the throne as emperor with the title *Xin* 新 'New' for his dynasty (*Han shu* 99a, 4099–100). Liu Xin became one of Wang's chief ministers, and provided intellectual support for his claim to be patterning his government on the model of antiquity. Despite his membership of the old imperial clan, Liu survived a number of political purges during the next three decades, including some involving his own children, until in 23 AD he was executed after the discovery of a plot to assassinate Wang and restore the Han to power.

17. Another example of this was Liang Qiuhe 梁 丘 賀, who was active in the period 73–47 AD and was appointed to (honorary?) military rank as a result of his ability in *xin ji* (*Han shu* 88, 3600).

18. 善為算能商功利…習於商功分銖之事.

19. 博士許商治尚書, 善為算, 能度功用.

20. The dates of individuals are given in the Appendix.

Before turning to Liu Xin's own writing, we may note in passing the significance of a project in which he was involved jointly with his father Liu Xiang 劉向. This involved the classification of all books in the imperial collection—some would even say the construction of at least some of those books—and the making of a systematic classified catalogue, preserved for us in abbreviated form in chapter 30 of the *Han shu*. Although most of the books mentioned in this listing have now been lost, there is still much to be learned from the book titles themselves, and also from the way they are classified. In particular, we may ask whether we see signs of the recognition of *suan* as an independent category of knowledge on which it would be appropriate to write books.

The answer is generally negative. At first one might be encouraged to find that a major portion of the catalogue is devoted to *shu shu* 數術, a term which one might be tempted to translate literally as 'numerical procedures/methods'. But immediately we look at the subheadings of the section and at the books listed therein, it is evident that *suan* is not the main or even a major topic here. As already hinted, *shu* 數 can refer to much more than 'numbers' in the sense of things one might count or measure with, and has strong connections with divinatory activity. Titles of the subsections of the *shu shu* section are as follows: *tian wen* 天文 'celestial patterns' (astrology), *li pu* 曆譜 'astronomical systems and listings' (calendrical astronomy), *wu xing* 五行 'five phases' (divinatory cosmology), *shi gui* 蓍龜 'milfoil and tortoise' (divination using these), *za zhan* 雜占 'miscellaneous divination', *xing fa* 形法 'the method of forms' (divination by visual inspection of landscapes, faces, and so on).

Within all this overwhelmingly divinatory material we can recognize the presence of *li*, *lü*, and *suan*, but the emphasis is not even. *Li* certainly dominates the second section, but *lü* appears in only three titles out of 31 titles in the third (we also find a book title referring to bells, which is probably also relevant to *lü*). *Suan* is tucked away at the end of *li pu* in the following two entries:

Xu Shang's *suan* procedures, 26 rolls
Du Zhong's *suan* procedures, 16 rolls.[21] (*Han shu* 30, 1766)

Xu Shang is almost certainly to be identified with the official mentioned above, but Du Zhong is otherwise unknown. Both books are now lost, and we have no known quotations from them in extant works.

What are we to make of this situation? One message is clear: *suan* was not a very important subject in the literary repertoire of elite knowledge at the time this listing of books was made. That much is clear from the fact that only two books refer to it. What we are to deduce from the fact that books on *suan* are

21. 許商算術二十六卷.
杜忠算術十六卷.

classified under 'astronomical systems and listings' is not obvious. While it is possible that the '*suan* procedures' they contain do relate to astronomical practice, the stated interests of Xu Shang do not support this, and it may be safer to assume that the two Lius, faced with the problem of where to put these books, simply placed them with others that contained calculations even if the 'family resemblance' was otherwise not a close one.

The main material by Liu Xin that we shall examine stems from a great conference of the learned summoned by Wang Mang in 5 AD, as part of his campaign to build himself into the image of an ideal Confucian minister and potential ruler before seizing the reins of power:

He summoned those in the empire who had a comprehensive knowledge of lost classical texts, ancient records, celestial patterns, *suan* [pertaining to] astronomical systems, bells and pitchpipes, 'elementary studies', historical compilations, recipes and procedures, pharmacognosy, and who professed the Five Classics, the Analects [of Confucius], the Classic of Filial Piety, and the Literary Expositor.[22] (*Han shu* 12, 359)

Liu Xin's report on the proceedings of this gathering is preserved in *Han shu* 21a, although the editors state that it has been subject to some censorship. The theme of the whole text is the role of numbers in the cosmic order. As Liu says:

Numbers are 1, 10, 100, 1000, 10,000. They serve to reckon affairs and things, and to accord with the patterns of inborn nature and fate. The '[Book of] documents' says: 'First one [performs] *suan* [to determine] fate'. The basis arises from the numbers of the Yellow Bell. Begin from one and triple it; pile up one tripling upon another until you have passed through the numbers of the 12 *chen*, [then with] 177,147 the five numbers are complete.[23] In the practice of *suan* one uses bamboo [rods] 1 *fen* in diameter, and of length 6 *cun*. 271 rods form a hexagon, which makes a bundle.[24] In diameter they image the 1 of the supernal pitchpipe Yellow Bell; in length they image the length of the chthonic pitchtube Forest Bell.[25] Their number comes from 50, the Great Expansion number of the *Changes*, of which 49 is used, and forms the 6 lines of the Yang, so as to attain to the image which runs through the 6 Voids.[26] In deriving astronomical systems, producing pitchpipes and making vessels, encompassing the circle and setting right the square, weighing the heavy and balancing

---

22. 徵天下通知逸經、古記、天文、曆算、鍾律、小學、史篇、方術、本草及以五經、論語、孝經、爾雅教授者.

23. The twelve *chen* are the cyclical signs that can represent such things as the double-hours into which the day is divided, the months of the year, or the compass directions. The first *chen* corresponds to unity, after which there are eleven multiplications by 3, giving 3^{11} = 177,147.

24. Each rod is a hexagonal prism; the reader may like to carry out the exercise of verifying that a bundle of 271 such rods will itself be such a prism.

25. The standard series of resonating tubes for musical pitch were theoretically generated by beginning from a tube of diameter one *cun* 寸 (approximately an inch) and length 9 *cun*. Subsequent lengths were defined by in effect multiplying lengths alternately by 2/3 and 4/3. The Yellow Bell Pipe and all other odd-numbered pipes in the sequence were labelled as cosmically *yang* 陽 and given the title *lü* 律, whereas the even-numbered pipes (of which the first was Forest Bell) were labelled as cosmically *yin* 陰 and given the title *lü* 呂.

26. The reference here is to the divinatory procedures using yarrow stalks set out in the 'Book of changes' (*Yi jing* 易經), an ancient divinatory text which by the Han period had become the basis for an elaborate numerical cosmology. For detailed explanations, see Nielsen (2003).

what is equal, levelling the [builder's] line and making fair the capacities, in 'Seeking the profound and drawing out the obscure, hooking up from the deep and attaining to the distant'—there is nothing in which [numbers] are without application. [By this means] in measuring the long and short one will not be out by a hair's-breadth, in estimating much and little one will not be out by a mere pinch, in weighing the light and heavy one will not be out by a tiny grain. [Numbers] take their guiding thread from one, unite in ten, grow in a hundred, become great with a thousand, and overflow into a myriad. The methods [for handling them] are found in *suan* procedures. These are found throughout the empire, and are taken as a pattern in elementary studies.[27] Responsibility for such matters lies with the Grand Clerk, and they are handled by the *Xi He* official.[28] (*Han shu* 21a, 956)

This introductory statement is played out in detail in the pages which follow, giving us a priceless insight into the view of numbers taken at the highest levels of the elite to which Liu Xin belonged—for the fact is that it was Liu Xin himself who held the office of *Xi He*, a role whose title had been chosen by Wang Mang to recall the names of the star-clerks charged with control of calendrical astronomy by the legendary Emperor Yao in high antiquity (*Han shu* 99b, 4103; 99a, 4090). It is to his skill in *li* that we owe the Triple Concordance astronomical system (*San tong li* 三統曆), the earliest preserved example of a system with a complete basic planetary theory.[29] Part of his essay is in fact devoted to an attempt to derive the basic constants of the Triple Concordance system from what Liu considers the 'basic theory' of the 'Book of changes'. But of *suan* itself Liu Xin says little directly. It is done with counting rods, and its practice is based on a repertoire of '*suan* procedures', which form part of 'elementary studies'—which we may recall was the department of learning named immediately after '*suan* [pertaining to] astronomical systems, bells and pitchpipes', namely *li* and *lü* in the syllabus of Wang Mang's conference.

If we continue our prosopographical search into the Eastern Han, some features already revealed in the Western Han persist. Thus the most conspicuous form of activity involving the use of numbers amongst the elite continues to be *suan* in relation to astronomical systems, *li suan*. Skill in *suan* alone is much less conspicuous. We do however find evidence that experts in *li* reflected on the extent to which *suan* procedures could be applied to it, as in this extract from a discussion by Jia Kui 賈逵 around 100 AD. In considering the changing position of the winter solstice, he stresses the necessity for periodic revisions of

27. 'Elementary studies' is my rendering of the term *xiao xue* 小學, which also refers to the schools a child traditionally entered at the age of eight. The *Han shu* listing of books under this heading contains works on writing, and the subsequent editorial discussion focuses solely on that topic. (*Han shu* 30, 1719–1721).

28. 數者，一、十、百、千、萬也，所以算數事物，順性命之理也．書曰：「先其算命．」本起於黃鐘之數，始於一而三之，三三積之，歷十二辰之數，十有七萬七千一百四十七，而五數備矣．其算法用竹，徑一分，長六寸，二百七十一枚而成六觚，為一握．徑象乾律黃鐘之一，而長象坤呂林鐘之長．其數以易大衍之數五十，其用四十九，成陽六爻，得周流六虛之象也．夫推歷生律制器，規圓矩方，權重衡平，準繩嘉量，探賾索隱，鉤深致遠，莫不用焉．度長短者不失豪氂，量多少者不失圭撮，權輕重者不失黍絫．紀於一，協於十，長於百，大於千，衍於萬，其法在算術．宣於天下，小學是則．職在太史，羲和掌之．

29. See Cullen (2004a) for a discussion of the Triple Concordance system and its applications.

astronomical systems, since none of them can stay permanently in step with the cosmos. His reasons for this are nothing to do with some mysterious ineffability of the heavens or limitations of human reason, but are purely technical, and to do with the way that *suan* treated fractional quantities:

The alignments and discrepancies of the way of heaven are not commensurable. There must be remainders, and those remainders [themselves] will themselves differ in ways that cannot be made commensurable through equalising. Those who manage *li* [arbitrarily] chop off a period of 76 years [for commensurability of day, lunation and solar cycle], so the fractional remainders gradually increase until they build up to a day.[30] (*Hou Han shu* 2, 3028)

The terms *qi* 齊 'commensurable' and *deng* 等 'equalize' form part of the technical vocabulary of the manipulation of fractional quantities in ancient Chinese *suan* practice. Two quantities are 'commensurable' in this sense if one can find a denominator that enables the fractional parts of both quantities to be represented precisely. As part of the process of arriving at this denominator in its simplest terms, one will seek for a common factor or *deng shu* 等數 'equalizing number'.[31] Jia Kui's point is that if one demands that all astronomical cycles have to be commensurable within some fixed period (such as the 76-year period used by systems of the type Jia Kui is discussing here), there are bound to be discrepancies between the real value of a quantity and its representation in simple fractional terms, and that in time these discrepancies will build up to a significant amount.

Nevertheless, there are some interesting changes in emphasis during this period, which I shall try to bring out in what follows. One unexpected feature revealed by searching for the word *suan* in the *Hou Han shu* (which uses the form 筭 rather than 算 for *suan*) is that the first people in Chinese history identifiable as forming a student/teacher pair in the subject of *suan* are women. They are the Empress Dowager Deng 鄧 and Ban Zhao 班昭. The Empress's father had given his daughter a literary education along with her brothers, to the displeasure of her mother, who reproached her for her lack of interest in 'women's work' by asking sarcastically 'Are you going in for a *bo shi* 博士 "doctorate"?' Ban Zhao was the sister of Ban Gu 班固, the scholar primarily responsible for the *Han shu*; on his death Ban Zhao was commissioned to finish the work, and wrote portions of it including the monograph on astrology *tian wen*. We are told in the Empress's biography:

From the time that the Empress entered the Palace Apartments, she took lessons from Her Excellency Madame Cao [namely Ban Zhao, who was the widow of a Mr Cao] in classical writings, together with *tian wen*, and the application of *suan* to numbers.[32] (*Hou Han shu* 10a, 424)

---

30. 天道參差不齊，必有餘，餘又有長短，不可以等齊．治曆者方以七十六歲斷之，則餘分消長，稍得一日.

31. For a discussion of how a *deng shu* is to be found, see below.

32. 太后自入宮掖，從曹大家受經書，兼天文、筭數.

Outside the palace there are also signs that *suan* was beginning to be seen as a topic of serious interest amongst the male master–pupil scholarly lineages that dominated classical studies. When Zheng Xuan 鄭玄 sought admission to the body of more than four hundred students gathered round the great scholar Ma Rong 馬融, he was for three years kept out of the core group of fifty who met Ma face to face—until Ma heard that he was *shan suan* 善筭 'good at *suan*' and allowed him in (*Hou Han shu* 35, 1207). Zheng was to become, like Ma, an immensely influential figure in the intellectual life of his time, and his commentaries on classical texts were to be cited for centuries to come; in effect, he became *the* Han explicator of the classics.

It is in this milieu that we first hear of a named book on *suan* that is still extant today. This is the *Jiu zhang suan shu* 九章算術. The title of this book has been rendered into European languages in many different ways. Although the most literal English rendering would in my view be something like 'methods/procedures for *suan*, under a ninefold division', the word *zhang* is most frequently rendered as 'chapters' in this context, so that English-speaking readers often call the book the 'Nine chapters' for short. A brother of Ma Rong, Ma Xu 馬續, is said to have studied this book as a youth around 110 AD, and it likewise formed part of the education of Zheng Xuan (*Hou Han shu* 24, 862; 35, 1207). We know that Ma Rong was in touch with Empress Deng and Ban Zhao, and that Ma Xu, like Ban Zhao, also worked on the *tian wen* monograph in the *Han shu* (*Hou Han shu* 10, 3215), so given all these contacts it is likely that Empress Deng and Ban Zhao would have known the 'Nine chapters' too.

My main interest in this discussion is the people who wrote and used texts rather than the texts themselves, so I shall not give a lengthy description of the 'Nine chapters' here. The material in the book is clearly from the same tradition as the *Suan shu shu* 筭數書 'Writings on reckoning' described earlier, and the problem-answer-method pattern of parts of the earlier collection is dominant throughout the later book. In terms of mathematical sophistication, the 'Nine chapters' goes somewhat further than its predecessor, mainly by adding material on problems that would nowadays be formulated as systems of linear equations in several unknowns, and on applications of what is called in the West the Pythagorean theorem. As its name implies, the book is divided into nine sections, but the contents of a given section do not always form a natural unity. In fact the main reason for the nine-fold division seems to be that the anonymous compiler of this work wanted to imitate what he or she believed was the pattern of mathematical education laid down by the sage rulers of antiquity. The 'Ritual of Zhou', *Zhou li* 周禮, is nowadays thought to be a text of the late Warring States or early imperial age, perhaps from the third century BC. But at the time when the 'Nine chapters' first appears in the historical record the *Zhou li* was thought to be an accurate description of government organization in the early Western

Zhou dynasty around 1000 BC, written by one of the dynastic founders, the Duke of Zhou, Zhou Gong 周公. In that work, the education of young aristocrats is said to include the *liu yi* 六藝 'Six Arts', which comprise 'the five rituals, the six musics, the five archeries, the five chariot-drivings, the six writings, and the *jiu shu* 九數 "nine numberings/reckonings"' (*Zhou li, Shi san jing zhu shu* ed. of 1815, 212–3). The *Zhou li* itself gives us no idea what the last of these might be, but in view of the likely sophistication of the students imagined in the text one would be surprised to find that this expression (of which there is no other attestation before the time when the 'Nine chapters' appears) meant much more than the multiplication table. The usual term for the multiplication table is *jiu jiu* 九九 'the nine nines', but in the *Guan zi* 管子 book—a collection of earlier material edited by Liu Xiang c 26 BC—it is referred to as *jiu jiu zhi shu* 九九之數 'the numbering/reckoning of the nine nines', a term suggestive of the *Zhou li* usage (Rickett 1998, 499). However from the Eastern Han onwards, all commentators gloss *jiu shu* with a list of the section headings of the 'Nine chapters', or some variant thereof. Presumably this was part of the Han scholars' project to 'reconstruct' a lost antiquity in as much detail as possible.

For all the importance later ascribed to the 'Nine chapters' in modern works on the history of mathematics, it is striking how little is said about it for the remaining century or so of the Eastern Han dynasty after its first appearance around 100 AD. Apart from the instances already cited, there is no contemporary evidence of links between this book and Eastern Han people who definitely possessed a high degree of numeracy. In the case of the polymath astronomer, maker of instruments, and litterateur Zhang Heng 張衡, we only have a reference in a commentary on the 'Nine chapters' written a century later, referring to 'Zhang Heng's *suan*' and '[Zhang] Heng's *shu* 術 "method/procedure"' in connection with a discussion of the volume of a sphere.[33] No direct connection between Zhang Heng and the 'Nine chapters' itself can be made. The most eminent practitioner of astronomical calculation later in the dynasty was undoubtedly Liu Hong 劉洪. According to the polymath and bibliophile Cai Yong 蔡邕, who worked with him closely on astronomical matters, Liu Hong *mi yu yong suan* 密於用筭 'was accurate in the use of *suan*' (*Hou Han shu* 3, 3083 commentary). The only connection with the 'Nine chapters' known in his case is that a Buddhist monk writing several centuries later referred to a book he called Liu Hong's *Jiu jing suan shu* 九京算術 *suan* 'procedures/methods of the Nine Capitals', which seems a probable copyist's error (京 ← 章) for *Jiu zhang suan shu* 九章算術 (Guo 1992, 128). But since there are no other attestations of a work under this title by Liu Hong, it seems unlikely on the basis of this late and garbled reference that such a book actually

---

33. Chemla and Guo (2004, 384–5). I see no reason to assume with these authors that 'Zhang Heng's *suan*' is a book title.

existed. We do hear of Liu Hong's astronomical methods being passed on by Xu Yue 徐岳 to Gan Ze 闞澤 (*Jin shu* 17, 503), but although later sources mention (presumably commented) versions of the 'Nine chapters' by both the latter two men, there seems to be no direct evidence that their learning in that area came from Liu Hong.[34]

As we have seen in following the story through from Zhang Cang at the beginning of the Han, through Empress Deng and Ban Zhao, and onwards through Zhang Heng to Liu Hong, during the four centuries of the Han *suan* in itself was an element whose presence in official and intellectual life was continual but not major. If the history of Han 'mathematics' is to be defined as the history of *suan*, then that history in itself is not a very rich one. We know relatively few people whose main claim to fame was their skill in *suan*, and we do not know the name of anyone responsible for a major innovation in the field. Apart from the single instance of Wang and Yang in the 'Writings on reckoning' we have no evidence of contention or competition in *suan*. Claiming to have originated a new piece of *suan*, or claiming that one's teacher had done so, was evidently not a worthwhile manoeuvre in Han intellectual life: the contrast with ancient Greece is striking. That certainly does not mean that early imperial society was innumerate, or that numbers did not enter into intellectual life. The skill-package we first identified in the case of Zhang Cang and saw fully played out in the case of Liu Xin—*li*, *lü*, and *suan* taken together—was certainly an intellectually significant element in the life of the Eastern Han, as in the Western Han. Its significance was not confined to scholarly writing alone, but could on occasion spill into the area of open debate, in which technical points would be argued out between opponents in front of an audience.[35] Certainly such debates would have been meaningless to those without skill in *suan*, even though the focus was on the results produced by its use in the area of *li* rather than on the process of *suan* in itself. As one such debater (Cai Yong 蔡邕) remarked around 180 AD:

When one begins dealing with *lü* and *li*, one takes *suan* with counting rods[36] as the basis, and uses the patterns of heaven as a check.[37] (*Hou Han shu* 3, 3083 commentary)

Numbers mattered a great deal in the early Chinese empire, even though the way they mattered was not always the same as in other areas of the ancient world.

---

34. I am not convinced by the conjecture to this effect by a Qing dynasty commentator; see Guo (1992, 128).

35. See Cullen (2007b). A large amount of material from memorials and oral debate on this topic is preserved in *Hou Han shu*, zhi 2.

36. Here the alternative word *chou* 籌 is used to refer to the counting rods themselves, which leaves *suan* to refer to the act of manipulating the rods. The motivation here is probably the literary one of wanting another two-character phrase to parallel *tian wen*.

37. 先治律曆，以籌算為本，天文為驗。

Given the fact that *suan* in the period we are discussing was an essential skill but does not appear to have been a major focus of intellectual attention in itself, it is not surprising that the extant Han dynasty written material on *suan* (which largely consists of the 'Writings on reckoning' and the 'Nine chapters') is almost entirely concerned with telling us what methods should be used to solve problems, but says little or nothing directly about why those methods should work, or how new methods might be constructed. The only example of material of the latter kind, which perhaps we might call meta-*suan*, from the Han dynasty comes in the form of a dialogue between two persons who are almost certainly fictional. The material in question is found in the scribally transmitted book known as the *Zhou bi* 周髀 'Gnomon of Zhou', which is mainly concerned with quantitative cosmography and calendrical astronomy. In part of this book, a part which I have argued is probably from the first century BC (Cullen 1996, 148–156), a teacher, Chen Zi 陳子, is represented as telling his student Rong Fang 榮方 how to learn *suan*. Neither of these characters is known to history.

Long ago, Rong Fang asked Chen Zi, 'Master, I have recently heard something about your Way. Is it really true that your Way is able to comprehend the height and size of the sun, the [area] illuminated by its radiance, the amount of its daily motion, the figures for its greatest and least distances, the extent of human vision, the limits of the four poles, the lodges into which the stars are ordered, and the length and breadth of heaven and earth?'[38] (after Cullen 1996, 176)

The key to all this, Chen Zi explains, is the knowledge of *suan*:

All these things can be attained to by *suan* procedures/methods. Your [ability] in *suan* is sufficient to understand such matters if you sincerely give reiterated thought to them.[39] (after Cullen 1996, 177)

For Chen Zi, according to the explanation he goes on to give, the essence of relating to numbers is the ability to *tong lei* 通類 'generalize categories'. From this follows the key ability in using *suan* effectively—*neng lei yi he lei* 能類以合類 'being able to categorize in order to unite categories'. To do this, *tong shu xiang xue tong shi xian guan* 同術相學同事相觀 'similar methods are studied comparatively, and similar problems are comparatively considered'. And thus one is enabled to understand *suan shu zhi shu* 算數之術 'procedures for applying *suan* to numbers' (Cullen 1996, 177). All this might stand as a user's guide to the 'Writings on reckoning',[40] in which similar problems and similar methods may be gathered together in the same short text. And the more systematic arrangement of the 'Nine chapters'

38. 昔者．榮方問于陳子．曰．今者竊聞夫子之道．知日之高大．光之所照．一日所行．遠近之數．人所望見．四極之窮．列星之宿．天地之廣袤．夫子之道．皆能知之．其信有之乎.
39. 此皆算術之所及．子之于算．足以知此矣．若誠累思之.
40. As noted above, in the label 'Writings on reckoning', 'reckoning' renders *suan shu* 算數, the expression which begins the phrase just quoted.

certainly carries the process further. We may note, however, that in telling his student how to study *suan*, Chen Zi says nothing about any kind of standard book to be studied; his instructions are a better fit with the earlier world of the 'Writings on reckoning', in which it is up to the student to assemble related material, than they are with the 'Nine chapters', in which the job has been done once and for all.

Liu Hui: a new departure

No-one in either the Western or Eastern Han dynasties seems to have written further on *suan* in any mode resembling Chen Zi's methodological reflections. After the end of the Han, in the third century AD we encounter a step change of discourse in which Chen Zi's programme (although not mentioned explicitly) is carried out with immense energy and in great detail. Somewhere around 263 AD, a man called Liu Hui 劉徽, otherwise unknown to history, wrote a lengthy and detailed commentary on the 'Nine chapters'. Alone of any other commentaries on this work from antiquity, this one has survived to our own day.[41] The commentator's preface is an entirely new departure in itself, since there we find not only the first recorded reflections of a real historical figure on *suan* in itself (as opposed to the package involving *li* and *lü*), but are also given a programme of explication and justification unlike anything attempted before. One of the first things we may choose to note from Liu Hui's essay is a confirmation of the remarks already made here about the lack of intellectual prominence of *suan* in the centuries before he wrote. Liu writes in tactful terms, but his meaning is quite clear:

Now *suan* was amongst the Six Arts,[42] and anciently those who were worthy and expert [in it] were promoted as retainers, and it was used in training the young of a state. Even though one [only] speaks of the Nine Numberings, its potential exhausts the arcane and enters into the subtle, and its ability to investigate is unlimited. When we come to consider the passing on of such methods, it is like the fact that the compass, trysquare,[43] and measures of length and capacity can be shared by all. There is nothing specially difficult in doing this. But those who favour [such studies] nowadays are few. Thus although there have been many of comprehensive talents and penetrating learning in our age, they have not yet always been able to take a full view of it.[44] (after Chemla and Guo 2004, 127)

He begins his account of his own work, however, by telling us of his own route to mastery of the 'Nine chapters':

---

41. Two translations of the 'Nine chapters' that include Liu Hui's commentary and give much additional materials are Shen, Crossley, and Lun (1999) and Chemla and Guo (2004); the latter, which includes Chinese text, is better based in sinological terms, but see also my essay review of some associated problems (Cullen 2006).

42. See the reference to the *Zhou li* above.

43. An L-shaped tool used to set out right angles by carpenters and other craftsmen.

44. 且算在六藝，古者以賓興賢能，教習國子．雖曰九數，其能窮纖入微，探測無方．至於以法相傳，亦猶規矩度量可得而共，非特難為也．當今好之者寡，故世雖多通才達學，而未必能綜於此耳．

When I was young I was drilled in the 'Nine chapters' and when I grew up I went over it again carefully. I looked into the breaking apart of Yin and Yang, took a comprehensive view of the basis of *suan* methods/procedures, and of the suppositions involved in seeking the unknown, and thus attained to realisation of its significance.[45] (see also Chemla and Guo 2004, 127)

The words used here by Liu Hui are significant. In his youth, he approached the 'Nine chapters' through *xi* 習, which describes a process of learning how to do something (such as reciting a text) through repetition and practice. In his maturity, he attained a different level of insight, described here as *wu* 悟, a word used in Buddhist discourse for the sudden break-through to enlightenment. What he aims to do in his book is to help other mature minds to make the same leap:

Therefore I have ventured to exert my meagre capacities to the utmost, and have selected from what I have seen in order to make a commentary. The categories under which the problems [treated herein fall] extend each other [when compared], so that each benefits [from the comparison]. So even though the branches are separate they come from the same root, and one may know that they each show a separate tip [of the same tree]. Now I have analysed principles with explanations, and dissected forms using diagrams, to make them simple and comprehensive, thorough and unconfusing, so that those who review them will find their ideas are more than half-way there.[46] (see also Chemla and Guo 2004, 127)

Essentially speaking, Liu Hui sets himself the task of convincing the reader of the 'Nine chapters' that its methods will work, and making it clear why they work. Let us take two simple but very different examples to demonstrate how he does this. The first is his explanation of the method of simplifying fractions prescribed in the 'Nine chapters'. The algorithm is given as follows:

Method for simplifying parts: What can be halved, halve them. As for what can not be halved, separately set out the numbers for the denominator and numerator. Subtract the lesser from the greater. Go on decreasing by subtracting from one another. This is seeking for equality. Simplify using this equal number.[47] (see also Chemla and Guo 2004, 156–7)

This is the mutual subtraction algorithm well-known from a number of cultures. But why does it work? Liu Hui tells us:

Simplifying by the equal number means dividing [by it]. The reason for the subtracting of one [number] from another is because they are both simply piled up accumulations of the equal number.[48] (see also Chemla and Guo 2004, 157)

---

45. 徽幼習九章，長再詳覽．觀陰陽之割裂，總算術之根源，探賾之暇，遂悟其意.
46. 是以敢竭頑魯，采其所見，為之作注．事類相推，各有攸歸，故枝條雖分而同本榦者，知發其一端而已．又所析理以辭，解體用圖，庶亦約而能周，通而不黷，覽之者思過半矣.
47. 約分術曰：可半者半之，不可半者，副置分母子之數，以少減多，更相減損，求其等也。以等數約之。
48. 等數約之，即除也．其所以相減者，階等數之重疊.

In other words, if two numbers are commensurable, then they can each be regarded as made up of whole numbers of some common subunit, which it was the object of the subtraction process to reveal.

In a more complex case, Liu Hui is faced with justifying to his reader the relation between what we would call in modern English the base, altitude, and hypotenuse of a right angled triangle—which are for Liu Hui respectively the *gou* 勾 'hook', *gu* 股 'leg', and *xian* 弦 'bowstring'.

> The *gou* multiplied by itself makes the red square, and the *gu* multiplied by itself makes the blue square. Let there be taking out, and putting in, and being made complete, each following its kind. Thus one reaches [a state where] the differences no longer are to be adjusted. Together they form the area of the *xian* square. Find the side of this square, and this is the *xian*.[49] (see also Chemla and Guo 2004, 704–5)

It is clear that Liu Hui assumes his readers can see a diagram (now lost) in which the two smaller squares on the *gou* and the *gu* are in some way dissected and reassembled to form the larger square. For him, no further explanation is necessary; we are not in an intellectual milieu where the structure of axiomatic-deductive proof is an end in itself.[50] Liu Hui does not take an uncritical approach to his text. Thus when the 'Nine chapters' informs us that to find the area of a circular cap of a sphere we should mutiply diameter and circumference, and divide by four, he begins his discussion by saying bluntly, *ci shu bu yan* 此術不驗 'This procedure does not check out', and goes on to explain why (Chemla and Guo 2004, 188–189). Further, he is prepared to start from a relatively simple feature of the 'Nine chapters', such as the assumption that the circumference of a circle is precisely three times the diameter, and launch himself into a lengthy and (as he acknowledges) completely original discussion in which he specifies the construction of polygons of an increasing number of sides within a circle, and envisages carrying the process as far as a 3072-gon (Chemla and Guo 2004, 187–193).

My object in placing this very brief glance at the nature of Liu Hui's work at the end of my prosopographical reconnaissance has simply been to point out how it is entirely unlike anything we can trace in the preceding four and a half centuries over which we have sketched the work of practitioners of *suan* in China. We are in the presence of something quite new. This should not be a surprise to us, because Liu Hui was in fact living in an intellectual world whose parameters were radically changed from those of the Han dynasty scholars whose work we have been discussing so far.

---

49. 句自乘為朱方。股自乘為青方。令出入相補，各從其類。因就其餘不移動也，合成弦方之幕。開方除之，即弦也。

50. The reader may enjoy pausing to try to make such a dissection, before turning to one possible and ingenious solution suggested by Don Wagner (1985), which may be more conveniently accessed through Wagner's website: <http://www.staff.hum.ku.dk/dbwagner/Pythagoras/Pythagoras.html>.

In what is commonly called the 'period of division' between the end of the Han Empire in 221 AD and the Sui 隋 dynasty reunification in 581 AD, the old land-marks of the Han synthesis were gone. The unified empire of the past four cen-turies was broken apart into competing kingdoms, and the confident imperial ideology that had underpinned it, with a version of Confucianism at its core, was no more. What moved into the place of that ideology was a very varied collection of intellectual and religious movements. Some of those movements looked back to points of departure in traditions of the Warring States period that had preceded the Qin unification, such as the *Lao zi* 老子 and *Zhuang zi* 莊子 books, and the writings of the followers of Mo Di 墨翟 (fifth–fourth centuries BC), some of which dealt with logical and mathematical definitions. Other movements flowed from the establishment in China of Buddhism, which not only supplied a new ration-ale for living through its religious concepts, but also brought with it a sophisti-cated philosophical package with Indian roots.[51] It has long been a common-place amongst Chinese scholars that Liu Hui's work must be interpreted in the context of the intellectual world in which he wrote (e.g., Horng 1983; Liu 1993, 71–75).

There is however an important converse to this need to set Liu Hui in his cul-tural context as a new voice on the 'Nine chapters' and the significance of *suan*. That converse is that writing on *suan* from the period before Liu Hui must be interpreted as far as possible without projecting backwards all the suppositions and patterns of thought that belong in his time, and emphatically do not belong in earlier centuries. If we do not heed this caution, we run the risk of obliterating all earlier evidence by viewing the past through the wrong historical spectacles. In the days when almost all the evidence we had of the practice of *suan* in the early Chinese empire was the 'Nine chapters', which is intimately associated with Liu Hui's commentary, there was some excuse for that error. But now that the 'Writings on reckoning' have come to light, the time is ripe for a reassessment of the numerate arts of the Han in the context of Han culture and history. This attempt at a prosopography is offered as a contribution to that project.

Appendix: Persons connected with *suan* in early Imperial China

The following table gives data on everybody during the Qin and Han periods who is said by available historical sources to have practised, or studied, or taught a dis-cipline that has an explicit connection with *suan*. I have not included here people who may be deduced on the grounds of their known activity to have had *suan*

---

51. For an accessible sketch of this new milieu from two well-informed scholars, Paul Demiéville and Timothy Barrett, see Twitchett, Loewe, and Fairbank (1986, 808–878).

Table 7.1.1 Persons connected with *suan* in early Imperial China

Name	Dates	Identity	Relevant expertise	Reference	Source
Zhang Cang 張蒼	c 250–152 BC	Statesman and expert in numerical arts	'Good at the use of *suan*, at *lü* and *li*'	蒼乃自秦時為柱下御史，明習天下圖書計籍，又善用算律曆，故令蒼以列侯居相府，領主郡國上計者．	HS 42, 2094
Sang Hongyang 桑弘羊	c 140–80 BC	Mental calculator; economic planner	Mental reckoning	弘羊，洛陽賈人之子，以心計，年十三侍中 Works on salt and iron: 於是以東郭咸陽孔僅為大農丞，領鹽鐵事；桑弘羊以計算用事，侍中 (SJ 30, 1428)	HS 24b, 1164
Luoxia Hong 落下閎	active c 104 BC	Unofficial expert in *li*	'Carried out *suan* to revise the *li*'	至今上即位，招致方士唐都，分其天部；而巴落下閎運算轉曆，然後日辰之度與夏正同．	SJ 26, 1260
Liang Qiuhe 梁丘賀	active under Emperor Xuan (73–47 BC)	military officer	Mental reckoning	梁丘賀字長翁，琅邪諸人也．以能心計，為武騎	HS 88, 3600
Du Zhong 杜忠	unknown	unknown; author of (lost) writings on *suan* procedures/methods in 16 rolls		杜忠算術十六卷．	HS 30, 1766
Geng Shouchang 耿壽昌	active 57–52 BC	Official; expert on *li* and observational astronomy	'Good at doing *suan*; able to evaluate the advantage of public works'	善為算能商功利 …習於商功分銖之事	HS 24a, 1141
Xu Shang 許商	active under Emperor Cheng (33–7 BC)	Official; author of (lost) writings on suan procedures/methods in 26 rolls	'Good at doing *suan*; able to plan the use of works'	博士許商治尚書，善為算，能度功用 Author of writings on suan: 許商算術二十六卷．(HS 30, 1766)	HS 29, 1688
Ma Yannian 馬延年	active under Emperor Cheng (33–7 BC)	Official	'Clear in reckoning and *suan*'	商，延年皆明計算，能商功利，足以分別是非，擇其善而從之，必有成功	HS 29, 1689

Name	Dates	Occupation	Description	Chinese text	Reference
Zhuo Mao 卓茂	d 28 AD, having studied in the time of Emperor Yuan (48–31 BC)	Official and scholar	*Suan* applied to *li*	元帝時學於長安，事博士江生，習詩、禮及歷筭，究極師法，稱為通儒	HHS 25, 869
Ban Zhao 班昭	c 45–c 117 AD	Female scholar and historian	The patterns of heaven; applying *suan* to numbers	See Empress Deng: as Mme Cao she instructs Empress in *tian wen*, 筭數. She also finishes the *Han shu tian wen zhi*: 兄固著漢書，其八表及天文志未及竟而卒，和帝詔昭就東觀藏書閣踵而成 (HHS 84, 2784)	HHS 10a, 424
Fan Ying 樊英	died c 129 AD aged over 70, so born c 60 AD	Diviner, recluse, and official	Stellar *suan*	少受業三輔，習京氏易，兼明五經．又善風角、星筭、河洛七緯，推步災異．	HHS 82a, 2721
Di Pu 翟酺	active in time of Emperor An (107–125 AD)	Diviner, shepherd, official	The patterns of heaven; *suan* applied to *li*	尤善圖緯，天文、歷筭．	HHS 48, 1602
Lang Zong 郎宗	active in time of Emperor An (107–125 AD)	Diviner	Stellar *suan*	學京氏易，善風角、星筭、六日七分，能望氣占候吉凶、常賣卜自奉．	HHS 30b, 1053
Wang Yi 王逸	Active 114–19 AD	Official, literary scholar, and editor of *Chu ci* anthology	*suan*	Made 'accounting clerk': 元初中舉上計吏；Studies *suan* with son: 王子山與父叔師到泰山從鮑子真學筭	HHS 80a, 2617 and comm.
Zhang Heng 張衡	78–139 AD	Polymath	*suan* applied to *li*	尤致思於天文、陰陽、歷筭．	HHS 59, 1897
Empress Deng 鄧皇后	80–121 AD	Empress	The patterns of heaven (*tian wen*); applying *suan* to numbers	太后自入宮掖，從曹大家受經書，兼天文、筭數．	HHS 10a, 424

Table 7.1.1 *(cont.)*

Name	Dates	Identity	Relevant expertise	Reference	Source
Ma Xu 馬續	c 100 AD	Scholar and historian; brother of Ma Rong	'Good at the *Jiu zhang suan shu*'	善九章筭術. He also works on the *Han shu*: 孝明帝使班固馴漢書，而馬續述天文志. (HHS zhi 10, 3215)	HHS 24, 862
Liu Yu 劉瑜	active c 165 AD	Official, expert in interpretation of apocryphal texts	The patterns of heaven; procedures/ methods for *suan* applied to *li*	少好經學，尤善圖讖，天文、歷筭之術.	HHS 57, 1854
Zheng Xuan 鄭玄	127–200 AD	Scholar, commentator on classics; student of the *Jiu zhang suan shu*	'Good at *suan*'	遂造太學受業，師事京兆第五元先，始通京氏易，公羊春秋，三統歷，九章筭術. Admitted to see Ma Rong on basis of skill in *suan*: 玄在門下，三年不得見，乃使高業弟子傳授於玄. 玄日夜尋誦，未嘗怠倦. 會融集諸生考論圖緯，聞玄善筭，乃召見於樓上. (HHS 35, 1207) Comments on *Qian xiang li*: 凡玄所注...乾象歷，又著天文七政論，...凡百餘篇萬言. (HHS 35, 1212)	HHS 35, 1207
He Xiu 何休	died 182 AD aged 55, so lived from 128 AD	official and recluse	*suan* applied to *li*	休善歷筭	HHS 79b, 2582–3
Liu Hong 劉洪	(c 135–210 AD)	official, mathematical astronomer	'Good at *suan*— unmatched in his age'	洪善筭，當世無偶 & 密於用筭	HHS 2, 3043 & HHS 3, 3083 (comm.)
Shan Yang 單颺	active c 195 AD	official, including Grand Clerk	The heavenly offices; procedures/ methods for *suan*	以孤特清苦自立，善明天官、筭術. 舉孝廉，稍遷太史令、侍中.	HHS 82b, 2733

expertise, but in connection with whom the word *suan* or compounds containing it are never mentioned.

The abbreviations SJ, HS, and HHS refer to the *Shi ji*, *Han shu*, and *Hou Han shu* respectively.

Bibliography

Chemla, Karine, and Guo Shuchun, *Les neuf chapitres: le classique mathématique de la Chine ancienne et ses commentaries,* Dunod, 2004.

Cullen, Christopher, 'Motivations for scientific change in ancient China: Emperor Wu and the Grand Inception astronomical reforms of 104 BC', *Journal for the History of Astronomy*, 24 (1993), 185–203.

Cullen, Christopher, *Astronomy and mathematics in ancient china: the* Zhou bi suan jing, Needham Research Institute Studies, 1, Cambridge University Press, 1996.

Cullen, Christopher, 'The birthday of the old man of Jiang county and other puzzles: work in progress on Liu Xin's *Canon of the ages*', *Asia Major*, 14 (2004a), 27–70.

Cullen, Christopher, *The* Suàn Shù Shū (筭數書) *Writings on reckoning: A translation of a Chinese mathematical collection of the second century BC, with explanatory commentary,* Needham Research Institute Working Papers, Needham Research Institute, 2004b.

Cullen, Christopher, 'Can we make the history of mathematics historical? The case of ancient China', *Studies in the History and Philosophy of Science*, 37 (2006), 515–525.

Cullen, Christopher, 'The *Suàn Shù Shū* (筭數書) *Writings on reckoning*: rewriting the history of early Chinese mathematics in the light of an excavated manuscript', *Historia Mathematica*, 34 (2007a), 10–44.

Cullen, Christopher, 'Actors, networks and "disturbing spectacles" in institutional science: 2nd century Chinese debates on astronomy', *Antiqvorvm Philosophia*, 1 (2007b), 237–268.

Guo Shuchun, *Gu dai shi jie shu xue tai dou Liu Hui* 古代世界数学泰斗刘徽 '*Liu Hui—an eminent mathematician of the ancient world*', Shandong Kexuejishu chubanshe, 1992.

Hacking, Ian, *The social construction of what?*, Harvard University Press, 2000.

Heath, T L, *A history of Greek mathematics*, 2 vols, Dover, 1921, repr 1981.

Ho Peng Yoke, *Li, qi and shu: An introduction to* Science and Civilization in China, 1985.

Horng Wann-sheng, 'Zhong shi zheng ming de shi dai 重視證明的時代 "An age that emphasised proof"', in Dai Liu (ed), *Zhong guo wen hua xin lun: ke ji pian* 中國文化新論:科技篇 '*New discussions of Chinese culture: science and technology*', Sanlian, 1983.

Lam Lay Yong and Ang Tian Se, *Fleeting footsteps: tracing the conception of arithmetic and algebra in ancient China*, World Scientific 1992.

Liu Dun, *Da zai yan shu* 大哉言数 '*What a great discourse on numbers*', Liaoning Educational, 1993.

Loewe, Michael, *Records of Han administration*, University of Cambridge Oriental Publications, 11–12, Cambridge University Press, 1967.

Major, John S, and Christopher Cullen, *Heaven and earth in early Han thought: Chapters three, four and five of the Huainanzi*, Suny Series in Chinese Philosophy and Culture, State University of New York Press, 1993.

Nielsen, Bent, *A companion to* Yi Jing *numerology and cosmology: Chinese studies of images and numbers from Han (202 BCE–220 CE) to Song (960–1279 CE)*, Routledge Curzon 2003.

Quine, Willard van Orman, 'Natural kinds', in *Ontological reality and other essays*, Columbia University Press, 1969.

Rickett, W A, *Guanzi: Political, economic and philosophical essays from early China*, vol 2, Princeton University Press, 1998.

Ruan Yuan, *Chou ren zhuan*, Taipei, 1981.

Shen Kangshen, Crossley, J N, and Lun, Anthony W C, *The nine chapters on the mathematical art: companion and commentary*, Oxford University Press, 1999.

Twitchett, Denis Crispin, Loewe, Michael, and Fairbank, John King, *The Cambridge history of China. Vol 1, the Ch'in and Han empires, 221 BC–AD 220*, Cambridge University Press, 1986.

Unguru, Sabetai, 'On the need to rewrite the history of Greek mathematics', *Archive for History of Exact Sciences*, 15 (1975), 67–114.

Wittgenstein, Ludwig, *Philosophical investigations: Philosophische Untersuchungen*, Blackwell, 1958.

Mathematics in fourteenth-century theology

Mark Thakkar

A ll would-be historians of medieval mathematics must ask themselves where to look for their subject matter. One obvious place to start would be in works with promising titles; approaching the Latin fourteenth century in this way, one might investigate Bradwardine's *Arithmetica speculativa*, *Geometria speculativa*, and *De proportionibus velocitatum in motibus*, Swineshead's *Liber calculationum*, Oresme's *De proportionibus proportionum*, and so on.[1] But this method, for all its initial merits, has limited scope. This chapter explores a less obvious source of material: commentaries on a theological textbook called the *Sententiae in quattuor libris distinctae*, 'Sentences divided into four books'.

The 'Sentences', a compilation of authoritative opinions from the Church Fathers and later theologians, was put together in the 1150s by Peter Lombard, a master at the cathedral school of Notre Dame.[2] Its originality lay solely in its

---

1. There is a brief *dramatis personae* at the end of this chapter; basic biobibliographical information on almost all of these characters can be found in Gracia and Noone (2003). On obviously mathematical works like those just mentioned, see the chapters by Mahoney (145–178) and Murdoch and Sylla (206–264) in Lindberg (1978), and the new studies in Biard and Rommevaux (2008).

2. Lombard divided the *Sententiae* into short chapters but in the 1220s it was divided thematically into larger sections called *distinctiones* 'distinctions' (Lombard 1971, I 137–144). The Latin text is edited in Lombard (1971); Books I and II are now translated in Lombard (2007; 2008); for an overview, see Rosemann (2004). On the *sententiae* genre, see Teeuwen (2003, 336–339). On the commentary tradition, see the studies in Evans (2002); for its development into the fourteenth century, see Friedman (2002b).

selective arrangement of extant material, but its importance for the history of Western thought can scarcely be overstated. It is not simply that it became an enormously popular textbook, or that it earned its author a portrayal as one of Beatrice's crowning lights in *Paradiso* X (106–108). It is rather that in the thirteenth century it was increasingly used by theologians as a matrix for their own lectures, giving rise to a prolific commentary tradition that lasted for over three hundred years.

Still, the reader might be forgiven for thinking that little of interest to historians of mathematics could possibly be found in commentaries, however original, on a theological textbook. It would be as well to address such misgivings with some preliminary remarks on the context in which such works were produced.[3]

First, theology students—secular ones, at least—were required to hold the wide-ranging degree of Master of Arts, which took around seven years to obtain. By the time they proceeded to the 'higher' faculty of theology, therefore, they were already trained in, among other things, the arts of logic, arithmetic, geometry, and astronomy. This was not a period of narrow specialization; indeed, the modern emphasis on interdisciplinary studies pales beside what has aptly been called the 'unitary character' of education in the medieval university (Murdoch 1975; Asztalos 1992; Marenbon 2007, 205–328).

Second, theology was regarded as the pinnacle of intellectual enquiry, and attracted many of the sharpest minds; commentaries on the 'Sentences' are certainly some of the meatiest intellectual products of the day. The arts faculty, by contrast, was regarded as inferior, not least because of its propaedeutic role. The Parisian arts master John Buridan, one of very few notable scholastics never to have moved on to theology, suggested that another factor was 'the wealth of those who profess in the other faculties' (Zupko 2003, 143, 338).[4] By this he could have meant either that those faculties gave greater financial reward and so were more attractive, or that their members were rich and ipso facto highly regarded. Either way it would not have been odd for men of a mathematical bent to become theologians.

Third, the remit of theology was broader than one might expect. The four books of the 'Sentences' dealt respectively with God, creation, the Incarnation, and the sacraments. Fourteenth-century commentaries tended to be weighted in favour of the first two books, and commentaries on Book II, in particular, involved matters which we would not now think of as theological, such as motion, perception, cosmology, and astrology (Murdoch 1975, 277–279; Grant 2001, 264–280).

Consequently, the 'Sentences' came to be used in the fourteenth century as a springboard for discussion of all kinds of topics. An extreme example is Roger

---

3. For an excellent introduction to the medieval intellectual world, see Grant (2001).
4. Quare autem nostra facultas sit infima? Potest dici quod hoc est propter divitias eorum qui alias profitent.

Roseth's *Lectura*, written probably in Oxford in around 1335, which bears no resemblance in structure, style, or content to Lombard's work. Roseth instead used five theological questions as pegs on which to hang discussions of, inter alia, the universality of logic, the relationship between a whole and its parts, and infinity and the continuum (Roseth 2005; Hallamaa 1998). His lectures were so divorced from the roots of the tradition that part of the first question even circulated as a separate treatise, *De maximo et minimo*, which certainly fulfils the naive search criterion suggested above.

This phenomenon became so widespread that in 1346 Pope Clement VI wrote a letter of complaint to the masters and scholars of Paris. Most theologians, he said, were ignoring the Bible and the writings of saints and other church authorities in order to waste time on 'philosophical questions, subtle disputations, suspect opinions and various strange doctrines' (Denifle and Chatelain 1889–97, II 588–589, §1125).[5] The tone of the letter is vaguely threatening: if the warning is not heeded, 'we will no doubt think of another remedy'. Twenty years later, the new Parisian university statutes included the following (Asztalos 1992, 434):

Those reading the 'Sentences' should not treat logical or philosophical questions or topics, except insofar as the text of the 'Sentences' demands or the solutions to arguments require; but they should pose and treat questions of speculative or moral theology that are relevant to the distinctions. Also, those reading the 'Sentences' should read the text in order, and expound it for the utility of the audience. (Denifle and Chatelain 1889–97, III 144, §1319)

The fourteenth-century context, then, was more conducive than one might have thought to the inclusion of technical material in theological commentaries. Still, it is hard to imagine where mathematics might have found a foothold. Without further ado, let us look at some examples.

Mathematics in theology

In distinction 24 of the first book of the 'Sentences', Lombard posed some questions about the ever-mysterious Trinity. What, for instance, is signified by the number three when we say that God is three persons? Here is Lombard's answer:

When we say three persons, by the term three we do not posit a numerical quantity in God or any diversity, but we signify that our meaning is to be directed to none other than Father and Son and Holy Spirit, so that the meaning of the statement is this: There

5. Plerique quoque theologi, quod deflendum est amarius, de textu Biblie, originalibus et dictis sanctorum ac doctorum expositionibus [...] non curantes, philosophicis questionibus et aliis curiosis disputationibus et suspectis oppinionibus doctrinisque peregrinis et variis se involvunt, non verentes in illis expendere dies suos [...] Alias autem, nisi nostris monitis hujusmodi utique salubribus vobisque multum expedientibus non obtemperaveritis cum effectu [...] cogeremur proculdubio de alio, sicut videremus expediens, remedio providere.

are three persons, or Father and Son and Holy Spirit are three, that is, neither the Father alone, nor the Son alone, nor the Father and Son alone are in the divinity, but also the Holy Spirit, and no one else than these. Similarly, it is not only this or that person who is there, or this one and that one, but this, that, and the other, and no one else. And Augustine sufficiently shows that this is the sense in which we must understand this, when he says that by that term 'the intention was not to signify diversity, but to deny singleness.' (Lombard 2007, 132)

Lombard went on to give a similarly unilluminating commentary on the phrase 'two persons'. Now, if we look up this same distinction in some fourteenth-century commentaries, we find something very different. Here the question is 'whether the Trinity is a true number', and the consensus seems to be that we must first ask what numbers are. The Franciscan William of Ockham, revising his Oxford lectures for publication in the early 1320s,[6] devotes thirty pages to this latter question, with no mention of the Trinity, before resolving the theological issue in one page. Using Ockham as a source, his confrère Adam Wodeham, lecturing at a seminary in Norwich, likewise devotes sixty-two pages to the general question and only three to its theological application. The figures for the Augustinian Gregory of Rimini are twenty and two respectively.[7]

One contentious issue was whether numbers had real existence outside the mind. The difficulty was that if they did, the existence of any two numbers would guarantee the existence of infinitely many objects, which most scholastics found metaphysically abhorrent. The proof has a distinctly mathematical flavour: given a pair of sticks and a pair of stones, we would ipso facto also have a pair of pairs, making three pairs all told; but this triple of pairs would itself be an object, so we would have four objects; and so on ad infinitum. The problem was resolved either by allowing numbers only mental existence, or by denying that they existed separately from the objects that they numbered.

The contrast with Lombard's discussion—brief, unquestionably theological, and devoid of mathematical interest even in the broadest sense—is astonishing. Nor is this an isolated instance. Later in the 'Sentences', Lombard wrote about God's omnipotence, asking such questions as whether He can sin, lie, walk, die, do something that He has not foreseen, and so on. Commenting on this passage, Gregory of Rimini instead asks 'whether God, through His infinite power, can produce an actually infinite effect' (I.42–44.4 in Gregory 1979–87, III 438–481).[8] This occasions over forty pages of discussion, during which he argues that God

---

6. Lectures were sometimes recorded by students in a set of notes called a *reportatio*. A lecturer could rework a *reportatio* into a more polished *ordinatio*, published at the university stationers. On university publication in Paris and Oxford, see Bataillon, Guyot, and Rouse (1988) and Parkes (1992).

7. These figures, intended only to give a rough and ready comparison, are based on the pagination of the critical editions of I.24.2: Ockham (1979, 90–121); Wodeham (1990, 346–411); Gregory (1979–87, III 34–58).

8. Utrum deus per suam infinitam potentiam possit producere effectum aliquem actu infinitum.

can indeed create an infinite multitude, an infinite magnitude, and an infinitely intense quality.

Gregory appeals in each case to the division of an interval into proportional parts, that is, parts that diminish successively by a fixed proportion. (A modern mathematician might think of this in terms of geometric series with common ratio $1/n$, such as $1/3 + 1/9 + 1/27 + \ldots = 1/2$.) For instance, if God creates an angel at the start of each successive proportional part of an hour—one at the start, one after say half an hour, one after three quarters of an hour, and so on—then by the end of the hour he will have created an infinite multitude of angels (Gregory 1979–87, III 443:3–12). This is a clever line for Gregory to take. He himself is perfectly happy to say that a continuum contains an actual infinity of parts, and that the existence of an infinite multitude is not absurd, either of which makes the question trivial.[9] But he knows that his serious opponents might allow only a 'potential' infinity, so that although further increase is always possible, infinite increase can never be completed. His construction of a 'supertask', a task consisting of an infinite number of accelerated subtasks (Thomson 1954–5), neatly sidesteps this objection. All must agree that, no matter how fast God works, the stars move still and the clock will strike.

Speaking of angels, in 'Sentences' II.2.iv Lombard asked where they were created; his answer was that they were created in the highest heaven, the empyrean, and not in the firmament. Gregory asks instead *utrum angelus sit in loco indivisibili aut divisibili* 'whether an angel is in an indivisible or a divisible place', prompting the general question *an magnitudo componatur ex indivisibilibus* 'whether a magnitude is composed of indivisibles', to which he devotes fifty-three pages (II.2.2 in Gregory 1979–87, IV 277–339; see also Cross 1998; Sylla 2005). His answer is negative: a magnitude is composed of, as one might put it, magnitudes all the way down. He gives surprisingly short shrift to the thesis of composition from infinitely many indivisibles, arguing erroneously that infinitely many indivisibles would yield an infinite magnitude. He is keener to discredit the 'more commonly' held thesis of composition from finitely many indivisibles, which he does with a barrage of nine mathematical and four physical arguments.

Gregory's mathematical arguments use simple geometrical constructions to deduce absurdities from the atomist thesis. The first, for instance, runs as follows. Draw a line of six points. Construct on this base an isosceles triangle with two sides of fifteen points, and draw lines from one side to the other, joining the thirteen pairs of opposite points. These lines must shorten towards the apex, so since the base consists of only six points, they soon become smaller than a point, which is *ex hypothesi* impossible (Gregory 1979–87, IV 279).

9. Gregory also gives the quicker answers to which his position entitles him (1979–87, III 441:19–28; 443:13–27). On the distinction between actual and potential infinity see, for example, Dewender (1999, 286–287).

Gregory was by no means alone in using Lombard's angelology as a pretext for a geometrical refutation of atomism. The tradition seems to have begun in the first few years of the century with John Duns Scotus's Oxford lectures, in which he asked *utrum angelus possit moveri de loco ad locum motu continuo* 'whether an angel can move from place to place in a continuous motion' (II.2.2.5 in Scotus 1973, 292–300; see also Murdoch 1962, 24–30; 1982, 579; Trifogli 2004). Indeed, two of Gregory's arguments are explicitly adapted from Scotus ('*Doctor subtilis*'), and three more are borrowed from Wodeham ('*unus doctor*').[10]

Geometry also gave rise, in the mid-fourteenth century, to some peculiar arguments concerning the relative perfection of different species. The source here was *Elements* III.16, where Euclid says that the curvilinear angle between a tangent and the circumference of a circle (the angle of contingence) is less than any acute rectilinear angle, while the remaining angle between the circumference and the diameter perpendicular to the tangent (the angle of the semicircle) is greater than any acute rectilinear angle.[11] In his Parisian 'Sentences' lectures of 1348–9 the Cistercian Peter Ceffons (Fig. 7.2.1) used this proposition, together with the idea that these angles could be increased or decreased by varying the size of the circle, to derive nineteen corollaries on the proportional excess of certain types of angles over others (Murdoch 1969, 238–246; 1982, 580–582). These results could be applied to 'theological' problems of the following sort: a man and an ass are both infinitely inferior to God, but a man, although of finite perfection, is infinitely superior to an ass.

A more obviously theological problem is that of human free will and divine judgement, but even this was not immune from mathematical intrusion. In his 'Sentences' commentary of 1331–3, the English Dominican Robert Holcot raised a difficulty based, like Gregory's divine supertask, on the proportional parts of an hour. Holcot did not specify a proportion, but let us take it to be a half. Now suppose that a man is meritorious over the space of half an hour, sinful over the next fifteen minutes, meritorious over the next seven and a half minutes, and so on, and suppose that he dies at the end of the hour. Then God cannot reward or punish him, because there was no final instant of his life that would determine whether he died a bad man or a good man.[12] Holcot followed this up with eight similar arguments based on the continuum (Murdoch 1975, 327 n101).

My final example of a theological problem that attracted mathematical speculation is the question of the eternity of the world. Theologians were obviously

10. Renowned scholastics acquired honorific titles like 'the Subtle Doctor' (Scotus) and 'the Venerable Inceptor' (Ockham, also known as 'the More Than Subtle Doctor'), but contemporary authors were usually alluded to indirectly as 'one doctor' or 'some people'. On fourteenth-century citation practices, see Schabel (2005).

11. The proposition is numbered III.15 in some editions of Euclid (Murdoch 1963, 248–249).

12. Holcot's scenario is similar, though ultimately not identical, to that of the 'Thomson's lamp' paradox, in which a lamp is switched on and off with increasing rapidity (Thomson 1954–5).

Figure 7.2.1 Peter Ceffons lecturing on the 'Sentences', from the sole manuscript that preserves his unedited commentary (Troyes BM 62 f. 1, c 1354). By permission of Médiathèque de l'Agglomération Troyenne, photo by Pascal Jacquinot

committed to the fact that the world had a beginning in time, but it was disputed whether this could be proved using reason alone. In his Parisian 'Sentences' commentary of the early 1250s, the Franciscan theologian Bonaventure (canonized in 1482) compiled a battery of six arguments to demonstrate that the notion of an eternal world was incoherent. Here I will mention only two. The first was that each passing day adds to the past revolutions of the heavens, and moreover the revolutions of the moon are twelve times as numerous as those of the sun; but one cannot add to or exceed the infinite because there is nothing greater than it. The fifth was that, given the permanence of species and the immortality of the soul, an eternal world would contain infinitely many rational souls; but it is impossible for infinitely many things to exist at the same time (II.1.1.i.2 in Bonaventure 1885, 20–22; Byrne 1964).[13]

Bonaventure's fifth argument explains why the subsequent debate was often conducted in terms of multitudes of souls, but the first one is more interesting for our purposes. In the fourteenth century two lines of response were developed.

---

13. These arguments ultimately came from the sixth-century Christian John Philoponus via the Islamic world, though the specific example of souls was introduced by the twelfth-century Muslim al-Ghazali (Davidson 1987, 117–134; Sorabji 1983, 214–226).

One was to deny, as Galileo was more famously to do in *Two new sciences* (1989, 40–41), that terms like 'equal to' and 'greater than' were applicable to the infinite. The other was to try to explain how these terms, and the terms 'part' and 'whole', behaved when they were applied to the infinite (Murdoch 1982, 569–573; Dales 1990; Friedman 2002a).

These, then, are some of the mathematical topics that one finds discussed in fourteenth-century theological works. It is hard to get a real sense of the territory from such an overview, though, so let us look more closely at two theologians writing in the early 1340s who disagreed on the question of infinite multitudes.

Infinite multitudes: Thomas Bradwardine

Thomas Bradwardine is known to historians of mathematics as one of the Oxford calculators, a group of technically-minded thinkers associated with Merton College in the second quarter of the fourteenth century.[14] It is in this context that we find him praising mathematics in his *Tractatus de continuo* 'Treatise on the continuum' as 'the revelatrix of all pure truth, which knows every hidden secret and bears the key to all subtle letters' (Murdoch 1969, 216 n1),[15] and quoting Boethius' remark from the *Institutio arithmetica* that 'whoever neglects mathematical studies has clearly lost all knowledge of philosophy' (Bradwardine 1961, 64).[16]

Bradwardine is also known for his later work as a theologian, and for holding the position of Archbishop of Canterbury for a month before succumbing to the Black Death in 1349. Unfortunately, his 'Sentences' commentary, which would have been written in around 1332, has not come down to us. Instead, we shall look at a theological work that has a substantial thematic overlap with such commentaries: his sprawling magnum opus of 1344, *De causa Dei contra Pelagium et de virtute causarum*, 'In defence of God against Pelagius, and on the power of causes', which he dedicated *ad suos Mertonenses* 'to his Mertonians' (Bradwardine 1618; Dolnikowski 1995).

The *De causa Dei* is essentially a polemic on divine freedom. Pelagius, a British monk active at the turn of the fifth century, had denied original sin and argued against Augustine that men were responsible for their own salvation (Pelikan 1971, 308–318). Bradwardine, perceiving such heretical tendencies among his contemporaries, took up the cudgels, stressing God's freedom to bestow grace wherever He saw fit. Bradwardine's stance on predestination was famous enough

---

14. On the calculators, see Snedegar (2006), North (2000), Sylla (1982), Kaye (1998, 163–199); on the Merton connection, see Martin and Highfield (1997, 52–62); on Bradwardine, see Leff (2004), Dolnikowski (2005).

15. Ipsa est enim revelatrix omnis veritatis sincere, et novit omne secretum absconditum, ac omnium litterarum subtilium clavem gerit.

16. testante Boethio, primo *Arithmeticae* suae: Quisquis scientias mathematicales praetermiserit, constat eum omnem philosophiae perdidisse doctrinam.

to be mentioned in the *Canterbury tales*.[17] Later, its resonance with Calvinism must account for the pedigree of the first printed edition: commissioned by the Archbishop of Canterbury, George Abbot, it was edited by the mathematical scholar and royal courtier Sir Henry Savile and printed in an unusual format at crippling expense by the King's Printer (Weisheipl 1968, 192; Vernon 2004; Wakely and Rees 2005, 484–487).

Savile warns the reader in his introduction that Bradwardine, 'since he was a first-rate mathematician, did not shrink from that art even in treating theological matters' (Bradwardine 1618).[18] Indeed, the *De causa Dei* is presented in so peculiarly Euclidean a style, proceeding from postulates to theorems and corollaries, that it has been described as having 'characteristics of a *Theologiae christianae principia mathematica* [mathematical principles of Christian theology]' (Molland 1978, 113; Sbrozi 1990). The deductive method cannot go very far unaided in such matters, though, as Savile observes: 'if in the lemmas and propositions he has not been able to attain such mathematical precision throughout, the reader will remember to impute this not to the author but to the subject matter of which he treats' (Bradwardine 1618).[19]

The subject matter of the *De causa Dei* turns out to be broader than its title suggests. In the fortieth and final corollary of the first chapter, Bradwardine fulminates at length against the Aristotelian doctrine of the eternity of the world, using what he calls *rationes quasi mathematicae* 'quasi-mathematical arguments' to deduce paradoxical consequences from the existence of actual infinities (Bradwardine 1618, 119–145). I will look at only a limited selection of Bradwardine's many arguments; some of the others are translated into French in Biard and Celeyrette (2005, 183–196).

Suppose we have an infinite multitude A of souls and an infinite multitude B of bodies, both arranged consecutively. Now 'let the souls be distributed [...] in this way: the first soul to the first body, the second to the second, and so on; when the distribution is complete, each soul will have a unique body, and each body a unique soul. So these [multitudes] jointly and severally correspond equally to one another' (Bradwardine 1618, 122A).[20] So far so good. But now instead:

let the first soul be given to the first body, the second to the third (or the tenth, or to one as distant as you please from the first), and the third soul to the body as distant from the

17. In the 'Nun's Priest's tale', where Chaucer rhymed 'Bradwardyn' with 'Augustyn' (lines 475–476).

18. ut huius libri genius Lectori melius innotescat, non abs re fuerit pauca praemonuisse: Primo *Thomam* nostrum, cum summus esset Mathematicus, ut ex praecedentibus apparet, etiam in Theologicis tractandis non recessisse ab arte.

19. Quod si in lemmatibus, & propositionibus non semper ἀκρίβειαν illam Mathematicam potuit usquequaque assequi, meminerit Lector non id Auctori imputandum, sed subiectae, quam tractat, materiae.

20. distribuantur animae per Dei omnipotentiam, vel per imaginationem hoc modo: Prima, primo corpori; secunda, secundo; et ita deinceps, qua distributione completa quaelibet anima unicum corpus habebit, et quodlibet corpus unicam animam. Haec igitur singillatim atque coniunctim mutuo sibi aequaliter correspondent.

second ensouled body as the latter is to the first, and so on until the whole distribution is completed in this way. This done, either all the individual souls have been distributed to bodies, or there are some souls left over. If all the individual souls have been distributed to such bodies, the whole multitude A jointly and severally corresponds equally to that part of B, and vice versa.[21] If any soul is left over, then since there are only finitely many between it and the first, the bodies already taken from the multitude B are the same in number and finite; so the whole multitude B—which was supposed to be infinite—is likewise finite.[22] (Bradwardine 1618, 122A)

Thinking of it another way, Bradwardine argues that we could instead assign a thousand souls to each body in turn, which leaves us with a similar problem. If we run out of souls, then A was finite after all, contrary to the supposition. But if on the other hand the distribution can be completed, then:

to every unit of B there correspond a thousand units of A—nay, even ten thousand, a hundred thousand, a thousand thousand, or as large a finite number as you like, as long as it is distributed to the former in the above manner [...] From all this it follows, clear as day, that multitude A is enough to ensoul multitude B, and double B, and four times B, and so on without end.[23] (Bradwardine 1618, 122A–B, 122D–E)

Bradwardine's first complaint is metaphysical: such sheer superfluity 'in no way befits God most wise, [...] does not fit with nature, and is detested by all philosophers' (Bradwardine 1618, 123A–B).[24] But he also takes issue with infinite multitudes from a mathematical point of view:

Many people in many ways have their hands full responding to arguments like this, for they are not even ashamed to deny that 'every whole is greater than its part', or to concede that a whole is equal to its part; so that if A is the whole infinite multitude of all souls, B just one of them, and C the whole remaining multitude, they say that A is not greater

21. As a modern mathematician might put it, there is a bijection between $\{a_i\}$ and $\{b_{ki}\}$ for any positive integer k.

22. detur prima anima primo corpori, secunda tertio, vel decimo, vel quantum volueris distanti a primo; et tertia anima corpori tantum distanti a secundo corpore animato, quantum illud a primo, et ita deinceps donec tota distributio huiusmodi compleatur. Quo facto vel singulae et omnes animae sunt huiusmodi corporibus distributae, vel sunt aliquae remanentes. Si singulae et omnes sunt corporibus talibus distributae, tota A multitudo illi parti B divisim et coniunctim correspondet aequaliter et e contra. Si aliqua anima remanet, cum ab illa ad primam sint tantum finitae, et omnia talia corpora praeaccepta B multitudinis sunt totidem et finita; quare et tota B multitudo similiter est finita, quae posita fuerat infinita.

23. dentur primo corpori mille animae, et secundo totidem, et deinceps quamdiu multitudo sufficit animarum. Vel ergo distributio ista alicubi desinet, vel ad singula et omnia corpora se extendet: Si alicubi desinet, cum inter illum locum seu corpus loci illius, et primum corpus sint corpora finita tantummodo, erunt et totidem; quare et finiti tantummodo millenarii omnium animarum, et tota A multitudo finita, quae posita fuerat infinita. Si autem distributio illa ad omnia et singula corpora se extendit, cuilibet unitati B correspondet unus millenarius unitatum A, imo et decem, et centum, et mille millenarii, et quantuscunque numerus finitus volueris, si tantus distribuatur in primis modo praedicto [...] Ex his quoque ulterius luculenter infertur, quod A multitudo sufficit animare B multitudinem, et duplam, et quadruplam, et deinceps sine termino, sine statu.

24. ut quid ibi superfluunt animae infinitae, et infinities infinitae, ut patet perspicue ex praemissis? [...] ut quid ergo ibi superfluunt corpora infinita, et infinities infinita, sicut ex prioribus clare patet? Hoc Deum sapientissimum nusquam decet, [...] hoc natura non convenit, hoc omnes Philosophi detestantur.

than *C* but equal to it—which, consequently, they must also have to say about any two infinite amounts compared to one another. But does not Euclid in book I of his *Elements* suppose it as a principle immediately known to anyone that 'every whole is greater than its part', [...] which all mathematicians and natural philosophers will unanimously acknowledge? And to which it seems anyone's mind, upon knowing the terms, freely consents; and which seems to be evident from the meanings of the terms? Surely one thing is greater than another if it contains it and more, or another amount beyond or outside it? Whose mind says otherwise?[25] (Bradwardine 1618, 132D)

The complaint that Bradwardine voices so strongly here is a natural one, and not even the modern mathematical theory of the infinite has entirely silenced it. Nonetheless, supporters of actual infinity did find ways of answering it. One particularly notable response was that of Gregory of Rimini, to whom we now turn.

Infinite multitudes: Gregory of Rimini

Gregory of Rimini was a powerful and careful thinker whose influence—especially on the topic of predestination, on which he held a view not unlike Bradwardine's— was felt right through to the seventeenth century. A member of the order of the Hermits of St Augustine, he studied theology at Paris in the 1320s before teaching in Bologna, Padua, and Perugia in the 1330s. His return to Paris in around 1342 to lecture on the 'Sentences' is now recognized as a crucial link in the transmission of novel ideas from Oxford to Paris by way of Italy. He was unanimously elected the Augustinians' Prior General in 1357, a year before his death.[26]

Gregory quoted Bradwardine's *De causa Dei* on two occasions in his lectures on Book II of the 'Sentences', as Savile proudly notes in his introduction, but sadly for our purposes neither was in the context of infinity.[27] In fact, the dating of the two works is so close, and the *De causa Dei* so long (just shy of nine hundred folio pages in Savile's edition), that Gregory may not have read the

---

25. Ad hoc autem et huiusmodi multi multipliciter satagunt respondere, quidem namque non verecundantur negare, Omne totum esse maius sua parte, neque concedere totum esse aequale suae parti; ut si *A* sit tota multitudo infinita omnium animarum, *B* vero una earum, *C* autem tota residua multitudo, dicunt quod *A* non est maior *C* sed aequalis, quod et dicunt, sicut et habent necessario dicere consequenter, de quibuslibet duobus quantis infinitis ad invicem comparatis. Sed nonne Euclides I. elementorum suorum supponit istud principium tanquam per se notum cuilibet, Omne totum est maius sua parte, [...] quod et omnes Mathematici atque naturales Philosophi concorditer profitentur? cui et videtur cuiuslibet animus sponte notis terminis consentire; quod et videtur clarere ex significationibus terminorum. Nonne illud est maius alio, quod continet illud et amplius, seu aliud quantum ultra vel extra? cuius animus contradicit?

26. Schabel (2007); on transmission, see Schabel (1998); on Gregory's 'Sentences' commentary, see Bermon (2002); on his views on predestination, and their relation to Bradwardine's, see Halverson (1998).

27. Gregory (1979–87, VI) quotes Bradwardine on free will and grace (131), and on sinful intentions (298–300).

passages on infinity before dealing with the same subject in his own lectures on Book I.[28]

In any case, Bradwardine's complaint that every whole is greater than its part was surely a common one, and Gregory tackles it head on. If this Euclidean maxim is supposed to be evident from the meanings of the terms, we must be clear about what those terms mean. What is a whole and what is a part, and what is it for one to be greater than the other? Gregory distinguishes two ways of answering both questions:

> I respond to the argument by making a distinction about 'whole' and 'part', for these can be taken in two ways, that is, generally and properly. (1) In the first way, everything that includes a thing—that is, everything which is a thing plus something else besides that thing and anything of that thing—is called a whole with respect to that thing; and everything included in this way is called a part of the thing that includes it. (2) In the second way, something is called a whole if it includes a thing in the first way and also includes more of a given amount than the included thing does (*includit tanti tot quot non includit inclusum*); conversely, such an included thing, not including as many of a given amount as the including thing (*non includens tot tanti quot includens*), is called a part of it.[29] (Gregory 1979–87, III 457:37–458:6)

In the context of multitudes, the general sense of 'part' clearly corresponds to the modern notion of a proper subset, and in this sense one infinite multitude can indeed be part of another. For instance, says Gregory, 'the multitude of proportional parts of one half of a continuum is a part of the multitude of parts of the whole continuum' (Gregory 1979–87, III 458:11–15).[30]

The additional condition for the proper sense is harder to understand. Gregory expands on it as follows: a proper whole includes 'more of a given amount (*tanti*)— that is, more of a particular quantity, for instance more pairs or triples—than the included multitude does' (Gregory 1979–87, III 458:16–19).[31] The pairs and triples here seem intended merely to indicate different ways of enumerating a multitude,

---

28. Two caveats are in order here. First, neither dating is certain. In particular, Bradwardine mentions at one point (1618, 559B) that he is writing in Oxford, leading one historian to argue that a major part of *De causa Dei* must have been written before 1335, the date of Bradwardine's move from Oxford to London (Oberman 1978, 88 n20). Second, ideas can of course circulate without being available in writing.

29. Secundo respondeo ad rationem distinguendo de toto et parte, nam haec dupliciter sumi possunt, scilicet communiter et proprie. Primo modo omne, quod includit aliquid, id est quod est aliquid et aliud praeter illud aliquid et quodlibet illius, dicitur totum ad illud; et omne sic inclusum dicitur pars includentis. Secundo modo dicitur totum illud, quod includit aliquid primo modo et includit tanti tot quot non includit inclusum, et econverso tale inclusum non includens tot tanti quot includens dicitur pars eius.

30. Et hoc modo una multitudo infinita potest esse pars alterius, sicut multitudo partium proportionalium unius medietatis continui est pars multitudinis partium totius continui, nam multitudo totius est omnes partes, sive omnia quorum quodlibet est pars, unius medietatis, et omnes partes alterius medietatis, quae sunt totaliter aliae ab illis. [I have altered the editors' punctuation a little. MT]

31. Secundo modo omnis multitudo includens aliam modo iam dicto et includens tanti tot, id est tot determinatae quantitatis, verbi gratia tot binarios vel tot ternarios, quot non includit multitudo inclusa, est totum respectu illius et illa econverso pars dicitur huius.

perhaps to allow the proper sense (like the general sense) to apply to wholes and parts that are not themselves multitudes. Understood in this way, a proper part is one that does not include as many units as the whole that includes it. In the proper sense, then, 'no infinite multitude is a whole or a part with respect to an infinite multitude, because none includes so many of a given amount without the other including as many' (Gregory 1979–87, III 458:19–21).[32]

Gregory's 'proper' sense is an odd way of understanding the terms 'whole' and 'part', but its additional condition has a more natural counterpart in his distinction between two senses of 'greater' and 'smaller':

Secondly, I make a distinction about 'greater' and 'smaller', although there would be no need if it were not that some people use them improperly. (i) For in one way they are taken properly, and in this way a multitude is called greater if it contains a given amount more times, and smaller if it contains it fewer times; or in another way, which comes to the same thing, that is called greater which contains one more times or [contains] more units, and that is called smaller which contains [one] fewer times or [contains] fewer [units]. (ii) In another way they are taken improperly, and in this way every multitude which includes all the units of another multitude and some other units apart from them is called greater than it, even if the former does not include more than the latter; and in this way to be a greater multitude than another is none other than to include it and to be a whole with respect to it, taking 'whole' in the first way.[33] (Gregory 1979–87, III 458:26–35)

The second sense of 'greater' and 'smaller' that Gregory identifies does indeed seem improper. In this sense, 'one infinite is greater than another, just as it is also a whole with respect to the other, taking "whole" in the first way' (Gregory 1979–87, III 458:37–459:1).[34] In the proper sense, however, 'greater and smaller are not said of infinites with respect to each other, but only of finites, or of infinites with respect to finites and vice versa' (Gregory 1979–87, III 458:35–37).[35]

Having established these definitions, Gregory explores the connections between them. Of course, anything that is a proper whole or a proper part is also

32. Et hoc modo nulla multitudo infinita est totum aut pars respectu multitudinis infinitae, quia nulla tot tanti includit quin tot tanti alia includat.

33. Secundo distinguo hos terminos 'maius' et 'minus', quamvis non oporteret nisi propter aliquos improprie illis utentes: Nam uno modo sumuntur proprie, et sic multitudo dicitur maior, quae tantundem pluries continet, illa vero minor, quae paucies; sive alio modo, et venit in idem, illa dicitur maior, quae pluries continet unum vel plures unitates, illa vero minor, quae paucies seu pauciores. Alio modo sumuntur improprie, et sic omnis multitudo, quae includit unitates omnes alterius multitudinis et quasdam alias unitates ab illis, dicitur maior illa, esto quod non includat plures quam illa; et hoc modo esse maiorem multitudinem alia non est aliud quam includere illam et esse totum respectu illius, primo modo sumendo totum.

34. Secundo vero modo unum infinitum est maius alio, sicut etiam est totum ad illud primo modo sumendo totum.

35. Primo modo maius et minus non dicuntur de infinitis ad invicem, sed de finitis tantum vel de infinitis respectu finitorum et econverso.

a general whole or a general part, but the converse does not hold. The more interesting comparison is between the two senses of 'greater' and 'smaller':

Not everything which contains more units than another contains those which the other contains, just as a group of ten men living in Rome includes more units than a group of six men living in Paris, but it does not include those [units]; and therefore not everything which is greater in the first way is greater in the second way. And not everything greater in the second way is greater in the first way, as is clear from one infinite multitude with respect to another infinite [multitude] which it includes.[36] (Gregory 1979–87, III 459:5–10)

It is clear from this that Gregory's two senses here correspond to the modern notions of (i) the size or 'cardinality' of a set, and (ii) the inclusion of a proper subset within a set.[37]

Finally, Gregory turns to the objection, which he has stated in the form of a dilemma: 'if there were an infinite multitude, either a part would not be smaller than its whole, or one infinite would be smaller than another' (Gregory 1979–87, III 459:13–14).[38] His response depends on how the terms are taken, and we may summarize his subsequent treatment of three of the four possibilities as follows:

(1.i) An infinite proper subset would indeed not have a lower cardinality than the set of which it is a part, but this is only to be expected; after all, it would not contain fewer things. (Here Gregory defuses Bradwardine's objection by showing the Euclidean maxim to be violated in a benign way; surely one multitude cannot be greater than another if it does not contain more things.)

(1.ii) An infinite set would indeed, as a proper subset, be 'smaller' in the improper sense than another infinite set; but the one infinite would not exceed the other, 'for nothing is properly said to be exceeded by another unless because it does not contain as many of a given amount as the other, which is not true of any infinite' (Gregory 1979–87, III 459:30–31).[39]

(2.i) An infinite proper subset is not a 'part' in the proper sense, that is, a part of lower cardinality. 'And this is the only sense in which it is absurd (*inconveniens*) to concede that a part is no smaller than its whole or that an infinite is smaller than another infinite' (Gregory 1979–87, III 459:35–36).[40]

---

36. Nam non omne, quod continet plures unitates quam aliud, continet illas, quas continet illud aliud, sicut denarius hominum existentium Romae plures unitates includit quam senarius existentium Parisius, non tamen includit illas; et ideo non omne, quod est maius primo modo, est maius secundo modo, item nec omne maius secundo modo est maius primo modo, sicut patet de multitudine una infinita respectu alterius infinitae, quam includit.

37. We need not be squeamish about using the terminology of set theory for the purposes of exposition; the essence of these notions was not an invention of the nineteenth century.

38. si esset aliqua multitudo infinita, vel pars non esset minor toto, vel unum infinitum esset minus alio.

39. nam nihil proprie dicitur excedi ab alio, nisi quod non continet tanti tot quot aliud; quod de nullo infinito est verum.

40. Et hoc modo tantum est inconveniens concedere partem non esse minorem toto aut infinitum esse minus esse infinito.

Surprisingly, Gregory does not deal with the fourth combination, (2.ii). But here he would again say, as in (2.i), that an infinite proper subset is not a 'part' in the proper sense, so that a fortiori it is not an example of a part that is smaller than its whole; and he would again say, as in (1.ii), that, as a proper subset, it would be 'smaller' in the improper sense.

Conclusion: the historiography of medieval mathematics

Historians of mathematics have traditionally said little about the scholastics, and what they have said has tended to be dismissive. There are, it must be admitted, whole areas of mathematics in which this judgement appears to be sound—algebra, for instance. But even a brief look at fourteenth-century debates over infinity should quash the notion that the scholastics either failed to notice the apparent paradoxes involved or simply put them aside. However, despite a surge in scholarly literature on the topic over the past forty years, this notion remains surprisingly widespread. Where not explicitly stated, it is often implicit in the following potted history: the Greeks abhorred the actual infinite, the medievals agreed, Galileo noticed that the integers could be paired off with their squares, Bolzano noticed the full extent of the phenomenon, and finally of course there was Cantor.[41]

Why has this misapprehension been so persistent? One undeniable factor is the deep-rooted conviction that the medieval period was one of pedantic stagnation; to see this, one need only look up 'medieval' or 'scholastic' in a dictionary.[42] But this cannot be the whole story, because even sympathetic writers have traditionally despaired of scholastic views on infinity; Bolzano, in his *Paradoxien des Unendlichen* 'Paradoxes of the infinite', claimed that the relationship between infinite sets and their proper subsets had previously been overlooked (Bolzano 1919, §20, 27; 2004, 16),[43] while Cantor, summarizing the history of the topic, wrote that 'as is well known, throughout the Middle Ages "infinitum actu non datur" [there is no actual infinite] was treated in all the scholastics as an incontrovertible proposition taken from Aristotle' (Cantor 1932, §4, 173–174).[44]

41. Any overview that fails to mention Gregory of Rimini is likely to give a similar story; in this regard, Zellini (2004) and Moore (2001) are commendable, though Moore (2001, 54) misrepresents Gregory's position on continua. The surge in scholarly literature is almost entirely due to the industry of John Murdoch.

42. For the roots of this prejudice in Renaissance self-congratulation, and its subsequent development, see Grant (2001, 283–355); a popular caricature of the debate on continua is discussed in Sylla (2005).

43. die man aber bisher zum Nachteil für die Erkenntnis mancher wichtigen Wahrheiten der Metaphysik sowohl als Physik und Mathematik übersehen hat.

44. Bekanntlich findet sich im Mittelalter durchgehends bei allen Scholastikern das 'infinitum actu non datur' als unumstößlicher, von Aristoteles hergenommer Satz vertreten.

Cantor's attachment to the scholastics is explored in Dauben (1979, 271–299) and the rather discursive Thiele (2005).

A second factor is the difficulty of gaining access to the relevant texts. For a long time they were available only in manuscripts or early printed editions, which demand far more time, patience, and training than can be expected of a casual researcher.[45] To make matters worse, medieval scholarship used to focus on the thirteenth century, which was thought to represent the zenith of scholasticism.[46] The past thirty or forty years have, however, seen critical editions of several important fourteenth-century 'Sentences' commentaries and an accompanying profusion of scholarly literature within the (inevitably and appropriately) broad field of medieval philosophy (Evans 2002). Much manuscript material remains to be edited, of course, but the textual situation is far happier than it once was, allowing a new appraisal of the quality of fourteenth-century thought.

Perhaps, though, the trouble lies also in the methodological question raised at the start of this chapter; from what we have seen above, it is entirely possible that some of the best mathematical brains of the time have wrong-footed historians of mathematics by working as theologians. For who, investigating treatments of the relationship between infinite sets and their proper subsets, would have thought to look in Thomas Bradwardine's polemic on divine freedom, or in Gregory of Rimini's commentary on a theological textbook? If there is some truth in this diagnosis, it may be worth repeating something that John Murdoch suggested to historians of science over thirty years ago:

in terms of the subject involved, the historian's search for accomplishments of significance should not be guided by the resemblance of the subject to some feature within modern science. Thus, I would submit that a good deal more of substance, of importance and of interest can be found, for example, in the medieval analysis of the motion of angels than in whatever astronomy occurs in Easter tables, or in the examination of the question of whether or not the infinite past time up to today is greater than the infinite past time up to yesterday than in the geometry of star polygons. One would discover more of importance because we would learn more of the whole tenor of late medieval thought. (Murdoch 1974, 73)

Acknowledgement

I would like to thank Rachel Farlie and Cecilia Trifogli for helpful comments on earlier drafts, and Olli Hallamaa for sending me material on Roger Roseth.

45. Williams (2003) gives a sympathetic overview of the textual issues facing medieval historians.
46. Two towering exceptions are Pierre Duhem (1956, 1985) and Anneliese Maier (1964).

Dramatis personae

Peter Lombard (c 1100–1160), Paris

Bonaventure of Bagnoregio (c 1217–1274), Paris; Franciscan

John Duns Scotus (c 1266–1308), Oxford and Paris; Franciscan

William of Ockham (c 1285–1347), Oxford and London; Franciscan

Francis of Marchia (c 1290–1344+), Paris and Avignon; Franciscan

Robert Holcot (c 1290–1349), Oxford, London, and Northampton; Dominican

John Buridan (c 1300–c 1360), Paris; secular cleric

Thomas Bradwardine (c 1300–1349), Oxford and London; secular cleric

Adam of Wodeham (d 1358), London, Norwich, and Oxford; Franciscan

Roger Roseth (fl c 1335), Oxford; Franciscan

Gregory of Rimini (c 1300–1358), Bologna, Padua, Perugia, Paris; Augustinian

Peter Ceffons (fl 1348–1349), Paris; Cistercian

Richard Swineshead (fl 1340–1355), Oxford

Nicole Oresme (c 1320–1382), Paris

Bibliography

Asztalos, Monika, 'The faculty of theology', in Hilde de Ridder-Symoens (ed), *A history of the university in Europe,* vol I: *Universities in the Middle Ages*, Cambridge University Press, 1992, 409–441.

Bataillon, Louis, Guyot, Bertrand, and Rouse, Richard (eds), *La production du livre universitaire au Moyen Âge: exemplar et pecia: actes du symposium tenu au Collegio San Bonaventura de Grottaferrata en mai 1983*, Éditions du Centre National de la Recherche Scientifique, 1988.

Bermon, Pascale, 'La *Lectura* sur les deux premiers livres des *Sentences* de Grégoire de Rimini O.E.S.A. (1300–1358)', in Evans (2002), 267–285.

Biard, Joël, and Celeyrette, Jean, *De la théologie aux mathématiques: l'infini au XIVe siècle*, Les Belles Lettres, 2005.

Biard, Joël and Rommevaux, Sabine (eds), *Mathématiques et théorie du mouvement (XIVe–XVIe siècles)*, Presses Universitaires du Septentrion, 2008.

Bolzano, Bernard, *Paradoxien des Unendlichen*, ed František Příhonský, 1851, reprinted with notes by Hans Hahn, Felix Meiner, 1919.

Bolzano, Bernard, *The mathematical works of Bernard Bolzano*, trans Steve Russ, Oxford University Press, 2004.

Bonaventure, *Commentaria in quatuor libros Sententiarum magistri Petri Lombardi: in secundum librum Sententiarum* (Opera Omnia, II), Collegium S. Bonaventurae, 1885.

Bradwardine, Thomas, *De causa Dei contra Pelagium, et de virtute causarum, ad suos Mertonenses*, ed Henry Savile, London, 1618.

Bradwardine, Thomas, *Tractatus proportionum seu de proportionibus velocitatum in motibus*, in H Lamar Crosby (ed), *Thomas of Bradwardine: his* Tractatus de proportionibus: *its significance for the development of mathematical physics*, University of Wisconsin Press, 1961.

Byrne, Paul, 'St. Bonaventure: selected texts on the eternity of the world', in St Thomas Aquinas, Siger of Brabant, St Bonaventure, *On the eternity of the world (de aeternitate mundi)*, trans Cyril Vollert, Lottie Kendzierski, and Paul Byrne, Marquette University Press, 1964.

Cantor, Georg, *Grundlagen einer allgemeinen Mannigfaltigkeitslehre: ein mathematisch-philosophischer Versuch in der Lehre des Unendlichen*, Leipzig, 1883, reprinted in Ernst Zermelo (ed), *Gesammelte Abhandlungen mathematischen und philosophischen Inhalts*, Springer, 1932, 165–208.

Cross, Richard, 'Infinity, continuity, and composition: the contribution of Gregory of Rimini', *Medieval Philosophy and Theology*, 7 (1998), 89–110.

Dales, Richard, *Medieval discussions of the eternity of the world*, Brill, 1990.

Dauben, Joseph, *Georg Cantor: his mathematics and philosophy of the infinite*, Harvard University Press, 1979.

Davidson, Herbert, *Proofs for eternity, creation and the existence of God in medieval Islamic and Jewish philosophy*, Oxford University Press, 1987.

Denifle, Heinrich and Châtelain, Emile, *Chartularium universitatis Parisiensis*, 4 vols, Paris, 1889–97.

Dewender, Thomas, 'Medieval discussions of infinity, the philosophy of Leibniz, and modern mathematics', in Stephen Brown (ed), *Meeting of the minds: the relations between medieval and classical modern European philosophy*, Brepols, 1998, 285–296.

Dolnikowski, Edith, *Thomas Bradwardine: a view of time and a vision of eternity in four-teenth-century thought*, Brill, 1995.

Dolnikowski, Edith, 'Thomas Bradwardine', in Thomas Glick, Steven Livesey and Faith Wallis (eds), *Medieval science, technology and medicine: an encyclopedia*, Routledge, 2005, 98–100.

Duhem, Pierre, *Le système du monde: histoire des doctrines cosmologiques de Platon à Copernic*, vol VII, Hermann, 1956.

Duhem, Pierre, *Medieval cosmology: theories of infinity, place, time, void, and the plurality of worlds*, trans Roger Ariew, University of Chicago Press, 1985.

Evans, Gillian, *Mediaeval commentaries on the Sentences of Peter Lombard: current research*, vol I, Brill, 2002.

Friedman, Russell, 'Francesco d'Appignano on the eternity of the world and the actual infinite', in Domenico Priori (ed), *Atti del primo convegno internazionale su Fr. Francesco d'Appignano*, Centro Studi Francesco d'Appignano, 2002a, 83–102.

Friedman, Russell, 'The *Sentences* commentary, 1250–1320: general trends, the impact of the religious orders, and the test case of predestination', in Evans (2002b), 41–128.

Galilei, Galileo, *Two new sciences: including centers of gravity and force of percussion*, trans Stillman Drake, Wall and Thompson, 2nd ed, 1989.

Gracia, Jorge, and Noone, Timothy (eds), *A companion to philosophy in the Middle Ages*, Blackwell, 2003.

Grant, Edward, *God and reason in the Middle Ages*, Cambridge University Press, 2001.

Gregory of Rimini, *Lectura super primum et secundum Sententiarum*, ed A Damasus Trapp, Venício Marcolino *et al*, 7 vols, Walter de Gruyter, 1979–87.

Hallamaa, Olli, 'Continuum, infinity and analysis in theology', in Jan Aertsen and Andreas Speer (eds), *Raum und Raumvorstellungen im Mittelalter*, Walter de Gruyter, 1998, 375–388.

Halverson, James, *Peter Aureol on predestination: a challenge to late medieval thought*, Brill, 1998.

Kaye, Joel, *Economy and nature in the fourteenth century: money, market exchange, and the emergence of scientific thought*, Cambridge University Press, 1998.

Kretzmann, Norman, Kenny, Anthony, and Pinborg, Jan (eds), *The Cambridge history of later medieval philosophy: from the rediscovery of Aristotle to the disintegration of scholasticism*, Cambridge University Press, 1982.

Leff, Gordon, 'Thomas Bradwardine (c.1300–1349)', *Oxford dictionary of national biography*, Oxford University Press, 2004.

Lindberg, David (ed), *Science in the Middle Ages*, Chicago University Press, 1978.

Lombard, Peter, *Sententiae in IV libris distinctae*, vol I, ed Ignatius Brady, Editiones Collegii S. Bonaventurae ad Claras Aquas, 3rd ed, 1971.

Lombard, Peter, *The Sentences, Book 1: the mystery of the Trinity*, trans Giulio Silano, Pontifical Institute of Mediaeval Studies, 2007.

Lombard, Peter, *The Sentences, Book 2: on creation*, trans Giulio Silano, Pontifical Institute of Mediaeval Studies, 2008.

Maier, Anneliese, 'Diskussionen über das aktuell Unendliche in der ersten Hälfte des 14. Jahrhunderts', 1947, reprinted in *Ausgehendes Mittelalter: Gesammelte Aufsätze zur Geistesgeschichte des 14. Jahrhunderts*, vol I, Edizioni di Storia e Letteratura, 1964, 41–85.

Marenbon, John, *Medieval philosophy: an historical and philosophical introduction*, Routledge, 2007.

Martin, Geoffrey, and Highfield, John, *A history of Merton College, Oxford*, Oxford University Press, 1997.

Molland, A George, 'An examination of Bradwardine's geometry', *Archive for History of Exact Sciences*, 19 (1978), 113–175.

Moore, Adrian, *The infinite*, Routledge, 2nd ed, 2001.

Murdoch, John, *'Rationes mathematice': un aspect du rapport des mathématiques et de la philosophie au Moyen Âge*, Palais de la Découverte, 1962.

Murdoch, John, 'The medieval language of proportions', in Alistair Crombie (ed), *Scientific change*, Heinemann, 1963.

Murdoch, John, *'Mathesis in philosophiam scholasticam introducta*: the rise and development of the application of mathematics in fourteenth century philosophy and theology', in *Arts libéraux et philosophie au Moyen Âge: actes du Quatrième Congrès International de Philosophie Médiévale, Université de Montréal, 27 août–2 septembre 1967*, Institut d'Études Médiévales, 1969, 215–249.

Murdoch, John, 'Philosophy and the enterprise of science in the later Middle Ages', in Yehuda Elkana (ed), *The interaction between science and philosophy*, Humanities Press, 1974.

Murdoch, John, 'From social into intellectual factors: an aspect of the unitary character of late medieval learning', in John Murdoch and Edith Sylla (eds), *The cultural context of medieval learning*, Reidel, 1975, 271–348.

Murdoch, John, 'Infinity and continuity', in Kretzmann, Kenny, and Pinborg (1982), 564–591.

North, John, 'Medieval Oxford', in John Fauvel, Raymond Flood, and Robin Wilson (eds), *Oxford figures: 800 years of the mathematical sciences*, Oxford University Press, 2000, 28–39.

Oberman, Heiko, 'Fourteenth-century religious thought: a premature profile', *Speculum*, 53 (1978), 80–93.

Ockham, William, *Scriptum in primum Sententiarum: ordinatio*, ed Girard Etzkorn and Francis Kelly, Franciscan Institute, 1979.

Parkes, Malcolm, 'The provision of books', in Jeremy Catto and T A Ralph Evans (eds), *The history of the University of Oxford*, vol II: *Late medieval Oxford*, Oxford University Press, 1992, 407–483.

Pelikan, Jaroslav, *The Christian tradition: a history of the development of doctrine*, vol I: *The emergence of the Catholic tradition (100–600)*, Chicago University Press, 1971.

Rosemann, Philipp, *Peter Lombard*, Oxford University Press, 2004.

Roseth, Roger, *Lectura super Sententias, quaestiones 3, 4 & 5*, ed Olli Hallamaa, Luther-Agricola-Seura, 2005.

Sbrozi, Marco, 'Metodo matematico e pensiero teologico nel "De causa Dei" di Thomas Bradwardine', *Studi medievali*, 31 (1990), 143–191.

Schabel, Christopher, 'Paris and Oxford between Aureoli and Rimini', in John Marenbon (ed), *Medieval philosophy* (Routledge History of Philosophy, III), Routledge, 1998, 386–401.

Schabel, Christopher, 'Haec ille: citation, quotation, and plagiarism in fourteenth-century scholasticism', in Ioannis Taifacos (ed), *The origins of European scholarship*, Franz Steiner, 2005, 163–175.

Schabel, Christopher, 'Gregory of Rimini', in Edward Zalta (ed), *Stanford encyclopedia of philosophy*, 2007. [online]

Scotus, John Duns, *Ordinatio: liber secundus: distinctiones 1–3* (Opera Omnia, VII), Typis Polyglottis Vaticanis, 1973.

Snedegar, Keith, 'Merton calculators (act. c.1300–c.1349)', *Oxford dictionary of national biography*, Oxford University Press, 2006.

Sorabji, Richard, *Time, creation and the continuum: theories in antiquity and the early middle ages*, Duckworth, 1983.

Sylla, Edith, 'The Oxford calculators', in Kretzmann, Kenny, and Pinborg (1982), 540–563.

Sylla, Edith, 'Swester Katrei and Gregory of Rimini: angels, God, and mathematics in the fourteenth century', in Teun Koetsier and Luc Bergmans (eds), *Mathematics and the divine: a historical study*, Elsevier, 2005, 249–271.

Teeuwen, Mariken, *The vocabulary of intellectual life in the Middle Ages*, Brepols, 2003.

Thiele, Rüdiger, 'Georg Cantor (1845–1918), in Teun Koetsier and Luc Bergmans (eds), *Mathematics and the divine: a historical study*, Elsevier, 2005, 523–547.

Thomson, James, 'Tasks and super-tasks', *Analysis*, 15 (1954–55), 1–13, reprinted in Wesley Salmon (ed), *Zeno's paradoxes*, Hackett, 2001, 89–102.

Trifogli, Cecilia, 'Duns Scotus and the medieval debate about the continuum', *Medioevo*, 24 (2004), 233–266.

Vernon, E C, 'Twisse, William (1577/8–1646)', *Oxford dictionary of national biography*, Oxford University Press, 2004.

Wakely, Maria, and Rees, Graham, 'Folios fit for a king: James I, John Bill, and the King's Printers, 1616–1620', *Huntingdon Library Quarterly*, 68 (2005), 467–495.

Weisheipl, James, 'Ockham and some Mertonians', *Medieval Studies*, 30 (1968), 163–213.

Williams, Thomas, 'Transmission and translation', in A Stephen McGrade (ed), *The Cambridge companion to medieval philosophy*, Cambridge University Press, 2003, 328–346.

Wodeham, Adam, *Lectura secunda in librum primum Sententiarum*, Rega Wood and Gedeon Gál (eds), Franciscan Institute, 1990.

Zellini, Paolo, *A brief history of infinity*, trans David Marsh, Penguin, 2004 (originally published in 1980 as *Breve storia dell'infinito*).

Zupko, Jack, *John Buridan: portrait of a fourteenth-century arts master*, University of Notre Dame Press, 2003.

Mathematics, music, and experiment in late seventeenth-century England

Benjamin Wardhaugh

The Scientific Revolution saw many subjects given new scrutiny, with attempts to use mathematical, mechanical, or experimental modes of explanation to gain understanding of them. One of those subjects was music. Already a tradition of mathematical study of musical intervals stretched back through the middle ages to ancient Greece, where the emphasis had been on ratios of the lengths of strings that formed particular musical intervals. In the seventeenth century there were new mathematical techniques and new kinds of mechanical explanation that could be applied instead (Wardhaugh 2006). There were also new experiments and experimental instruments. In this chapter I will discuss those instruments, in the particular context of late seventeenth-century England.

The Royal Society, founded in 1660, provided a meeting-place for diverse approaches to music, and was a potential source of legitimization for the few studies of music that incorporated experiments. I will discuss below some of the musical experiments performed by the Society: they included the use of a very long string to find the absolute frequency of musical vibrations; the use of a short string to display relationships between the lengths and tensions of strings and their musical pitch; the use of a vibrating glass to display patterns of standing waves; the use of a toothed wheel to demonstrate the effect of particular ratios of frequency; and finally an experimental musical performance using specially modified musical instruments. None of these was an 'experiment' in the more modern sense of producing knowledge or testing a theory; each in fact displayed quantitative knowledge which some or all of those present already possessed.

In this chapter I will consider four instruments close to the boundary between 'mathematical instruments' and 'musical instruments'. As well as throwing that boundary into relief, these instruments illustrate a range of issues that arose when scholars attempted to make sense in the new seventeenth-century context of the mathematical musical tradition they had inherited.

The Musical Compass

Fig. 7.3.1 shows an instrument, of sorts, called the Musical Compass. It is a paper instrument, of the type known as a 'volvelle': that is, it consists of two sheets of paper attached together at a single place so as to rotate. The upper sheet is circular, and both are printed. Only one copy of the 'compass' now survives, as the final page of a pamphlet printed in London in 1684. The pamphlet is anonymous but internal evidence strongly suggests that its author was Thomas Salmon, an Oxford-trained clergyman who had studied with the mathematician John Wallis and who had a lifelong interest in music theory.

This device dates from a period when Salmon was trying out different strategies for musical tuning. It was surely modelled on similar devices for navigation or astronomy, which were sometimes made of paper but more often of brass. Volvelles, such as those by Fernandez (1626) and Cantone (1668), were also sometimes used in non-mathematical music theory at this period, but I do not know of an English example, and it seems relatively unlikely that Salmon would have seen either of these (from Lisbon and Turin respectively).

Whatever its inspiration, this is a clear example of an 'instrument' that embodies a theory, containing a considerable amount of information in a small space. How does it work? The rotating paper disc lists the string lengths for each note of a one-octave scale, for a string of length 1000 units, so that 1000 units represents the lowest note of the octave and 500 the highest. On the page beneath are printed the letter-names of notes, and a second set of string lengths. Rotating the disc allows the mobile string lengths to be variously aligned with the fixed note names. The 'compass' was probably intended to be used for building or modifying a musical instrument, to facilitate the placing of frets on a set of equal-length strings tuned to different pitches. To fret a D string, for example, the user would line up the 1000.00 on the volvelle with the D on the page, and read off the string lengths of other notes from the device. It would also be necessary to multiply these lengths by a constant factor depending on the actual length of the string in question: a table printed beside the 'compass' illustrates this for a string of length thirteen units.

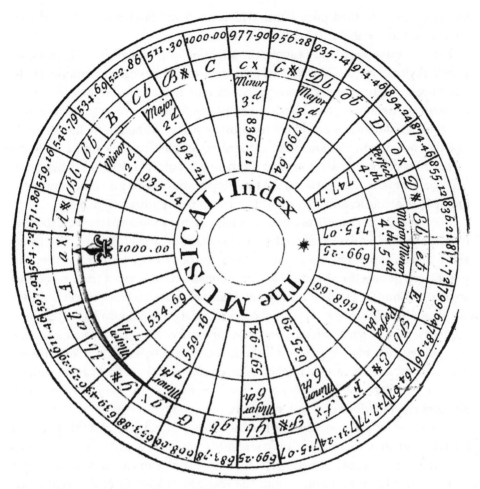

Figure 7.3.1 The Musical Compass. By permission of the British Library

Also marked on the moving page are the names for the relative positions of notes within the key: major second, major third, perfect fourth, and so on; and when the device is rotated these move to correlate with the fixed letter names. In D, for example, the major second will be E and the perfect fourth G. So the 'compass' also illustrates the variable relationship of the major or minor scale with respect to absolute pitches, an emerging feature of music theory more generally at this time.

Inspection of the string lengths on the 'compass' reveals that they divide the octave into thirty-one equal parts. A 31-pitch octave had been proposed in the fourteenth century by Marchetto of Padua, and had made occasional appearances since then: in this scheme, which was also used by Christiaan Huygens, the tone and semitone were taken as 5/31 and 3/31 of an octave respectively, good approximations to their pure values (Huygens 1986; Marchetto 1985; Herlinger 1981, 193; Barbour 2004, 117–121).

The 'Compass' apparently caught the attention of the English mathematician Brook Taylor, who was the first to provide a mathematical description of the vibrating string by (in effect) solving its differential equation. Among his papers are found, dismantled, two 'musical compasses'. They are more elaborate than Salmon's, each with not one but three moving parts.[1] Although we do not know exactly when or how Salmon's work came to Taylor's notice, their similarity to the earlier 'musical compass' is too great for them to be independent of it. They probably date from the 1720s, around the time Taylor was working on other aspects of mathematical music, and after his 1714 analysis of the vibrating string.

The first of Taylor's musical compasses has circular moving parts with, starting from the middle, scales based on the division of the octave into twelve, fifty-three, and nineteen apparently equal parts. These divisions are not arbitrary; they can result from quite natural criteria for scale construction derived from successive approximations for the relationship between the octave and the perfect fifth (see Wardhaugh 2006, 71–80). Taylor's second musical compass also has three moving parts, each of which divides the octave into fifty-three. Each whole compass is about ten centimetres across. All the markings are by hand, unlike Salmon's printed compass. There are a total of eighteen spare compass parts, most with no markings at all, probably reflecting an intention to make further compasses with different combinations of octave divisions.

Taylor's musical compasses are hard to interpret, but it is clear that he was interested in constructing devices consciously based on Salmon's musical compass and which would demonstrate the variable alignment of different letter names and relative positions within the scale, in at least three different equal divisions of the octave. There is no evidence that Taylor was interested in a correlation with string lengths, although these could have appeared on a lost page to which the moving discs and rings would have been fixed.

Salmon's compass illustrates both the use of decimals to quantify pitch and the precision with which that quantification could be realized. He gave string lengths to five figures, a degree of precision wholly unrealizable with seventeenth-century instruments and which scattered references in the sources tell us was known to be wholly imperceptible by at least the conscious faculty of hearing. Taylor's compasses illustrate mathematical music in a purer form: they appear to be not a tool for instrument building but an aid to the conceptualization of pitch and the comparison of different musical scales which were in discussion among theorists.

These are instruments of display: they display first the belief that music was a mathematical art, and second the mathematical description of a specific musical

---

1. Cambridge, St John's College Library, Classmark U 19: Brooke Taylor Papers (unfoliated).

tuning or group of tunings. The constraints of a rotating circular device were perhaps uniquely suited to theories of tuning in which the octave was equally divided, in which the depiction of the scale would have a high degree of rotational symmetry.

Other paper instruments for the mathematics of music existed. One was the 'Grand Scale', produced and demonstrated by the musician and theorist John Birchensha in the 1660s. Founded on a very elaborate table of string lengths, this instrument appears to have had a primarily didactic function, as a way to teach Birchensha's music students about his personal theory of musical tuning. Intermediate between Salmon's aids for instrument building and Taylor's apparently private tools for musical theorizing, the 'Grand Scale' would have formed the centrepiece of a whole musical treatise, which was unfortunately never completed (Wardhaugh 2006, 117–118, 273–292; Field and Wardhaugh 2009). An idea of what the 'Scale' might have looked like is given by a table of thirty-four pitches made by the English mathematician John Pell in 1665, possibly in response to a request from Birchensha: each pitch has a ten-digit integer string length, its prime factorization, a reference number, and up to three further factorizations indicating its relationships with other notes.[2]

Long strings

While these paper instruments displayed already-established (or assumed) relationships between pitch and number, there were also instruments whose function was more ambivalent. One of the most prominent was the long string.

It had been known since antiquity that when simple ratios were realized in the lengths of vibrating strings, the pitches produced by those strings formed consonant musical intervals. Vincenzo Galilei extended this observation to the thickness and tension of the strings (Galilei 1589; Cohen 1984, 82–83).

His son Galileo Galilei quoted and discussed these results, but said that 'it is quite impossible to count the vibrations of a sounding string, since it makes so many of them' (translated in Galilei 1974, 144). The relationship between the dimensions of the string and its frequency, important for understanding the physical cause of consonance, therefore remained a matter of surmise, or of argument from analogy. (The latter approach was pursued in particular by Galileo.) The first person to lengthen the string so that its vibrations could be counted was apparently Marin Mersenne, a French member of the order of Minim friars, who recorded in 1636 that a string making the pitch G '*bat 168 fois l'air…* *dans le temps d'une seconde minute*', 'strikes the air 168 times…in a second'.

---

2. London, British Library, Add MS 4388, ff. 14r–37v: John Pell, notes and calculations on music, f. 37v.

This was based on extrapolation from the absolute frequency of a string of 67½ feet (about 22 m), which he estimated at two cycles per second when tensed by a mass of half a pound (about 245 g).[3] When this string was reduced in length to five inches (13 or 14 cm) it made the G in question (Mersenne 1636, III 169; Dostrovsky 1974–75, 198). The figure of G = 168 implies A ≈ 377, which is too low to be credible. But Mersenne was relatively uninterested in absolute frequencies or in numerical precision: he was keener to encourage the reader to experiment for himself, and to show that frequency could in principle be measured.

The reason that these apparently relatively simple experiments had not been performed before was probably to do with the theories of sound that prevailed. Although it would not have been difficult, at any time from Pythagoras onwards, to slow down a musical string by lengthening or slackening it, the incentive to do so was absent because the string's vibrations were not held to be directly associated with the sound's pitch. Of the classical sources only one, the Euclidean *Sectio canonis*, described frequency as essential to pitch: most ancient writers apparently thought of frequency as only accidentally associated with a sound's pitch, both frequency and pitch being caused by the particular speed of the sound's travel through the air (Barker 1989, 190–208, esp 192 and n2; also 107 n39). Even in the late seventeenth century the fairly astute observer Claude Perrault argued that the visible vibrations of a string were larger in size and much slower in frequency than the invisible vibrations which constituted sound (Perrault 1680, 40–41, 62, 78–84, 113–7).

Long-string experiments were also performed by Walter Charleton, the English physician and natural philosopher, which he described in his *Physiologia Epicuro-Gassendo-Charletoniana* (1654), a work mainly concerned with the presentation of his controversial atomism. He began by using a slack string to show that vibrations of very low frequency were individually visible but not audible, those somewhat faster became invisible but produced 'a certain dull stridor', and vibrations of sufficient frequency created a pitch (Charleton 1654, 222). Next he established the relationship between frequency ratios, as opposed to string length ratios, and intervals: 'Fasten a long Lute-string at one extreme on a hook nayled to a wall, and suspend a small weight at the other; then strike the string at convenient distance above the weight.' If the initial vibrations were slow enough, it could be confirmed that halving the length of the string doubled their frequency. Turning to a string whose vibrations were audible as pitches, the same exercise of halving the string's length would produce a rise in pitch of an octave, 'and thence you cannot but concede, that the Acuteness of this half of the sonant chord, above that of the whole sonant chord, is caused only by the doubly

3. These equivalents assume that Mersenne used the Paris pound and foot.

more frequent Percussions of the Aer, and proportionate strokes of the Sensory'. Similar arguments would associate the other musical intervals with their proper ratios (Charleton 1654, 222–223). He extended his observations to the effect of changing the string's tension, and went on to explain how the coincidence of strokes could produce consonances. But he did not attempt to establish the absolute frequency of particular musical notes, as Mersenne had.

Pierre Gassendi, Charleton's source for the long string experiments as for the *Physiologia* generally, referred to Mersenne's work on the speed of sound (Gassendi 1649, 418–419; see also Gassendi 1658, 38–39, although this text would not have been known to Charleton in 1654). And when the experiments were repeated in 1664 at the Royal Society of London, Charleton was present. There is therefore a plausible line of transmission of the long string experiments from Mersenne their originator, through Gassendi and Charleton to the Royal Society. Since Charleton was also named by the Stationers' Register as the translator of Descartes' *Compendium musicae* into English in 1653, he emerges as quite substantially responsible for bringing musical science to England in the 1650s and 60s, an aspect of Charleton's career which has not previously been recognized (van Otegem 1999, 199; *Transcript*, I 402).

The long string experiments of the Royal Society were stimulated by contact with the musician and theorist John Birchensha (see above), who appeared at a meeting later that summer. The person who actually set up the string, built the monochord, and performed the experiments was Robert Hooke: he was the Society's Curator of experiments at the time, and his role in the musical experiments is also implied by a brief reference in a letter to Robert Boyle (Boyle 2001, II 292). Here 'G Sol. Re. Ut.' is the name of a single note: the G which, in modern notation, lies in the top space of the bass clef stave.

[6 July 1664] An experiment was made to measure the velocity of a sounding string, or to determine how quick the vibrations thereof are in a certain space of time. There was taken a brass wire of 136 foot long, of 1/32 of an inch diameter; and weighing this string, extended by a weight of 3¾lb. + 1lb. 10 ounces, and being made to vibrate in the middle, its vibrations were found to be half seconds. Then being stopped in the middle, and the half of that made to vibrate in the middle, was found twice as swift, or to vibrate quarter seconds: whence the length and vibrations appeared reciprocal.... Then farther stopping the wire within one foot of the end, and striking that short part, it was guessed to give a note of G. Sol. Re. Ut; which was to be experimented by a pipe at the next meeting. So that it seemed, that the velocity of the vibration of a string tuned to G. Sol. Re. Ut. is two hundred seventy-two times in a second (Birch 1756-7, I 446).

A similar account appeared in a letter of Moray to Christiaan Huygens two weeks later, adding that a mass of 4 lb 7 oz was initially used and then 'adjusted' so that half-second vibrations would result (Huygens 1888–1950, V 95). But the figure of 272 cycles per second for G is inconsistent with any contemporary pitch standard. For the long string described one would expect on theoretical grounds a frequency of about 0.82 cycles per second, not 1, and for the short string 112 (or 224, correcting for the fact that the Society took 1 cycle to contain 2 strokes). The latter is somewhat more plausible than 272 cycles per second as a late seventeenth-century G: the most likely source of error seems the estimation of 'half seconds' (Charleton took his pulse as a standard for seconds in his experiment), although the Fellows' estimation of G cannot be relied upon (Morse 1948, 84, 169; Dostrovsky 1974–5; Haynes 2001).[4]

The next week the experiment was repeated, and the results agreed with those of the first trial (Birch 1756–7, I 449). A week later again, a monochord was set up 'to know the diversity of notes by':

[20 July] The brass wires were extended upon a long square box of four feet long, one with a weight, the other with a pin, till they became unisons. Then the one being stopt in the middle with a moveable bridge, the two halves on either side were unisons to one another, and one of them an eighth higher than the other, which was not stopt. (Birch 1756–7, I 451)

There followed divisions in the ratios 1:2, producing a fifth with the unstopped string, and 1:4, producing a double octave between the two stopped parts. The next week the effect of weights was investigated:

[27 July]…one wire being extended by five pounds weight, the other was tuned to an unison with it; and then the same string being stretched with a weight of twenty pounds, it was found just an octave higher: which shews, that the weight is in a duplicate proportion to the sound or vibration (Birch 1756–7, I 456).

The final experiments before Birchensha's appearance are the most interesting:

[3 August] Two strings being tuned unisons, one of them was stopt at one third, and the lower end of it gave a fifth, and the shorter end was an eighth higher than the longer. Then one of the strings was so stopt, as to make it a note [that is, a whole tone] higher than the whole; and the proportion of the shorter to the whole was found less than 9 to 10. Then the string was stopt a third higher, and the proportion was found as 3 to 4. This was estimated so by the ear. (Birch 1756–7, I 456)

The ear became increasingly important during the series of experiments. On 6 July the ear was used to 'guess' the absolute pitch produced by a given string, but distrust of the ear was expressed by the promise to check it using a pipe at

---

4. Frequency $v = (\sqrt{T/\rho\sigma})/2l$ where T = tension, ρ = density, σ = cross-sectional area and l = string length: here l = 136 feet = 41.45 m, T = 19.67 N (= wg where w = 5lb 6 oz. = 5.375 lb = 2.006 kg, and g = 9.80665), ρ = 8600 kg/m^3 and σ = 4.948 × 10^{-7} m^2 (= πr^2 where r = 1/64 in. = 3.969 × 10^{-4} m). For the wire described the nonlinearity due to stiffness is only a few percent.

the next meeting. The implication on 13 July—'the experiments of determining the velocity of the vibrations of a brass wire, to afford a certain sound, was prosecuted, and agreeing with what was made at the last meeting …'—was either that that pipe had not been supplied, or that it had confirmed the ear's judgement.

On 20 and 27 July the ear was asked to confirm that certain string-length ratios indeed produced the expected musical intervals. This did not place any great reliance on the ear, particularly since the intervals involved were limited to the octave, fifth, and double octave. It is possible that the pipe referred to earlier was in fact used to check these intervals also, although a question would have arisen about the accuracy of the pipe's realization of musical intervals: since the pipe was not mentioned, it probably did not feature.

Finally, on 3 August, the procedure was reversed: instead of measuring out ratios and checking their musical result, two musical intervals were set up 'by the ear' and the corresponding ratios measured. This use of vibrating strings, to move from musical interval to mathematical ratio, featured in the Pythagorean legends of the discovery of harmonious ratios, and was mentioned as an unrealized possibility by Ptolemy, but this is a very unusual instance of its actual performance at any period.

The ratios that this experiment produced are also of interest, although it is frustrating that the Society did not go on to generate more of them. They suggest, broadly speaking, that the intervals the Fellows of the Society produced 'by ear' were close to those of the 'mean tone' tuning, and relatively distant from the Pythagorean values which most mathematical theorists of the period might have expected the ear to prefer. At the end of this experiment the Fellows again expressed distrust of their own ears, asking Silas Taylor to bring John Birchensha to the next meeting: the outcome showed that what the Society wanted from Birchensha was to make use of his musical ear.

[10 August] Mr Birchinsha being accordingly called in, tuned the string by his ear, to find how near the practice of music agreed with the theory of proportions.

This was exactly what they had been trying to find out the previous week: the emphasis was on the superior ability of Birchensha's ear to do it.

The effect was, that he could not by his ear distinguish any difference of sounds (upon the moving of the bridge) above half an inch, especially in the fourths, thirds, and tones.

Assuming they were still using the four-foot monochord of 20 July, moving the bridge by half an inch from any of the intervals named would produce an error of between 20 and 25 cents, that is up to a quarter of a semitone. This is not a huge amount, but it should have been audible to a competent musician.

Whereupon it was resolved, that a virginal should be as exactly tuned, as could be done by the ear, and then the monochord examined by it. (Birch 1756–7, I 457)

So Birchensha's ear was rejected in favour of an instrument, which presumably would be more reliable. Given an accurately tuned set of virginals all the experimenters' ears would have to do would be to judge when certain notes produced by the monochord were unisons with notes on the musical instrument. Unfortunately the fatal flaw was implied as soon as this strategy was mentioned: the virginal itself must be tuned 'by the ear', and it would therefore provide no more reliability than the ear itself. It is not recorded that anyone pointed this out, and the proposed trial was never performed in any case. Birchensha's own suggestion, to use a bass viol, hinted that the brass-stringed monochord had made it hard 'to distinguish the musical notes': but although the viol was accepted, no (experimental) work was apparently done with it (Birch 1756–7, I 460).

The long string provides an intriguing example of an experimental instrument which was clearly conceptually derived from musical instruments but was kept distinct from them in practice. It was an instrument which could be operated satisfactorily by the Fellows of the Royal Society, who made no special claim to specifically musical skills (although many of them may well have possessed such skills, as Penelope Gouk (1999, 23–65) has documented). It generated an experimental programme in which distrust of the ear was a key problem, mentioned explicitly on more than one occasion: a programme which failed when it was demonstrated that even the ear of a professional musician could not (though perhaps due to the limited abilities of the individual concerned) provide the level of mathematical precision the Fellows desired.

Depictions of a vibrating string, or even the strings themselves, had long been didactic tools for the display of theories about tuning: the depiction of the monochord and the discussion of that depiction were the standard means to display such theories during the Middle Ages and Renaissance. The conceptual reversal of the string so as to determine, rather than display, the relationship between pitch and string length, is a striking development.

Comparable experiments with other instruments are elusive. Later in 1664 the Royal Society obtained, at Huygens' suggestion, a flat plate of bell metal 'for the trial of the vibrations of hard bodies sounding'. Their intention was to obtain several such plates, of different sizes, to determine how their sound depended on their size. But the specimen produced was 'found useless for the experiments' (perhaps it was cracked), and the project did not proceed (Birch 1756–7, I 460, 475). The plate was found among Hooke's possessions at his death in 1703 (Hunter and Schaffer 1989). Attempts to find absolute frequency using sounding pipes were rare until the work of Joseph Sauveur in Paris in the final decade of the seventeenth century (Sauveur 1984). Huygens attempted such an experiment using organ pipes—he calculated frequency from pipe length and the speed of sound, which he found experimentally—but I know of no English example (Dostrovsky 1974–5, 201).

The toothed wheel

One other device that was used by Hooke and others, specifically to display the mechanical cause of consonance, was the toothed wheel, the teeth of which were made to strike against a fixed metal plate as the wheel turned. It has been linked to Francis North's *Philosophical essay of musick*, which contained a diagram plotting pulses at various frequencies along a horizontal 'time' axis, clearly displaying on the page the various rates of coincidence among different pairs of frequencies (North 1677 (unpaginated plate variously placed in different copies); Kassler 2004). Roger North reported in his biography of his brother Francis that Hooke converted this diagram 'into clockwork':

and made wheels, with small ligulae, in the manner of coggs, which moving each upon its pinn, as the wheel turned, struck upon an edg, one after another equably; the wheel turning slow the pulses were distinguishable, and had no other vertue; but then turning swifter, the distinction ceased, and a plain musicall tone emerged. This for one[;] then, another wheel was contrived to strike 3 to 2 (for instance) and as the distinction begun to fail, and continuation took place, one might hear a consort 5th coming on, and setling in the manifest accord so named (North 1995, 250; Chan, Kassler, and Hine 1999, 73).

Another description emphasized coincidences of strokes:

some wheels should strike together puls for puls, and others in proportions, as 1/2, 3/2, etc.... He would begin to turne slow, and so long the pulses were distinct, and he could discerne them, as smiths at anvill, without any other idea; but then coming to a mighty swiftness, the consonance called fifth (for instance,) which is 3/2.... which sort of demonstration of the nature of musicall accords is irrefragable [irrefutable].[5]

And another had more specific details of the ratios available:

The ingenious Mr. Hook, made an engin of wheels that made pulses in any musical proportion, as 2, 3, 4, 5, or 6 to 1 and so 3 to 2 and the like.[6]

For Roger North, this device illustrated that a continuous sensation in general resulted from a series of separate events too frequent to be distinguished. It also confirmed the already-known relationship between particular frequency ratios and musical intervals.

---

5. London, British Library, Add MS 32537, ff. 66–109: Roger North, 'Hasty essay', quote ff. 91v–92r, see Chan, Kassler, and Hine (1999, 49–170); Kassler (2004, 72–3).

6. London, British Library, Add MS 32546, ff. 33–90: Roger North, essay 'The world', quote ff. 33r–v, see Chan, Kassler, and Hine (1999, 72).

Hooke was already working on sound wheels in March 1676, and his diary records that he received material from Francis North only the following November; North's book was published the next year (Hooke 1968, 223). Conceivably Hooke had seen the sheets of North's *Essay* as they came from the press, but I think it more likely that Hooke's development of the musical wheel preceded his contact with North's ideas (Chan, Kassler, and Hine 1999, 73; Gouk 1999, 210 and n62).

In 1681 the device was shown to the Royal Society:

an experiment of making musical and other sounds by the help of teeth of brass wheels, which teeth were made of equal bigness for musical sounds, but of unequal for vocal sounds. (Birch 1756–7, IV 96)

Although incorporating perhaps the earliest attempt at sound synthesis, this demonstration did not include what might now seem the most obvious use of the brass wheel: to establish the absolute frequency of specific notes. To turn the wheel at a known speed would have been quite easy using a gearing mechanism, and the sound could then have been matched with a pitch from a musical instrument. This would certainly have been more accurate than the use of the progressively shortened string for the same purpose.

Another use of the brass wheels could have been to falsify the 'coincidence theory' of consonance (the term is not a seventeenth-century one, though it catches the essence of the idea). This theory proposed that the aural experience of consonance, the 'blending' of certain pairs of pitches, could be accounted for by conceiving pitched sounds as associated with regular series of pulses, the frequency of which was linked to the sound's pitch. If the frequencies of two sounds formed a simple ratio of whole numbers, many pulses in the two series would coincide. This persuasive explanation had first been proposed in letters of 1563 and a publication of 1585 by the Italian music theorist Giovanni Battista Benedetti (Benedetti 1585; Cohen 1984, 75). Though seriously flawed because of its failure to consider what in modern terms would be called the phase of the two sets of pulses, the theory was considered unproblematic by writers in the later seventeenth century, when it became a routine set-piece at the beginning of treatises on mathematical music. It could support a commitment to the founding of music theory on a mechanical basis, but did not necessarily do so (contrast North 1677 and Holder 1694).

Robert Hooke's apparatus could have shown easily that matching of phase was unnecessary for the production of consonance, thereby falsifying the coincidence theory. But although only one of Roger North's three descriptions explicitly mentioned the two sounds' being in phase, we have no positive evidence that Hooke ever considered trying out unphased sounds.

It has been repeatedly suggested that Hooke's sounding wheel demonstrated for the first time the correctness of the identification of musical interval with relative frequency (Gouk 1999, 208; Dostrovsky 1974–5, 199). It is not clear whether it did any more in this respect than the long strings of Mersenne, Gassendi, Charleton, and the Royal Society already had: indeed, it still left open the possibility that, as Perrault was to suggest in the 1680s, visible vibrations and their frequency were only accidentally related to pitch, the vibrations which actually constitute sound being much smaller and faster (Perrault 1680). A remark made by Hooke to Christopher Wren and William Holder (another mathematical music theorist) in 1676 that 'the vibrations of a string were not Isocrone [of constant period] but that the vibration of the particals was' hints that he, too, was considering a similar possibility (Hooke 1968, 211). This is ambiguous, but it certainly suggests that although Hooke considered sound a series of strokes of some kind (at this date, almost certainly not waves) with a definite frequency, he did not necessarily identify this frequency with that of the sounding body's visible vibrations. This might account for his apparent uninterest in establishing the absolute frequency of visible vibrations in the case of the wheel. From other references in his diary Hooke seems to have been developing his theories of sound throughout 1676, but unfortunately we have very little more information about their content (see Kassler and Oldroyd 1983; Gouk 1980).

Huygens, on the other hand, did use a toothed wheel to measure absolute frequency, in about 1682. He drove a small toothed wheel from a larger wheel using a driving belt, and calculated the frequency of the sound produced as 547 cycles per second. He judged the pitch to be the same as the D on his harpsichord (Huygens 1888–1950, XIX, 375–376; Dostrovsky 1974–5, 199–201). This is a plausible figure: it implies A = 410, which is well within the range of pitches in use at the time (pitches around A = 400 were normal for instruments used at home) (Haynes 2001). It is unlikely that this appearance of a toothed-wheel experiment shortly after Hooke's was a coincidence: Huygens had been in touch with the Royal Society via Robert Moray, and it is possible that lost letters in that correspondence transmitted the idea for the experiment.

Somewhat later Brook Taylor, whose 'musical compasses' I discussed earlier, performed similar toothed-wheel experiments (as with 'G. Sol. Re. Ut' above, 'A la mi re' is the name of a single note, in this case the A that falls within the treble clef stave in modern notation):

6 March 1712/13

I applied a quill to the crown wheel of my chamber clock, and making it fast to one of the [pillars] of the clock, I let the works run down for 7 minutes and by my Harpsichord I found the quill to sound A la mi re in alt:, and by the works of the clock the quill struck 766 teeth per second.[7]

---

7. Cambridge, St John's College Library, classmark U 19: Brooke Taylor papers (unfoliated); see Cannon and Dostrovsky (1981, 19).

He also matched this pitch, two octaves lower, with that of a wire whose frequency he was able to calculate from its length, tension, and density. The fact that the prediction matched the frequency observed in the wheel he seems to have taken as confirmation of his analysis of the vibrating string, which he had presented to the Royal Society the previous year (Taylor 1713).

The experiment implies that Taylor's harpsichord had A = 383, a very low value, but one which Taylor's careful experiment and the agreement with his theoretical prediction (which is correct for the length, tension, and density he gives) oblige us to take seriously.[8] Of course, matching the wheel's pitch on a harpsichord involved rounding to the nearest semitone and therefore introduced an error of up to half a semitone. Three days later Taylor repeated the experiment with a different frequency and pitch, and found that this result was consistent with the first.[9]

The toothed-wheel apparatus differs from the long string in the crucial respect that it is not derived from a musical instrument, and its operation is consequently still further from the exercise of distinctively musical skills (and the sound produced by a cog striking a brass 'edge', though pitched, is unlikely to have been a recognizably 'musical' one). The only role for the ear here was in recognizing when two sounds were at the same pitch.

The diversity of the uses of this apparatus by different experimenters is therefore striking: for Hooke this was another instrument for theory display, but for Huygens and Taylor it was a more genuinely experimental apparatus, capable of producing data which in Taylor's case constituted a meaningful check on a quantitative theory. That theory, though, dealt with the vibration of strings in general: it was not a theory of music. Next I turn to a rare and valiant attempt at experimental verification of a specifically musical theory.

Modified viols

Thomas Salmon, whom we met above as the maker of the first musical compass, pursued an interest in mathematical music theory throughout his life. In 1672–3 he was involved in a controversy about the reform of musical notation: his short book on the subject, *An essay to the advancement of musick*, was violently attacked in print by Matthew Locke, organist to the Queen's chapel (Salmon 1672a; Locke 1672). Two later volumes in the dispute were more concerned with trading insults than resolving strictly musical questions (Salmon 1672b; Locke 1673). Although

---

8. He gives length = 12.3 inches, weight = 12 ounces, and density = 1 grain per foot, which implies frequency = 382.4 Hertz (correcting for Taylor's terminology which introduces an extra factor of two in frequencies).

9. Cambridge, St John's College Library, classmark U 19: Brooke Taylor papers (unfoliated); see Cannon and Dostrovsky (1981, 19).

a storm in a teacup, the dispute gave both Salmon and Locke the opportunity to explain at length their respective ideas about the nature of musical knowledge and the proper way(s) to acquire it.

Strikingly, Salmon incorporated public 'experiments' into the study of music. 'I don't know', his 'publisher' wrote, 'what to request more advantageous for [the scheme's] acceptance, than an Experiental tryal' (Salmon 1672a, 'epistle', A4r). After Locke's challenge he appealed to experiment again as a way to establish the superiority of his scheme: 'which I have experienc'd before several judicious persons' (Salmon 1672a, 61). This recalls the Royal Society's semi-public trials and its appeals to a group of reliable observers. Locke, by contrast, was not interested in public experiments: when he did refer to experience it was to individual, private experience.

The point of Salmon's trials was to persuade hearers of the excellence of the scheme and thereby result in modifications to musical practice: 'surely, 'twere well worth the while for Instruments to be contriv'd accordingly... for the excellency of Musick' (Salmon 1672b, 20). A review in the *Philosophical Transactions* endorsed the scheme, and again 'recommended it to publique practise' (Anon 1671–2, 3095). Perhaps the emphasis on experiment had in part been a strategy to arouse the Society's interest.

Musical hearing had an important role in Salmon's project: it was supposed to be able to recognize harmonic excellence reliably. Salmon believed that 'God hath created a peculiar faculty of hearing, to receive harmonious sounds, clearly different from that by which we perceive ordinary noises', although he declined to speculate on whether this faculty was physiological or resided in the soul (Salmon 1672a, 2). A difference between the ordinary and musical hearings was their sensitivity to small deviations from pure tuning:

The Keys of an *Harpsochord* [sic] are now tuned in a common diluted proportion...though a vulgar ear may not be able to judg the difference...yet there will be a dissatisfaction, though it be not evident in what particular to complain. (Salmon 1672b, 20)

Salmon felt that, unlike his opponents, he was concerned with 'the true nature of music', and often returned to the desire to unite theory and practice (Salmon 1672b 'epistle', [i], 12–13). But he always assumed that mathematical theory led practice: music 'consists in proportions', it 'is a combination of sounds as they are proportioned in numbers'; therefore it is 'part of the *Mathematicks*' (Salmon 1672a, 2; Locke 1673, 7). For Salmon (as for his Greek sources, most notably Ptolemy), aural recognition of a musical interval and intellectual computation of a ratio were two different ways of apprehending the same thing. Eventually, Salmon explicitly outlined a whole method for musical science; he would:

(1)   establish the mathematical givens, supposedly by doing experiments with strings, but in reality by study of other theorists;

(2) use these to produce a mathematical division of the octave into smaller musical intervals;

(3) relate that division to practical knowledge by establishing how these intervals correspond to steps of the scale;

(4) invite the reader to check these intervals by measuring them out on a string and comparing them with his expectations;

(5) organize persuasive public performances using his scheme; and finally,

(6) the scheme would be widely adopted in practice. (Salmon 1672b, 248)

Although experiment was prominent here, it was not meant to test but to persuade. The checking of the harmonic intervals by the reader was meant to allow the reader to persuade himself the scheme was correct, not to allow Salmon to correct it if it were found wrong: in fact, Salmon did not even say that he had performed the checks himself. And the point of the public demonstration was to persuade others, not to check whether the mathematics had worked (Salmon 1672b, 6). (One might say that the reader and later the audience were invited to become witnesses to the correctness of Salmon's mathematics and its applicability to music: a strategy which arguably owed something to writings of Boyle and others which Salmon may have read.)

By contrast, here is Matthew Locke's statement about how to acquire musical knowledge.

All Creatures that have Ears are apprehensive of Sounds, but not of distinguishing them; those, whose Ears Nature hath prepared for Practical Music, by dividing and sub-dividing a String (for Example) come to experience their difference and distances; and from thence, by comparing them, to Tones, which (the Ear having distinguished into Consonants and Dissonants) they Arithmetically divide to the greatest quantity Practicable...and thence...advance to That we call *Composition*, the Mother of all Vocal and Instrumental Musick.

...

More of the Mathematicks than this, Sir, (excepting what belongs to the Mechanical Part thereof for the Making Instruments) signifies nothing to us.... You have...quitted the Field of *Practical Musick*, and run for shelter to the *Nature and Causes of Sounds*, which properly belongs to Philosophy. (Locke 1673, 15–16)

Salmon's career after the dispute with Locke was less explosive (partly, perhaps, because Locke died in 1677). I discussed above the pamphlet probably by him, 'The Musicall Compass', about musical tuning and notation, which appeared in 1684. In 1688 he wrote a book on tuning. He corresponded with John Wallis, and there are manuscripts of treatises by Salmon very similar to his 1688 tuning theory in both London and Cambridge: the Cambridge manuscript is among Newton's papers (Salmon 1688; Salmon 1705).[10] The 1688 book was in fact endorsed by Wallis.

All this is by way of prelude to, and to shed some light on, my fourth example of a mathematical musical instrument, a different type of modified viol. In each of these tuning texts Salmon described in detail a method for realizing his preferred tuning on stringed instrument. By contrast with a normal instrument, on which straight 'frets' govern the position of the fingers across all of the strings, Salmon proposed a scheme which placed corresponding frets at slightly different positions on each string (Salmon 1688, foldout). At the end of his book were scale diagrams for the system, which could in principle be transferred to the relevant part of an instrument, the fingerboard, quite easily: Fig. 7.3.2 shows an example. An advertisement in the *London Gazette* in 1689 offered to modify lutes according to Salmon's scheme (see Tilmouth 1961, 8).

Salmon continued to promote his new fretting at the Royal Society until shortly before his death, and in 1705 a musical 'experiment' was finally performed at a meeting of the Society under his direction:

Two Viols were Mathematically set out, with a particular Fret for each String, that every Stop might be in a perfect exactness: Upon these, a Sonata was perform'd by those two most eminent Violists, Mr *Frederick* and Mr *Christian Stefkins*, Servants to his Majesty; whereby it appear'd, that the Theory was certain, since all the Stops were owned by them to be perfect. And that they might be prov'd agreeable to what the best Ear and the best Hand performs in Modern practice, the famous *Italian*, Signior *Gasperini*, plaid another Sonata upon the Violin in Consort with them, wherein the most compleat Harmony was heard.[11] (Salmon 1705, 2069)

The ear did not produce knowledge here; it was simply asked to assent to knowledge already possessed. If the musical compass was an instrument for the visual display of musical theories, the modified viols enabled their aural 'display'. It is not clear what would have happened if the assembled company had failed to judge the harmonies they heard to be excellent: perhaps Salmon would have blamed the performers or listeners rather than revised his mathematical theories; but it is hard to imagine he would have allowed the trial to go ahead if he were not confident of its result.

One of the oddities of Salmon's tuning scheme was that it contained at least five different sizes of semitone, whose detailed arrangement depended on the key in which the performer wished to play. Salmon suggested, in fact, that several

10. Oxford, Bodleian Library, MS Eng Lett C 130 ff. 27–8: letters, Thomas Salmon to John Wallis, 31 December 1685; Wallis to Salmon, 7 January 1685/6; Cambridge, University Library, Add MS 3970, ff. 1–11: Thomas Salmon, 'Division of a monochord' (copy); London, British Library, Add MS 4919, ff. 1–11: [Thomas Salmon?], 'The practicall theory of musick …' (copy: diagrams only are in Salmon's hand).

11. See also Royal Society Journal Book X, 97, 102.

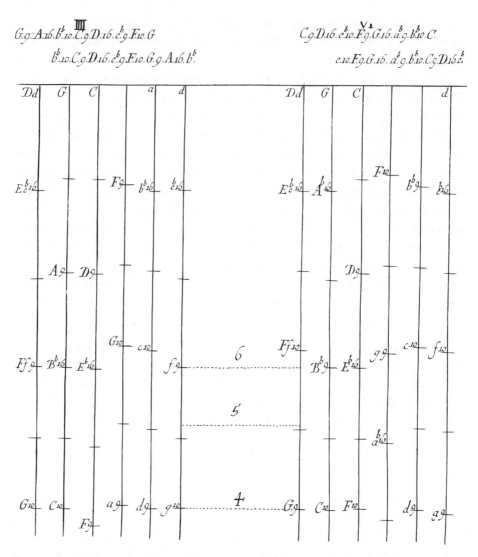

Figure 7.3.2 Where to place the frets on a viol (Salmon 1688, foldout). By permission of the British Library

different fingerboards be constructed, and a different one attached to the instrument each time a piece of music in a new key was to be played. This raises obvious questions about the feasibility of using his scheme to play real music, where key changes within continuous sections of music were hardly unusual by this period. But the problems of producing satisfactory intonation from a viol in a variety of keys were not trivial, and unusual fretting schemes are by no means unknown in viol technique: slanted, curved, even split frets (Crum and Jackson 1989, 159–163). What is revealing about Salmon's scheme is its avoidance of such practical and workable compromises, in favour of embodying in a musical instrument a mathematical perfection which may not even have been fully translated into sounds: the details of how the fingers are placed on the strings in relation to the frets can also be used to modify the pitch, and Penelope Gouk (1982, 230–231) has suggested that in Salmon's trial the performers may have used such a technique to cancel the effect of his unusual placement of frets. Although he achieved contact of a kind with the world of professional musical performance, he seems to have remained ignorant of the interplay between playing and listening by which good tuning could be achieved even with theoretically imperfect instruments, and indeed of the level of ad hoc negotiation which may be involved in such a thing as 'good tuning' for both players and listeners.

Apart from this failure of communication—a complex and recurring issue in the mathematical music theory of the period—perhaps the most remarkable feature of Salmon's theory of music is that both he and ultimately the Royal Society apparently thought that a meeting of the Society was an appropriate place to demonstrate it. Salmon's modified viols differ from the other experimental instruments which I have discussed by being very closely related to ordinary musical instruments: even after their mathematical modifications they remained real musical instruments which would be played, for preference, by professional musical performers. They were not pieces of experimental apparatus to be operated by the Curator of experiments. But the very closeness of Salmon's scheme to real musical performance, and the fact that musical skill was required to judge it, caused problems of interpretation which perhaps account for the fact that Salmon's musical experiment was entirely unique in the early history of the Royal Society.

Conclusion

These four very different instruments used for the mathematical study of music do not by any means exhaust the contents, or the idiosyncrasies, of mathematical music in late seventeenth-century England. It raised distinctive and probably unique problems about the relationship of mathematical knowledge to the senses and to instruments and, during this particular period, writers about the mathematics of music also laboured to achieve a workable relationship between musical practice and the new experimental practices of early modern science. One conclusion which these four instruments illustrate is that those writers failed to achieve a consensus.

The relationship of music to the mathematizing impulse which arose in the seventeenth century was complicated by the fact that music was already considered a branch of mathematics. Each of these instruments reflected a belief that music was inherently mathematical, a programme of study which aimed to update the mathematical basis of music, and uncertainty about how that was to be achieved.

These instruments illustrate how, in the context of the new theories of knowledge of the seventeenth century, mathematical music was exceptional, particularly in its relationships with mathematics, with the ear, and with musical practice. Questions about the reliability of the ear and its ability to produce knowledge were complex in the seventeenth century. Some musical scientists assigned the ear as nearly as possible no role at all, working mathematically in terms of pure reason and simply asserting that their results corresponded to real music: the musical compass illustrates such an approach, though not in its most extreme form. Others permitted the ear the role of recognizing mathematical excellence when it was rendered audible: as in Salmon's demonstration with modified viols. Others tried to find a role for the ear in producing knowledge, as fleetingly occurred during the long-string experiments at the Royal Society, but which proved problematic.

If the object of study in musical science was seriously considered to be musical sound, it was necessary for it to rely on musical practitioners, who alone could produce musical sound. Any experiments or experimental instruments used were liable also to rely on the skills of musicians and musical instrument makers. Musical knowledge was difficult to embody except in musical instruments and performances.

But this was not a situation in which the kind of objectivity sought in other sciences could be attained or even envisaged. A scientific instrument is typically supposed to make an experiment more 'objective', since 'an instrument cannot be prejudiced or passionate', but it is far from obvious that this could ever be said of a musical instrument or its operator (Hankins and Silverman 1995, 229). This is why Salmon's performance with modified viols was so exceptional, with its close dependence on real musical instruments and musical skills both of production and of judgement.

The relationship of mathematical theory to musical practice was problematic also because that theory remained prescriptive rather than descriptive. The few attempts to create mathematical descriptions of contemporary practice were hampered by the lack of suitable instruments and by lack of confidence in the ear's ability to turn sounds into numbers. The perception by practitioners that mathematical theory was therefore irrelevant to them may well have contributed to theorists' failure to build a relationship with practitioners which might have solved these problems.

Bibliography

Anon, 'An Accompt of some Books', *Philosophical Transactions of the Royal Society*, 6 (1671–2), 3088–3095.

Anon [Thomas Salmon?], *The musical compass,* [London, 1684].

Anon, *Transcript of the register of the worshipful company of stationers; from 1640–1708,* London: private printing, 1913–14.

Barbour, J Murray, *Tuning and temperament: a historical survey,* Dover publications, 2004 (reprint).

Barker, Andrew (ed), *Greek musical writings II: harmonic and acoustic theory,* Cambridge University Press, 1989.

Benedetti, Giovanni Battista, *Diversarum speculationum mathematicarum et phusicarum liber,* Turin, 1585.

Birch, Thomas, *A history of the Royal Society of London,* London, 1756–7.

Boyle, Robert, *Correspondence,* M Hunter, Antonio Clericuzio, and Lawrence M Principe (eds), Pickering and Chatto, 2001.

Caldwell, John, 'Music in the faculty of arts', in T H Aston (ed), *The history of the university of Oxford,* III, *The collegiate university,* James McConica (ed), Oxford University Press, 1986, 201–212.

Cannon, John T, and Dostrovsky, Sigalia, *The evolution of dynamics: vibration theory from 1687 to 1742,* Springer, 1981.

Cantone, *Armonica gregoriana,* Turin, 1668.

Chan, Mary, Kassler, Jamie C, and Hine, Janet, *Roger North's writings on music c. 1704–ca 1709: digests of the manuscripts by Mary Chan and Jamie C Kassler with analytical indexes by Janet D Hine,* University of New South Wales Press, 1999.

Charleton, Walter, *Physiologia Epicuro-Gassendo-Charletoniana: or a fabrick of science natural, upon the hypothesis of atoms,* London, 1654.

Christensen, Thomas Street (ed), *The Cambridge history of western music theory,* Cambridge University Press, 2002.

Cohen, H Floris, *Quantifying music: the science of music at the first stage of the scientific revolution, 1580–1650,* D Reidel, 1984.

Crum, Alison, and Jackson, Sonia, *Play the viol: the complete guide to playing the treble, tenor, and bass viol,* Oxford University Press, 1989.

Dostrovsky, Sigalia, 'Early vibration theory: physics and music in the seventeenth century', *Archive for the History of Exact Sciences,* 14 (1974–5), 169–218.

Fend, Michael, 'The changing functions of *senso* and *ragione* in Italian music theory of the late sixteenth century', in Charles F Burnett, Michael Fend, and Penelope Gouk (eds), *The second sense: studies in hearing and musical judgement from antiquity to the seventeenth century,* Warburg Institute, 1991, 199–221.

Fernandez, António, *Arte de musica…,* Lisbon, 1626.

Field, Christopher D S, and Wardhaugh, Benjamin (eds), *John Birchensha: writings on music,* Ashgate, 2009.

Galilei, Galileo, *Discorsi e dimostrazioni matematiche, intorno a due nuove scienze,* Leiden, 1638, translated by Stillman Drake as *Two new sciences including centers of gravity and force of percussion,* University of Wisconsin Press, 1974.

Galilei, Vincenzo, *Discorso intorno alle opere de Gioseffo Zarlino et altri importanti particolari attenenti alla musica* [Discourse concerning the work of Gioseffo Zarlino and other important particulars pertaining to music], Venice, 1589.

Gassendi, Pierre, *Syntagma philosophiae Epicuri…,* The Hague, 1649.

Gassendi, Pierre, *Manductio ad theoriam musices,* in Gassendi, *Opera omnia,* Lyon, 1658, V 633–658; translated into French as *Initiation à la théorie de la musique, texte de la 'Manuductio'* [Introduction to music theory: text of the 'Manuductio'] by Gaston Guieu, Edisud, 1992.

Gibson, Strickland (ed), *Statuta antiqua universitatis oxoniensis*, Oxford University Press, 1931.

Gouk, Penelope M, 'The role of acoustics and music theory in the scientific work of Robert Hooke', *Annals of Science*, 37 (1980), 573–605.

Gouk, Penelope M, 'Music in the natural philosophy of the early Royal Society', unpublished PhD thesis, University of London, 1982.

Gouk, Penelope M, *Music, science and natural magic in seventeenth-century England*, Yale University Press, 1999.

Hankins, Thomas L, and Silverman, Robert J, *Instruments and the Imagination*, Princeton University Press, 1995.

Haynes, Bruce, 'Pitch' in L Macy (ed), *Grove music online*, http://www.grovemusic.com (2001). Accessed on 18 February 2004.

Herlinger, Jan W, 'Marchetto's Division of the Whole Tone', *Journal of the American Musicological Society*, 34 (1981), 193–216.

Holder, William, *A treatise of the natural grounds, and principles of harmony*, London, 1694.

Hooke, Robert, *The diary of Robert Hooke 1672-1680*, Henry Robinson and Walter Adams (eds), Wykeham Publications, 1968.

Hunter, Michael, and Schaffer, Simon (eds), *Robert Hooke: new studies*, Boydell, 1989.

Huygens, Christiaan, *Le cycle harmonique (1691) and Novus cyclus harmonicus (1724)*, Rudolf Rasch (ed), Diapason Press, 1986.

Huygens, Christiaan, *Oeuvres complètes* , The Hague: Martinus Nijhoff, 1888–1950.

Kassler, Jamie C, *The beginnings of the modern philosophy of music in England: Francis North's A philosophical essay of musick (1677) with comments of Isaac Newton, Roger North and in the Philosophical transactions*, Ashgate, 2004.

Kassler, Jamie C, and Oldroyd, David R, 'Robert Hooke's Trinity College "Musick scripts", his music theory and the role of music in his cosmology', *Annals of Science*, 40 (1983), 559–595.

Locke, Matthew, *Observations upon...An essay to the advancement of musick by Thomas Salmon*, London, 1672.

Locke, Matthew, *The present practise of musick vindicated against the exceptions and new way of attaining musick, lately publish'd by Thomas Salmon MA*, London, 1673.

Marchetto da Padova, *Lucidarium*, Jan W Herlinger (ed), University of Chicago Press, 1985.

Mersenne, Marin, *Harmonie universelle...*, Paris, 1636.

Morse, Philip M, *Vibration and sound*, McGraw Hill, 1948.

Moyer, Ann E, *Musica scientia: musical scholarship in the Italian renaissance*, Cornell University Press, 1992.

North, Francis (Anon), *A philosophical essay of musick, directed to a friend*, London, 1677.

North, Roger, *The life of the Lord Keeper North, edited with an introduction, notes and appendices by M Chan*, Mellen, 1995.

Otegem, Matthijs van, 'Towards a Sound Text of the *Compendium musicae*, 1618–1683, by Rene Descartes (1596-1650)', *Lias*, 26 (1999), 187–203.

Perrault, Claude, *Essais de physique, ou receuil de plusieurs traitez touchant les choses naturelles II: 'De la bruit'*, Paris, 1680, 1688.

Salmon, Thomas, *A vindication of an essay to the advancement of musick, from Mr Matthew Locke's Observations. By enquiring into the real nature, and most convenient practise of that science*, London, 1672a.

Salmon, Thomas, *An essay to the advancement of musick, by casting away the perplexities of different cliffs. And uniting all sorts of musick...in one universal character*, London, 1672b.

Salmon, Thomas, *A proposal to perform musick, in perfect and mathematical proportions*, London, 1688.

Salmon, Thomas, 'The theory of musick reduced to arithmetical and geometrical proportions', *Philosophical Transactions*, 24 (1705), 2072–2077, 2069 [misnumbered].

Sauveur, Joseph, *Collected writings on musical acoustics (1700–1713)*, Rudolf Rasch (ed), Diapason Press, 1984.

Taylor, Brook, 'De motu nervi', *Philosophical Transactions of the Royal Society*, 28 (1713), 26–32.

Tilmouth, Michael, 'A Calendar of References to Music in Newspapers Published in London and the Provinces (1660–1719)', *Royal Musical Association Research Chronicle*, 1 (1961), 1–107.

Wardhaugh, Benjamin, 'Mathematical and mechanical studies of music in late seventeenth-century England', unpublished DPhil thesis, University of Oxford, 2006.

Wardhaugh, Benjamin, 'Musical logarithms in the seventeenth century: Descartes, Mercator, Newton', *Historia mathematica*, 35 (2008), 19–36.

Wardhaugh, Benjamin, *Music, experiment, and mathematics in England, 1653–1705*, Ashgate, 2008.

Modernism in mathematics

Jeremy Gray

There were dramatic changes in several branches of mathematics between 1880 and 1920. Among the best documented are the rise of modern algebra, associated with the work of Emmy Noether and her followers and which she traced back to the work of Richard Dedekind; Henri Lebesgue's axiomatic theory of the integral; the introduction of algebraic topology by Henri Poincaré; and the new axiomatic geometry first developed by the Italian school around Guiseppe Peano and later by David Hilbert.[1] This chapter considers the extent to which these developments can be regarded as a 'modernism' similar to the rise of modernism in cultural spheres such as painting, music, and literature. The first section sets out some of the general issues in making a claim of this kind, followed by arguments that put the case for a modernist period in algebra, analysis, and geometry.

Modernism in mathematics?

The term 'modernism' in the arts is well understood to refer to the kind of paintings that Pablo Picasso did, or the kind of music that Arnold Schoenberg wrote.

---

1. There were comparable changes in some other branches of mathematical analysis, notably in the theory of metric spaces due to Fréchet and the creation of functional analysis in the wake of Hilbert's contributions, but there is not space here to consider them.

James Joyce's *Ulysses* is an exemplary modernist novel, and we can also speak of modernist architecture. Beyond the general agreement that something major happened in the practice of those disciplines there is, inevitably, much disagreement. When did it start, and when did it end? Was it a complete change in attitudes, or at best a partial success? Some of this disagreement is apparent even with the names just cited: many more people look at Picasso's painting with pleasure than listen to Schoenberg's music with enjoyment. We must accept that modernism refers, at best, to a sprawling, overlapping set of activities, loosely centred on a decade or so around 1900 that has at its core some characteristic features. This chapter argues that we can equally well apply the term 'modernism' to changes in mathematics, in particular that there is a similarly identifiable core, and that it can be useful to do so. But first we should look to see if there is any chance of success.

In fact, it is almost a commonplace that mathematics underwent a major transformation in the years around 1900. Various branches of mathematics, most notably modern algebra, wear the adjective 'modern' on their sleeve to this day, and until recently the terms 'modern geometry' and 'modern analysis' summoned up recognizable lists of topics that dated more or less from that period. But when one attempts to characterize that transformation in any precise way, problems arise and multiply. What indeed is involved in portraying this transformation as a modernist one? What are the obstacles in our path?

One is simple chronology. What period of time counts as the decisive one? There should presumably be quite a short period of time, a matter of two or three decades at the most, when modernism was novel and fighting for acceptance, although outliers can be accepted, and then a possibly longer period where it was the new orthodoxy, if still exciting. The arrival should be roughly synchronous with those of the artistic modernisms if the name is to have any real value. In point of fact there is a period from about 1880 to 1914, which covers much of the work of Dedekind, Georg Cantor, and Hilbert, the heyday of logicism, and numerous other developments, that is a strong candidate. Modern algebra, which is particularly associated with the work of Noether and her school in the 1920s and early 1930s, would then belong to the period of acceptance, as would the rise to dominance of Nicholas Bourbaki and the structuralist movement in the 1930s and 1950s.

Two more obstacles are the problem of deciding what modernism is, and of adapting it to a mathematical setting. The danger here, as Corry (forthcoming) has pointed out, is that of firing the arrow first and then drawing the bull's eye round it. In other words, one might identify some features of mathematics around 1900 as coherent, then look around for a suitable exponent of modernism (there are many, and one is free to join in) and then announce that these mathematical features are examples of that kind of modernism. At its worst, such an exercise would be devoid of value, its conclusions artifacts of the construction process and liable to collapse as soon as the terms of the analysis were altered. While it is easy to avoid such naked gerrymandering of the debate, more sophisticated variants of it are harder to prevent. Corry's own proposal, that the processes of mathematical change in some period be analysed and only be agreed to be somehow modernist if they seem to be close to the processes involved in the cultural modernist transformations, is a good one and will be adopted here, but it does not guarantee immunity from the charge of drawing the bull's eye last.

Another obstacle is that abstraction has long been a characteristic of certain kinds of mathematics. If one is to lay stress on the formal and abstract character of modern painting, music, and architecture with the associated rejection of classical conventions of meaning and beauty, then one cannot argue that mathematics went the same way at the same time. The most that can be said is that it went from being abstract one way to being abstract in a new way, albeit a deeper, more significant way.

Finally, two smaller but still significant issues. Mathematics is held by its practitioners to remarkable standards of rigour. You cannot simply say what you like, as you can in painting or (perhaps to a lesser extent) music. This cognitive aspect distinguishes it from the arts and puts it with the sciences, whatever its differences from science may be. Connected with this is the social structure of the mathematics profession, which is closer to that of architects, who must also obey a number of external criteria, than to that of novelists or painters. So if one's criteria for modernism include political ones (for example, a rejection of the bourgeoisie and all they stand for) it is unlikely that one will find modern mathematicians. And since one does not find anti-bourgeois professors, can one reject that aspect of modernism without, once again, drawing the bull's eye last?

If all these obstacles can be overcome, one still has to redeem the effort. Labelling a certain batch of mathematical activities as modernist will only be worthwhile if some features of mathematics and mathematical life in the period stand out more clearly as a result, if certain useful analogies can be drawn, if new and better questions can be asked. We have to know something we did not know before, and know it in a useful and insightful way. This, rightly, is the lure that has drawn a number of authors starting with Mehrtens (1990), and including Everdell (1997), Epple (1999), and myself (Gray 2004; 2008). It is right to hold such people to high standards, because they wish to have their work regarded as important. If they are to start a trend we need some assurance that they are heading in a good direction.

Here the present state of the historiography of mathematics needs to be considered. Over the past thirty years, much of the nineteenth century has been well studied, the largest single exception being the vast topic of partial differential equations with its deep connection to physics. The history of logic and mathematical logic in the period has also been well worked over. There is a feeling among historians that much has been done, that it would be convenient if it could be drawn together, and that perhaps some substantial general claim has been missed. (For what it is worth, there is presently no articulate exponent of the view that such pattern drawing is wrong and good history, like the devil, is all in the details, but the point should be remembered). Much less is known about the period after 1918, for the simple reason that the mathematics is much harder, and some historical accounts are little more than a trudge through the major relevant papers. Mathematicians have contributed a number of useful surveys, historians of mathematics have made inroads, but gaps of all kinds make generalizations almost impossible. That in turn contributes to the feeling that some kind of an assessment can be made of the period to 1914, but not for any later date.

Let us now see what can be made to push against these obstacles. It will be helpful to consider mathematical developments under three headings: algebra, analysis, and geometry. If it is agreed that the label 'modernist' can only be used about developments starting at roughly the same time as modernism began in the artistic spheres, that could narrow the field to a period from about 1900 to 1920. But it must be allowed, of course, that there is no agreed starting date for any kind of modernism. If one takes a hard line, with say Picasso and Schoenberg, then 1900 is a good round figure. If one admits Stéphane Mallarmé and Paul Cézanne, then an earlier starting point is allowed. It does not seem unreasonable, therefore, to begin by considering the mathematics of the later decades of the nineteenth century.

Fishing in these well-trawled waters brings up the following much studied items. In algebra, the work of Dedekind in the 1870s and 1880s produced a structural theory of algebraic number theory, one often juxtaposed to the alternative contemporary version due to Leopold Kronecker, which was overtly computational and algorithmic. Dedekind's version was the one picked up by Hilbert in his influential *Zahlbericht* 'Report on the theory of numbers' in 1897, his report on the state of research in number theory. From there it passed after World War I to Noether. There is also the emergence of abstract group theory and the explicit recognition of a new mathematical object, the group. In analysis there is the creation of measure theory and the Lebesgue integral (1902), Maurice Fréchet's theory of metric spaces (1906), and the emergence of point-set and algebraic topology. There is also the construction of the real numbers by Dedekind, Cantor, Charles Méray, and Heinrich Heine in the 1870s and 1880s. In geometry there is Riemannian differential geometry and the end of the hegemony of Euclidean geometry, and abstract axiomatic geometry, most notably but not only on display in Hilbert's *Grundlagen der Geometrie* 'Foundations of geometry' (1899). We should also note the gradual acceptance of the idea that set theory is the proper foundation for mathematics, although space does not permit a discussion of that in this chapter (Ferrierós 1999).

This trawl generates some problems. We have to settle on a manageable catch and not let it become impossibly too large to survey, but the items caught must also make the case, so we have to see what is potentially modernist about these fish. Evidently there is not room in this chapter to describe each of these items adequately and then to argue that they display convincing and characteristic modernist features. Besides, we must address the bull's eye problem, and as yet we have not set out a definition of modernism.

There is, however, a well-grounded consensus about what these works contain and what is most significantly novel about them. Since any claim to novelty involves a contrast (it used to go like *that* but now it goes like *this*) it will help to set out briefly the way things were before, so that a sharp contrast can be made. Mathematics in the eighteenth century, for both the philosopher Immanuel Kant and the mathematician Jean le Rond d'Alembert, was about quantity: counting and measuring. On this view, which may be called naive abstractionism, mathematicians have privileged access to certain primitive concepts, such as number and length: geometry abstracts from the world its key mathematical features and is the idealized study of the space around us. In the eighteenth century there was no sharp divide between mathematics and science: the calculus spanned much of high level mathematics and almost all of theoretical science. This view persisted through the nineteenth century, although it became more sophisticated as mathematics and the sciences deepened and diversified. The supposedly transparent quality of mathematical abstraction permitted mathematicians to claim their conclusions are true, and that their proofs can be seen to be without error. It also limited the applications of mathematics to domains that are broadly amenable to similar acts of abstraction. It would be hard, for example, with this philosophy of geometry to study spaces of dimensions greater than three. Part of the claim made for mathematical modernism is that these views collapsed and were replaced by something very different.

We now have our candidates for mathematical modernism and criteria for comparing them with what went before, so it is time to confront the bull's eye problem. The only way round it is to propose a definition of modernism that is fair to the cultural movements commonly reviewed under that heading, to see that it can be adapted to the context of mathematics, and that when the case has been made there is as little contrivance about the conclusions as possible. It is not necessary for an author to drag in every piece of evidence that supports his or her case; enough may well be enough. Perhaps items left undiscussed will further support the case. More importantly, there should not be extensive passages in the history of mathematics in the period that refute the case.

The definition of modernism that is proposed here is this: modernism is an autonomous body of ideas, having little or no outward reference, placing considerable emphasis on formal aspects of the work and maintaining a complicated, indeed anxious, rather than naive relationship with the day-to-day world; further, it is the de facto view of a coherent group of people, such as a professional or discipline-based group who have a high sense of the seriousness and value of what they are trying to achieve.

There is not the space here to argue that this, surely somewhat simplified, description captures the essential features of a modernism. But it is at least consistent with the view of one early practitioner and astute critic, Guillaume Apollinaire, who in 1912, speaking of many young painters said:

These painters, while they still look at nature, no longer imitate it, and carefully avoid any representation of natural scenes which they may have observed, and then reconstructed from preliminary studies. Real resemblance no longer has any importance, since everything is sacrificed by the artist to truth, to the necessities of a higher nature whose existence he assumes, but does not lay bare. The subject has little or no importance any more. [...] Thus we are moving towards an entirely new art which will stand, with respect to painting as envisaged heretofore, as music stands to literature. It will be pure painting, just as music is pure literature. [...] This art of pure painting, if it succeeds in freeing itself from the art of the past, will not necessarily cause the latter to disappear; the development of music has not brought in its train the abandonment of the various genres of literature, nor has the acridity of tobacco replaced the savoriness of food. (Translated by Lionel Abel and cited in Chipp 1968, 222—223)

Similar remarks could be drawn from the writings of Schoenberg about his twelve-tone system, or other works that define modernism in music, as these words by Charles Rosen attest:

With Schoenberg and Webern, [...] and with Stravinsky (starting with the Rite of Spring) we must generally begin with a dispassionate understanding of the art and an appreciation of the technique in order to comprehend the emotional content [...]. In those works of the modernist movement considered difficult or hermetic, in short, the content is partially withheld from us until we have understood the technique. (Rosen 1999, 44)

The next three sections will demonstrate how closely this definition of modernism permits us to speak of the transformation of mathematics in these terms. But just as no-one who speaks of modern art or modern music requires *every* painter or composer to work in the modernist way, and merely requires that enough did, so too the claim that there was a distinctly mathematical modernism requires only that many mathematicians worked that way, not that all of them did. How many, and to what extent, will occupy us in the final section.

Modernist algebra

Dedekind's algebraic number theory is a paradigm case, and the comparison with Kronecker's work done at exactly the same time helps to point up crucial aspects of its novelty (Reed 1994; Corry 2003; Avigad 2006).

For most of his life, Richard Dedekind worked on algebraic number theory, and much of his work took the form of extensive commentaries on the work of Dirichlet, which itself was a working and re-presentation of the work of Carl Friedrich Gauss. Both Dedekind and Kronecker saw themselves as the inheritors of a major tradition that had started with the publication of Gauss's *Disquisitiones arithmeticae* (see Goldstein, Shappacher, and Schwermer 2007). According to this tradition number theorists were occupied with certain algebraic generalizations of the familiar integers (the numbers 0, 1, 2,... and also -1, -2,...) to other objects with integer-like properties. The objects might be the roots of polynomial equations with integer coefficients, or (Kronecker's view) the polynomials themselves. These properties include one number object dividing another, or being a prime, and so forth. The problem that Dedekind and Kronecker grappled with most intently was not merely to extend these concepts but to do so in the 'right' way, which makes it all the more interesting that their answers differed markedly.

Kronecker employed Gauss's idea of congruences, with which the *Disquisitiones arithmeticae* opens. Kronecker noted that to define the negative integers one can introduce a letter (an unknown or variable) x and subject polynomials in x with positive integer coefficients to the congruence $x + 1 \equiv 0$. Similarly, imaginary numbers can be defined via the congruence $x^2 + 1 \equiv 0$. In fact, Kronecker's general treatment of number theory started with the integers as described, the rational numbers, and so on, using arbitrary congruences with an arbitrary but finite number of unknowns (see Neumann 2007). As his obituarist, Heinrich Weber, famously remarked, it was Kronecker's opinion that 'God made the integers, all else is the work of Man' (Weber 1891–2, 19). Numerical expressions of this kind met his standards for what could be truly known, whereas the real numbers could only be understood, in Kronecker's opinion, as sequences of better and better approximations to them by rational numbers. Kronecker felt that his essentially algorithmic approach was constructive, and he placed non-constructive arguments outside the realm of strict mathematics.

Dedekind's successive reformulations of number theory began with an issue that was also visible in Kronecker's theory but which went back to the work of Kronecker's mentor Eduard Kummer, who had studied numbers called cyclotomic integers. These are polynomial expressions in powers of roots of unity, and Kummer had come up with a divisibility theory for them. He had found that it was possible to define irreducible numbers of this kind: they are numbers that cannot be written in a nontrivial way as a product of two cyclotomic integers. It is even possible to define primes in this context: a cyclotomic integer is prime if whenever it divides a product it divides one of the factors. But, rather to his surprise, Kummer found that such prime numbers can lack some of the usual properties of primes: specifically, there were cases when the uniqueness of factorization into prime numbers failed for the cyclotomic integers (Edwards 1977).

This phenomenon is time consuming to illustrate with the cyclotomic integers, but it is easy to demonstrate that, in certain circumstances, irreducible numbers can fail to be primes, which already shows that some of the familiar properties of the usual integers will fail to generalize. Consider 'integers' of the form $m + n\sqrt{-5}$, where m and n are ordinary integers. We have $6 = 2 \times 3 = (1 + \sqrt{-5})(1 - \sqrt{-5})$, and it is easy to show that, in this system, while 2, 3, $(1 + \sqrt{-5})$, and $(1 - \sqrt{-5})$ are irreducible, none is prime.

In Kummer's case, to cope with the extensive calculations that can arise, he developed a test to show when a cyclotomic integer is divisible by a prime. Faced with the problems non-uniqueness of prime factorization would cause, he restored uniqueness by allowing that a cyclotomic integer might have ideal factors, in which case its prime divisors (the 'numbers' that made it up as a product) would not all be cyclotomic integers. Dedekind's first objection was that it was wrong to have a test for a new number being prime which gave a negative answer, and then not to say what the prime factors are. So Dedekind began with the question of divisors in mind, and was moved to create still newer kinds of numbers (called, reasonably enough, 'ideals') to rescue unique factorization by concretely defining the new divisors.

After a long struggle to give his new objects the 'right' definitions, Dedekind defined his ideals as infinite sets of numbers—a point worth noting—and created an arithmetic for them. Interestingly, it was only on his fourth formulation of the theory that he was able to base the theory on multiplication of ideals and derive division as the inverse of multiplication. More significantly still, on a number of occasions he disdained to use Kronecker's explicit methods, because he believed that they placed too much emphasis on the representation of algebraic numbers and not enough on what they really are. It seems likely this distinction between an object and its representation is one that he had learned to appreciate from his friend Riemann, who had based his dramatically novel theory of complex functions on exactly this distinction in the 1850s. Neither man suggested that objects cannot be studied via their representations, but both believed that one must be vigilant to ensure that one establishes properties of the objects themselves and not the properties of merely this or that representation, and to this end it was best to avoid explicit representations whenever possible.

When Hilbert published his *Zahlbericht* in 1897 it was clear that he preferred Dedekind's way of working to Kronecker's. From there, abstract ring theory passed eventually to Noether, who more or less eliminated polynomial methods in her papers of the 1920s. Her inspiration too came from Dedekind, and indeed the essential first step into algebraic modernism had been taken by him. Since then, the objects of algebraic number theory have been sets, usually infinite sets. An abstract arithmetic of these sets is defined, and the fundamental properties of algebraic number theory are said to be those of (to be anachronistic) commutative ring theory in the number-theoretic setting. It is a long way from, or underneath, the familiar theory of the natural numbers, and it makes no appeal to experience as a source of knowledge about the objects under study.

Modernist analysis

To this day, modern analysis—the rigorous theory of the calculus—is a topic that separates mathematicians from engineers and even physicists. Not only does it seem abstract and formal but, and this is the problem it presents to any thesis about mathematical modernism, so much had already been done by Cauchy in the 1820s that one can wonder what was left to be done. Cauchy, after all, had given recognizably modern definitions of continuity, differentiability, and inte-grability along with theorems connecting them, couched in the abstract language of ε, δ methods. If no further transformations happened in mathematical ana-lysis then the modernist thesis will fall because of chronological difficulties or, if Cauchy's achievements are magicked away, it will fall into the bull's eye problem. However, the second half of the nineteenth century saw radical innovations in a number of key areas, and here four will be briefly discussed: the idea of a general function, the concept of the real number, new ideas about integration, and the emergence of algebraic topology.

In a major paper of 1854 on the limitations of the method of Fourier series, Bernhard Riemann made a distinction between what he regarded as functions that arise in nature—that is, in the study of natural phenomena—and others that arise outside the physical sciences, for example in the number theory he was studying. He argued that the need for clarity and rigour in the principles of the infinitesimal calculus could not be met until these new functions were prop-erly understood. Could there be functions to which the calculus could not apply? He produced explicit examples of functions that failed in various ways to agree with their Fourier series, or failed to have a Fourier series at all. In Riemann's opinion, these functions were not only entirely arbitrary, they were the exclusive property of mathematics because they do not occur in nature. He was quite clear that mathematics went beyond objects one might say were (possibly idealized) abstractions from things in the 'real' world.

The problem is particularly acute because one might say that there is an elem-ent of nonlinearity at work. Riemann's functions could not be defined without the appropriate mathematical techniques; they were a product of the techniques used to analyse them. Mathematics then, for Riemann, was about concepts created by the mathematician and an important problem was to evaluate the methodo-logical tools by which functions are created. It was the artificial world of objects created by mathematicians that might baffle them, not the ingenuity of nature.

The second issue, the construction of the real numbers, is much better known, and I can be brief. Problems with Fourier series made it seem worthwhile to define the real numbers rigorously. Dedekind's celebrated cuts, first published in 1872, defined a real number as a division of the rational numbers into a pair of disjoint sets, say L and R, where every member of L is less than every member of R. If L contains a greatest element, or R a least element, the cut reduces to that element, but it defines a new, irrational, number if no such element exists. Dedekind went on to define an arithmetic for these numbers and to show that repeating the construction gives nothing new, so the set of rational and irrational numbers is complete (it has no 'gaps'). Cantor (like Heine and, independently, Méray) started from the idea that an irrational number is usefully approximated by a sequence of rational numbers that 'converges' to it, and defined the irrational numbers as equivalence classes of Cauchy sequences of rational numbers.

So mathematicians now constructed the real numbers, and no longer simply presumed that they existed. This programme, which Klein (1895) called the arithmetization of analysis, in a sense reduced the real numbers to the integers, but at the high price of admitting infinite sets into the foundations of mathematics. The measuring numbers, as they were sometimes called, which were among the fundamental objects of classical mathematics, changed from idealized abstractions obtained from thinking about length and measurement to infinite sets of integers, which could then be proved to have the well-known properties required for measurement along lines. Once this was done the way was open, and eventually taken, to describe objects in a variety of settings that had only some of the familiar properties of the real numbers. At the same time, the requirement that properties of the real numbers be properly established changed from being a matter of common sense to something requiring proof, and the way properties of the real numbers entered into the proofs of theorems changed accordingly as well.

The third example is Lebesgue's theory of the integral as he presented it in 1903. This illustrates another characteristic feature of modern mathematics, one that had been introduced a few years earlier by Hilbert in his work on geometry (to be discussed below): the use of axiomatic formulations. In 1902—3 Lebesgue was invited to give the prestigious Cours Peccot Lectures at the Collège de France. He found the opportunity stimulating, and the published version, his *Leçons sur l'intégration et la recherche des fonctions primitives* 'Lessons on integration and the search for primitive functions', carries a fine axiomatization of the idea of integration. Lebesgue showed that there was essentially only one good definition of the integral that satisfied certain natural axioms, and that his definition of the integral exemplified them. The axioms are:

1. $$\int_a^b f(x)\,dx = \int_{a+h}^{b+h} f(x-h)\,dx;$$

2.
$$\int_a^b f + \int_b^c f + \int_c^a = 0;$$

3.
$$\int_a^b (f+g) = \int_a^b f + \int_a^b g;$$

4.
If $f(x) \geq 0$ and $b > a$; then $\int_a^b f \geq 0$;

5.
$$\int_0^1 1 = 1;$$

6.
If $f_n(x) \uparrow f(x)$ then $\int_a^b f_n \uparrow \int_a^b f$.

The axioms look unproblematic, even to be the very minimum that anyone would require of a theory of the integral. But in fact the sixth is not true of the Riemann integral, and Lebesgue had to work hard to show that there was a way of defining the integral (known, of course, today as the Lebesgue integral) that satisfied all six properties. More importantly for present purposes, the axiomatic approach conceals another novelty. The axioms specify what the integral is intended to do. They do not start from an idea that the integral is about, say, area, or any other primitive concept. It is necessary to show that there is a model of these axioms, but once that is done it is at least possible to prove properties of the integral directly from the axioms and without reference to any model of them. The axioms are sometimes said to define their object implicitly, or to create it. There is no reference to a primitive concept available via abstraction from the natural world.

Lebesgue's theory of the integral was intimately tied to his definition of the measure or size of a set. That too was defined axiomatically, and axiomatically presented material can be used in ways that concrete examples cannot: to show that some things are incompatible with any theory of the indicated kind. This was dramatically illustrated when Felix Hausdorff published what became known as his paradox, which established that on any plausible definition of the measure of a set there must be non-measurable sets (Hausdorff 1915).

Mathematicians now had three choices. They could either agree that there are non-measurable sets on any definition of measure that obeys Lebesgue's axioms; or they must reject the concept of measure and look for another concept altogether—but Lebesgue's six axioms are very natural ones; or they must scrutinize the proof and hope to find a flaw. The French mathematician Émile Borel, who had been close to some of the ideas that informed Lebesgue's theory of measure, reacted in the third way. In the second edition of his *Leçons sur la théorie des fonctions* 'Lessons on the theory of functions' he rejected the paradox on the grounds that the sets in Hausdorff's paradox were not properly defined, because they had been 'constructed' using the axiom of choice. 'If one scorns precision and logic', he wrote, 'one is led to contradictions' (Borel 1914, 256).[2] Borel was not alone at this time in doubting the axiom of choice. Hausdorff, on the other hand, was not at all bothered that any definition of the area of a set is inherently imperfect, even though this was a conclusion that could never have been dreamt of by researchers a generation before.

The nascent field of topology displays the most obviously modernist objects, even if they can also be presented very informally (see Epple 1999). Here, one example must stand for many. In 1895 Poincaré began to publish a series of papers on *analysis situs* 'analysis of position', as he called topology, that were intended to open up the field. The opening paper began: 'The geometry of *n* dimensions has a real subject; no-one doubts this today. The objects of hyperspace can be given precise definitions just like those of ordinary space, and if we cannot represent them we can conceive of them and study them' (Poincaré 1895, 1).[3] In fact, even precise definitions were hard to give that could also be made to yield results. Poincaré knew very well that surfaces can be made out of polygons by glueing pairs of edges together. Analogously, one of the ways in which he constructed three-dimensional 'spaces' that had then to be classified was by glueing pairs of faces of a solid polyhedron together. He then asked: what aspect of the study of curves and surfaces in ordinary space generalize, what features are entirely new?

He found it was a productive way forward to take an idea that had proved useful in the classification of surfaces, and to consider loops in the three-dimensional 'space'. Two loops are considered equivalent if one can be deformed into the other. For example, every loop drawn in a solid ball can be shrunk to a point, so all loops are equivalent in this 'space'. But consider a solid ball from which an unknotted solid tube has been removed. Now there are several kinds of in-equivalent loop one can draw, which vary according to how they are wrapped around the tube (it is still more complicated if the tube is knotted). Poincaré studied the different types of loop one can draw in the spaces he was interested in, and exploited the idea that if all the loops start and finish at the same point in the 'space' then one can follow one loop by another and obtain a third loop. In this way (after making some technical refinements) he was able to regard his loops as elements of a group. For the unknotted solid tube, the group obtained in this way is the integers, for the solid tube knotted like a torus, the group is more complicated. For the spaces Poincaré was interested in, he found that he could only guess, but not prove, that the solid three-sphere (the locus $x^2 + y^2 + z^2 + w^2 = 1$ in R^4)

2. Si l'on fait si de la précision et de la logique, on est conduit à des contradictions.
3. La Géométrie à *n* dimensions a un objet réel; personne n'en doute aujourd hui. Les êtres de l'hyperespace sont susceptibles de définitions précises comme ceux de lespace ordinaire, et si nous ne pouvons pas nous les représenter, nous pouvons les concevoir a étudier.

can be distinguished from other spaces in this group-theoretic way. This is the origin of the famous Poincaré conjecture, which was only solved over a hundred years after his comments, in work for which the Russian mathematician Grigory Perelman was awarded a Fields Medal in 2006. Such are the difficulties of this branch of mathematics.

The modernist shift here is from the world of genuine three-dimensional objects to artificially defined 'spaces' studied via the groups that describe some of their cruder geometric features. At stake was the question of how to extend the methods of geometry to problems involving three or more variables, and thus was born algebraic topology.

Modernist geometry

Modern geometry may be said to have started in Germany and Italy, but was pursued in each country with different aims. This explains the different fortunes of the two schools: Hilbert's became famous, but after World War I the Italian contribution was largely forgotten. The most radical and influential among the Italians was Mario Pieri. His work is characterized by the complete abandonment of any intention to formalize what is given in experience. Instead he treated projective geometry 'purely deductively and abstractly [...independently of] any physical interpretation of the premises' (Pieri 1895; here Pieri 1980, 13).[4] Primitive terms, such as line segments, 'can be given any significance whatever [...] in harmony with the postulates that will be successively introduced' (Bottazzini 1988, 276; Marchisotto and Smith 2007).[5] In Pieri's (1898) presentation of plane projective geometry nineteen axioms were put forward (typically: any two lines meet).

Initially it was the Italian work, rather than Hilbert's, which travelled best. Pieri's ideas were taken up by Louis Couturat (1905), Alfred North Whitehead (1906), and John Wesley Young (1911). It seems that in the early years of the twentieth century Pieri's ideas met with a greater degree of acceptance than is commonly recognized today. But the Italians' work was limited in two ways: they missed the potential for creating novel geometries, and they failed to see the broader significance of the axiomatic method. Gino Fano (1892), for example, worked his way through the axioms of projective geometry, listing them successively and at each stage testing for independence by finding a model that satisfies all the previous axioms but not the new one. For example, given three points on a line, is there a fourth point on the line that is the fourth harmonic point of the previous three? Fano saw that the answer is evidently not if there are only three points on a line,

---

4. Puramente deduttivo e astratto [...] ogni interpretazione fisica delle premesse.
5. si può attribuire qualsivoglia significato [...] in armonia coi postulate che saranno man mano introdotti.

and he produced a geometry in which there *are* precisely three points on a line. Therefore an axiom is needed to ensure that there are at least four points on a line. Fano was certainly not treating points and lines as abstractions obtained from the real world. But he also made it clear that a geometry with only three points on a line is to be excluded, not embraced and studied. The creative aspect of the axiomatic method was passed over, and with it the chance to promote the same method in other branches of mathematics.

The Fano plane, as this geometry is called today, was neglected because of the pedagogic mission of the Italians. Their intention was to spell out, once and for all, what elementary geometry was for the purpose of educating future school teachers. Novel geometries and research in this area were not on the agenda.

Matters were very different with Hilbert. Novel geometries had a certain interest for him, although Adolf Hurwitz's insight was undoubtedly right when he wrote to Hilbert to say: 'You have opened up an immeasurable field of mathematical investigation which can be called the 'mathematics of axioms' and which goes 'far beyond the domain of geometry' (Toepell 1986, 257).[6] Hilbert's presentation of geometry in the *Grundlagen der Geometrie* sold the axiomatic idea very powerfully, and in his lectures over the next decade he promoted the method with varying degrees of success in various branches of mathematics and physics.

Hilbert's axiomatic message would not have been so clear if he had not also dramatically changed the whole approach to geometry. Hilbert's presentation of geometry is based on the introduction of successive families of axioms (the first three families concern incidence, order, and congruence), together with an exploration of what can and cannot be done with the collection of axioms at any stage. So Hilbert was the first to prove what others had already begun to suspect, that while the axioms of projective geometry in three dimensions do permit one to prove Desargues' theorem, the axioms of projective geometry in two dimensions do not. This has the remarkable consequence that there are plane projective geometries that cannot be embedded in any three-dimensional projective space.

Hilbert found his way to these results through an astute use of what he called segment arithmetic, the geometric manipulation of segments that allows them to be added and multiplied. Different axiom systems generate different segment arithmetics, which in turn determine the sorts of numbers that can be admitted as coordinates. Desargues's theorem, it transpired, is true only if the coordinates are drawn from a field, and will be false in a geometry in which the coordinates belong to a non-commutative ring.[7] Geometry in this setting

---

6. Sie haben da ein unermeßliches Feld mathematischer Forschung erschlossen, welches als 'Mathematik der Axiom' bezeichnet werden könnte und weit über das Gebiet der Geometrie hinausreicht.

7. Desargues theorem says that if two triangles ABC and $A'B'C'$ are in perspective from a point O, so that the points O, A, A', the points O, B, B', and the points O, C, C' are collinear, and if the lines AB and $A'B'$ meet

is assuredly modern, abstract, and axiomatic, and it can seem unduly so, as the Italians seem to have thought.

Hilbert reserved a special place for the investigation of the Archimedean axiom. In geometric terms this says that given two segments a and b where a is smaller than b there is an integer n such that na is greater than b. The axiom is instinctively felt to be true, and so non-Archimedean geometry is a challenge to our fundamental beliefs about geometry. Before Hilbert, the existence of non-Archimedean geometry had been disputed by a small group of Italian mathematicians, of whom Giuseppe Veronese was the only one Hilbert cited. His work was difficult to follow, and that by Rodolfo Bettazzi was altogether sharper. But only Hilbert took seriously the question of how geometry would fare in the presence or the absence of the Archimedean axiom, and this question was then taken up masterfully by his student Max Dehn, who connected it to non-Euclidean geometry.

By the 1880s mathematicians were well aware that there were two distinct geometries with a claim to be the geometry of space: the non-Euclidean geometry discovered by János Bolyai and Nikolai Lobachevskii and made rigorous by Riemann, Eugenio Beltrami, and Poincaré, and ordinary Euclidean geometry. One might argue that they were slow to recognize there were many other possibilities, and a few alternatives were in fact canvassed in the 1890s. But the widely accepted view was that there were just these two physically plausible two-dimensional geometries. (Geometry on the surface of a sphere, was ruled out on the grounds that in it lengths cannot be indefinitely extended.)

These geometries were distinguished by the angle sums of triangles. In Euclidean geometry the sum is always two right angles; in Bolyai–Lobachevskii geometry it is always less; and on the sphere it is always greater. Likewise, given a line l and a point P not on it, in Euclidean geometry there is a unique line through P not meeting l; in Bolyai–Lobachevskii geometry there are infinitely many lines through P not meeting l; and in spherical geometry there are none.

Dehn looked at this tidy trichotomy and saw that the theorems which established it had often and naturally invoked the axiom of Archimedes. Following Hilbert's example, he investigated what would happen if the axiom was dropped and found two more possibilities. There is a non-Archimedean geometry in which the angle sum of a triangle is greater than two right angles but given a line l and a point P not on it there are infinitely many lines through P not meeting l; and there is a non-Archimedean geometry in which the angle sum of triangle is equal to two right angles but given a line l and a point P not on it there are infinitely many lines through P not meeting l. This last one struck Hilbert as particularly unexpected and remarkable.

at R, the lines BC and $B'C'$ meet at P, and the lines CA and $C'A'$ meet at Q, then the points P, Q, and R are collinear.

The work of Hilbert, Dehn, and others showed that there was a valuable difference between axiomatizing in the manner of Euclid and in the new way. The suggestion that there was no logical choice in the matter had been blown away by the discovery of non-Euclidean geometry. As a result, the idea that axioms codify what we know was weakened, although it was not clear what the philosophical status of the 'wrong' geometry was. Hilbert's work showed that the axiomatic method, backed up by the construction of suitable models, was creative, and the gap between pure and applied geometry became so wide that it was not even clear that Euclidean geometry could be taken to be true.

It was in this context that Poincaré (1902) put forward his philosophy of geometric conventionalism, which said that there was no way a logical decision could be made as to whether Euclidean or non-Euclidean geometry was true. Poincaré asked his readers to consider an experimental test that showed that the sum of angles in a triangle was less than two right angles. The researcher may say either that space is Euclidean but that light no longer travels in straight lines, or that light travels along the straight lines (geodesics) in a non-Euclidean space. This choice, said Poincaré, can only be made by convention. He suggested that we should always choose the simpler hypothesis, which was Euclidean geometry. This argument does not seem to have met with widespread acceptance, but to have provoked a debate about which concepts belong to physics and which to geometry.

Resistance to change

The mathematicians who were least touched by these changes were the British, trained as they had been at the University of Cambridge with its strong tradition of applied mathematics. Only Bertrand Russell and Whitehead can be counted as exceptions, and Russell was a philosopher while Whitehead, regarded in Cambridge as an applied mathematician, was on his way to philosophy (of a different kind from Russell's). In the period 1900 to 1914, G H Hardy and J E Littlewood were emerging into the front rank of mathematical analysis and number theory. In 1911 Hardy teamed up with Littlewood for a collaboration that lasted thirty-five years. From 1913 they worked also with the Indian mathematician Ramanujan until he died in 1920. These three were pure mathematicians, the first of international stature in Britain to establish a lasting school, and their influence on the growth of British mathematics was profound, but they did not adopt a modernist perspective.

It was Hardy's view that 'mathematical reality lies outside us, that our function is to discover to observe it, and that the theorems which we prove, and which we describe grandiloquently as our "creations", are simply our notes of our observations' (Hardy 1941, 123–124). Little in Hardy's view, neither its uncompromising stance, nor its implication that it is stating the obvious, nor its lack of philosophical sophistication, separates it from the opinion of many a mathematician from ancient Greek times to today. It is plainly not modernist.

One can go further. Much of real and complex analysis to this day is deep and valuable without being particularly modern. The same may be said of much work in ordinary differential equations and dynamics. The place of topology can be disputed. There is no question that when it was first introduced topology was abstract. It was a largely axiomatically defined study of sets of particular kinds. But in a climate where naive set theory serves as a foundation, a standard course in point set topology can seem no more than a repository of techniques. Just as one might argue that biology does not change significantly with the advent of better microscopes, one can argue that analysis has absorbed topology and is in many ways unaltered. That said, the opposite case can be argued: biology *is* different now that cells can be looked at in detail, topological thinking *has* transformed analysis. We return to the open question mentioned at the start of this chapter: what constitutes a major change is something every writer and reader has to decide for themselves.

One can go further still. Many emerging branches of applied mathematics seem less tied to the creations of modern mathematics and more directly linked to the natural world (or worlds, one should say, of biology, sociology, economics, and so on). Even the modernist orthodoxies of Bourbaki (Corry, Chapter 6.4 in this volume) look a little less inevitable these days, a little bit more like historical events with historical causes. On reflection that is reassuring. It might be possible even now to write a Hegelian history in which ideas evolve according to some set of natural laws, and the mathematics of this era gives rise to the mathematics of the next. But we are more comfortable these days with accounts that stress human agency, circumstances, and some degree of chance.

Conclusion

The cultural modernists were often deeply immersed in the art from which they sprang, with a thorough grounding in traditional technique. A considerable portion of Schoenberg's work can be fairly labelled late romantic; Joyce's *Ulysses* was preceded by much more conventional narratives; Picasso made his modernist turn earlier but retained a great respect for earlier, classical, styles; and so on. These people felt a need to be composers, novelists, artists, but to be so, eventually, in radically new ways. Those ways can be thought about at many levels, among them artistic intention (for example, the claim that art should no longer be 'beautiful'), technique, form, and content. All of these were of vital interest to the composers, novelists, and artists of previous generations, but at some period around 1900 leading figures moved to definitively new positions.

The driving forces at work among writers, painters, and composers included a strong emphasis on novelty of form and on new criteria for appreciation, which were much more internal. The same can be said of the mathematical moderniz-ers. As modern algebra emerged, it was based on naive set theory. So too was ana-lysis, insofar as it rested on a rigorized theory of the real numbers. Elementary geometries of various kinds and even the theory of the integral were given novel axiomatic foundations. These new foundations were accompanied by new modes of proof, appropriate to such formal and non-intuitive concepts. The whole ques-tion of mathematical intuition and the relation of mathematical truths to natural or scientific truths was considerably widened.

These novel ways of doing mathematics made it inevitable that only math-ematicians could judge the technical quality of a mathematical paper, not just for the traditional reason that doing high level mathematics requires training and practice, but because only a mathematician had the training in such mat-ters as naive set theory and abstract axiomatics. More importantly, the novelties ensured that only a mathematician could pronounce upon the value or import-ance of such work. The grey area between truly important work and mere puzzles moves at various times in the history of mathematics, it may well vary to some extent between mathematicians, but it became much harder to navigate as the separation between mathematics and physics grew. To recall just one example from those mentioned above, why should Dedekind's creation of structural alge-braic number theory have been a worthwhile thing to do? Only a mathematician can say; no scientific implication was even suggested.

How, then, should the changes in mathematics around 1900 be regarded? Do they constitute enough of a change, and a change of the right type, to be charac-terized as mathematical modernism? Answers to that question return us to the spectrum of responses discussed at the start of this chapter. Art critics and art historians have learned not to over-state the novelty of modern painting and to see continuities and influences while nonetheless holding on to a sense of the strong element of change and difference in the principal works of modern art. The modern novel presents a more complicated case, because it did not become impossible to do first-rate work in the manner of the nineteenth century. There were changes, to be sure, but the element of continuity has proved to be much stronger. Just so in mathematics the insistence of novelty in form and content (here: concept) should not blind us to the strong continuities, some of which were sketched above.

That said, these continuities presented themselves in a changed mathemat-ical world. Changed in its very foundations, because naive abstractionism had by 1900 been replaced by naive set theory and that in turn was to be made more sophisticated as the paradoxes began to bite (Ferreirós 1999). Changed in the nature of the objects it dealt with, which were not only removed from daily life (even daily scientific life) but more abstract even for the professional mathematician. Changed, accordingly, in its methods: abstract, axiomatically defined objects can only be handled by methods that emphasize the formal over the intuitive; thus the axiomatic method came to be adopted in geometry, in algebra and number theory, and in the theory of integration. Changed by the growing recognition that there was a definite subject called, variously, pure mathematics, conceptual mathematics, or even (Cantor's preferred term) *freien Mathematik* 'free mathematics' (Cantor 1932, 182). This new subject had its own sense of worth as being more fundamental than the older more seamless blend of mathematics and physics. Modern mathematics was the hegemonic discipline to which all questions about the validity of any mathematical innovation would, most likely, have to submit, and its leading exponents controlled the process by which difficult decisions were ratified. The new generation of mathematicians around 1900, most noticeably in Göttingen, were masters of what they surveyed.

Insofar as this new picture of modern mathematics convinces it brings with it a coherence to the developments in the field, it makes certain features stand out more clearly and shows that they were not particular to this or that branch of the subject but were widespread and characteristic. It invites questions, some of which I have dealt with elsewhere but some of which will be new: can one see a new relationship between mathematics and physics as a result? Will the presently imperfectly understood history of partial differential equations fit or contradict this account? What large-scale accounts of the history of mathematics are we willing to accept? I believe that if it is accurate and helpful to see a modernist transformation in painting or music then it is also accurate and helpful to see it in mathematics. There is the same reconfiguration of the field, the same novel sets of priorities and aesthetic criteria which have to be learned, the same emphasis on the acquisition of methods and techniques without which understanding is impossible.

Bibliography

Avigad, Jeremy, 'Methodology and metaphysics in the development of Dedekind's theory of ideals', in José Ferreirós and Jeremy J Gray (eds), *The architecture of modern mathematics*, Oxford University Press, 2006, 159–186.

Borel, Émile, *Leçons sur la théorie des functions*, Paris, 1914.

Bottazzini, Umberto, 'Fondamenti dell'aritmetica e della geometria', in Paolo Rossi (ed), *Storia della scienza moderna e contemporanea*, Unione Tipografico Editrice Torinese, 1988, 253–288.

Cantor, Georg, *Grundlagung einer allgemeinen Mannigfaltigkeitslehre. Ein mathematisch-philosophischer Versuch in der Lehre des Unendlichen*, Teubner, 1883; in *Gesammelte Abhandlungen mathematischen und philosophischen Inhalts*, Berlin, 1932, 165–209.

Chipp, Herschel B, *Theories of modern art: a source book by artists and critics*, University of California Press, 1968.

Corry, Leo, *Modern algebra and the rise of mathematical structures*, Birkhäuser, 1996; 2nd ed, 2003.

Corry, Leo, 'How useful is the term Modernism for understanding the history of early twentieth-century mathematics?', in Moritz Epple and Falk Mueller (eds), *Modernism in the sciences, ca. 1900–1940*, Akademie Verlag (forthcoming).

Couturat, Louis, *Principes des mathématiques*, Paris, 1905.

Edwards, Harold M, *Fermat's last theorem*, Springer, 1977.

Epple, Moritz, *Die Entstehung der Knotentheorie*, Vieweg, 1999.

Everdell, William, *The first moderns*, University of Chicago Press, 1997.

Fano, Gino, 'Sui postulati fondamentali della geometria proiettiva', *Giornale di Matematiche*, 30 (1892), 106–131.

Ferreirós, José, *Labyrinth of thought*, Birkhäuser, 1999; 2nd, revised, ed, 2007.

Goldstein, Catherine, Schappacher, Norbert, and Schwermer, Joachim (eds), *The shaping of arithmetic*, Springer, 2007.

Gray, Jeremy J, 'Anxiety and abstraction in nineteenth-century mathematics', *Science in Context*, 17 (2004), 23–48.

Gray, Jeremy J, *Plato's ghost: the modernist transformation of mathematics*, Princeton University Press, 2008.

Hausdorff, Felix, 'Bemerkung über den Inhalt von Punktmengen', *Mathematische Annalen*, 75 (1915), 428–433.

Hardy, Godfrey Harold, *A mathematician's apology*, Cambridge University Press, 1941; 2nd ed, 1969.

Hilbert, David, 'Die Theorie der algebraischen Zahlkörper (Zahlbericht)', *Jahresbericht der Deutschen mathematiker Vereinigung*, 4 (1897), 175–546; in *Gesammelte Abhandlungen*, 3 vols, Berlin, 1935, I 63–363.

Klein, C Felix, 'Ueber Arithmetisirung der Mathematik', *Nachrichten der Königlichen Gesellschaft der Wissenschaften zu Göttingen*, (1895), 82–91; in *Gesammelte Mathematische Abhandlungen*, 3 vols, Berlin, 1921, II 232–240. English translation in *Bulletin of the American Mathematical Society*, 2 (1896), 241–249.

Marchisotto, Elena A, and Smith, James T, *The legacy of Mario Pieri in geometry and arithmetic*, Birkhäuser, 2007.

Mehrtens, Herbert, *Moderne Sprache, Mathematik: eine Geschichte des Streits um die Grundlagen der Disziplin und des Subjekts formaler Systeme*, Suhrkamp, 1990.

Neumann, Olaf, 'Divisibility theories in the early history of commutative algebra and the foundations of geometry', in Jeremy J Gray and Karen H Parshall (eds), *Episodes in the history of modern algebra*, American and London Mathematical Societies, 2007, 73–106.

Pieri, Mario, 'Principii della geometria di posizione, composti in sistema logico deduttivo', *Memorie della Reale Accademia delle Scienze di Torino*, (2) 48 (1898), 1–62.

Pieri, Mario, 'Sui principi che reggono la geometria di posizione', *Torino Atti* 30 (1895), 607–641; 31 (1896), 381–399, 457–470, in *Opere sui fondamenti della matematica*, Unione Matematica Italiana, 1980.

Poincaré, Henri, 'Analysis situs', *Journal de l'École Polytechnique*, (2) 1 (1895), 1–123.

Poincaré, Henri, 'L'expérience et la géométrie', in Henri Poincaré, *La science et l'hypothèse*, Paris, 1902, 77–94; *Oeuvres*, Paris, 1953, VI 193–288.

Reed, David, *Figures of thought: mathematics and mathematical texts*, Routledge, 1994.

Riemann, Bernhard, 'Über die Darstellbarkeit einer Function durch einer trigonometrische Reihe', *Nachrichten der Königlichen Gesellschaft der Wissenschaften zu Göttingen*, 13 (1854), 87–132, in *Gesammelte Mathematische Werke, Wissenschaftliche Nachlass und Nachträge, Collected Papers*, Rarhavan Narasimhan (ed), Springer, 1990, 259–303.

Rosen, Charles, 'Mallarmé the magnificent', *New York Review of Books*, 20 May 1999, 42–46.

Toepell, M-M, *Über die Entstehung von David Hilberts 'Grundlagen der Geometrie*, Vandenhoeck and Ruprecht, 1986.

Weber, Heinrich L, 'Kronecker', *Jahresbericht der Deutschen Mathematiker-Vereinigung*, II 5–23.

Whitehead, Alfred North, *The axioms of projective geometry*, Cambridge, 1906.

Young, John Wesley, *Lectures on fundamental concepts of algebra and geometry*, London, 1911.

8. Mathematical

CHAPTER 8.1

The transmission of the *Elements* to the Latin West: three case studies

Sabine Rommevaux

The history of the text of Euclid's *Elements* during the Middle Ages and the Renaissance is now well known and documented.[1] The twelfth-century translation with the largest number of surviving manuscripts is that attributed to Robert of Chester (Busard and Folkerts 1992). In fact it exists in several different versions. Initially it consisted of nothing more than itemizations of the definitions, principles, and propositions, compiled from one or more Arabic translations. Later, authors added proofs in the margins, frequently in abbreviated form. Finally, copyists inserted these proofs, sometimes reworked, into the text itself. At the end of the thirteenth century Robert of Chester's version was supplanted by that of Campanus de Novare, which served as the standard text until the Renaissance. Compiled in the 1260s, it was not a translation based on Greek or Arabic manuscripts but rather a rewritten version of the *Elements*, based on other versions, notably one of the texts of Robert of Chester's version, and also on other texts such as the *Arithmetic* of Jordanus de Nemore, from the first half of the thirteenth century, and a commentary on the *Elements* by an-Nayrīzī, translated into Latin by Gerard of Cremona. The Campanus version was first published in Rome

---

1. All the most important translations of the twelfth and thirteenth centuries have now been published, principally by Hubert Busard. For an overview of the various texts, see Busard (1998) and also the introduction to Busard (2005, I 1–40). For the principal Renaissance editions see Murdoch (1971, IV 437–459).

in 1482, and there were numerous re-editions throughout the Renaissance. We will take it as the starting point for our study of the reception and assimilation of the Euclidean treatise in the Latin West. It was recently published in a critical edition (Busard 2005), and we shall refer to that edition throughout this chapter.

We will examine how the mathematicians of the Middle Ages and the Renaissance reacted to the Euclidean treatise and what they did with the theories that they found there. To do this, we will turn our attention to three aspects: the study of pyramids and prisms; the theory of the irrationality of magnitudes; and the theory of ratios and proportions.

Prisms and pyramids

A first example of the erroneous reception of Euclid's *Elements* in the Middle Ages concerns the study of pyramids and prisms in Book XII. When we compare the medieval versions with Heiberg's edition of the Greek text (Heiberg 1883–8), we find certain important differences.[2]

In the Greek text, according to definition 12 of Book XI, 'A pyramid is a solid figure contained by planes, which is constructed from one plane to one point' (Heath 1956, III 261); in other words it consists of a solid with a polygonal base, and triangular faces with a common apex. According to definition 13, 'A prism is a solid figure contained by planes, two of which, namely those which are opposite, are equal, similar and parallel, while the rest are parallelograms' (Heath 1956, III 261). For example, parallelepipeds are prisms. The study of these two families of solids is presented in propositions XII.3 to XII.9 (Heath 1956, III 378–400). Propositions 3–5, and 7–9 concern only pyramids and prisms with triangular bases. In XII.5 it is demonstrated that triangular pyramids of equal height are proportional to their bases; and in XII.6 this result is generalized to pyramids based on any polygon. In XII.7, it is proved that triangular prisms can be divided into three triangular pyramids (see Fig. 8.1.1, where the prism ABCDEF, based on the triangle ABC, is divided into the pyramid ABCD with base ABD and summit C, the pyramid EBCD with base EBC and summit D, and the pyramid ECFD with base ECF and summit D).

In a corollary, this result is extended to prisms based on any polygon: a pyramid based on any given polygon is a third part of a prism based on the same polygon and of the same height. Finally, in XII.8 it is shown that triangular pyramids that are similar (that is, whose sides are in proportion) are in a triplicate

2. In this chapter Heiberg's edition of the *Elements* is taken as the standard Greek version. However, as Saito (Chapter 9.2 in this volume) shows, Heiberg's text itself is not without its problems.

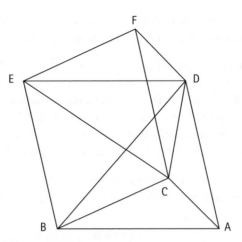

Figure 8.1.1

ratio of their sides.[3] In a corollary this result is extended to pyramids on any polygonal base. Thus we have a series of results for triangular pyramids and prisms, which are generalized to prisms and pyramids on all polygonal bases.

In the twelfth century Arabic–Latin translations, the definition of the prism is replaced by that of a *corpus serratile* (or sometimes *figura corporea servata*)[4] which is a solid with five faces, of which two, opposite to each other, are triangles, connected by parallelograms. It is therefore a triangular prism. There is no mention of polygonal prisms, which were nonetheless used in some proofs of the *Elements*, as Christoph Clavius pointed out with some surprise in his early seventeenth-century commentary (Clavius 1611–12, I 480). For example, in XII.10 of the Greek text, during the proof, we are asked to take a square and construct a prism based on it. In Robert of Chester's version,[5] the construction of such a prism is reduced to the construction of two *corpore serratilia* by dividing the square along its diagonal, forming two triangles as bases (Busard and Folkerts 1992, I 213–216). Campanus treats the proposition in the same way.

Moreover, XII.6 and the corollaries of XII.7 and XII.8 are missing from the twelfth-century versions. Thus, these versions contain a study of triangular

3. If ABCD and EFGH are two pyramids with summits D and H, the ratio of their volume is, for example, the triplicate ratio of the side AB to the side EF, that is $\left(\dfrac{AB}{EF}\right)^{3}$.

4. Adelard of Bath in his version defines a '*figura corporea servata*' (Busard 1983a, 299, def. vii); Gerard of Cremona discusses a '*figura corporea serratilis*', (Busard 1983b, c.337, def. vi). In the versions by Robert of Chester and Campanus, it is a '*corpus serratile*' (Busard and Folkerts 1992, I 265, def. vii; Busard 2005, I 389, def. vii).

5. In this version the proposition is XII.9 (Busard and Folkerts 1992, I 300).

pyramids and prisms only. This led Campanus to make the following remark after the demonstration of his XII.6 (corresponding to XII.7 of the Greek text):[6]

Since Euclid does not propose to demonstrate anything at all on the subject of prisms, with the sole exception of those whose bases are triangles, in order that we can sufficiently draw all possible knowledge from the elements he supplies we judge that it is not useless to add certain results to the demonstrations given here. For Euclid, by contenting himself only with the elements, omits many things which, even though they are consequences, do not appear without difficulty to students.[7] (Busard 2005, I 436)

Campanus's remark makes sense only in the light of the corrupted text he had inherited from Robert of Chester: Euclid certainly does deal with pyramids based on any polygon, as we have seen.

Campanus then presents a series of five propositions, by way of extension to XII.6. In the first he shows that if two solids, one of which is a prism and the other a triangular pyramid, are constructed on the same base or on two equal triangles, or the prism on a quadrilateral and the pyramid on a triangle which is half the quadrilateral base of the prism, and if the two solids have equal heights, then the prism is the triple of the pyramid. This result corresponds to the proof corollary to XII.7 of the Greek text, even if Campanus's result is less general. In the second proposition, he proves that pyramids on any bases are equal so long as their bases and heights are the same. He generalizes this in his fifth proposition, corresponding to XII.6 of the Greek text: there he proves that pyramids on any base having equal heights are proportional to their bases. His third and fourth propositions serve to demonstrate his fifth. Thus, in his third proposition, he shows that triangular pyramids of equal height are proportional to their bases (this has already been shown in XII.5 but it is found again here); and in his fourth proposition, he proves that if we have two pyramids of the same height, one with a triangular base and the other a base of any other polygon, the pyramids are proportional to their bases.[8]

Proposition XII.6 of the Greek version, along with the corollary of XII.7, are thus to be found in the Campanus version as additions to his XII.6 (corresponding to XII.7 in the Greek). As for the corollary of XII.8 in the Greek, missing

6. In framing this proposition Robert of Chester and Campanus did not specify that the prism is triangular, but it is so by definition (Busard and Folkerts 1992, I 298; Busard 2005, I 436). On the other hand this specification was made by Adelard (Busard 1983a, 339) and Gerard of Cremona (Busard 1983b, c.377).

7. Quoniam autem Euclides nihil demonstrandum proponit de piramidibus lateratis exceptis solis hiis quarum sunt bases triangule ut omnium cognitionem ex elementis, que ponit, sufficienter elicere possimus, quedam arbitramur non inutile demonstrationibus hic positis adiungere. Solis enim elementis contentus Euclides multa ex eis pretermittit que quamvis ex eis consequantur, non tamen sine difficultae patent studentibus.

8. Note that Campanus's demonstration contains a circularity: the proof of proposition 3 makes use of proposition 1, which itself is demonstrated from XII.6. But XII.6 is deduced from XII.5, which is also his proposition 3, where we started.

Table 8.1.1 Propositions XII.5–XII.8 of Campanus compared with propositions from the Greek text

Campanus's version	Greek text (Heiberg)
XII.5	XII.5
XII.6	XII.7
Addition 1 to XII.6 (generalization of XII.6)	Corollary to XII.7
Addition 2 to XII.6 (special case of addition 5)	
Addition 3 to XII.6 (special case of addition 5)	XII.5
Addition 4 to XII.6 (special case of addition 5)	
Addition 5 to XII.6 (generalization of XII.5)	XII.6
XII.7	XII.9
Addition to XII.7 (generalization of XII.7)	
Proposition XII.8	XII.8
Addition 1 to XII.8 (generalization of XII.8)	Corollary to XII.8
Definition of *columna laterata* (prism)	Definition XI.13
Addition 2 to XII.8 = addition 5 to XII.6 for the *columna laterata*	
Addition 3 to XII.8 = addition 1 to XII.6 for the *columna laterata*	
Addition 4 to XII.8 = XII.7 for the *columna laterata*	
Addition 2 to XII.8 = addition 1 to XII.8 for the *columna laterata*	

from Robert of Chester's version, it is to be found in the first proposition of an addition to Campanus's proposition XII.8. Here he shows that if we have two similar pyramids, the ratio of one to the other will be the triplicate ratio of their sides. In a second proposition added to XII.8, Campanus introduces what he calls a *columna laterata* and explains that this is a solid whose base and top are equal polygons, and that the faces joining them are parallelograms. We recognize here the Greek definition of a prism on any polygonal base, which, as we saw above, was missing from Robert of Chester's version. In further propositions added to XII.8, Campanus demonstrates for any prism a set of results earlier demonstrated for pyramids (for details see Table 8.1.1). The proofs rely on the division of polygonal bases into triangles and use similar results on *corpore serratilia* and triangular prisms.

Finally, in Campanus's XII.7 (corresponding to XII.9 in the Greek), it is shown that if two triangular pyramids are equal then their bases are inversely proportional to their heights, and vice versa. Campanus generalizes this result to pyramids based on any polygon, a generalization not found in the Greek. We therefore find in Campanus's XII.5–XII.8, and in additions to certain of these propositions, a study of pyramids and prisms on any polygon which is absent from the Arabic–Latin versions of the twelfth century, but which was partially present in the Greek.

The irrationality of magnitudes

In Book X of Euclid's *Elements*, which contains the theory of commensurable and incommensurable magnitudes, the erroneous transmission goes beyond the loss of information that we have just seen in the study of prisms and pyramids. Now the theory itself is put into question, allowing an arithmetical reading of results in Book X, which Euclid treated geometrically.

In the first definitions of Book X, Euclid introduced, for magnitudes, an initial pair of concepts: *summetra* 'commensurable' and *assumetra* 'incommensurable'; and for straight lines and surfaces a second pair of concepts: *rétè* 'expressible' and *alogoi* 'irrational'.[9] Thus we say that two straight lines are commensurable (we often add 'in length') if a same line can be used to measure both; otherwise we say that they are incommensurable. And we say that straight lines are commensurable in square only, when they themselves are incommensurable but the squares constructed on them are commensurable, that is to say when there exists a surface that can be used to measure both squares. If we now consider a straight line *E* which we call 'expressible' (for Euclid, this is in general one of the elements of whichever figure is being considered), we say that straight lines that are commensurable with *E*, either in length or in square only, are also expressible, while other straight lines are irrational. We can sum this up in Fig. 8.1.2.

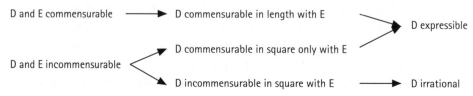

Figure 8.1.2

Let us take as an example the diagonal and the side of a square. The square constructed on the diagonal is twice the magnitude of the original square, so these two squares are commensurable. Taking the side of the original square as our reference line, we can therefore say that the diagonal is expressible, even though it is incommensurable with the side.

Notice that these two pairs of notions are not parallel: expressibility does not imply commensurability. This misled many readers of Book X. Thus, we can read in Pappus of Alexandria's commentary from the fourth-century:

---

9. The terms 'expressible' and 'irrational' preserve the asymmetry of the Greek terms *rétè* and *alogoi*. In his English translation, Heath talks of 'rational' and 'irrational' lines, picking up, as we shall see, the Latin terminology (Heath 1956, III 10).

Euclid, on the other hand, calls the line which is commensurable with the rational [expressible] line, however commensurable, rational, without making any stipulation whatsoever on that point: a fact which has been a cause of some perplexity to those who found in him some lines which are called rational [expressible], and are commensurable, moreover, with each other in length but incommensurable with the given rational [expressible] line (Pappus 1930, 81).

In the medieval versions, the definitions of commensurable and incommensurable magnitudes, as well as the definitions of straight lines commensurable in square only and incommensurable, are like those in the Greek. In contrast, in the Arabic then Latin versions,[10] and therefore particularly in the versions of Robert of Chester and Campanus, the definition of expressible (or rational) lines is abbreviated. Neither specifies that rational straight lines must be commensurable with the reference line either in length or in square, but only that they must be commensurable.[11] We can interpret their definition as saying that the adjective 'commensurable' refers to length only, so that only straight lines commensurable in length are to be called rational. Thus we have the situation in Fig. 8.1.3.

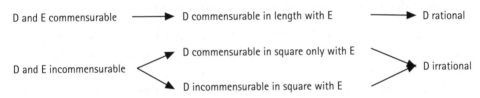

Figure 8.1.3

We thus have a strict parallel between rationality (or expressibility) and commensurability. This is how Tartaglia, for example, interprets Campanus's definition (Tartaglia 1565, 175v).

The parallelism between commensurability and rationality (or expressibility) is also amplified by the introduction in the propositions of non-Euclidean notions of lines rational in length and lines rational in square only, which correspond to commensurability in length and in square only, in relation to a reference line (Rommevaux 2001, 101–105). Thus we have the diagram shown in Fig. 8.1.4.

---

10. The Greek terms *summetra* and *assumetra* are translated into Arabic by the terms *muštarak* 'commensurable' and *ghayr muštarak* 'non commensurable' or *mutabāyin* 'separate' and the terms *rétè* and *alogos* by the terms *munêaq* 'expressible' and *ghayr munêaq* 'non expressible' or *aṣamm* 'surd. In Latin, we have on the one hand *communicantes* or *commensurabiles* and *incommensurabiles* or *incommunicantes* and on the other *rationalis* and *irrationalis* or *surde* (Rommevaux, Djebbar, and Vitrac 2001, 259).

11. The definition in the Greek text is as follows (definition X. 3): '[…] Let then the assigned straight line be called rational, and those straight lines which are commensurable with it, whether in length and in square or in square only, rational […]' (Heath 1956, III 10). Campanus has: 'Omnis autem linea cum quo ratiocinamur posita vocetur rationalis. Lineeque ei communicantes dicuntur rationales' (And let any given line we reason with be called rational. And the lines commensurable with it we say are rational.) (Busard 2005, 306).

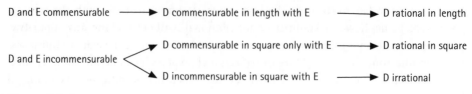

Figure 8.1.4

Returning to the example of the diagonal and side of a square, the diagonal is here rational in square only, relative to a side supposed to be rational.

Thomas Bradwardine, in his treatise on ratios of speeds in 1328, spoke of commensurable and incommensurable magnitudes or quantities (see Crosby 1955). He thereby established a parallel between quantity and ratio (Crosby says 'proportion'):

Rational differs from irrational proportion, moreover, in that the former exists only between commensurable or rational quantities, whereas the latter is found to exist only between incommensurable or irrational quantities (Crosby 1955, 67).

We see then that the terms 'rational' and 'irrational', which for Campanus referred to straight lines that are commensurable or incommensurable with a reference line, are here applied to quantities. This helps to reinforce the parallelism between commensurability and rationality that we have already noted. And this parallelism includes ratio: ratios between rational quantities are rational, and those between irrational quantities are irrational. Thus, returning to the example of the diagonal and the side of the square, and again taking the side as the reference line, we would hold that the diagonal is irrational (whereas for Euclid it was expressible, and for Campanus it was rational in square only). Further, the ratio of the diagonal to the side is also said to be irrational. After Bradwardine, we find this same terminology for quantities and ratios in the works of fourteenth- and fifteenth-century authors such as Nicholas Oresme and Blasius of Parma.

For Euclid, the reference line that is said to be expressible or rational was often one of the elements of the figure under discussion. In the medieval versions, on the other hand, and in particular in Campanus's version, this line is posited a priori, without any link to the figure. Let us examine XIII.6, for example. In the Greek version, it is demonstrated that any expressible line divided in extreme and mean ratio is an *apotome*. Dividing a line AB in extreme and mean ratio means dividing it at a point C in such a way that AB : AC = AC : BC. An *apotome* is a line D such that $D = D_1 - D_2$ where D_1 and D_2 are expressible, and commensurable with each other in square only. The concept of an *apotome* is therefore relative to a reference line, since it is in relation to this that the lines D_1 and D_2 are said to be expressible. If we examine the proof of the proposition in the Greek version, we see that the reference line is actually the line that is to be divided and which is called 'expressible' in the description of the proposition. In Campanus's version of this proposition he specifies in a commentary that the first part of the proof, which is the same as that in the Greek, holds only if the reference line is rational in length or in square. Thus Campanus implicitly posits a reference line, external to the problem, to which the line to be divided is said to be rational in length or in square.

This change in status of the reference line is important. Positing a priori a reference line external to the given conditions of the problem is the same as introducing a unit to which all lines will be compared. This is what the majority of the commentators of Euclid do. Thus we find in certain Greek manuscripts an addition at the beginning of definition X.3, which describes the posited expressible line as the one from which measures may be taken, for example, a cubit, a palm, a finger, or a foot (Euclid 1990–2001, III 34, n48). The expressible line is also interpreted as a unit of measure by certain Arab commentators, like an-Nayrīzī and Ibn al-Haytham. Clavius, on the other hand, who in the sixteenth century rediscovered the Euclidean text as we know it today, explained that the reference line is called *rétè* in Greek and 'rational' in the Latin texts, because it is *semper certa et nota* 'always known and given' (Clavius 1611–12, I 396). He did not see it as a unit of measure.

The proportionality of numbers and magnitudes

We have seen how the faulty transmission of Book X of the *Elements* altered the theory it contained. With Books V and VII, it was the understanding of the theory of proportionality which was brought into question, leading Campanus to introduce a non-Euclidean concept, that of the denomination of ratio.

A ratio is defined as a quantitative relationship between two magnitudes of the same kind.[12] At the same time, it is not the case that any two magnitudes have a ratio. They do so only if the smaller one, multiplied as often as necessary, can be made to exceed the larger. Euclid sets out this condition at the start of Book V,[13] but it is absent from the versions of Robert of Chester and Campanus. We should note, however, that it does appear in the proof of proposition X.1. Campanus also uses it in his commentary to this same proposition, where he argues that the angle of contact between a straight line and a circle can never be multiplied enough times to exceed a right angle, and therefore there is no ratio between them.

---

12. Definition V.3 in the versions of Robert of Chester and Campanus.
13. Definition V.4 in Heiberg's edition.

As for Definition V.5, on the proportionality of magnitudes and their equimultiples,[14] it is replaced by two different definitions in the versions of both Robert of Chester and Campanus. First, there is a definition of continuous proportion:

Quantities which are said to be in continuous proportion are those for which the equimultiples are either equal to, or exceed or fall short of, each other in the same way, without interruption.[15] (Busard 2005, I 161)

Then the definition of being 'in the same ratio':

Quantities which are said to be in the same ratio, the first relative to the second and the third to the fourth, are those for which the equimultiples of the first and the third are similar,[16] whether in excess or in deficit or in equality, to the equimultiples of the second and of the fourth, if they are taken in the same order.[17] (Busard 2005, I 164)

The second definition introduces the idea of similarity of equimultiples, which therefore needs to be defined. This is what Campanus does in his commentary:

And the similarity in excess or in deficit is to be understood here [...] not as excess relative to quantity, but as excess relative to the ratio.[18] (Busard 2005, I 164–165)

He goes on to make the idea more precise:

Four quantities are not continually proportional and the ratio of the first to the second is as that of the third to the fourth, when having taken equimultiples of the first and the third, and in the same way equimultiples of the second and the fourth, the ratio of the multiple of the first to the multiple of the second is as that of the multiple of the third to the multiple of the fourth.[19] (Busard 2005, I 165)

In the end Campanus brought the definitions of proportionality of quantities back to the proportionality of equimultiples. But he admitted that his argument was

14. Definition V.5: Magnitudes are said to be in the same ratio, the first to the second and the third to the fourth, when, if any equimultiples whatever be taken of the first and the third, and any equimultiples whatever of the second and fourth, the former equimultiples alike exceed, are alike equal to, or alike fall short of, the latter equimultiples respectively taken in corresponding order. (Heath 1956, II 114).

15. Quantitates que dicuntur habere continuam proportionalitatem sunt, quarum eque multiplicia equa sunt aut eque sibi sine interruptione addunt aut minuunt.

16. Busard chooses the wording *simul*, which approaches the idea of simultaneity and which is to be found in the Greek definition (Heath 1956, II 120); but older manuscripts have *similes* which seems to me closer to Campanus's commentary to these two definitions. Likewise, the majority of the manuscripts used by Busard and Folkerts for their edition of Robert of Chester's version have *similes* (Busard and Folkerts 1992, II 540; Rommevaux 2007).

17. Quantitates que dicuntur esse secundum proportionem unam, prima ad secundam et tertia ad quartam, sunt, quarum prime et tertie multiplicationes equales multiplicationibus secunde et quarte equalibus fuerint simul vel additione vel diminitione vel equalitate eodem ordine sumpte.

18. Similitudo autem in addendo aut diminuendo intelligatur hic [...] non quantum ad quantitatem excessus, sed quantum ad proportionem.

19. Incontinue proportionales sunt 4 quantitates et proportio prime ad secundam sicut tertie ad quartam cum sumptis eque multiplicibus ad primam et tertiam, itemque eque multiplicibus ad secundam et quartam erit proportio multiplicis prime ad multiplex secunde sicut multiplicis tertie ad multiplex quarte.

circular, defining 'the same by the same' (Busard 2005, I 165). Most Renaissance mathematicians were to take up the same mistaken interpretation.

Murdoch (1963, 251–261) has underlined the fact that this misunderstanding of the role played by equimultiples in the definition of proportionality led Campanus to introduce the notion of the denomination of a ratio. It appears in his long commentary following the definitions in Book V. Campanus begins by admitting his perplexity when faced with the use of equimultiples in the definition of proportionality. He next explains that if all ratios were rational, we would have a clear understanding of the equality or non-equality of ratios, because ratios of equal denominations would be equal. He then refers to the *Arithmetic* of Jordanus, where we find the definition of the denomination of a ratio at the start of Book II:

What we call the denomination of a ratio, at least of a smaller number to a greater, is the part or parts that the smaller is of the greater; and of a greater number to a smaller, the number by which it contains it and the part or parts of the smaller that remain in the greater.[20] (Busard 1991, 75)

Thus, in the case of a ratio between a and b where a is smaller than b (we say in this case that there is a ratio of lesser inequality), the denomination is the fraction p/q if a is p qth parts of b. In the case of a ratio between c and d where c is greater than d (a ratio of greater inequality), we determine how many whole times c contains d, say n, then the denomination of the remainder in relation to d, say k/l; the denomination is then $n + k/l$. Thus the denomination is a quantity associated with the ratio that allows us to see how the terms of the ratio relate to each other. The denomination also allows us to compare ratios: one ratio is larger than another if its denomination is larger.

We find an almost identical definition of denomination at the beginning of Book VII in Campanus's version. There the concept of denomination was no doubt introduced to overcome a difficulty in the version of Robert of Chester, who defined the proportionality of numbers like this:

Numbers are proportional where the first is in the second in the same way as the third is in the fourth, or where the second is in the first in the same way as the fourth is in the third.[21] (Busard and Folkerts 1992, 187)

---

20. Denominatio dicitur proportionis minoris quidem ad maiorem pars vel partes quote illius fuerit, maioris vero ad minus numerus secundum quem eum continet et pars vel partes minoris que in maiore superfluunt.

21. Numeri proporcionales sunt, quorum primus in secundo tamquam tercius in quarto aut in primo secundus tamquam in tercio quartus.

This differs from the definition in the Greek texts: 'Numbers are proportional when the first is the same multiple, or the same part, or the same parts, of the second that the third is of the fourth' (Heath 1956, II 278).

The concept of denomination allows us to specify precisely how one number is 'in' another. Campanus also adds to the beginning of Book VII a definition of numerical ratio (absent from the Greek):

What we call the ratio of a number to a number, at least of a smaller number to a greater, is that by which [*in eo quod*] it is a part or parts of the greater; and of a greater number to a smaller, that by which [*secundum quod*] it contains it, and its part or parts.[22] (Busard 2005, I 230)

We cannot fail to see the similarity of this definition to that of denomination given earlier. Thus, Campanus's introduction of the concept of denomination at the beginning of Book VII underlies the idea of numerical proportion (Rommevaux 1999, 89–106).

Campanus's definition of denomination can be related to the nomenclature of numerical ratios coined by Nicomachus (around 100 AD) and transmitted to the Latin world by Boethius in his *Arithmetic* (around 500 AD). Numerical ratios are separated in the first place into ratios of equality (between a and b where $a = b$), ratios of greater inequality (where $a > b$) and ratios of lesser inequality (where $a < b$). Ratios of greater inequality are further separated into five types: multiple ratios of denomination n, called doubles, triples, and so on; superparticular ratios of denomination $1 + 1/k$, called sesqui-k-ian (sesquialternate, sesquitertian, sesquiquartian, and so on); superpartient ratios of denomination $1 + h/k$, called super-h-partient k-ian (for example, the supertripartient quartian, between 7 and 4, with denomination $1 + 3/4$); multiple superparticular ratios of denomination $n + 1/k$; and multiple superpartient ratios of denomination $n + h/k$; the latter are named by combining the name of the whole multiple and of the superparticular or superpartient ratio. The relationships between the names of ratios and their denominations is presented in this way by Oresme in his treatise on ratios of ratios written between 1351 and 1360 (Oresme 1966, 210) and again in Clavius' commentary to his edition of Euclid's *Elements*, in the second edition of 1589 (Clavius 1611–12, 176, translated in Rommevaux 2005).

Another definition of denomination is to be found in two little treatises on proportion from the thirteenth century,[23] whose subject is the study of the relations that we can deduce from an initial relation of the type $(a{:}b) = (c{:}d){\cdot}(e{:}f)$. In these treatises denomination is defined simply as the result of the division of the antecedent or first term of the ratio by the consequent or second term (Busard 1971, 205, 213). It is not specified that the denomination is of the form $n + k/l$. The introduction of this notion serves to define composition or, if one prefers, the

---

22. Numeri ad numerum dicitur proportio minoris quidem ad maiorem in eo quod maioris pars est aut partes. Maioris vero ad minorem secundum quod eum continet et eius partem vel partes.
23. Busard (1971) attributes these to Jordanus and Campanus, but the attribution is doubtful.

multiplication of proportions, by the multiplication of their denominations, and the division of ratios by the division of their denominations.

We therefore have two definitions of the denomination of a ratio: in the first the denomination reveals the type of relationship (the larger term contains the smaller term a whole number of times together with some parts of the smaller term); in the second the denomination is simply the result of division of the terms of the ratio, without specifying the form in which the result should be presented. Under the first definition, the denomination is distinct from the fraction that can be made from the terms of the ratio. Thus, for the ratio between 6 and 4, the denomination is $1 + \frac{1}{2}$, whereas the corresponding fraction is 3/2.

It is tempting to compare the second definition of denomination to the Greek notion of *pélikotès*, which conveys the idea of quantity.[24] This comparison was indeed made by Renaissance mathematicians (see, for example, Fine 1551, 87–88; Clavius 1611–12, 176; Rommevaux 2005, 143). The notion of *pélikotès* appears in the *Elements*, in definition VI.5: 'a ratio is said to be compounded of ratios when the magnitudes (*pélikotétès*) of the ratios multiplied together make some (? ratio, or size)' (Heath 1956, II 189). Renaissance translators interpreted *pélikotès* by the Latin term *quantitas*. Thus, definition VI.5 leads us to associate ratios—which are relationships—with quantities, and the multiplication of these quantities give the quantity of the composite ratio. The definition is missing from most of the Latin translations of the twelfth century (it is today considered a late interpolation into the Greek text), and in particular from the versions of Robert of Chester and Campanus. Where the definition does appear, *pélikotès* is rendered by *quantitas* and not by *denominatio* (see Busard 1987, 125; Busard 1983b, c 137).

The concept of *pélikotès* is also to be found in Eutocius' sixth-century commentary on Archimedes' treatise *On the sphere and cylinder*:

So it ought to be recalled how a ratio is said to be composed of ratios. For as in the *Elements*: 'when the quantities [*pélikotétès*] of the ratios multiplied, produce a certain quantity,' where 'quantity' clearly stands for 'the number' whose cognate is the given ratio [...] which is the same as saying: 'the number which, multiplied on [by] the consequent term of the ratio, produces the antecedent as well'. (Netz 2004, I 313)

Eutocius refers to definition VI.5 of the *Elements* and he does not specify the form the *pélikotès* should take. Elsewhere, he explains that the *pélikotès* gives its name to the ratio. It is likely that he was referring to Nicomachus' nomenclature, but he does not specify it. Eutocius's commentary was translated into Latin along with some of Archimedes' treatises by Guillaume de Moerbecke in 1269. He too translated *pélikotès* as *quantitas*.

---

24. The interrogative adverb *pélikos* means 'What size?' In Arabic *pélikotès* was translated by *kammiya*, which also evokes the idea of quantity.

From our current understanding, it is difficult to establish an incontestable textual ancestry between the concepts of denomination and *pélikotès*, even if the two concepts are mathematical neighbours. And if the concept of denomination does indeed derive from that of *pélikotès*, it remains to be explained when and why Latin scholars rendered *pélikotès*, which contains the idea of the quantification of a ratio, by the term *denominatio*, which strictly speaking is a name rather than a quantity, but which brings to the fore a quantity associated with a ratio.

It also remains to find the origin of the definition of denomination that appears in Jordanus's *Arithmetic*, and subsequently in Book VII of Campanus's *Elements*, and which allows us to understand how the terms of ratio relate to each other. As far as we know, there is no Greek or Arab text containing such a definition. I would like to propose a hypothesis as to its origin. In Boethius's *Arithmetic*, the noun *denominatio* and the verb *denominare* appear several times in relation to the parts of a number (see Boethius 1995, 17–19): the number 3, for example, gives its name, or 'denominates', a third part. Campanus takes up this notion of denomination of a part at the beginning of Book VII: 'The number denominating is that by which a part is taken in its whole' (Busard 2005, I 230).[25] Campanus also adds a criterion concerning the similitude of the parts: 'We call similar parts those which are denominated by the same number' (Busard 2005, I 230).[26] We can hardly fail to compare these definitions with those of the denomination of proportions, and the similitude of ratios, that Campanus puts forward in this same set of definitions at the beginning of Book VII.

The denomination allows us to associate a quantity with a rational ratio. Was this enough for medieval and Renaissance mathematicians to *identify* the ratio with this quantity? The answer is no. Mathematicians of this period always maintained a strict distinction between a ratio and its denomination, continuing to insist that ratio is a relationship and not a quantity. Thus they held firmly to Euclidean orthodoxy. By way of example, let us look at Blasius of Parma's reflections on ratio, and the possibility of considering ratio as a quantity by way of its denomination. Blasius taught logic, natural philosophy, and consequently mathematics, in some of the universities of northern Italy at the turn of the fourteenth and fifteenth centuries. It was no doubt while engaged in this that he wrote *Questions on Master Thomas Bradwardine's treatise on ratios*, two versions of which have survived. It is the second that interests us here (Blasius of Parma 2005), which contains twelve questions concerning the theory of ratios and its applications to the study of movement.

In question 2, Blasius asks what a ratio is. Taking up the Euclidean definition he explains that in a strict sense it consists of a relation between two quantities

---

25. Denominans est numerus secundum quem pars sumitur in suo toto.
26. Similes dicuntur partes que ab eodem numero denominantur.

of the same kind. According to Blasius, a ratio, as a relationship, is therefore nothing but the terms of which it is composed.[27] Thus the double ratio is nothing other than 2 and 1; and the ratio of equality between a soul and itself is nothing other than the soul itself. This being so, the double ratio of 2 to 1 and the subdouble ratio of 1 to 2 are the same, since they are nothing other than 2 and 1 (Blasius of Parma 2005, 64). At the same time, a mathematician needs to distinguish these two ratios, and also to be able to work on ratios as objects independently of the terms of which they are composed. So in question 5, Blasius introduces what he calls the formal aspect of a ratio, by which we consider how the terms of the ratio relate to each other:

I note that on the subject of ratio we express it in two ways. First, we can speak of a ratio as things compared to each other. Second, we can speak of a ratio as things compared according to the aspect by which they are called equal or unequal in extension or in perfection, etc.[28] (Blasius of Parma 2005, 90)

In both cases, the ratio is nothing other than things considered in some relation; there is therefore no essence which would exist independently of the terms of which the ratio is composed. At the same time, in the second case it can be studied for itself, by a certain way of conceiving which signifies that formal aspect. And this is how the mathematician considers it, so that he can draw a formal distinction between the double ratio and the subdouble ratio. It is according to its formal aspect that the ratio can be considered as a quantity:

The ratio can be considered in another way, according to its formal aspect. And thus, strictly speaking, we say that a ratio is not a quantity but has the nature of a quantity. And that comes to the same thing as saying that its formal aspect is explained in terms of a category of quantity and not otherwise.[29] (Blasius of Parma 2005, 92)

What gives away the quantitative character of the ratio is precisely its denomination:

First conclusion: every ratio is a certain quantity or has the nature of a quantity. This is obvious because every ratio has a denomination according to which it is called a ratio of equality or inequality, and consequently according to which this ratio is said to be equal or unequal to another. And since this is a property of quantity, any ratio will be a certain quantity.[30] (Blasius of Parma 2005, 91)

---

27. To see the status in which Blasius of Parma holds relations, see Biard (2003, 390–396).

28. Noto quarto quod de proportione potest esse duplex sermo. Uno modo potest esse sermo de proportione tamquam de rebus invicem comparatis. Alio modo potest esse sermo de proportione tamquam de rebus comparatis secundum rationem secundum quam dicuntur equales vel inequales in extensione vel perfectione, et sic de aliis.

29. Alio modo potest considerari proportio secundum eius formalem rationem. Et sic loquendo proprie, dico quod proportio non est quantitas sed habet rationem quantitatis. Et hoc est dicere quod ratio eius formalis explicatur per terminos de predicamento quantitatis et non aliter.

30. Prima conclusio: omnis proportio est quedam quantitas vel habet rationem quantitatis. Patet quia omnis proportio habet denominationem secundum quam dicitur proportio equalitatis vel inequalitatis,

Thus, since ratios can be compared according to their denominations, they can be considered as quantities, because a property of a quantity is that we can say it is larger or smaller than another. A ratio remains nonetheless a relationship: a quantitative relationship.

Ratios of ratios

The transmission of Euclid's *Elements* gave rise to modifications to its theories, as we have seen in the three examples above. But the reception of the *Elements* was also an opportunity for novel extensions, and the theory of ratios of ratios is one of them.

We have seen that we can name numerical ratios, and also rational ratios (of magnitudes), according to their denominations. But what about irrational ratios? Bradwardine, in his treatise on ratios, explained that irrational ratios are denominated by means of rational ratios, which themselves are denominated by a number. He gave the example of the diagonal and the side of a square, which he called 'half the double ratio' and that of half a musical tone, which he called 'half the sesquioctave ratio'. He went no further and it was left to Oresme to devise a mathematical theory that gave meaning to these denominations.

To do this, he had to define ratios between ratios. How to do this is not obvious, since a ratio is by definition a relationship between two magnitudes. If we are to define a ratio between ratios we need to be able to consider ratios themselves as magnitudes. What are the essential properties of such magnitudes? First, they must be infinitely divisible and further, they must satisfy the property known as the Archimedean Axiom, that is, the smaller of two magnitudes can be taken as many times as necessary until it exceeds the larger. Oresme showed that ratios of greater inequality can be considered in the same way as magnitudes. Actually, he showed that any ratio can be infinitely divided by the insertion of means between the terms of the ratio: thus the quadruple ratio, 4 to 1, is divided into two double ratios by the insertion of 2 between 4 and 1, and so on (the means do not need to be whole numbers). If we compose, or multiply by itself, a ratio of greater inequality as often as necessary, it becomes larger than any given ratio of greater inequality (for example, if we take the double and sextuple ratios, we see that the ratio composed of the double ratio three times is the octuple ratio, which is larger than the sextuple ratio). Thus, ratios of greater inequality can be considered as magnitudes, subject to the definitions of Book V of the *Elements*. We can therefore talk of a ratio between ratios of greater inequality. On the other hand

et per consequens secundum quam ista proportio dicitur esse equalis vel inequalis alteri. Et quia hoc est proprium quantitati, ideo omnis proportio erit quedam quantitas.

we cannot talk of a ratio between a ratio of equality or of lesser inequality and a ratio of greater inequality. For example, the subtriple ratio, 1 to 3, multiplied by itself as often as we like, always gives a ratio of lesser inequality, which is smaller than any ratio of greater inequality.

Denomination of irrational ratios is therefore founded on an additive interpretation of the composition of ratios. If a, b, and c are three magnitudes, the ratio $a : c$ is said to be composed of the ratios $a : b$ and $b : c$, in the same way that a line segment AC is composed of segments AB and BC, where B lies between A and C. Let us consider composition as an addition and suppose that the magnitudes a, b, and c are continuously proportional, that is to say that the ratios $a : b$ and $b : c$ are equal. Since the two ratios compose the ratio $a : c$, each of them will be called the 'half' of this ratio. So, for example, the ratio of the diagonal to the side of a square is called 'half the double ratio', because the ratio composed of it twice is the double ratio. We can speak in the same way about the irrational ratio 'two-thirds of the triple ratio', for example: it consists of the ratio which composed three times yields the duplicate of the triple ratio, that is to say the nonuple ratio.

In this way we can name irrational ratios, which are parts of rational ratios. Bradwardine seemed to think that all irrational ratios were like this. Oresme had an intuition that this was not the case, and there could be irrational ratios which are not parts of a rational ratio, but he gave no example (Oresme 1966, 160–162). We now know that the ratio of the circumference of a circle to its diameter is such a ratio, but this was not proved until the nineteenth century.

Conclusion

The text of the *Elements* that reached the Latin West in the twelfth century diverged in several ways from the Greek text that we now know. The variations can seem, on the face of it, to be of little significance. However, attentive study of the texts, with particular attention to the vocabulary used by the translators and commentators, and a precise examination of their claims and demonstrations, shows that the divergences sometimes change the *Elements* and its theories in important ways.

In the study of the prisms and pyramids, we were able to present what became a concern of many commentators, in particular Campanus: to generalize the results presented by Euclid. Faulty transmission had led Campanus to believe that Euclid had considered only pyramids and prisms with triangular bases. He therefore generalized the propositions that he found in Robert of Chester's version, to pyramids and prisms on any polygonal bases.

The study of irrationality offers an example of subtle divergences which profoundly modified the theory. A corrupt definition allowed medieval writers to put in place a formal parallelism between the notions of commensurability and of rationality, a parallel that does not exist in Euclid. Further, while Euclid referred to an expressible or rational line that was part of the problem under consideration, medieval writers posited a priori a unit of measure against which all other lines could be compared. Thus we have the conditions for the arithmetization of Book X. We should emphasize that Campanus did not himself proceed to this arithmetization: for him, the choice of reference line remained implicit. And he did not invoke the aid of radical numbers to translate the different concepts of measure, nor did he interpret the claims in terms of algebraic operations on these numbers.

Finally, the study of the proportionality of magnitudes and numbers shows how Campanus was led to introduce the non-Euclidean concept of the denomination of a ratio, in order to put the theories of proportionality in Books V and VII onto solid foundations. Thus a ratio, which initially was only a relationship, came to have a quantity associated with it. It was not until the seventeenth century, however, that ratios became irrevocably assimilated with quantities, ready to give birth later to the theory of real numbers.

These three examples show how faulty transmission of the *Elements* led to readjustments and important modifications to the theories. Campanus played a central role in the reception of the *Elements* in the Middle Ages and in its transmission to the Renaissance, and the theories he put in place nourished medieval mathematics and its applications. Thus, the theory of irrationality and the concept of denomination of a ratio have an important place in the mathematical theories devised by Bradwardine and developed by Oresme, in the context of calculations of the speed. Bradwardine explained that if we have two bodies moved by two movers, the ratio between the speeds of their movements is the ratio between two other ratios: the ratio of the power of the first mover to the resistance of the first body and the ratio of the power of the second mover to the resistance of the second body. To give meaning to this formulation, Oresme developed a theory of ratios of ratios which applies to the theories in Books V, VII, and X of the *Elements* in the modified form of Campanus, of which we have given certain elements (see also Rommevaux, in press). This theory of ratios was to be taken up and criticized until the sixteenth century, and we find an echo of the debates it inspired in Clavius's commentary on the *Elements* (Rommevaux 2005, 69–72). Oresme's theory of ratios is also to be found at the origin of certain logarithmic constructions in the seventeenth century, for instance, that of Kepler.

In the Middle Ages, theories of ratio and of irrationality found many applications beyond the theory of movement, notably in music and architecture. In these fields, studies remain to be done to analyse which theory of ratios is being used and to gauge what place is given to the modifications made by Campanus.

The impact of Campanus's edition, however, was felt beyond the Middle Ages. Although mathematicians of the Renaissance returned to the Greek text of

Euclid as we know it, they still drew on Campanus's commentary and on those of other medieval mathematicians, who sometimes helped them to understand difficult concepts, or to complete the Euclidean treatise where it seemed incomplete. Thus, the non-Euclidean concept of the denomination of ratio crops up often in Renaissance editions of the *Elements* to explain what ratio and proportionality are. We also find numerous references to Campanus, explicit or otherwise, in sixteenth- and even seventeenth-century editions of the *Elements*. These borrowings sometimes come with criticisms, especially that Campanus misunderstood Euclid. But we should be aware that such criticisms are sometimes baseless: our study has shown that the same term can hide different concepts, and that the definitions Campanus had to use were not the Euclidean definitions. Renaissance editors of the *Elements* did not always take this into account, sometimes even intentionally, because they needed to show that Medieval commentators had distorted the *Elements* and that it was necessary to return to the Greek sources.

Bibliography

Biard, Joël, 'Mathématiques et philosophie dans les *Questions* de Blaise de Parme sur le *Traité des rapports* de Thomas Bradwardine', *Revue d'histoire des sciences*, 56 (2003), 383–400.

Blaise de Parme, *Questiones circa tractatum proportionum magistri Thome Braduardini*, Introduction et édition critique de Joël Biard et Sabine Rommevaux, Vrin, 2005.

Boethius, *Institution arithmétique*, ed and trans Jean-Yves Guillaumin, Les Belles Lettres, 1995.

Busard, Hubert L L, 'Die Traktate *De proportionibus* von Jordanus Nemorarius and Campanus', *Centaurus*, 15 (1971), 193–227.

Busard, Hubert L L, *The first Latin translation of Euclid's Elements commonly ascribed to Adelard of Bath*, Pontifical Institute of Medieval Studies, 1983a.

Busard, Hubert L L, *The Latin translation of the Arabic version of Euclid's Elements commonly ascribed to Gerard of Cremona*, New Rhine Publishers, 1983b.

Busard, Hubert L L, *The medieval Latin translation of Euclid's Elements made directly from the Greek*, Franz Steiner, 1987.

Busard, H L L, *Jordanus de Nemore, De Elementis Arithmetice Artis. A medieval treatise on number theory*, Franz Steiner Verlag, 1991.

Busard, Hubert L L, 'Über den lateinischen Euklid im Mittelalter', *Arabic Sciences and Philosophy*, 8 (1998), 97–129.

Busard, Hubert L L, *Campanus of Novara and Euclid's Elements*, 2vols, Franz Steiner Verlag, 2005.

Busard, Hubert L L, and Menso Folkerts, *Robert of Chester's (?) redaction of Euclid's Elements, the so-called Adelard II Version*, 2 vols, Birkhaüser Verlag, 1992.

Clavius, Christoph, *Christophori Clavii…Opera mathematica*, 5 vols, Mainz, 1611–12.

Crosby, H Lamar, *Thomas Bradwardine: his Tractatus de proportionibus. Its significance for the development of mathematical physics*, University of Wisconsin Press, 1955.

Euclide, *Les Éléments*, traduction et commentaires par Bernard Vitrac, 4 vols, Presses Universitaires de France, 1990–2001.

Finé, Oronce, *In sex priores libros geometricorum elementorum Euclidis...*, 3rd ed, Paris, 1551.

Heath, Thomas L, *The thirteen books of Euclid's Elements*, 3 vols, Dover publications, 1956.

Heiberg, I L, *Euclidis opera omnia*, 8 vols, Teubner, 1883–1916.

Murdoch, John, 'The medieval language of proportions: elements of the interaction with greek foundations and the development of new mathematical techniques', in Alistair Cameron Crombie (ed), *Scientific change*, Heinemann, 1963, 237–271.

Murdoch, John, 'Euclid: transmission of the *Elements*', in Charles C Gillepsie and F L Holmes (eds), *Dictionary of scientific biography*, 16 vols, Charles Scribner's sons, 1970–1981, IV 437–459.

Netz, Reviel, *The works of Archimedes: vol I: The two books on the sphere and the cylinder*, Cambridge University Press, 2004.

Oresme Nicole, *De proportionibus proportionum and Ad pauca respicientes*, ed and trans Edward Grant, University of Wisconsin Press, 1966.

Pappus, *The commentary of Pappus on Book X of Euclid's Elements*, Arabic text and translation by William Thomson, Harvard University Press, 1930.

Rommevaux, Sabine, 'La proportionnalité numérique dans le livre VII des *Éléments* de Campanus', *Revue d'histoire des mathématiques*, 5 (1999), 83–126.

Rommevaux, Sabine, 'Rationalité et exprimabilité: une relecture médiévale du livre X des *Éléments* d'Euclide', *Revue d'histoire des mathématiques*, 7 (2001), 91–119.

Rommevaux, Sabine, *Clavius: une clé pour Euclide au XVIe siècle*, Vrin, 2005.

Rommevaux, Sabine, *Théorie des rapports (XIIIe – XVIe siècles): Réception, appropriation, innovation*, forthcoming.

Rommevaux, Sabine, Djebbar Ahmed, Vitrac Bernard, 'Remarques sur l'histoire du texte des *Éléments* d'Euclide', *Archive for History of Exact Sciences*, 55 (2001), 221–295.

Rommevaux, Sabine, 'La similitude des équimultiples dans la définition de la proportion non continue de l'édition des *Éléments* d'Euclide par Campanus: une difficulté dans la réception de la théorie des proportions an Moyen Âge', *Revue d'histoire des mathématiques*, 13 (2007), 181–202.

Tartaglia, Niccolo, *Euclid...diligentemente rassettato et alla integrita ridotto...con una ampla espositione...*, Venice, 1565.

'Gigantic implements of war': images of Newton as a mathematician

Niccolò Guicciardini

T hat Newton was a great mathematician was evident in the early 1670s to
those very few who had privileged access to his early mathematical discoveries.[1] By the 1680s Newton was regarded by his contemporaries as an outstanding
mathematician. It was in order to find an answer to a mathematical question concerning the shape of planetary orbits, for example, that Edmond Halley travelled
to Cambridge to see Newton in August 1684. There he found that Newton had
already broached and answered that difficult question. When in 1687 Newton's
answer appeared in *Philosophiae naturalis principia mathematica* 'Mathematical
principles of natural philosophy', not a few had reservations about his cosmology
based on gravitational action at a distance, but even critics of gravitation theory
were impressed by the depth and scope of the mathematical structure Newton
had developed.

Since then the image of Newton as one of the greatest mathematicians in history
has remained unshaken. But the nature of his contribution to mathematics has
always been difficult to define. Evaluations, most of them eulogistic, have stressed
different aspects of Newton's work. To some he has appeared as the initiator of

---

1. The reader seeking information on Newton's early mathematical work can turn to Westfall (1980,
105–139), and for a deeper analysis to Whiteside's editorial commentary in Newton (1967–81, I).

new and powerful algorithms, while others have been captivated by his geometrical style reminiscent of ancient Greece. There has been, as we will see in this chapter, considerable disagreement between those who have tried to delineate Newton's mathematical legacy, which has been constructed and reconstructed again and again. The process of reinvention of Newton as a mathematician has been pursued for different purposes and with different agendas in mind; and in some cases, as we shall see, with the purpose of downgrading and criticizing him. Newton's mathematical works have always proved puzzling, and his mathematics has engendered reactions from frustration to awe. His style appeared to many as obsolete, or rather, endowed with an aura of a past golden age impossible to emulate. Further, in too many cases Newton seemed to be hiding his method of discovery, or even to be obfuscating proofs by failing to disclose a number of crucial passages.

I begin my chapter by discussing the priority dispute which broke out in the first decade of the eighteenth century and which divided Newton's supporters from those of Leibniz. In their confrontation, Newtonians and Leibnizians rendered explicit some shared values that worked as tacit assumptions in normal mathematical practice. The priority dispute, therefore, played an important role in the making of Newton's image as a mathematician. We will then consider the reception and evaluation of two of his works, the *Principia*, and the somewhat less celebrated *Enumeratio linearum tertii ordinis*. We conclude the chapter with a section devoted to criticisms that emerged after a century of successful applications of Leibniz's differential and integral calculus.

The *Commercium epistolicum*

The circumstances surrounding the controversy between Newton and Leibniz have been analysed in detail by Rupert Hall (1980) and D T Whiteside (Newton 1967–81, VIII 469–538). In broad outlines let us recall a few bare facts. Newton formulated his method of series and fluxions between 1665 and 1669.[2] Leibniz had worked out his equivalent algorithm, the differential and integral calculus, around 1675 and printed it in a series of papers from 1684, and it is clear from manuscript evidence that he arrived at his results independently from Newton. It was only in part in Wallis's *Algebra* in 1685 and *Opera* in 1693 and 1699, and in full in an appendix to his *Opticks* in 1704, however, that Newton's method appeared in print. In 1708 a British mathematician, John Keill, stated in the *Philosophical Transactions of the Royal Society* that Leibniz had plagiarized Newton (Keill 1708). After Leibniz's protest a committee of the Royal Society secretly guided by

---

2. From the 1690s Newton denoted 'fluent quantities' by letters such as x, y, and their 'fluxions', or rates of change, by $\dot{x}$, $\dot{y}$.

its President, Isaac Newton, produced a publication, the so-called *Commercium epistolicum* 'Exchange of letters', which maintained that Newton was the 'first inventor' and that '[Leibniz's] Differential Method is one and the same with [Newton's] Method of Fluxions' ([Newton] 1712, 121). It was also suggested that Leibniz, after his visits to London in 1673 and 1676, and after receiving letters from Newton's friends and from Newton himself, had gained sufficient information about the fluxional method to allow him, after changing the symbols, to publish the calculus as his own discovery. Most notably, Newton addressed two letters, the so-called *epistola prior* 'first letter' and *epistola posterior* 'later letter', to Leibniz in 1676 through Henry Oldenburg, secretary of the Royal Society. Only after the work of twentieth-century historians such as Hall and Whiteside do we have proof that Keill's accusation was unjust: Newton and Leibniz arrived at equivalent results independently and following different paths of discovery.

The *Commercium epistolicum* can be considered as Newton's last mathematical work. Based on archival material (letters, manuscripts, excerpts from printed works) it was printed in late 1712 and distributed in January and February 1713. Its purpose was to reply to Leibniz's demand, addressed in December 1711, that the Royal Society, of which Leibniz was a Fellow, should protect him from the 'empty and unjust braying' of such an 'upstart' as Keill (Newton 1959–77, V 207).[3] Formally the *Commercium* was the work of an independent committee. Materially, as the manuscripts, edited by Whiteside, now show, it is a work carefully drafted and engineered by Newton himself, who honed and perfected every detail of it (Newton 1967–81, VIII 539–560). Not a word passed into print without his supervision.

The *Commercium epistolicum* has been considered puzzling by many commentators. Its purpose was to prove Leibniz's plagiarism of the calculus, but to many readers it failed to do so since it dealt with topics which are deemed to be only loosely related to the calculus. Most notably, the two letters that Newton addressed to Leibniz in 1676, which constitute the main proof that crucial information was passed to Leibniz, are often described as lacking this very evidence.

With hindsight we know that Newton could have provided documents that would have been considered more convincing to our critical readers. For instance, he did not use the manuscript treatise, composed in 1671, known as 'Tractatus de methodis serierum et fluxionum' 'Treatise of the method of series and fluxions', which includes algorithms and rules equivalent to Leibniz's differential calculus, and their application to the calculation of tangents and curvatures to plane curves (Newton 1967–81, III 74–81, 116–194). Why did Newton not use excerpts from these problems during the controversy with Leibniz? Why, instead, did he amass information on topics that can hardly be considered as proof of his transmission of the calculus to Leibniz? The easy answer, that Newton was just

3. vanae et injustae vociferationes […] cum homine docto, sed novo.

dishonest, does not capture the complexity of his position. Even if one wants to concede dishonesty in Newton's handling of the quarrel with Leibniz, it would have been simply stupid not to provide evidence which was as strong as possible in the *Commercium epistolicum*. In this section I will expand on these themes to show that in the priority dispute Newton had a different view of the nature and importance of the discovery which, he was convinced, Leibniz had stolen from him.

While Leibniz insisted on the importance of his discovery and publication of the algorithm for differentiation, Newton and his acolytes focused instead on methods of quadrature (namely integration) via power series expansions. These methods were developed by Newton in an early treatise entitled *De analysi per aequationes numero terminorum infinitas* 'On analysis through equations with an infinite number of terms', composed in 1669 (Newton 1967–81, II 206–247), which was reproduced in the *Commercium epistolicum*. Newton divided his method of fluxions into the 'direct method', which is equivalent to differentiation, and the 'inverse method', which is equivalent to integration. When one analyses the mathematical examples adduced in the *Commercium epistolicum* it emerges that Newton, and his acolytes who were slavishly editing it, referred to the inverse method of fluxions applied to quadratures (for an example, see Fig. 8.2.1).

Leibniz protested, of course: this was not his point. He claimed to have discovered the rules of the differential calculus independently from Newton, but the *Commercium epistolicum* had nothing to say about this. In his rebuttal, the *Historia et origo calculi differentialis* 'The history and origin of the differential calculus', which was to remain unpublished until the nineteenth century, he wrote of himself in the third person:

> They have changed the whole point of the controversy, for in their publication [...] one finds hardly anything about the differential calculus; instead every other page is made up of what they call infinite series [...] This is certainly a useful discovery, for by it arithmetical approximations are extended to the analytical calculus; but it has nothing at all to do with the differential calculus. They use this sophism, that whenever his adversary works out a quadrature by addition of the parts by which a figure is gradually increased, at once they hail it by the use of the differential calculus (as for instance on p. 15 of the *Commercium epistolicum* [see Fig. 8.2.1]) [...] Since therefore his opponents, neither from the *Commercium epistolicum* that they have published, nor from any other source brought forward the slightest bit of evidence whereby it might be established that his rival used this calculus before it was published by our friend; therefore all the things that they have reported may be rejected as extraneous to the matter. They have made recourse to the skill of ranters with the purpose to divert the attention of judges from the matter on trial to other things, namely to infinite series.[4] (Leibniz 1849–63, V 393, 410)

---

4. Mutarunt etiam statum controversiae, nam in eorum scripto [...] de calculo differentiali vix quicquam (invenitur): utramque paginam faciunt series, quas vocant, infinitae. [...] Utile est inventum, et appropinquationes Arithmeticas transfert ad calculum Analyticum, sed nihil ad calculum differentialem. [...] Cum ergo

Longitudines Carvarum invenire.

Sit ADLE circulus cujus arcus AD longitudo eft indaganda. Duĉto
tangente DHT, & completo indefi-
nite parvo rectangulo HGBK, & po-
fito AE = 1 = 2AC. Erit ut BK five
GH, momentum Bafis AB(*x*), ad HD
momentum Arcus AD :: BT : DT
:: BD ($\sqrt{x - xx}$) : DC ($\frac{1}{2}$) :: 1 (BK):

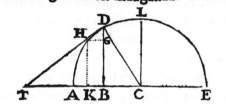

$\frac{1}{2\sqrt{x-xx}}$ (DH). Adeoque $\frac{1}{2\sqrt{x-xx}}$ five $\frac{\sqrt{x-xx}}{2x-2xx}$ eft momentum Arcus AD.

Quod reductum fit $\frac{1}{2}x^{-\frac{1}{2}} + \frac{1}{4}x^{\frac{1}{2}} + \frac{3}{16}x^{\frac{3}{2}} + \frac{5}{32}x^{\frac{5}{2}} + \frac{35}{256}x^{\frac{7}{2}} + \frac{63}{512}x^{\frac{9}{2}}$ &c.
Quare, per regulam fecundam, longitudo Arcus AD eft

$x^{\frac{1}{2}} + \frac{1}{6}x^{\frac{3}{2}} + \frac{3}{40}x^{\frac{5}{2}} + \frac{5}{112}x^{\frac{7}{2}} + \frac{35}{1152}x^{\frac{9}{2}} + \frac{63}{2816}x^{\frac{11}{2}}$ &c.

five $x^{\frac{1}{2}}$ in $1 + \frac{1}{6}x + \frac{3}{40}x^2 + \frac{5}{112}x^3 + \frac{35}{1152}x^4 + \frac{63}{2816}x^5$, &c.
Non fecus ponendo CB effe *x*, & radium CA effe 1, invenies Arcum
LD effe $x + \frac{1}{6}x^3 + \frac{3}{40}x^5 + \frac{5}{112}x^7$, &c.
Sed notandum eft quod unitas ifta quæ pro momento ponitur eft Super-
ficies cum de Solidis, & linea cum de fuperficiebus, & punĉtum cum de
lineis (ut in hoc exemplo) agitur.
Nec vereor loqui de unitate in punĉtis, five lineis infinite parvis, fi
quidem proportiones ibi jam contemplantur Geometræ, dum utuntur me-
thodis Indivifibilium.
Ex his fiat conjeĉtura de fuperficiebus & quantitatibus folidorum, ac de
Centris Gravitatum.

Figure 8.2.1 'An example of calculation by moments of fluents ' from *De analysi*
(1669) reproduced in the *Commercium epistolicum*. First Newton sets *AB* = *x*, and
considers the circle with diameter *AE* = 1. The ratio of the 'moment' of the arc
AD, to the 'moment' of the abscissa *AB* (that is, *HD/GH*) is $\left(\sqrt{x - x^2}\right)/(2x - 2x^2)$.
Expanding this as an infinite power series and integrating term by term, he obtains
the arc-length *AD* of the circle as: $x^{1/2}(1 + x/6 + 3x^2/40 + 5x^3/112 + \cdots)$. Then,
setting *CB* = *x* and the radius *CA* = 1, Newton observes that the arc *LD* is given by:
$x + x^3/6 + 3x^5/40 + 5x^7/112 + \cdots$, which we now recognize as the *arcsin* series.
([Newton] 1712, 15)

The policy followed in the *Commercium epistolicum* surely and bitterly appeared
to Leibniz as a clever way of shifting the level of discourse in order to avoid a
fair confrontation. But what we know about Newton's intellectual trajectory as a

adversarii neque ex Commercio Epistolico, quod edidere, neque aliunde vel minimum indicium protulerint,
unde constet aemulum tali calculo usum ante edita a nostro; ab his allata omnia ut aliena sperni possunt. Et
usi sunt arte rabularum, ut judicantes a re de qua agitur ad alia diverterent, nempe ad series infinitas.

mathematician, what we know about his statements before the controversy, tell us that there is something deeper. On the one hand we have Leibniz who focuses on algorithm and the basic rules of the 'simple' differential calculus. On the other there is Newton for whom the prowess of a mathematician is measured by the hard 'inverse' problems he is able to solve. Newton's rules, on which much was said in the *Commercium epistolicum*, were the rules of the inverse method of fluxions.

The fact that Newton and his acolytes focused on the use of series and the inverse method (integration), rather than on the direct one (differentiation), is supported by the attribution, encountered repeatedly in their writings, of the direct method to Barrow rather than to Newton or Leibniz. Wallis, for instance, who in 1693 was the first to print the fluxional notation and algorithm (Wallis 1693–9, 2, 391–396), stated in his commentary, which was cited in the *Commercium epistolicum*, that

akin to this method (the direct of fluxions) there is on the one hand the method of Leibniz, and on the other hand that method, older than either, which Dr Isaac Barrow has expounded in his *Lectiones geometricae*: and this is acknowledged in the *Acta Lipsica* in January 1691.[5] ([Newton] 1712, 98)

The *Acta Lipsica* were the *Acta eruditorum* (published in Leipzig), and the reference is to a paper where Jacob Bernoulli, one of Leibniz's most prolific acolytes, claimed that the differential calculus was the same as Barrow's.[6] David Gregory was of the same opinion. Much impressed by the mine of mathematical results that Newton showed him in his chambers in May 1694, once back in Oxford he jotted down a widely circulated little treatise entitled *Isaaci Newtoni methodus fluxionum; ubi calculus differentialis Leibnitij, et methodus tangentium Barrovij explicantur.*[7] In the opening lines, Gregory made it clear that both Leibniz's and Newton's methods 'differ only in name' and 'flow easily from Barrow's Method of Tangents treated in the 10th chapter of his Lectiones geometricae'.[8]

This interpretation seems not to have bothered Newton too much. Quite the contrary, in the *Commercium epistolicum*, a text that was drafted under Newton's careful supervision, one repeatedly finds the direct method attributed to Barrow, James Gregory, and René François de Sluse. In commenting on Leibniz's letter

---

5. Huic Methodo affinis est tum Methodus differentialis Leibnitii tum utraque antiquior illa quam D^r Is. Barrow in Lectionibus Geometricis exposuit. Quod agnitum est in *Actis Leipsiensis* (Anno 1691, mense Jan.) à quodam qui methodum adhibet Leibnitii similem.

6. Jacob Bernoulli's paper was to arouse Leibniz's wrath. An English translation of his draft of a long reply to Bernoulli can be found in Leibniz (2005, 11–20). What provoked Leibniz's anger was peacefully accepted by Newton.

7. Personal contact with Newton was the procedure one had to follow, with due deference, in order to obtain private disclosure on the new analysis. The translation of Gregory's title is 'Isaac Newton's method of fluxions; in which the differential calculus of Leibniz, and Barrow's method of tangents are explained'.

8. Calculus Differentialis Leibnizij et Methodus Fluxionum Newtoni tantum nomine tenus differant [...] et facile fluunt ex Methodo Tangentium Barrovij Lect: 10. Geom: tradita. (Christ Church, Oxford, MS B.13. cxxxI1r)

of 11/21 June 1677,[9] a very important document in which the notation and rules of the differential calculus were first displayed to Newton, the editors of the *Commercium epistolicum* note that 'Barrow did the very same thing [...] and by a very similar calculus' ([Newton] 1712, 88).[10] John Keill in his answer to Leibniz, presented in the *Philosophical Transactions* for 1711 and reproduced near the end of the *Commercium epistolicum*, stated that Sluse, James Gregory, and Barrow had methods for drawing tangents to curves 'which do not differ too much from the method of fluxions' ([Newton] 1712, 112).[11]

These statements, which appeared with Newton's approbation, have often been considered counterproductive. For instance, in the second edition of Montucla's *Histoire des mathématiques* (1802) one reads:[12]

Is it not contradictory to say that Leibniz's method, described in the letter which we are considering [Leibniz to Oldenburg, 11/21 June 1677], is just that of Barrow, and that it is the same as the one that Newton had communicated in 1669, which is claimed to be his method of fluxions? Because from this it follows that Newton's method is equivalent to Barrow's, excepting the notation.[13] (Montucla 1799–1802, III 107)

It is highly unlikely that Newton allowed the attribution of the direct method to Barrow because of a careless editing of the *Commercium epistolicum*. Quite the contrary, we know from the many manuscripts that Newton left that he supervised its publication with an almost obsessive attention to detail.

The attribution of the direct method to Barrow, Sluse, and James Gregory would have appeared incomprehensible to Leibniz: his elegant algorithm, he frequently claimed, had many advantages over previous tangent methods, the most important being that it could be applied to irrational quantities. In tangent methods prior to Newton and Leibniz one had to eliminate irrational quantities first. Leibniz's 'remarkable calculus', by contrast, was 'impeded neither by fractional nor by irrational quantities', a claim which appeared in the title of his very first paper on the calculus, published in 1684.[14] Newton also praised the fact that his own direct method of fluxions could deal with irrationals 'without taking away surds'. But for Newton, overcoming

---

9. During the seventeenth century England continued to follow the 'old style' calendar, so that 21 June in France or Germany was 11 June in England.

10. Idem fecit Barrow in ejus Lect.10 Anno 1669 impressa, idque calculo consimili.

11. quae a Fluxionum methodo non multum abludebat.

12. When Montucla died, pages 1–336 of volume III of his new edition of the *Histoire* were already printed. The rest was revised by de Lalande. We do not know how much Montucla's text was changed, especially after page 336, but it is fair, I surmise, to attribute to Montucla quotations that appear before that.

13. N'y a-t-il pas de la contradiction à dire que la méthode de Leibnitz, décrite dans la lettre é dont nous parlons, n'est que celle de Barrow, et qu'elle est la même que celle que Neuton avoit communiquée dès 1669, qu'on prétend être son calcul des fluxions. Car il suivroit delà que la méthode même de Neuton ne seroit que celle de Barrow à la notation près.

14. *Nova methodus pro maximis et minimis, itemque tangentibus, quae nec fractas nec irrationales quantitates moratur, et singulare pro illis calculi genus* 'A new method for maxima and minima, as well as tangents, which is impeded neither by fractional nor irrational quantities, and a remarkable type of calculus for this'.

this hurdle, at least in the most elementary cases, was a simple application of the binomial expansion that he had communicated to Leibniz in the *epistola prior*: namely the fluxion of x^n was easily calculated by expanding $(x + o)^n$, where o is a small increment of the fluent x, even when n was a fraction. As he explained in the *Tractatus de quadratura curvarum* 'Treatise on the quadrature of curves' in 1704:

Suppose the quantity x flows uniformly and the fluxion of the quantity x^n is to be found. In the time that the flowing quantity x comes to be x + o, the quantity x^n will come to be $(x + o)^n$, that is, (when expanded) by the method of infinite series

$$x^n + nox^{n-1} + \frac{1}{2}(n^2 - n)o^2 x^{n-2} + \cdots.$$

The increases o and $nox^{n-1} + \frac{1}{2}(n^2 - n)o^2 x^{n-2} + \cdots$ are to one another as 1 to $nx^{n-1} + \frac{1}{2}(n^2 - n)ox^{n-2} + \cdots$. Now let those increases to vanish and their last ratio will be 1 to nx^{n-1}; consequently the fluxion of the quantity x is to the fluxion of the quantity x^n as 1 to nx^{n-1}.[15] (Newton 1967–81, VIII 126–129)

As in the inverse method of fluxions, the key to the direct method is in power series development, the main topic of the *Commercium epistolicum*.

While Newton put little importance on the direct method for drawing tangents, he was very sensitive about his techniques for the inverse method. A study of his mathematical correspondence reveals that he did not easily share his methods of quadrature. His disclosures reveal a complex strategy, most notably that as the years passed Newton allowed more and more of his precious inverse methods to be known. The progress in integration techniques achieved in the 1690s by John Craig, David Gregory, Leibniz, and Jacob and Johann Bernoulli began eroding Newton's conviction that he had an unbridgeable advantage over his contemporaries. To summarize, one can say that from the early 1670s he considered the quadrature methods of the *De analysi* to be in the public domain: for instance, he revealed them to Barrow, Collins, and Wallis, and to Leibniz in the two *epistolae*. But he showed a remarkable jealousy until the mid-1690s in revealing his more sophisticated techniques (now recognized as integration by substitution and by parts), which are to be found in the concluding pages of the 1671 'Treatise of series and fluxions', and which constitute the main body of the *De quadratura curvarum*. When, in his letter of 17/27 August 1676, Leibniz informed Newton about his transmutation method applied to circle quadrature (Newton 1959–77,

---

15. *Fluat quantitas x uniformiter et invenienda sit fluxio quantitatis x^n.* Quo tempore quantitas x fluendo evadit $x + o$, quantitas x^n evadet $\overline{x + o}|^n$, id est per methodum serierum infinitarum $x^n + nox^{n-1} + \frac{1}{2}(nn - n)oox^{n-2} + \&c.$; Et augmenta o et $nox^{n-1} + \frac{1}{2}(nn - n)oox^{n-2} + \&c.$ sunt ad invicem ut 1 et $nx^{n-1} + \frac{1}{2}(nn - n)ox^{n-2} + \&c.$ Evanescant jam augmenta illa, et eorum ratio ultima erit 1 ad nx^{n-1}; ideoque fluxio quantitatis x est ad fluxionem quantitatis x^n ut 1 ad nx^{n-1}.

II 57–64), Newton replied to Collins (on 8 November) that he himself had a much more powerful technique. He boasted that:

I say there is no such curve line but I can in less then half a quarter of an hower [sic] tell whether it may be squared or what are ye simplest figures it may be compared with, be those figures Conic sections or others'. (Newton 1959–77, II 179)

But after this tantalizing hint at his method of quadratures Newton added:

This may seem a bold assertion [...] but it's plain to me by ye fountain I draw it from, though I will not undertake to prove it to others. (Newton 1959–77, II 180)

In the 1690s Newton, challenged by his younger contemporaries, allowed some of his quadrature techniques, his secret 'fountain', to be printed in Wallis's *Opera* (1693–9, II 391–396).

The authorial and publication strategies adopted by Newton, as well as the policy he followed during the controversy with Leibniz in editing the *Commercium epistolicum*, reveal something about his agendas and priorities. For us 'the calculus' is a deductive theory based on definitions (of limit, derivative, differential, and so on) and basic rules for differentiation. The crucial questions for many historians have often been: who was the first to discover these rules? Who was the first to publish them? As a matter of fact, it is easier to find the rules of the differential calculus in Leibniz's *Nova methodus* (the first publication of his calculus in 1684) than in any of Newton's papers. But these questions do not address what was crucial for Newton. A formal theory and its basic rules were not a matter of great interest for him. Rather, he was concerned with a method for solving geometrical problems, most notably problems of quadrature, and this was the whole issue of the *Commercium epistolicum*. In his opinion, his method of series and fluxions showed its power only when tested against hard problems in squaring curves (integration) or in what was known as the 'inverse method of tangents' (integration of differential equations).[16] Thus he saw himself as the discoverer not of simple rules for finding tangents, but of a secret 'fountain' that allowed him to solve such inverse problems.

The following remark taken again from the third volume of Montucla's *Histoire* can be cited as an example of how, from a different perspective from Newton's, the *epistola posterior*, one of the chief evidential documents of the *Commercium epistolicum*, can be viewed as defective:

Here we note that after having read and re-read this letter [*epistola posterior*], we find the method of fluxions described only with regard to its consequences and advantages, but not with regard to its principles.[17] (Montucla 1799–1802, III 103)

---

16. An inverse tangent problem requires the determination of a curve when its tangent is known at every point.

17. Nous remarquons ici qu'après avoir lu et relu cette lettre, nous y trouvons seulement cette méthode décrite, quant à ses effets et ses avantages, mais non quant à ses principes.

But Newton, unlike Leibniz, had always presented his method of fluxions as a panoply of successful problem-solving techniques: its 'consequences' and its 'advantages' were his priority. When asked to show his prowess, he preferred to deliver results rather than methods; Leibniz on the other hand was concerned to establish himself as the discoverer of a successful algorithm, and was happy to leave the burden and merit of its applications to others. A typical statement that Newton anonymously circulated in 1717 can be cited as further evidence of his viewpoint:

In the year 1684 Mr Leibnitz published only the Elements of the Calculus differentialis & applied them to questions about Tangents & Maxima & Minima as Fermat Gregory & Barrow had done before, & shewed how to proceed in these Questions without taking away surds, but proceeded not to the higher Problems. The *Principia mathematica* gave the first instances made publick of applying the calculus to the higher Problems. (Newton 1967–81, VIII 513)

Newton's pronouncement concerning the use of calculus in the *Principia* is, however, extremely problematic. The image of Newton as a mathematician that emerges from his *magnum opus* is complex and in a way contradictory. In the next section I will therefore turn to this image as it was construed on the basis of the evidence provided in the *Principia*. This construal, of course, played an important role in the controversy with Leibniz, since proof of the use of calculus in the *Principia* would have tipped the balance in Newton's favour.

The *Principia*

Contrary to Newton's own pronouncement quoted above, the *Principia* is written almost entirely in geometric form: it is hard to find 'instances made publick' of applications of 'the calculus to the higher Problemes' in its pages, which are heavily adorned with geometrical diagrams. However, in a number of propositions algebraic reasoning and infinite series occur. Further, some of the propositions open with a rather intriguing statement: the reader is asked to concede that a 'method for squaring curvilinear figures' is available. In Book I Proposition 41, for instance, which is concerned with determining an orbit under a central force, Newton reduces the problem to the quadrature of a curve, in Leibnizian terms to an integration. In the corollaries that follow, the results are obtained thanks to previously acquired knowledge of the calculation of the curvilinear area, but the reader is given not the slightest hint on how such crucial calculations are to be carried out.[18] Clearly, a strategy of non-disclosure of heuristic techniques is

18. Corollary 3 is particularly interesting as it deals with the difficult problem of determining the trajectories traced by a 'body' accelerated by an inverse-cube force.

at work here, and this frustrated all the competent readers of the *Principia*, like Christiaan Huygens, who complained about Newton's secrecy.

The plurality of methods and complexity of authorial strategies employed in the *Principia* are the main causes for the diversity of judgements that have been given over the centuries. A passage from the preface to l'Hôpital's *Analyse des infiniment petits pour l'intelligence des lignes courbes* (1696) has remained justly famous since Newton himself quoted it in the 1710s:

> Furthermore, it is a justice due to the learned M. Newton, and that M. Leibniz himself accorded to him: That he has also found something similar to the differential calculus, as it appears in his excellent book entitled Philosophiae Naturalis *Principia* Mathematica, published in 1687, which is almost entirely about this calculus.[19] (l'Hôpital 1696, xiv)

This passage, now believed to have been written by Bernard le Bovier de Fontenelle, shows us how Newton's mathematical natural philosophy was already perceived by some late seventeenth-century natural philosophers: as based on modern techniques, geometrical limits, or infinitesimals, and therefore ready to be translated into the language of fluxional, or differential and integral, algorithms. The geometry of the *Principia* thus appeared to many as disguised calculus.

A completely different evaluation was given in 1837 by William Whewell, who wrote in *History of the inductive sciences*:

> The ponderous instrument of [geometric] synthesis, so effective in [Newton's] hands, has never since been grasped by one who could use it for such purposes; and we gaze at it with admiring curiosity, as on some gigantic implement of war, which stands idle among the memorials of ancient days, and makes us wonder what manner of man he was who could wield as a weapon what we can hardly lift as a burden. (Whewell 1837, 167)

A few decades later Maximilien Marie seems to reply to Fontenelle, writing:

> In the *Principia* one finds excellent infinitesimal geometry, but I could not find any infinitesimal analysis: I add that to those who wish to see the calculus of fluxions in the *Principia,* one could then also show the differential calculus in Huygens's *Horologium* (Marie 1883–8, VI 13).[20]

While Fontenelle stresses the modernity of Newton's mathematical methods in the *Principia*, underlining their equivalence with the new calculus, Whewell and Marie are impressed by the distance that separates Newtonian mathematical natural philosophy from modern analytical mechanics. Marie disagreed completely

---

19. C'est encore une justice dûë au sçavant M. Newton, & que M. Leibniz lui a renduë lui-même: Qu'il avoit aussi trouvé quelque chose de semblable au calcul différentiel, comme il paroît par l'excellent Livre intitulé *Philosophiae Naturalis Principia Mathematica*, qu'il nous donna en 1687, lequel est presque tout de ce calcul.

20. On trouve dans les *Principes de Philosophie naturelle* d'excellente Géométrie infinitésimale, mais je n'y ai pas découvert d'Analyse infinitésimale: j'ajoute qu'à celui qui voudrait voir le calcul des fluxions dans le *Livre des Principes,* on pourrait aussi bien montrer le calcul différentiel dans l' *Horologium*.

with Fontenelle, seeing Newton's method as closer to the 'infinitesimal geometry' of Huygens than to the differential and integral calculus of Leibniz. To understand the reasons for the enigmatic character of Newton's mathematics as presented in the *Principia* one needs to consider the aims and values of its author, which in turn shaped his publication policy.

We have to step back to the early 1670s, when Newton began lecturing in Cambridge as a young Lucasian Professor, and had a clear agenda which would continue to inform his intellectual life. He saw himself as a mathematician who could inject certainty into natural philosophy through the use of geometry. Newton positioned himself against the Baconian inductivism in vogue at the Royal Society as much as against Cartesian rationalistic hypotheticism. From his many pronouncements on method one can infer that he considered both Bacon's 'bottom to top' experimental procedure and Descartes' 'top to bottom' method as doomed to yield only probability. Bacon's proposed bottom line, the patient collection of experimental results, was not sufficient to guarantee certainty. Descartes' 'top', consisting of clear and distinct ideas, was not necessary if it was just to deliver, as Newton wrote, 'little better then a romance' (CUL MS Add 3970, f. 480v).[21]

In Newton's time many virtuosi of the Royal Society fostered a moderate scepticism that eschewed the arrogance of certainty and instead aimed at probable truth reached through patient experimentation. A passage such as the following, taken from the lectures on optics that Newton delivered in about 1670, was anathema to eminent fellows of the Royal Society like Henry Oldenburg, Robert Hooke, and Robert Boyle. Newton stated:

I hope to show—as it were, by example—how valuable is mathematics in natural philosophy. I therefore urge geometers to investigate nature more rigorously, and those devoted to natural science to learn geometry first. Hence the former shall not entirely spend their time in speculations of no value to human life, nor shall the latter, while working assiduously with an absurd method, perpetually fail to reach their goal. But truly with the help of philosophical geometers and geometrical philosophers, instead of the conjectures and probabilities that are being blazoned about everywhere, we shall finally achieve a science of nature supported by the highest evidence.[22] (Newton, 1984, 86–89, 438–439)

21. 'But if without deriving the properties of things from Phaenomena you feign Hypotheses & think by them to explain all nature you may make a plausible systeme of Philosophy for getting your self a name, but your systeme will be little better then a Romance.'

22. spero me quasi exemplo monstraturum quantum Mathesis in Philosophia naturali valeat; et exinde ut homines Geometras ad examen Naturae strictius aggrediendum & avidos scientiae naturalis ad Geometriam prius addiscendam horter: ut ne priores suum omnino tempus in speculationibus humanae vitae nequaquam profuturis absumant, neque posteriores operam praepostera methodo usque navantes, a spe sua perpetuo decidant: Verum ut Geometris philosophantibus & Philosophis exercentibus Geometriam, pro conjecturis et probabilibus quae venditantur ubique, scientiam Naturae summis tandem evidentijs firmatam nanciscamur. Translation from Latin by A Shapiro.

Ideas very similar to these were to occur in the Preface to the second edition of the *Principia* (1713) written by Roger Cotes—but under Newton's careful supervision—some forty years afterwards.[23]

In order to defend his somewhat isolated position, Newton had to explain why mathematics could be considered as a source of 'highest evidence' in natural philosophy. A tension between his philosophical agenda and his mathematical practice soon emerged, since his early works on the method of series and fluxions were based on procedures that lacked secure mathematical foundations. Most notably, he deployed infinitely small magnitudes ('moments'), and handled infinite series on the basis only of analogies and extrapolations. He very soon came to the conclusion that what is nowadays considered his greatest mathematical discovery, the calculus, could be seen only as a heuristic tool, to be discarded in published work. The calculus could *not* claim to be the mathematical method that allowed natural philosophers to become geometrical philosophers by overcoming uncertainty and probability.

Newton, therefore, turned to geometry and began reading ancient texts, especially Federico Commandino's 1588 translation of Pappus' *Collectio mathematica* (a compilation of Greek geometry from the fourth century AD). In the seventh book Newton read that the ancients possessed a method of discovery, the so-called 'method of analysis', which—so he interpreted Pappus' text—they kept secret, preferring to publish their results according to the rigorous synthetic method. Many Early Modern mathematicians were fascinated by Pappus' words. Some of them attributed knowledge of algebra to the ancients: this would have been their hidden method of discovery. Newton, after a long and tortuous exegesis of the works of Apollonius and Pappus, concluded that the ancient hidden analysis was a form of projective geometry. In fact several of Pappus' results were expressed in terms of the invariance of cross-ratios. Newton's results in this field, developed thanks to a creative reading of Pappus' text, align him with Blaise Pascal, Gérard Desargues, and Philippe de la Hire as one of the great geometers of the seventeenth century.

As the years passed by, the method of fluxions became less and less important in Newton's eyes. He relegated algebraic methods to the role of heuristic tools which were to be discarded in published proof. In his writings on mathematical method Newton frequently deployed the archaic Pappusian concepts of analysis (or resolution) and synthesis (or composition): the former was a method of discovery, the latter a truly demonstrative method. In the 1690s he stated:

if a question be answered [...] that question is resolved by the discovery of the equation [...], but it is not solved before the construction's enunciation and its

23. Alan Shapiro (2004) has taught us to appreciate the pervasive role of the quest for certainty in Newton's thought. We learn from him that Newton did indeed change his ideas concerning this matter, especially as far as the possibility that philosophical geometers and geometrical philosophers actually have of reaching absolute certainty.

complete demonstration is, with the equation now neglected, composed. Hence it is that resolution so rarely occurs in the Ancient's writings outside Pappus' collection.[24] (Newton, 1967–81, VII 307)

The Ancients, according to Newton:

accomplished [the solution of problems] by certain simple propositions, judging that nothing written in a different style was worthy to be read, and in consequence they were concealing the analysis by which they found their constructions.[25] (Newton, 1967–81, IV 277)

Consequently, Newton's policy was to shun disclosure of algebraic techniques:

careful considerations should be given to fabricating a demonstration of the construction which as far as permissible has no algebraic calculation, so that the theorem embellished with it may turn out worthy of public utterance.[26] (Newton, 1967–81, III 279)

Historians of Newton's mathematics cannot avoid feeling disconcerted when they realize that most of the mathematical discoveries achieved by Newton in the late 1660s and early 1670s were printed only decades later. These discoveries, especially those concerning infinite series and fluxions, were so innovative that late seventeenth-century European mathematics would have been very different if Newton had been more prompt in publishing some of his early manuscripts. Not least, the priority dispute with Leibniz would have been avoided.

Several psychological explanations have been proposed for Newton's delayed publication of calculus and the absence of calculus from the *Principia*. These focus on Newton's obsessive fear of disputes and criticisms; they contain more than a grain of truth and have to be considered by any serious biographer. But they risk underestimating the rationale of Newton's publication policy. While he was adamant about the heuristic validity of his early mathematical methods, he felt acutely that their status was controversial and concluded that they were not 'worthy of public utterance'.

One should also note that Newton's readers in 1687 would have found it impossible to follow a text that proposed not only new mechanics and cosmology, but also a new mathematical language. In 1687 the language of geometry was the most obvious choice in a work devoted to natural philosophy. However, the geometrical demonstrative structure of the *Principia* could not hide the presence of statements whose demonstration was lacking. These gaps can only be filled with

---

24. si quaestioni per constructionem aequationis alicujus respondeatur, quaestio illa resolvitur per inventionem aequationis, componitur per constructionem ejusdem, sed non prius solvitur quam constructionis enunciatio ac demonstratio tota componitur, aequatione neglecta. Hic est quod resolutio in veterum scriptis extra Pappi collectanea tam raro occurrat.

25. At illi rem peregerunt per simplices quasdam Analogias, nihil judicantes lectu dignum quod aliter scriberetur, & proinde celantes Analysin per quam constructiones invenerunt.

26. de constructionis demonstratione consulendum est, quacum sine Computo Algebraico quantum liceat contexta ornetur Theorema ut evadat publicae notitiae dignum.

the aid of algebra, series, higher-order derivatives, and integration, but very little detail of these algorithmic techniques is given in the printed text. Both Newton's British acolytes and his fierce critics in continental Europe were interested in the missing steps. Newton might easily have added details on the use of algebraic equations, for example: these were certainly well known techniques to mathematically trained readers in 1687. He could also have given indications on the use of more advanced techniques, such as series and quadratures, in an appendix, for instance. He actually considered this option in the 1690s when he began pondering a revised second edition. But the *Principia* remained, even in its second (1713) and third (1726) editions, in a geometrical language which was increasingly perceived as obsolete by younger mathematicians, a geometrical cloak that hid what seemed mathematically more interesting.

If we take into consideration Newton's explicit policy of 'neglecting the equation' from the printed page, the *Principia*'s mathematical style is perhaps explained. In many propositions Newton made recourse to integration techniques (quadratures, as he would say). We know this both because Newton, somewhat mysteriously, states in the *Principia* that he is using a 'method for squaring curvilinear figures', and because he communicated these algorithmic techniques to his acolytes when they turned to him for explanations (see Fig. 8.2.2).

Other readers of the *Principia* were left with the gaps in its demonstrative structure: much to their frustration the author was hiding what seemed to be advanced expertise in integration techniques. Until the mid-1690s Newton used correspondence and controlled circulation of his manuscripts to divulge knowledge of his algebra and calculus. This double communication code (public for geometry, private for algorithmic techniques) was evident to all competent readers of the *Principia*. As Montucla wrote:

As a matter of fact, even though his *Principia* offers many examples of the ancient way, in general the calculus surfaces through a concealment with which Newton hides it; a drawback which is common to many books delivered according to the ancient method, and which are nothing but a concealed algebra.[27] (Montucla 1799–1802, III 6)

The complex stratification between text and subtext in the *Principia* has given rise to diverging evaluations of its mathematical methods. To some, after the development of analytic mechanics by men such as Johann Bernoulli and Leonhard Euler, its language seemed simply archaic and obsolete. Others argued that Newton was capable of writing the whole work in terms of calculus. Recent research has shown that both positions miss the point (Guicciardini 1999). Newton did use calculus techniques, but he did so sporadically and unsystematically. The second

27. En effet, quoique ses principes nous offrent en bien des endroits des exemples de ce tour ancien; en général le calcul y perce à travers le déguisement dont Newton l'a couvert, espèce de défaut, commun à bien des livres donnés pour écrits suivant la méthode ancienne, et qui ne sont que de l'algèbre déguisée.

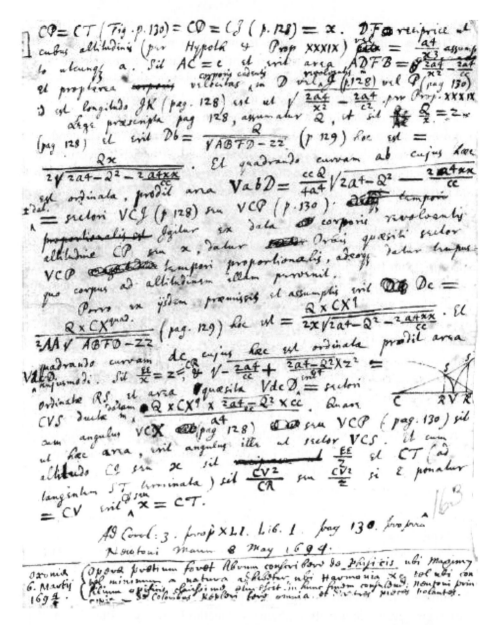

Figure 8.2.2 Letter sent by Newton to David Gregory in 1694, an example of how information about the analytic subtext of the *Principia* circulated amongst Newton's acolytes. It contains details of the quadrature required in Book 1, Proposition 41, Corollary 3 of the *Principia*, where the centripetal force is inverse cube. Newton denotes the distance of the body from the centre of force by x, and sets F, the centripetal force, inversely proportional to the cube of the distance, equal to a^4/x^3, where a is a constant. This allows the geometrical proportions occurring on pages 128 and 129 of the *Principia* to be translated as fluxional (differential) equations, which Newton solves for Gregory's benefit. For a detailed discussion see Brackenridge (2003) (Gregory MS, f. 163, by kind permission of the Royal Society)

point needs to be stressed. Newton never conceived the calculus as a top-down theory which could allow a formulation of the laws of dynamics. The method of fluxions was conceived by him as a heuristic tool, to be deployed in the resolution, or analysis, of problems, in order to overcome obstacles that block a much preferred geometrical route.

The enigmatic character of the mathematical methods of the *Principia* was to play an important role in the Newton–Leibniz controversy. According to Johann Bernoulli, one of Leibniz's staunchest supporters, the *Principia* was proof of two weaknesses in Newtonian mathematized natural philosophy. First, that Newton was unable to deploy higher-order infinitesimals. Second, that he showed no sign of being able to formulate and integrate differential equations of motion. As any reader of Newton's mathematical manuscripts knows, the former criticism is false. The latter is more problematic. It is difficult to believe, as Newton famously stated, speaking of himself in the third person, that:

By the help of this new Analysis Mr Newton found out most of the Propositions in his *Principia Philosophiae*. But because the Ancients for making things certain admitted nothing into Geometry before it was demonstrated synthetically, he demonstrated the Propositions synthetically that the systeme of the heavens might be founded upon good Geometry. And this makes it now difficult for unskillful men to see the Analysis by wch those Propositions were found out. (Newton 1967–81, VIII 598–599)

As I have remarked above, we have good reason to believe that, in certain propositions, Newton did employ quadrature techniques (integrations, in Leibnizian terms), which barely surfaced in the printed text. That the *Principia* shows evidence of the method of fluxions was repeated again and again by the acolytes who set themselves the task of rebuking Leibniz's challenge (see Fig. 8.2.3).

The *Enumeratio*

The *Principia* was not the only work that aroused puzzlement and diverging interpretations because of Newton's publication policy. Another interesting case is the *Enumeratio linearum tertii ordinis* 'Enumeration of curves of third order', the slim treatise on the classification of cubic curves that appeared in 1704 as an appendix to the *Opticks*.

Newton developed most of his results on cubics in the 1670s, while he was also actively working on Apollonian geometry with a view to recovering the methods of the ancients. He systematized them in the 1690s, when he was at the peak of his fascination with ancient mathematicians and philosophers. Notwithstanding his vehement anti-Cartesianism and his admiration for the Ancients, Newton achieved most of the results on cubics through application of Cartesian analytic geometry, or even through the use of infinite series. Thus, in the late 1670s there

Figure 8.2.3 Engraving from William Jones' 1711 edition of Newton's mathematical treatises, *Analysis per quantitatum series, fluxiones, ac differentias,* printed during the dispute with Leibniz. The image expresses an idea cherished by both Newton and his acolytes: that the fluxional methods revealed in these treatises constituted the hidden analysis of the *Principia.* The mythological characters display scrolls and shields bearing diagrams from some key propositions of the *Principia.* From left to right are several diagrams from Book 1: Proposition 94 (on the ground) on the motion of refracted light corpuscles; Proposition 66 (held by a putto) on the three-body problem; Propositions 32 and 43 (both on the shield) on fall accelerated by an inverse-square force, and on precession of orbits; Proposition 1 (on the ground) on the law of areas; Proposition 91, Corollary 2 (on the ground) on the attraction exerted by an oblate ellipsoid. The message addressed to Leibnizians like Johann Bernoulli could not be clearer

emerged a tension between Newton's mathematical practice, and the views on method that he elaborated in his writing on the use of analysis and synthesis by the Ancients (Newton 1967–81, IV 274–335).

This tension led Newton to structure his published work on cubics, the *Enumeratio,* in a way that does not render the use of 'common analysis' (algebra) and 'new analysis' (calculus and series) wholly explicit. In many cases Newton provided just a hint, or no trace at all, of demonstrations of his statements. His way of presenting his results is often declaratory rather than argued. His readers often complained about this, and tried to obtain clarification from manuscript sources or oral communication. The *Principia,* as we have seen, is a rich repertoire of such mysteries. In the *Enumeratio* we encounter quite a number of extraordinary unproved statements.

One of the most disconcerting aspects of the *Enumeratio* is precisely that no proof is given of most of its propositions. One of its most important results is that, just as the conic sections can be regarded as projections of a circle, so every (nondegenerate) cubic can be generated as a projection of one of five particular cubic curves called 'divergent parabolas'. Nowhere, however, does the reader find proof of this fact. Indeed, Newton's Section 5, devoted to the generation of curves as shadows, is so concise that we can quote it in full:

If the shadows of curves caused by a lumininous point, be projected on an infinite plane, the shadows of conic sections will always be conic sections; those of curves of the second genus [that is, of third order] will always be curves of second genus; those of the third genus will always be curves of the third genus; and so on *ad infinitum*.

And in the same manner that the circle, projecting its shadow, generates all conic sections, so the five divergent parabolas, by their shadows, generate all other curves of the second genus. And so some of the more simple curves of other genera might be found, which would form all curves of the same genus by the projection of their shadow on a plane. (Talbot 1861, 25)

This is all Newton has to say. How did he achieve this profound result? This is the question that was invariably asked by readers of the *Enumeratio*, and before the publication of Newton's *Mathematical papers* it was not clear how to find an answer.

Sections 3 and 4 of the *Enumeratio*, devoted to a long classification of cubics, are perhaps less mysterious, but here again Newton limited himself to the classification and gave the reader little hint about how it was achieved. Each curve is meticulously drawn, creating a world of strange and beautiful objects made up of ovals and branches extending to infinity (see figure 8.2.4), but there is no instruction on how to construct the curves. It was James Stirling who, in *Lineae tertii ordinis neutonianae* 'Newton's curves of third order' (1717), provided a commentary on the *Enumeratio* where the algebraic character of Newton's work, and his use of infinite series, was spelled out. Revealing the analysis behind Newton's published proofs had by now become a highly esteemed practice among his acolytes.

Newton's obscure style soon generated complaints and comments, which oscillated between frustration and reverence; Newton was seen as a man who flew so high that he did not need or care to bow down towards mortals who want to be told how curves can be constructed and theorems proved. Jean Paul de Gua de Malves wrote with veneration:

This geometer, whose works are characterized by a unique sublimity, especially in this one seems to have elevated himself to an immense height, to which all other minds less penetrating and strong would have attempted in vain to attain. But the path he has followed in such a difficult enterprise escapes the sight of those who marvel at the degree of elevation to which he has arrived. The exception are a few light traces that he cared to leave in places which would have deserved that he would have stopped there for a much longer interval of time. These places, moreover, are almost always very far one from the

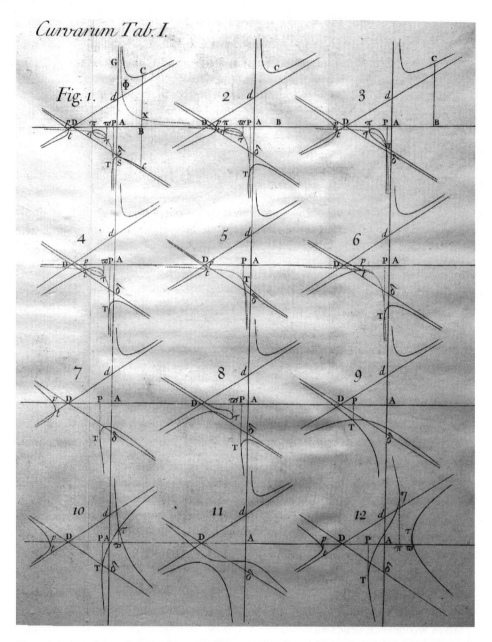

Figure 8.2.4 One of the tables that adorn Newton's *Enumeratio*. When such diagrams were published it was not easy to divine Newton's methods for plotting them in such fine detail. (Newton 1704, Table 1, facing page 162)

other. If one desires to follow the same route, one is compelled to guide oneself along such distant intervals.[28] (de Gua de Malves 1740, xi–xii)

Gabriel Cramer instead openly showed a critical attitude, not devoid of moral reproach:

It is deplorable that Newton was satisfied in displaying his discoveries without adding the demonstrations, and that he has preferred the pleasure of being admired to that of providing instructions.[29] (Cramer 1750, viii–ix)

C R M Talbot, who translated and commented on the *Enumeratio* in the nineteenth century, came to Newton's defence. But although he found 'the criticisms of the French mathematicians [...] ill founded' he had to admit that 'some explanation and illustration is wanted' (Talbot 1861, vii). W W Rouse Ball, who wrote on the *Enumeratio* in the 1890s, observed more bluntly that in Newton's treatise 'no proofs of the propositions are given' (Ball 1891, 105).

These are typical attitudes to Newton's published works, which arouse astonishment at their results, and criticisms for their opacity. Such reactions are evident even in Newton's contemporaries, and are thus a sign that Newton was following a strategy that seemed peculiar even to those of his own time. In fact publication practices in mathematics were undergoing deep changes during Newton's lifetime. The foundation of scientific academies and scientific journals after the 1660s rendered the double register of the printed book and scribal publication much less acceptable.

Further, as in the case of the *Commercium epistolicum* and the *Principia*, what Newton wanted to deliver in the *Enumeratio* were results not methods. A famous episode illustrates this attitude well. In January 1697, Johann Bernoulli proposed the brachistochrone problem,[30] as a challenge *acutissimis qui toto orbe florent mathematicis* 'to the sharpest mathematicians in the whole world'.[31] Newton's solution appeared anonymously in the *Philosophical Transactions*. He had probably found it through a fluxional (differential) equation similar to that employed (but not published) for the solid of least resistance in the *Principia*;[32]

---

28. Ce géomètre dont tous les ouvrages portent un caratctère singulier de sublimité, paroit en particulier dans celui-ci s'être élevé à une hauteur immense, à laquelle toute autre génie moins pénétrant et moins fort que le sien, auroit tenté vainemment d'atteindre: mais la route qu'il a tenue dans une enterprise si difficile, se dérobe aux yeux de ceux qui apperçoivent avec étonnement le degré d'élévation auquel il est parvenu. On doit en excepter quelques legères traces qu'il a eu soin de laisser sur son passage, aux endroits qui avoient mérité qu'il s'y arrétât plus long tems. Ces endroits, au reste, sont presque toujours assez distants les uns des autres. Si l'on se propose donc de suivre la même carrière, on est obligé se guider soi-même dans de long intervalles.

29. Il est facheux que M. Newton se soit contenté d'étaler ses découvertes sans y joindre les Démonstrations, et qu'il ait préféré le plaisir de se faire admirer à celui d'instruire.

30. Given two points *A* and *B* in a vertical plane, what is the curve traced out by a point acted on only by gravity, which starts at *A* and reaches *B* in the shortest time?

31. Bernoulli's challenge, which had already appeared in the *Acta eruditorum* for June 1696, circulated as a broadsheet. Newton's copy is held at the Royal Society of London. A transcript can be found in Newton (1967–81, VIII 80–85).

32. Book II Proposition 35 (34 in the second edition).

in his paper there is a geometrical construction of the required curve (a cyc-loid), but not the fluxional analysis. Newton's inclination to present himself as the deliverer of results rather than methods distanced him from the agendas that were to dominate in the eighteenth century, a period in which universality, method, and justification were deemed more important than achievements in problem solving. Newton's biographers have often reported a story recollected by his niece Catherine Barton: Newton, coming back home after a tiresome day at the Mint, found Johann Bernoulli's defiant brachistochrone problem on his desk and spent just a few hours solving it. The somewhat inaccurate story continues by describing the amazement of Johann Bernoulli, who immediately recognized the anonymous author 'as the lion from its claw'. This episode celebrates Newton the problem solver, whose published paper betrays a genius who idiosyncratically hides his method of discovery.

The nineteenth century

The mathematical legacy Newton left his followers was complex (Guicciardini 1989). Newton devoted much effort to the development of algebra and calculus. But he also conveyed to his disciples the idea that the Greek classics were superior to modern mathematics, and that the Ancients possessed hidden geometric tools based on projective geometry, which could be recovered by patient analysis of the surviving texts.

We conclude this chapter by considering how this complex and somewhat enig-matic legacy was re-evaluated in a moment of rupture in the history of mathem-atics, namely during the first half of the ninenteenth century. This was a period in which the calculus, Newton's and Leibniz's great discovery, underwent a deep change, often dubbed as 'rigourization of analysis', and epitomized by Cauchy's *Cours d'analyse* (1821). The calculus moved from being a mathematical algorithm based on often uninterpretable algebraic manipulations (as was characteristically the case with Euler and his contemporaries) to become a carefully crafted theory based upon definitions of convergence, continuity, limit, and derivative, and it was unavoidable that a change in the perception of its history also occurred. In particular, the confrontation between Newton's and Leibniz's algorithms and the 'metaphysics' of the calculus attracted renewed interest.

At the beginning of the nineteenth century there was also renewed interest in Newton's work and biography. This began with disclosure of information on Newton's theological and alchemical work, acquired by somewhat hurried sur-veys of the Portsmouth Papers (then in private custody at Hurstbourne Castle) by Samuel Horsley, David Brewster, and Stephen P Rigaud. Newton's image as a rational and pious scientist began to be questioned. Francis Baily's edition of

Flamsteed's papers (1835) raised doubts about Newton's moral conduct. Biot's biographical work (1822) defended Laplace's claim that Newton's interest in theology was the result of senility, or even mental derangement, and paved the way to de Morgan's scathing and irreverent writings, which rehabilitated Leibniz in the priority controversy and dealt at length with the Catherine Barton affair (highly embarrassing in Victorian England). Finally the Biot–Lefort edition of the *Commercium epistolicum* (Newton 1856) had an enormous impact in reshaping the image of Newton's role in the priority dispute. The overall effect of these writings was to cast a shadow upon Newton's dealings during the controversy with Leibniz. But more interesting for us is that in the meantime general scepticism was growing in France, Germany, and Italy about the effectiveness of his methods, a scepticism which soon crept into his native Britain. The Newtonian legacy appeared to many as defective in two fields: integration and mathematical astronomy, exactly those advanced fields Newton boasted about as the sign of his superiority over Leibniz.

Montucla's evaluation is typical in this respect:

The reader should not conclude that Newton resolved the problem [of the integration of real valued functions] completely; this would not fit well with what we have said above. Newton's method only delivers the sought relationship [between the independent and the dependent variables] as an infinite series [...] This is why geometers, reserving Newton's method for the most desperate cases, have sought means, both for integrating in finite terms, when this is possible, and for separating the indeterminates.[33] (Montucla 1799–1802, III 165)

Newton's method of integration by infinite series thus appeared to Montucla inferior to integration in finite terms. Lacroix expressed very similar ideas, also stressing the simplicity of Leibniz's notation compared to Newton's:

The continental geometers did not neglect the theory of infinite series; but they did not go so far as abusing it, as instead did the English geometers of second rank, who have often applied series to problems which admit a solution in finite terms [...] The school of Leibniz had a marked superiority over that of Newton, due perhaps more to the superiority of the former's methods than to the genius of his disciples, the Bernoullis [...] When Newton's writings were circulated on the Continent, one could see that he was in possession of the method of fluxions well before Leibniz had invented his differential calculus; but while it was possible for Newton's genius to deduce everything from his method that

---

33. Le lecteur ne doit cependant pas en conclure que Neuton ait résolu le problème en entier ; cela s'accorderoit mal avec ce qu'on a dit plus haut. La méthode de Neuton donne seulement le rapport cherché en série infinie. Content de cette solution générale, Newton n'a pas poussé plus loin ses recherches. [...] C'est pourquoi les Géomètres, réservant la méthode de Neuton pour les cas désespérés, ont recherché des moyens, soit pour intégrer en termes finis, lorsque cela se peut, soit pour séparer les indéterminées.

Leibniz could deduce from his own, the latter could be applied much more easily than the former.[34] (Lacroix 1797, xviii–xix)

In the second decade of the nineteenth century, the ideas of Montucla and Lacroix were to be adopted as a manifesto by the Analytical Society of Cambridge, a short lived but influential group of reformers of the Cambridge mathematical curriculum, who promoted an algebraic approach to the calculus in the hope of overcoming the stagnation of British mathematics.

But the Newtonian achievement in mathematical astronomy also came under attack. The opinions of John Playfair, a highly reputable Scottish mathematician, are interesting in this regard. One of Playfair's most influential papers was a review, published in 1808, of Laplace's *Mécanique céleste*, often praised as one of the first attempts to awaken the interest of British mathematicians in the works of the French school. Playfair not only commented on the contents of the first four volumes of Laplace's masterpiece but also placed them in the context of the development of eighteenth-century astronomy and, by way of conclusion, added several considerations on the reasons for the inferiority of British achievements. He concluded his account with a query:

In the list of the mathematicians and philosophers, to whom that science [mathematical astronomy], for the last sixty or seventy years, has been indebted for its improvements, hardly a name from Great Britain falls to be mentioned. What is the reason of this?

Playfair went on to describe the situation in England and Scotland:

a man may be perfectly acquainted with every thing on mathematical learning that has been written in this country, and may yet find himself stopped at the first page of the works of Euler and D'Alembert. He will be stopped, not from the difference of the fluxionary notation, (a difficulty easily overcome), nor from the obscurity of these authors, who are both very clear writers, especially the first of them, but from want of knowing the principles and the methods which they take for granted as known to every mathematical reader. If we come to works of still greater difficulty, such as the *Méchanique* [sic] *Céleste,* we will venture to say, that the number of those in this island, who can read that work with any tolerable facility, is small indeed. If we reckon two or three in London and the military schools in its vicinity, the same number at each of the two English Universities, and perhaps four in Scotland, we shall not hardly exceed a dozen; and yet we are fully persuaded that our reckoning is beyond the truth. (Playfair 1808, 279–281)

---

34. Les Géomètres du continent ne négligèrent point la théorie des Suites; mais ils n'allèrent pas jusqu'à en abuser, comme firent les Géomètres Anglois du second ordre, qui appliquèrent souvent les séries à des problémes dont on pouvoit avoir la solution par équations finies [...] L' école de Leibnitz avoit sur celle de Newton, une supériorité décidée, due peut-être autant à la simplicité de la Méthode du premier, qu'au génie des Bernoulli ses disciples [...]Lorsque les écrits de Newton furent répandus dans le continent, on vit qu'il avoit été en possession de la Méthode des fluxions, long-tems avant que Leibnitz eût découvert le Calcul différentiel; mais quoiqu'il fut possible au génie de Newton de tirer de sa méthode tout ce que Leibnitz pouvoit déduire de la sienne, l'une étoit d'une application bien moins facile que l'autre.

Playfair's review is typical of the pessimism towards the Newtonian mathematical legacy shared by several early nineteenth-century British mathematicians. Many mathematical reformers, including Robert Woodhouse, John Toplis, Charles Babbage, John Herschel, and George Peacock, denounced the lack of communication between continental Europe and Britain. Their self criticisms reveal the chasm separating the British and the Continental schools. Newton's image as a mathematician was lessened in many accounts that were produced at the beginning of the nineteenth century because of the superiority of the developments of the integral calculus and the *mécanique céleste* on the Continent when compared with the achievements of the British school. Now eulogy could take Whewell's somewhat lame form: Newton's 'gigantic implements of war', namely his geometrical methods, seemed to lie useless on the battlefield of cutting-edge mathematical research.

But a new turn in mathematics was again to tip the balance. A renewed interest in projective, or synthetic geometry, opened the way for a re-evaluation of still another current of the Newtonian legacy, that stemming from Newton's protracted attempts to 'divine' Euclid's lost books of *Porisms*. It was Michel Chasles who in 1837 referred with veneration to Newton and his heirs, Robert Simson and Matthew Stewart, who, working along the lines initiated by Pascal, Desargues, la Hire, and Newton himself, had developed elegant geometrical results based on cross-ratio invariance. Once again Newton's image as a mathematician was reshaped according to agendas that were polarizing debates amongst practising mathematicians.

Conclusion

Newton produced his mathematical work in a period of tumultuous change in the history of mathematics, a change which might well be seen as an aspect of a more general 'crisis' in European thought described by Paul Hazard (1935) and Basil Willey (1940). His philosophical agendas elaborated in opposition to Cartesianism and Baconianism, his preoccupations concerning mathematical method and certainty, the tensions which so worried him between his mathematical practice and the ancient exemplars, are all rooted in seventeenth-century English mathematical culture. A deep reorientation took place in the first half of the eighteenth century: algebraic language acquired independence from geometry, and the Greek mathematical classics became models of rigour useful for pedagogic aims in the University curriculum but which had little to teach to the practising mathematician. In his mature years Newton himself was aware of a

shift towards algebra which was slowly isolating him from the younger generation of mathematicians. In the late 1710s, referring to the *Principia*, he wrote:

> To the mathematicians of the present century, however, versed almost wholly in alge-
> bra as they are, this [i.e. the *Principia*'s] synthetic style of writing is less pleasing,
> whether because it may seem too prolix and too akin to the method of the ancients,
> or because it is less revealing of the manner of discovery. And certainly I could have
> written analytically what I had found out analytically with less effort than it took me
> to compose it. I was writing for Philosophers steeped in the elements of geometry,
> and putting down geometrically demonstrated bases for physical science.[35] (Newton
> 1967–81, VIII 451)

While the 'philosophers' in 1687 were steeped in geometry, the younger mathematicians, trained at the Bernoulli's school in Paris and Basel, who began their studies in higher mathematics reading l'Hôpital's *Analyse des infiniment petits* (1696),
found the *Principia* obscure. Newton's mathematical work became interesting
more for the unresolved problems it contained rather than for its methods, which
appeared more and more obsolete. When, in 1739–42, T Le Seur and F Jacquier
produced their richly annotated edition of the *Principia* (Newton 1739–42) they
provided eighteenth-century readers with long explanatory footnotes written in
the language of the differential and integral calculus: these footnotes replaced
Newton's enigmatic geometric text. As A R Hall writes:

> The *Principia* was to remain a classic fossilized, on the wrong side of the frontier between
> past and future in the application of mathematics to physics. (Hall 1958, 301)

Newton's text, with its idiosyncratic style, and its concealments due to the author's
publication policy, became opaque even to well-trained mathematicians.

The status of Newton as the prime mover of the scientific revolution remained,
however, impossible to disown. The readers who turned to his celebrated texts
had to translate them, as Le Seur and Jacquier had done, into more familiar language, and they had even to fill the gaps that were so frequent in Newton's printed
works. This process of interpretation and interpolation was indeed a process of
reinvention that generated a plurality of images of Newton as a mathematician:
we have considered a few of them in this chapter.

The plurality that we have encountered in our brief survey of the reception
of Newton's mathematical work is part of the same phenomenon that has been
studied by historians of his natural philosophy and theology, who have made us

---

35. Mathematicis autem hujus saeculi qui fere toti versantur in Algebra, genus hocce syntheticum
scribendi minus placet, seu quod nimis prolixum videatur & methodo veterum nimis affine, seu quod
rationem inveniendi minus patefaciat. Et certe minori cum labore potuissem scribere Analytice quam ea
componere quae Analytice inveneram: sed propositum non erat Analysin docere. Scribebam ad Philosophos
Elementis Geometriae imbutos & Philosophiae naturalis fundamenta Geometrice demonstrata ponebam.

aware both of the significance of Newton's influence on eighteenth-century culture and of the plurality of eighteenth-century 'Newtonianisms' (Schofield 1978; Schaffer 1990; Mandelbrote 2002). The Newtonian heritage in natural philosophy and theology branched into many different schools and styles of thought, which frequently developed in directions distant from the great master's intentions. The same holds true for mathematics. Present-day historians of mathematics are thus facing two challenges. The former is to attempt to go beyond the plurality of images of Newton in order to relocate his mathematical works, printed and manuscript, into the context in which they were produced. The latter is to study the complex process of reception, which spans from adulation to rejection, of Newton's mathematical heritage.

Bibliography

Baily, Francis, *An account of the Revd. John Flamsteed, the first astronomer-royal: compiled from his own manuscripts, and other authentic documents, never before published: to which is added his British catalogue of stars*, London: Printed by order of the Lords Commissioners of the Admiralty, 1835.

Biot, Jean-Baptiste, 'Newton, Isaac', in *Biographie universelle,* vol. 31, Paris, 1822, 127–194.

Brackenridge, Bruce, 'Newton's easy quadratures omitted for the sake of brevity', *Archive for history of exact sciences,* 57 (2003), 313–336.

Cauchy, Augustin-Louis, *Cours d'analyse de l'Ecole Royale Polytechnique*, Paris, 1821.

Chasles, Michel, *Aperçu historique sur l'origine et le développement des méthodes en géométrie, particuliérement de celles qui se rapportent à la géométrie moderne*, Bruxelles, 1837.

Cramer, Gabriel, *Introduction a l'analyse des lignes courbes algébriques*, Genève, 1750.

de Gua de Malves, Jean Paul, *Usages de l'analyse de Descartes pour découvrir, sans le secours du calcul différentiel, les proprietées, ou affections principales des lignes géométriques de tous les ordres*, Paris, 1740.

Guicciardini, Niccolò, *The development of Newtonian calculus in Britain, 1700–1800,* Cambridge University Press, 1989.

Guicciardini, Niccolò, *Reading the Principia: the debate on Newton's mathematical methods for natural philosophy from 1687 to 1736*, Cambridge University Press, 1999.

Hall, A Rupert, 'Correcting the Principia', *Osiris,* 13 (1958), 291–326.

Hall, A Rupert, *Philosophers at war: the quarrel between Newton and Leibniz,* Cambridge University Press, 1980.

Hazard, Paul, *La crise de la conscience européenne: 1680–1715,* Paris, 1935.

l'Hôpital, Guillaume F A de, *Analyse des infiniment petits pour l'intelligence des lignes courbes,* Paris, 1696.

Keill, John, 'Epistola ad clarissimum virum Edmundum Hallejum geometriae professorem savilianum, de legibus virium centripetarum', *Philosophical Transactions of the Royal Society,* 26 (1708), 174–188. [The volume for 1708 was printed in 1710]

Lacroix, Sylvestre F, *Traité du calcul différentiel et intégral,* 3 vols, Paris, 1797.

Leibniz, Gottfried W, 'Nova methodus pro maximis et minimis, itemque tangentibus, quae nec fractas nec irrationales quantitates moratur, et singulare pro illis calculi genus', *Acta eruditorum,* (1684), 467–473.

Leibniz, Gottfried W, *Leibnizens mathematische Schriften,* 7 vols, Berlin, 1849–63; reprinted Olms, 1971.

Leibniz, Gottfried W, *The early mathematical manuscripts of Leibniz,* Dover publications, 2005.

Mandelbrote, Scott, 'Newton and eighteenth-century Christianity', in I Bernard Cohen and George E Smith (eds), *The Cambridge companion to Newton,* Cambridge University Press, 2002, 409–430.

Marie, Maximilien, *Histoire des sciences mathématiques et physiques,* 12 vols, Paris, 1883–88.

Montucla J E, *Histoire des mathématiques,* 4 vols, Paris, 1799–1802. [vols 3 and 4 'Achevé et publié par Jérôme de La Lande', assisted by a number of other scholars]

Newton, Isaac, *Opticks, or, A treatise of the reflexions, refractions, inflexions and colours of light. Also two treatises of the species and magnitude of curvilinear figures,* London, 1704.

Newton, Isaac, *Analysis per quantitatum series, fluxiones, ac differentias: cum enumeratione linearum tertii ordinis,* London, 1711.

Newton, Isaac, *Philosophiae naturalis principia mathematica,* 3 vols, Genève, 1739–42.

Newton, Isaac, *The correspondence of Isaac Newton,* 7 vols, Cambridge University Press, 1959–77.

Newton, Isaac, *The mathematical papers of Isaac Newton,* 8 vols, Cambridge University Press, 1967–81.

Newton, Isaac, *The optical papers of Isaac Newton,* Cambridge University Press, 1984.

[Newton, Isaac, *et al*], *Commercium epistolicum D. Johannis Collins, et aliorum de analysi promota: jussu Societatis Regiae in lucem editum,* London, 1712.

[Newton, Isaac *et al*], *Commercium epistolicum J. Collins et aliorum de analysi promota, ou correspondance de J. Collins et d'autres savants célèbres du XVII siècle: relative à l'analyse supérieure: réimprimée, completée et publiée par J B Biot et F Lefort,* Paris, 1856.

Pappus, *Pappi alexandrini mathematicae collectiones a Federico Commandino urbinate in latinum conversae, et commentariis illustratae,* Pesaro, 1588.

Playfair, John, 'Traité de mechanique celeste', *The Edinburgh Review,* 22 (1808), 249–284.

Rouse Ball, W W, 'On Newton's classification of cubic curves', *Proceedings of the London Mathematical Society,* 50 (1891), 104–143.

Schaffer, Simon, 'Newtonianism', in R C Olby, G N Cantor, J R R Christie, and M J S Hodge (eds), *Companion to the history of modern science,* Routledge, 1990, 610–626.

Schofield, Robert E, 'An Evolutionary Taxonomy of Eighteenth-Century Newtonianisms', *Studies in Eighteenth Century Culture,* 7 (1978), 175–192.

Shapiro, Alan, 'Newton's "Experimental Philosophy"', *Early Science and Medicine,* 9 (2004), 185–217.

Stirling, James, *Lineae tertii ordinis neutonianae, sive illustratio tractatus D.Neutoni de enumeratione linearum tertii ordinis, cui subjungitur, solutio trium problematum,* Oxford, 1717.

Talbot, Christopher R M (ed and translator), *Sir Isaac Newton's enumeration of lines of the third order: generation of curves by shadows, organic description of curves, and construction of equations by curves,* London, 1861.

Wallis, John, *A treatise of algebra, both historical and practical,* London, 1685.

Wallis, John, *Opera mathematica,* 3 vols, Oxford, 1693–9.

Westfall, Richard S, *Never at rest: a biography of Isaac Newton*, Cambridge University Press, 1980.

Whewell, William, *History of the inductive sciences,* Cambridge and London, 1837.

Willey, Basil, *The eighteenth century background: studies on the idea of nature in the thought of the period*, Chatto and Windus, 1940.

From cascades to calculus: Rolle's theorem

June Barrow-Green

Rolle's theorem is a simple but important result, familiar to anyone who has moved just beyond elementary calculus into the beginnings of analysis. Essentially it tells us that if a differentiable function has equal values at *a* and *b*, then somewhere between those two points it must have a local maximum or a local minimum (Fig. 8.3.1).

A more formal statement of the theorem, typical of those given in modern textbooks, is as follows.

Let *f* be a function that is continuous on the closed interval [*a*, *b*] and differentiable on the open interval (*a*, *b*). If *f*(*a*) = *f*(*b*), then there exists a point *c* in (*a*, *b*) for which $f'(c) = 0$.

It is clear from the language of functions and derivatives that the theorem is now presented as a theorem of calculus. Its importance lies in the fact that it is needed in the proof of the mean value theorem and for establishing the existence of Taylor series.[1] When Michel Rolle (1652–1719) made the first statement of this theorem in 1690, however, Taylor series had not yet been discovered and

---

1. The mean value theorem states that if *f* is a function that is continuous on the closed interval [*a*, *b*] and differentiable on the open interval (*a*, *b*), then there exists a point *c* in (*a*, *b*) for which $f'(c) = \dfrac{f(b) - f(a)}{b - a}$.

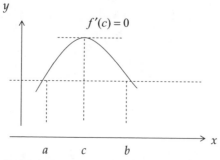

Figure 8.3.1

calculus itself was in its infancy. Moreover, Rolle was deeply suspicious of its methods. His theorem first appeared not in the context of calculus at all but of equation solving. It is paradoxical, therefore, that the name of a man renowned for his opposition to the infinitesimal calculus should end up attached to one of the fundamental theorems in the subject. I began to investigate how this might have happened after I came across a translation of Rolle's statement of the theorem in David Eugene Smith's *Sourcebook* (1959, 253–260). Florian Cajori, the translator, wrote an article (Cajori 1911) in which he gave a (not very clear) account of the theorem, and a chronology of sources in which it had later appeared, but with little indication of their content. I therefore decided to study Rolle's original work, to follow up some of Cajori's references and to seek others, to see how the theorem was transformed from a theorem in algebra to a theorem of calculus.

Rolle and his work

Rolle was born in Ambert, a small French town in the Auvergne, in 1652. At the age of 23, with sufficient education to become a teacher, he moved to Paris. There he became increasingly interested in mathematics and in particular in algebra. His first public success, in 1682, was the solution to an indeterminate problem posed by Jacques Ozanam:[2] to find four numbers the positive difference of any two of which is a square, and the sum of any two of the first three is also a square. Ozanam believed that the smallest of the four numbers would have at least fifty figures but Rolle found solutions in which the smallest numbers each had only

---

2. Ozanam, while not a research mathematician, was a successful teacher and writer of several texts. He is best remembered today for his *Récréations mathématiques et physiques* (1694), which went through many editions.

seven figures.[3] Rolle's result was published in the *Journal des sçavans*,[4] where Rolle was described as a 'Professeur d'Arithmetique', indicating that he was by this time a teacher of elementary mathematics ([Rolle] 1682). His achievement brought him to the attention of Jean-Baptiste Colbert, the minister of finance and original 'architect' of the Paris Academy of Sciences, and whose support was decisive for Rolle's future career. In 1685, Rolle was elected to the Academy as an *élève astronome* 'pupil astronomer', responsible for preparing experiments, and in 1699 he was promoted to the salaried position of *pensionnaire géomètre* 'stipendiary geometer'.[5] This was a distinguished position—of the seventy members of the Academy only twenty were *pensionnaires*—and one he retained until 1719 when, only a few months before he died at the age of sixty-seven, he became a *pensionnaire veteran*.

Around 1690 Rolle also enjoyed the patronage of the Marquis de Louvois. For a time he taught his fourth son, to whom he dedicated his *Traité d'algebre* 'Treatise of algebra' in 1690 and the *Démonstration* 'Demonstration' that followed it in 1691. The Marquis de Louvois was Louis XIV's secretary of state for war and it was through this connection that Rolle briefly held an administrative position in the ministry of war (de Fontenelle 1719, 96).

In the opening years of the eighteenth century, Rolle gained notoriety for his involvement in a lively dispute in the Academy on the validity of infinitesimal methods of the Leibnizian calculus as set out in l'Hôpital's *Analyse des infiniment petits* 'Analysis of the infinitely small' (1696) (Mancosu 1989). Rolle, who attacked the infinitesimal calculus for its lack of rigour and, as he believed, its propensity for error, initially pitted himself against Pierre Varignon, an established Leibnizian, the debate taking place within the confines of the Academy. The two men locked horns until the end of 1701 when the Academy itself called a silence on their disputations. The following year Rolle took up the cudgels again, this time engaging in a more public battle through the pages of the *Journal des sçavans*. So determined had Rolle been to reopen the debate that he had managed to have a special issue of the journal published—it was dated Thursday instead of the usual Monday—in which to fire his opening salvo: a paper on tangent problems (Rolle 1702).[6] It drew a direct response from Joseph Saurin (1702), a protégé of l'Hôpital, who himself had a special issue of the journal published, and battle lines were drawn. At the beginning of 1706, after four years of acrimony, with even Leibniz himself being drawn into the fray, the Academy found a way to

3. For a description of Rolle's solution of Ozanam's problem, see Dickson (1920 [1999], 447).
4. The *Journal des sçavans* (later renamed *Journal des savants*), began in January 1655, and was the first scientific journal in Europe.
5. At this time the term *géométrie* was used for what would now be described more generally as mathematics.
6. I am grateful to Jeanne Peiffer for alerting me to the special nature of this issue.

put an end to the fight and Rolle finally desisted, having been 'asked to conform better to the regulations of the academy' (Mancosu 1989, 243). Mancosu (1989) has convincingly argued that Rolle's retreat was a decisive factor in the victory of the infinitesimal calculus in France, but it seems that Rolle himself never fully accepted the legitimacy of the new methods.

Rolle also published extensively on topics in Cartesian geometry. In *De l'evanoüissement des quantitéz inconnuës dans la géométrie analytique* 'Of the vanishing of unknown quantities in analytic geometry' (1709) he was, according to Boyer (1956 [2004], 155), the first to use the term 'analytic geometry' in the modern sense.

Though he first became known in Paris for his proficiency in indeterminate problems in arithmetic, Rolle preferred algebra, and in particular working on equations of the kind that are now known as polynomial. In 1690 he published what is now his most famous work, his *Traité d'algebre*. Its most significant feature was Rolle's 'method of cascades' for finding roots (see below). The method is effective, if cumbersome, for equations in which all the roots are real and distinct. It seems that Rolle was criticized, however, for not justifying why it worked. To satisfy his critics, he almost immediately published his *Démonstration* (Rolle 1691),[7] written to prove that the method is infallible. And it was in this work, almost incidentally, that he established the theorem for which he is now famous.

Unlike the *Algebre*, the *Démonstration* appears to have had only a small circulation.[8] It disappeared from view almost completely during the eighteenth and nineteenth centuries: the only reference to it during this period seems to be a cursory mention by Montucla, who gave its date of publication as 1692, which suggests that he never actually saw it (1799–1802, II, 167). At the beginning of the twentieth century, the Swedish historian of mathematics Gustaf Eneström had never seen it either, but correctly conjectured that it was the book in which Rolle's theorem was to be found (Eneström 1906, 301–302).

Rolle's method of cascades and statement of the theorem

Rolle's statement of his theorem appears as part of his justification for his method of cascades. Rolle was attempting to find what he called '*hypotheses*', more usually known as 'limits' of the roots of an equation. Using modern notation,

---

7. Although the book was published in 1691, permission for printing was granted by the Academy on 30 December 1690 and Rolle (1699, Preface, 3) gave its date as 1690.

8. I am aware of only four copies: three in Paris, as listed by Cajori (1911, 300), and one in the New York Public Library.

given a polynomial $P(z)$ and two (real) numbers a and b such that $P(a)$ and $P(b)$ are of opposite sign, then between a and b there is a number c for which $P(c) = 0$. The numbers a and b are then the 'limits' of the root.[9] Rolle's method assumes that all the roots are real and distinct. His first step was to 'prepare' the equation using the transformation $x = h - z$, where h is taken large enough to ensure that all the roots x are positive. To find a suitable value for h, which he called the *grande hypothese* 'largest limit' for the roots of the equation, Rolle took the negative coefficient $-g$ of the original equation with largest absolute value, divided g by the coefficient of the highest power of the unknown, added 1 to the quotient, and rounded up to the nearest integer.[10] Since the equation now has all roots positive, 0 is the lower bound, or *petite hypothese* 'smallest limit', for the roots of the equation.

Suppose the prepared equation is $P(x) = p - qx + rx^2 - \ldots = 0$. Now a 'cascade' is formed. The method Rolle used throughout the *Algebre* was to multiply each term of $P(x)$ by its own exponent, divide the result by x, and equate the result, $-q + 2rx - 3sx^2 + \ldots$, to 0. This equation was what Rolle called a cascade. The process can be repeated until an equation of first degree is obtained. It is easily seen that it is equivalent to taking successive derivatives, but that is not how Rolle described it. As he explained it, he was multiplying each term by the corresponding term in the arithmetical progression 0, 1, 2, 3,.... This technique had first been devised by Johann Hudde, though Rolle made no reference to him.[11]

Here Rolle was using particular arithmetic progressions to generate the cascades, namely those whose first term is zero and whose difference is one. This is so that each cascade will be one degree lower than the previous one, the reduction in degree being essential for the method to be of use in finding roots. Although Rolle always used this construction to generate cascades in the *Algebre*, it is clear from the *Démonstration*, as we will see below, that his concept of a cascade was actually more general, and that a cascade could in fact be generated by any arithmetic progression (as in Hudde's method).

The cascades allow one to find the limits of the roots, and once that is done the roots themselves are found by continuous halving of the intervals between the limits. Rolle used the following equation as an example (Rolle 1690, 133).[12]

$$v^4 - 24v^3 + 198v^2 - 648v + 473 = 0$$

9. This notion of limit implicitly assumes the intermediate value theorem, which was rigorously proved only in the nineteenth century.

10. Rolle did not prove that this method for finding h did in fact give an upper bound; this was done later for cubics by Colin Maclaurin (1748, 172–174).

11. In 1659 Hudde had used the technique to find any double roots of a polynomial equation. For an explanation of his method, see Edwards (1979, 127–128).

12. For a more detailed example of the method of cascades, see Shain (1937, 25–27).

The cascades are:

$$v^4 - 24v^3 + 198v^2 - 648v + 473 = 0 \tag{4}$$
$$4v^3 - 72v^2 + 396v - 648 = 0 \tag{3}$$
$$6v^2 - 72v + 198 = 0 \tag{2}$$
$$4v - 24 = 0 \tag{1}$$

Note that here the cascades are in modern notation and in the opposite order to Rolle's.

Cascade (1) has just one root, $v = 6$. This root is now one of the limits of (2). The other limits of (2) are 0 (the *petite hypothese*) and 13 (the *grande hypothese*). An approximation to the root between 0 and 6 is found by repeated interval bisection, and yields 4. (Rolle approximated only to the nearest integer.) A similar process yields 7 as an approximation to the root between 6 and 13. Since 4 and 7 are approximate roots of (2), the limits of (2) are 0, 4, 7, and a new *grande hypothese*, 163. A similar calculation to that for (2), gives 3, 6, and 9 as approximate roots of (3). The limits of (4) are then 0, 3, 6, 9, and 649. Using these limits in a similar way leads to 1 as an exact root of (4), and 6, 8, and 10 as approximate roots.

Today, interpreting Rolle's 'cascades' as 'derivatives' it is not difficult to understand why his method works. Rolle, however, neither used nor trusted calculus. Considering his method algebraically, it is not at all obvious what is happening. In the *Algebre* Rolle gave no clue as to any theoretical underpinning and introduced the idea of multiplying by an arithmetic progression without giving any reason for it. Nor did he prove that the roots of each cascade are limits for the previous equation. The latter in particular is not easy to see—it relies on some clever algebraic manipulation—and the fact that it was hidden from the reader in the *Algebre* is one of the reasons that Rolle realized the necessity of bringing out his *Démonstration*.

In the *Algebre* Rolle states the following *regle* 'rule' for moving from a cascade of lower degree to the next one *upwards*, of higher degree: 'The roots of each cascade are taken for the mean limits of the following cascade' (Rolle 1690, 127).[13] However, in the *Démonstration* his statement of the theorem, given in Corollary III, moved in the other direction, from an equation or cascade of higher degree to the next one *downwards*, of lower degree.

Corollary III. It is also clear that the roots are intermediate numbers lying between the limits, and consequently if the roots are substituted in an equation whose roots are these limits, the substitution must give results that are alternately positive and negative, or negative and positive. For example:

$(y - 6)(y - 21)(y - 30)$ Roots of the equation

$(y - 0)(y - 12)(y - 26)$ Roots of the cascade[14]

---

13. Les racines de chaque cascade seront prises pour les hypotheses moyennes de la cascade saivante.
14. Here the original equation is $y^3 - 57y^2 + 936y - 3780 = 0$ and the cascade, before dividing by y, is $3y^3 - 114y^2 + 936y = 0$ or $y^3 - 38y^2 + 312y = 0$.

where one can see that 6 being substituted in the roots of the cascade will give, following Articles III & IV, results whose product is positive, that 21 will give results whose product is negative, and that 30 will give results whose product is positive; and consequently the roots 6, 21, 30, each being substituted separately into the cascade must give alternatively + and –, according to Article II. Thus the roots are themselves limits of their own limits.[15] (Rolle 1691, 20–21)

The significant part of the Corollary is the last sentence, *Ainsi les racines sont hypoteses des hypoteses mêmes* 'Thus the roots are themselves limits of their own limits.' In other words, roots and limits alternate in a regular way. In particular, two consecutive roots will give rise to a limit on the level of the next cascade downwards, that is, they will produce a root on that level. The first part of Rolle's sentence, 'Thus the roots are themselves limits', is what today we call Rolle's 'theorem'. In this example, however, Rolle gives priority to his 'rule', namely, that roots of an upper cascade, if they exist, lie between roots of the lower cascade, that is, they are 'limits of their own limits'. Thus, in effect, Rolle gives his 'theorem' as a corollary to his 'rule'. Nevertheless, his proof of the theorem, as given in the *Demonstration*, is independent of that assumption.

An outline of the proof, given in modern notation, is as follows.[16] Given a monic polynomial $P(z)$ of degree n, with positive, distinct, real roots $a > b > c > \ldots > m$, multiply $P(z)$ term by term by the arithmetic series $y, y + v, y + 2v, \ldots,$ $y + nv$. We now need to show that the resulting cascade $C(z)$ takes positive and negative (or negative and positive) values for the two consecutive roots a, b of $P(z)$, thus making a, b limits (*hypotheses*) of a root of $C(z)$.

First we may write $P(z) = (z - a)(z - b) K(z)$, where $K(z)$ is a monic polynomial of degree $(n - 2)$. Now multiply $(z - a)(z - b) = ab - (a + b)z + z^2$ term by term by $y, y + v, y + 2v$. This gives the 'partial cascade'

$$C^*(z) = aby - (a + b)z(y + v) + (y + 2v)z^2$$
$$= aby - ayz - byz - avz - bvz + yz^2 + 2vz^2.$$

Now $C^*(a) = (a - b)va$ and $C^*(b) = (b - a)vb$. Note that $C^*(a)$ and $C^*(b)$ depend only on v, not on y, and (although Rolle did not explicitly make this point) this

---

15. Corollary III. Il est clair aussi que les racines sont des nombres moyens entre les hypoteses, & par consequent les racines estant substituées dans l'égalité que renferme les hypoteses, leur substitution doit donner des resultats alternativement positifs & negatifs, ou negatifs & positifs. En voicy un exemple.

$$y - 6.y - 21.y - 30 \text{ Rac. de l'ég.}$$
$$y - \theta.y - 12.y - 26 \text{ Rac. de la Cascade.}$$

Où l'on peur voir que 6 estant substitué dans les racines de la Cascade, donnera suvant les Arti. 3 & 4 des resultats dont le produit est positif. Que 21 donnera des resultats dont le produit est negatif, & que 30 donnera des resultats dons le produit est positif; & par consequent ces racines 6.21.30 estant substituées chacune séparément dans la Cascade, elles doivent donner alternativement + & – selon l'Article 2. Ainsi les racines sont hypoteses des hypoteses mêmes.

16. I am grateful to Reinhard Siegmund-Schultze for enlightening discussions about Rolle's proof.

is the reason that his theorem holds for multiplication of $P(z)$ by any arithmetic progression. Note too that since a and b are positive, $C^*(a)$ and $C^*(b)$ will be of opposite sign.

If instead of starting from $(z - a)(z - b)$ we start from $(z - a)(z - b)z^n$ it is easily seen that the partial cascade $C^*(z)$ now has values $C^*(a) = (a - b)va^{n+1}$ and $C^*(b) = (b - a)vb^{n+1}$. The final stage is to use the polynomial $P(z)$ itself, to obtain the full cascade $C(z)$. In this case we have $C(a) = (a - b)vaK(a)$ and $C(b) = (b - a)vbK(b)$.

We now note that $K(a) \neq 0$ and $K(b) \neq 0$. Further $K(a)$ and $K(b)$ are of the same sign, because $K(z) = (z - c)\ldots(z - m)$ and, according to the original assumption, a and b are both greater than any root of $K(z)$. On the other hand, since a and b are both positive, $(b - a)b$ and $(a - b)a$ are of opposite sign. Therefore $C(a)$ and $C(b)$ must also be of opposite sign. What has been shown here for a and b is, as Rolle (1691, 29–30) showed, easily adapted for any other pair of consecutive roots. Thus any two consecutive roots of $P(z)$ produce opposite signs in the cascade $C(z)$ and therefore a root of $C(z)$ lies between them. The theorem is thus proved for equations with (positive) real and distinct roots (Rolle 1691, 30–31).

This outline proof follows Rolle in spirit if not to the letter. Rolle's proof was considerably longer, with details of the term by term multiplication worked in full. At the same time it was less explicit, because he did not write down the crucial co-factors of $(b - a)$ and $(a - b)$, namely $vaK(a)$ and $vbK(b)$, but said only that $C(b)$ is 'measured' by $(b - a)$. Clearly Rolle knew that his theorem was true for general arithmetic series, regardless of the values of y and v. In this sense his original theorem was more general than its modern counterpart, because it was not restricted only to derivatives. At the same time it was also more restricted because it applied only to polynomial functions. Rolle gave no indication that he considered this theorem to have any more importance than his other results, despite the fact that it is clearly a cornerstone of his method of cascades (see Fig. 8.3.2).

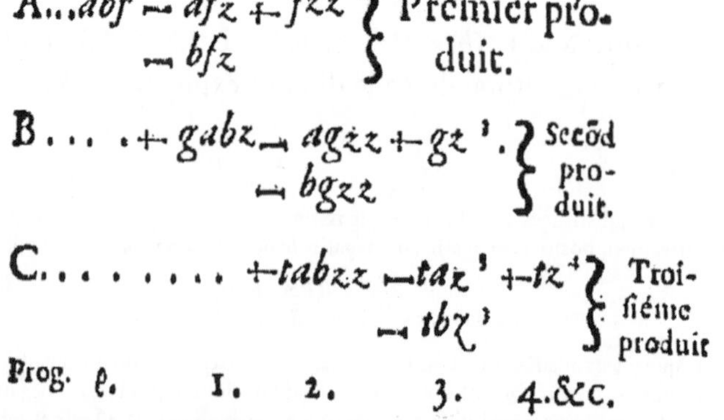
A...abf − afz + fzz } Premier produit.
 − bfz
B... + gabz − agzz + gz³. } Secõd produit.
 − bgzz
C......... + tabzz − taz³ + tz⁴ } Troisiéme produit
 − tbz³
Prog. e. 1. 2. 3. 4.&c.

Figure 8.3.2 Multiplication by an arithmetic series, from Rolle (1691, 27)

The assumption (which Rolle did not initially state or prove) that the roots of each equation are the limits of the next cascade *downwards* (of lower degree) is crucial for the validity of the method. Translated into the language of the calculus, this says that between two consecutive (real) roots of an equation $f(x) = 0$ there exists *at least* one (real) root of the 'derived' equation $f'(x) = 0$. This is of course Rolle's 'theorem'.

But the 'rule' originally given in the *Algebre* stated the converse: that the roots of each cascade are the limits for the next equation *upwards* (of higher degree). In modern terms: between two consecutive (real) roots of an equation $f(x) = 0$ there lies *at most* one (real) root of the equation $f(x) = 0$ from which $f'(x) = 0$ is derived. In this case the roots of $f'(x) = 0$ serve as limits for the root of $f(x) = 0$. This is what Cajori (1911, 301) called Rolle's 'corollary', though in fact it was Rolle's original 'rule'. Its importance lies in the fact that it guarantees, providing all the roots are real, that the method of cascades finds *all* the roots of the original equation. It is easily deduced from Rolle's 'theorem', a fact that caused later authors who had seen it in the *Algebre* but who had not seen the *Démonstration* to question whether Rolle's theorem was after all really due to Rolle.

One further remark about Rolle's method seems appropriate. His argument was applicable only to polynomial functions. Rolle did not transcend the latter to include other functions for which his theorem is also valid. This can perhaps be seen as a consequence of his difficulties with the new methods of infinitesimal analysis, which are hinted at in the following remark in the *Démonstration*:

There are nevertheless several mathematicians who do not always take care in this respect [in constructing their proofs], and one finds even today some whose frequent mistakes would make one believe that in them is formed the habit of being deceived about the idea of the infinite.[17] (Rolle 1691, Preface, 2–3)

Though effective, Rolle's method is lengthy. It was enthusiastically received by Rolle's contemporaries, who were eager to seize upon any new technique for finding roots, but it did not enjoy an enduring success. By the end of the eighteenth century it had fallen into disuse, largely due to its length (Lagrange 1798, Note VIII, 1) and to Rolle's rather inelegant style and idiosyncratic notation (Montucla 1799–1802, 168).

It is natural to ask why Rolle did not forestall criticism by including the *Démonstration* in his *Algebre*. Here one must remember that he worked in a mathematical environment that favoured methods that could be shown to work, regardless of their theoretical foundation or lack of it. He had provided a description of the method of cascades, and he had demonstrated its efficacy through numerous

---

17. Il y a neanmoins plusieurs Mathematiciens qui n'ont pas toûjours égard à ce rapport, & il s'en trouver meme aujourd'huy donts les frequentes méprises seroient croire qu'il se forme en eux une habitude à se tromper sur l'idée de l'infini.

examples. For most readers, that would probably have been sufficient and perhaps Rolle initially thought so too. But the mathematicians gathered around the Academy were as strong (and as combative) as any in Europe at the time and it was natural for them to put Rolle, who was still making a name for himself, under closer scrutiny.

Rolle's theorem from 1691 to 1910

Rolle's theorem, or statements equivalent to it, was published several times during the eighteenth century, although it was not always attributed to Rolle. Its first appearance seems to have been in Charles-René Reyneau's *Analyse demontrée ou la methode de résoudre les problêmes des mathematiques* 'Analysis demonstrated or the method of solving the problems of mathematics' (1708), the first advanced account of the integral calculus (Greenberg 1986, 66). Reyneau was a member of the group of mathematicians assembled around the Cartesian philosopher Nicolas Malebranche in Paris. He would certainly have been familiar with Rolle's work because he had taken an active interest in the Rolle–Varignon calculus debates a few years earlier (Mancosu 1989, 230). Reyneau's book contained an account of the method of cascades, and in the preface he acknowledged it as Rolle's:

In the sixth book we explain and demonstrate a method for finding the magnitudes of the limits of the values of the unknown in numerical equations of all degrees (Mr Rolle is the author of this method); and we give several ways to find, by the [arithmetical] mean of these limits, the values of the unknowns of these numerical equations differing as little from the exact values as we require.[18] (Reyneau 1708, xii)

Reyneau formulated the theorem itself as follows, replacing Rolle's 'hypotheses' by the more familiar 'limits'.

Corollary VII. Which is fundamental.
    The roots of a first equation are none other than the limits of the roots of a second, the roots of the second are none other than the limits of the roots of the first. Consequently if one multiplies each of the terms of any equation in which all the roots are real, positive and distinct by the number which is the exponent of that term, and the last term by zero, the roots of the equation which result from this multiplication are the limits of the roots of the original equation.
    For example, suppose that $x^4 - nx^3 + pxx - qx + r = 0$ represents an equation where all the roots are real, positive and distinct. If one multiplies each term by the exponent of the degree of the unknown of that term, and the last term by zero, one will have

---

18. On explique et l'on démontre dans le sixiéme livre la methode de trouver les grandeurs qui sont les limites des valeurs de l'inconnue dans les équations numeriques de tous les degrés; (*Monsieur Rolle est l'auteur de cette méthode;*) & l'on donne plusieurs manieres de trouver par le moyen de ces limites, les valeurs des inconnues des équations numeriques aussi peu differentes des valeurs exactes qu'on le peut desirer.

$4x^4 - 3nx^3 + 2pxx - qx = 0$; then dividing by x, $4x^3 - 3nxx + 2px - q = 0$, the roots of this last equation are the limits of the original equation.[19] (Reyneau 1708, 290)

Reyneau was already describing the theorem as 'fundamental'. After Rolle's disputes in the Academy, it is ironic that this first restatement of his theorem should have appeared in a textbook on the Leibnizian calculus, though still presented firmly in the context of algebra.

In Scotland, Colin Maclaurin worked on methods for finding 'impossible' (complex) roots of equations, proving the following theorem in a paper published in 1729:

THEOREM III. In general, the roots of the equation $x^n - Ax^{n-1} + Bx^{n-2} - Cx^{n-3}$ & c. = 0, are the *limits* of the roots of the equation $nx^{n-1} - (n-1)Ax^{n-2} - (n-2)Bx^{n-3} +$ & c. = 0, or of any equation that is deduced from it by any arithmetical progression $l \mp d, l \mp 2d, l \mp 3d$ & c. and conversely the roots of this new equation will be the *limits* of the roots of the proposed equation $x^n - Ax^{n-1} + Bx^{n-2} - Cx^{n-3}$ & c. = 0. (Maclaurin 1729, 88)

Maclaurin did not use Rolle's method of cascades. His proof was based on two Lemmas, the first of which involved transforming the equation $x^n - Ax^{n-1} + \ldots = 0$ by putting $x = e + y$ and showing that the coefficient of each term y^r of the new equation could be deduced from the coefficient of the previous term y^{r-1} by multiplying each part of the coefficient by its e exponent and dividing the product by re. The second Lemma gave the conditions for two quantities to be limits of one or more *real* roots of the equation. These results were then used to show that the roots of the equation are limits of the equation $nx^{n-1} - (n-1)Ax^{n-2} \ldots = 0$. Arithmetic progressions were invoked only at the end of the proof when they were used to produce (by 'multiplication' with the original equation) an equation whose roots are the limits of the original equation.

Like Reyneau, Maclaurin made no explicit connection to the methods of the calculus. Nor did he make any direct reference to Rolle. However, in 1725 a copy of Rolle's *Algebre* had been donated to the University Library of Aberdeen, where Maclaurin was professor of mathematics.[20] Reyneau's text was also known to Maclaurin,[21] so it is virtually certain that Maclaurin was aware of Rolle's work.

---

19. Corrollaire VII. *Qui est fundamental*. Mais les racines d'une premiere équation ne sçauroient êtres les limites des racines d'une seconde, que les racines de la seconde ne soient aussi les limites des racines de la premiere; par consequent si on multiplie les termes d'une équation quelconque, dont tout les racines sont réelles, positives & inégales, chacun par le nombre qui est l'exposant de l'inconnue de ce terme, & le dernier terme par zero, les racines de l'équation qui vient de cette multiplication, sont les limites des racines de l'équation proposée. Par example supposant que $x^4 - nx^3 + pxx - qx + r = 0$ représente une équation dont tous les racines sont réelles, positives & inégales, si on multiplie chaque terme par l'exposant du degré de l'inconnue de ce terme, & le dernier terme par zero, l'on aura $4x^4 - 3nx^3 + 2pxx - qx = 0$, ou bien divisant par x, $4x^3 - 3nxx + 2px - q = 0$, les racines de cette dernier équation sont les limites des racines de la proposée.

20. The book was donated by James Fraser, a well known bookseller and court librarian. Maclaurin was professor of mathematics in Aberdeen in 1721–2 and again in 1724–5.

21. Reyneau's text is cited in Campbell (1727, 517), a paper that Maclaurin discussed with James Stirling in a letter of 11 February 1728; see Mills (1982, 185–188) for a transcript.

During the eighteenth century the theorem lived on as a theorem in algebra, appearing in a form comparable to that given by Maclaurin. A typical example is the version presented by the Italian mathematician Odoardo Gherli (1771, 151). But while mathematicians like Gherli still adhered to the language of algebra, a radical change had already taken place elsewhere. In 1755 Leonhard Euler had published a version of the theorem in his *Institutiones calculi differentialis* 'Foundations of differential calculus' (1755, 657–660). The theorem still appeared in the context of equation solving, but was now expressed in the language of calculus.

Suppose we have an equation $x^n - Ax^{n-1} + Bx^{n-2} - \ldots = 0$ with distinct roots $p < q < r < \ldots$. Euler considered the general function $z = x^n - Ax^{n-1} + Bx^{n-2} - Cx^{n-3} + \ldots$ and allowed x to increase from '$-\infty$' through its range of values:

… continuing to place larger values in place of x, it is clear that z will take values greater than zero or less than zero, but it does not vanish before we put $x = p$; in which case $z = 0$. As the values of x are increased beyond p, the values of z become positive or negative, until we arrive at the value $x = q$; in which case again $z = 0$. Therefore it is necessary, since the values of z move from 0 to 0 again that, in between, z will have either a maximum or minimum value.[22] (Euler 1755, 657–658)

Euler had earlier shown that the values of x which make the function z a maximum or minimum are roots of the differential equation $\dfrac{dz}{dx} = nx^{n-1} - (n-1)Ax^{n-2} + \ldots = 0$, and so could go on to argue:

Since between any two real roots of the equation $z = 0$ there holds one of the cases, that the function z becomes maximum or minimum, it follows that if the equation $z = 0$ has two real roots, then the equation $\dfrac{dz}{dx} = 0$ necessarily has one real root. Equally, if the equation $z = 0$ has three real roots, then the equation $\dfrac{dz}{dx} = 0$ certainly has two real roots. And in general if the equation $z = 0$ has m real roots, the equation $\dfrac{dz}{dx} = 0$ necessarily has at least $(m - 1)$ real roots.[23] (Euler 1755, 660–661)

Euler's presentation of the theorem was thus fundamentally different from that of his predecessors. With the calculus at his fingertips he had no need for Rolle's

22. …continuo maiores in locum ipsius x collocari; perspicuumque est z nacturum hinc esse valores vel nihilo maiores vel nihilo minores, neque prius esse evaniturum, quam ponatur $x = p$; quo casu sit $z = 0$. Augeantur valores ipsius x ultra p, atque valores ipsius z vel affirmativi vel negative fient, donec perveniatur ad valorem $x = q$; quo casu iterim erit $z = 0$. Necesse ergo est, ut cum valores ipius z ab 0 iterum ad 0 accesserint, interea z habuerit valorem vel maximum vel minimum.

23. Quia inter binas quavis aequationis $z = 0$ radices reales datur unus casus, quo functio z sit maximum vel minimum; sequitur si aequatio $z = 0$ necessario unam radicem habituram esse realem. Pariter si aequatio $z = 0$ tres habeat radices reales, tum aequatio $\dfrac{dz}{dx} = 0$ certo duas habebit radices reales. Atque generatum si aequatio $z = 0$ habeat m radices reales, necesse est ut aequationis $\dfrac{dz}{dx} = 0$ ad minimum sint $m - 1$ radices reales.

method of cascades so for the first time, albeit still in the context of polynomial equations, the theorem resembles its modern counterpart. Whether or not Euler had read Rolle is not known. Characteristically, he makes no reference to any earlier work.

Lagrange, like Euler, discussed the theorem in the context of equation solving but in the language of the calculus. Unlike Euler, however, he specifically referred to Rolle and his method of cascades:

The need to deal with this problem [finding the limits of roots of equations] was recognized before the end of the seventeenth century, and once it had been found that the equation formed by multiplying each term of the given equation by the exponent of its unknown contains the conditions for the equality of roots of the original equation, it was not long before it was discovered that the roots of this same equation thus formed were the limits of those of the primitive equation. It is known that Hudde is the author of the first of these important discoveries, and I believe that the second is due to Rolle, who gave it in his *Algebre*, printed in 1690, and which has as its base his method of cascades.[24] (Lagrange 1798, Note VIII, 1)

The version Lagrange refers to is only Rolle's 'rule', as given in the *Algebre*. Lagrange had not seen the *Démonstration*, but he referred to the theorem itself (which of course he would have seen (unattributed) in Euler's *Institutiones calculi differentialis*) in his subsequent discussion:

It is first clear that the equation $F(x) = 0$ of degree m will have m real roots and that the derived equation $F'(x) = 0$ of degree $m - 1$ will necessarily have $m - 1$ real roots, since between two consecutive real roots of the equation $F(x) = 0$, there is always a real root of the equation $F'(x) = 0$.[25] (Lagrange 1798, Note VIII, 12)

By the 1830s the theorem appeared in a number of textbooks on the theory of equations, but was still not associated with Rolle. The person who appears to have been the first to call it 'Rolle's theorem' was Wilhelm Drobisch (Cajori 1911, 309). Drobisch, was professor of mathematics at the University of Leipzig, and noted for his clarity and precision (Heinze 1904, 81), published 'Rolle's Sätze' in a textbook (1834, 179). He cited Lagrange (1798, Note VIII) as his source despite the fact that Lagrange himself had not actually associated the theorem with Rolle. Drobisch, having noticed the symbiotic nature of the theorem and the rule,

---

24. On a senti avant la fin du xviie siècle la nécessité de s'occuper de ce problème, et, dès qu'on eut trouvé que l'équation formée en multipliant chaque terme d'une équation donnée par l'exposant de son inconnue renferme les conditions de l'égalité des racines de la proposée, on découvrit bientôt que les racines de cette même équation ainsi formée étaient les limites de celles de l'équation primitive. On sait que Hudde et l'auteur de la première de ces deux importantes découvertes, et je crois que le seconde et due à Rolle, qui l'a donnée dans son Algèbre, imprimée en 1690, et qui en a fait la base de sa méthode des cascades.

25. Il est d'abord évident que l'équation $F(x) = 0$ of degré m aura m racines réelles et que l'équation dérivée $F'(x) = 0$ du degré $m - 1$ aura aussi nécessairement $m - 1$ racines réelles, puisque, entre deux racines réelles consécutives de l'équation $F(x) = 0$, il tombe toujours une racine réelle de l'équation $F'(x) = 0$.

advertised it thus in the table of contents: 'Rolle's theorems on the limitation of the real roots of the derived equation by the real [roots] of the original and vice versa' (Drobisch 1834, xxvi).[26]

Six years later, François Moigno, a teacher of mathematics at the College Sainte Geneviève in Paris, included 'Rolle's theorem' in the title of an article (Moigno 1840), as did the French mathematician and historian of mathematics, Orly Terquem (1844), shortly afterwards. Joseph Liouville (1864, 84) proved an extension of the theorem to complex roots, describing it as 'le théorème celebre de Rolle'. Nevertheless, according to Cajori (1911, 309), it was not until 1868, when the theorem was attributed to Rolle in a German edition of Joseph Serret's influential *Cours d'algèbre supérieure*, that the association with Rolle became more widely known (Serret 1868a, 216). The theorem was then included, with its own subheading, in the next (fourth) French edition of Serret's text:

If a and b are two consecutive roots of the equation $f(x) = 0$, such that the equation has no other root between a and b, the equation $f'(x) = 0$, obtained by equating to zero the derivative of $f(x)$, has at least one root between a and b, and if it has several, then the number of these roots is odd.[27] (Serret 1877, 271)

In the latter half of the nineteenth century the theorem underwent its second significant change. From being a useful result in the theory of equations it was transformed into a fundamental theorem in analysis. Although it is now used to prove the mean value theorem, the two theorems had originally existed separately. Cajori (1911, 310) conjectured that the first person to bring the two theorems together was Pierre-Ossian Bonnet, a French mathematician better known for his work on differential geometry. Bonnet's derivation of the mean value theorem from Rolle's theorem was reported by his colleague Joseph Serret in his calculus textbook (1868b, 19), although Serret neither mentioned Rolle's name nor provided a reference for Bonnet's work.

In 1873 Charles Hermite used the theorem in his *Cours d'analyse* 'Course of analysis' in the context of the theory of Taylor series, now clearly attributing it to Rolle:

When a continuous function is zero for two values x_0 and X, the derivative, if it is itself continuous, vanishes for a value between x_0 and X.

This last proposition, that is to say Rolle's theorem, together with the rules of arithmetic established so easily in algebra for the formation of derivatives of sums, products and powers of functions, is sufficient to establish the existence of Taylor series.[28] (Hermite 1873, 48)

---

26. Rolle's Sätze von der Begrenzung der reellen Wurzeln der derivirten Gleichung durch die reellen der ursprünglichen, so wie dieser durch jene.

27. Si a et b désignent deux racines consécutives de l'équation $f(x) = 0$, en sorte que cette équation n'ait aucune autre racine comprise entre a et b, l'équation $f'(x) = 0$, obtenue en égalant à zero la dérivée de $f(x)$, a au moins une racine comprise entre a et b, quand elle en a plusieurs, le nombre de ces racines est impair.

28. Lorsqu'une fonction continue est nulle pour deux valeurs x_0 et X, la dérivée, si elle est elle-même continue, s'annule pour une valeur comprise entre x_0 et X.

Note that this formulation is not Rolle's theorem in the generality we know it today. Not only does Hermite seem to assume that continuous functions are always differentiable (although he may have known from his close correspondent Weierstrass that this was not true)[29] but, more importantly, he includes the unnecessary condition that the derivative has to be continuous. After giving a proof of the existence of Taylor series, Hermite concluded:

> But the very principle of this simple proof, which rests on Rolle's theorem, belongs to Homersham Cox, as one can see in the work of Todhunter (*A Treatise on the differential Calculus*).[30] (Hermite 1873, 49–50).

Rolle's name does not appear in any of the works to which Hermite referred (Homersham Cox 1851a, 80–81; 1851b, 37; Todhunter 1855, 68),[31] but Hermite would have been familiar with the origins of the theorem from his knowledge of the theory of equations. Hermite was the leading French analyst of his generation and his *Cours d'analyse*, which ran to four editions, was extremely influential in France during the latter part of the nineteenth century. His unequivocal association of the theorem with Rolle was thus decisive for future writers.[32]

From the mid-1870s onward the attribution became standard: during the years 1876 to 1888 the subject index of the *Royal Society catalogue* listed six publications under the subheading 'Rolle's Theorem' (1908, 167).[33]

There was yet another turn in the theorem's history at the beginning of the twentieth century when some mathematicians doubted that the theorem was due to Rolle at all.[34] This was because they had assumed that it was in his *Algebre,* but had only been able to find there the weaker 'upward' result, 'Rolle's Corollary'. It was only after Eneström (1906, 301–302), having read Reyneau (1708), suggested that the *Démonstration* was the place to look that the doubts were laid to rest.

C'est cette dernière proposition, c'est-à-dire le théorème de Rolle, jointe aux règles de calcul établies si facilement en Algèbre pour la formation des dérivées de sommes, de produits et de puissances de fonctions, qui nous suffira pour établir la série de Taylor.

29. Weierstrass had given an example of a nowhere-continuous differentiable function in a lecture in 1872.

30. Mais le principe même de cette démonstration si simple, qui repose sur le theoreme de Rolle, appartient a M Homersham Cox, ainsi qu'on peut le voir dans l'Ouvrage de M Todhunter (*A Treatise on the differential Calculus*).

31. The first edition of Isaac Todhunter's calculus textbook (1852) does not contain a statement of Rolle's theorem nor a reference to Homersham Cox's publications.

32. Edmond Laguerre, for example wrote: ... *sur la théoreme de Rolle (voir notamment Cours d'Analyse...) M. Hermite 1867–68)* '... on Rolle's theorem (see notably *Cours d'analyse*...of Hermite 1867–68)' (Laguerre 1880, 230).

33. Cajori (1911, 309) incorrectly gives 1876 as the date of the earliest article in the Index with Rolle's orem in its title, but Moigno (1840), Terquem (1844), and Liouville (1860) are all listed.

34. Cajori (1911, 301) mentioned Alfred Loewy, Alfred Pringsheim, and Anton von Braunmühl.

Conclusion

The theorem's transition from algebra to analysis, begun in 1755 by Euler and completed more than one hundred years later by Serret and Hermite, mirrors the increasing interest in foundational aspects of the calculus over the period. Concomitant with this transition was the growing recognition of Rolle as the original formulator of the theorem. As the theorem itself grew in status, so too the association with Rolle became more widespread. That the attribution became a subject of discussion and doubt at the end of the nineteenth century serves only to highlight the theorem's increasing significance. Had the few surviving copies of the *Démonstration* been lost, the truth would have been hard to confirm and the story might have had a very different ending.

That Rolle's theorem, one of the fundamental theorems in analysis, should have begun its life as an unexceptional theorem in algebra, now seems rather remarkable. Given Rolle's own views about the calculus it is perhaps less surprising. Nevertheless it is now so firmly entrenched in analysis that for some it seems inconceivable that it could ever have been otherwise. As an egregious example of such misguided judgement consider the Wikipedia entry for Rolle's theorem, which claims that 'A version of the theorem was first stated by the Indian astronomer Bhaskara in the 12th century. A proof of the theorem had to wait until centuries later when Michel Rolle in 1691 used the methods of the differential calculus'.[35] Without vouching for the correctness or otherwise of the first claim, we can be unequivocal in condemning the second.

Bibliography

Boyer, Carl, *History of analytic geometry*, Dover publications, 2004.

Cajori, Florian, 'On Michel Rolle's book "Methode pour résoudre les egalitez" and the history of "Rolle's Theorem"', *Bibliotheca mathematica*, 11 (1911), 300–313.

Campbell, George, 'A method for determining the number of impossible roots in adfected equations', *Philosophical Transactions of the Royal Society*, 35 (1727), 515–531.

Cox, Homersham, 'A demonstration of Taylor's Theorem', *The Cambridge and Dublin Mathematical Journal*, 6 (1851a), 80–81.

Cox, Homersham, *A manual of the differential calculus, with simple examples*, Cambridge, 1851b.

Dickson, Leonard Eugene, *History of the theory of numbers*, 3 vols, New York, 1919; reprinted AMS Chelsea Publishing, 1999.

Drobisch, Moritz Wilhelm, *Grundzüge der Lehre von den höheren numerischen Gleichungen*, Leipzig, 1834.

35. http://en.wikipedia.org/wiki/Rolle's_theorem, accessed 5 November 2007.

Edwards, Charles Henry, *The historical development of the calculus*, Springer-Verlag, 1979.

Eneström, Gustaf, 'Kleine Mitteilungen', *Bibliotheca mathematica*, 7 (1906), 282–309.

Euler, Leonhard, *Institutiones calculi differentialis*, St Petersburg, 1755.

de Fontenelle, Bernard, 'Eloge de M. Rolle' *Histoire de l'Acadèmie royale des sciences*, Paris, 1719, 94–100.

Gherli, Odoardo, *Gli elementi teorico-practici della matematiche pure*, Modena, 1771.

Greenberg, John L, 'Mathematical physics in eighteenth-century France', *Isis*, 77 (1986), 59–78.

Heinze, Walther, 'Drobisch, Moritz Wilhelm' in *Allgemeine Deutsche Biographie*, Band 48, Nachträge bis 1899, Leipzig: Historische Commission bei der Königl. Akademie der Wissenschaften, 1904, 80–82.

Hermite, Charles, *Cours d'analyse de l'École Polytechnique*, Paris, 1873.

Itard, Jean, 'Michel Rolle' in *Dictionary of scientific biography*, Charles Scribner and Sons, 1975, XI 511–513.

Lagrange, Joseph Louis, *Traité de la résolution des équations numériques de tous les degrés*, Paris, 1798.

Laguerre, Edmond, 'Sur quelques propriétés des équations algébriques qui ont toutes leurs racines réelles', *Nouvelles Annales de Mathématiques*, 1880 (19), 224–239.

Liouville, Joseph, 'Extension du théorème de Rolle aux racines imaginaires des équations', *Journal de Mathématiques Pures et Appliquées* (2), 9 (1864), 84–88.

Maclaurin, Colin, 'A second letter…concerning the roots of equations', *Philosophical Transactions of the Royal Society of London*, 36 (1729), 59–96.

Maclaurin, Colin, *A treatise of algebra, in three parts*, London, 1748.

Mancosu, Paolo, 'The metaphysics of the calculus: a foundational debate in the Paris Academy of Sciences, 1700–1706' *Historia mathematica*, 16 (1989), 224–248.

Mills, Stella, *The collected letters of Colin MacLaurin*, Shiva Publishing Limited, 1982.

Moigno, François, 'Note sur la détermination du nombre des racines réelles ou imaginaires d'une équation numérique comprises entre des limites données. Théorèmes de Rolle, de Budan ou de Fourier, de Descartes, de Sturm et de Cauchy', *Journal de mathématiques Pures et Appliquées*, 5 (1840), 75–94.

Montucla, Jean Étienne, *Histoire des mathématiques*, 4 vols, Paris, 1799–1802.

Ozanam, Jacques, *Récréations mathématiques et physiques*, Paris, 1694.

Reyneau, Charles, *Analyse demontrée ou la methode de résoudre les problêmes des mathematiques*, Paris, 1708.

[Rolle, Michel], 'Probléme résolu par le Sieur Rolle' *Journal des Sçavans*, 31 August 1682, 284–286.

Rolle, Michel, *Traité d'algebre ou principes généraux pour resoudre les questions de mathématique*, Paris, 1690.

Rolle, Michel, *Démonstration d'une Methode pour resoudre les égalitez de tous les degrez; suivie de deux autres Methodes, dont la premiere donne les moyens de resoudre ces mêmes égalitez par la Geometrie, et la seconde, pour resoudre plusieurs questions de Diophante qui n'ont past encore esté resoluës*, Paris, 1691.

Rolle, Michel, *Methodes pour resoudre les questions indéterminées de l'algebre*, Paris, 1699.

Rolle, Michel, 'Règles et remarques, pour le problême général des tangents' *Journal des Sçavans*, 13 April 1702, 239–254.

Rolle, Michel, 'De l'evanoüissement des quantitéz inconnuës dans la géometrie analytique' *Histoire et Mémoires de l'Académie Royales des Sciences*, 1709, 419–450.

Royal Society of London, *Catalogue of Scientific Papers 1800–1900. Subject Index, Volume 1, Pure Mathematics*, Cambridge University Press, 1908.

Saurin, Joseph, 'Réponse à l'écrit de M. Rolle de l'Ac. R. Des Sc. Inseré dans le Journal du 13 Avril 1702 sous le titre Règles & remarques pour le problême generale des tangentes' *Journal des Sçavans*, 3 August 1702, 519–534.

Serret, Joseph-Alfred, *Cours d'algèbre supérieure*, Paris, 1849.

Serret, Joseph-Alfred, *Handbuch der Höhere Algebra*, Leipzig, 1868a.

Serret, Joseph-Alfred, *Cours de calcul différentiel et intégral*, Paris, 1868b.

Serret, Joseph-Alfred, *Cours d'algèbre supérieure*, 4th ed, Paris, 1877.

Shain, Julius, 'The method of cascades', *American Mathematical Monthly*, 44 (1937), 24–29.

Smith, David Eugene, 'Rolle's theorem' in David Eugene Smith (ed), *A source book in mathematics*, New York: Dover, 1959 (original work published 1929), 253–260.

Terquem, Orly, 'Théorèmes de Descartes, de Rolle, de Budan et Fourier, de Mm. Sturm et Cauchy deduits d'un seul principle', *Nouvelles Annales de Mathématiques*, 3 (1844), 188–194, 209–213, 555–565, 577–580.

Todhunter, Isaac, *A treatise on the differential calculus*, Cambridge, 2nd ed, 1855.

Abstraction and application: new contexts, new interpretations in twentieth-century mathematics

Tinne Hoff Kjeldsen

During the twentieth century mathematics grew on all fronts. Many universities were built, universal education became a widespread political goal, technology advanced with a speed never seen before. Worldwide, mathematics was the most frequently taught school subject, at all levels. The number of PhDs in mathematics rose, as did the number of research mathematicians. Mathematics expanded into fields such as economics, biology, physiology, and psychology. Many new mathematical disciplines saw the light of day. In the twentieth century mathematics thrived. To get an idea of the volume of research mathematics produced in the twentieth century, and the development and expansion of the field, one can compare the Mathematics Subject Classification (MSC) of 2000 with the one used at the beginning of the century. In 1900 the MSC had twelve headings, forty-one subheadings and forty-two sub-subheadings, occupying just a few pages of the *Jahrbuch über die Fortschritte der Mathematik*. By 2000 there were ninety-eight headings, more than three thousand subheadings, and a huge number of sub-subheadings. The whole scheme, as downloaded from the internet, takes up sixty-nine pages. Philip Davis and Reuben Hersh (1981, 24) estimated that two-hundred thousand mathematical proofs were produced annually, calling the twentieth century 'the golden age for mathematical production'.

Just listing the new developments that took place during this period would exceed the limits of this chapter. But it is possible to single out two trends in twentieth-century mathematics that most historians and mathematicians would recognize and agree upon: (1) a trend towards abstraction, and (2) a trend to mathematization that went beyond the physical sciences, resulting in the emergence of new disciplines of applied mathematics, a trend that was greatly accelerated by World War II and the advent of the computer. Both led to radical new interpretations of pieces of mathematics that had previously been considered unimportant and which now became central. This phenomenon will be illustrated through two in-depth case studies: the emergence of the theory of convex sets and the creation of mathematical programming in the wake of World War II. But first I give a brief introduction to the approach used here.

A multi-perspective approach to the history of mathematics

In the paper 'Where did twentieth-century mathematics go wrong?' Chandler Davis, Editor-in-Chief of *The Mathematical Intelligencer*, uttered his frustration with the picture that twentieth-century mathematicians painted of mathematics: 'Most 20th-century mathematicians talk as if they had a subject-matter outside of time and space. No wonder they seem snooty to others!' (Davis 1994, 132). Behind the attitude that Davis opposes lies a strongly Platonic view of mathematics as an autonomous science with an unchanging eternal subject-matter to be gradually uncovered over the course of time. At present the general opinion among historians of mathematics is that such a view of mathematical concepts as time-, place-, and context-independent is not very fruitful if one wants to understand the historical development of mathematics. A much more rewarding approach is to focus on concrete practices of mathematics, acknowledging that, despite its universal character, mathematical knowledge is produced by mathematicians who live, interact, and communicate in concrete social settings.

Through mathematicians' activities, mathematical ideas and knowledge emerge and develop at local places and in specific intellectual contexts and times.[1] Problems, concepts, definitions, and proofs emerge, develop, and change through mathematical activities. To understand how this happens, it is productive on the one hand to investigate how and why mathematicians have decided to discuss and work on particular problems, how and why they have introduced certain concepts and definitions, and employed particular strategies of proof; and on the other hand to identify changes in the understanding of mathematical entities, notions, and approaches. Such investigations provide a foundation for answering

---

1. In the history of science there has been a trend to emphasize the local nature of research practices (e.g., Buchwald and Franklin 2005, 1). Recently this approach has also been taken up by historians of mathematics (e.g. Epple 2000; 2004). For methodological discussions in history of mathematics see also Corry (2004); Kjeldsen *et al* (2004).

questions like: what influenced the development of mathematics? What driving forces can be identified behind its development, and what are they dependent on? What kind of factors and actions has modified the course of mathematics?

One way of answering such questions is to analyse concrete mathematical activities and episodes from several perspectives—or points of observation.[2] In the following two case studies I use such a multi-perspective approach to study the emergence of the theory of convex sets and the rise of nonlinear programming. I analyse the sources from one or more of the following points of view: mathematicians' motivations and goals, their perceptions of the objects involved, their methods and techniques of investigation, the mathematical context, and the institutional context. The choice of perspectives has been dictated by the historical research questions investigated and by the available sources. The consequent analysis shows that a variety of decisive factors, scientific as well as extra-scientific, enter into the explanation of developments and changes in mathematics—illustrating that its subject-matter is not outside of time and space, while casting light on how local differences matter in the history of mathematics.

The emergence of the theory of convex sets

The modern theory of convex sets emerged at the turn of the twentieth century. In some textbooks on convex sets introductory historical sketches explain that Hermann Brunn was the first to study geometrical objects which were characterized only by the property of convexity. The theory was then developed further by Hermann Minkowski, who introduced fundamental new concepts to reveal some of its many applications (e.g., Bonneson and Fenchel 1934; Klee 1963). The emergence of the theory of convexity is portrayed as a homogenous linear historical process that began with Brunn's first papers in 1887–9, which were succeeded by Minkowski's works, published in 1896 to 1903. In these historical sketches Brunn's and Minkowski's work are interpreted from the perspective of later theoretical developments. By contrast, the following presentation of their work is based on a reading and interpretation from various perspectives which emphasize the differences between Brunn's and Minkowski's mathematical practices, differences that will highlight some of the changes in mathematicians' conceptions of the subject-matter of mathematics that became characteristic of twentieth-century mathematics.

Karl Hermann Brunn (1862–1939) grew up and lived most of his life in Munich. He wrote an autobiography (Brunn 1913), which is the source of most of what is known about him: for example, that besides publishing mathematical works he

---

2. This is partly inspired by the work of the Danish historian Bernard Eric Jensen (2003).

also wrote poetry. Brunn was intrigued by what he called the elementary geometry of figures. This fascination is reflected in Brunn's inaugural thesis from the University of Munich, *Über Ovale und Eiflächen* 'On ovals and egg-surfaces' (1887), an investigation of geometrical properties of these special kinds of figures. The title he chose reflects the emphasis on the figures themselves: ovals and egg-surfaces refer to shapes of things we encounter in physical space.

Brunn defined his objects in the first chapter of the thesis (Brunn 1887, 1). By an oval he understood a closed curve with the property that every intersecting straight line in the plane of the curve intersects the curve in two and only two points (Fig. 8.4.1). He used the term egg-surface to denote the corresponding three-dimensional object, where the word 'curve' is replaced by the word 'surface' in the definition. Today we would call such curves and surfaces boundaries of closed convex sets in the plane and in space. Brunn also introduced the terms *volles Oval* 'full ovals' and *volles Eigebilde* '(full) egg-bodies', by which he understood the bodies one gets by considering ovals and egg-surfaces taken together with their inner points.

In the thesis, Brunn investigated what he could prove about his newly defined objects when he applied basic geometrical notions like curvature, area, and volume. A characteristic feature of Brunn's work was his strong opinions on proper methodology in geometry. In the introduction he emphasized that his study belonged to the category he called *elementargeometrischen Untersuchungen* 'elementary geometrical investigations' (Brunn 1887, Vorwort), where he restricted himself to the synthetic method in geometry—that is, in the Euclidean style. He explicitly stated that he had refrained from presenting these egg-forms analytically, that is describing—or replacing—them by algebraic formulae.[3] This preference was a major issue for Brunn, as he clearly stated in his autobiography:

I was not entirely satisfied with the geometry of that time, which strongly stuck to laws that could be presented as equations, quickly leading from simple to fuzzy figures that have no connection with common human interests. I tried to treat plain geometrical forms in general definitions. In doing so I leaned primarily on the elementary geometry that Hermann Müller, an impressive character with outstanding teaching talent, had taught me in the Gymnasium, and I drew on Jakob Steiner for stimulation.[4] (Brunn 1913, 40)

Jakob Steiner (1796–1863) had been one of Germany's leading geometers. In Steiner's time geometers had been engaged in a controversy concerning the proper way of reasoning, with the synthetic geometers on one side and the analytical

3. For historical discussions of analytical and synthetic geometry, see Kline (1972, chapter 35); Daston (1986); Epple (1997).

4. Von der damaligen Geometrie, die sich stark an die in Gleichungsform darstellbaren Gesetzmäßigkeiten hielt und vom einfachen rasch zu krausen, dem allgemein menschlichen Interesse fern stehenden Gestalten führte, war ich nicht ganz befriedigt. Ich versuchte einfachste geometrische Formen in allgemeinerer Definition zu behandeln. Dabei stützte ich mich wesentlich auf die Elementargeometrie, die mich im Gymnasium Hermann Müller, eine imponierende Persönlichkeit von hervorragendem Lehrtalent, gelehrt hatte, und schöpfte Anregung aus Jakob Steiner.

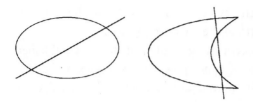

Fig. 8.4.1 Examples of an oval and a curve which is not an oval

geometers on the other. Steiner belonged to the synthetic camp, insisting that Euclidean descriptive or projective approaches were the only acceptable ones. His commitment to the intuitive synthetic method was so great that he literally hated analytical geometry. Steiner died only a year after Brunn was born and soon the controversy died out, as can be inferred from Felix Klein's description of the differences between the two approaches in 1872:

The difference between recent synthesis and recent analytical geometry has no longer to be considered an essential one, since the ways of reasoning on both sides have gradually evolved into quite similar forms. (cited by Epple 1997, 181)

But even though these purist methodological views were no longer strongly present in geometry when Brunn wrote his thesis, he chose to work very much in Steiner's spirit, using synthetic methods. He argued from the figures and their cross-sections themselves and did not translate the figures into equations because he felt—as he expressed it in his autobiography—that manipulations of the equations lead to figures that have no connection with common human interests. Here Brunn was repeating one of the main arguments in the earlier controversy about analytical methods: that the transformations of the geometrical figure corresponding to the algebraic manipulations are difficult to keep track of, and that the analysts 'often lost themselves in blind calculations, devoid of any geometric representation', as Klein put it (1908 [1939], 56).

Why did Brunn work on these forms? What was his motivation? Only towards the end of the thesis did he reflect a little on his work. Here it appears that the driving force had been to generalize theorems involving the length of lines in the two-dimensional plane to areas and surfaces in three-dimensional space, replacing the notion of lines with oval-shaped curves and egg-shaped surfaces. He finished the thesis with the following remark:

However, the complete work [which Brunn had not yet provided] is intended to show that geometrical figures constructed from a few, unusual and specialized laws also enable one to make statements that are not quite obvious.[5] (Brunn 1887, 42)

---

5. Die arbeit im Ganzen aber möchte zeigen, dass sich auch über geometrische Gebilde von ungemein wenig spezialisiertem Bildungsgesetz immerhin noch einiges nicht ganz auf der Hand liegenden aussagen lässt.

In short, no particular unsolved problem lay behind Brunn's thesis; he was simply trying to build a (synthetic) geometrical theory of egg-forms.

Whereas Brunn introduced the concept of egg-forms in his first published work, Hermann Minkowski's (1864–1909) concept of convex bodies was developed through three different phases in his mathematical practice (Kjeldsen 2008). In the first phase Minkowski used a geometrical method to treat the minimum problem for positive definite quadratic forms in n variables:

$$f(x_1,\ldots,x_n) = \sum_{h,k=1}^{n} a_{h,k} x_h x_k \quad a_{h,k} \in \mathfrak{R}, \; a_{h,k} = a_{k,h}$$

where $f(x_1,\ldots,x_n) > 0$ for all $(x_1,\ldots,x_n) \neq (0,\ldots,0)$. The problem of finding the smallest number N that can be represented by f for integer values (not all zero) of the variables $x_1,\ldots x_n$ is called the minimum problem. Minkowski was interested in the minimum problem because it was closely related to the theory of reduction for positive definite quadratic forms. In the second phase he investigated the geometrical method used in the first phase, introducing the notions of *nirgends konkave Körper mit Nullpunkt* 'nowhere-concave bodies with middle point', *Eichkörper* 'gauge-bodies', that functioned as a kind of measuring tool, and *nirgends konkave Körper* 'nowhere-concave bodies'. Finally, in the third phase Minkowski began to investigate convex bodies—or nowhere-concave bodies as he then called them—for their own sake, investigations that led to the twentieth-century theory of convex sets.

Therefore, to understand the historical development of Minkowski's notion of convex bodies the point of departure is his geometrical treatment of the minimum problem.[6] From the preface to *Geometrie der Zahlen* (Minkowski 1896) as well as from Minkowski's papers, it is clear that his work was inspired primarily by two sources. The first was the letters from Hermite to Jacobi published in *Crelle's Journal* (Hermite 1850, 263), in which Hermite proved that for a positive definite quadratic form $f(x)$ in n variables $x = (x_1,\ldots,x_n)$, there exist integer values $(x_1,\ldots,x_n)$ (not all zero) for the variables such that:

$$f(x) \leq (4/3)^{1/2(n-1)} D^{1/n}$$

where D is the determinant of the form. The second source was a paper by Dirichlet (1850) published in the same volume, in which Dirichlet used a geometrical interpretation of positive definite quadratic forms in three variables to study their reduction.[7]

6. The purpose of this case study is to understand the emergence of the concept of convex bodies in Minkowski's work. Schwermer (2007) discusses Minkowski's work in relation to the history of the reduction theory of quadratic forms.

7. This geometrical interpretation actually goes back to Gauss, who in 1831 had indicated the interpretation of binary and ternary positive definite quadratic forms as point systems in the plane and space, respectively. See (Gauss 1863).

In a rectangular (x, y) coordinate system the level curves $f(x, y) = \lambda$ for a positive definite quadratic form in two variables:

$$f(x, y) = ax^2 + 2bxy + cy^2$$

form ellipses. Through a transformation, new coordinates (u, v) can be found such that the level curves in the (skew) (u, v) coordinate system form circles. The vectors $(1, 0)$ and $(0, 1)$ in the (x, y) plane are represented by (m_1, m_2) and (n_1, n_2) respectively, where:

$$m_1^2 + m_2^2 = a, \quad n_1^2 + n_2^2 = c$$

and the angle φ between (m_1, m_2) and (n_1, n_2) is given by:

$$\cos\varphi = \frac{b}{\sqrt{ac}}.$$

The unit squares in the (x, y)-system are mapped to (fundamental) parallelograms in the (u, v) plane. The lattice points are the corners of the fundamental parallelograms and they correspond to integer values of x and y. The determinant of the quadratic form f is the square of the area of the fundamental parallelogram, and for integers x_0 and y_0, $f(x_0, y_0)$ is the square of the distance from the corresponding lattice point to the origin (Fig. 8.4.2).

Through this geometrical construction, a positive definite quadratic form can be presented by a lattice. Through geometrical reasoning about the lattice, number-theoretical problems can be solved. For example, the problem of whether there exist integer values of x and y for which a given number N can be represented by f transforms into the question of whether a circle with radius $\sqrt{N}$ and centre in the origin passes through a lattice point or not. In particular, the minimum problem transforms into the problem of determining the square of the smallest distance between two points in the lattice associated with the quadratic form.

Minkowski did not publish his geometrical way of thinking about the minimum problem until 1891, but he had combined Hermite's result with Dirichlet's geometrical image several years earlier. One of its first appearances can be found in the probationary lecture he gave for his habilitation in Bonn on 15 March 1887 (Schwermer 1991). In the manuscript Minkowski explains how an upper bound for the minimum problem for a positive definite quadratic form in three variables can be reached by using a very elegant and intuitive geometrical argument about the corresponding lattice. He mentioned that the same could be applied to lattices of any dimension. In a letter to Hilbert dated two years later, Minkowski gave an explicit formulation of the minimum result for quadratic forms in n variables, stating that:

In a positive quadratic form with n (≥ 2) variables and determinant D one can always assign integer values to the variables such that the form becomes $< nD^{1/n}$. For the

$$f(x,y)=2x^2 + 6xy + 5y^2$$

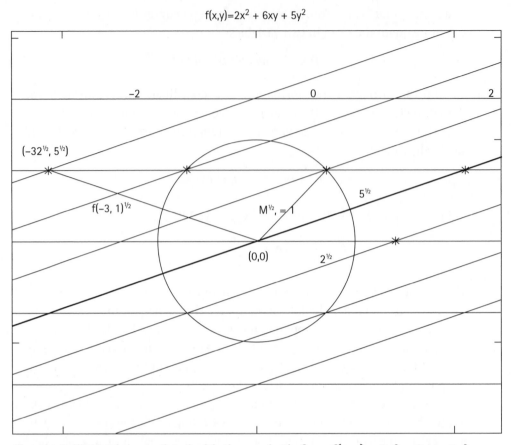

Fig. 8.4.2 The lattice associated with the quadratic form $f(x, y) = 2x^2 + 6xy + 5y^2$. The smallest distance $\sqrt{M}$, between two lattice points is $\sqrt{M} = 1$. The two coordinate axes are the bold lines

coefficient n Hermite had only $(4/3)^{\frac{1}{2}(n-1)}$, which obviously, in general, is a much larger limit.[8] (Minkowski 1973 [1889], 38)

Minkowski published the first proof of this statement in 1891, acknowledging his debt to Hermite and to the geometrical interpretation of forms of three variables (Minkowski 1911, I [1891a], 246). His proof was purely geometrical and indeed 'almost obvious': the positive definite quadratic form f in n variables was associated with a lattice. Around each lattice point Minkowski imagined a n-dimensional (hyper-) cube, with the lattice point as the centre and edges equal in length to $\dfrac{1}{\sqrt{n}}\sqrt{M}$, where $\sqrt{M}$ denotes the smallest distance in the lattice. He

8. In einer positiven quadratischen Form von der Determinante D mit n (≥ 2) Variabeln kann man stets den Variabeln solche ganzzahligen Werthe geben, dass die Form $< nD^{1/n}$ ausfällt. Hermite hat hier für den Coefficienten n nur $(4/3)^{\frac{1}{2}(n-1)}$, was offenbar im Allgemeinen eine sehr viel höhere Grenze ist.

visualized the hyper-cubes organized in a parallel pattern and argued that the hyper-cubes on the one hand have no inner points in common and on the other hand do not fill out the whole space. Minkowski then concluded that the volume of one of the hyper-cubes is smaller than the volume of the fundamental parallelotope and thereby smaller than the square root of the determinant D of the form, which means that:

$$\left(\frac{1}{\sqrt{n}}\sqrt{M}\right)^{n} < \sqrt{D}$$

or

$$M < n\sqrt[n]{D}.$$

Since the construction of the lattice ensures that if $\sqrt{M}$ denotes the smallest distance between two points in the lattice then the square, $(\sqrt{M})^{2}$, represents the smallest number different from zero that can be represented by f with integer variables, Minkowski had reached an upper bound that, as he mentioned in the letter to Hilbert as well as in the paper, is better than the one given by Hermite. By repeating the argument using n-dimensional (hyper-) spheres of radius $\frac{1}{2}\sqrt{M}$ instead of the hyper-cubes, Minkowski reached an even better upper bound for the minimum because the hyper-spheres circumscribe the hyper-cubes (Minkowski 1911, I [1891a] 255–256).

As first noticed by Joachim Schwermer (1991, 50), Minkowski's argument in the 1891 paper can be interpreted as containing the core of what is known today as Minkowski's convex body theorem about how big a bounded convex (or nowhere-concave) body with the origin as middle point has to be in order to contain a nonzero lattice point. But the theorem Minkowski proved in 1891 is about the minimum of positive definite quadratic forms in n variables, not about nowhere-concave bodies with middle point. How did a notion of convexity enter into this discussion? And how did the minimum question transform into the question of the volume of a nowhere concave body with middle point?

It is hard to tell from the published sources exactly when Minkowski became aware of the significance of nowhere-concave bodies with middle point, but he mentioned them in print for the first time in a resume of a talk he gave in Halle in 1891 with the title *Über Geometrie der Zahlen* 'On the geometry of numbers' (Minkowski 1911, I [1891b], 264). Until this point he had only talked about the lattice in connection with studies of quadratic forms, but here he introduced the lattice of integer points in ordinary Euclidean space with perpendicular axes. The lattice and corresponding objects were the key terms under investigation; quadratic forms were not mentioned.

Nowhere-concave bodies with middle point are discussed in the resume first as 'very general kind of bodies that are constructed in such a way that in a certain manner they encompass a specific point of the lattice—the origin' and then as a category of bodies 'having the origin as middle point and whose boundary outward is nowhere concave' (Minkowski 1911, I [1891b], 264–265).[9] This way of describing the objects shows that they were not generally familiar to mathematicians. They did not yet have their own name. According to the resume, in the talk Minkowski presented the key theorem, later known as the convex body theorem, for three dimensions as follows:

The first category of bodies consists of all those bodies that have the origin as middle point and whose boundaries towards the outside are nowhere concave, and the property in question for this category is: If the volume of a body from this category is $\geq 2^3$ then this body necessarily contains additional lattice points besides the origin.[10] (Minkowski 1911, I [1891b], 264–265)

To understand the significance of the convex body theorem and why Minkowski was led to introduce the nowhere-concave bodies with middle point we must go back and analyse more closely that proof of the minimum theorem for positive definite quadratic forms from Minkowski's 1891 paper.

As we saw, Minkowski interpreted the positive definite quadratic form f geometrically and considered the corresponding lattice. In the (three dimensional) lattice the equation $f = \lambda$ forms a sphere with the origin as centre and radius $\sqrt{\lambda}$. In the rectangular three-dimensional coordinate system, $f = \lambda$ represents an ellipsoid and here the lattice points are the points that correspond to integer values of the variables. Hence, finding the minimum value of f for integer values of the variables is similar to determining how big the ellipsoid has to be in order for it to contain a nontrivial lattice point. The result, Minkowski announced in the talk, states that in a three-dimensional rectangular coordinate system an ellipsoid with volume greater than 2^3 will contain a point other than the origin with integer coordinates.

This explains why the notion of volume became essential; but what about the nowhere-concave bodies with middle point? Returning to Minkowski's proof for the minimum question, we can see that the argument depends on the fact that the spheres ($f = \frac{1}{4} M$) with radius $\frac{1}{2}\sqrt{M}$ which Minkowski placed around each lattice point do not have any inner points in common. Minkowski did not explain this further in the 1891 paper, but it is clear that he had already realized that the significant properties of these bodies were their symmetry about the centre and

9. welche im Nullpunkte einen Mittelpunkt haben, und deren Begrenzung nach aussen hin nirgends konkav ist.

10. Die erste Kategorie von Körpern besteht aus allen denjenigen Körpern, welche im Nullpunkte einen Mittelpunkt haben, und deren Begrenzung nach aussen hin nirgends konkav ist; und die fragliche Eigenschaft für diese Kategorie lautet: Wenn der Inhalt eines Körpers dieser Kategorie $\geq 2^3$ ist, so schliesst der Körper notwendig noch weitere Punkte des Zahlengitters ausser dem Nullpunkte ein.

their nowhere-concave shape. In Minkowski's *Collected works* there is a paper of a talk that was read at the International Mathematical Congress in Chicago in 1893, from which it follows that he had already realized that such bodies could be used to measure distances. He introduced what he called a radial distance function and its associated *Eichkörper* 'gauge-body'. Minkowksi pointed out that every nowhere-concave body with the origin as middle point is the Eichkörper of what he called an *einhellig* 'reciprocal' distance function. Today we would call such a radial distance function a metric and think of the associated Eichkörper as a unit ball (Minkowski 1911, I [1893], 272–273).

So far we have seen how this concept of nowhere-concave bodies with middle point emerged out of Minkowski's work on the minimum problem for positive definite quadratic forms, leading to his convex body theorem. Further evidence for this interpretation of Minkowski's path to the general concept of convex bodies can be found in Minkowski's good friend and colleague David Hilbert's commemorative speech of 1909. Here Hilbert praised Minkowski's geometrical proof of the minimum for positive definite quadratic forms and emphasized that:

Minkowski's proof led, by the generalization to forms with n variables, to a more natural and a far smaller upper bound for the minimum M, than the one Hermite had found. But still more important than that was that the essential thought in Minkowski's argument only used the property of the ellipsoid that it is a convex body and has a middle point, and therefore could be transferred to any convex figure with middle point. This fact led Minkowski to realise for the first time that *the concept of a convex body* is a fundamental concept in our science and belongs amongst its most fruitful research methods.[11] (Hilbert 1909 in Minkowski 1911, I XI)

Hilbert referred to Minkowski's geometrical proof of the minimum problem as *eine Perle Minkowskischer Erfindungskunst* 'a pearl of the Minkowskian art of inventions'. It is indeed a very powerful theorem, which connects the number theoretical property of the existence of lattice points in a nowhere-concave body with middle point with the geometrical property of its volume. Through this theorem Minkowski was able to translate number theoretical problems into geometrical problems and solve them without burdensome arithmetic calculations.

Even though Minkowski used analytical methods in *Geometrie der Zahlen* there is no doubt that his intuition came from spatial geometrical considerations.

---

11. Bei der Verallgemeinerung auf Formen mit n Variablen führt der Minkowskische Beweis auf eine natürlichere und weit kleinere obere Schranke für jenes Minimum M, als sie bis dahin Hermite gefunden hatte. Noch wichtiger aber als dies war es, dass der wesentliche Gedanke des Minkowskischen Schlussverfahrens nur die Eigenschaft des Ellipsoides, dass dasselbe eine konvexe Figur ist und einen Mittelpunkt besitzt, benutzte und daher auf beliebige konvexe Figuren mit Mittelpunkt übertragen werden konnte. Dieser Umstand führte Minkowski zum ersten Male zu der Erkenntnis, dass überhaupt der *Begriff des konvexen Körpers* ein fundamentaler Begriff in unserer Wissenschaft ist und zu deren fruchtbarsten Forschungsmitteln gehört.

In a publisher's advertisement from 1893, Minkowski wrote about the forthcoming book:

I have chosen the title *Geometry of numbers* for this work because I reached the methods that give the arithmetical theorems by spatial intuition. Yet the presentation is analytic throughout, which was necessary for the reason that I consider manifolds of arbitrary order right from the beginning.[12] (Minkowski 1910, V)

In the announcement of the 1896 edition he again emphasized this geometrical intuition:

I have reached my theorems through spatial intuition [...]. But here I have prepared a purely analytical presentation because the limitation to a three-dimensional manifold seemed impossible. But I aim to use expressions that are suitable for evoking the geometrical imagination.[13] (Minkowski 1910, VI)

After the publication of *Geometrie der Zahlen* Minkowski followed up on the ideas he had developed and began to work out a theory of convex bodies, as he soon renamed the nowhere-concave bodies, detached from number theory, thus initiating the modern theory of convex sets.

By approaching the works of Brunn and Minkowski from the perspectives of their motivations, goals, and how they perceived the objects involved it is possible to get insights into the differences of their mathematical practices and the different impact of their work.

Brunn wanted to generalize theorems involving the length of lines to theorems about areas and surfaces for oval- and egg-shaped figures. For him the study of the properties of these objects was a goal in itself. This was not the case for Minkowski. His primary goal was to deal with problems in number theory, *not* to investigate properties of nowhere-concave bodies with middle point for their own sake. At the outset, he only did so because he wanted to use them as a tool for answering number-theoretical questions. As we have seen, these different motivations had an impact on how Brunn and Minkowski developed their theories. Minkowski's n-dimensional bodies were dictated by positive definite quadratic forms with n variables—a need that was not determined by the nowhere-concave bodies with middle point but by the quadratic forms. For Brunn no such need was present.

---

12. Geometrie der Zahlen habe ich diese Schrift betitelt, weil ich zu den Methoden, die in ihr arithmetische Sätze liefern, durch räumliche Anschauung geführt bin. Doch ist die Darstellung durchweg analytisch, wie dies schon durch den Umstand geboten war, dass ich von Anfang an eine Mannigfaltigkeit beliebiger Ordnung betrachte.

13. Ich bin zu meinen Sätzen durch räumliche Anschauungen gekommen [...]. Weil aber die Beschränkung auf eine Mannigfaltigkeit von drei Dimensionen unthunlich erschien, so habe ich die Darstellung hier rein analytisch gefasst, nur befleissige ich mich des Gebrauchs solcher Ausdrücke, die geeignet sind, geometrische Vorstellungen wachzurufen.

One of the premises for the historical sketches that illustrate the emergence of convex sets as a homogenous linear historical process is that both Brunn and Minkowski worked on objects that we today recognize as convex sets. But if we take a closer look at their work from the perspective of the objects a more diffuse picture emerges, revealing that the objects they studied were in practice quite different. Brunn was considering geometrical objects in two or three dimensions. He gave them concrete names and probably considered them as quasi-empirical objects abstracted from forms in our physical space. Minkowski's objects were significantly different: they were not abstractions of empirical objects but abstract mathematical entities that 'lived' in n-dimensional manifolds.

Brunn's and Minkowski's work impacted very differently on the development of twentieth century mathematics. Whereas Minkowski's work laid the foundation for what became the theory of convexity, Brunn's work only became known and acknowledged after Minkowski read and criticized Brunn's thesis around 1893–4 and began citing it (Kjeldsen forthcoming). The conclusion to be drawn from this is that at the outset convex objects were not mathematically interesting in themselves, but placed into a particular context by Minkowski they emerged as an effective tool to solve important problems in number theory, and consequently became mathematically interesting. This is reflected in the third phase in Minkowski's mathematical practice, leading to the modern theory of convexity, where he began to investigate the geometry of nowhere-concave bodies, or convex bodies as he renamed them, for their own sake. Before he suddenly died of a ruptured appendix in 1909 he had published four papers (a fifth unpublished manuscript was found after his death) in which he initiated a systematic study of the geometry of convex sets, thereby laying the foundation for the twentieth-century discipline of convexity in mathematics. His work inspired other mathematicians, in the beginning especially Ernst Steinitz and Constantin Carathéodory, who extended Minkowski's work for different purposes and in different directions.

Nonlinear programming: a consequence of World War II?

Nonlinear programming is one of the new disciplines of applied mathematics that materialized during the second half of the twentieth century. It concerns finite-dimensional, inequality-constrained optimization where the problem is to minimize a function f defined on $\mathbf{R}^n$ where the variables have to fulfil some constraints given as inequalities. It quickly expanded into other areas, all collected under the common heading of mathematical programming, which developed into a major field in applied mathematics of importance in fields such as economics and operations research. The main theorem in nonlinear programming is known today as the Kuhn–Tucker theorem. It gives the necessary conditions—now called the Kuhn–Tucker conditions—for the existence of a minimum for a nonlinear programming problem where the functions involved are differentiable. It was derived by two Princeton mathematicians, Harold W Kuhn and Albert W Tucker, who presented the result to the mathematical community in 1950. The following year the theorem was published in the paper 'Nonlinear Programming' (Kuhn and Tucker 1951), now considered a classic in mathematical programming and the founding paper of nonlinear programming. The result itself, as well as the creation of the new research field, made Kuhn and Tucker famous in the mathematical community.

Later, Kuhn (1991) became aware that the Kuhn–Tucker theorem was not a new result in 1950. Two other mathematicians, William Karush and Fritz John, had already proved the same theorem in 1939 and 1948 respectively, but each time the result had gone almost unnoticed. This raises two interesting questions: why were these seemingly identical results perceived so differently? And what enabled Kuhn's and Tucker's derivation of the result to create a new research field in 1950?[14] A Platonic view of mathematical concepts as independent of time, place, and context cannot provide answers to these questions, and neither can a limitation of one's analysis to the mathematical ideas alone. The following case study examines the historical events and the mathematical practices of Karush, John, and Kuhn and Tucker from several perspectives, which provide an account of the story that can give explanatory answers to these questions.

In 1939 William Karush handed in his master's thesis, 'Minima of functions of several variables with inequalities as side conditions', to the mathematics department at the University of Chicago (Karush 1939). People familiar with mathematical programming will immediately recognize the title as a nonlinear programming problem. But as this term did not exist in 1939 Karush's thesis was not about nonlinear programming. So what then was it considered to be about? What problem did Karush work on in his thesis and why?[15]

The mathematics department at Chicago, where Karush had studied, was often referred to as the Chicago School of Calculus of Variation. Its high point came in 1910, after which it gradually declined, especially during the years of Gilbert Ames Bliss's leadership in the 1930s. In retrospect people tend to explain the decline by inbreeding and a too narrowly defined focus on 'the study of local interior minimum points for certain prescribed functionals given by integrals of a special form' (Duren 1976, 245; Browder 1989; Mac Lane 1989; Stone 1989). Karush's thesis was fostered in this local milieu. He set out to investigate necessary and sufficient

---

14. The significance of World War II for the emergence of new mathematical disciplines, as well as the history of nonlinear programming and the Kuhn–Tucker theorem, has been discussed in several papers. See for example Kuhn (1991); Kjeldsen (1999; 2000a; 2000b; 2003; 2006).

15. For further details and arguments for the claims made in this section, see Kjeldsen (1999; 2000a).

conditions for the existence of a relative minimum for a function $f(x_1,\ldots,x_n)$ in the set of points $x = (x_1,\ldots,x_n)$ satisfying the inequality conditions $g_\alpha(x) \geq 0$, $\alpha = 1,2,\ldots,m$ where the functions f and g_α are subject to certain continuity and differentiability conditions. In section three of the thesis he formulated and proved what later became known as the Kuhn–Tucker theorem (Karush 1939, 13). The problem had been proposed by Karush's supervisor Lawrence M Graves. At first sight it might seem strange to propose such a finite-dimensional problem when the interesting case, from the perspective of calculus of variation, is the infinite-dimensional case. Karush himself did not explain the relevance of examining the finite-dimensional case, but he did refer to a work by Bliss published the year before, which treated a similar finite-dimensional problem in which the constraints were given as equalities instead of inequalities (Bliss 1938, 365–367). Bliss carried out this analysis because he was interested in the significance of the notions of normality and abnormality for the calculus of variations, an important tool in the investigation of which was to clarify the significance of the notion of normality for the finite-dimensional problem. After referring to this work by Bliss, Karush presented the problem Graves had put to him, as described above (Karush 1939, 1). This juxtaposition suggests that Graves's motivation could have been that a finite-dimensional analysis might provide a corresponding insight into calculus of variation problems with inequality constraints.

Nearly a decade later, in 1948, Fritz John published a paper with the title 'Extremum problems with inequalities as subsidiary conditions' in a collection of papers put together in honour of Richard Courant's sixtieth birthday (John 1948). John had been a student of Richard Courant in Göttingen. He moved to the USA in 1935, where he was employed—except for two years of war work at Aberdeen Proving Ground—by the University of Kentucky, until 1946 when he accepted a position at the Courant Institute at New York University (Reid 1976, 131–132, 154–155). John worked in various areas of mathematics but at the time of Courant's birthday volume more than half of his publications had been on the theory of convexity, and as we shall see below, this is also the context in which to understand John's 1948 paper on extremum problems. As with Karush's masters thesis, from the title alone the paper would later have been considered to be about nonlinear programming. But again, nonlinear programming did not yet exist in 1948. So the same questions that were asked about Karush's work can be asked again: since John's paper did not belong to the discipline of mathematical programming what then can it be said to be about? Why did John investigate this problem? Just two years later Kuhn's and Tucker's paper resulted in the emergence of a new mathematical theory; why didn't John's paper do just that?

John's paper is divided into two parts. The first deals with the theoretical result that, except for the absence of the so-called constraints qualification, is what became known as the Kuhn–Tucker conditions, namely necessary conditions for the existence of a minimum. The second part is devoted to two geometrical applications of the theoretical result to problems concerning convex sets. I will not go through the technical details of his paper (for which see Kjeldsen 1999; 2000a) but only list the conclusions of the analysis. Even though the structure of John's paper and the title he chose suggest that the theoretical part is the important one, several circumstances indicate that geometrical applications were the main focus of his attention, while the theoretical result in the first part was simply a tool developed for handling those geometrical problems. Instead of being perceived as a contribution to optimization theory, John's paper should rather be thought of as belonging to the theory of convexity.

The first application concerns the minimal closed sphere containing a bounded set in R^m. The problem was to find the sphere of least positive radius that enclose a given bounded set S in R^m. By interpreting spheres in R^m as points in R^{m+1} where the first m coordinates represent the centre and the last coordinate is the square of the radius, John reformulated the problem as an inequality constraint optimization problem. The constraints ensure that the minimum is sought among spheres that enclose S. In the second application John needed the theorem on the necessary conditions for the existence of a minimal ellipsoid containing a bounded set in R^m to prove a theorem in the theory of convexity that seems to have been his main interest. John later revealed in letters to Kuhn that the background for his work was an attempt to show that the boundary of a compact, convex set S in R^m lies between two homothetic ellipsoids of ratio $1/m$ (Kuhn, pers. comm. 1998). Indeed, this is the result of John's second application (John 1948, 202). The goal was not to find minimum points but rather to derive general theorems about closed convex sets and their relations to ellipsoids.

In short, the result that two years later became known as the Kuhn–Tucker conditions has the status of a tool in John's paper. He was not in general interested in extremum problems subject to inequality constraints. In the introduction to the paper he had mentioned, as a kind of 'defence' for the theoretical part, some tools or theories that could be developed further, but he never took this point up again and nor did he actually develop any of them. He used his result exclusively as a tool to derive other general results about convex sets. Also John's formulation of the result was clearly dictated by the problems in the applications. It was formulated in such a way that it could be used directly in the two geometrical applications. This also explains why the constraint qualification is absent from John's version of the theorem: the problem which the constraint qualification takes care of does not show up in either of the applications that John treated in the second part.

Just two years after Fritz John's paper appeared in the publication marking Courant's birthday, Harold Kuhn and Albert Tucker wrote their names into the 'Hall of Fame' of operations research. The explanation lies not in the result itself; rather, the significance, interpretation, and importance of the result were dependent on the (local) context in which it emerged. The question then is: what was the context that allowed Kuhn and Tucker to launch a new research field in which the Kuhn–Tucker theorem became important?

As already mentioned, Kuhn and Tucker presented their result at the Second Berkeley Symposium in 1950 and published it in the proceedings of the meeting a year later, where they noted that their work had been sponsored by the Office of Naval Research's (ONR) logistics programme (Kuhn and Tucker 1951). This programme originated in 1948 as a result of George B Dantzig's work with so-called programming planning problems in the US Air Force during and after World War II. In October 1947 Dantzig visited John von Neumann, in von Neumann's capacity as a consultant to the Air Force, to discuss the possibility of solving such an Air Force problem. At this point Dantzig and his group had built a mathematical model for the problem, a model they first called 'programming in a linear structure' but soon to become known as a linear programming problem. Von Neumann, who had just completed the first book on game theory with Oskar Morgenstern (von Neumann and Morgenstern 1944), suggested that Dantzig's programming problem was equivalent to a finite two-person zero-sum game. This connection to game theory provided the linear programming problem with a mathematical foundation in the theory of systems of linear inequalities and convexity.[16]

The ONR decided to set up a separate logistics branch with its own research programme to further support this kind of research. Mina Rees, then head of the mathematics division of ONR, later recalled how this programme originated:

…when, in the late 1940s the staff of our office became aware that some mathematical results obtained by George Dantzig [...] could be used by the Navy to reduce the burdensome costs of their logistics operations, the possibilities were pointed out to the Deputy Chief of Naval Operations for Logistics. His enthusiasm for the possibilities presented by these results was so great that he called together all those senior officers who had anything to do with logistics, as well as their civilian counterparts, to hear what we always referred to as a 'presentation'. The outcome of this meeting was the establishment in the Office of Naval Research of a separate Logistics Branch with a separate research program. This has proved to be a most successful activity of the Mathematics Division of ONR, both in its usefulness to the Navy, and in its impact on industry and the universities. (Rees 1977, 111)

---

16. For further literature on this story, the involvement of mathematicians in the US during World War II, and details on the history of von Neumann's work on two-person zero-sum games, see Dantzig (1982; 1991); Owens (1989); Leonard (1992); Kjeldsen (1999; 2000a; 2000b; 2001; 2002; 2003; 2006).

The project started out as a university-based trial in the summer of 1948. Tucker seems to have become involved by chance. He met Dantzig at one of the latter's meetings with von Neumann at Princeton, where he showed an interest in mathematical problems concerning linear programming. He was asked by the ONR whether he would take on the job of principal investigator for the project. Tucker hired two of the department's graduate students, Harold W Kuhn and David Gale, and together they developed the mathematical theory of linear programming, published the first rigorous proof of the important duality result for linear programming, and showed its equivalence with a two-person zero-sum game. For any linear programming problem, that is the minimization/maximization of a linear form in n variables subject to linear inequality constraints, another linear programming problem, called the dual programme, can be formulated on the same set of data. If the original programme, called the primal programme, is a minimization problem then the dual is a maximization problem and vice versa. The duality result states that if either the primal or the dual programme has a finite optimal solution then so does the other one, and the optimal solutions have the same value. This duality theorem, a key result in linear programming, caught Tucker's attention. He chose to continue working on the topic together with Kuhn (Albers and Alexanderson 1985; Kjeldsen 1999; 2000a; 2006).

During the fall of 1949 Tucker was on sabbatical at Stanford University, where he continued thinking about the duality question. He and Kuhn began to work on an extension of the duality theorem to quadratic programming problems, at some point during 1949 changing the focus from the quadratic to the general nonlinear case. Their joint paper 'Nonlinear programming', which contained the Kuhn–Tucker theorem, was the result of this work. Kuhn and Tucker did not succeed in deriving a duality result for nonlinear programming, but an analysis of the mathematics in their paper shows that this was in fact their point of departure. Their formulation of the Kuhn–Tucker theorem is different from both Karush's and John's. Kuhn and Tucker reformulated the extremum problem as a saddle-value problem, an approach that was clearly dictated by the duality result for linear programming. That in turn had been suggested by the duality of optimal strategies for two-person zero-sum games. Kuhn's and Tucker's 'Nonlinear programming' paper stimulated further work in linear and nonlinear programming, fields that rapidly expanded into many others which are now collectively referred to as mathematical programming (Kjeldsen 2000a).[17]

The ONR programme that Tucker headed was an essential component of the Kuhn–Tucker theorem's success. Another was the inclusion of linear and nonlinear programming in the toolbox of operations research, along with the

17. The first duality result for nonlinear programming was proved by Werner Fenchel (1953).

establishment of operations research as an academic discipline in the post-war period. The connection with game theory, which had become a major mathematical research area at the RAND Corporation just after the war, also secured further (military) funding for mathematical programming.[18] Finally, the construction of the first computer in 1946, with the promise that solutions to this kind of problem could actually be calculated, also played a part in the establishment of nonlinear and eventually mathematical programming in the last half of the twentieth century.

Karush's priority was recognized in 1975 by the mathematical programming and the operations research communities. At this time Kuhn was preparing a historical overview of nonlinear programming for an American Mathematical Society (AMS) symposium. As he explained in a letter to Karush, Kuhn had become aware of Karush's thesis through his reading of Takayama's book *Mathematical economics* (1974), whereupon he requested a copy of Karush's thesis (Kuhn 1976, 10). From this unpublished letter, dated 4 February 1975, it is clear that Kuhn wanted to use the AMS symposium to call attention to Karush's work:

First let me say that you have clear priority on the results known as the Kuhn–Tucker conditions (including the constraint qualification). I intend to set the record as straight as I can in my talk.

Kuhn also quoted Richard Cottle, the organizer of the AMS symposium who is supposed to have said of Karush when confronted with Kuhn's plan, '"you [Karush] must be a saint" not to complain about the absence of recognition'. Kuhn had a short correspondence with Fritz John too, whose paper he also became aware of through Takayama's book (Kuhn 1991). As a result of Kuhn's record-straightening historical talk, both John's paper and Karush's became generally known to the mathematical programming community and from then on both were considered classics in nonlinear programming.

The significance by then attached to the Kuhn–Tucker conditions is illustrated by an unpublished letter to Kuhn from his colleague Richard Bellman on 11 February 1975:

I understand from Will Karush that you will try and set the record straight on the famous Kuhn–Tucker condition. I applaud your effort. Fortunately there is enough credit for everybody. It would certainly be wonderful if you wrote it as the Kuhn–Tucker–Karush condition.

Like many important results, it is not difficult to establish, once observed. That does not distract from the importance of the condition.

---

18. The research and development centre Project RAND was established in Sante Monica, California in the spring of 1946. In 1948 it became a non-profit free-standing corporation.

It follows that Bellman thought of this result as important in its own right. One gets the impression that importance is an intrinsic property of a theorem, something that can be objectively ascribed to a mathematical result in itself. But the fact that Karush's thesis was never published, and that nobody at the time apparently encouraged him to do so, shows that this is not the case. Whether a result is important or not cannot necessarily be judged outside of the mathematical and institutional practice or context in which it was developed.

Two questions concerning the emergence of nonlinear programming were posed in this second case study: why were Karush's, John's, and Kuhn's and Tucker's seemingly identical result perceived so differently by the mathematical community? Why was Kuhn's and Tucker's derivation of the result able to create a new research field in 1950? Analyses of the protagonists' concrete mathematical activities, from several perspectives, provide answers to these questions.

Karush did his work in a narrowly defined mathematical milieu of calculus of variation. Regarded from within that context his result was relatively unimportant, just a finite dimensional simplification of the 'real' thing. And viewed from inside that world it is not at all surprising that nobody was able to anticipate the future interest in that result. By contrast, the proper context in which to analyse Fritz John's work is the theory of convexity. From that perspective his result can be understood as a tool developed to prove general results about convex sets. Viewed from inside the mathematical context of convex problems it became transparent that the inequality constraint optimization was not John's main focus of interest.

Kuhn's and Tucker's work, on the other hand, belonged to the newly developed field of linear programming, and the military promotion of new scientific disciplines in the post-war USA. These different settings offer a variety of decisive explanatory factors: the internal mathematical context of linear programming; the internal scientific context of operations research; the development of the computer; the sociological context of the establishment of operations research as a scientific discipline; and the more global sociological context of World War II and the consequent shaping and supporting of science in the post-war USA.

Conclusion

How and why do these two case-studies illustrate how mathematics was interpreted in the twentieth century? The first illustrates how mathematicians' conception of the subject matter of mathematics went through fundamental changes in the period 1870–1940, changes that transformed mathematics from a science that examined numbers and geometrical objects to the much more abstract twentieth-century mathematics of investigating structures. Not only did the emergence of the modern theory of convex sets happen in the middle of this period, to a certain extent it also exemplifies this transformation. Situating this story within the larger changes mathematics went through at the turn of the twentieth century, we have seen that Brunn perceived of mathematics as a science investigating quasi-empirical objects, whereas Minskowski studied the geometry of n-dimensional entities in mathematical spaces, work that clearly exhibited features of twentieth-century mathematics. Like Brunn, Minkowski was inspired by 'spatial intuition' but his notion of *Eichkörper* and his new concept of distance, for example, were detached from empirical experiences in physical space (Kjeldsen 2008). These entities belong to the modern side of what Jeremy Gray (1992) has called the ontological revolution in geometry. Brunn's and Minkowski's works were rooted on either side of the fracture line that divided the twentieth century's abstract mathematics from the mathematics of earlier times.

In the second case, one of the characteristics of twentieth-century mathematics is its migration into a wide variety of non-physical sciences. The emergence of nonlinear programming was the result of such an interaction and exemplifies the twentieth-century dynamic between mathematics and its non-physical areas of application. Both case studies examine new mathematical disciplines that emerged and were established within the subject matter of mathematics during the twentieth century and as such they are also exemplary of the huge expansion of mathematics that happened in this period. But they also illustrate how pieces of mathematics accrue value through the context and language in which they are used, and the meanings, values, and uses that mathematicians give to them. In the right circumstances, the irrelevant, trivial, or uninteresting can be reinvented as key theorems and concepts in new mathematical disciplines.

Bibliography

Albers, D J, and Alexanderson, G L (eds), *Mathematical people: profiles and interviews*, Birkhäuser, 1985.

American Mathematical Society, *Classification*, http://www.ams.org/msc/classification.pdf accessed on 10 October 2006.

Bellman, R, *Letter to Harold W Kuhn*, 11 February 1975 (unpublished).

Bliss, G A, 'Normality and abnormality in the calculus of variations', *Transactions of the American Mathematical Society*, 43 (1938), 365–376.

Browder, F E, 'The Stone Age of mathematics on the Midway', in P Duren (ed), *A century of mathematics in America*, vol 2, American Mathematical Society, 1989, 191–193.

Bonnesen, T, and Fenchel, W, *Theorie der konvexen Körper*, Springer, 1934.

Brunn, H, 'Ueber Ovale und Eiflächen', Inaugural-Dissertation, Munich: Akademische Buchdruckerei von F Straub, 1887.

Brunn, H, 'Brunn, Karl Hermann, Dr. Phil', in *Geistiges und Künstlerisches München in Selbstbiographien*, Max Kellers Verlag, 1913, 39–43.

Buchwald, J Z, and Franklin, A, *Wrong for the right reasons*, Springer, 2005.

Corry, L, 'Introduction: the history of modern mathematics: writing and rewriting', *Science in Context*, 17 (2004), 1–21.

Dantzig, G B, 'Reminiscences about the origins of linear programming', *Operations Research Letters*, 1 (1982), 43–48.

Dantzig, G B, 'Linear programming', in Jan Karel Lenstra, Alexander H G Rinnooy Kan, and Alexander Schrijver (eds), *History of mathematical programming: a collection of personal reminiscences*, North-Holland, 1991, 19–31.

Daston, L J, 'The physicalist tradition in early nineteenth-century French geometry', *Studies in History and Philosophy of Science*, 17 (1986), 269–295.

Davis, C, 'Where did twentieth-century mathematics go wrong?', in S Chikara, S Mitsuo, and J W Dauben (eds), *The intersection of history and mathematics*, Birkhäuser, 1994, 129–142.

Davis, P J, and Hersh, R, *The mathematical experience*, Birkhäuser, 1981.

Dirichlet, G L, 'Über die Reduction der positiven quadratischen Formen mit drei unbestimmten ganzen Zahlen', *Journal für die reine und angewandte Mathematik*, 40 (1850), 209–227.

Duren, A L, 'Graduate student at Chicago in the twenties', *The American Mathematical Monthly*, 83 (1976), 243–248.

Epple, M, 'Styles of argumentation in late 19th century geometry and the structure of mathematical modernity', in Michael Otte and Marco Panza (eds), *Analysis and synthesis in mathematics*, Kluwer Academic Publishers, 1997, 177–198.

Epple, M, 'Genies, Ideen, Institutionen, mathematische Werkstätten: Formen der Mathematikgeschichte', *Mathematische Semesterberichte*, 47 (2000), 131–163.

Epple, M, 'Knot invariants in Vienna and Princeton during the 1920s: epistemic configurations of mathematical research', *Science in Context*, 17 (2004), 131–164.

Fenchel, W, *Convex cones, sets, and functions*, Lecture Notes, Department of Mathematics, Princeton University, 1953.

Gauss, C F, *Gesammelte Werke* (eds Ernst Schering and Richard Dedekind), 12 vols, Springer, 1863–1933.

Gray, J, 'The nineteenth-century revolution in mathematical ontology', in Donald Gillies (ed), *Revolutions in mathematics*, Clarendon Press, 1992, 226–248.

Gray, J, 'Anxiety and abstraction in nineteenth-century mathematics', *Science in Context*, 17 (2004), 23–47.

Hermite, C, 'Extraits de letters de M Ch Hermite a M Jacobi sur différents objects de la théorie des nombres', *Journal für die reine und angewandte Mathematik*, 40 (1850), 261–315.

Hilbert, D, 'Hermann Minkowski. Gedächtnisrede', 1909, in H Minkowski, *Gesammelte Abhandlungen*, vol 1, Teubner, 1911, V-XXXI.

Jensen, B E, *Historie—livsverden og fag*, Gyldendal, 2003.

John, F, 'Extremum problems with inequalities as subsidiary conditions', in K O Friedrichs, O E Neugebauer, and J J Stoker (eds), *Studies and essays, presented to R Courant on his 60th birthday, January 8, 1948*, Interscience, 1948, 187–204.

Karush, W, 'Minima of functions of several variables with inequalities as side conditions', Dissertation, Department of Mathematics, University of Chicago, 1939 (unpublished).

Karush, W, *Letter to Harold W Kuhn*, 10 February 1975 (unpublished).

Kjeldsen, T H, 'En kontekstualiseret matematikhistorisk analyse af ikke-lineær programmering: udviklingshistorie og multipel opdagelse', Dissertation, IMFUFA text 372, Roskilde University, 1999.

Kjeldsen, T H, 'A contextualized historical analysis of the Kuhn–Tucker Theorem in nonlinear programming: the impact of World War II', *Historia Mathematica*, 27 (2000a), 331–361.

Kjeldsen, T H, 'The emergence of nonlinear programming: interactions between practical mathematics and mathematics proper', *The Mathematical Intelligencer*, 22/3 (2000b), 50–54.

Kjeldsen, T H, 'John von Neumann's conception of the Minimax Theorem: a journey through different mathematical contexts', *Archive for History of Exact Sciences*, 56 (2001), 39–68.

Kjeldsen, T H, 'Different motivations and goals in the historical development of the theory of systems of linear inequalities', *Archive for History of Exact Sciences*, 56 (2002), 469–538.

Kjeldsen, T H, 'New mathematical disciplines and research in the wake of World War II', in B Booss-Bavnbek and J Høyrup (eds), *Mathematics and war*, Birkhäuser, 2003, 126–152.

Kjeldsen, T H, 'The development of nonlinear programming in post-war USA: origin, motivation, and expansion', in H B Andersen, F V Christiansen, K F Jørgensen, and V Hendricks (eds), *The way through science and philosophy: essays in honour of Stig Andur Pedersen*, College Publications, 2006, 31–50.

Kjeldsen, T H, 'From measuring tool to geometrical object: Minkowski's development of the concept of convex bodies', *Archive for History of Exact Sciences*, 62 (2008), 59–89.

Kjeldsen, T H, 'Egg-forms and measure-bodies: different mathematical practices in the early history of the modern theory of convexity', *Science in Context* forthcoming.

Kjeldsen, T H, Pedersen S A, and Sonne-Hansen, L M, 'Introduction', in T H Kjeldsen, S A Pedersen, and L M Sonne-Hansen (eds), *New trends in the history and philosophy of mathematics*, University Press of Southern Denmark, 2004, 11–27.

Klee, V (ed), *Convexity: Proceedings of symposia in pure mathematics*, vol 7, American Mathematical Society, 1963.

Klein, F, *Elementary mathematics from an advanced standpoint* (English translation of the third German edition from 1908), The Maximillan Company, 1939.

Kline, M, *Mathematical thought from ancient to modern times*, Oxford University Press, 1972.

Kuhn, H W, *Letter to H Karush*, 4 February 1975 (unpublished).

Kuhn, H W, 'Nonlinear programming: a historical view', *SIAM-AMS Proceedings*, 9 (1976), 1–26.

Kuhn, H W, 'Nonlinear programming: a historical note', in Jan Karel Lenstra, Alexander H G Rinnooy Kan, and Alexander Schrijver (eds), *History of mathematical programming: a collection of personal reminiscences*, North-Holland, 1991, 82–96.

Kuhn, H W, Personal interview with H W Kuhn, Princeton University, Princeton, New Jersey, 23 April 1998.

Kuhn, H W, and Tucker, A W, 'Nonlinear programming', in J Neyman (ed), *Proceedings of the Second Berkeley Symposium on Mathematical Statistics and Probability*, University of California Press, 1951, 481–492.

Leonard, R J, 'Creating a context for game theory', in E Roy Weintraub (ed), *Towards a history of game theory*, Duke University Press, 1992, 29–76.

MacLane, S, 'Mathematics at the University of Chicago: a brief history', in P Duren (ed), *A century of mathematics in America*, vol 2, American Mathematical Society, 1989, 127–154.

Minkowski, H, 'Über die positiven quadratischen Formen und über kettenbruchähnliche Algoritmen' (1891a), in H Minkowski, *Gesammelte Abhandlungen*, vol 1, Teubner, 1911, 243–260.

Minkowski, H, 'Über Geometrie der Zahlen' (1891b), in H Minkowski, *Gesammelte Abhandlungen*, vol 1, Teubner, 1911, 264–265.

Minkowski, H, 'Über Eigenschaften von ganzen Zahlen, die durch räumliche Anschauung erschlossen sind' (1893), in H Minkowski, *Gesammelte Abhandlungen*, vol 1, Teubner, 1911, 271–277.

Minkowski, H, *Geometrie der Zahlen*, Teubner, 1896.

Minkowski, H, *Geometrie der Zahlen*, Teubner, 1910.

Minkowski, H, *Briefe an David Hilbert*, Rüdenberg & Zassenhaus, 1973.

Owens, L, 'Mathematicians at war: Warren Weaver and the Applied Mathematics Panel, 1942–1945', in D E Rowe, and J McCleary (eds), *The history of modern mathematics, vol II: institutions and applications*, Academic Press, 1989, 287–305.

Reid, C, *Courant in Göttingen and New York: the story of an improbable mathematician*, Springer, 1976.

Rees, M S, 'Mathematics and the government: the post-war years as augury of the future', in D Tarwater (ed), *The bicentennial tribute to American mathematics, 1776–1976*, Mathematical Association of America, 1977, 101–116.

Schwermer, J, 'Räumliche Anschauung und Minima positive definiter quadratischer Formen', *Jahresbericht der Deutschen Mathematiker-Vereinigung*, 93 (1991), 49–105.

Schwermer, J, 'Reduction theory of quadratic forms: towards *Räumliche Anschauung* in Minkowski's early work', in C Goldstein, N Schappacher, and J Schwermer (eds), *The shaping of arithmetic after C F Gauss's Disquisitiones Arithmeticae*, Springer, 2007, 483–504.

Stone, M H, 'Reminiscences of mathematics at Chicago', in P Duren (ed), *A century of mathematics in America*, vol 2, American Mathematical Society, 1989, 183–190.

Takyama, A, *Mathematical economics*, The Dryden Press, 1974.

von Neumann, J, and Morgenstern, O, *Theory of games and economic behavior*, Princeton University Press, 1944.

9. Historical

Traditions and myths in the historiography of Egyptian mathematics

Annette Imhausen

How frequently it happens that books on the history of mathematics copy their assertions uncritically from other books, without consulting the sources! How many fairy tales circulate as 'universally known truths'. (van der Waerden 1954, 6)

This statement of van der Waerden's rather surprised me. It was not that I did not agree with it wholeheartedly, since I have encountered just that situation many times myself. Rather, it was the fact that it came from one of the authors I blame as the source of some of the very same 'universally known truths' about Egyptian mathematics that I consider to be wrong.

A close look at the variety of widely-held myths about Egyptian mathematics reveals that van der Waerden's criticism is correct, but it is not the only one that could be made. There are at least two further reasons for the origin and longevity of these 'myths'. The first is the modern popular perception of ancient Egypt, originating at least partly in its climate and geography, which account for the pattern of survival and destruction of its historical sources. The second is an obsolete style of historiography, which is particularly associated with van der Waerden, Otto Neugebauer, and others whose names have become inextricably linked to popular knowledge of Egyptian mathematics. In what follows, I will outline the specific geographical conditions and the resulting source situation in more detail, then give examples of some of the more prominent myths and explain what makes them mythical. Of course, while many of these myths have prevailed for a long time now, there are also authors who have managed to avoid falling for them.

Modern scholars traditionally divide ancient Egyptian history into three separate phases known as 'kingdoms', each followed by an 'intermediate period'.[1] Thus the Old Kingdom (2686–2160 BC), Middle Kingdom (2055–1650 BC), and New Kingdom (1550–1069 BC), are each followed by the First (2160–2055 BC), Second (1650–1550 BC), and Third (1069–664 BC) Intermediate Periods. The kingdoms were times when Egypt was unified under the rule of one king only. They are associated with periods of stability and cultural activity. The intermediate periods, in contrast, were times when two political centres fought for supremacy, for example Memphis and Thebes during the First Intermediate Period. Within all of these periods, dynasties of kings are distinguished, often indicating succession of the throne within one family. Pharaonic history (c 3100–332 BC) also includes the end of the Predynastic Period (c 5300–3000 BC) followed by the Early Dynastic Period (3000–2686 BC). The Late Period (664–332 BC), during which Egypt struggled to remain independent, constitutes the end of the Pharaonic Period. It was followed by the Greco-Roman Period (332 BC–395 AD) when Egypt was ruled by a succession of Ptolemies and finally became a province of the Roman Empire.

Although mathematical works are extant only from the time of the Middle Kingdom (2055–1650 BC) on, the earliest evidence of writing already includes numerical information. The mace-head of king Narmer (c 3000 BC) proves that the Egyptian decimal system existed, fully developed, even before the unification of Egypt into a single political entity under the pharaohs. Egyptian writing took two separate forms throughout pharaonic history: hieroglyphs were mostly used for monumental inscriptions on stone, whereas a cursive form of writing, now known as hieratic, was used on papyrus and pot sherds for everyday purposes such as letters, administrative documents, and literature. During the Late Period, hieratic evolved into an even more cursive form, which is known as demotic.[2]

Mathematical texts fall into two groups, those written in hieratic and those in demotic. The first date almost exclusively to the time of the Middle Kingdom (for example the Moscow Mathematical Papyrus, pMoscow), with the exception of the Rhind Mathematical Papyrus (pRhind), which was written during the Second Intermediate Period but claims to be a copy of an older document. The second group (for example the Cairo Mathematical Papyrus on the back of the legal Codex Hermopolis) originated during the Greco-Roman Period.

---

1. Years in this outline follow Shaw (2000), where detailed accounts of cultural and political events of each period can be found.

2. For an overview of the various stages of Egyptian language see Loprieno 1995. Parkinson 1999 provides an up-to-date account of the decipherment of the Rosetta stone, plus a catalogue and essays about various stages of Egyptian writing.

Egypt's climate and geography

For the western world pharaonic Egypt has always held an unrivalled fascination, which even today can be traced by the success of travelling exhibitions of ancient Egyptian artefacts[3] and the use of ancient Egypt in modern movies.[4] The exhibitions often focus on royal evidence found in tombs or temples, while the cinema makes regular use of the myths that surround them in forms of curses, magical objects, and the like.

What does the popular perception of ancient Egypt have to do with Egyptian mathematics? It creates certain expectations in the modern reader. On the one hand, amazing buildings such as pyramids, temples, and other monuments left by Egyptian culture have made such a deep impression on western visitors that they are inclined to credit the ancient Egyptians with the invention of many modern concepts. This is one of the origins of myths around ancient Egyptian knowledge. On the other hand, the exhibits, which often consist of tomb decorations and other objects meant to secure the afterlife, may project an image of Egyptian culture exclusively focused on death and the afterlife. Both of these expectations are reinforced by an uncritical reading of ancient Greek historians such as Herodotus (fifth century BC), who credited the Egyptians with the invention of geometry but also described them as 'religious beyond measure, more than any other people' (Herodotus, *Histories* 2.37, 109).[5] Corinna Rossi (2004, xv) has summarized the negative influence that this sort of infatuation can have on modern research:

Egypt, with its impressively oversized architectural remains, its legendary wealth, its obscure and fascinating writing, seems to be the ideal candidate to hide the key of a lost wisdom. Even if the ancient Egyptians would have been flattered by this attitude, the results of this kind of speculation have, unfortunately, little to do with the actual historical and archaeological remains.

Instead, the preponderance of religious artefacts in the archaeological record of ancient Egypt is simply the outcome of the country's specific geographical and climatic conditions. The consequences for the preservation of objects can skew our perception of its historical culture. The main topographical feature of Egypt, in ancient and in modern times alike, is the Nile, which runs through the whole country, creating a fertile strip of land next to it some fifteen kilometres wide (Butzer 1976). Only in this area are agriculture and urban life possible, so that

---

3. E.g., *Egypt's dazzling sun/Amenophis III: le pharaon soleil*: Cleveland, Fort Worth, Paris 1992–3 (Kozloff et al 1993); *The quest for immortality: treasures of ancient Egypt*: North America 2002–7 (Hornung and Bryan 2002); and *Tutankhamun and the golden age of the pharaohs*: London 2007, Dallas 2008 (Hawass 2005).

4. E.g., *Stargate* (1994), *The Mummy* (1999), *The Mummy Returns* (2001).

5. Literature with a critical assessment of Herodotus' views is plentiful, for example, Hartog (1988).

ancient as well as modern cities have been located almost exclusively along the Nile, or on major oases. This settlement pattern has two consequences for modern archaeology. First, only areas that are not currently occupied by modern habitation can be excavated. Second, excavations in those areas within the Nile valley will reveal only that which has survived millennia of damp. Perishable, organic materials such as papyrus are swiftly destroyed under such conditions.

Only objects that came to be left outside this narrow strip of moist and fertile land, in the desert or on its margins—the location of ancient temples and tombs—stood a good chance of being preserved for the long term. Hence most of the papyrological evidence for pharaonic culture, but also for Greek and Roman civilization, originates from Egypt's deserts. This situation very obviously favours the preservation of archaeological and textual evidence from funerary and ritual contexts. So, on the one hand, given the humidity of the Nile Valley we are lucky that any ancient Egyptian papyri have survived at all. On the other hand, only about fifteen of them contain mathematical texts—since mathematics was used in the business of life, not of death—and these are dispersed across two periods separated by over a thousand years. It is exactly this lack of evidence that has enabled some myths in the historiography of Egyptian mathematics to flourish.[6] And rather than accepting this situation and working with it, trying to establish as many positive statements as possible while acknowledging the limits of the evidence, many historians and mathematicians have tried to exploit the scraps of evidence to try to prove ancient Egyptian knowledge or lack thereof of specific (modern) mathematical concepts. As Rossi (2004, xv) puts it, such 'theories do not necessarily provide any useful information about the ancient culture to which they are supposed to refer', but rather provide evidence for 'the culture and the historical period that produced them—that is, Europe in the last two centuries'.

Moreover, this very uneven preservation of original sources helps to explain why Mesopotamia has fared so much better than Egypt in popular accounts of the history of mathematics. Mesopotamian scribes wrote their mathematics on clay tablets, while Egyptian scribes used papyrus. Both cultures presumably produced massive amounts of written text, including works describing, performing, and explaining mathematical operations. Clay tablets turned out to be much more resistant than papyrus to long-term decay, in a wider range of climatic and geographical conditions. Hence, while there are fewer than ten copies of Egyptian tables, for instance, literally thousands have survived from ancient Mesopotamia. Likewise, there are many hundreds of published mathematical problems from

---

6. As indicated by Mott Greene (1992, 26), the contrast of the impressive achievements of ancient Egypt and the lack of evidence showing how these were attained may also have contributed to the creation of certain myths: 'A corollary to the postulate of Egyptian wisdom has always been that it was esoteric, arcane, hidden. This assumption is not too difficult to concede but even the most enthusiastic supporters of the Egyptians have been more or less driven to the postulation of hidden wisdom by the yawning gap between the primitive character of their techniques and the great beauty, size and precision of their architectural executions.'

Mesopotamia but only about a hundred from Egypt (the majority of which originate from just one source, the Rhind mathematical papyrus). A comparison of the two cultures will always be biased because of this situation.[7]

An outmoded historiography of (ancient) mathematics

Since the 1990s, the aims and methodology of ancient Mesopotamian, Egyptian, Greek, and Roman mathematics have been undergoing radical change, as part of larger developments in the history of mathematics (see for example Bottazzini and Dalmedico 2001). For much of the twentieth century, Egypt and Mesopotamia were perceived as the cradle of (modern) mathematics, and hence were often given pride of place in the introductory chapters of general textbooks and overviews. Such chapters typically attempted to describe the sparse roots these civilizations had laid down for whatever branch of mathematics was under discussion, from π (Beckmann 1971) to trigonometry (Maor 1998). As Jim Ritter (1995, 44–45) has noted, historians of ancient mathematics also held a peculiar place within the estimation of their colleagues:

Thus it is that the few historians who work on the earliest traces of mathematics are generally considered by their colleagues to be exotic specimens, content with childish babblings long since surpassed and quite rightly forgotten by both working mathematicians and those who study them.

This positioning of ancient Egypt at the opening of grand historical narratives is based on the assumption that there is only one mathematics, which continues to develop as time progresses and—apart from minor aberrations—inexorably leads to current mathematical concepts. It has now been recognized that this view of mathematics is rather simplistic. It ignores above all that mathematics is a cultural and social activity, and hence dependent on the societies and groups in which it originates.[8]

Jens Høyrup (1996) has identified several phases in the historiography of Mesopotamian mathematics: an initial 'heroic era' (1930–40) during which the heroes (Otto Neugebauer and Francois Thureau-Dangin) first translated what were then known as Babylonian mathematical texts and established their algebraic interpretation; this was followed by 'the triumph of translations'

---

7. That a comparison can still be achieved, and can usefully bring out specific characteristics of each of the two cultures, has been demonstrated by Jim Ritter (1995; 2004).

8. The best-known article against the universality of mathematics is Unguru (1975). It and the reactions it elicited have now been reprinted by Christianidis (2004). For Mesopotamia, the work of Jens Høyrup (2002) has identified a distinct mathematical culture. For a method of analysing procedures of Mesopotamian and Egyptian mathematics that brings out the characteristic elements of each, see Ritter (2004).

(1940–1975), characterized by a tendency to replace the Babylonian source texts with their modern mathematical equivalents. From 1971 onwards, 'a fresh start from the sources through new approaches' was made, resulting in a periodization of Mesopotamian mathematics and a better understanding of some of its individual phases and social background.[9] The historiography of Egyptian mathematics followed a similar path, with the addition that, compared to Mesopotamia, Egyptian mathematics was perceived as more primitive, more accessible to the modern reader (thanks to fewer extant sources and the Egyptian decimal system), but ultimately less interesting (as no new manuscripts came to light). The first translations of ancient Egyptian mathematics, most of which were made in the early twentieth century, thus became accepted as 'the sources'. Over time, it was often forgotten that 'the sources' were originally written in Egyptian, not in English or German. The results, not surprisingly, were sometimes theories that could not be substantiated by the source material they were supposedly based on. This holds especially for the numerous theories concerning the creation of the $2 \div n$ table, as discussed further below.

The move towards cultural context in the historiography of ancient mathematics has improved the interpretation of Egyptian mathematical writings. It is now recognized that it is no longer adequate simply to re-express their mathematical content in modern terms. When instead the formal features and cultural context of a text are taken into account, a whole new range of interesting questions can be asked (Ritter 1995; 2000; Rossi 2004). In order to assess the sources fully, a range of expertise is required. That includes an ability to read the Middle Egyptian language and hieratic script, an understanding of ancient Egyptian history and culture, and a knowledge of mathematics (ancient and modern). From this broad base the aim is to overcome the traditional gap between the humanities and the sciences and to provide a study that satisfies readers from both groups.[10]

Despite these changes, certain earlier historians' conclusions, which do not withstand critical re-examination, have proved surprisingly resistant to revision. Over time these myths have acquired the status of truths, whether because they accommodate a widely perceived (but false) notion that ancient Egyptian culture was overwhelmingly religious, or because it is easier to look at Egyptian mathematics from a misleadingly modern point of view. In some cases the evolution of the myth can be traced over time; in other instances, an initial, carefully-phrased

9. Since then, he and Eleanor Robson have produced studies that have taken our knowledge of Mesopotamian mathematics to a new level (e.g., Robson 1999; Høyrup 2002). For an overview of the historiography of ancient mathematics in general see also Netz (2003).

10. However, individual researchers do not have to be versed in *all* of the necessary skills to make useful contributions to understanding and interpreting ancient mathematics: see, for example Abdulaziz (2008) for a recent mathematician's analysis of the $2 \div n$ table. Likewise, teams of researchers can work collaboratively together. In such a team, however, a certain critical assessment of each other's theories and methodologies is indispensable.

observation became simplified and thereby falsified over time. The following section presents examples of such myths, and explains why they are considered obsolete in current historiography of ancient science.

Myth no. 1: Egyptian π

Consider the following statements:

It is certain from repeated examples, certain too from its rather good applicability, for it corresponds to a value of

$$\pi = \left(\frac{16}{9}\right)^2 = 3.1604\ldots$$

for the ratio of the circumference to the diameter, which is far from the worst a mathematician has ever made use of.[11] (Cantor 1880, I 50)

This means a quadrature of the circle by $\pi\, r^2 = \pi\left(\frac{d}{2}\right)^2 = \left(\frac{8}{9}d\right)^2$ wherefrom the excellent approximation $\pi = \left(\frac{16}{9}\right)^2 = 3.1605$. The error is only 0.0189. (Engels 1977, 137)

In the Egyptian Rhind Papyrus, which is dated about 1650 BC, there is good evidence for

$$4 \times \left(\frac{8}{9}\right)^2 = 3.16 \text{ as a value for } \pi.[12]$$

Simply by experimenting, mathematicians in early civilizations must have figured out that a rope wound around the periphery of a circle equaled just over three lengths across its diameter. With more accurate measurements, they probably discovered the value of the additional bit of rope at more than one-eighth of a length and less than one-fourth. The earliest known record of this ratio was written by an Egyptian scribe named Ahmes around 1650 BCE on what is now known as the Rhind Papyrus. (Blatner 1997: 7–8)

This, as well as the relatively accurate value 3.16 for π resulting from the above formula, gave Egyptian geometry a lead over the corresponding arithmetical achievements. (Neugebauer 1969, 78; cf Neugebauer 1929, 14; 1934, 123)

---

11. Gesichert ist sie durch wiederholtes Auftreten, gesichert ist auch ihre ziemlich gute Anwendbarkeit, denn sie entspricht einem Werte

$$\pi = \left(\frac{16}{9}\right)^2 = 3.1604\ldots$$

für die Verhältniszahl der Kreisperipherie zum Durchmesser, der weitaus nicht der schlechteste ist, dessen Mathematiker sich bedient haben.

12. School of Mathematics and Statistics, University of St Andrews, 'A history of pi', *The MacTutor History of Mathematics archive*, http://www-groups.dcs.st-and.ac.uk/~history/HistTopics/Pi_through_the_ages.html. Accessed on June 15, 2007.

But the Babylonians and the Egyptians knew more about π than its mere existence. They had also found its approximate value. By about 2000 BC, the Babylonians had arrived at the value $\pi = 3\frac{1}{8}$ and the Egyptians at the value $\pi = 4\left(\frac{8}{9}\right)^2$. (Beckman 1971, 12)

Modern mathematical treatment of circles is based on the constant π, the ratio of circumference to diameter. We calculate the area of a circle as the product of π and the squared radius, and its circumference as the product of π and the diameter. The number π is transcendental, as proved by Ferdinand von Lindemann (Lindemann 1882). When it was believed that mathematics was universal, ancient calculations of the circle were assessed by their exactness compared to modern formulae. In this view, a comparison of modern formulae with the ancient Egyptian procedure for calculating the area of a circle produced a value for the 'Egyptian π' of 3.16 or $\frac{256}{81}$.

The first of the quotations listed above is a statement by Moritz Cantor, made just a few years after the first decipherment of the Rhind Papyrus. Over time it has acquired the status of a truth that does not need reassessment: statements about Egyptian calculations of the area of a circle move from describing a procedure that is equivalent to assuming a value for π of $\frac{256}{81}$ or 3.16... to the assertion that Egyptian mathematics not only knew the concept of π but had also established a comparatively good value for it.

However, the ancient Egyptian procedure for calculating the area of a circle, found explicitly in pRhind problems 41–43 and 50, and implicitly in 48, is to subtract $\frac{1}{9}$ of the diameter from it and then to square the remainder:

pRhind, 50	*numerical representation*	*symbolic representation*
Method of calculating a circular area of diameter 9 ḥt.	9	D_1
What is its amount as area?		
You shall subtract $\frac{1}{9}$ of it, namely 1,	(1) $\frac{1}{9} \times 9 = 1$	(1) $\frac{1}{9} \times D_1$
remainder 8.	(2) $9 - 1 = 8$	(2) $D_1 - (1)$
You shall multiply 8 8 times; it shall be 64.	(3) $8 \times 8 = 64$	(3) $(2) \times (2)$

Consistent with all other known examples, this procedure uses the diameter as its point of departure. It then comprises three steps, which use only the constant $\frac{1}{9}$ and the value of the diameter. In order to arrive at the 'Egyptian value for π', one would have to express this procedure as a modern formula:

$$A_{circle} = \left(d - \frac{1}{9}d\right)^2 = \left(\frac{8}{9}d\right)^2$$

and compare it to the modern one:

$$A_{circle} = \pi\, r^2.$$

Since a circle is a circle is a circle,[13] this exercise yields $\pi = \frac{256}{81} = 3.16....$ While this may be a sound way to establish how accurate the Egyptian procedure was, it remains historiographically incorrect to state that the Egyptians used an approximation of π which was $\frac{256}{81}$. The Egyptians did not use π. As can be seen from the symbolic representation of the text, above, the constant used to calculate the area of a circle was $\frac{1}{9}$, which was clearly not an extremely bad approximation of π, but rather the constant appropriate to the ancient Egyptian method of mathematizing circles.

It is remarkable that secondary literature is more concerned with the fact that Egyptian mathematics arrived at such a good approximation for π than with the actually rather striking observation that Egyptian mathematics used a procedure that did not involve π but resulted in a comparatively accurate result.

Another Egyptian π can be found in studies of the pyramids.[14] One of the best known achievements of ancient Egypt is its pyramids, the three most famous of which are located at Giza. Did you notice anything in this last sentence? It should have read either 'were located at Giza' or better 'the remains of the three most famous...', since most of their casing is gone. It is therefore impossible to recover their exact dimensions.

From the measurements of what is left, the following observations have nevertheless been made: dividing the perimeter of the Great Pyramid of Giza by its height results in 3.1399667 or in ancient units, assuming that the Great Pyramid did indeed measure 440 cubits on a side and 280 cubits in height, 3.1428571—that is, (an approximation of) π.[15] Strangely enough, however, the measurements of the two neighbouring pyramids fail to conform to the same ratio. Further, as shown above, the calculations of the area of circles in the Rhind Papyrus make no use of this supposed π 'found' in the Great Pyramid at Giza. Consequently, it seems far more probable that the value obtained by carrying out an arbitrary, modern arithmetical operation using guesses of measurements of the remains of an ancient building are nothing but that: the result of a modern calculation.[16]

---

13. For a more sophisticated view of the circle in various cultures see Goldstein (1995).

14. For a detailed discussion of this π see Hollenbeck (1997); Rossi (2004, 200–202).

15. Depending on what measurements were used, and how the author decided to round, various similar values for π have also been obtained (Hollenbeck 1997, 62).

16. For a reliable overview of architectural practices in ancient Egypt see Arnold (1991). For anachronistic, coincidental 'discoveries' of π or φ in ancient Egyptian architecture see for example Robins and Shute (1985; 1990, 78).

Myth no. 2: the Horus eye fractions

One of ancient Egypt's many appeals for modern students of that culture is its writing. Not only is hieroglyphic one of the world's earliest attested scripts, but it comes in the form of little pictures which simultaneously look pleasing and convey a sense of secrecy resembling a code. While readers of histories of Egyptian mathematics make little direct contact with the actual Egyptian script—indeed Egyptian mathematical texts were not even written in hieroglyphs but in the much less attractive hieratic—there are two groups of hieroglyphs that are encountered regularly in modern works on the subject: the hieroglyphic signs for Egyptian numbers and the signs supposedly used to represent fractions of the basic capacity unit, the *heqat* (approx. 4.8 litres). The latter are commonly known as the 'Horus-eye-fractions' because the individual symbols (◁○ ⌐◁ ⌐◁ ◊) can apparently be assembled to form a very Egyptian-looking eye (👁). Further, this composite 'eye' has been connected with the myth of the Eye of Horus, which was destroyed by his evil brother Seth and then restored by the ibis god Thoth.[17] The alleged use of these symbols to represent parts of the *heqat* then confirms the impression originally voiced by Herodotus that the Egyptians were 'religious beyond measure'.

This historiographical myth originated, as Jim Ritter (2002) has shown, in 1911 with the publication of Georg Möller's hieratic palaeography.[18] It then made its way into Alan Gardiner's influential *Egyptian Grammar* (1927), and has—apart from few doubts arising as early as 1923—since been accepted as a truth. However, as Ritter has argued in detail, the signs commonly referred to as 'Horus-eye fractions' were not originally associated with the *heqat* capacity unit. There is evidence for hieroglyphic versions of the hieratic signs, which are clearly not parts of the eye of Horus, representing the subunits of the *heqat* from as early as the Second Dynasty and then further evidence from the Old Kingdom. In addition, the later, New Kingdom evidence that was originally used by Möller for his identification is far from conclusive (Ritter 2002, 307–311).

---

17. The most explicit reference can be found in the Middle Kingdom version of spell 17 from the Book of the Dead: 'I have filled the eye when it was injured on this day of the conflict of the two rival gods. What is that, the conflict of the two rival gods? That is the day on which Horus fought with Seth when he (Seth) wounded the face of Horus, when Horus seized the testicles of Seth. It was Thoth who did this with his fingers' (Griffiths 1960, 29).

18. The relationship between hieroglyphs and hieratic is in some respects similar to that of our printed writing and handwriting. Just as some hands today are hard to read, deciphering hieratic can be more or less complicated and constitutes the first step in working on an Egyptian papyrus. Editions of hieratic texts therefore usually include a hieroglyphic version of the text, to indicate how the editor read the hand of the ancient scribe. Hieratic writing also changes over time, hence the need for palaeographies which list typical forms of hieratic signs and their hieroglyphic counterparts.

In the original ancient Egyptian documents, the submultiples of the *heqat* are always written in their hieratic form, and look nothing like the parts of an eye. They do not lend themselves to the association with the myth of Horus, Seth, and Thoth. Rather, they are no more and no less than what they appear to be at face value: specific signs to indicate a system of subunits of the basic grain measure.

Myth no. 3: rope stretching, right angled triangles, and Pythagoras

We can imagine, albeit without any justification yet, that the Egyptians could have known that three sides of length 3, 4, 5 form a triangle with a right angle between the two smaller sides.... A period which had developed the theory of angles so far that it computed the Seqt [*seked*], we can also imagine capable of knowledge of the right-angled triangle with sides 3, 4, 5. This will have been gained substantially through experience, without thinking of a strict geometrical proof in our modern sense of the word.[19] (Cantor 1880, I 57)

Another reason for supposing that the Egyptians knew of the 3,4,5 triangle is based on the proportions of the fourth dynasty pyramid of Khafre (Chephren) at Giza and of many of the later Old Kingdom pyramids. The same proportions occur in some pyramid problems included in the Rhind Mathematical Papyrus (nos. 56–59). (Shute 2001, 350)

Claims about the supposed Egyptian use of Pythagorean triplets (especially 3–4–5) have also spread over from mathematics into architecture (Rossi 2004, 216–221). Certainly, what we call today a 'right angled triangle' was not unknown in Egyptian mathematics, as demonstrated by problem 7 of the Moscow Mathematical Papyrus:

pMoscow, 7	*numeric procedure*	*symbolic procedure*
Method of calculating a triangle.		
If you are told: A triangle of area 20	20	D_1
and the *ideb* of $2\frac{1}{2}$.	$2\frac{1}{2}$	D_2
You shall double its area.	(1) $2 \times 20 = 40$	(1) $2 \times D_1$
40 shall result.		
Calculate times $2\frac{1}{2}$.	(2) $40 \times 2\frac{1}{2} = 100$	(2) (1) $\times D_2$

19. Denken wir uns, gegenwärtig allerdings noch ohne jede Begründung, den Aegyptern sei bekannt gewesen, dass die drei Seiten von der Länge 3, 4, 5 zu einem Dreiecke verbunden ein solches mit einem rechten Winkel zwischen den beiden kleineren Seiten bilden, (…). Einer Zeit, welche die Winkellehre so weit ausgebildet hatte, dass sie den Seqt berechnete, können wir auch die Kenntnis des rechtwinkligen Dreiecks von den Seiten 3, 4, 5 zutrauen, die wesentlich erfahrungsmässig gewonnen worden sein wird, ohne dass irgendwie an einen strengen geometrischen Beweis in unserem heutigen Sinne des Wortes gedacht werden müsste.

100 shall result.

Calculate the square root.	(3) $\sqrt{100} = 10$	(3) $\sqrt{2}$

10 shall result.

Divide 1 by $2\frac{1}{2}$.	(4) $1 \div 2\frac{1}{2} = \frac{1}{3}\frac{1}{15}$	(4) $1 \div D_2$

What results is $\frac{1}{3}\frac{1}{15}$.[20]

Calculate for 10.	(5) $\frac{1}{3}\frac{1}{15} \times 10 = 4$	(5) (4) $\times$ (3)

4 shall result.

It is 10 as length to 4 as width.

This triangle is proven to be a right-angled triangle by the calculations carried out to solve the problem; its designation as *sepedet* 'triangle' at the start of the text does not indicate this special property. However, the technical terms used here also include the word *ideb*, which is used to designate the ratio of two sides which encompass a right angle.

The area of the triangle and the ratio of the two sides encompassing the right angle are given. The length of these sides is to be calculated. The procedure transforms the triangle into a rectangle (*step 1*), and then into a square whose base is identical to the longer side of the triangle (*step 2*). Then the length of that side is calculated (by extracting of the square root in *step 3*), and finally the length of the other side of the triangle is found by multiplying the side with the inverse of the *ideb* (*step 5*).

The same procedure is also used in problems 6 and 17 of the Moscow Papyrus, and a similar one can be found in the Lahun fragment UC32162 (Imhausen and Ritter 2004, 79–80). There is no evidence for the use of the hypotenuse, or for its identity as the square root of the sum of the squares of the two shorter sides of the triangle, in any of these problems. However, this does not necessarily mean that Egyptian mathematicians were unaware of this property; there simply is no evidence of it within the extant hieratic mathematical texts.[21]

The pyramid problems (pRhind, 56–59), cited by Charles Shute (2001, 350) in the second quotation above, are all concerned with the *seked*, the ancient Egyptian measure of a sloping surface which indicated the horizontal displacement of the sloping face for a vertical drop of one cubit. It was always indicated in palms (and, if necessary, fingers), where 1 (royal) cubit (approx. 52 cm) = 7 palms = 28 fingers. Problems 56–59 all deal with the relationship between the side, height, and *seked* of pyramids: from two of these quantities the third is calculated. For instance, in problem 56 a pyramid of side 360 and height 250 is given, from which the *seked* is determined as $5\frac{1}{25}$ palms. While these problems all involve right angled triangles

---

20. Egyptian fractions are discussed further below.
21. This situation had changed by the Graeco-Roman period. Problems that involve Pythagorean triplets can be found in pCairo JE 89129/89137/89139 (Parker 1972, 35–40).

(through the concept of the *seked*), neither their 'hypotenuses' (the length of the sloping side of the pyramid from bottom to top) nor the Pythagorean rule are involved in their solution.[22] Therefore, while it cannot be excluded that Egyptian mathematics and architecture might have used Pythagorean triplets, most notably 3–4–5, it must be kept in mind that our actual 'evidence' for this is based only on measurements of the remains of buildings, which—as we have already seen— may well be misleading.

Myth no. 4: Egyptian fractions were restricted to unit fractions

Ahmes did not use fractions in the most general sense of the word, i.e., implied divisions, in which the numerator, like the denominator, can be of arbitrary size, but only unit fractions—i.e., those which have an integral denominator and unity as numerator, and which he indicated by writing the number of the denominator and putting a small dot over it.[23] (Cantor 1880, I 21)

As the technique of calculation developed, the set of fractions was extended to include the unit fractions (fractions which have 1 as a numerator). The Egyptian was not able to write any other fraction, except those which have been mentioned. (van der Waerden 1954, 19)

The Egyptians could not possibly get beyond linear equations and pure quadratics with one unknown, with their primitive and laborious computing technique. (Van der Waerden 1954, 29)

To some extent Egyptian mathematics has had some, though rather negative, influence on later periods. Its arithmetic was widely based on the use of unit fractions, a practice which probably influenced the Hellenistic and Roman administrative offices. (Neugebauer 1969, 72)

The primitive, strictly additive, Egyptian way of computing with unit fractions had a detrimental effect throughout, even on Greek astronomy. (Neugebauer 1975, 559)

Another myth often encountered in historians' assessments of the ancient Egyptians is that their cumbersome fraction reckoning, which was restricted to unit fractions, prevented them from advancing in their mathematics, for instance to produce mathematical astronomy, as their neighbours in Mesopotamia did. This myth consists of two elements: first, that Egyptian mathematics was 'restricted' to unit fractions; and second, that Egyptian fraction reckoning was so cumbersome that it prevented the further development of mathematical techniques.

---

22. For a distinction between the Pythagorean theorem and the Pythagorean rule see Høyrup (1999).

23. Ahmes benutzt nämlich nicht Brüche in dem allgemeinsten Sinne des Wortes, d.h. angedeutete Theilungen, wobei der Zähler wie der Nenner von beliebiger Grösse sein können, sondern nur Stammbrüche, d.h. solche, die bei ganzzahligem Nenner die Einheit als Zähler haben und die er dadurch anzeigte, dass er die Zahl des Nenners hinschrieb und ein Pünktchen darüber setzte.

Table 9.1.1 Egyptian common fractions

Hieratic	Hieroglyphic	Modern
𝄃	⌓	$\frac{2}{3}$
⊃	⌐	$\frac{1}{2}$
╱	⌓	$\frac{1}{3}$
✗	⊗	$\frac{1}{4}$

There are two types of Egyptian fractions. The most frequently used were what are now called the 'common fractions', $\frac{2}{3}$, $\frac{1}{2}$, $\frac{1}{3}$, and $\frac{1}{4}$, which were written with specific signs in hieratic and hieroglyphic writing (see Table 9.1.1). All other fractions were written by placing a 'fraction marker' (a dot in hieratic and the symbol ⌒ in hieroglyphic) above what constitutes, in current terminology, the denominator. This type of fraction corresponds to modern fractions of the kind $\frac{1}{n}$, or, unit fractions.

Hence, from a modern point of view, ancient Egyptian fraction reckoning was 'restricted' to fractions with numerator 1, unit fractions. However, a look at the Egyptian notation for fractions described above reveals rather that they do not have a numerator at all. It is possible (though not currently provable) that the concept of fractions, or maybe better inverses, developed through the generalization of the common fractions to the idea of the inverse of any counting number, for which a general notation (dot/⌒) was created. From this basic general concept, fractions of values $\frac{m}{n}$ (in modern notation) were created by selecting those fractions that would add up to this value and juxtaposing them in order of magnitude. For instance, $\frac{5}{6}$ would be written in hieratic as ╱ ⊃ $\frac{1}{2}$ $\frac{1}{3}$ (note that hieratic is written from right to left). While such a comparison with modern notation makes Egyptian fractions seem like unit fractions, it is more accurate to think of them as inverses,[24] and to keep in mind that the Egyptian fractional notation did not encompass a numerator. This was recognized by Otto Neugebauer (1926), the first historian to work on Egyptian fraction reckoning, who created a transcription for Egyptian fractions which respected their character as inverses. It rendered the dot/⌒ as a bar over the number ('denominator'), for instance

---

24. *Pace* Greene (1992, 36): 'The concept of fraction as a part of a thing rather than the reciprocal of an integer was tied to its origins in measurement, from which it was never subsequently freed and abstracted.' While I agree that the origin of fractions is most probably to be sought in dividing and apportioning (Greene 1992, 35), I believe that the notation for fractions, and their uses within the mathematical texts, indicate that the general concept of a fraction was exactly that of an inverse. For Egyptian technical terms relating to fractions, see pRhind, 61b.

$\overline{\underline{\text{\tiny III}}} = \overline{5}$. Neugebauer also used this notation for the common fractions (with $\overline{\overline{3}}$ representing 2/3), thus providing a systematic transcription for all Egyptian fractions, but obliterating the notational differences between the common fractions and the inverses.

It is evident, then, that the Egyptians did not 'restrict' themselves to unit fractions; rather, their concept of fractions did not include a numerator. Hence a different method was used to express fractional values of the (modern) form m/n.[25] Did this cause problems for Egyptian arithmetic? Can this concept of fractions be blamed for having 'prevented' Egyptian mathematics from developing further? What evidence is there to help answer these questions? The mathematical texts themselves give two clues. First, the texts come in two varieties: mathematical problems (and their solutions), such as the examples from the Rhind papyrus given here, and mathematical tables. Extant tables are for fraction reckoning, most notably the $2 \div n$ table (found at the beginning of the Rhind papyrus and in the Lahun fragment UC 32159) and metrological tables for the conversion of measures (for instance, pRhind, 81). Jim Ritter (2000, 129) has argued that tables were created in order to help overcome technical difficulties with calculation. According to the evidence, fraction reckoning was indeed a tricky area of Egyptian mathematics. However, with the aid of tables such as the $2 \div n$ table, which lists the result of divisions $2 \div$ (odd) n in form of Egyptian fractions, these technical difficulties could to a certain degree be overcome. A look at the problems of the Rhind papyrus that involve fractions does not reveal obvious pitfalls for a competently numerate scribe. Second, another argument in favour of Egyptian fractions may be that they are the one characteristic element of Egyptian mathematics that spread and survived beyond pharaonic Egypt. One reason for their continued popularity may have been their obvious practicality in assessing the size of a fraction. Expressing $\frac{5}{19}$ as $\frac{1}{6}$ $\frac{1}{12}$ $\frac{1}{114}$ $\frac{1}{228}$ (pRhind, 32) may look cumbersome to us (especially if expected to perform further mathematical operations on it). However, this representation has (at least) one advantage over $\frac{5}{19}$: its magnitude is immediately apparent. The modern decimal representation of fractions works much in the same way. As for calculations, the accuracy needed could be chosen as well through including or ignoring as many elements of the fraction as desired, thereby facilitating some operations.

---

25. It is often stated that the demotic mathematical papyri show the use of fractions with a numerator. However, as David Fowler (1999, 259–262) argued, these should rather be interpreted as unfinished divisions.

Myth no. 5: Egyptian 'algebraic equations'

At the zenith of these exercises stand the *Hau* calculations, whose contents are no different to those which today we call algebraic equations of the first degree in one unknown.[26] (Cantor 1880, 32)

These *aha*-calculations are quite like our linear equations in one unknown. (van der Waerden 1954, 27)

These eleven problems deal with the methods of solving equations in one unknown of the first degree. (Gillings 1972, 154)

Thus, some 3,500 years before the creation of modern symbolic algebra, the Egyptians were already in possession of a method that allowed them, in effect, to solve linear equations. (Maor 1998, 4)

Egyptian 'algebraic equations' were at the centre of one of the most heated debates in the historiography of Egyptian mathematics during the first half of the twentieth century. While scholars were in agreement that sections 24 to 34 of the Rhind papyrus constituted problems that are expressed today as algebraic equations, there was at first much disagreement about the way they were solved and whether one could indeed credit ancient Egyptian mathematical practitioners with a knowledge of algebra. The question was whether the problems designated with the Egyptian technical term *aha* 'quantity' were solved as moderns would solve algebraic equations or by the method of false position.

Note above how the similarity between *aha* problems and algebraic equations in the quotations from 1880 and 1954 had become 'a method to solve algebraic equations' by the quote from 1998. An example of such an 'equation' can be found in problem 26 of the Rhind papyrus:

pRhind, 26	*numeric procedure*	*symbolic procedure*
A quantity whose $\frac{1}{4}$ is added to it	$\frac{1}{4}$	D_1
becomes 15.	15	D_2
Calculate with 4.[27]	(1) $[*\frac{1}{4}] = 4$	(1) $[*D_1]$
You shall calculate its $\frac{1}{4}$ as 1,	(2) $\frac{1}{4} \times 4 = 1$	(2) $D_1 \times (1)$
sum 5.	(3) $4 + 1 = 5$	(3) $(1) + (2)$
Divide 15 by 5. 3 shall result.	(4) $15 \div 5 = 3$	(4) $D_2 \div (3)$
Calculate 3 times 4. 12 shall result.	(5) $3 \times 4 = 12$	(5) $(4) \times (1)$

26. An der Spitze dieser Aufgaben stehen die Hau-Rechnungen, die dem Inhalte nach nichts anderes sind als was die heutige Algebra Gleichungen ersten Grades mit einer Unbekannten nennt.

27. This instruction cannot be expressed unambiguously as an arithmetic operation. It may result from calculating the inverse of the first datum, but an explicit instruction to calculate it is not given.

From a modern point of view, the assessment of these problems as predecessors of 'algebraic equations' seems quite straightforward. As the Egyptologist Thomas Eric Peet (1923, 62) commented in his edition of this problem, 'The equation here solved is $x + \frac{1}{4}x = 15$'. Egyptian 'equations' (of course) lack the use of symbols, most notably the variable x. In line with the Egyptian concept of mathematics as a systematic collection of procedures, the statement of the problem describes an operation and its numerical outcome. The quantity manipulated in the operation is an unknown, and the problem is to determine it from the number given as the result.

The historiographical problem here is similar to that faced above with Egyptian fractions. Egyptian mathematics has a sort of 'equivalent' in a modern mathematical concept; however, it is different enough to require a careful examination of its characteristic features. In the case of the Egyptian 'equations' another indicator of the inadequacy of modern assessments may be the argument that raged about the method of their solution. As already mentioned, scholars adhered to one of two groups: the first was adamant that these problems were solved by manipulating an equation; the second, objecting to this, suggested the method of false position instead.[28] For instance, the first group interpreted problem 26 of the Rhind papyrus in the following way:[29]

$$x + \tfrac{1}{4}x = 15$$

can be rewritten as

$$\tfrac{4}{4}x + \tfrac{1}{4}x = 15$$

therefore

$$\tfrac{5}{4}x = 15$$

hence

$$\tfrac{1}{4}x = 3$$

and so

$$x = 12.$$

28. For a summary of the beginning of this debate see Peet (1923, 60). Later contributions were made by Wieleitner (1925), Vogel (1930), and Neugebauer (1931). The controversy was never really resolved. When this group of problems appears in more recent publications, only one side of the argument is cited without mentioning the other: see for example Couchoud (1993, 97): manipulations of algebraic equations; Caveing (1994, 364–368): false position.

29. Cf. Eisenlohr (1877, 63): 'Die Division von 15 durch 5 oder was dasselbe ist, die Multiplikation von 5, bis 15 erreicht ist, wird zuerst vorgenommen und dann der Quotient 3 mit 4 multiplicirt = 12; der Hau ist also 12, zu welchem ¼ addirt, 15 geben muss. Text und Rechnung sind selbstverständlich.' [The division of 15 by 5 or, which is the same, the multiplication of 5 until 15 is reached, is done first and then the quotient 3 is multiplied by 4 = 12; the *aha* is thus 12, to which 1/4 is added, to give 15. Both text and calculation are obvious.]

The alternative interpretative strategy, of false position, can also be easily demonstrated. Using this example, 4 is chosen as the suitable trial number, giving 5 as the false result. Division of 15 by 5 determines the correction factor of 3, which is multiplied by the trial number to obtain the result. Supporters of this latter analysis have the advantage that their 'method' is a strategy that can be moulded into various formal expressions. That is, even if the interpretation of the *aḥa* problems as equations is given up in favour of interpreting them as procedures (which now seems historiographically more appropriate), the underlying strategy of some of those procedures may be that of the false position.

Within the mathematical papyri, there are fifteen *aḥa* problems in total (Rhind, nos. 24–34, Moscow, nos. 19 and 25, UC32134A, pBerlin6619, 1). Individual problems can be assigned to this type based on their usage of the technical term *aḥa* 'quantity', found in thirteen of them, or based on their position within the Rhind papyrus (as in the case of problems 28 and 29, which occur between problems 24–27 and 30–34). An analysis of the procedures they use reveals that they can be assigned to several groups, each of which applies a distinct strategy in its method of solution (Imhausen 2002, 155–158). The order of the *aḥa* problems in the Rhind papyrus reflects these groups. But only the problems of one group use the strategy of false position—which is therefore not a distinguishing feature of *aḥa* problems within ancient Egyptian mathematics.

Conclusion

Egyptian mathematics did not use π, equations, or 'general fractions'. However, to assess it as 'primitive' is historically misleading and based on a comparison with modern mathematics more than 2000 years later. If there is little to link the two, it points to the inadequacy of describing this ancient mathematical culture in terms of modern categories like algebra, trigonometry, and so on. The meagre evidence indicates that in Egypt, as in Mesopotamia, mathematics constituted a key element in the education of scribes (cf. Robson, Chapter 3.1 in this volume, Imhausen 2003b). Numerous administrative texts (accounts) throughout pharaonic history indicate the use of mathematical techniques. Egypt produced a mathematical culture that predates most others worldwide. It was motivated by practical needs, but did not remain limited to practical applications (e.g., pRhind, 79). The limited sources available restrict us in what we can learn about it; however, we will be able to get a better understanding if we stop evaluating it according to modern criteria, and instead aim for careful, detailed description that is sensitive to its cultural context.

Bibliography

Abdulaziz, Adulrahman, 'On the Egyptian method of decomposing 2/n into unit fractions', *Historia Mathematica*, 35 (2008), 1–18.

Arnold, Dieter, *Building in Egypt: pharaonic stone masonry*, Oxford University Press, 1991.

Beckmann, Petr, *A history of π*, St Martin's Press, 1971.

Blatner, David, *The joy of π*, Penguin, 1997.

Bottazzini, Umberto, and Dalmedico, Amy Dahan, 'Introduction', in Umberto Botazzini and Amy Dahan Dalmedico (eds), *Changing images in mathematics: from the French revolution to the new millennium*, Routledge, 2001, 1–14.

Butzer, Karl W, *Early hydraulic civilization in Egypt*, University of Chicago Press, 1976.

Cantor, Moritz, *Vorlesungen über Geschichte der Mathematik*, Leipzig: Teubner, 1880.

Caveing, Maurice, *Essai sur le savoir mathématique dans la Mésopotamie et l'Egypte anciennes*, Presses universitaires de Lille, 1994.

Christianidis, Jean, *Classics in the history of Greek mathematics*, Boston Studies in the Philosophy of Science, 240, Kluwer, 2004.

Couchoud, Sylvia, *Mathématiques Égyptiennes*, Editions le Léopard d'Or, 1993.

Eisenlohr, August, *Ein mathematisches Handbuch der alten Ägypter: Papyrus Rhind des British Museums*, Leipzig: Hinrichs, 1877.

Engels, Hermann, 'Quadrature of the circle in ancient Egypt', *Historia Mathematica*, 4/2 (1977), 137–140.

Fowler, David, *The mathematics of Plato's Academy: a new reconstruction*, 2nd ed, Clarendon Press, 1999.

Gillings, Richard J, *Mathematics in the time of the pharaohs*, MIT Press, 1972.

Goldstein, Catherine, 'Stories of the circle', in Michel Serres (ed), *A history of scientific thought*, Blackwell, 1995, 160–190.

Greene, Mott T, *Natural knowledge in preclassical antiquity*, Johns Hopkins University Press, 1992.

Griffiths, J Gwyn, *The conflict of Horus and Seth from Egyptian and classical sources*, Liverpool University Press, 1960.

Hartog, François, *The mirror of Herodotus: the representation of the other in the writing of history*, University of California Press, 1988.

Hawass, Zahi, *Tutankhamun and the golden age of the pharaohs: official companion book to the exhibition*, National Geographic Books, 2005.

Hollenbeck, George M, 'The Myth of Egyptian Pi (π)', *Skeptic,* 5 (1997), 58–62.

Hornung, Erik, and Bryan, Betsy M, *The quest for immortality: treasures of ancient Egypt*, National Gallery of Art, 2002.

Høyrup, Jens, 'Changing trends in the historiography of Mesopotamian mathematics: an insider's view', *History of Science*, 34 (1996), 1–32.

Høyrup, Jens, 'Pythagorean "rule" and "theorem": mirror of the relation between Babylonian and Greek mathematics', in Johannes Renger (ed), *Babylon: Focus mesopotamischer Geschichte, Wiege früher Gelehrsamkeit, Mythos in der Moderne*, Harrassowitz, 1999, 292–407.

Høyrup, Jens, *Lengths, widths, surfaces: a portrait of Old Babylonian algebra and its kin*, Springer, 2002.

Imhausen, Annette, 'The algorithmic structure of the Egyptian mathematical problem texts', in John M Steele and Annette Imhausen (eds), *Under one sky: astronomy and mathematics in the ancient Near East*, Ugarit, 2002, 147–166.

Imhausen, Annette, 'Egyptian mathematical texts and their contexts', *Science in Context,* 16 (2003a), 367–389.

Imhausen, Annette, 'Calculating the daily bread: Rations in theory and practice', *Historia Mathematica,* 30 (2003b), 3–16.

Imhausen, Annette, and Ritter, Jim, 'Mathematical fragments', in Mark Collier and Stephen Quirke (eds), *The UCL Lahun Papyri: religious, literary, legal, mathematical and medical,* Archaeopress, 2004, 71–96.

Kozloff, Arielle P *et al, Amenophis III, le pharaon-soleil,* Réunion des musees nationaux, 1993.

Lindemann, Ferdinand, 'Über die Zahl π', *Mathematische Annalen,* 20 (1882), 212–225.

Loprieno, Antonio, *Ancient Egyptian: a linguistic introduction,* Cambridge University Press, 1995.

Maor, Eli, *Trigonometric delights,* Princeton University Press, 1998.

Netz, Reviel, 'Introduction: the history of early mathematics—ways of re-writing', *Science in Context,* 16 (2003), 275–286.

Neugebauer, Otto, *Die Grundlagen der ägyptischen Bruchrechnung,* Springer, 1926.

Neugebauer, Otto, *Über vorgriechische Mathematik,* Teubner, 1929.

Neugebauer, Otto, 'Arithmetik und Rechentechnik der Ägypter', *Quellen und Studien zur Geschichte der Mathematik, Abteilung B: Studien* 1 (1931), 301–380.

Neugebauer, Otto, *Vorlesungen über Geschichte der antiken mathematischen Wissenschaften, Vol 1: Vorgriechische Mathematik,* Springer, 1934.

Neugebauer, Otto, *The exact sciences in antiquity,* Dover, 1969.

Neugebauer, Otto, *A history of ancient mathematical astronomy,* 3 vols, Springer, 1975.

Parker, Richard A, *Demotic mathematical papyri,* Brown University Press, 1972.

Parkinson, Richard B, *Cracking codes: the Rosetta Stone and decipherment,* British Museum, 1999.

Peet, Thomas E, *The Rhind mathematical papyrus,* Hodder & Stoughton, 1923.

Ritter, Jim, 'Measure for measure: mathematics in Egypt and Mesopotamia', in Michel Serres (ed), *A history of scientific thought: elements of a history of science,* Blackwell, 1995, 44–72.

Ritter, Jim, 'Egyptian mathematics', in Helaine Selin (ed), *Mathematics across cultures. The history of non-Western mathematics,* Kluwer, 2000, 115–136.

Ritter, Jim, 'Closing the eye of Horus', in John Steele and Annette Imhausen (eds), *Under one sky: astronomy and mathematics in the ancient Near East,* Ugarit, 2002, 297–323.

Ritter, Jim, 'Reading Strasbourg 368: a thrice-told tale', in Karine Chemla (ed), *History of science, history of text,* Kluwer, 2004, 177–200.

Robins, Gay, and Shute, Charles, 'Mathematical bases of ancient Egyptian architecture and graphic art', *Historia Mathematica,* 12 (1985), 107–122.

Robins, Gay, and Shute, Charles, 'The 14 to 11 proportion in Egyptian architecture', *Discussions in Egyptology,* 16 (1990), 75–78.

Robson, Eleanor, *Mesopotamian mathematics, 2100–1600 BC: technical constants in bureaucracy and education,* Clarendon Press, 1999.

Rossi, Corinna, *Architecture and mathematics in ancient Egypt,* Cambridge University Press, 2004.

Shaw, Ian, *The Oxford history of ancient Egypt,* Oxford University Press, 2000.

Shute, Charles, 'Mathematics', in Donald B Redford (ed), *The Oxford encyclopedia of ancient Egypt,* vol 2, Oxford University Press, 2001, 348–351.

Unguru, Sabetai, 'On the need to rewrite the history of Greek mathematics', *Archive for History of the Exact Sciences,* 15 (1975), 67–114.

Vogel, Kurt, 'Die Algebra der Ägypter des Mittleren Reiches', *Archeion,* 12 (1930), 126–162.

van der Waerden, B L, *Science awakening,* Noordhoff, 1954.

Wieleitner, H, 'Zur muslimischen und aegyptischen Gleichungsauflösung', *Archivio di storia della scienza,* 6 (1925), 46–48.

Reading ancient Greek mathematics

Ken Saito

Since the history of mathematics tells narratives about past mathematics, it is obvious that all that it narrates is based on one or more source documents, for nobody among us has ever met Archimedes, or even Gauss. However, the source documents are not in themselves history, and the fundamental task of the historian is to make a history from those documents. There are particular features and difficulties in the interpretation of sources for the history of mathematics, which depend on the period and area treated. In this chapter I illustrate some of the problems surrounding the sources for ancient Greek mathematics, many of which are relevant to other times, places, and cultures too.

The Archimedes palimpsest

The most dramatic twentieth-century story about sources for the history of mathematics is that of the Archimedes palimpsest, a unique Byzantine manuscript containing the *Method*, an attractive work in which Archimedes expounds his procedure for finding areas and volumes. The manuscript was made in the tenth century AD but much of it was reused to copy a prayer book some three centuries later, partially erasing the original text. It was rediscovered in the mid-nineteenth century but its contents remained unknown until 1906, when Johannes Heiberg, the great philologist of Greek mathematics, consulted it in Istanbul. Heiberg discovered that in the *Method*, Archimedes introduced and elaborated an audacious technique for determining the volume of a solid. He cut the solid into an infinite number of parallel planes and put each of them on a virtual balance to evaluate the volume of the whole solid. Archimedes' use of a virtual balance was already known from his *Quadrature of the parabola*, but no historian had ever imagined that he had applied the technique so spectacularly.

Heiberg's discovery was already an extraordinary event, but was only the beginning of an equally breathtaking saga. The manuscript went astray in the aftermath of World War I and reappeared some decades later in the possession of a family in Paris, who put it up for auction at Christie's in 1998. An anonymous purchaser bought the palimpsest for two million dollars, then generously deposited it with the Walters Arts Museum in Baltimore for study and conservation.

The manuscript was in a miserable condition. Evidently it had been badly looked after for most of the twentieth century. Worst of all, somebody had faked religious images on four of the pages, so that the text underneath them was completely hidden. However, thanks to modern technology some of the pages that had been illegible even to Heiberg, an expert paleographer of Greek mathematics, could be read for the first time in centuries.[1] In 2001, it was discovered that Archimedes affirmed that two sets of the plane sections infinite in number were *isai plēthei* 'equal in multitude', a clear reference to real infinity (not the potential infinity allowed by Aristotle) by one of the greatest mathematicians of antiquity, and something that nobody had expected (Netz, Saito, and Tchernetska 2001; 2002).

Later, a palimpsest page which contained the beginning of the *Stomachion* was examined and a hitherto unreadable word was read. By intriguing coincidence, it was again the word *plēthos* 'multitude'. This led to a tentative hypothesis that the *Stomachion* does not describe a game of making various figures from the fourteen pieces obtained by splitting a square—the standard interpretation which had thus dismissed this work as unimportant. Rather, Archimedes was trying to find the number of possible ways to make up the square from these fourteen pieces. (Netz, Acerbi, and Wilson 2004; Netz and Noel 2007, chapter 10). If that

---

1. The reading of the Palimpsest has been greatly aided by enhanced computer-generated images, as explained in detail with illustrations on the website http://www.archimedespalimpsest.org. The ink of the Archimedean text reflects red light better than the prayer book text that overlies it. As it looks bright in ordinary photos and very dark in ultraviolet photos, the Archimedean text is revealed by combining the red channel from the former and the blue channel from the latter. It then appears in red, clearly distinguished from the overtext and background. X-rays can also be used to distinguish the Archimedean text from the overwritten prayer book and the forged images. However, the discovery of 2001 described below was made before these marvellous innovations.

interpretation is right, the *Stomachion* is a work of combinatorics and thus key to understanding an important aspect of Archimedes' mathematical activity that has hitherto been neglected because the evidence is scattered sparsely throughout his writings. In the *Sand reckoner*, for instance, he enumerates the thirteen semi-regular solids, presenting his own notation for very large numbers; in the *Cattle problem* he presents a set of indefinite equations whose integer solution he surely did not know, for it is over two hundred thousand digits long. All these were known, but just as minor achievements of Archimedes unrelated to one another. Now his concern with combinatorics in the *Stomachion*, as suggested by the new reading of the palimpsest, would present Archimedes as a great calculator, as well as a superlative geometer and engineer.

The dramatic example of the Archimedes palimpsest illustrates the importance of primary sources, where close attention to only one word may profoundly change our view of the past. If, however, discovery of hitherto unknown works were the only important type of event in the history of mathematics, scholars of Greek mathematics would have been completely redundant between 1906 and 1998. Of course this was not the case. Even without new sources, historians always have something to do. Almost all Greek mathematical manuscripts available today were already known in the second half of the sixteenth century. However, they still offer challenging problems to historians. In a sense, the sources give us questions, not solutions, for they require interpretation. What kind of process is interpretation? Why is it necessary (and inevitable)? And why can a definitive interpretation not put an end to this process?

In the following we will take the start of the *Elements*, by far the most frequently read and studied source of Greek mathematics for many centuries. We shall see why this mathematically simple text has always required interpretative effort since ancient times, and explore what that effort consists of. We will then discuss the diagrams in manuscripts which have been silently altered in modern print editions, and try to see why diagrams have attracted scholars' attention only very recently.

Sources and interpretation: the case of the *Elements*

Anyone who has ever glanced at the *Elements* knows that the work has a very particular style. At the beginning there are several definitions to which we will return. Then the reader is asked to consent to some obvious statements without demonstration, called postulates and common notions. After that the propositions begin. They are demonstrated one after another, using only the initial postulates, common notions, and the propositions that have already been demonstrated. That is all that the *Elements* contains. Euclid gives mathematical proofs, but never speaks of other matters, even about mathematics: the purpose of the work, the reason behind the choice of some particular definition or style of proof, why he proves certain apparently useless propositions, and so on. In the *Elements*, meta-mathematics is strictly prohibited.[2] This particularly austere style has always made a strong impression on a certain type of reader: for them, being mathematical entails looking like the *Elements*, while mathematicians are those who are moved by reading Euclid and who are able to develop arguments in the Euclidean manner.

The distinctive style of the *Elements* (and of Greek mathematics in general) leads to interpretative problems. Though Euclid writes nothing but mathematical reasoning—rather, because he writes nothing else—readers try to reconstruct what Euclid did not write. Given that it is extremely difficult to resist such temptations when reading the *Elements*, how should one fill the void that Euclid left unspoken?

For a long time, philosophy was the language for talking about what Euclid did not say. A huge number of commentaries have been written on the *Elements* from a philosophical point of view. Proclus, a neo-Platonic philosopher of fifth century AD, wrote *A commentary on the first book of Euclid's Elements*, which is an influential representative of this tendency (though this work is not the most important of his numerous and voluminous philosophical works). Indeed, his approach has been so predominant that this sort of commentary forms an important strand of the history of mathematical philosophy.

Because mathematics is constantly changing and growing, Euclid's readers inevitably bring with them some knowledge of the mathematics contemporary to them. If something in the *Elements* seems superfluous or defective, they assume that some philosophical reason must have given rise to such imperfection. This was, in principle, the predominant approach to Greek mathematical sources until the 1970s. Since there was no neat distinction between mathematicians and philosophers in ancient times, and since philosophy and mathematics have always influenced each other, this seems a legitimate approach. However, one risks assuming that the development of ancient mathematics was stunted because mathematicians were haunted by philosophical concerns.

A typical example of this approach can be seen in the prevalent interpretation of the particular style in which the *Elements* treats length, area, and volume. All the arguments concerning such geometrical magnitudes are expressed in the language of ratio and proportion, without assigning any numerical values. For example, a formula like 'the area of a rectangle is given by the product of the

---

2. Authors like Archimedes and Apollonius put this kind of meta-mathematical discussion in the preface, away from the main text where only mathematical reasoning appears. Even so, the meta-mathematical information provided in prefaces tends to be sparse, sporadic, and sometimes even enigmatic.

base and the height' is never found in the *Elements*.[3] Instead, the corresponding proposition is: 'triangles and parallelograms which are under the same height are to one another as their bases' (VI.1). Though this avoidance of numbers seems folly for us moderns, we should remember that the Greeks did not have numbers capable of expressing geometrical magnitudes, so that it was natural not to assign numerical values to them. Moreover, Euclid is very careful not to assign concrete values even to integers in the arithmetical books (VII to IX) of the *Elements*. As we will see later, this particular style can be attributed to a concern with the generality of proof.

The prevalent interpretation of this style has been that the absence of numbers in geometry is the result of a scandalous collapse of the Pythagorean principle 'all is number', caused by the discovery of incommensurability in the fifth century BC. I do not agree with this interpretation, but it is not my point here.[4] Rather, this example shows that the *Elements* necessitates interpretations all the more because of its particular style. And as we will soon see, any interpretation, however objective and disinterested one endeavours to be, is subject to the environment—not only mathematical and philosophical, but more generally intellectual and even technological—in which one lives. As times change, environments change with them, and new interpretative problems arise. Thus we cannot decide which interpretation is best, but we may say that an attitude toward the sources is better if it enables us to pose fruitful interpretative problems: that is, those which reveal our ignorance and improve our insight.

Now let us examine some of the propositions of the *Elements*, the main part of the work, constructed from logical chains of demonstrations. The first proposition is a problem, to construct an equilateral triangle on a given line.

Let the given line be AB, and draw a circle BΓΔ with centre A, distance (i.e., radius) AB (Postulate 3). Similarly, draw a circle AΓE with centre B, distance BA. Let the circles cut each other at Γ, and AΓ, BΓ are joined (Postulate 1).[5] Then, it is proved that AB, BΓ and Γ equal one another.

---

3. Heron's *Metrica*, by contrast, is full of this kind of formulae expressed verbally, accompanied by numerical examples, including the famous one for the area of a triangle widely known by his name—in modern algebraic terms, the area of a triangle is $\sqrt{\{s(s-a)(s-b)(s-c)\}}$, a, b, c being the three sides and s the half of the sum $a + b + c$ —though an Arabic source attributes it to Archimedes.

4. For the alleged impact of incommensurability on Greek mathematics, see Christianidis (2004, part 3).

5. The first three postulates ask for the possibility of basic geometrical constructions: to draw a straight line joining two points (Postulate 1), to prolong a given straight line (Postulate 2) and to draw a circle with given centre and distance (radius). Árpád Szabó (1978 [1969]) argued that such obvious postulates reflect a conscious effort to escape the Eleatic criticism against motion. The Eleatic philosophical school began with Parmenides in the early fifth century BC. The Eleatics argued, elaborating their founder's thesis that all is one, that motion (and any change) is impossible. The well-known Zeno's paradox, as in Achilles and tortoise, is a typical example of their arguments. Szabó argued that it was Eleatic philosophy that gave birth to deductive mathematics based on postulates and axioms. Though it is difficult to accept any thesis that ascribes the birth or invention of deductive mathematics in Greece to one factor, the Eleatic influence on Greek mathematics seems undeniable. At the very least, both Eleatic philosophy and early deductive mathematics flourished in the same intellectual milieu which made much of logical consistency. It is indeed curious that the *Elements* seldom makes explicit

Figure 9.2.1 Diagram of Proposition I.1

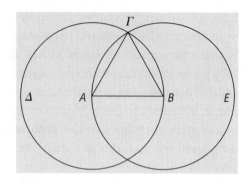

The style of Greek mathematics is concentrated in this simple proposition. First, Euclid does not explain why he has chosen this problem to begin the work. Second, each construction and deduction is strictly based on postulates or (in later propositions) on the results of preceding propositions.

The strictly logical style, however, makes the reader hyper-sensitive to any logical flaw. In this proposition, Euclid takes it for granted that the two circles he has drawn have a common point (more precisely, two). This unproved assumption has long been the target of severe attacks, some of which are mentioned in Proclus' commentary. For us moderns, this is evidence of Euclid's dependence on a spatial intuition which he could not entirely dispose of. But this criticism is rather beside the point, for it was not possible to construct geometrical objects that were independent of spatial intuition until the purely axiomatic construction of real numbers in the nineteenth century.

We should rather note that Euclid defined the line not as a set of points, but based on a kind of intuition of continuity (Definition I.2). His definition of the circle is similar: the circle is not the set of points equally distant from a given point, but it is a line with the property of being equidistant from some point which is called the centre (Definitions I.15–16). We can better understand this idea if we suppose that for the Greeks the line was an entirely different geometrical entity from the point, for the point does not have the line's most basic property, namely extension. In this interpretation the notorious (pseudo-) Definition I.2, 'the line is length without breadth', makes some sense.

Proposition 2 is also a problem (Fig. 9.2.2).[6] Given a point A and straight line BΓ, it is required to construct a straight line beginning at point A and equal to BΓ.[7] First, construct equilateral triangle ABΔ on AB (Prop. 1), prolong ΔA and ΔB

recourse to motion. Apart from the first three postulates, Euclid only speaks of rotation of the semicircle, triangle, and rectangle when he defines the sphere, cone, and cylinder respectively in Book XI.

6. We provide diagrams redrawn from the Vatican manuscript P (see below), together with images from Heiberg's edition, except for Proposition I.1, whose diagram is consistent. The differences between Heiberg's diagrams and those in the manuscript are discussed in a later section.

7. For the sake of simplicity, we now omit the general statement (protasis or enunciation) found at the beginning of the proposition, and explain its content with labelled objects.

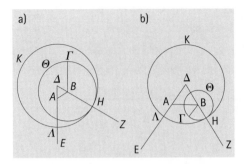

Figure 9.2.2 (a) Heiberg's diagram of Proposition I.2; (b) the same diagram, redrawn from Codex P

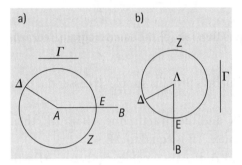

Figure 9.2.3 (a) Heiberg's diagram of Proposition I.3; (b) the same diagram, redrawn from Codex P

(Post. 2) to E and Z, draw a circle with centre B and distance BΓ (Post. 3; though Euclid does not say so explicitly, its intersection with ΔZ is H). Now with centre Δ, distance ΔH, draw a circle (Post. 3). Let Λ be the intersection of this circle and ΔE, then AΛ is proved to be equal to BH (in justifying this, common notion 3 is used).

One thing is evident from this argument. Euclid did not think that Postulate 3 ('to describe a circle with any centre and radius') permitted him to draw a circle with centre A and distance BΓ, for such a construction would have made the whole proposition unnecessary. Though Euclid is extremely taciturn about his understandings and considerations concerning each of the expressions in the *Elements*, sometimes we can make out his intention clearly.

Let us move to Proposition 3 (Fig. 9.2.3). Given a line AB and another line Γ, it is required to cut off a line equal to Γ from AB. Drawing a circle with centre A and distance equal to Γ (Prop. 2), one can cut off a line AΔ on AB, equal to Γ. This means that you can 'carry' any line segment onto any other line, without changing its length.

So far we have witnessed a typical chain of logical demonstrations: Proposition 1 used to prove Proposition 2, which in turn is necessary for Proposition 3. Proposition 4, however, causes a problem. It asserts the side-angle-side test for congruence (Fig. 9.2.4). If two triangles have two sides equal to two sides, and the angles contained by equal sides are equal, then the two triangles are congruent. In two triangles ABΓ and ΔEZ, it is supposed that AB = ΔE and AΓ = ΔZ and that

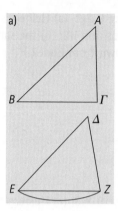

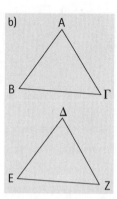

Figure 9.2.4 (a) Heiberg's diagram of Proposition I.4; (b) the same diagram, redrawn from Codex P

the angle at A is equal to the angle at Δ. Euclid's proof is simple. Its essential part is as follows.

If the point A be placed on the point Δ, and the straight line AB on ΔE, then the point B will also coincide with E, because AB is equal to ΔE.

Then, from the equality of angle A and Δ, if AB coincides with ΔE, AΓ will coincide with ΔZ, and since AΓ=ΔZ, the point Γ will fall on Z. Then, BΓ will also coincide with EZ,[8] and the whole triangle ABΓ coincides with the whole triangle ΔEZ.

In this argument Euclid apparently does not use the preceding three propositions. In fact, Proclus says, 'The proof of this theorem, as anyone can see, depends entirely on the common notions and grows naturally out of the very clarity of the hypotheses' (Proclus [1970], 240). Then a question arises: if this proposition depends on nothing but some of the common notions, why did Euclid decide to prove the preceding three propositions before it?

Proclus seems to have been aware of this question, for he says that Euclid had to include the first three propositions, which are classified as problems (not theorems), to show that the triangle exists, and that equal lines exist, and so on. In Proclus' interpretation the constructions in the first three propositions serve as a kind of proof of existence. Proclus' idea was prevalent until very recently. Zeuthen (1896) maintained that the constructions in the *Elements* are 'proof of existence', and this thesis remained unchallenged until Wilbur Knorr's rebuttal (Knorr 1983).[9] In short, historians were all under the influence of Proclus' concern for ontology until just 25 years ago.

---

8. Otherwise there would be two lines connecting Δ and Z, which is absurd. However, Common Notion 4, which explicitly states its impossibility, is certainly a later addition, probably inspired by this argument.

9. Knorr shows Heiberg's arguments (that go back as far as Proclus) do not fit the practice of Greek mathematicians. First, Knorr examines the chains of constructions in the *Elements* which lead to some complicated figure, such as the construction of the five regular polyhedra in Book XIII. However, Euclid had already

The idea of proof of existence is indeed attractive in some situations, for example, when Euclid suddenly shows how to construct a square (I.46) just before the theorem of Pythagoras (I.47). However, as Knorr convincingly argued, the problem of existence is not always related to construction, and nor does construction always serve to prove existence. And it seems out of place to assume existential concern on Euclid's part in the first three propositions of the *Elements*.

Then why do they exist? The simplest answer ought to be that they somehow serve Proposition 4. As we have seen, in the proof of this proposition points and lines are moved and placed onto other points and lines. This is made possible by Propositions 2 and 3, although Euclid never says so explicitly.[10] The point A is moved onto another point Δ, then the line AB is moved and placed onto another given line ΔE. It is natural to assume that Euclid implicitly resorts to these propositions when he moves the sides of the triangles in Proposition 4. Strangely, this seems to be a rather recent interpretation. As far as I know, Beppo Levi (1949) was the first to have this idea, while Bernard Vitrac (1990–2001, 202), also suggested it, apparently independently.

This interpretation and that of Proclus are quite different, but they share a common motivation. Both try to explain the reason why Propositions 1–3 appear before Proposition 4. For Proclus, it was an existential issue, while for Levi and Vitrac it was Euclid's concern about the possibility of motion. If the latter interpretation is right, and if Euclid was indeed conscious of the Eleatic criticisms against motion, then the basic philosophical issue had undergone considerable change between Euclid's and Proclus' time. Thus Proclus was projecting his own concern, that of existence, onto Euclid. It is not important whether Proclus was right or wrong. My point is simply that interpretative effort on the text of the *Elements* was already necessary in Proclus' time, for he no longer had access to the intellectual environment in which the *Elements* had taken shape. This point is equally true for mediaeval scholars (Rommevaux, Chapter 8.1 in this volume), early modern mathematicians, and today's historians: the *Elements* has continued to pose different interpretative questions in different eras, and

---

named them as the pyramid, cube, octahedron, icosahedron, and dodecahedron in their definitions in Book XI. With these names he shows no concern about their existence: for it is not known a priori that a regular polyhedron whose surface comprises regular pentagons exists, still less that it is a 'dodecahedron', a (solid with) twelve surfaces. So Euclid's construction of polyhedra cannot be interpreted as proof of existence. On the other hand, the existence of a geometric figure equal to the one given (e.g., a straight line equal to a given circumference), or satisfying some proportion (e.g., x satisfying the proportion $a:b = c:x$ for given a, b, c) is often simply assumed in the context of quadrature of figures. Thus the construction problems show, according to Knorr, not the existence of the figure constructed, but the possibility of constructing that figure using only some specific postulates of construction (in the case of the *Elements*, only the use of the straight line and the circle is permitted). Knorr concludes that it is rather philosophical concern (ancient and modern) that has sustained the thesis of construction as proof of existence. See also Harari (2003).

10. Euclid never explicitly refers to the propositions he uses. Therefore Euclid's silence cannot be construed as evidence that he did not intend to use Propositions 2 and 3 here.

readers always have to face such questions, no matter whether their answers are right or wrong.

Textual criticism: do we really read what the ancients wrote?

So far I have been treating the text of the *Elements* as firmly established, as if we still had Euclid's own autograph copy. However, this is far from the case. The *Elements*, like all ancient Greek works, is available only through medieval copies which differ from each other. If we seriously want to argue about the content of the *Elements*, we should first try to establish the original text through such manuscripts as we possess; and if we cannot establish it, we should at least recognize the extent to which the surviving text resembles the original. This task of establishing the original text from extant copies, whether for poetry, philosophy, or history, is called textual criticism (see West 1973). I shall briefly explain some basic principles of textual criticism, and the particular situation with regard to the *Elements*.

When more than one manuscript of a classical work survives, their texts invariably differ. There may be anything from two or three to more than a dozen different readings (variants) on one page of a print edition, amounting to thousands of variants in one work. There are two main causes of variants. First any copyist inevitably errs, omitting some words, mistaking one word for another, or inadvertently skipping one or more lines between the same word(s). Second, a scribe may consciously 'correct' or 'improve' the text; for example, a rare word or expression may be taken as an error made by a previous scribe. Some may write a gloss in the margin, which is later incorporated into the text when the manuscript is recopied. The cornerstone of textual criticism is that every time a work is copied, errors are inherited and more are added. (For the sake of simplicity, I use the word 'error' to denote any difference between a copy and its original.)[11] Comparison of the errors in individual manuscripts can help to determine whether one surviving manuscript is a copy of another.

We can thus draw a diagram, or stemma, to represent the relationships between the manuscripts of a work (Fig. 9.2.5). A copy of another extant manuscript tells us nothing new about the author's original autograph work, so it is neglected in the editorial reconstruction of the text. Sometimes all the extant manuscripts turn out to be direct or indirect copies of a single extant manuscript (as in Figures 9.2.5a–9.2.5.d, if codex A is extant). This is the case for the *Conics* of Apollonius and the *Collection* of Pappus, so to edit those works it suffices to read

---

11. Some simple errors may be corrected by conjecture. The errors we speak of are limited to irrecoverable ones, such as gaps of one or more lines.

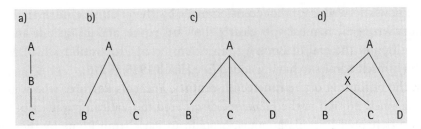

Figure 9.2.5 (a–d) Possible relationships between different manuscripts of the same text

and transcribe the source manuscript as carefully as possible.[12] More often, we have several copies that share some errors but which each have particular errors of their own. They are thus independent copies of a lost original, or archetype. For example, in the stemma of Fig. 9.2.5b, if A is no longer extant we are in this situation. Our task is, of course, to restore the text of A from the extant manuscripts B and C.

Let us examine some possible cases more in detail. Suppose A is lost and there is another copy D of A (Fig. 9.2.5b). For the restoration of the archetype A, the two situations in Figs. 9.2.5c and 9.2.5d are significantly different. If we have three independent copies B, C, D of the archetype A (Fig. 9.2.5d), then it would be exceptional for the three copies to give three different readings at any one place in the text. So at least two of the copies should always agree, thus giving the reading of A. So, except in rare cases, we can establish the text of A. On the other hand, if we find that B and C share some common errors which are not found in D, then B and C must have derived from an intermediate copy X of A, which is the cause of those common errors (Fig. 9.2.5d). In this case, even if the reading of B and C agree against D, the majority rule does not apply, for the agreement of B and C shows only the reading of a single copy X, which is worth no more than D. If one of the two is clearly corrupt and makes no sense, we can choose the other as the reading of the archetype A. But if both traditions give readable and plausible texts it can be difficult to decide between the two.

The situation of the *Elements* is the latter, more difficult one. Let us briefly look back at how the critical edition of the Greek text of the *Elements* was made. The first print edition of the *Elements* was published in 1533 by Simon Grynaeus, based on two manuscripts, which are far from the best of those known today. However, this edition was the source of all subsequent editions and translations until the early nineteenth century. We now understand that all the Greek manuscripts

12. Even in this type of case, there are often other witnesses, that is, citations, translations, and commentaries (in mathematical works, commentaries often take the form of lemmas to facilitate the understanding of difficult arguments), which inform us of different textual traditions. Those traditions must then be evaluated, in a similar manner, in order to establish the text.

then known derived from the recension made by Theon of Alexandria in the 4th century AD. Some manuscripts clearly show his name, and all include an extra proposition at the end of Book VI (an extension of Proposition VI.33), which Theon himself admits to having added (see Heath 1925, II 276).

At the beginning of the nineteenth century, François Peyrard, who was preparing a new edition of the *Elements*, discovered the Vatican manuscript Grec. 190 (later named P in his honour), brought to Paris from Rome by the scholars who followed Napoleon's army. This manuscript does not name Theon as editor, lacks the proposition added by Theon,[13] and in many respects provides good readings. He recognized that this manuscript belongs to a different tradition from all known 'Theonian' manuscripts. Acknowledging the importance of this manuscript, and probably inspired by its discovery, Peyrard made a thorough examination of the manuscripts of Euclid's *Elements* and *Data* possessed by the Bibliothèque Imperiale (today's Bibliothèque Nationale in Paris), but did not dare to change the accepted text a great deal in his edition (Euclide 1814–18). Following the standard procedure of his time, he was generally content to add a critical apparatus in which he reported the variants in the Vatican manuscript compared to David Gregory's edition of 1703.

It was only when Johannes Heiberg edited the *Elements* in the 1880s that this non-Theonian manuscript was properly used to establish the text. Heiberg's edition is still considered the standard edition of the *Elements*. Changing standards in nineteenth-century classical scholarship made it mandatory for Heiberg to investigate as many manuscripts as possible to construct his new edition. He was indeed able to consult the principal manuscripts held in European cities, thanks to the railway. In the description at the beginning of the first volume of his Greek edition of the *Elements*, Heiberg presents the six manuscripts he used to establish the critical text, of which three are in Italy (P in the Vatican, F in Florence, and b in Bologna). The other three are in Oxford (B), Paris (p), and Vienna (V). In 1880 he consulted V in his home city of Copenhagen (thanks to a generous loan of the Vienna manuscript) and Book 1 of p in Paris. The next year he went to Italy to consult Books 4 to 9 of P in Rome (the first three books were examined by colleagues). In Florence he was lucky enough to consult F alongside b, loaned from Bologna municipal library. Then in 1882 he went to Oxford to examine B and then consulted the Parisian manuscript p again, which this time was loaned to the Royal Library of Copenhagen. The first volume of his edition, containing Books 1 to 4 with a critical apparatus detailing all the different readings of these

13. More precisely, Theon's addition to VI.33 is written in the margin by another hand. This shows that it was added later, by someone who had access to another manuscript. This is an example of the phenomenon known as contamination, through which a manuscript inherits two traditions, making the reconstruction of the stemma difficult or impossible when content from another tradition has entered the text in subsequent copies.

codices and distinguishing the different hands in them, appeared in 1883. By 1885, all four volumes containing the thirteen books of the *Elements* had come out. Finally, in 1888 Heiberg published the very thick volume 5 which contains the additional books 14 and 15 (not actually by Euclid, but which traditionally accompany some manuscripts) and numerous scholia, prefaced by over a hundred pages of critical introduction, editorial notes, and studies of previous editions.

It is evident that without the railway network Heiberg's edition would have been impossible to put together (just remember Goethe's *Italian journey* to imagine the difficulty of travelling before the railway age). At the same time, one cannot but notice that this was an incredibly rapid work by a great but still young scholar. As he himself admits in the preface, he did not collate all the extant Greek manuscripts of the *Elements* (an impossible task even now, given the enormous number of them) and the first volumes in particular have the flavour of a work in progress. In fact he honestly admits, 'I preferred to publish this now, whatever it may be, than to prolong the work to infinity'.[14] Indeed, he reviewed his edition, and modified some of his choices between variants, in the critical introduction of 1888, and later published an article in which he collected newer results and discoveries, such as that of a papyrus fragment which proved one whole Proposition I.40 to be spurious (Heiberg 1903).

In short, Heiberg published a provisory edition of the *Elements* in the 1880s and never ceased improving on it in later life. Some of those results are duly included in his successors' translations (so that Proposition I.40 is always bracketed these days), but most of his notes and modifications remain unnoticed, and even today the text of the 1880s passes as 'definitive', despite Heiberg's intentions. Now I turn to Heiberg's edition and point out some problems with it, with no intention of criticizing him. If the Greek edition we use today is problematic, that is due to later historians who have simply accepted Heiberg's edition as definitive, sometimes overlooking even Heiberg's own later remarks.

Heiberg used Peyrard's Vat. Grec. 190 and several other manuscripts (all belonging to Theonian tradition), of which most important was F. But F is badly damaged, so other manuscripts were also indispensible. Though none of these manuscripts is a copy of another, the tradition is so complicated that it was impossible to reconstruct a stemma for them.[15] But at least they all derive from Theon's version (let us name it T), which is reconstructible from extant manuscripts, if not with absolute certainty in every passage. So we have a stemma like Fig. 9.2.6. Note that even though P is free of Theon's editorial work, it is a later manuscript, of the ninth century, so it cannot be the exemplar that Theon worked with.

---

14. …haec, qualiacunque sunt, nunc edere malui, quam opus in infinitum differre: *Elementa* (Heiberg) I VII-VIII; *Elementa* (Heiberg-Stamatis) I VIII.

15. *Elementa* (Heiberg) 5:XLIX, (Heiberg-Stamatis) 5:XXXVII. See note 13 for one reason why.

Figure 9.2.6 Stemma of manuscripts assumed in Heiberg's editorial work

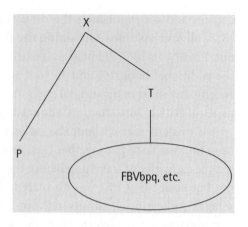

Figure 9.2.7 (a) Heiberg's diagram of Proposition III.31; (b) the same diagram, redrawn from Codex P

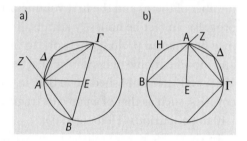

Our task is to restore the archetype X. It is obvious that with the discovery of P we can go back beyond T, coming closer to the original. However, it is not guaranteed that the archetype of X can be reconstructed, for we have only two testimonies of X (the situation in Fig. 9.2.3b), and efforts to reconstruct X often leave open questions where P and T do not agree. What is worse is that mathematical text is easy to alter, delete, or add to, without leaving any visible traces. It is not a poem, nor is it an inimitable masterpiece of prose. It is a chain of logical inferences constructed from extremely limited vocabulary and expressions, so that anyone with good mathematical sense and some training could try to restore corrupt text or add explanations.

Here is an example. In Proposition III.31 (Fig. 9.2.7), it is argued that the angle in a semicircle is a right angle, that in a greater segment is less than a right angle, and that in a smaller segment is greater than a right angle. In the figure, ΒΓ is the diameter. After showing that angle ΒΑΓ (the angle in the semicircle) is right and that the angle ΑΒΓ (the angle in segment ΑΒΓ, the segment greater than the semicircle) is less than a right angle, Euclid goes on to show that the angle ΑΔΓ, an angle in segment ΑΔΓ, is greater than a right angle. I quote the text:

(a) Next, since ΑΒΓΔ is quadrilateral in a circle, (b) and the opposite angles of quadrilaterals in circles are equal to two right angles (III.22),[16] [(c) therefore, the angles ΑΒΓ, ΑΔΓ

---

16. The reference to the number of the proposition used in this reasoning is not contained in the manuscripts.

are equal to two right angles,] (d) while the angle ABΓ is less than a right angle; (e) therefore the angle AΔΓ which remains is greater than a right angle. (after Heath 1925, II 62)

Now, phrase (c) between the brackets was expunged by Heiberg (and so does not appear in Heath's translation), because in P this phrase was added in the margin by a more recent hand, while it appears in all the Theonian manuscripts. However, there is also the possibility that the scribe of P (or a scribe of some manuscript between X and P) accidentally omitted phrase (c). This is quite plausible, for both (b) and (c) end with the same expression 'equal to two right angles' so that a scribe may have jumped from its first occurrence to the second. This is a common phenomenon, known as *homoioteleuton*. If an omitted phrase is difficult to replace by conjecture, it is almost certain that the text containing that phrase (in this case, T) is to be preferred. However, phrase (c) says the same thing as phrase (b), with the addition of the names of the angles. Any commentator could have inserted it. Therefore, it seems that neither interpretation can easily be excluded. The passage may have been added in T, but it may just as well have been omitted in P.[17] As I have already argued, when there are only two independent traditions coming down from an archetype, textual criticism cannot give us a definite answer about which to choose. In this particular case, Heiberg's decision can probably be justified by other omissions of similar intermediate (and relatively trivial) steps of proofs in codex P found in Propositions I.36, III.7, IV.7, etc. If such arguments are regularly omitted in codex P, they are less likely to be chance omissions in the tradition to which P belongs, than additions of obvious intermediate arguments by Theon or some other commentator. From a different point of view, phrase (b), which appears in both traditions, is not above suspicion either. Euclid does not usually cite general enunciations. Thus it is not altogether excluded that the original text also lacked this part, so that it contained only phrases (a), (d), and (e) of the cited passage.

In this hypothetical argument, I introduced another criterion for deciding the authenticity of the text: the question of style was tacitly invoked in Heiberg's decision to expunge phrase (c). Let us examine this criterion further, for scholars today are increasingly sensitive to the stylistic deviations found even in the best manuscripts of Euclid. Scholars agree that classical Greek mathematical works, including Euclid's *Elements*, have certain stylistic features, though there have always been disagreements over whether certain specific features are Euclidean or not.

Euclid's most reliable feature is that he never goes backwards. He prepares everything necessary beforehand, and the proof goes always forward in this manner. He argues: A, therefore B; and it is proved that C, so that D. It is not usual

---

17. See Heath (1925, I 52–53) for further discussion on the possible explanations for disagreements between P and Theonian manuscripts.

for Euclid to argue, for example: B, for A holds; so that D; this is because of C. Recently this feature has turned out to be by far more consistent in Euclid (and consequently powerful and compulsory for identifying spurious parts of the *Elements*) than it was thought to be. Wilbur Knorr (1996) pointed out that all the lemmas and corollaries of Book XII, as well as dozens of additional explanations that come after particular statements, are all absent from the Arabo-Latin tradition. This discrepancy had already been raised by Martin Klamroth in 1881, but Heiberg had replied that the Arabo-Latin tradition was a shortened version, while the Greek manuscripts preserved the genuine form. But Knorr persuasively showed that the contrary is much more plausible: that the Greek manuscripts P and T both suffered later editorial work (but this must be pre-Theonian, for P agrees with T in this respect), while the Arabo-Latin tradition is exempt from this alteration, therefore preserving the better text.[18]

Let us examine an example of a justification following a statement, which Knorr judged to be spurious. In XII.2, in which it is proved that two circles are (in the same ratio) as the squares (drawn) on their diameters, Euclid inscribes a square EFGH in the circle EFGH, and says:

then the inscribed square is greater than the half of the circle EFGH, *inasmuch as (epeidēper), if through the points E, F, G, H we draw tangents to the circle,*[19] *the square EFGH is half the square circumscribed about the circle, and the circle is less than the circumscribed square; hence the inscribed square EFGH is greater than the half of the circle EFGH.*

I have italicized the passage which explains why the inscribed square is greater than the half of the circle. This passage, introduced by the Greek conjunction *epeidēper*, appears in all the extant Greek manuscripts (except one, see note 18), leading Heiberg to include it in his edition. However, it is absent from the mediaeval Latin translations from Arabic. It, and other similar passages, are now regarded as later additions, invented by those who wanted to supplement Euclid's concise reasoning. Without going into the details of Knorr's arguments, I should add that the use of the subject in the first person plural in this passage ('we draw tangents...') is exceptional in Euclid, and should be regarded as a sign of interpolation. Thus, at least for this part of the *Elements*, the authenticity of codex P has collapsed, and we should imagine a stemma like in Fig. 9.2.8.[20]

18. Knorr also found that the Theonian manuscript b contains a text which is very close to that of Arabo-Latin tradition in this section. (But in other books of the *Elements*, its text is not particularly better than other Theonian manuscripts.) The peculiarity of b had already been noticed by Heiberg, who gave up trying to record its variants in footnotes (for the text was too different). Instead he put the last part of Book XI and the whole of Book XII (XI.36–XII.18) from manuscript b into an appendix. However, he seems to have been unaware of its importance. Manuscript b also implies that the tradition of the *Elements* is so complicated that a manuscript may not necessarily be homogeneous but can contain passages from different traditions.

19. The tangents are not drawn in the figure.

20. Around the end of Book XII (especially in proposition XII.17), the text of b deviates from that of Arabo-Latin tradition, showing the complexity of the traditions. Knorr concludes that the Arabo-Latin

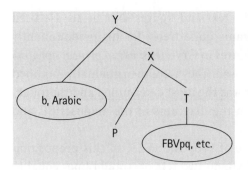

Figure 9.2.8 Stemma of manuscripts of the *Elements* XI. 36–XII.18, with Arabic sources included

Stimulated by Knorr's study, Fabio Acerbi (2003a) showed that in Book V too the Arabo-Latin tradition preserves the better text. However, the situation for Book X has turned out to be more complicated (Djebbar, Vitrac, and Rommevaux 2001), and we are now aware that there is no one tradition which consistently preserves a better text than any other. In some books one tradition provides the better text; in other books, some other tradition is preferable; and in many places we simply do not know which tradition is best. We are sure that we are closer to Euclid's text of the *Elements* than when we ingenuously believed that Heiberg's edition was 'definitive', but we are now much more aware of the distance (and sometimes unreachability) of Euclid's original text from the documents we possess (see also Vitrac 2004).

Diagrams

So far I have discussed the text of the *Elements*. No overall re-examination has yet been carried out of Heiberg's ground-breaking comparative study, though scholars have begun to realize that there is room for improvement to his edition, thanks to documents from the Arabo-Latin tradition and new stylistic methodology. In this section, I will look at the manuscripts from a different point of view, examining the diagrams they contain. The diagrams are, of course, indispensable constituents of the mathematical text. Except for some very simple cases, it is impossible to understand a proposition without its diagram. However, there seem to have been no criteria for editing diagrams comparable to those for editing text.

All modern translations of the *Elements* reproduce the diagrams in Heiberg's edition, which in turn copies those in the edition of E F August (Euclid 1826–9).[21]

tradition preserves the oldest form, while b shows an intermediate stage between the Arabo-Latin text and other Greek manuscripts. However, Vitrac, in his French translation of the *Elements* (1990–2001, IV 364–371), poses reasonable questions about Knorr's somewhat hasty conclusion with regard to the tradition of XII.17. Indeed, the Arabo-Latin text of this proposition does not seem better from a mathematical point of view, and the question is still open.

21. In the arithmetical books (Books VII to IX), Heiberg restored the presentation of diagrams with lines, while August represented numbers by dotted lines or even by some particular numerical value.

Did August consult manuscripts for the text and figures? Not at all. He used Peyrard's text, and modified the diagrams quite freely for his predominantly educative purpose. The fact is that the figures in Grynaeus' *editio princeps* of 1533 were fairly faithful to the manuscript he used. They were then gradually modified by subsequent editors, of whom August was the most determined and thorough. What, then, are the differences between the diagrams in the manuscripts and those in Heiberg's edition?

Let us begin with the simple example of Proposition I.7. In this proposition Euclid shows that given a straight line AB, no two sets of straight lines equal to each other AΓ, ΓB, and AΔ, ΔB (AΓ = AΔ and ΓB = ΔB) can be constructed on the same side of the line AB. This is an obvious lemma for the next Proposition, I.8, where two triangles having three sides equal to three sides are congruent to each other. Euclid proceeds by reductio ad absurdum, assuming that such two sets of lines have been constructed.

Fig. 9.2.8 shows an image of codex P and a diagram redrawn from it.[22] Only the redrawn figures are shown for other manuscripts, for they are all faithful enough to the originals for present purposes. In all the manuscripts containing Arabic and Latin translations that I have seen, the diagram is roughly symmetrical and AB is parallel to ΓΔ. Now, since no other condition is assumed, there is no particular reason for the figure to be symmetrical, or for AB to be parallel to ΓΔ. In Heiberg's edition, however, the diagram is deliberately drawn so that it is not symmetrical.

A similar example is found in Proposition I.43, where Euclid proves that when the diagonal AΓ of a parallelogram ABΓΔ is drawn, the complements about the diameter, that is, the two small parallelograms BΓEΔ and ΔZKH, are equal to each other. In Heiberg's edition the parallelogram is not rectangular, nor is the point K the midpoint of the diagonal. However, all the manuscripts show a rectangle (that is, a special parallelogram) and in many the point K bisects the diagonal. First, we show the reproduced image of codex B, then the redrawn figures.

In these examples, the tendency of the manuscripts is clear. If two sides of a triangle can be equal, they tend to be drawn as equal. If an angle can be right, it

---

22. Redrawn diagrams from the principal manuscripts of all 48 propositions in Book I of the *Elements* are available in Saito (2006); that is, for the Greek manuscripts PBVb, and for two Latin manuscripts of Gerard's translation. The diagrams are copied and reproduced by a simple computer programme developed for this purpose. It works as follows. The coordinates of the principal points—usually the points to which labels are attached—are recorded by clicking on them. Then one registers how the points are joined, either by straight lines or the arc of a circle. Thus, for example, from the image of Proposition I.7 in manuscript P (adequately trimmed) one gets the following data: A(303, 607), B(1199, 641), Γ(510, 208), Δ(1009, 218); and the lines connecting these points are the following: AB, BΔ, AΓ, ΓA, AΔ, BΓ (as the line BΓ does not exactly start from the point B, the coordinates of that starting point are also registered). From this set of information alone, a figure can be drawn which is faithful to the original image if one is ready to give up some particularities such as widths of lines, etc. The reason for this fidelity lies in the fact that scribes used only ruler and compass to construct their diagrams.

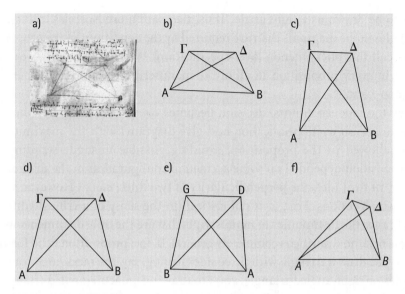

Figure 9.2.9 (a) The diagram of Proposition I.7 in Codex P; (b) the same diagram, redrawn from Codex P; (c) the same, from Codex F; (d) the same, from Codex V; (e) the same, from Bruges 521 (Gerard's translation); (f) Heiberg's diagram of Proposition I.7

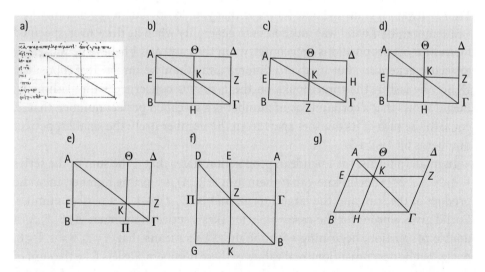

Figure 9.2.10 (a) The diagram of Proposition I.43 in Codex B; (b) the same diagram, redrawn from Codex B; (c) the same, from Codex P; (d) the same, from Codex b; (e) the same, from Codex V; (f) the same, from Bruges 521 (Gerard's translation); (g) Heiberg's diagram of Proposition I.43

tends to be drawn as a right angle. Thus, there are more isosceles triangles and rectangles in the diagrams than are required by the conditions of the proposition. Let us call this phenomenon 'over-specification'. We can find similar examples almost in every proposition in all of the manuscript traditions, whether Greek or Arabo-Latin.

Now the modern editors' decision becomes comprehensible: they thought it better to banish over-specification and give diagrams with the maximum generality allowed in the proposition, avoiding possible misunderstandings that the proposition depended on some accidental configuration in the diagram (the equality of two sides, the perpendicularity of two lines, etc.).[23] However, we may raise two objections. First, it is not critical (in the sense of a critical edition) to silently alter the testimonies of manuscripts that are the only documents we possess. Second, no drawing is completely general. If, in a proposition valid for all triangles, you draw a triangle with three sides having, say, the ratio 7:6:5 to avoid an isosceles or right-angled triangle, you are drawing an acute-angled triangle and thus excluding obtuse triangles and right-angled triangles for which the theorem is also valid. Even if you draw all three types of triangle, it does not mean that you have drawn all possible triangles for which the proposition is valid. Therefore, from a strictly logical point of view, even if you make the effort to draw a triangle with no conspicuous feature (non-isosceles, non-right-angled), the triangle you draw is always specific in some sense, so that you do not gain any generality.[24]

I do not think Euclid was indifferent to generality when he drew over-specified figures for his propositions on geometry. On the contrary, I believe he was eager to secure generality, but in a very different way than we might. In Proposition IX.36, the last in the three books on the *Elements* concerned with arithmetic, Euclid shows that a certain type of number is a so-called perfect number, that is, equal to the sum of its divisors apart from the number itself; the smallest perfect number is $6 = 1 + 2 + 3$.

In modern terms, the Euclidean proposition states that if the sum of the series $1, 2, 2^2, \ldots, 2^{p-1}$ (Euclid knew that their sum is $2^p - 1$) is a prime number, then the product of the sum and the largest term, that is $(2^p - 1)2^{p-1}$, is a perfect number. Euclid gives a proof for the case $p = 5$. He takes a series of numbers A, B, Γ, Δ in double proportion, beginning with the unit. This means that $A = 2$, $B = 4$, $Γ = 8$, $Δ = 16$, but he never mentions real values of these numbers. He lets E be the sum of them (including the initial unit), so that $E = 31 = 2^5 - 1$. And the product of E and Δ is represented by ZH (= 496), which he is going to prove to be a perfect number.

Then he takes another series of numbers in double proportion ΘK, Λ, M (62, 124, 248). The choice of two letters ZH and ΘK for certain numbers is due to

---

23. It should be added that Heiberg was much more faithful to manuscript diagrams in his later work, e.g., the edition of Theodosius' *Spherics* (Heiberg 1927).

24. For a consideration of generality in Greek mathematics, see also Netz (1999, chapter 6).

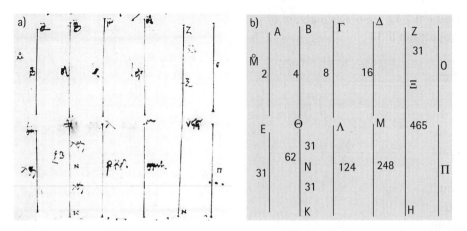

Figure 9.2.11 (a) The diagram of Proposition IX.36 in Codex B; (b) the same diagram, redrawn

the necessity of dividing them into the sum of two (or more) numbers. Then, he shows that the divisors of ZH (except ZH itself) are only E, A, B, Γ, Δ and ΘK, Λ, M, and these make up ZH itself when added together. Throughout the argument he consistently pretends not to know the values of the each of the numbers A, B, Γ, Δ, etc. (except that A = 2), and that the perfect number ZH at issue is 496. (Of course, he did know all of these values, and he deliberately chose the case $p = 5$ in which the sum of the first series $1 + 2 + 4 + 8 + 16 = 31$ really is a prime number, and in manuscripts later commentators wrote the value of each number beside the line.)

Euclid's argument reveals that he was concerned with the generality of his proofs, whether or not we judge that Euclid argued successfully.[25] What is interesting about this proposition is that in the diagram beside the text, all the numbers are represented by line segments whose lengths are almost the same (Fig. 9.2.11), though in reality they range from 2 to 496 and Euclid was surely aware of that. Here the lengths of the lines are not intended to express quantitative relationships between numbers, just as the lines in over-specified geometrical diagrams are not meant to express equality or inequality between them. The only function of the line segments in the arithmetical books is to show that the sum of two line segments corresponds to the sum of two numbers represented by those segments, just as the diagrams in the geometrical books are supposed to be a schematic representation of the figures.

---

25. If it is difficult to understand Euclid's concern for generality, it is because we use expressions like 2^p-1 which automatically guarantee the generality of the argument. Here again, it is the change in our intellectual environment that necessitates the intepretative effort.

Figure 9.2.12 Heiberg's diagram of Proposition III.13

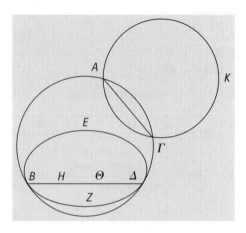

It was probably wise of ancient mathematicians not to charge diagrams with too much meaning, for one could not expect the figures to be copied exactly, keeping every metrical relationship between their components. By contrast, the persistence of the apparent generality of the modern editors' figures may be partly due to the fact that print technology enabled their drawings to be reproduced correctly in thousands of identical copies. Thus attitudes towards the sources (in this case diagrams) seems to have been partly influenced by the means and technologies available.

Another intriguing example of a diagram is in Proposition III.13, where Euclid proves that no two circles touch each other at more than one point, whether internally or externally. Euclid proceeds by reductio ad absurdum, his usual method for proving that something cannot happen, so that he had to draw two circles touching at two points, a nonexistent figure! However, Heiberg draws two circles cutting each other at two points, which is copied in modern translations. I shall explain the difficulty in understanding Euclid's arguments with Heiberg's figures (Fig. 9.2.12).

The impossibility of two circles touching internally at two points is based on the preceding Proposition III.11, which shows that if two circles touch each other internally, the straight line joining their centres, when produced, passes through the point of contact.[26] Euclid assumes that the circle ABΓΔ and circle EBZΔ touch at two points B and Δ, and lets their centres be H and Θ respectively. Then, the straight line joining H and Θ must pass both B and Δ, so that BH = HΔ and BΘ = ΘΔ, leading at once to a contradiction.

Then Euclid proceeds to the case of two circles touching externally. Here the key to the argument is a simple proposition, III.2, which guarantees that any chord of a circle falls inside it. Now, let the circles ABΓΔ and AΓK touch externally

---

26. It is taken for granted that the centres of the two circles are different points, for if the circles touching each other had the same point for their centres, they would coincide completely.

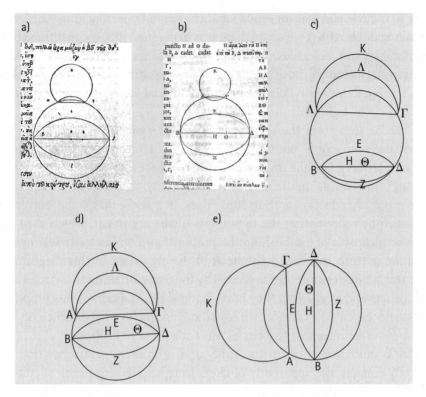

Figure 9.2.13 (a) The diagram of Proposition III.13 from Grynaeus' edition (1533); (b) the same diagram, from Gregory's edition (1703); (c) the same, redrawn from Codex P; (d) the same, from Codex B; (e) the same, from Codex V

at A and Γ. So, by Proposition III.2, the chord AΓ falls inside both circles. Then Euclid states, 'but it fell inside the (circle) ABΓΔ, [therefore it must fall] outside the (circle) AΓK'.[27] The reader is slightly bewildered by this sudden affirmation. The line AΓ indeed falls inside the circle ABΓΔ. Then why does this entail that it falls outside AΓK? This deduction must depend on the understanding of 'touch externally'. According to Euclid's definition, 'circles are said to touch one another which, meeting one another, do not cut one another' (III.def.3). So, since circles ABΓΔ and AΓK touch externally, circle AΓK is entirely outside the circle ABΓΔ, and the points inside ABΓΔ are necessarily outside AΓK.

However, this is not explicitly stated and, what is worse, in the figure the line AΓ is inside both circles! It is precisely here that Heiberg's figure diverges in an important and substantial way from those given by the manuscripts. Fig. 9.2.13 shows the diagrams in early editions and in the principal manuscripts.

27. The original text is: *alla tou men ABΓΔ entos epesen, tou de AΓK ektos.* As readers who know Greek will notice, the word 'circle' in parenthesis is not in the text but it is usual and even necessary to add it; the words between brackets are additions based on my interpretation.

The first circle is ΑΒΓΔ (or ΑΒΔΓ) and the second one, touching ΑΒΓΔ at two points Α and Γ, is ΑΓΚΛ which, in diagrams in manuscripts, is constituted by two arcs of circles and looks like a crescent moon (a shape called a lunule).[28] All the manuscripts, including those belonging to Arabo-Latin tradition (as far as I have seen), represent the second circle by a lunule, so that we may safely attribute this 'lunule as circle' to Euclid.

As can be seen from the reproduced image, Grynaeus' edition of 1533 still follows the manuscripts, but Gregory's edition of 1703 has two intersecting circles, as in Heiberg (see the images of figures of these editions). Of course this is a 'wrong', 'impossible' figure, but the impossibility comes from the hypothesis of the proof by reductio ad absurdum. Thus, in a sense this is the 'correct' figure, faithfully representing the hypothesis of the argument, which then turns out to be incapable of establishing the proposition. Euclid clearly distinguishes the intuitive truth from the hypothesis of the argument in this diagram. It is ironic that modern editors are so fixated by the common sense that circles do not have corners and angles that they have changed the diagram of this proposition without documented evidence, while criticizing Euclid's dependence on intuition when he assumes that two circles meet at a point in Proposition I.1.

There is much room for further study of diagrams in manuscript traditions. Strangely enough, the apparently obvious fact that Heiberg's critical edition of Euclid (critical in the sense that the text is established by the procedure of textual criticism) does not faithfully reproduce the diagrams has only recently been noticed, along with the realization that their study may reveal something about ancient Greek mathematicians' sense of generality.

Today we can consult the manuscript on the screen of a computer, comparing it with any other one with ease. Until very recently—and still today in some cases—one had to sit in front of microfilm reader in a library or, in the best case, have them printed on thermal paper, which gave forth a strange smell. Heiberg made his edition of the *Elements* in the early 1880s when microfilms were not available, not to mention computers.[29] We have seen that the railway and the inter-library loan of manuscripts both facilitated Heiberg's work. However, he could not work as we do today. When he read a manuscript in the Bodleian Library in Oxford, he did not have to hand another important manuscript from the Vatican. He had to copy the text and marginal notes with great care (by hand, of course), but it was not easy to register every particularity of all the figures without photography. In short, his conditions were very unfavourable for studying diagrams, and it was an understandable choice to use the diagrams of a previous edition in order

28. In several manuscripts the point Λ is lacking in the text, and in codex P it has been added by a later hand. The Λ may thus be a later addition to indicate the second circle with less ambiguity.

29. When Heiberg studied the Archimedes palimpsest in 1906, he had many of its pages photographed and continued his study with the photos (now preserved in the Royal Library of Copenhagen).

to publish his new edition quickly. The rise of interest in the diagram tradition among today's scholars is dependent on the development of the computer, which enables us to consult and compare figures with ease and, if we wish, to calculate every parameter in each figure.[30]

Conclusion

Sources in themselves do not represent facts. They require interpretation on the part of those who read them. In other words, sources do not teach us facts but pose questions. The questions are always different according to the knowledge, interests, and attitudes of the reader and the means at their disposal. For Proclus, philosophical interests such as the existence of figures, or the construction of the regular polygons (as cosmic figures), prevailed. After the Renaissance the aim of many scholars was to restore ancient mathematics in a form understandable to contemporary mathematics, and occasional redundancy and idiosyncracies were attributed to the (sometimes alleged) philosophical concerns of Greek mathematicians. Today, we believe we are trying to understand Greek mathematical texts in their proper context, but it is almost certain that coming generations will criticize us for imposing some interests of our own, which were not those of the ancients.

We also know that no Greek mathematical text is an autograph and are thus interested in textual traditions. If we are more sensitive to this aspect now, it is not only because we are open-minded (although we would like to think that we really are), but also because we can more easily obtain and examine more sources (Greek, Arabic, and Latin manuscripts) with the aid of modern technology, which also supports the recently renewed interest in diagrams. Thus, with the inevitable changes of interest and available tools for study, no interpretation can remain complete nor definitive, and the sources will not cease to pose new questions to historians. Thus the history of mathematics is no less a dialogue between the past and the present than other branches of history.

Bibliography

Acerbi, F, 'Drowning by multiples. Remarks on the fifth book of Euclid's *Elements*, with special emphasis on Prop. 8', *Archive for History of Exact Sciences,* 57 (2003a), 175–242.
Acerbi F, 'On the shoulders of Hipparchus: a reappraisal of ancient Greek combinatorics', *Archive for History of Exact Sciences,* 57 (2003b), 465–502.

30. For example, it is easy to calculate every angle in a figure. Then we can examine how exactly right angles are drawn; in some manuscripts they range from 87° to 93°, while in another where the scribe was much more meticulous most right angles are between 89° and 91°. This shows that the scribe of the latter used some sort of tool to obtain a right angle.

Busard, H L L (ed), *Campanus of Novara and Euclid's Elements*, 2 vols, Steiner, 2005.

Christianidis, J (ed), *Classics in the history of Greek mathematics*, Kluwer Academic Press, 2004.

Djebbar, A, S Rommevaux, and Vitrac, B, 'Remarques sur l'histoire du texte des Eléments d'Euclide', *Archive for History of Exact Sciences,* 5 (2001), 221–295.

Drake, S, *Galileo at work: his scientific biography*, University of Chicago Press, 1978.

Euclide (Euclid), *Les Oeuvres en Grec, en Latin et en Français par François Peyrard*, 3 vols, Patris, 1814–8.

Euclid, *Euclidis elementa ex optimis libris in usum tironum. Graece edita ab Ernesto Ferdinando August*, 2 vols, Berolini, 1826–9.

Euclide (Euclid), *Les Éléments. Traduction et commentaires par Bernard Vitrac*, 4 vols, Presses Universitaires de France, 1990–2001.

Giusti, E, *Euclides reformatus: La teoria delleproporzioni nella scuola galileiana*, Bollati Boringhieri, 1993.

Harari, O, 'The concept of existence and the role of constructions in Euclid's *Elements*', *Archive for History of Exact Sciences*, 57 (2003), 1–23.

Heath, T L, *The thirteen books of the Elements*, 3 vols, 2nd ed, Cambridge University Press, 1925. Reprint, Dover, 1956.

Heiberg, J L, 'Paralipomena zu Euklid', *Hermes*, 38 (1903), 46–74, 161–201, 321–356.

Heiberg J L (ed), *Theodosius tripolites sphaerica*, Weidmann, 1927.

Heiberg, J L, and Menge, H (eds), *Euclidis opera omnia*, 8 vols and a supplement, Teubner, 1883–1916. (The first 5 volumes, published between 1883 and 1888, are the *Elements*.)

Heiberg J L, and Stamatis, E S (eds), *Euclidis Elementa*, 5 vols, Teubner, 1969–77. (Reprint, without Latin translation, of the first 5 volumes of Heiberg's edition of Euclid's *Elements*.)

Knorr, W R, 'Construction as existence proof in ancient geometry', *Ancient Philosophy,* 3 (1983), 125–128. (Reprinted in Christianidis [2004, 115–137].)

Knorr, W R, 'The wrong text of Euclid: on Heiberg's text and its alternatives', *Centaurus*, 38 (1996), 208–276.

Netz, R, *The shaping of deduction in Greek mathematics*, Cambridge University Press, 1999.

Netz, R, and Noel, W, *The Archimedes codex*, Weidenfeld & Nicolson, 2007.

Netz, R, Saito, K, and Tchernetska, N, 'A new reading of Method proposition 14: preliminary evidence from the Archimedes Palimpsest', *SCIAMVS*, 2 (2001), 9–29; 3 (2002), 109–125.

Netz, R, Acerbi, F, and Wilson, N, 'Towards a reconstruction of Archimedes' Stomachion', *SCIAMVS*, 5 (2004), 67–99.

Proclus, *A commentary on the first book of Euclid's Elements. Translated with introduction and notes by Glenn R Morrow*, Princeton University Press, 1970. (Paperback edition with a foreword by Ian Mueller, 1992.)

Saito, K, 'A preliminary study in the critical assessment of diagrams in Greek mathematical works', *SCIAMVS*, 7 (2006), 81–144.

Szabó, Á, *The beginnings of Greek mathematics* (trans A M Ungar), Reidel, 1978 (original work published 1969).

Vitrac, B, 'A propos des demonstrations alternatives et autres substitutions de preuves dans les Éléments d'Euclide', *Archive for History of Exact Sciences,* 59 (2004), 1–44.

West, M L, *Textual criticism and editorial technique: applicable to Greek and Latin texts*, Teubner, 1973.

Zeuthen, H G, 'Die geometrische Construction als "Existenzbeweis" in der antiken Geometrie', *Mathematische Annalen*, 47 (1896), 222–228.

Number, shape, and the nature of space: thinking through Islamic art

Carol Bier

Islamic art, when observed through the western lens of art history, is often seen as decorative and ornamental. This view sets up an opposition between Islamic art and general trends identified in western art since the Renaissance: the focus on representations of the human form, pictorial narrative, and spatial perspective. The treatment of pattern, which evolved and flourished from the tenth century onwards, is at once both universal within the Islamic world and sufficiently distinct from other cultural traditions that it may justifiably be characterized as Islamic. Just what led to the extraordinary proliferation of geometric ornament within the Islamic world, extending from Spain to India and beyond, and why geometric ornament proved to be both fascinating and enduring as an expression of Islamic art to the present, is not yet fully understood. Islamic art's emphasis on geometry and surface, with the effects of light on form, has not yet received the attention it deserves from historians of art, and even less from historians of mathematics, science, and Islamic philosophy. One might argue in the latter case that the primacy given to the textual tradition and its transmission has precluded sufficient consideration of the experiential understanding of crafts and their technologies in relation to both the production of art and the production of knowledge.

From another perspective, the very characteristics which may be seen as decorative and ornamental in Islamic works—from architecture and carpets to book illumination, ceramics, metalwork, ivory-carving, and other media—express abstract mathematical ideas and principles such as linear symmetries (border patterns) and symmetries of the plane (field patterns), or geometry and algorithms (units and their repeats)—in short, relationships among numbers, shapes, and space in one, two, and three dimensions. From the perspective of higher mathematics in the twenty-first century, the existence of such relatively elementary mathematics in the visual arts may be readily recognized—and easily dismissed. It may be seen on the one hand as mathematically trivial, and on the other hand as inherent in the processes of artistic production. But what is thereby dismissed is the recognition of the presence in the arts of mathematics at a time when this mathematics was newly created. What is yet difficult to determine is whether the abstract forms, geometric relationships of shapes, and visual algorithms are preferred in Islamic art solely for aesthetic reasons, or whether they are intentionally expressive of emergent mathematical ideas in the context of their creation.

Debate continues within the field of Islamic art history over whether architects and artisans were sufficiently familiar with the work of contemporary mathematicians to apply such thinking, assuming that the arts represent applications of knowledge arrived at independently, or, whether mathematicians may have been directly involved in the production of architectural forms of ornament (Grabar 1992; Özdural 1995). The presence of five-fold symmetries and aperiodic tilings on Islamic monuments of the eleventh and twelfth centuries raises questions about the sophistication of mathematical knowledge and its application to architecture nearly a millennium before Roger Penrose described similar two-dimensional quasi-crystalline patterns (Makovicky 1992; Bonner 2003; Lu and Steinhardt 2007). Sheila Blair and Jonathan Bloom (2006), for instance, attribute tile patterns of stars and polygons, the most complicated of which comprise pentagons and ten-pointed stars, to traditional artisans' 'tricks of the trade' without recourse to 'higher mathematics' (Fig. 9.3.1a). Apart from this recent explanation, there has been little acknowledgment of empirical play with forms and patterns in the production of artistic works, which are so richly endowed with geometric possibilities, and even less consideration of art as a possible source for the recognition and development of abstract mathematical ideas (Bier 2005a).

More than a century ago, Owen Jones isolated ornament as a subject of inquiry removed from cultural origins. His seminal work, *The grammar of ornament* (1856), which promoted decorative schemes in architecture, flourished for decades as a source of inspiration in the development of architecture and the decorative arts in the West. Jones' notion that ornament is devoid of meaning has persisted to the present. If, however, one delves into the discourse of classical Islamic philosophy and theology, and the history of mathematics, one finds meaningful intersections of forms of visual expression with both mathematics

a)

b)

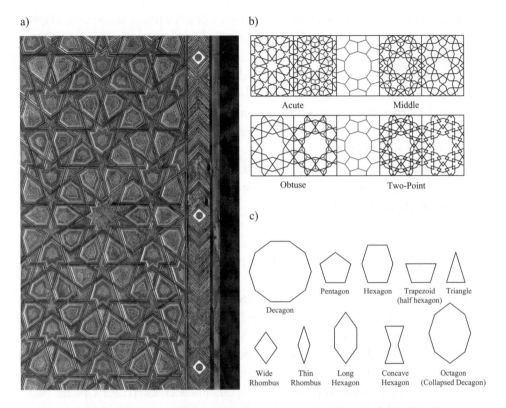

Acute   Middle

Obtuse   Two-Point

c)

Decagon   Pentagon   Hexagon   Trapezoid   Triangle
                                   (half hexagon)

Wide   Thin   Long   Concave   Octagon
Rhombus   Rhombus   Hexagon   Hexagon   (Collapsed Decagon)

Figure 9.3.1 a) Wood door, Iran, seventeenth century. David Collection inv. no. 35/2000, detail (photo by Pernille Klemp). Cat. no. 41, front, detail (Blair and Bloom 2006, 111). b) Diagram demonstrating the generation of the four principal pattern families from common five-fold polygonal sub-grid (Bonner 2003, fig. 11; courtesy of Jay Bonner). c) Standard sub-grid elements that comprise the five-fold system of geometric pattern generation (Bonner 2003, fig. 14; courtesy of Jay Bonner)

(Chorbachi 1989) and metaphysics (Akkach 2005). The proposition of a mathematical and metaphysical basis for Islamic art has perhaps been eclipsed by disciplinary attention in the arts, on the one hand, to primacy of the figural tradition and perspective, and in mathematics, on the other, to the modernist focus on quantification, symbolism, and abstraction from the real world (see Gray, Chapter 7.4; Corry, Chapter 6.4; Kjeldsen, Chapter 8.4 in this volume).

Efforts to relate geometry and Islamic art are extensive, although somewhat diffuse; different approaches have emerged to suit different purposes. They include efforts to interpret the meaning, or meanings, of ornament; to analyze geometry in historical works; to understand how Islamic-style geometric patterns can be generated today; to explore Islamic aesthetics; and to relate the history of Islamic mathematics to art. All of these approaches are in one way or another interdisciplinary. In what follows, I present synopses of several bodies of recent work. I then try to frame the question of whether geometry is inherent in

Islamic art simply because of the relationships of juxtaposed shapes, or whether the mathematical aspects of Islamic art are themselves visually expressive forms, selected purposefully and intentionally to articulate new mathematical ideas. Key issues concern pattern making by repetition using different media, and the contemporaneity of such visual forms with the development of new ways of thinking. Finally, I address the crucial issue of technologies of transfer: how was relevant, useful knowledge transmitted across time and space? The goal of this chapter is to suggest new points of departure and lines of exploration for the study of intersections of art and mathematics generally, and Islamic art and the history of mathematics, philosophy, and metaphysics in particular. The interrelationships between texts, artifacts, and monuments are the nexus of such future studies.

Influential approaches to the study of pattern in Islamic art

Jones, in *The grammar of ornament* (1856), divorced ornament from its original cultural contexts and meanings because he sought to derive principles of design that would have universal application. This treatment was reinforced in both the nineteenth and early twentieth centuries through the influence of Islamic design on architecture and the arts in Europe and North America, as well as by the active promotion by cultural entrepreneurs such as Arthur Upham Pope, who sanctioned the collection of Persian arts as offering a 'symphony of pure form' (Wood 2000). Literally until the fourth quarter of the twentieth century, it was deeply unfashionable to treat Islamic art as anything but decorative. A few die-hard scholars, such as J M Rogers (1973), sought to position the academic understanding of Islamic arts and architecture in relation to local cultural meanings. But for the most part, the paradigm established by Jones held fast. Ever since publication of *The grammar of ornament*, geometric pattern in Islamic art has been characterized in the West as decorative and ornamental, rather than as systems of signification with cultural associations and contextual meanings.

Within the study of Islamic art history, by contrast, several scholars have attempted to understand ornament and its meaning in cultural context. Richard Ettinghausen approached pattern as *syntactic*, looking at the whole in relation to its parts, rather than addressing individual elements as separate isolated entities. He considered *Gestalt* 'form' as an integrated totality and he sought to determine 'principal methods by which the artisans handled the extensive combinations of patterns so as to avoid bare areas which it would seem were aesthetically unsatisfactory' (Ettinghausen 1979, 15). He defined an aesthetic 'which managed to overcome in a pleasant fashion the *horror vacui*, yet did not create the impression' of overcrowding.

Figure 9.3.2 *Hazār bāf* ('thousand weavings'), Tomb of the Samanids, Bukhara, tenth century AD. (Photo by Ruth Harold, fh–bukhara49, courtesy of Frank and Ruth Harold, http://depts.washington.edu/silkroad/)

Taking a different stance, Lisa Golombek (1984) ascribed a 'textile mentality' to Islamic lands, based on the extent to which patterned textiles were used and appreciated, including the use of metaphors such as the Persian term *hazār bāf* 'thousand weavings' to describe brickwork (Fig. 9.3.2). She hypothesized that this might account for what she identified as a 'textile aesthetic' that seems to inform Islamic arts and architecture in ways that distinguish it from the arts of other cultures. Endorsed by the influential Oleg Grabar (1992), Golombek's view has held sway in recent decades. Grabar, however, sought to explore ornament as mediation. He expressed puzzlement over why geometric pattern attracted such extraordinary focus in Islamic cultures when it was so patently accessible to all. He identified a series of what he called intermediaries, which included calligraphy, geometry, floral elements, and architectural motifs. Margaret Olin (1993) described Grabar's approach as 'systematically investigating ornamental syntax'.

She emphasized that ornament in architecture comprises precisely those elements that are not direct expressions of structure and materials. In my own critique of Grabar's work I called him to task for using the term 'mediation of geometry', instead of 'mediation of pattern', implying the lack of recognition of an emphasis on the rhythmic repetition of forms rather than the geometry that is emergent in the relationships among shapes when they are repeated (Bier 1994).

Gülru Necipoğlu (1995) and Yasser Tabbaa (2001), by contrast, both attribute political motivation to the use of particular geometric patterns: in their works historical circumstances are considered primary factors in the choices and use of patterning. Necipoğlu explains eleventh-century geometric patterns as expressions of dynastic identity, functioning as a form of propaganda (Necipoğlu 1995, 96–99, 108–109, 192). Tabbaa focuses on the early Islamic development of proportional styles of Arabic calligraphy, in particular the geometrical regularization of *naskh* script that was put forward by the calligrapher Ibn Muqla around the year 1000. Tabbaa identifies the role of these new conventions as a tool for the restoration of orthodoxy in eleventh-century Baghdad, capital of the Abbasid Empire, after a period of relative religious liberalism (Tabbaa 2001, 141–145). More recently, Blair and Bloom (2006) have offered a new appellation, *cosmophilia*, 'love of ornament', as an explanation for the phenomenon of pattern in Islamic art, ascribing its purpose to providing visual pleasure.

Outside the art historical literature, there is a growing body of work that addresses mathematics in Islamic art. A major exhibition of Islamic arts at the Hayward Gallery in London, held in conjunction with the World of Islam Festival Trust in 1976, generated a spate of publications that sought to explain the geometry of Islamic art analytically, approached through a metaphysical lens and linked to a cosmological world view (Critchlow 1976). The underlying analyses of geometric structures reveal careful mathematical thinking and understanding (El-Said and Parman 1976 [1988]). Issam El-Said (1993) lays out three systems of design which underlie two-dimensional patterns in Islamic art, the first based on the square root of 2 (the diagonal of a square unit), the second the square root of 3 (the altitude of half an equilateral triangle), and the third the square root of 5 (based on the golden mean). Other works of the time proffered a more mystical interpretation, leaning heavily on contemporary Sufi ideas (Bakhtiyar 1976; Bakhtiyar and Ardalan 1973). For this group of essentialists, spirituality is an essential component in the construction of pattern. Ismail Al-Faruqi and Lois Lamya al-Faruqi (1986) begin to explore aspects of Islamic art, such as abstraction, modularity, rhythm and repetition, which they relate to expressions of *tawḥīd*, the doctrine of unity and oneness of God.

Nearly a hundred years earlier, Jules Bourgoin had prepared technical drawings with geometric constructions documenting patterns he saw on monuments in Cairo and Damascus; his classic work was reissued in 1973 and is still in print today. More recent analytical studies of geometry by Jay Bonner, Craig Kaplan, Reza Sarhangi, Slavik Jablans, Jean-Marc Castéra, Lynn Bodner, and Chris Palmer have been presented at Bridges Conferences (Mathematical Connections in Art, Music, Science and Culture). Often with applicability to contemporary design, their works reckon with systems of repeat and the modularity of Islamic patterns; these are published (with related references) in the annual Bridges proceedings.[1]

In addition to the approaches of art history, mathematics, and design to the study of geometric pattern in Islamic art, another body of work addresses more broadly the phenomenon of ornament in relation to abstraction and the psychological dimension of perception. Ernst Gombrich (1979) sought to establish theoretical categories for the functions of ornament, which he identified as framing, filling, and linking. But he never considered ornament *qua* ornament as the subject of representation or artistic form. Other theories of ornament also tend to focus on formal aspects of ornament and the abstraction of forms found in nature, thereby deriving meaning from visual forms and our perception of them by means of seeing (Freedberg 1989; Brüderlein 2001; Trilling 2001; 2003; Summers 2003). This approach depends upon what is optically recognizable. Discussions of patterning in Islamic art that focus on the figures themselves explore literal aspects of geometry, rather than its potential meanings on different levels.

However, the questions I think we should be asking are not about decoration and ornament, but about surfaces and the plane, about units and repeats, and about circles and the nature of two-dimensional space. It has not been adequately recognized that the forms contained within geometric patterns, no matter how basic they are understood to be today, were new to the artistic vocabulary of pattern at the time they were created. They seem to represent intensive efforts to explore the mathematical properties of space in two and three dimensions. Is pattern non-representational? Or is it representational in the deepest meaning of the word: a visual metaphor of relationships, of existence, of the cosmos, an expression of realities beyond that which can be merely seen?

Models and forms

By taking pattern as our subject, we may return to the study of Islamic art with mathematical approaches in mind. This may prove a useful exercise in better understanding relationships expressed through repetition of forms, using a variety of traditional technologies on a wide range of available materials including clay, metal, wood, ivory, leather, fiber, stone, and glass. Relationships between

---

1. See <http://www.bridgesmathart.org/>.

arithmetic and geometry, as explored in Islamic intellectual discourse from the eighth to the eleventh century, may allow us to consider a new theory of Islamic ornament that treats patterns not as representations or symbols but as expressive forms—expressive of newly emergent mathematical ideas (Bier 2007a). Meaning is not so much evident in the forms themselves as it is inherent in the processes of formation. According to this line of thinking, at some point between the eighth and the eleventh century, Islamic ornament and its formal expression became connected to abstract ideas articulated in contemporary philosophy, mathematics, and religion. If this identification of the origin of a paradigm shift is correct, the effect on artistic production was dramatic. A new aesthetic of pattern came to depend upon the direct relationship with the new mathematical idea of algorithms, and found deliberate articulation in repeat patterns in textiles as well as the expression of patterns in all media.

What is meant by this new aesthetic of pattern? Let us begin by considering carpets, even though surviving examples are from the Saljuq period (thirteenth century) and later. Typically, traditional Oriental carpets (a Western category used to encompass carpets produced in a variety of weaving cultures of the Islamic world) exhibit a multiplicity of patterns. This profusion of patterns contributes to the complexity and intricacy which so characterize the appearance of Oriental carpets. Various systems of repeats are displayed within the oblong central field, framed by the borders. This arrangement results in a contrast between field patterns (designs repeated in the plane) and border patterns (designs repeated in linear fashion). The presence of both symmetry and symmetry-breaking—creating an expectation of symmetry that is not met—yields patterns that both please the eye and tease the mind, offering delight as they confound. In contrast, patterns that rely solely on symmetry are boring because they are so readily perceived; such patterns do not actively engage the mind. As we often see in Oriental carpets, it is a playfulness with symmetry that results in intriguing patterns. In nature, symmetry is imperfect although mathematicians treat it as an ideal. The evidence of Islamic art suggests a deep appreciation of the visual impact of the approximation of symmetry, rather than its precision, to create patterns perceived as beautiful (Bier 2005b).

Patterns in carpets (Fig. 9.3.3) are carried by the pile, but the underlying fabric structure relies first upon the rectilinearity of the loom, a structure designed to secure the warps (longitudinal elements) and keep them taut during weaving (see Brezine, Chapter 5.4 in this volume). The woven structure of a carpet results from the interlacing of the warps with wefts (transverse elements), which are inserted as weaving progresses. The interlacing of warp and weft creates a mathematical grid that underlies the patterns, which are carried by colored yarn segments secured in the weave to form the pile. These yarn segments are inserted between two warps and wrapped around another one or two warps (or pairs of warps) to form

Figure 9.3.3 Turkmen carpet. Fine Arts Museums of San Francisco 2000.118.5 The Caroline and H McCoy-Jones Collection, gift of Caroline McCoy-Jones

what are called knots. The knots are arranged by color in horizontal rows, each row of knots inserted above a weft yarn as weaving progresses. Patterns result from counting and repeating sequences of knots of the same or different colors. Field patterns and border patterns in all handmade Oriental pile carpets rely upon the weaver's counting and repeating sequences of knots. Patterns emerge based upon the weaver's choices of color, and the execution of repeated sequences of selected designs. The pile projects from the foundation; after weaving, it is cut to a uniform height. In the completed carpet, only the pile yarns are visible on the surface, since they hide the warp and weft. It is the pile that carries the colors, designs, and patterns. Thus, Oriental carpets are two-dimensional in the appearance of surface design; they are, however, three-dimensional in structure.

The technology involved in carpet-weaving is very simple; it is much simpler than that of other weaving technologies. Because of the technology involved, and the counting and repeating of sequences of knots, the two-dimensional patterns in carpets bear many features that may be explored mathematically. Indeed, carpets bear inherently significant mathematical concepts that their makers may have understood intuitively or empirically. Not only are the arithmetic systems of addition, subtraction, multiplication, and division rendered visible, but so are halves and wholes, multiples and fractions, squares and square roots, grids and tessellations, and points, lines, angles, and shapes in the plane; in short, the fundamental elements of geometry. Not immediately apparent to the untrained eye, the sharp contrast of finite borders enclosing an infinitely repeating pattern in the central field renders visible the representation of infinity and ambiguity, features also encountered frequently in Islamic art in other media. But field patterns and border patterns in carpets also relate to algorithms, number theory, topology, sometimes knot theory, even fractals, which nowadays may all be taught using carpets. Weaving at the simplest logical level—interlacing wefts with warps in a sequence over-one-under-one—can be related to linear algebra, while more complicated weave structures render visible geometrical combinatorics (Brezine, Chapter 5.4 in this volume). The question remains as to whether these fundamental mathematical principles as we understand them today were understood at a theoretical level by those who designed and wove carpets in traditional weaving cultures.

The fascination with color and pattern in Islamic art, however, is not restricted to carpets. Rather, it is expressed in all media in many areas of the Islamic world, from the western Mediterranean to Indonesia, where two-dimensional repeat patterns provide the primary means for organizing color and space. Ornamentation of architecture, ceramics, metalwork, book illumination, and textiles all rely upon the repetition of complex or simple designs, creating overall surface patterns of apparent complexity and intricacy, which may be reduced through analysis to the identification of symmetries and symmetry-breaking (Bier 2005b).

In looking at these artifacts and monuments from a mathematical perspective, it is useful to keep in mind that a pattern depends upon three characteristics—a unit, repetition, and an organizing principle. To relate patterns to the processes of pattern formation links both craft and technology to the concept of an algorithm. Today an algorithm is understood to be a finite set of steps, which when repeated, always yields the same result. It functions in the manner of a recipe or formula. For example, when you create an eight-pointed star and then repeat it successively (Fig. 9.3.4), adding one star to the next, the result is a pattern comprising a square grid with centers all in alignment forming parallel and perpendicular sets of axes, intersecting at right angles to create an orthogonal grid (Bier 2007c). As you line up the stars, negative space emerges within the spatial configuration and new relationships are formed within the set of visual elements.

By participating in the processes of pattern making, an artisan may quickly become familiar with the principles of applied geometry without ever needing to know or understand the symbolic or formulaic expression of repetitions, or the geometric relationships they represent. Art that relies upon the principles of pattern making forces the artist or pattern maker to engage in processes over which the individual has little control. Making a pattern, the maker is engaged

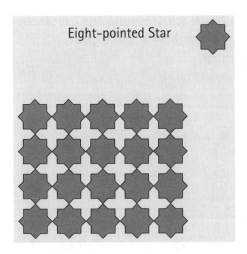

Eight-pointed Star

Figure 9.3.4 Eight-point stars arranged additively to form a square grid

in a process that is itself limited by the laws of symmetry. Symmetry allows for repetition, but only according to certain principles (Bier 1996–2007). Decisions are made at the level of design; but once a decision is taken to manipulate that design to form a pattern, the laws of symmetry prevail until the pattern is accidentally or deliberately broken. The artist thus participates in a process by which number and pattern, unit and shape, are integrated into visual forms. The artistic production of repeated patterns may become deeply meaningful as both a meditative and contemplative exercise. The artist may experience an engagement whose meaning lies beyond optical perception. It is an engagement with process, rather than product.

Other fundamentals of mathematics are experienced through pattern making. Circles, when tight-packed, naturally combine in two ways. One method yields centers that form a square grid (Fig. 9.3.5); the other method yields a more tightly packed triangular grid. By connecting their centers with straight lines, regular polygons result. By highlighting sections of curves or segments of lines, or by adding flourishes, many complex patterns may be generated from simple algorithms. Tight-packed circles in a square grid are typical of the patterns of eighth and ninth-century silk textiles woven in regions north of Iran. Such textiles quickly became the rage throughout the known world—they were presented as gifts from ambassadors (as depicted at Afrasiyab, offering a bolt of patterned silk) (Fig. 9.3.6), preserved in European church reliquaries wrapping the relics of saints, excavated in the tombs of nomadic chieftains in the Caucasus, treasured in the emperor's repository at shrines in Japan (Bier 2004).

To give another example, by selecting centers in a triangular grid, one may establish either a rhombic grid or a hexagonal grid. The triangular grid lends itself to six-pointed stars and hexagons; the square grid often yields eight-pointed stars and cross forms. Further connecting (and obliterating) lines may produce

Figure 9.3.5 Dirhams laid out tangentially to form a square grid

interlaced patterns. Once a generating unit and its mode of repetition have been determined, it is the process of pattern formation that carries the craftsman from conception to completion. Such basic mathematical considerations underlie the play of pattern in Islamic art.

The shift in focus from viewer to pattern maker is particularly useful for understanding the processes by which geometric patterns are made. Apart from the steps involved in constructions using a compass and straight edge, the process of pattern formation is radically different for each medium. For ceramic tile the process may entail incising, cutting, and glazing, but for a pile carpet the patterns are effected by counting and repeating sequences of knots (pieces of colored yarns, wrapped around pairs of warps). On an illuminated page of a manuscript, patterns may be outlined using a compass and straight edge, then filled in with a pen or brush, using pigments and various gilding techniques. For carved stucco, the pattern may be first constructed using a compass and straight edge, then carved and painted. Alternatively, templates may be carved then the stucco molded or stamped with them in a manner similar to the printing of patterns on cloth. For pattern-woven textiles, an entirely different technique is utilized, whereby a model is established to effect mechanical repetition at the loom by calculated manipulation of warp and weft. Each technology affects the process by which relationships of number and shape produce what may be perceived on completion as a geometric pattern. The resulting visual effect in a work of art is far removed from—and yet integral to—the temporal processes of pattern formation.

Figure 9.3.6 Wall painting of ambassadors at Afrasiyab, Uzbekistan, c seventh century AD; central figure offers a bolt of patterned silk

Mathematical aspects of Islamic art inform a beauty of form, pattern, and structure. Within the Islamic world, pattern making has served a primary function in the organization of two- and three-dimensional space, both in architecture and in arts of the object in all media. Seemingly complicated patterns may be analyzed to identify a unit and its repeat according to the principles of symmetry, exemplifying pattern making as a process that is at once unitary and systemic. The physical and visual effects of pattern making as a process of reiteration thus connect the maker to an implication of infinity. By examining relationships among numbers and shapes, units and repeats, circles and centers, with a view towards understanding the processes of pattern making, we may recognize meaning in the fascination with pattern that is evident in classical Islamic art. Several monuments in particular exhibit extraordinary play with the elements of

Figure 9.3.7 Tomb tower at Kharraqan, Iran, 1093 AD. Photograph courtesy of Ann C Gunter

pattern, giving evidence of the artisans' efforts to explore the extraordinary possibilities of two-dimensional space.

Intersecting polygons executed in cut brick project to catch sunlight on the eight sides of each of two Saljuq tomb towers at Kharraqan in northern Iran, one built in 1067–68 AD and other in 1093 AD (Fig. 9.3.7) (Bier 2002). In the tomb towers at nearby Maragha (Fig. 9.3.8), constructed about a century later, color is added to highlight even more complex patterns of illusionary interlace (Makovicky 1992). At both Kharraqan and Maragha, units and repeats relate to

Figure 9.3.8 Gunbad–i Kabud (tomb tower) at Maragha, Iran, dated 1196–97 AD (AH 593), western and northwestern sides, detail. Courtesy of the Freer Gallery of Art and the Arthur M Sackler Gallery Archives, Smithsonian Institution (photo by Hans C Seher-Thoss, Lantern slide C156)

number and shape, creating a multiplicity of patterns that cover the plane. The patterns themselves are conceivably infinite, made finite by arbitrary endings at the edge of a side, or bounded by the form of a niche or engaged pillars. In contrast, the field patterns in carpets are arbitrarily cut off by borders. In all such cases, the patterns express algorithms, sequences of steps through which units are repeated according to particular organizing principles.

We thus shift our vision from the object itself to its role as a medium, its service as the bearer of something else that is more important than its physicality. Think of the example of a book, the contents of which are distinct from the materials of the book and its binding. Ideas are portrayed in words, which are separate and distinct from the physicality of the book. Can we consider the notion that ideas are portrayed in patterns, distinct from the medium in which they are carried?

In so formulating our thoughts, we move into the realm of philosophy. The discourses of mathematics and philosophy had a long history before Islam, evident not only in works of classical Greek philosophy and Hellenistic mathematics, and in the works of neo-Platonist philosophers of later Antiquity, but also in the reconstructed thought of Pythagoras and others involved in seeking to understand the nature of number, space, and the universe. The Abbasid court in Baghdad, capital of the Islamic empire from the middle of the eighth century, sponsored translations of much of the ancient Greek scientific corpus, with its emphasis on geometry and philosophy, and hosted resident scholars and delegations of scholars from India who shared their knowledge of arithmetic and astronomy (Gutas 1998; Brentjes, Chapter 4.1 in this volume).

Surviving Arabic texts provide evidence that discussions of the one and many, limits and infinity, center and circumference, and signs of God, were pertinent to many areas of theoretical exploration in philosophy that may also have relevance for the arts. In Abbasid Baghdad, mathematicians and philosophers with court patronage advanced our collective human understanding of what today we call algebra and algorithms. They studied the significance of ideas such as zero, infinity, and limits or boundaries, and explored notions of space (see also Thakkar, Chapter 7.2 in this volume). Most obviously, the modern word 'algorithm' originates in a thirteenth-century Latin corruption, *algorismus*, of the name of Muḥammad ibn Mūsā al-Khwārizmī, a ninth-century Abbasid scholar who set forth his mathematical problems in *Kitāb al-mukhtaṣar fī ḥisāb al-jabr wa-al-muqābala* 'The condensed book of calculation by restoring and balancing', from which we also get the word algebra. For al-Khwārizmī's algebra did not only give definitions, theorems, and proofs in the ancient style. It also gave clear instructions on solving algebraic problems, addressed directly to the reader, as follows:

What must be the square which, when increased by ten of its own roots, amounts to thirty-nine?

The solution is this: you halve the number of roots, which in the present instance yields five. This you multiply by itself; the product is twenty-five. Add this to thirty-nine; the sum is sixty-four. Now take the root of this, which is eight, and subtract from it half the number of the roots, which is five; the remainder is three. This is the root of the square which you sought for; the square itself is nine. [...]

The proceeding will be the same if the instance be, [...] what must the amount of a square, the half of which when added to the equivalent of five of its roots, is equal to twenty-eight?[2] (Rosen 1831, 5; Berggren 2007, 543–544).

Thus the idea of following a carefully controlled set of procedures in order to achieve consistent results is exemplified mathematically, along with the conditions under which those procedures are valid. Furthermore, al-Khwārizmī conceived of numbers themselves as an infinitely repeating pattern of identically constructed entities:

When I considered what people generally want in calculating, I found that it always is a number. I also observed that every number is composed of units, and that any number may be divided into units. Moreover, I found that every number, which may be expressed from one to ten, surpasses the preceding by one unit; afterwards the ten is doubled or tripled, just as before the units were; thus arise twenty, thirty, etc., until a hundred; then the hundred is doubled and tripled in the same manner as the units and the tens, up to a thousand; then the thousand can be thus repeated [...]; and so forth to the utmost limit of numeration.[3] (Rosen 1831, 3; Berggren 2007, 543)

Parallel to these developments in mathematical ideas were theological discussions that centered on the nature of God, and the relationships of man and the cosmos. These subjects were debated throughout the Islamic empire in intellectual circles at a time of passionate, vibrant, and sponsored explorations of ideas. A group of philosophers known as the *Ikhwān al-Ṣafā'* 'Brethren of Purity' wrote a series of fifty-two *Rasā'il* 'epistles' or 'letters' covering a vast array of subjects (Netton 1982). They categorized numbers in this theological context, focusing on the number one and the additive function, adding one to each number in succession. Their epistles probably contain significant information not yet brought to bear on the study of Islamic art. For instance, the following passage from the first *risāla* is reminiscent of al-Khwārizmī's infinitely repeating integers, giving them cosmological significance in that 'the whole scheme of creation and generation resembled the generation of numbers from one':

Know, O brother, that the first thing the Creator originated and invented from the light of his Unity was a simple essence, called 'The Active Intellect' just as he produced two

2. ayy māl idhā zidta ʿalayhi mithla ʿashara ajdhār balagha dhālika kulluhu tisʿa wa thalāthīn? Fa-qiyāsuhu an tanṣufu al-ajdhār wa hiya fī hādhihi al-masʾala khamsa fa-taḍrubuhā fī mithalihā fa-yakūnu khamsa wa ʿashrīn fa-tazīduhā ʿalā al-tisʿa wa al-thalāthīn fa-yakūnu arbaʿa wa sittīn fa-taʾkhudhu jadhrahu wa-huwa thamāniyya fa-tanṣufu minhu niṣf al-ajdhār wa huwa khamsa fa-yabqā thalātha wa huwa jadhr al-māl alladhī turīdu wa al-māl tisʿa.

3. Wa innī lammā naẓartu fī-mā yaḥtāju ilayhi al-nās min al-ḥisāb wajadtu jamīʿ dhālika ʿadadan wa wajadtu jamīʿ al-aʿdād innamā tarakkabat min al-wāḥid wa al-wāḥid dākhilun fī jamīʿ al-aʿdād. Wa wajadtu jamīʿ mā yulfaẓu bi-hi min al-aʿdād mā jāwaza al-wāḥid ilā al-ʿashara yakhruju makhraj al-wāḥid thumma tathannā al-ʿashara wa tathallathat kamā fuʿila bi-al-wāḥid fa-yakūnu minhā al-ʿishrūn wa al-thalāthūn ilā tamām al-miʾa thumma tathannā al-miʾa wa tathallathat ka-mā fuʿila bi al-wāḥid wa bi al-ʿashara ilā al-alf thummā kadhālika yuraddadu al-alf...ilā ghāyati al-mudrak min al-ʿadad.

from one by repetition. Then, he created the 'Universal Celestial Soul' from the light of the intellect, just as he created three by the addition of one to two. Then he created prime matter from the movement of the Soul, just as he created four by the addition of one to three.[4] (Netton 1982, 34)

At the same time, the decades between the eighth and eleventh centuries were an extraordinary period of formation and transformation in Islamic art and architecture (Grabar 1973; Rogers 1973; Grabar 1992; Necipoğlu 1995; Tabbaa 2001). From the tenth century on, Islamic art seems to express a new or renewed understanding of the plane, exploring many aspects of two-dimensional space. The visual effects of such explorations reverberated throughout the arts in the Middle East, Africa, and India in the following centuries. An argument may be advanced to suggest that these experiments eventually stimulated explorations in the Renaissance that led to the discovery of perspective.

Sometimes inscriptions on buildings and objects are indicative of the close relationship between architecture and the arts, and the contemporary intellectual tradition. We now know that this is the case for major monuments such as the Dome of the Rock in Jerusalem, Alhambra in Andalusia, and the Taj Mahal in Agra. My own work suggests that this may also be the case for the tomb towers at Kharraqan (Bier 2002). The Qur'anic passage 59:21 is inscribed on both towers:

Had we sent down this Qur'an on a mountain, verily thou would have seen it humble itself and cleave asunder for fear of God. Such are the similitudes which we propound to men that they reflect.[5] (trans Ali 1978, 1527–1528)

One is tempted to associate the *amthāl* 'similitudes' with the actual patterns depicted on the monuments, and thereby consider the patterns to call for our reflection both literally and metaphorically. The dramatic number of patterns present in the early tower at Kharraqan are pushed to an even higher number in the later tower, built less than a generation later, by means of dividing up the two-dimensional space enclosed by the arches on each face of the monument. Clearly, there is a degree of experimentation evident here, with the range of forms and patterns suggesting new understandings of two-dimensional space, its boundedness and extent. At Kharraqan, two-dimensional patterns in three-dimensional forms interact and combine to inform an aesthetic that depends upon repetition, patterns, and structures, resulting in standards of beauty that soon spread throughout the Islamic world.

---

4. I'lam, yā akhī, anna al-bārī jalla thanā'uhu awwal shay'in ikhtara'ahu wa abda'ahu min nūr waḥdāniyyatihi jawhar basīṭy uqālu lahu al-'aql al-fa'āl kamā ansha'a al-ithnayn min al-wāḥid bi al-tikrār. Thumma ansha'a al-nafs al-kulliyya al-falakiyya min nūr al-'aql kamā ansha'a al-thalātha bi-ziyādati al-wāḥid 'alā al-ithnayn. Thumma ansha'a al-hayūlā al-ūlā min ḥarakati al-nafs, kamā ansha'a al-arba'a bi-ziyādati al-wāḥid 'alā al-thalātha.

5. law anzalnā hādhā al-qur'ān 'alā jabalin la-ra'aytahu khāshi'an mutaṣaddi'an min khashyati allāh wa tilka al-amthālu naḍribuhā li al-nās la'allahum yatafakkarūna.

As we have seen, the timeframe of that broad proliferation of geometric patterns in Islamic art and monuments from Spain to India coincides almost precisely with significant advances in mathematical knowledge. This raises important questions pertaining to the transmission of mathematical knowledge in relation to the arts across expanded trade routes in the eighth to tenth centuries (Bier 2007b). Further, we may recognize that the same geometric patterns executed in cut brick at Kharraqan can be reproduced using other technologies to form patterns in other media, a recognition that will lead us once again to distinguish an ontology of pattern that is distinct from the medium in which it appears. That is to say that the category of brick, as the medium in this instance, is separate from the patterns carried by the bricks. Let us explore further how this may relate to the transmission of knowledge.

Technologies of transfer and the transmission of knowledge

By engaging in pattern play, we may begin to recognize aspects of intricacy and complexity that are, in essence, simple. They may result from the reiterative manipulation of a single module, with only one or two variables of form, color, orientation, or placement. The question remains as to how or whether artisans understood mathematically what they were doing when they engaged in pattern play. The transmission of mathematical knowledge has been treated as part of the textual tradition. In eighth and ninth-century Baghdad, there was an effort to gather all knowledge of the world and to translate it into Arabic, often with commentary and new understandings. Among the works translated was Plato's *Timaeus* (Bury 1929, 1–254), one of the late dialogues, in which Plato introduces through Timaeus the notion of two ideal triangles—a right isosceles triangle (45°-45°-90°) and half an equilateral triangle (30°-60°-90°). Timaeus proceeds to use these triangles to create three-dimensional forms that came to be called Platonic solids (tetrahedron, cube, octahedron, icosahedron; the dodecahedron formed otherwise using pentagonal faces). The same triangles, representative of $\sqrt{2}$ and $\sqrt{3}$, respectively, underlie the structure of many if not most two-dimensional patterns in Islamic art (El-Said 1993; El-Said and Parman 1988). The question of the possible relationship of Plato's ideal triangles to Islamic art and newly emergent mathematical ideas of the eighth and ninth centuries of our era should not be dismissed, although research has not yet provided clear evidence of direct influence (Bier 2006b).

Another technology of transfer for the transmission of knowledge is instruction. The tenth-century work of Abu l-Wafā' Buzjānī on *Kitāb fīmā yaḥtāju ilayhi al-ṣāni min al-a'māl al-handasiya* 'On the geometric constructions necessary for the artisan' tells that this was also the case in classical Islam (Özdural 1995; 1996;

2000). Until at least the seventeenth century, across the Islamic world, there were regular seminars where mathematicians and craftsmen gathered to share knowledge. That a delegation of Indian astronomers was present in Baghdad is also documented, and travel has been recognized as relevant both to instruction and to the publication of books to transmit knowledge.

All of these technologies of transfer were recognized in the fourteenth century by the Arab historian, essayist, and encyclopedist Ibn Khaldūn in his *Muqaddimah* 'Introduction (to history)' (Rosenthal 1967). Ibn Khaldūn addresses the question of how knowledge is transferred among individuals, across generations, and to different cultures. Of the means to generate and transfer of knowledge, he acknowledges books, observation, travel, and instruction. To these he adds the crafts, recognizing the value of experiential learning. He says that writing 'enables the innermost thoughts of the soul to reach those who are far and absent' and enables the 'intention (of one person) to be carried to distant places, and, thus, the needs (of that person) may be executed without his personally taking care of them. It enables people to become acquainted with science, learning, with the books of the ancients, and with the sciences and information written down by them'. Ibn Khaldūn links book learning to instruction: 'The transformation of writing in man from potentiality to actuality takes place through instruction' (Rosenthal 1967, 327; Bier 2007b).[6]

With the benefit of hindsight, and highlighting the contributions of the Greeks and Persians of pre-Islamic times, Ibn Khaldūn recounts several sources of learning and methods of instruction. Among the latter, he highlights the importance of hands-on exercises in addition to book learning, identifying crafts as a source for knowledge. He emphasizes that the individual human being cannot exist without the cooperation of others, for 'to make all the things he needs, a man by himself would require longer than the time he can keep alive without them. The ability to think... enables human beings to cooperate'. As for crafts, he says that 'The mind does not cease transforming all kinds of (crafts)... from potentiality into actuality through the gradual discovery of one thing after another, until they are perfect. This is achieved in the course of time and of generations' (Rosenthal 1967, 314).[7] While emphasizing that the crafts require teaching, he also notes that 'the crafts and their habit always lead to the acquisition of scientific norms, which result from the habit. Therefore, any experience provides intelligence' (Rosenthal 1967, 331).[8] He argues that the crafts result from man's natural ability to think

6. fa-hiya taṭlaʿu ʿalā mā fī al-ḍamāʾir wa-tata'addā bihā al-aghrāḍ ilā al-balad al-baʿīd fa-taqḍī al-ḥājāt wa-qad dufiʿat muʾnatu al-mubāshara la-hā wa yaṭlaʿu bihā ʿalā al-ʿulūm wa al-maʿārif wa ṣuḥaf al-awwalīn wa mā katabūhu min ʿulūmihim wa akhbārihim wa hiya sharīfa bi-hādhihi al-wujūh wa al-manāfiʿ wa khurūjiha fī al-insān min al-quwwa ilā al-fiʿl innamā yakūnu bi al-taʿlīm.

7. wa lā yazāl al-fikr aṣnāfahā...min al-quwwa ilā al-fiʿl bi al-istinbāṭ shayʾan fa-shayʾan ʿalā al-tadrīj ḥattā takmalu wa lā yaḥsulu dhālika dafʿatan wa innamā yaḥsulu fī azmān wa ajyāl.

8. wa al-ṣanāʾiʿ abadan yuḥsalu ʿanhā wa ʿan malakatihā qānūn ʿilmī mustafād min tilka al-malaka fa li-hādhā kānat al-ḥunka fī al-tajriba tufīdu ʿaqlan.

and to determine outcomes, the causes and effects of reality. If one considers, for example, what happens in the laying of bricks one by one in the construction of masonry, one may discover by doing, or by observation and analysis, the formulas by which a unitary process is at the same time systemic.

There is an additional, potentially significant, technology for the transmission of mathematical knowledge across time and space: the embodiment of mathematical knowledge in the woven patterns and physical structures of textiles (Bier 2007b). Both physical structures and patterning are visual rather than verbal technologies of transfer; both involve three-dimensional construction and surface appearance. When considering textiles as a means of transfer, most often cited are the transmission of styles and the transfer of cultural forms of expression through designs and motifs—what is *represented* visually, rather than what is *expressed* physically or through visual means. This significant aspect of their role in cultural transmission should not be underestimated. But at the same time, textiles may also have conveyed other information that is at once both tangible and abstract, concerning the relationships between number, shape, and the nature of space in the formal relationships of patterns that embody a unit and its reiteration according to particular principles of organization. This information, conveyed within the textile medium, is distinct from the physical form of the textile itself and ontologically different from the style or content of its decoration. Just as a book is a 'technology of transfer' of information, so may textiles have served to impart information different from their physical or visual qualities. Textiles, through their patterns, 'embody' mathematical knowledge just as bricks through their patterns embody mathematical knowledge; but bricks are not so easily transported across great distances, nor in a format that preserves patterns. As such, textiles—in the plural—may have played a significant role in the transmission of mathematical knowledge and other cultural values from the central lands of the Islamic world in all directions.

Conclusion

Islamic art, with its focus on abstract floral forms and geometry, is often considered to be nonrepresentational in contrast to the figural tradition more familiar in the West. The study of art history through the figural tradition has privileged the arts of western Europe and North America. With expanded appreciation of diversity and more global perspectives in the late twentieth century, the discipline of the history of art expanded to include consideration of nonwestern arts, often addressing the arts of India and China, which are nonetheless figural in their representational aspects. Less well studied in the western world, the arts of Islam have more recently attracted broad attention. Yet even within the study of Islamic art as a subject in the art history curriculum, more publications have treated the subjects of book illustration (often called miniatures), exploring aspects of style, attribution, the work of individual artists, and pictorial narrative, than the more abstract aspects of geometric pattern and floral ornament.

By the middle of the nineteenth century Islamic geometric pattern had entered European visual consciousness through the publication of several sumptuous volumes (Murphy 1815; Jones 1856). By the end of the century the graphic documentary works of Prisse d'Avennes (1877) and Jules Bourgoin (1879) had appeared, today available as Dover Reprints. Such designs, predominantly drawn from the Alhambra in Spain and Mamluk monuments in Cairo, were initially perceived as exotic, decorative, and opulent, non-figural and two-dimensional—in short, counter in nearly every way to the European artistic idiom. Even M C Escher, so profoundly influenced by tessellations at the Alhambra, sought to enliven his patterns with figural forms.

Reconsidering aspects of Islamic art according to a paradigm of algorithmic patterns, we may begin to understand the integral relationship between the history of mathematics and Islamic art within the context of contemporary discourse in philosophy, religion, and metaphysics. Further study of the role of algorithms in association with the processes of pattern making in Islamic art may advance our understanding of both the origins and development of algorithms and the significance of reiterations in the arts. Through analysis of the relationships of crafts and technologies, units and repeats, numbers and shapes, algorithms and geometry, we may approach recognition of two-dimensional patterns in three-dimensional forms. This recognition is the inverse of the acknowledged achievement of Renaissance artists and theoreticians whose experimentation and analysis resulted in the phenomenon of perspective, representing three-dimensional space in the two-dimensional picture plane. In this sense, the decorative and ornamental qualities of Islamic art are geometric forms of expression of great cultural significance worthy of further exploration.

Bibliography

Abas, Syed Jan, *Symmetries of Islamic geometrical patterns*, World Scientific, 1995.
Akkach, Samer, *Cosmology and architecture in pre-modern Islam: an architectural reading of mystical ideas*, State University of New York, 2005.
Ali, A Yusuf, *The holy Qur'an: text, translation and commentary*, The Islamic Center, 1978.
Allen, Terry, *Islamic art and the argument from academic geometry*, Solipsist Press, 2004.
Bakhtiar, Laleh, *Sufi: expressions of the mystic quest*, Thames & Hudson, 1976.
Bakhtiar, Laleh, and Ardalan, Nader, *The sense of unity: the Sufi tradition in Persian architecture*, University of Chicago Press, 1973.

Behrens-Abouseif, Doris, *Beauty in Arabic culture*, Markus Wiener Publishers, 1999.

Berggren, J Lennart, 'Mathematics in medieval Islam', in Victor J Katz (ed), *The mathematics of Egypt, Mesopotamia, China, India, and Islam: a sourcebook*, Princeton University Press, 2007, 515–675.

Bier, Carol, 'Elements of plane symmetry in Oriental carpets', *The Textile Museum Journal*, 31 (1992), 53–70.

Bier, Carol, 'Ornament and Islamic art: Oleg Grabar, *The mediation of ornament* (Princeton University Press, 1992) and Alois Riegl, *Problems of style: foundations for a history of ornament* (Princeton University Press, 1992)', *Middle East Studies Association Bulletin*, 28 (1994), 28–30.

Bier, Carol, *Symmetry and pattern: the art of Oriental carpets*, The Math Forum @ Drexel University and The Textile Museum, http://mathforum.org/geometry/rugs 1996–2007.

Bier, Carol, 'Mathematical truth and beauty: the case of Oriental carpets', *Institute for the Humanities Notes* (September), University of Michigan, 1998.

Bier, Carol, 'Geometry and the interpretation of meaning: two monuments in Iran', in Reza Sarhangi (ed), *Bridges: mathematical connections in art, music, and science*, Tarquin 2002, 67–78 Winfield, KS

Bier, Carol, 'Pattern power: textiles and the transmission of mathematical knowledge', in Carol Bier (ed), *Appropriation, acculturation, transformation: proceedings of the 9th biennial symposium of the Textile Society of America 2004*, Textile Society of America, 2005a, 144–153.

Bier, Carol, 'Symmetry and symmetry-breaking: an approach to understanding beauty', in Reza Sarhangi and Robert V Moody (eds), *Renaissance Banff-Bridges: mathematical connections in art, music, and science*, Tarquin, 2005b, 219–226.

Bier, Carol, 'Islamic art at Doris Duke's Shangri La: playing with form and pattern' (with David Masunaga), in Reza Sarhangi and Carlo Sequin (eds), *Bridges: mathematical connections in art, music, and science*, Tarquin 2005c, 251–258.

Bier, Carol, 'Islamic art: an exploration of pattern', in Reza Sarhangi and John Sharp (eds), *Bridges: mathematical connections in art, music, and science*, Tarquin, 2006a, 525–532.

Bier, Carol, 'Number, shape and the nature of space: an inquiry into the meaning of geometry in Islamic art', in James Boyd White (ed), *How to talk about religion*, Erasmus Institute publications, University of Notre Dame Press, 2006b, 246–277.

Bier, Carol, 'From textiles to algorithms: revising an Islamic aesthetic paradigm', paper presented at College Art Association, Annual Meetings, New York, February 2007a.

Bier, Carol, 'Patterns in time and space: technologies of transfer and the cultural transmission of mathematical knowledge across the Indian Ocean', *Ars Orientalis*, 34 (2007b), 174–196.

Bier, Carol, 'From folding and cutting to geometry and algorithms: integrating Islamic art into the mathematics curriculum', in Reza Sarhangi and J Barallo (eds), *Bridges Donostia: mathematics, music, art, architecture, culture*, Tarquin 2007c, 453–458.

Bixler, Harry, 'A group theoretic analysis of symmetry in two dimensional patterns from Islamic art', PhD thesis, New York University, 1980.

Blair, Sheila S, and Bloom, Jonathan M, *Cosmophilia: Islamic art from the David Collection, Copenhagen*, McMullen Museum of Art, Boston College, distributed by the University of Chicago Press, 2006.

Bonner, Jay, 'Three traditions of self-similarity in fourteenth and fifteenth century Islamic geometric ornament', in Javier Barrallo *et al* (eds), *Meeting Alhambra: ISAMA-Bridges Conference Proceedings*, University of Granada, 2003, 1–12.

Bourgoin, J, *Arabic geometrical pattern and design*, Dover, 1973 [first published 1879].

Brüderlein, Markus (ed), *Ornament and abstraction: the dialogue between non-Western, modern, and contemporary art*, Fondation Beyeler and DuMont, 2001.

Bury, R G, *Plato, IX: Timaeus, Critias, Cleitophon, Menexenus, Epistles*, Loeb Classical Library, 1929.

Chorbachi, W K, 'In the tower of Babel: beyond symmetry in Islamic design', *Computers and mathematics with applications*, 17 (1989), 751–789.

Critchlow, Keith, *Islamic patterns*, Thames & Hudson, 1976.

al-Faruqi, Ismail R, and Lamya al-Faruqi, Lois, *The cultural atlas of Islam*, Macmillan, 1986.

Field, Robert, *Geometric patterns from Islamic art & architecture*, Tarquin Publications, 2004.

Freedberg, David, *The power of images: studies in the history and theory of response*, University of Chicago Press, 1989.

Golombek, Lisa, 'The draped universe of Islam', in Priscilla P Soucet (ed), *Content and context of visual arts in the Islamic world*, Pennsylvania State University Press, 1984, 25–49.

Gombrich, E H, *The sense of order*, Cornell University Press, 1979.

Gonzalez, Valérie, *Beauty and Islam: aesthetics in Islamic art and architecture*, Tauris, 2001.

Grabar, Oleg, *The formation of Islamic art*, Yale University Press, 1973.

Grabar, Oleg, *The mediation of ornament*, Princeton University Press, 1992.

Grabar, Oleg, and Robinson, Cynthia (eds), *Islamic art and literature*, Markus Wiener Publishers, 2001.

Grünbaum, Branko, and Shepherd, G C, 'Interlace patterns in Islamic and Moorish art', in Michele Emmer (ed), *The visual mind: art and mathematics*, MIT Press, 1993, 147–155.

Grünbaum, Branko, and Shepherd, G C, *Tilings and patterns*, Freeman, 1987.

Grünbaum, B, Grünbaum, Z, and Shepherd, G C, 'Symmetry in Moorish and other ornaments' *Computers and Mathematics with Applications*, 12B (1986), 641–653.

Gutas, Dimitri, *Greek thought, Arabic culture: the Graeco-Arabic translation movement in Baghdad and early 'Abbasid society (2nd–4th/8th–10th centuries)*, Routledge, 1998.

Hillenbrand, Robert, *Islamic architecture: form, function, meaning*, Columbia University Press, 1994.

Jones, Owen, *The grammar of ornament*, London: Bernard Quaritch, 1868 [1856].

Leaman, Oliver, *Islamic aesthetics: an introduction*, University of Notre Dame Press, 2004.

Lee, A J, 'Islamic star patterns', *Muqarnas*, 4 (1987), 182–197.

Lu, Peter J, and Steinhardt, Paul J, 'Decagonal and quasi-crystalline tilings in medieval Islamic architecture', *Science*, 315 (February 22, 2007), 1106–1110.

Makovicky, Emil, '800-year old pentagonal tiling from Maragha, Iran, and the new varieties of aperiodic tiling it inspired', in I Hargittai (ed), *Fivefold symmetry*, World Scientific, 1992, 67–86.

Murphy, James Cavanah, *The Arabian antiquities of Spain*, London: Cadell & Davies, 1815.

Nasr, Seyyed Hossein, *Islamic art and spirituality*, State University of New York, 1987.

Necipoğlu, Gulru, *The Topkapi scroll: geometry and ornament in Islamic architecture*, Getty Trust Publications, 1995.

Netton, Ian Richard, *Muslim Neoplatonists: an introduction to the thought of the brethren of purity (Ikhwan at-Safa')*, George Allen & Unwin, 1982.

Olin, Margaret, 'Review of Oleg Grabar, *Mediation of ornament*,' *Art Bulletin*, 75/4 (1993), 728–731.

Özdural, Alpay, 'Omar Khayyam, mathematicians, and *conversazioni* with artisans', *Journal of the Society of Architectural Historians*, 54/1 (1995), 54–71.

Özdural, Alpay, 'On interlocking similar or corresponding figures and ornamental patterns of cubic equations', *Muqarnas*, 13 (1996), 191–211.

Özdural, Alpay, 'Mathematics and arts: connections between theory and practice in the medieval Islamic world', *Historia Mathematica*, 27 (2000), 171–201.

Prisse d'Avennes, *L'art arabe d'après les monuments du Kaire: depuis le VIIe siècle jusqu'à la fin du XVIIIe*, Paris: A Morel et Cie, 1877.

Rogers, J M, 'The 11th century—a turning point in the architecture of the Mashriq?', in Donald S Richards (ed), *Islamic civilization 950–1150. Papers on Islamic history III*, Cassirer, 1973, 211–249.

Rosen, Frederic (ed and trans), *The algebra of Mohammed ben Musa*, London: the Oriental Translation Fund, 1831.

Rosenthal, Franz (trans), *Ibn Khaldūn, the* muqaddimah, *an introduction to history,* (edited and abridged by N J Dawood), Princeton University Press, 2nd ed, 1967 [originally published 1958].

el-Said, Issam, *Islamic art and architecture: the system of geometric design* (ed Tarek El-Bouri), Garnet Publishing, 1993.

el-Said, Issam and Parman, Ayse, *Geometric concepts in Islamic art*, Scorpion, 1988.

Seher-Thoss, Hans C, and Seher-Thoss, Sonia P, *Design and color in Islamic architecture: Afghanistan, Iran, Turkey*, Smithsonian Institution, 1968.

Summers, David, *Real spaces: world art history and the rise of western modernism*, Phaidon, 2003.

Tabbaa, Yasser, *The transformation of Islamic art during the Sunni revival*, University of Washington Press, 2001.

Trilling, James, *The language of ornament*, Thames & Hudson, 2001.

Trilling, James, *Ornament: a modern perspective*, University of Washington Press, 2003.

Wood, Barry D, 'A great symphony of pure form: the 1931 International Exhibition of Persian Art and its influence', in Linda Komaroff (ed), *Exhibiting the Middle East: collections and perceptions of Islamic art*, Ars Orientalis, 30 (2000), 113–130.

The historiography and history of mathematics in the Third Reich

Reinhard Siegmund-Schultze

After Hitler became chancellor of the German Reich on 30 January 1933 and the Nazi regime consolidated its power in March 1933, serious effects for science and mathematics became visible.[1] Mathematics was, by and large, less important to the regime than biology, chemistry, the technical sciences, or history. It could not be used for propaganda like biology and historiography and it did not have immediate importance for armament and war production although it was applied 'indirectly' in many areas. One can even say that most mathematicians (though not all as we will see) were spared from difficult moral dilemmas, at least as far as their work was concerned. They had 'only' to cope with the general moral problems of ordinary German citizens under the Nazis, an increasing subliminal feeling of guilt, which often forced those citizens into solidarity with the regime. Nevertheless, there were specific political effects of Nazi rule and Nazi ideology on mathematics as a field of research, teaching, and application. These effects were so clearly palpable that the history of mathematics in the Third Reich can be divided (by analogy with other professions) into 'three periods of coordination

---

1. In this chapter the terms National Socialist German Workers Party, NSDAP, and Nazi party are interchangeable.

and legitimization', directly parallel and connected to political events (Mehrtens 1989a; Siegmund-Schultze 1993b).

(1) 1933–35 Political coordination, and expulsion of Jews from the state-related mathematical professions.

(2) 1936–39 Relative stability and internal success of the regime; economic, scientific, and cultural self-isolation; expulsion of Jews from non state-related mathematical occupations, journals, and societies (Remmert 1999; 2004).

(3) 1939–45 War research; changes in the profession (diploma); strategies for occupation and subordination of European mathematics; mathematics in concentrations camps (Ebert 1998, Epple and Remmert 2000; Mehrtens 1996; Siegmund-Schultze 1986).

There were, of course, political factors acting continuously during all three periods, namely defamation, emigration, and pressure for political conformity.

In spite of the book by Sanford Segal (2003), which introduces research up to 1995 to English readers, a monograph on the history of mathematics in the Third Reich, devoted to all aspects of the problem, is still a desideratum. There is not enough space in the present article to give a full picture of the various topics that have been dealt with in the recent historiography of mathematics in the Third Reich, let alone of those that are still unexplored. The table in the Appendix provides an overview of themes that have been treated, at least in part.

In this chapter I will give a short, general, and cursory review of recent developments in research on the history of mathematics during the Third Reich and of some major results, particularly with respect to the political, sociological, and institutional aspects of mathematics. In particular I will reflect on the potential and demonstrable influences of Nazi ideology on mathematics itself as theory and practice, based on discussion of conditions before 1933. Finally I will briefly outline one of the most important effects of the Third Reich for global mathematics: mass emigration of mathematicians from Germany.

Historical research on mathematics in the Third Reich

The historiography of mathematics in the Third Reich dates back only to the 1980s.[2] This late beginning was due to a multitude of self-restrictions and taboos

2. It was basically German work by Mehrtens and Richter (1980) that started the historiography of science and mathematics under National Socialism. Since Mehrtens focused on mathematics, so in a sense the

in the historiography of science and mathematics in Germany. It suffices to mention widespread consideration for still active and influential scholars from the period under scrutiny (1933–45) and for their immediate students, a restraint which only began to weaken as a consequence of the student movement of the late 1960s.

At the same time, general historical research on the Third Reich was still in the midst of serious methodological controversies about the role of elites in the Third Reich, the specific and particular interests and influences of the state bureaucracy, military, and industry, and the role of university professors and students. Again there were political taboos at work in both West and East German research. In both countries there was a lack of critical will to investigate the part their ideological 'ancestors' (the bourgeois parties of the Weimar Republic on the one hand, and the Communists on the other) played in allowing Hitler to seize power in 1933. In the East, research was hindered by the undeniable deficit of democracy in both the Third Reich and the German Democratic Republic, with their parallel doctrines of the 'leading role of the party'. Besides, (self-proclaimed) 'Marxist' historiography was less interested in the 'grey zone' of the political behaviour of scientists, who could not usually be easily related to either 'proletarian resistance' or to the 'interests of finance capital'.[3] In the West the failure of 'de-Nazification' and the continued reliance on the elites of the Third Reich (stronger than in the East) was a source of embarrassment at home and abroad. Many university professors with Nazi connections, among them no few who had moved westwards in the last months of the war, were reinstalled and uncritically and unconditionally used in rebuilding society in the years of the *Wirtschaftswunder* 'economic miracle'.[4]

Research into the history of mathematics during the Third Reich can be roughly grouped under two headings: (I) historical and sociological, and (II) mathematics-related. However, the transitions are fluid and to some extent both areas have to be addressed at the same time to obtain a full picture. Research of type (I) is concerned with general conditions for intellectual activity and education in the Third Reich, including policies directed towards and responsible for public opinion (journalism), arts, and, in particular, science, technology, and mathematics. Research of type (II) asks for the distinctive features of mathematics that created specific conditions for mathematics compared to other sciences, allowing

historiography of mathematics had a pioneering role for the history of science and mathematics during the Third Reich.

3. For the same reason, historiographical explanation of the Nazi 'dogma of anti-Semitism' created insurmountable difficulties for East German scholars; see Siegmund-Schultze (1999).

4. Former Nazi scientists were employed for pragmatic reasons in the East too, but on a much smaller scale than in the West, not least due to lack of availability. The use of the number theorist Helmut Hasse by East German politicians is described by Siegmund-Schultze (1999, 68–70), although Hasse was not a 'culprit' in any extreme sense; see also Segal (1980).

its potential either to flourish or to be curtailed. Within this, one can specify different functions of 'mathematics', beginning principally with a tripartite division into research, teaching, and applications. To type (II) belongs everything concerning the methods and cognitive structure of mathematics, for instance, its symbolic language, its widely recognized rigour and generality, its special modes of communication, and its particular relationships to other sciences and engineering. The most visible but maybe not the most momentous field of contact and conflict between mathematics and the Third Reich on level (II) were undoubtedly the various racist theories of mathematics such as the *Deutsche Mathematik* 'German mathematics' of Ludwig Bieberbach (see below).

Obviously, problems of type (II) cannot be approached without knowledge of problems of type (I). One can go a step further and say that one must be aware of the distinctive features of the Third Reich compared to political systems before 1933 (the Weimar Republic) and after 1945 (the two German republics and Austria), and how the population at large, and Jews in particular, fared under these conditions. This latter, more general problem goes beyond the historiography of mathematics and has to be taken for granted, even though there is still controversial research in this area too.

Much of the research from the last twenty-five years on the history of mathematics during the Third Reich belongs at least partly to type (I), making it historiography of science in a 'social' or sociological sense. There were various reasons for this initial focus, some of them circumstantial (as for the delay in a general history of science in the Third Reich) and some more substantial. Among the latter is the fact that the most visible effects of National Socialist rule on mathematics, namely expulsion and emigration, and the accompanying racist theories of *Deutsche Mathematik*, had more to do with National Socialist ideology, particularly anti-Semitism, than with the ideology and research values of mathematics itself. Thus one has to recognize the particularly heavy influence of 'abnormal' social conditions in German science and mathematics between 1933 and 1945. But regardless of whether one is occupied with historiography of type (I) or (II), the aim of the historian of mathematics is to compare how mathematics is done generally and beyond the Third Reich, using the latter as an experimental testing ground, albeit a somewhat extreme, even abnormal one.

The basic consequences of National Socialist rule for official educational policies can be summarized by the following four doctrines, which of course had immediate influence also on the sciences and mathematics, not least through indoctrination in the school and university systems (Nyssen 1969):

(1) *Anti-Semitism*: a non-negotiable political taboo (in the sense that no discussion was allowed), and an overall aim of education. It implied the superiority of an 'Aryan' race and their right to rule.

(2) *Elitism*: the superiority of certain *Volksgenossen* 'people's comrades', which led both to the preservations of the traditional, hierarchical, four-level education system (elementary school, middle school, Gymnasium/Realschule, university), and to the foundation of *Napola* 'National Political Educational Institutions'.

(3) *Führer-Gefolgschaft* 'ideology of leader-follower': indisputable obedience to authority; subordination of the teacher to the school director, or the professor to the university rector; the influence of political organizations at schools and universities (*Hitlerjugend* 'Hitler youth', National Socialist student organizations, and so on); rejection of democracy.

(4) *Volksgemeinschaft* 'ideology of people's community': social-demagogic rejection of class differences, based on race theory; subordination of individual to community.

One of the more general conclusions of research to date has been that uncritical talk about an alleged 'misuse of science and mathematics' by regimes such as the Nazi dictatorship misses the point, given the eagerness of scientists to take advantage of certain extraordinary conditions given to them only under Nazi rule (see, for instance, Mehrtens 1996). Other work has studied the extent to which the after-effects of Germany's defeat in World War I, and long held anti-Semitic prejudices, shaped the ideology of mathematicians during the Nazi years and brought them into conformity with the regime.

There has been much discussion on how 'resistance' by scientists to political regimes like the Third Reich can be defined and where the line has to be drawn between the moral obligation to resist and the individual human right to conformity. In this respect existing research has justly been much more critical of postwar apologia by mathematicians than of their actual conformity with the regime. In particular the myth cultivated by some mathematicians after the war, that resistance to racist theories such as *Deutsche Mathematik* already constituted resistance to the regime, has been rejected by most historians.

One particular and repeated question in this context has been: to what extent can membership in the Nazi party be considered as a criterion of guilt. For mathematics, this question will be illustrated by a short study, based on the results of biographical work.

Research on the political behaviour of mathematicians has confirmed what has also been shown for other social groups in the Third Reich: membership of the Nazi party *cannot* be considered a decisive criterion of personal guilt. This is somewhat contrary to much of the older literature, and to the practice of de-Nazification after the war which stipulated the dismissal of party members. Moreover, it has become very clear that two assets, or social markers, (i) holding an established position in the university system (for example, a professorship) and (ii) acknowledged (mathematical) expertise were, particularly after the first tumultuous years, generally sufficient to avoid a third social marker (iii) political activity in its most visible forms, such as membership of the Nazi party. The possession of one of these two assets alone, however, was usually not strong enough to exempt an individual from the need to join political organizations such as the Nazi Party, at least formally. Biographical research on mathematicians such as Ludwig Bieberbach, Gustav Doetsch, Helmut Grunsky, Heinrich Heesch, Ernst Mohr, Oskar Perron, Erhard Schmidt, Kurt Schröder, Wilhelm Süss, and Oswald Teichmüller has revealed different patterns of behaviour (politically active, politically passive, and anti-Nazi, the latter often necessarily hidden), which, together with the main social markers (expertise, academic status, and Party membership) are shown in Table 1 (see Bigalke 1988; Mehrtens 1987; Litten 1996; Remmert 1999; Schappacher and Scholz 1992; Siegmund-Schultze 1989; 2004).

One recognizes a variety of combinations, in particular the fact that these mathematicians had to meet at least two of the three first and most visible social criteria in order to secure a career in Nazi Germany. In this respect, of course, older professors, who had been established before the regime, were in a much better position. By contrast, Grunsky and Schröder typify competent younger mathematicians (of whom there were many) who had to join the Nazi party against their own conviction in order to secure even a modest career outside the university system. Mohr received only an associate professorship. The equally young Heesch of four-colour-conjecture fame was too 'aloof': he met only one of the three main criteria and did not flourish. Mohr, Doetsch, and Süss appear with contradictory political attributes in the table. This is either because they violated

Table 9.4.1 Mathematicians' engagement for or against the Nazis in relation to their academic positions

Name	Expertise	Professorship	Member of the Nazi Party	Active Nazi	Passive Nazi	Anti-Nazi
Bieberbach	X	X	X	X		
Doetsch	X	X		X	X	
Grunsky	X		X		X	
Heesch	X				X	
Mohr	X	X			X	X
Perron	X	X				X
Schmidt	X	X			X	
Schröder	X		X		X	
Süss		X	X	X	X	
Teichmüller	X		X	X		

taboos (Mohr was accused of enemy propaganda during the war and was barely saved from execution), or because they felt they had to over-compensate for their political positions before 1933 (Doetsch had been a pacifist), or because it is still too difficult to label them unambiguously (as is the case with Süss). Teichmüller was the one brilliant young mathematician in the table who was also an ardent Nazi. This rather exceptional combination cost him his life when he volunteered for the Russian front during the war, a fate from which Bieberbach, an older mathematician of almost the same 'combination', was saved. Süss, who was president of the *Deutsche Mathematiker-Vereinigung* 'German Mathematicians' Association', or DMV, from 1937 until 1945, had to expend extra political effort to make up for his lack of research prowess.

A comparison with East Germany after the war sharpens the eyes, not only in the discussion of party membership, though with all due caution as to the incomparably more criminal character of the Nazi regime than the Communist. Schröder became rector of Humboldt University in East Berlin without being forced to join the Communist party: in the 1950s and 1960s he was in the position of a professor with expertise, as Schmidt had been twenty years before him.

Mathematics before Nazi rule

A general problem of historiography still under discussion is the following: to what extent were scientific careers discouraged by the general irrational ideological atmosphere in Nazi Germany? This leads into the complex of more specific problems of type (II) which I will now address.

Generally, in order to recognize the potential *within* mathematics for the discipline to be influenced by the Third Reich in a way *specific* to it, one has to go back into the history of German and international mathematics before 1933. But even taking this broader view we will see that the 'loose ends' in this very complicated picture of assumed or psychologically probable 'threads of influence' of Nazi ideology on mathematics have to be tied up, once again, by more general considerations of institutional and political developments. That is, only if one considers the political system of the Third Reich, including the policies of mathematicians themselves, can one understand which historically existing *potentials* for oppression or stimulation of particular mathematical disciplines were most likely to become *reality*. This, of course, refers back to problems of type (I), that is, problems on a more general social level than mathematics. It also means that some discussion of the mathematics-specific effects of Nazi ideology and Nazi rule has to be episodic and somewhat speculative, because many of these effects came about by intimidation or stimulation of individuals, and were moderated or not by personal ability or inability to cope with political conditions under the regime.

During the entire history of mathematics, the development of its ideas has depended on social and political conditions, and on structures of communication, which were usually based on dominating centres and dependent peripheries. The emergence of nation states (most notably after the French revolution), for example, had clear ramifications for the mathematical topics discussed in different places in Europe. The rise of Göttingen as the dominant mathematical centre in Germany around 1900 had thematic consequences both within the country and globally. It consolidated German mathematics as the most versatile and modern mathematical culture in the world, including applications and new developments in teaching under the guidance of Felix Klein. But even then German mathematics did not and could not cover all promising developments in an ever expanding discipline. It was, for instance, lacking parts of the French theories of real functions (developed by Baire, Borel, and Lebesgue), and of differential equations and celestial mechanics (Poincaré), of Italian algebraic geometry, of British and Scandinavian mathematical statistics, or of American work in the logical foundations of geometry. One of the American workers in geometry, Oswald Veblen, had the following to say about German mathematics when he visited Göttingen in 1913:

I am beginning to have definite impressions of Germany. Mathematically, even more than politically, it is a monarchy. The mathematical situation is well illustrated by a remark of Landau's when I asked him whether there was any interest in Abelian functions and the like (more than one variable): No one in Germany is interested in anything Hilbert has not worked with. They are only mildly interested in what is going on elsewhere, unless it touches pretty directly on their own work.[5]

It was fortunate for German mathematics around 1913 that the 'monarch', David Hilbert of Göttingen, one of the last mathematical universalists, was so broad in his research (see Gray 2000) and that the communication network developed by Felix Klein was so functional, with many foreign students flocking to Göttingen. But in the years after World War I (see Parshall, Chapter 1.4 in this volume), and with the rise of several national mathematical cultures, in the US, Soviet Russia, Poland, Austria, the Netherlands, and some Scandinavian countries, German mathematics was increasingly unable to cover the whole of mathematics. One may think, for example, of Polish functional analysis and point set topology, and of American, Dutch, and Austrian topology. The 'envoy' of Rockefeller philanthropy, Augustus Trowbridge, who came to Göttingen in 1926 with the leading American mathematician George David Birkhoff, had the following to say in his report:

Birkhoff thinks that the mathematics group to be [sic] a little too self-satisfied and needs to be shocked out of that state—at least all that A.T. overheard of the mere technical

---

5. Veblen to George David Birkhoff, December 25, 1913, from the Birkhoff Papers (Harvard University Archives, HUG 4213.2, box 7, handwritten, 4 pages). Thanks go to June Barrow-Green who kindly shared this document with me.

points B. was making pointed to B.'s evident desire to make the group feel that they needed to strengthen their position in mathematics by means of men rather than by means of equipment. (Siegmund-Schultze 2001, 153)

Even though German mathematical culture, with Göttingen in the lead, probably remained the most internationalized in the world until 1933, the dominance of Göttingen did not remain unchallenged. It met with envy and resistance even within Germany by mathematicians such as Bieberbach and Richard von Mises in Berlin. The arguments of these two, and of others such as Schmidt (also in Berlin), were partly political in the sense of promoting German nationalism or of general distrust of international mathematical relations. These sentiments were stirred by Germany's defeat in World War I and by the policies of the victorious nations, for instance, the boycott against German science in the years immediately after the War, which led to exclusion from mathematical congresses and actions against the dominance of German mathematical literature (Mehrtens 1987; Siegmund-Schultze 2001).

But there was also an early racist, anti-Semitic, dimension to the discussion, although it was certainly not promoted by mathematicians of Jewish descent like von Mises, even though he partly identified with the German/Austrian cause. Anti-Semitism had its early proponents in mathematicians such as Theodor Vahlen of Greifswald, who later became an influential politician in the Third Reich (see Siegmund-Schultze 1984), and in Walther von Dyck from Munich. When in 1925 the editors of the internationally leading German journal *Mathematische Annalen* controversially discussed the collaboration of French authors in an issue devoted to Bernhard Riemann's hundredth birthday, one member of the board, von Dyck, wrote to another, Bieberbach, to query the opinions of a third, Einstein:[6]

I would like to know Einstein's opinion about it. To be sure, he has declared, on the occasion of his memorable stay in Paris, that he is no German! By the way, one rather believes one is in Hungary, Poland, Russia and Bulgaria if one reads the names of those who now publish in the *Annalen*!

Already in 1923 Vahlen published a talk entitled 'Wert und Wesen der Mathematik', 'The value and nature of Mathematics', in which he called mathematics a 'Spiegel der Rassen', 'mirror of the races' (Vahlen 1923, 22). He thus anticipated theses that Bieberbach was to defend after 1933 (see below). Between 1924 and 1927 Vahlen was *Gauleiter* 'leader' of the Nazi party in the district of Pomerania. His nationalist colleagues, such as von Dyck, tried in vain to prevent his dismissal by the Prussian

6. 'Es wäre mir lieb, ganz bestimmtes zu erfahren, wie sich Einstein geäußert hat. Freilich hat er ja schon bei jenem denkwürdigen Aufenthalt in Paris erklärt, daß er kein Deutscher sei! Übrigens glaubt man auch eher in Ungarn, Polen, Rußland u. Bulgarien zu sein, wenn man die Namen derer liest, die jetzt in den Annalen publizieren!', Munich, 19 January 1925, from the Estate of Bieberbach Oberaudorf, handwritten, copy with Menso Folkerts, Munich. All translations from German to English are by the author.

government in 1927 for his nationalist and anti-Republican activities. Max Dehn's scathing and somewhat one-sided criticism (Dehn 1905) of Vahlen's book *Abstrakte Geometrie* 'Abstract geometry' (1905), apparently increased Vahlen's anti-Semitic resentments and contributed to his move to applied mathematics in the following years.[7] In retrospect, Vahlen explained his personal shock in a speech to the Prussian Academy of Sciences in 1938 in the following manner:[8]

In the university analytical topics prevailed. My destiny for the intuitive parts of mathematics only became evident after my studies. My *Abstract geometry* marks the change. After its completion I was attracted by the natural, concrete thinking of our race. (Vahlen 1938; 98)

Vahlen's talk about the 'concrete thinking of our race' was, of course, pure cliché and it was neither confirmed nor justified by the mathematical predilections of German-Jewish mathematicians of his time, among whom were many applied mathematicians such as Richard Courant, von Mises, and Felix Bernstein.

The name of Bernstein, director of the Göttingen Institute for Mathematical Statistics, conjures up another dimension to the ideological atmosphere of the 1920s which has to be taken into account in order to fully understand the conditions under which Nazi ideology was able to act after 1933: namely, eugenic thinking supported by mathematics. Bernstein had become famous in 1924 for discovering the mechanism of heredity in blood groups. In the following years, including the time of his enforced emigration to the United States after 1933, he did much statistical work on racial genetic markers. This was respectable research, stimulated by internationally shared concerns about the genetic degradation of the human race (Schappacher 2006b). As late as 1956 Bernstein argued in a letter to the *New York Times* (24 June 1956, page E8) that there were three types of mathematical aptitude: geometrical, algebraic-combinational, and purely logical, and that these were not uniformly distributed geographically.[9]

Taking into consideration developments and attitudes in German mathematics during the 1920s, which could only briefly be described here, one is somewhat better prepared to understand the events of the year 1933 and the backlash they entailed for mathematics under Nazi rule.

7. In the spring of 1935 Vahlen, now an official in the Nazi ministry of education, contributed to Dehn's early dismissal from his professorship in Frankfurt, although Dehn was then still officially exempted from these measures.

8. 'Auf der Universität überwogen die analytischen Fächer, meine Bestimmung für das Anschauungsmäßige trat erst nach beendetem Studim hervor. Meine 'Abstracte Geometrie' bezeichnet den Wandel, nach ihrer Vollendung zog es mich zu der natürlichen, konkreten Denkart unserer Rasse.'

9. Bernstein did not claim that one type of aptitude was superior to the others. However, he went as far as saying that a student's particular types of aptitude should be taken into account in teaching mathematics, though, unlike Bieberbach (see below), without implications for the employment of teachers. Even von Mises' ironic reaction to Bieberbach's racist *Deutsche Mathematik* in 1934 did not question the existence of different mathematical talents in different human races (Siegmund-Schultze and Zabell 2007).

Mathematics under Nazi rule

The theoretically best informed approach to the history of mathematics in Nazi Germany (and, incidentally, to the history of science as a whole in that period) is that of Mehrtens (1990a), who embeds the discussion of mathematics into larger political and philosophical controversies around the 'production of mathematics as a language' between the late nineteenth century and 1945. On the social level, partly inspired by Bourdieu, he considers mathematics as a system that exchanges the products of knowledge, plus political loyalty, for material resources and social legitimacy.

As to the influences of Nazi rule and Nazi ideology specifically on mathematics, that is, for questions of type (II), one can expect from the outset two main lines of influence: (a) intrusion, interference, and possibly stimulation on the one hand; and (b) neglect, restriction, and isolation on the other. Point (b), particularly international isolation due to the repercussions of general Nazi policies was, once again, a feature which mathematics partly shared with other sciences in the Third Reich. One can nevertheless look for the specific effects of that isolation in different subdisciplines, although little research has been done so far in this direction.

After the war, in 1947, the German topologist Kurt Reidemeister, who had been penalized by the Nazis in 1933 for his lack of compliance with the official ideology, wrote:

The general consequence of the infiltration of Nazi ideology was that one gradually abandoned the truth. The people in the National Socialist state in no way believed the officially sanctioned tenets taught in so many indoctrination courses and in so many papers. But they were willing to use the official vocabulary thoughtlessly.[10] (Reidemeister 1947, 35–36)

Reidemeister referred to the general influence of Nazi ideology, of course, not just in mathematics.

It is impossible to judge how many pupils and students were discouraged from taking on the 'rationalistic' and 'aloof' subject of mathematics, which had little support in the official political climate. On the other hand, even in the most blatant and absurd utterances of Nazi ideology with respect to mathematics there was, if not a rational core, then at least some faint reflection of real developmental problems within German mathematics. When the *Deutscher Verein zur Förderung des mathematischen und naturwissenschaftlichen Unterrichts*

10. 'Die allgemeine Folge der Weltanschauungspflege war, daß man sich allmählich der Wahrheit entwöhnte. Die Menschen im nationalsozialistischen Staat waren weit davon entfernt, die staatlich sanktionierte Lehre, die auf so vielen Schulungskursen gelehrt, in so vielen Schriften vorgetragen wurde, für bare Münze zu nehmen. Aber sie fanden sich bereit, sie einzukassieren und in das Kleingeld des eigenen Geredes umzuwechseln.'

'German Association for the Advancement of Mathematical and Natural Science Instruction', usually called *Förderverein*, cheerfully associated itself with the Third Reich in April 1933, it not only accepted the 'Aryan paragraph' of the Nazi 'Law on the Restoration of the Civil Service' (7 April 1933) in its bylaws. The *Förderverein* also stressed in its resolutions the *Unerbittlichkeit mathematischen Denkens* 'mercilessness of mathematical thinking', and connected it to the tasks of the military and of *Befreiung des Volkskörpers von erblicher Minderheit* 'freeing people's bodies from hereditary inferiority' (Lorey 1938, 105–106). The idea of the 'mercilessness of mathematical thinking' had been used in a somewhat less radical formulation (for instance as 'absolute rigour') in science and school policies prior to the Third Reich.

When, after 1933, the internationally renowned function theorist Bieberbach rather unexpectedly presented his infamous, racist, and partly intuitionist[11] 'Deutsche Mathematik' (Bieberbach 1934), he connected his points to discussions before 1933 on the foundations of mathematics. Indeed, his talk of an *Entmenschlichung der Mathematik* 'de-humanization of mathematics' as a result of modern axiomatics (Bieberbach 1934, 240) partly reflected concerns about an increasingly abstract mathematics, allegedly losing contact with classical problems of the nineteenth century and with applications. Some of these concerns were shared by other mathematicians before 1933, even by the applied mathematician Richard Courant of Göttingen, who by Nazi definition was Jewish. Under Nazi rule, Bieberbach now distorted the discussion towards an anti-Semitic context of 'mathematical styles'. He claimed 'intuition' and 'organic and fruitful' thinking for German and 'Nordic' mathematicians, while the non-German type of mathematician allegedly 'beamed his autistic thought into reality' (Mehrtens 1987, 230). Bieberbach never contested the correctness of foreign or Jewish mathematics, and seldom called into question the importance of their problems or solutions, but declared the mode of discovering those results 'un-German'.

It was a Jewish emigrant from Germany, the statistician Emil Julius Gumbel, who pointed out in 1938, that mathematics, unlike physics, did not exhibit a real antagonism of subdisciplines or methods, the existence of which had been a starting point for the idea of 'Deutsche Physik' promoted by the conservative experimental physicists and Nobel prizewinners Johannes Stark and Philipp Lenard (Gumbel 1938, 254). Though equally racist and demagogic, 'Deutsche Physik', was in this sense slightly more rational than Bieberbach's 'Deutsche Mathematik'.

In fact, few among Bieberbach's colleagues, usually the more careerist ones, were prepared to follow him. It has been shown that roughly the same 'arguments', accusing foreigners (French or Jewish in Bieberbach's mind) of lack of 'intuition'

---

11. That is, on the intuitionist side of the quarrel over the foundations of mathematics, following the ideas of the Dutch mathematician Luitzen Egbertus Jan Brouwer.

and of lacking 'relation to reality', had been used by conservative French philosophers like Pierre Duhem before and during World War I against German mathematics.[12] In fact, almost every propagandist of racist theories in mathematics had a different conception of what 'un-German' or *artfremd* 'alien' mathematics was, and the understanding was never independent of his own research interests. In this way representatives of applied mathematics tried to prove the futility of pure mathematics (Tietjen 1936), while others wanted to play 'German' geometry off against 'Jewish' number theory and analysis. Bieberbach himself, being a representative of pure mathematics, saw even doing mathematics for its own sake, as in some parts of number theory, a manifestation of German nature. At times he was concerned that his own racist theories could be misunderstood and misused for purposes that were too utilitarian. He remarked in 1934:

To prove the importance of mathematics for the people one refers quite often to the applications which figured prominently in Klein's reforms. It seems to me ... that mathematics is an emanation of our racial qualities too and anything which reveals our national character in a forceful manner requires no additional justification.[13] (Bieberbach 1934, 243)

Several mathematicians ridiculed Bieberbach's theories, at least in private. Some, like the Dane Harald Bohr, and the Germans Hermann Weyl, von Mises, and Oskar Perron also did so in public, though all but Perron from abroad (see Mehrtens 1989a for Bohr, and Siegmund-Schultze and Zabell 2007 for von Mises). They pointed out that Bieberbach's typology (based on theories of the Nazi psychologist Erich Jaensch) rarely connected individual mathematicians to certain styles of doing mathematics. In particular, many leading German mathematicians of the past (also living ones such as Hilbert) could easily be classified as 'foreign' or 'un-German' if Bieberbach's theories had been taken seriously. It comes as no surprise that some students of the great Jewish mathematician Emmy Noether in Göttingen tried in 1933 (if in vain) to prevent her dismissal, referring to the 'Aryan character' of her abstract algebraic works which were, as is well-known, much inspired by Richard Dedekind (Schappacher 1987, 351). Hermann Weyl, who like Noether had fled to the US, wrote in his obituary of her:

Her affinity with Dedekind, who was perhaps the most typical Lower Saxon among German mathematicians, proves by a glaring example how illusory it is to associate in a schematic way race with the style of mathematical thought. (Weyl 1935, 218)

---

12. Mehrtens (1990b, 116) argues, based partly on work by Andreas Kleinert, that much of Bieberbach's 'Deutsche Mathematik' was 'copied' from publications of Duhem, but this time (twenty years later) with a reversed direction of accusation. However, Duhem connected 'intuition' to French 'common sense' while Bieberbach, as a representative of geometric function theory, connected German 'intuition' to geometry.

13. 'Als Beleg für die Volksnotwendigkeit der Mathematik führt man namentlich gerne die Beziehung auf die Anwendungen an, die ja auch in den Kleinschen Reformplänen eine hervorragende Rolle spielt. Mir scheint aber...auch die Mathematik ein Betätigungsfeld völkischer Eigenart zu sein, und alles, worin sich unser Volkstum kraftvoll offenbart, scheint mir keiner ausführlichen Rechtfertigung mehr zu bedürfen.'

There was, it is true, some criticism, not only within Germany but also abroad, of the extreme way (maybe one can call it 'style') of presenting mathematics largely without motivation and in a purely logical order, a method for which the Jewish analyst and number theorist Edmund Landau was known.[14] But when Bieberbach took Landau as an example of the 'un-German' style of doing mathematics, the political aspect was immediately apparent. In one of the politically most explicit passages of his pamphlets on mathematical styles, Bieberbach openly supported a National Socialist student boycott against Landau with the following words:

A few months ago differences with the Göttingen student body put an end to the teaching activities of Herr Landau.... This should be seen as a prime example of the fact that representatives of overly different races do not mix as students and teachers.... The instinct of the Göttingen students was that Landau was a type who handled things in an un-German manner.[15] (Bieberbach 1934, 236)

Thus the main political function of Bieberbach's 'Deutsche Mathematik' was support for expulsions by the Nazis. It provided 'arguments' and pretexts for other mathematicians, often of lesser calibre than Bieberbach himself, to make careers for themselves.

Bieberbach, who remained influential in Berlin as Dean of the University's Philosophical Faculty, exerted further influence on mathematics based on conservative mathematical positions backed up by Nazi ideology. Beside promoting the careers of less important but politically 'reliable' mathematicians (Siegmund-Schultze 1989), and barring careers for promising ones (see the cases of Schröder and Grunsky referred to above), he had a conservative influence on the German system of mathematical reviewing. The old-fashioned *Jahrbuch über die Fortschritte der Mathematik* was kept alive until the end of World War II, despite the creation of the more modern *Zentralblatt für Mathematik* in 1931. The latter was more market-oriented and internationalist, and, above all, much more rapid. But it was also less systematic in publishing reviews than the *Jahrbuch*. For Bieberbach and other conservative German mathematicians there seems to have existed a subliminal connection between the systematic collecting function of the *Jahrbuch* and the foundational, rigour-providing function of axiomatics. Bieberbach approved of this last function of axiomatics, but he was suspicious of its creative, expansive function, as well as of uncontrolled and unsystematic mathematical reviewing (Siegmund-Schultze 1993a, 91–97). Here again there was some real problem in mathematics, in this case

14. Weyl belonged to those critics: Landau's style was very different from Noether's even though both were strongly focused on rigour and abstraction; they exhibited different strands of mathematical modernity.

15. 'Vor einigen Monaten haben Differenzen mit der Studentenschaft dem Lehrbetrieb des Herrn Landau ein Ende bereitet.... Man hat darin...ein Musterbeispiel dafür zu sehen, daß Vertreter allzu verschiedener menschlicher Rassen nicht als Lehrer und Schüler zusammenpassen.... Der Instinkt der Göttinger Studenten fühlte in Landau einen Typus undeutscher Art, die Dinge anzupacken.'

both in its reviewing system and in exaggerated use of axiomatics, to which Nazi ideology could demagogically refer.

Bieberbach's 'Deutsche Mathematik' and similar ad hoc racist theories in mathematics were gradually disregarded once they had fulfilled their function of providing a rationale for expulsions, and when pragmatic mathematical work was actually needed in preparation for the war. Ironically, the flight of many promising young scientists from the politically oppressive universities and technical colleges was more often than not to the benefit of institutions for armaments research, which some scientists perceived as oases of freedom of opinion.

It was indicative of Bieberbach's waning influence that in 1939 Perron, mathematics professor in Munich, could publish within Germany an ironic allusion to Bieberbach's 'Deutsche Mathematik'. He did so in the second edition (1939) of his book *Irrationalzahlen* 'Irrational numbers' (1920), when he alluded in the preface to Bieberbach's preference for the Cantor–Méray theory of real numbers. Georg Cantor, the German founder of set theory, was generally considered to be Jewish, while Charles Robert Méray, was obviously French and thus also 'un-German'. But in 1924, before his sympathies for the Nazis had developed, Bieberbach had criticized the first edition of Perron's book for basing its definition of real numbers on Dedekind's 'cuts'. Now, in 1939, Perron alluded to that review and 'defended' his own preference for the undoubtedly 'German' Dedekind, who, as mentioned, happened to be the true spiritual mentor to the 'Jewish' algebraist Emmy Noether, with the following words:

I believe a German who has the choice between a German product and an equally beautiful and valuable product of foreign origin should be allowed to follow his heart and to prefer the German one because it is German.[16] (Perron 1939, vi)

Judging the impact of 'neglect' and 'isolation' on a scientific discipline meets almost insurmountable methodological questions of a counterfactual, hypothetical nature. There is no doubt, however, that mathematics in the Third Reich was indeed neglected and isolated on several levels, both domestic and international. The international nature of mathematics offered some support to resistance against the Nazi regime, as the affairs around the political coordination of the DMV in 1934 showed, when consideration of foreign members led to a certain moderation of the Nazi policies spearheaded by Bieberbach (Mehrtens 1989a). But similar consideration was not given to matters that concerned applied or school mathematics, as the much 'smoother' political coordination (with respect to Nazi interests) of the respective organizations showed (Mehrtens 1989a).

---

16. '...glaube ich, daß es einem Deutschen, der die Wahl zwischen einem deutschen Erzeugnis und einem an sich ebenso schönen und wertvollen Erzeugnis fremden Ursprungs hat, immer erlaubt ist, der Stimme des Herzens zu folgen und das Deutsche vorzuziehen, weil es deutsch ist.'

At primary and secondary schools, teachers found it difficult to retain the value of mathematics as an intellectual subject, given the strong emphasis on physical and purely ideological education. Additional problems were caused by the Nazi doctrine of inheritance of personal traits and talents, as indicated in the following quotation by the leading school mathematician, Walther Lietzmann:

Didactics in mathematics has been fighting for decades against the assumption that mathematical talent is a pre-condition of any education at school. Now given that mathematical talent is inherited, is not a continuation of this fight doomed to failure? *What task remains for mathematical education* under these circumstances?[17] (Lietzmann 1935, 363)

In pure mathematical research, personal prejudices and mathematical predilections would occasionally get political support, as when the influential geometer Wilhelm Blaschke alluded to the 'grey' (meaning dubious and abstract) theory of real functions (Blaschke 1941, 44). Workers in probability and statistics stuck to old notions, avoiding measure theory, mainly due to a lack of 'German traditions' in the field and because of interrupted communication with the French and the Russians.[18] The German mathematician Eberhard Hopf, on returning from the US, said the following in his pioneering book *Ergodentheorie* which treated a topic neglected in Germany:

 Statistics is measure theory…Mathematicians who have no taste for the 'almost all' or the 'except for set of measure zero' must realise that in nature only that which happens 'as a rule' can be interpreted mathematically.[19] (Hopf 1937, iii)

Modern mathematical notions connected to Cantor's set theory were often made suspect in the context of Bieberbach's 'Deutsche Mathematik'. Prejudices against modern mathematical techniques based on Cantor's theories, such as Lebesgue's integral, though they also existed in other countries, gained additional political power under the peculiar political circumstances of the Third Reich.

Given that mathematics requires international communication it is clear from the outset that disturbances to such communication are bound to hamper the development of the discipline. However, to what extent and in what specific ways this damage occurred during the Third Reich are questions that remain largely

17. 'Die mathematische Methodik hat ein paar Jahrzehnte lang gegen das Axiom von der mathematischen Begabung gekämpft, ohne die mathematisches Verständnis auf der Schule nicht möglich sei. Müssen wir das nicht…als einen verfehlten Kampf ansehen, weil eben die mathematische Begabung erblich bedingt und also entweder da oder nicht da ist? Was hat unter solchen Umständen überhaupt noch *mathematische Erziehung für Aufgaben?*'

18. This is a particularly promising topic which the author is currently exploring. Ironically, German stochasticians remaining in Germany after 1933 mostly preferred the predominant 'German tradition', namely the dated frequency approach to probability introduced and defended by the Jewish émigré von Mises since 1919.

19. 'Statistik ist Maßtheorie…Den Mathematikern, die dem "fast alle" oder "bis auf eine Nullmenge" keinen Geschmack abgewinnen können, sei entgegnet, daß sich nur das, was in der Natur "in der Regel" sich ereignet, mathematisch interpretieren läßt.'

unanswered. Of course, many German mathematicians remaining in Germany after 1933 kept in contact with their expelled colleagues, even if those contacts were overshadowed by misunderstandings, disappointment, and self-censorship. Germans tended to send their letters from occasional trips abroad for fear of Nazi control. Heinrich Behnke accepted many invitations, especially to France and Switzerland as explained by Behnke (1978), which also describes the atmosphere of suspicion and denunciation in Germany after 1933. We have extensive correspondence between German mathematicians and Courant, who emigrated first to Cambridge (England) and then to New York, and there was continued collaboration between him and Kurt Friedrichs (Brunswig) on the second volume of 'Methods of Mathematical Physics' ('Courant–Hilbert') which finally appeared in 1937.

On the one hand, there was early Nazi interference in international work on mathematical publications, as in the case of the second edition of the *Enzyklopädie der Mathematischen Wissenschaften* 'Encyclopaedia of mathematical knowledge' published by Teubner in Leipzig. Bieberbach's partly racist journal *Deutsche Mathematik* (which paralleled and partly propagated the racist theories of the same name) appeared in 1936. It was published in Gothic letters which, together with its political dimension, contributed to its international isolation and to an underestimation of several valuable mathematical publications in it.[20] On the other hand, the mathematical publication system dominated by Springer, and mathematical reviewing by *Zentralblatt*, edited by Otto Neugebauer from Copenhagen after 1934, secured considerable international contacts for German mathematicians, at least until 1938.

The Nazi course towards economic autarchy in the years to come, the shortage of foreign currency, and finally the war in 1939, created an atmosphere of increasing self-isolation, palpable at the International Congress of Mathematicians in Oslo 1936. Alluding to the three invited lectures by the German mathematicians Helmut Hasse, Erich Hecke, and Carl Siegel, the *Führer* 'leader' of the German delegation, Walther Lietzmann, could still boast about German mathematics at that time, claiming that: 'The leading position of Germany in number theoretic research has been maintained from Gauss to the present'[21] (Siegmund-Schultze 2002, 343). But at the same time and in the same report, Lietzmann wrote the following to the Nazi ministry, which could perhaps be understood as a slight criticism of the existing ideological interference in mathematics in Germany: 'There is one thing that foreigners have a real problem understanding: our notion of the

---

20. For instance, Oswald Teichmüller's work on quasi-conformal mappings, whose importance was recognized only after the war (Schappacher and Scholz 1992).

21. 'Die führende Stellung Deutschlands in der zahlentheoretischen Forschung ist von Gauss bis auf den heutigen Tag erhalten geblieben.'

national peculiarity of science, regardless of the universal validity of its results'[22] (Siegmund-Schultze 2002, 343).

In the years to come, German participation in the international congresses for pure and applied mathematics was increasingly hampered by the Nazi dogma of anti-Semitism (Siegmund-Schultze 2002). Even though the communication structure of mathematics within Germany stabilized after the shockwave of expulsions, international communication, especially with western countries and the Soviet Union, deteriorated. This was mainly for two reasons which in a certain sense were intertwined: the insurmountable dogma of anti-Semitism and preparations for war. The affair around the dismissal of the Italian-Jewish editor of the *Zentralblatt für Mathematik*, Tullio Levi-Civita, in 1938 and the ensuing foundation of *Mathematical Reviews* in the United States have been much discussed and will not be repeated here (see Reingold 1981; Siegmund-Schultze 1993a). Suffice it to mention that even mathematicians such as Hasse, who were no *alte Parteigenossen* 'old comrades' of the Nazis, would support Nazi measures to discontinue reviewing of 'German authors' by Jewish mathematicians. Like Bieberbach in 1934, Hasse promoted an apartheid notion of 'internationalism', as in his letter (in English) to American Marshall Harvey Stone on 15 March 1939:

Looking at the situation from a practical point of view, one must admit that there is a state of war between the Germans and the Jews. Given this, it seems to me absolutely reasonable and highly sensible that an attempt was made to separate within the domain of the *Zentralblatt* the members of the two opposite sides in this war. I do not understand why the American mathematicians found it necessary thereon to withdraw their collaboration in bulk. I do not know whether it was the intention, but it certainly has the appearance of taking decidedly and emphatically one of the two sides, and thus deviating from a truly impartial and hence genuinely international course. (see Siegmund-Schultze 1993a, 164; 2002, 341)

The Nazis' distorted view of 'internationalism' did not help the development of mathematics. There were, for instance, attempts at international mathematical contact under German hegemony, particularly during the war. Several German mathematicians and politicians reached out for influence over southeast-European mathematics and tried to induce French mathematicians under German occupation after 1940 to collaborate. With the agreement of the Nazi ministry, Harald Geppert, Süss, and other German mathematicians tried to use the mutually traumatic experiences of French and German mathematicians in World War I to assist with their policies in World War II (Siegmund-Schultze 1986; 2002).

---

22. 'Eines ist allerdings für Ausländer ausserordentlich schwer zu begreifen: unsere Auffassung von der nationalen Eigenart einer Wissenschaft, unbeschadet der internationalen Bedeutung jeder Forschung.'

The emigration of mathematicians

The emigration of mathematicians from Europe after 1933 and the ensuing shift of the world centre of mathematics from Europe to the United States (where most of the refugees went) is arguably the most important historical consequence of Nazi rule for mathematics (Siegmund-Schultze 1998/2009).[23] Many of today's hotly debated problems in mathematics, in the US and worldwide (such as the relative advantages of various educational and school systems; the need for a classical European background in analysis; communication systems in research) cannot be discussed without reference to the historical event of emigration.

The discussion in the existing literature is largely restricted to German-speaking emigration, which was by far the most extensive, but was partly matched by Polish, and, to a lesser extent, French and Italian. Emigrant mathematicians from Germany included Emil Artin, Richard Courant, John von Neumann, Emmy Noether, Richard Brauer, Richard von Mises, Issai Schur, Hermann Weyl, and many others. Weyl and the American mathematician Oswald Veblen were leading figures at the Institute for Advanced Study in Princeton after 1933, and they played an outstanding role in the reception of refugees.

The aspects of emigration that can be discussed are both quantitative and qualitative. In the latter respect the different political and economic causes of emigration, and the conditions under which emigrants were received in various countries, remain to be investigated. The cognitive dimensions of the process include the consequences for the world-wide reception of abstract algebra and applied mathematics. Some results of research on emigration of German-speaking mathematicians are summarized in the following paragraph.

Nazi politicians gave no consideration to the international fame of their mathematicians and made no effort to keep them in the country, because the dogma of anti-Semitism was all-pervasive and allowed no exceptions. Although some Americans also held anti-Semitic prejudices, combined with concerns for the academic job market in the US as a result of immigration, the realization of the benefits for American mathematics finally prevailed. The emigration of European mathematicians had considerable importance for bringing classical European background knowledge to the attention of (particularly) American students. As far as German-speaking emigration was concerned, these were fields such as number theory, applied mathematics, and the history of mathematics. Although mathematics was one of the most important immigration gains for the US, it had to be organized by private initiatives. Immigrants to the US went mostly to minor institutions or to newly built ones which specialized in research rather than in teaching. Public

23. It is comparable in importance only with the reorientation of mathematics towards applications due to the war, both in Germany and abroad, which, in part, was also promoted by emigration. This aspect of influence of the Nazi rule and the war could not be discussed in this article.

recognition or support for mathematics was minimal until the end of the war. It was only after the war that growing state support for fundamental science in the US also contributed to a second wave of emigration from Germany. Some 'German' traditions in mathematical research were less represented in the emigration and became therefore gradually less visible internationally.[24] Important members or adherents of the 'Noether school' of abstract algebra (van der Waerden, Deuring, Witt, Hasse, Artin until 1937) remained in Germany, and the internationally acclaimed influence of the Noether school in the 1930s was exerted at least as much by Noether's American, French, and Polish followers as by the German emigrants, who mostly belonged to the Berlin school of Issai Schur.[25] Some modern mathematical trends of the 1930s which had had their origin partly in Germany (for example, the structural method of Bourbaki, see Corry, Chapter 6.4 in this volume) had to be re-introduced into Germany after the war due to the relative isolation of German mathematical culture in that period. For most emigrants, emigration was final; very few (Siegel, Artin, Reinhold Baer) returned after the war. But personal relations between Germans and emigrants such as Courant served to reintroduce international trends into German mathematics after the war.

Conclusion

This chapter had room to address only a few of the problems relating to mathematics in the Third Reich. One important aspect in particular, war research in mathematics, which has recently been given promising attention (Mehrtens 1996; Epple/Remmert 2000; Eckert 2006), could not be discussed here for reasons of space. But those developments that have been discussed, like the expulsion of about one quarter of German mathematicians after 1933, or the temporary rise of racist theories like 'Deutsche Mathematik', were unique events in the history of mathematics, as were the monstrous crimes of the Holocaust to which these events would partially contribute.

Science and mathematics have changed considerably since the 1930s with regard to their degree of professionalization, their social role, and the nature of their problems. 'Little science' was far more typical in the Nazi period than the few beginnings of 'big science', in rocket research, physics, and chemistry. This is paralleled by deep differences between then and now in mathematics, in

24. For instance, quasi-conformal mappings as in Teichmüller's work and the Leibnizian tradition of logic, represented by Heinrich Scholz.

25. For the Schur School and combinatorial group theory see Chandler and Magnus (1982). Schappacher (2006a) argues, from the example of Max Deuring and his pursuit of the arithmetic paradigm of the Hasse school, that some modern algebraic research in Germany remained isolated from the mainstream of algebraic geometry, which was influenced by Italian work on the theory of geometrical correspondences and its reception by André Weil and others.

the divide between the pre-computer age and the modern period of information technology. The indisputable fact that the Nazi regime and the ensuing war put an end to the 'German age' of science and mathematics should rule out simple-minded analogies to the situation today. However, as this chapter has shown, the history of science in National Socialist Germany, and of mathematics in particular, remains an inexhaustible source for discussion of burning questions about the social role of science and mathematics now.

Appendix: research on mathematics in the Third Reich

The following table lists the main themes that have so far been treated, together with the most relevant research (in English where it exists) and printed sources (memoires of contemporaries). The importance of archival material from university, ministerial and institutional files,[26] from the papers of emigrants (such as Richard Courant), emigrant committees (like the Emergency Committee in New York City), and foreign mathematicians helping immigrants (such as the Oswald Veblen Papers in the Library of Congress in Washington, DC) is of course all pervading. The unpublished sources can be identified from the relevant research articles. There is also useful information in biographical dictionaries such as Röder and Strauss (1983) and Tobies (2006).

Table 9.4.2

Topic	Research	Published Sources
Political Coordination 1933 and later, particularly DMV	Mehrtens 1989a; Remmert 1999; 2004; Schappacher/Kneser 1990	Behnke 1978; Thullen 2008
Expulsion and emigration	Siegmund-Schultze 1998/2008	Pinl 1965; 1969–72; Pinl/Dick 1974–76
Emigration and reception abroad	Danneberg et al. 1994; Reingold 1981; Rider 1984; Siegmund-Schultze 1998/2009; Thiel 1984	Thullen 2008; Pinl 1969/72; Gumbel 1938
Resistance	Litten 1996; 2000	Thullen 2008; Thomsen 1934; Perron 1939
Profiteers and Nazi activists	Hochkirchen 1998; Schappacher/Scholz 1992; Mehrtens 1987; Remmert 1999	Draeger 1941; Bieberbach 1934

26. Among the latter are the files of the *Deutsche Mathematiker-Vereinigung* (DMV) 'German Mathematicians' Association', which have been accessible only since 1997. They throw new light, for instance, on the expulsions of Jews from the DMV (Remmert 1999; 2004) but were not used by Segal (2003).

Table 9.4.2 (*Continued*)

Topic	Research	Published Sources
School mathematics	Mehrtens 1989b; Guntermann 1992; Genuneit 1984; Radatz 1984	Lorey 1938; Dorner 1935
Mathematics students in National Socialism	Segal 1992	
Mathematical research during Third Reich (with relevant information on National Socialism)	Chandler/Magnus 1982; Hochkirchen 1998; Hoehnke 1986; Lemmermeyer/Roquette 2006	Siegel 1965
Biographical and autobiographical books	Dawson 1997; Ebbinghaus 2007; Georgiadou 2004; Litten 2000; Menzler-Trott 2007; Bigalke 1988; Segal 2003; Reid 1976; Sigmund 2001	Behnke 1978; Fraenkel 1967
Biographical and autobiographical articles (only if focused on NS)	Segal 1980; 1992; Litten 1996; Peckhaus 1994; Remmert 1999; Soifer 2004–05; Siegmund-Schultze 2004; Sigmund 2004	Thullen 2008; Menger 1994
Development of the system of publication	Siegmund-Schultze 1993a; Knoche 1990	
International connections, congresses	Siegmund-Schultze 2002; Schappacher 2006a	Behnke 1978
Industry, including insurance mathematics	Tobies 2007; Petzold 1992	
War research	Mehrtens 1996; Epple and Remmert 2000	
Occupation, including Austria	Dalen 2005; Einhorn 1985; Ulam 1976; Weil 1992; Siegmund-Schultze 1986	
Mathematics in concentration camps	Ebert 1998	
Post-war policies, de-Nazification, and remigration	Rammer 2002; Litten 1996; 2000	Rudin 1997
Broader perspective, including comparison with Fascist Italy	Epple, Karachalios and Remmert 2005; Guerragio and Nastasi 2005	

Bibliography

Abbreviations

JDMV: Jahresbericht der Deutschen Mathematiker-Vereinigung
NTM: International Journal of the History of Science, Technology and Medicine
ZDM: Zeitschrift für Didaktik der Mathematik

Becker, H, Dahms, H-J, and Wegeler, C, *Die Universität Göttingen unter dem Nationalsozialismus. Das verdrängte Kapitel ihrer 250jährigen Geschichte*, Saur, 1987.

Begehr, H (ed), *Mathematik in Berlin. Geschichte und Dokumentation*, 2 vols, Shaker, 1998.

Behnke, H, *Semesterberichte. Ein Leben an deutschen Universitäten im Wandel der Zeit*, Vandenhoeck and Ruprecht, 1978.

Bieberbach, L, 'Persönlichkeitsstruktur und mathematisches Schaffen', *Unterrichtsblätter für Mathematik und Naturwissenschaften*, 40 (1934), 236–243.

Bigalke, H-G, *Heinrich Heesch. Kristallgeometrie, Parkettierungen, Vierfarbenforschung*, Birkhäuser, 1988.

Blaschke, W, 'Mauro Picone: Appunti di Analisi Superiore', (Review), in *JDMV*, 51 (1941), II, 44–45.

Brieskorn, E (ed), *Felix Hausdorff zum Gedächtnis, Band I. Aspekte seines Werkes*, Westdeutscher Verlag, 1996.

Butzer, P, and Volkmann, L, 'Otto Blumenthal (1876–1944) in retrospect', *Journal of Approximation Theory*, 138 (2006), 1–36.

Chandler, B, and Magnus, W, *The history of combinatorial group theory: a case study in the history of ideas*, Springer, 1982.

Cohen, Y, and Manfrass, K (eds), *Frankreich und Deutschland. Forschung, Technologie und industrielle Entwicklung im 19. und 20. Jahrhundert*, Beck, 1990.

van Dalen, D, *Mystic, geometer, and intuitionist: the life of L. E. J. Brouwer 1881–1966, 2, Hope and disillusion*, Clarendon Press, 2005.

Danneberg, L, Kamlah, A, and L. Schäfer L (eds), *Hans Reichenbach und die Berliner Gruppe*, Vieweg, 1994.

Dawson, J W, *Logical dilemmas: the life and work of Kurt Gödel*, A K Peters, 1997.

Dehn, M, 'K. Th. Vahlen, Abstrakte Geometrie' (Review), *JDMV*, 14 (1905), 535–537.

Dithmar, R (ed), *Schule und Unterricht im Dritten Reich*, Luchterhand, 1989.

Dorner, A (ed), *Mathematik im Dienste der nationalpolitischen Erziehung*, Moritz Diesterweg, 1935.

Draeger, M, 'Mathematik und Rasse', *Deutsche Mathematik*, 6 (1941), 566–575.

Ebbinghaus, H D, in cooperation with Peckhaus, V, *Ernst Zermelo, an approach to his life and work*, Springer, 2007.

Ebert, H, '"Häftlingswissenschaftler" im Einsatz für die SS 1944/45' in Begehr, 1998, II 219–242.

Eckert, M, *The dawn of fluid dynamics: a discipline between science and technology*, Wiley, 2006.

Einhorn, R, *Vertreter der Mathematik und Geometrie an den Wiener Hochschulen 1900–1940*, 2 vols, VWGÖ, 1985.

Epple, M, and Remmert, V, '"Eine ungeahnte Synthese zwischen reiner und angewandter Mathematik." Kriegsrelevante mathematische Forschung in Deutschland während des II.Weltkrieges', in, Doris Kaufmann (ed), *Geschichte der Kaiser-Wilhelm-Gesellschaft im Nationalsozialismus. Bestandsaufnahme und Perspektiven der Forschung*, 2 vols, Wallstein 2000, I 258–295.

Epple, M, Karachalios, A, and Remmert, V, 'Aerodynamics and mathematics in National Socialist Germany and Fascist Italy, a comparison of research institutes', in C Sachse and M Walker (eds), 'Politics and science in wartime, comparative international perspectives on the Kaiser Wilhelm Institutes', *Osiris,* (2) 20 (2005), 131–158.

Forman, P, and Sánchez-Ron, J M (eds), *National military establishments and the advancement of science and technology*, Kluwer, 1996.

Fraenkel, A A, *Lebenskreise. Aus den Erinnerungen eines jüdischen Mathematikers*, Deutsche Verlagsanstalt, 1967.

Genuneit, J, 'Mein Rechenkampf', in, *Stuttgart im Dritten Reich, Anpassung, Widerstand, Verfolgung*, Stuttgart, 1984, 205–236.

Georgiadou, M, *Constantin Carathéodory, mathematics and politics in turbulent times*, Springer, 2004.

Gray, J, *The Hilbert challenge*, Oxford University Press, 2000.

Guerraggio, A, and Nastasi, P, *Italian mathematics between the two world wars*, Birkhäuser, 2005.

Gumbel, E J (ed), *Freie Wissenschaft. Ein Sammelbuch aus der deutschen Emigration*, Brant, 1938.

Guntermann, U, *Walther Lietzmann und die Mathematikdidaktik im Nationalsozialismus*, Schriftliche Hausarbeit Mathematik, University of Wuppertal, 1992.

Hochkirchen, T, 'Wahrscheinlichkeitsrechnung im Spannungsfeld von Mass- und Häufigkeitstheorie-Leben und Werk des "Deutschen" Mathematikers Erhard Tornier (1894–1982)', *NTM*, 6 (1998), 22–41.

Hoehnke, H J, '66 Jahre Brandtsches Gruppoid', *Wissenschaftliche Beiträge Martin Luther Universität Halle*, 47 (M43) (1986), 15–79.

Hopf, E, *Ergodentheorie*, Springer, 1937.

Knoche, M, 'Wissenschaftliche Zeitschriften im nationalsozialistischen Deutschland' in Estermann and M Knoche (eds), *Von Göschen bis Rowohlt. Beiträge zur Geschichte des deutschen Verlagswesens*, Harrassowitz, 1990, 260–281.

Lemmermeyer, F, and Roquette, P (eds), *Helmut Hasse und Emmy Noether. Die Korrespondenz 1925–1935*, Göttingen Universitätsverlag, 2006. [Also available at <http://univerlag.uni-goettingen.de/hasse-noether/>]

Lietzmann, W, 'Die geistige Haltung des Mathematikers, Vererbung oder Erziehung', *Zeitschrift für mathematisch-naturwissenschaftlichen Unterricht*, 66 (1935), 361–365.

Litten, F, 'Ernst Mohr—das Schicksal eines Mathematikers', *JDMV*, 98 (1996), 192–212.

Litten, F, *Mechanik und Antisemitismus, Wilhelm Müller (1880–1968)*, Institut für Geschichte der Naturwissenschaften, 2000.

Lorey, W, *Der Deutsche Verein zur Förderung des mathematischen und naturwissenschaftlichen Unterrichts e.V. 1891–1938*, Salle, 1938.

Mehrtens, H, 'Ludwig Bieberbach and "Deutsche Mathematik"', in E R Philips (ed), *Studies in the history of mathematics*, Washington, 1987, 195–241.

Mehrtens, H, 'The Gleichschaltung of Mathematical Societies in Nazi Germany', *The Mathematical Intelligencer,* 11.3 (1989a), 48–60.

Mehrtens, H, 'Mathematik als Wissenschaft und Schulfach im NS–Staat', in R Dithmar (ed), 1989b, 205–216.

Mehrtens, H, *Moderne, Sprache, Mathematik. Eine Geschichte des Streits um die Grundlagen der Disziplin und des Subjekts formaler Systeme*, Suhrkamp, 1990a.

Mehrtens, H, *Der französische Stil und der deutsche Stil. Nationalismus, Nationalsozialismus und Mathematik, 1900–1940*, in Cohen and Manfrass (eds), 1990b, 116–129.

Mehrtens, H, 'Mathematics and war, Germany, 1900–1945', in P Forman, and J M Sánchez-Ron (eds), 1996, 87–134.

Mehrtens, H, and Richter, S (eds), *Naturwissenschaft, Technik und NS-Ideologie. Beiträge zur Wissenschaftsgeschichte des Dritten Reiches*, Suhrkamp, 1980.

Menger, K, *Reminiscences of the Vienna Circle and the Mathematical Colloquium*, eds L Golland, B McGuinness, and A Sklar, Kluwer, 1994.

Menzler–Trott, E, *Logic's lost genius: the Life of Gerhard Gentzen*, American Mathematical Society and London Mathematical Society, 2007.

Neuenschwander, E, 'Felix Hausdorffs letzte Lebensjahre nach Dokumenten aus dem Bessel–Hagen–Nachlass' in Brieskorn (ed), *Felix Hausdorff zum Gedächtnis, Band I. Aspekte seines Werkes*, Westdeutscher Verlag, 1996, 253–270.

Nyssen, E, *Schule im Nationalsozialismus*, Heidelberg, 1969.

Olff-Nathan, J (ed), *La science sous le Troisième Reich. Victime ou alliée du nazisme?*, Seuil, 1993.

Parshall, K H, and Rice, A C (eds), *Mathematics unbound, the evolution of an international mathematical research community, 1800–1945*, American Mathematical Society, 2002.

Peckhaus, V, 'Von Nelson zu Reichenbach, Kurt Grelling in Göttingen und Berlin' in Danneberg *et al* (eds), 1994, 53–86.

Perron, O, *Irrationalzahlen*, de Gruyter, 2nd ed, 1939.

Petzold, H, *Moderne Rechenkünstler. Die Industrialisierung der Rechentechnik in Deutschland*, Beck, 1992.

Pinl, M, 'In memory of Ludwig Berwald', *Scripta mathematica*, 27.3 (1965), 193–203.

Pinl, M, 'Kollegen in einer dunklen Zeit', part I: *JDMV*, 71 (1969), 167–228; part II: *JDMV*, 72 (1971), 165–189; part III: *JDMV*, 73 (1971/72), 153–208.

Pinl, M, and Dick, A, 'Kollegen in einer dunklen Zeit. Schluß', *JDMV*, 75 (1974), 166–208, Nachtrag und Berichtigung, *JDMV*, 77 (1976), 161–164.

Pinl, M and Furtmüller, L, 'Mathematicians under Hitler', *Yearbook Leo Baeck Institute*, 18 (1973), 129–182.

Radatz, H, 'Der Mathematikunterricht in der Zeit des Nationalsozialismus', *ZDM*, 16.6 (1984), 199–206.

Rammer, G, 'Der Aerodynamiker Kurt Hohenemser', *NTM*, 10 (2002), 78–101.

Reid, C, *Courant in Göttingen and New York: the story of an improbable mathematician*, Springer, 1976.

Reidemeister, K, *Über Freiheit und Wahrheit*, Berlin, 1947.

Reingold, N, 'Refugee mathematicians in the United States of America 1933–1941', *Annals of Science*, 38 (1981), 313–338.

Remmert, V, 'Mathematicians at war. Power struggles in Nazi Germany's mathematical community, Gustav Doetsch and Wilhelm Süss', *Revue d'histoire des mathématiques*, 5 (1999), 7–59.

Remmert, V, 'Die Deutsche Mathematiker–Vereinigung im "Dritten Reich"', *Mitteilungen der Deutschen Mathematiker–Vereinigung*, 12 (2004), 159–177, 223–245.

Rider, R, 'Alarm and opportunity, Emigration of mathematicians and physicists to Britain and the United States, 1933–1945', *Historical Studies in the Physical Sciences*, 15 (1984), 107–176.

Röder, W, and Strauss, H, (eds), *International biographical dictionary of Central European emigrés 1933–1945*, II, *The arts, sciences, and literature*, Sauer, 1983.

Rudin, W, *The way I remember it*, American and London Mathematical Societies, 1997.

Schappacher, N, 'Das Mathematische Institut der Universität Göttingen 1929–1950', in Becker, Dahms and Wegeler (eds), 1987, 345–373.

Schappacher, N, 'The Bourbaki Congress at El Escorial and other mathematical (non) events of 1936', *Madrid Intelligencer*, edited by F Chamizo and A Quirós, Springer, 2006a, 8–15.

Schappacher, N, 'De Felix Bernstein à Siegfried Koller, des implications politiques des statisticiens' in C Bonah, A Danion–Grilliat, J Olff–Nathan, and Norbert Schappacher (eds), *Nazisme, science et médecine*, Paris, 2006b, 65–91, 291–295.

Schappacher, N, and Kneser, M, 'Fachverband — Institut — Staat' in G Fischer, F Hirzebruch, W Scharlau, and W Toernig (eds), *Ein Jahrhundert Mathematik 1890–1990*, Festschrift zum Jubiläum der DMV, Vieweg, 1990, 1–82.

Schappacher, N, and Scholz, E (eds), 'Oswald Teichmüller—Leben und Werk', *JDMV*, 94 (1992), 1–39.

Segal, S, 'Helmut Hasse in 1934', *Historia mathematica*, 7 (1980), 46–56.

Segal, S, 'Ernst August Weiss, mathematical pedagogical invention in the Third Reich' in S Demidov *et al*, *Amphora. Festschrift für Hans Wussing zu seinem 65. Geburtstag*, Basel, 1992, 693–704.

Segal, S, *Mathematicians under the Nazis*, Princeton University Press, 2003.

Siegel, C L, *Zur Geschichte des Frankfurter Mathematischen Seminars*, Frankfurt, 1965.

Siegmund-Schultze, R, 'Theodor Vahlen—zum Schuldanteil eines deutschen Mathematikers am faschistischen Mißbrauch der Wissenschaft', *NTM*, 21.1 (1984), 17–32.

Siegmund-Schultze, R, 'Faschistische Pläne zur 'Neuordnung' der europäischen Wissenschaft. Das Beispiel Mathematik', *NTM*, 23.2 (1986), 1–17.

Siegmund-Schultze, R, 'Zur Sozialgeschichte der Mathematik an der Berliner Universität im Faschismus', *NTM*, 26.1 (1989), 49–68.

Siegmund-Schultze, R, *Mathematische Berichterstattung in Hitlerdeutschland. Der Niedergang des Jahrbuchs über die Fortschritte der Mathematik (1869–1945)*, Vandenhoeck and Ruprecht, 1993a.

Siegmund-Schultze, R, 'La légitimation des mathématiques dans l'Allemagne fasciste, trois étapes' in Olff–Nathan (ed), 1993b, 91–102.

Siegmund-Schultze, R, 'The shadow of National Socialism' in K Macrakis and D Hoffmann (eds), *Science under Socialism, East Germany in comparative perspective*, Harvard University Press, 1999, 64–81, 315–317.

Siegmund-Schultze, R, *Rockefeller and the internationalization of mathematics between the two World Wars*, Birkhäuser, 2001.

Siegmund-Schultze, R, 'The effects of Nazi rule on the international participation of German mathematicians: an overview and two case studies', in K H Parshall and A Rice (eds), 2002, 335–357.

Siegmund-Schultze, R, *Mathematiker auf der Flucht vor Hitler. Quellen und Studien zur Emigration einer Wissenschaft*, Vieweg, 1998; to appear in an extended English version, *Mathematicians fleeing from Nazi Germany: individual fates and global impact*. Princeton University Press, 2009.

Siegmund-Schultze, R, 'Helmut Grunsky (1904–1986) in the Third Reich, a mathematician torn between conformity and dissent', in Oliver Roth and Stephan Ruscheweyh (eds), *Helmut Grunsky, Collected Papers*, Lemgo: Heldermann, 2004, xxxi–l.

Siegmund-Schultze, R, and Zabell S, 'Richard von Mises and the "problem of two races": a statistical satire in 1934', *Historia Mathematica* 34 (2007), 206–220.

Sigmund, K, 'Kühler Abschied von Europa'—Wien 1938 und der Exodus der Mathematik', Catalogue of an exhibition Vienna September 2001, Österreichische Mathematische Gesellschaft, 2001.

Sigmund, K, 'Failing Phoenix, Tauber, Helly, and Viennese Life insurance', *The Mathematical Intelligencer*, 26.2 (2004), 21–33.

Soifer, A, 'In Search of Van der Waerden', *Geombinatorics*, 14 (2004–5), 21–40, 72–102, 124–161.

Thiel, C, 'Folgen der Emigration deutscher und österreichischer Wissenschaftstheoretiker und Logiker zwischen 1933 und 1945', *Berichte zur Wissenschaftsgeschichte*, 7 (1984), 227–256.

Thomsen, G, 'Über die Gefahr der Zurückdrängung der exakten Naturwissenschaften an den Schulen und Hochschulen', *Neue Jahrbücher für Wissenschaft und Jugendbildung*, (1934), 164–175.

Thullen, P, 'Erinnerungsbericht für meine Kinder', *Exil*, 20.1 (2000), 44–57; English translation in Siegmund–Schultze 2009.

Tietjen, C H, *Raum oder Zahl?*, Brandstetter, 1936.

Tobies, R, *Biographisches Lexikon in Mathematik promovierter Personen an deutschen Universitäten und Technischen Hochschulen WS 1907/08 bis WS 1944/45*, Dr Erwin Rauner Verlag, 2006.

Tobies, R, 'Zur Position von Mathematik und Mathematiker/innen in der Industrieforschung, am Beispiel früher Anwendung von mathematischer Statistik in der Osram G.m.b.H.', *NTM*, 15 (2007), 241–270.

Ulam, S, *Adventures of a mathematician*, Scribner, 1976.

Vahlen, T, *Abstrakte Geometrie*, Leipzig, 1905.

Vahlen, T, *Wert und Wesen der Mathematik*, Greifswalder Universitätsreden, 1923.

Vahlen, T, *Erwiderung, Sitzungsberichte Preußische Akademie der Wissenschaften*, 1938, 98–99.

Vogt, A, 'Die Berliner Familie Remak—eine deutsch–jüdische Geschichte im 19. und 20.Jahrhundert', in M Toepell (ed), *Mathematik im Wandel*, I, *Mathematikgeschichte und Unterricht*, Franzbecker, 1998, 331–342.

Weil, A, *The apprenticeship of a mathematician*, Birkhäuser, 1992.

Weyl, H, 'Emmy Noether', *Scripta mathematica*, 3 (1935), 201–220.

ABOUT THE CONTRIBUTORS

David Aubin
Institut de mathématiques de Jussieu, Université Pierre et Marie Curie, Paris

David Aubin teaches the history of the mathematical sciences at the Université Pierre et Marie Curie in Paris. Following his work on the history of chaos theory and dynamical systems, he started exploring the history of observatory sciences and techniques in the nineteenth century. He has edited a special issue of the *Cahier François Viète* (2006) about the transits of Venus of 1874 and 1882. In a volume co-edited with Charlotte Bigg and Otto Sibum, *The heavens on earth* (2008), he has established a project for a social and cultural history of the observatory. He is also currently working on individual scientific practices in non-academic settings.

Markus Asper
New York University

Markus Asper is Assistant Professor of Classics at New York University. His work focuses on two areas: Hellenistic poetry and the literature of ancient Greek science and technology. He has published a study of Callimachus' poetological metaphors, *Onomata allotria* (1997), a critical edition of his works, *Kallimachos' Werke* (2004), and a monograph on Greek science writing, *Griechische Wissenschaftstexte* (2007). Currently, he is working on *Writing science in antiquity*, an edition of and commentary upon a selection of Hippocratic treatises, and a study on Alexandrian poets and Ptolemaic identity.

June Barrow-Green
Open University

June Barrow-Green is a lecturer in the history of mathematics at the Open University. Her research centres on nineteenth- and early twentieth-century European and American mathematics. She is currently researching the role of British mathematicians during World War I, the contribution of George Birkhoff to the development of the theory of dynamical systems, and aspects of the history of the three-body problem. Recent publications include 'Burnside's Applied Mathematics' in *The Collected papers of William Burnside* (2004) and, with Jeremy Gray, 'Geometry at Cambridge, 1863–1940' in *Historia Mathematica* (2006). She is assistant editor of the forthcoming *Princeton companion to mathematics*, editor of *Historia Mathematica*, and Librarian of the London Mathematical Society.

Kate Bennett
Independent scholar

Kate Bennett did a BA and a DPhil in English Literature at Oxford University where she held a Junior Research Fellowship at Christ Church, and subsequently a British Academy Postdoctoral Fellowship. She then held a fellowship in English at Pembroke College, Cambridge, which she resigned in 2005 to take an extended maternity break and to concentrate on her research. She is completing a new, two-volume, annotated critical edition of John Aubrey's *Brief lives* for the Clarendon Press, and has published articles on Aubrey and his circle, in particular on Aubrey's relationship with the Royal Society and Early Modern science.

Carol Bier
The Textile Museum, Washington, DC

Carol Bier is Research Associate at The Textile Museum in Washington, DC, where she served as Curator for Eastern Hemisphere Collections from 1984 to 2001. Her award-winning on-line exhibition, *Symmetry and Pattern: the Art of Oriental Carpets* is hosted by The Math Forum@Drexel University, http://mathforum.org/geometry/rugs/. She has taught courses on Islamic arts and culture for Advanced Academic Programs at Johns Hopkins University, and undergraduate courses at the Maryland Institute College of Art, San Francisco State University, and Mills College. Currently, she is President of the Textile Society of America and serves on the editorial board of the *Journal of Mathematics and the Arts*.

Sonja Brentjes
Department of Philosophy, Logic and History of Science, University of Seville

Sonja Brentjes studied mathematics at Dresden Technical University and Arabic and Near Eastern Civilizations at Martin Luther University, Halle. She completed graduate studies at Leipzig and Dresden and has since held posts in Berlin, Frankfurt, London, Munich, Oklahoma, Princeton, and Cambridge Mass. Her research interests include cross-cultural transfer of knowledge, cultural history of the mathematical sciences in Islamic societies, and history of mathematics and cartography in Islamic societies. Recent publications include 'Pride and Prejudice: the invention of a "Historiography of Science" in the Ottoman and Safavid Empires by European travellers and writers of the sixteenth and seventeenth centuries' in *Religious values and the rise of science in Europe,* Istanbul, 2005, and 'Mapmaking in Ottoman Istanbul between 1650 and 1750: a domain of painters, calligraphers, or cartographers?' in *Frontiers of Ottoman studies,* London, 2005.

Carrie Brezine
Harvard University Department of Anthropology

Carrie J Brezine is a weaver and spinner with experience in both European and Andean textile construction. Her undergraduate work in mathematics at Reed College continues to inspire her research in fabric structure and ethnographic weaving. From 2002 to 2005 she was Database Administrator for the Harvard Khipu Database project, an effort to decipher the knotted-cord communication devices of the Inka empire. She is presently a PhD candidate in the Archaeology Program, Department of Anthropology, Harvard University.

Stephen Chrisomalis
Department of Anthropology, McGill University, Montreal

Stephen Chrisomalis was educated at McMaster University (BA, 1996) and McGill University (PhD, 2003). Following a postdoctoral fellowship at the University of Toronto, he is currently a Faculty Lecturer in the Department of Anthropology, McGill University (Montreal). His research focuses on cognitive anthropology and archaeology, the anthropology of science, and literacy studies, with a particular interest in numeration and numerals and the transmission and adoption of pre-modern knowledge systems. His recent publications include 'The Egyptian origin of the Greek alphabetic numerals' (*Antiquity*, 2003) and 'A cognitive typology for numerical notation' (*Cambridge Archaeological Journal*, 2004). His book, *A comparative history of numerical notation*, is forthcoming.

Christopher Cullen
Needham Research Institute, Cambridge

Christopher Cullen is Director of the Needham Research Institute, and a Fellow of Darwin College Cambridge. He is the Honorary Professor of the History of East Asian Science, Technology, and Medicine in the University of Cambridge. He edits the *Science and civilisation in China* series, and the Needham Research Institute monograph series. Currently he works mainly on the history of astronomy and mathematics in China. His monograph publications include *Astronomy and mathematics in ancient China: the Zhou bi suan jing* (1996), and *The Suàn shù shū* (筭數書) '*Writings on reckoning': a translation of a Chinese mathematical collection of the second century BC, with explanatory commentary* (2004).

Leo Corry
Tel Aviv University

Leo Corry is Director of the Cohn Institute for History and Philosophy of Science and Ideas, Tel Aviv University, and editor of the journal *Science in Context*. His current research interests include the history of computation in number theory. His publications include *Hilbert and the axiomatization of physics (1898–1918): from 'Grundlagen der Geometrie' to 'Grundlagen der Physik'*, (2004).

Mary Croarken
University of Warwick

Mary Croarken gained her PhD from the University of Warwick in 1986 and published her thesis as *Early scientific computing in Britain* in 1990. Since then she has continued to publish journal articles on mathematical tables, table makers and human computers. She co-edited *The history of mathematical tables* (2003) and is working on a co-edited volume on the mathematical history of Greenwich. She has held several research fellowships and has also worked in research management. She is currently a Visiting Research Fellow at the University of Warwick.

John Denniss
British Society for the History of Mathematics, Mathematical Association, and Textbook Colloquium

John Denniss taught mathematics in secondary schools before becoming a lecturer at Worcester Training College and then Head of Mathematics at Brentwood College of Education, now part of Anglia Ruskin University. He was invited to assist with the British Council Primary Mathematics Projects in Indonesia and India and visited both countries several times. Over the years he has acquired what is now a substantial collection of antiquarian mathematics books and manuscripts. Now retired, he is an active member of the *BSHM* and the *Textbook Colloquium*, contributing talks and articles to both organizations, particularly on arithmetic. He is currently researching the work of the seventeenth-century French mathematician, Pierre Hérigone.

Jeremy Gray
Open University and the University of Warwick

Jeremy Gray studied mathematics at Oxford and at Warwick University, where he took his doctorate in 1980. He has taught at the Open University since 1974. His most recent books include *The architecture of modern mathematics*, a collection of essays on the history and philosophy of modern mathematics, edited with José Ferreirós, (2006), *Worlds out of nothing; a course on the history of geometry in the nineteenth century* (2006), and *Plato's ghost – mathematics and modernism at the end of the nineteenth century* (2008).

William L Hanaway
University of Pennsylvania

William L Hanaway was born in New York City and educated at Amherst College (BA, History, 1951) and Columbia University (PhD, Iranian Studies, 1970). After teaching briefly at Columbia, he moved to the University of Pennsylvania where he taught classical and modern Persian, and Persian literature, for twenty-five years. His research interests include classical Persian poetry and prose, Persian epigraphy, oral narrative, medieval Persian popular literature and Persian epic poetry. His publications include *Reading Nasta`liq: Persian and Urdu hands, 1500 to the present*, with Brian Spooner (1995), *The PreSafavid Persian inscriptions of Khurasan, I* (1977), and numerous articles and reviews.

Dimitri Gouzévitch
Centre d'Etudes des Mondes russe, caucasien et centre-européen, EHESS, Paris

Dmitri Gouzévitch is a Doctor in Engineering Sciences (history of technology, Russia), and a DEA in History and Civilization (EHESS, France). From 1988 he researched the history of engineering, at the Institute of History of Science and Technology (Russian Academy of Sciences), and in 2001 joined the Centre des Mondes Russe, Caucasien et Est-Européen, Ecole des Hautes Etudes en Sciences Sociales (Paris). His general research is on the circulation of knowledge and technological transfer between Russia and Western Europe in the eighteenth and nineteenth centuries. Themes of particular interest are: comparative studies in the history of the engineering, technological change between different cultural and geographical areas, history of scholarship and professional migration. Recent publications include *La Russie et la culture technique français: quelques exemples de la circulation des idées*, with Irina Gouzévitch (2004).

Irina Gouzévitch
Centre Alexander Koyré, EHESS, Paris

Irina Gouzévitch is a Doctor in History of Technology, University of Paris VIII (2001). She has been a historian since 1988, when she became a researcher at the Institute for the History of Science and Technology (Moscow/St-Petersburg). From 1998 she has been a member of the Centre Alexandre Koyré (EHESS, Paris). Her general research focuses on the history of circulation, confrontation, and legitimization of scientific and technical knowledge in the eighteenth and nineteenth centuries, with particular regard for the comparative history of special professional education in European countries. Recent publications include *La formation des ingénieurs en perspective: modèles de référence et réseaux de médiation: XIIIe-XXe siècles*, edited with A Grelon and A Karvar (2005), and *The editorial policy as a mirror of Petrine reforms: textbooks and their translators in the early eighteenth-century Russia* (2006).

Niccolò Guicciardini
Università di Bergamo, Italy

Niccolò Guicciardini teaches the history of science at the University of Bergamo. He is the author of *The development of Newtonian calculus in Britain, 1700–1800* (1989) and *Reading the Principia: the debate on Newton's mathematical methods for natural philosophy from 1687 to 1736* (1999). His current research concerns Johann Bernoulli's role in the polemic between Newton and Leibniz.

Annette Imhausen
Johannes Gutenberg University, Mainz, and Cambridge University

Annette Imhausen studied mathematics (including history of mathematics) and Egyptology at the universities of Mainz, Heidelberg, and Berlin. She did her doctoral research under the supervision of David Rowe (Mainz) and Jim Ritter (Paris) and was awarded her doctorate *summa cum laude* in 2000. She was a postdoctoral fellow at the Dibner Institute for History of Science and Technology and a visiting fellow in the Department for History of Science at Harvard (2000–2) and Junior Research Fellow at Trinity Hall Cambridge (2002–5). She currently holds a Junior professorship at Mainz University where she is teaching courses in history of mathematics and Egyptology. Her research interests focus on Egyptian science, its cultural and social context as well as influences from neighbouring cultures.

Catherine Jami
REHSEIS, Université Denis Diderot et CNRS, Paris

Catherine Jami is Senior Researcher at the French CNRS (REHSEIS, Université Paris-Diderot). Starting with her book *Les méthodes rapides pour la trigonométrie et le rapport précis du cercle (1774). Tradition chinoise et apport occidental en mathématiques* (1990), she has published extensively on seventeenth- and eighteenth-century Chinese mathematics, as well as on the Jesuits and on the reception of the sciences they introduced in late Ming and early Qing China. She is currently completing another book, on the imperial appropriation of mathematics and other branches of Western learning during the Kangxi reign.

Tinne Hoff Kjeldsen
Institut for Natur, Systemer og Modeller, Roskilde University, Denmark

Tinne Hoff Kjeldsen is Associate Professor at NSM, Roskilde University. Her work includes contextualized historical analyses of the emergence and development of nonlinear programming in the wake of World War II in the USA, John von Neumann's minimax theorem in game theory, and the development of the modern concept of convex sets. Besides this, her main interests include history and philosophy of mathematical modelling, historiography, and didactics of mathematics (modelling competencies and the use of history in mathematics and science education). Recent publications include 'From measuring tool to geometrical object: Minkowski's development of the concept of convex bodies', in *Archive for History of Exact Science* (2008).

Snezana Lawrence
Simon Langton Grammar School for Boys, Canterbury, UK

Snezana Lawrence's main research interests lie in exploring the manifold relationships between the historical development of mathematics and the educational systems of Europe since 1800, with emphasis on France, England, and the countries of Eastern Europe and the Balkans. She has been involved in a number of national initiatives to promote the use of the History of Mathematics in Mathematics Education, most recently through her Gatsby Teacher Fellowship and in cooperation with the National Centre for Excellence in the Teaching of Mathematics. She is the Education Officer of the British Society for the History of Mathematics. Her recent publications include 'Alternatives to teaching space – teaching of geometry in nineteenth-century England and France', *Revue d'histoire des mathématiques* (2008).

G E R Lloyd
Needham Research Institute, Cambridge

For most of his academic career G E R Lloyd has been based in Cambridge. He was a Fellow of Kings College 1957–89, then Master of Darwin 1989–2000. Since retiring in 2000 he has been based at the Needham Research Institute, Cambridge. His current main research interests include the comparisons and contrasts between ancient Greek and Chinese philosophy and science. He has published nineteen books, various of which have been translated into eleven different languages, and has edited four more. His most recent publications include *Ancient worlds, modern reflections* (2004), *The delusions of invulnerability* (2005), and *Cognitive variations: reflections on the unity and diversity of the human mind* (2007).

Massimo Mazzotti
University of Exeter

Massimo Mazzotti teaches history and sociology of science at the University of Exeter. He is research associate at Egenis, the ESRC Centre for Genomics in Society, and visiting professor of history of science at the University of Bologna. He has been a postdoctoral fellow at the Dibner Institute (MIT), and the 2002 Kenneth May Fellow in the History of Mathematics. His research interests include the historical sociology of mathematics and the politics of science and technology in the age of Enlightenment. He is the author of articles in several international journals including *Isis, The British Journal for the History of Science, Technology and Culture, Actes de la Recherche en Sciences Sociales,* and of the book *The world of Maria Gaetana Agnesi, mathematician of God* (2007).

Karen Hunger Parshall
Departments of History and Mathematics, University of Virginia

Karen Hunger Parshall is Professor of History and Mathematics at the University of Virginia and Chair of the International Commission for the History of Mathematics. Her research interests lie in the history of nineteenth- and twentieth-century mathematics, with particular emphasis on the historical development of modern algebra. She has written *The emergence of the American mathematical research community, 1876–1900: J J Sylvester, Felix Klein, and E H Moore* (1994) (with David E Rowe), *James Joseph Sylvester: life and work in letters* (1998), and *James Joseph Sylvester: Jewish mathematician in a Victorian world* (2006). She is a co-editor of *Experiencing nature* (1997), *Mathematics unbound: the evolution of an international mathematical research community, 1800–1945* (2002), and *Episodes in the history of modern algebra (1800–1950)* (2007).

Kim Plofker
Department of Mathematics, Union College

Kim Plofker is currently Visiting Assistant Professor of Mathematics at Union College in Schenectady, NY, USA. Her research focuses on the history of mathematics and astronomy in India and its connections with science in the Islamic world. Her recent publications include the chapters 'Mathematics in India' in *The mathematics of Egypt, Mesopotamia, China, India and Islam,* edited by Victor J Katz (2007) and 'The problem of the sun's corner altitude and convergence of fixed-point iterations in medieval Indian astronomy' in *Studies in the history of the exact sciences in honour of David Pingree,* edited by Charles Burnett (2004). She has recently completed a book on the history of Indian mathematics for Princeton University Press (2008).

Volker R Remmert
AG Geschichte der Mathematik und der Naturwissenschaften, Institut für Mathematik, Universität Mainz

Volker R Remmert trained as mathematician and historian. He is Assistant Professor of History of Science and History of Mathematics at the University of Mainz. He works on the history of mathematics in the nineteenth and twentieth centuries and on the history of science, art, and culture in Early Modern Europe. His recent publications include a book on the role of frontispieces in the scientific revolution, *Widmung, Welterklärung und Wissenschaftslegitimierung: Titelbilder und ihre Funktionen in der Wissenschaftlichen Revolution* (2005). A study on mathematical publishing in Germany, *Eine Disziplin und ihre*

Verleger: Formen, Funktionen und Initiatoren mathematischen Publizierens in Deutschland, 1871–1949 (with Ute Schneider) is to be published in 2010.

Eleanor Robson
Department of History and Philosophy of Science, Cambridge

Eleanor Robson is a Senior Lecturer in the Department of History and Philosophy of Science at the University of Cambridge and a Fellow of All Souls College Oxford. She is the author of *Mesopotamian mathematics, 2100–1600 BC* (1999), *Mathematics in ancient Iraq: a social history* (2008), and many articles on the socio-intellectual history of the cuneiform world. Together with Steve Tinney she runs an AHRC-funded research project called *The geography of knowledge in Assyria and Babylonia* <http://cdl.museum.upenn.edu/gkab>, which studies the contents and contexts of four scholarly libraries of cuneiform tablets from the seventh to second centuries BC.

David Gilman Romano
University of Pennsylvania Museum of Archaeology and Anthropology and Department of Classical Studies, University of Pennsylvania

David Gilman Romano has worked as an archaeologist in Greece for more than thirty years. He is a specialist in Greek and Roman cities and sanctuaries, the ancient Olympic Games, ancient surveying and modern cartographic and survey techniques to reveal and study ancient sites. He is the author of three books and numerous articles on these subjects. His book *Athletics and mathematics in Archaic Corinth: the origins of the Greek stadion*, was awarded the John Frederick Lewis Prize for the best publication of the American Philosophical Society in 1993. He is a collaborator with Lothar Haselberger on *Mapping Augustan Rome*, 2002 and co-author with Irene Romano on the *Catalogue of the classical collections of the Glencairn Museum*, of 1999.

Sabine Rommevaux
CNRS, Centre d'études supérieures de la Renaissance, Tours

Sabine Rommevaux is a mathematician and a researcher on history of mathematics and natural philosophy in the Middle Ages and the Renaissance. She has published papers and books on the reception of Euclids *Elements*, on the application of proportion theory to the measure of motion, and on continuity. Recent publications include *Clavius: une clé pour Euclide à la Renaissance* (2005) and, with Joël Biard, editions of Blaise de Parme's *Questiones circa Tractatum proportionum magistri Thome Braduardini* (2005) and *Mathématiques et théorie du mouvement (xive- xvie siècles)* (2007).

Corinna Rossi
Collegio di Milano

Corinna Rossi graduated in Architecture in Napoli (Italy) and specialized in Egyptology with Barry J Kemp at Cambridge University, where she obtained an MPhil and a PhD. Part of her research focuses on the relationship between ancient Egyptian architecture and mathematics; on this subject, she has published *Architecture and mathematics in Ancient Egypt* (2004), plus several articles in specialist journals. In parallel, she is the co-director of the North Kharga Oasis Survey, an archaeological project on a chain of late-Roman military installations in Egypt's Western Desert, and is the co-author of regular field reports on the subject.

Ken Saito
School of Humanities and Social Sciences, Osaka Prefecture University

Ken Saito is associate professor at Osaka Prefecture University (Japan). He studied history of science at the University of Tokyo and Università degli studi di Roma la Sapienza. His research field and his method are best described in his article: 'Phantom theories of pre-Eudoxean proportion' in *Science in Context* (2003). He is participating in *Progetto Maurolico*, www.maurolico.unipi.it, whose aim is the publication of the complete works of the sixteenth-century mathematician and editor Federico Maurolico. He is managing editor of the journal *SCIAMVS*, which is dedicated to primary source materials for premodern exact sciences.

Reinhard Siegmund-Schultze
University of Agder, Kristiansand, Norway

Reinhard Siegmund-Schultze is Professor of the history of mathematics at the University of Agder. He was awarded a diploma in mathematics at Halle in 1971, and a PhD in the history of functional analysis at Leipzig in 1979. Since 1983 he has researched and published on mathematics in the Third Reich after 1983. He has also published on the history of mathematical reviewing (1993), on refugee mathematicians from Hitler's Germany (1998) and on Rockefeller Foundation support for mathematics (2001). He has written many articles on mathematics, the history of analysis, probability, and applied mathematics. He moved to Norway in 2000, and is currently working on the applied mathematician Richard von Mises.

Brian Spooner
University of Pennsylvania

Brian Spooner teaches anthropology at the University of Pennsylvania. He has conducted research in Iran, Afghanistan, and Pakistan since the 1950s, focusing on traditional economies, social organization, language and society, and religion. His publications include 'Weavers and dealers: the authenticity of an oriental carpet', in *The social life of things* (1986), *Ecology and development* (1984), and articles in the *Encyclopaedia Iranica* on Anthropology, Baluch, Ethnography, Desert, and Irrigation.

Jacqueline Stedall
The Queen's College, Oxford

Jacqueline Stedall is a Lecturer in history of mathematics and a Fellow of The Queen's College, Oxford. Her research focuses on European mathematics from the sixteenth century to the eighteenth, with a special interest in the development of algebra. Recent publications include *Mathematics emerging: a sourcebook 1540–1900* (OUP 2008) and *The 'Magisteria magna' of Thomas Harriot, and constant difference interpolation in the seventeenth century*, with Janet Beery (EMS 2008). She is also Editor of the *BSHM Bulletin*, journal of the British Society for the History of Mathematics.

Mark Thakkar
Balliol College, Oxford

Mark Thakkar read Mathematics and Philosophy at Balliol College, Oxford, before taking an MA in Cultural and Intellectual History 1300–1650 at the Warburg Institute, London, and an MSc in the History of Science at Imperial College, London. His MSc dissertation on a topic in fourteenth-century natural philosophy was published in 2006 as 'Francis of Marchia on the Heavens'. He also works on the philosophy of language, and is currently writing a DPhil thesis on Peter Auriol (c 1280–1322) and the logic of statements about the future.

Gary Urton
Department of Anthropology, Harvard University

Gary Urton is the Dumbarton Oaks Professor of Pre-Columbian Studies in the Archaeology program of the Department of Anthropology of Harvard University. His research focuses on a variety of topics in pre-Hispanic and early colonial intellectual history in the Andes drawing on materials and methods in archaeology, ethnohistory, and ethnology. He is the author of numerous articles and books on Andean/Quechua cultures and Inka civilization, including *At the crossroads of the earth and the sky* (1981), *The history of a myth* (1990), *The social life of numbers* (1997), *Inca myths* (1999), and *Signs of the Inka khipu* (2003). He is Director of the Khipu Database Project at Harvard University.

Alexei Volkov
Center for General Education and Institute of History, National Tsing Hua
University, Taiwan

Alexei Volkov obtained his PhD in History of Mathematics from the Academy of Sciences of the USSR (Moscow) in 1989. His research interests include the history of mathematics and of mathematics education in China and Vietnam, and the history of interaction between science and religion in China. His representative publications include *Notions et perceptions du changement en Chine,* (1994) (co-edited with V Alleton) and special issues of the *Extrême-Orient Extrême-Occident* (1994), entitled *Sous les nombres, le monde,* and of the *Taiwanese Journal for Philosophy and History of Science* (1997), entitled *Science in 14th century China,* which he published as Guest Editor.

Benjamin Wardhaugh
All Souls College, Oxford

Benjamin Wardhaugh holds degrees in mathematics, music, and history. He is currently doing postdoctoral work in Oxford, where he also teaches the history of mathematics. His current research focuses on the uses of mathematics in seventeenth- and eighteenth-century England; his doctoral work examined mathematical theories of music from that period. He is the author of *Music, experiment and mathematics in England, 1653–1705* (2008) and co-editor of *John Birchensha: Writings on music* (2008).

INDEX